Renewable Energy in the Service of Mankind Vol I

Ali Sayigh
Editor

Renewable Energy in the Service of Mankind Vol I

Selected Topics from the World Renewable Energy Congress WREC 2014

Springer

Editor
Ali Sayigh
Chairman
World Renewable Energy Congress
Brighton
United Kingdom

ISBN 978-3-319-17776-2 ISBN 978-3-319-17777-9 (eBook)
DOI 10.1007/978-3-319-17777-9

Library of Congress Control Number: 2015947131

Springer Cham Heidelberg New York Dordrecht London

Printed on acid-free paper

Springer International Publishing AG Switzerland is part of Springer Science+Business Media (www.springer.com)

Contents

Contributors

Ismaeil K. Abdou Egyptian Petroleum Research Institute, Nasr City, Cairo, Egypt

Ghanim Kadhim Abdulsada Department of Mechanical Engineering, Al-Mustansiriyah University, Baghdad, Iraq

Jorge Aguilera Tejero School of Architecture, University of Malaga, Málaga, Spain

Sheikh Saeed Ahmad Rawalpindi, Pakistan

Jong-Bo Ahn Smart Distribution Research Center, Korea Electrotechnology Research Institute, Changwon, Republic of Korea

Amela Ajanovic Vienna University of Technology, Vienna, Austria

Mounir Aksas Energetics Applied Physics Laboratory, Department of Material Physics, Faculty of Sciences, Hadj Lakhdar University, Batna, Algeria

Khaled A. Al-Sallal Department of Architectural Engineering, UAE University, Al-Ain, United Arab Emirates

K. Amalevičiūtė Institute of Agriculture, Lithuanian Research Centre for Agriculture and Forestry, Kėdainiai district, Lithuania

Lennart Andersson Volvo Group Trucks Technology, Advanced Engineering & Research, Gothenburg, Sweden

Ronnie Andersson Chalmers University of Technology, Gothenburg, Sweden

Fredrik Andreasson Alfa Laval AB, Lund, Sweden

Xuereb Annalise University of Malta, Msida, MSD, Malta

A. R. M. Ariffin Department of Architecture, University of Malaya, Kuala Lumpur, Malaysia

Francisco Eduardo Arruda Rodrigues Curso de Licenciatura em Química, Instituto Federal da Paraíba—Campus Sousa, Fortaleza, Brazil

Eyjolfur Ingi Asgeirsson School of Science and Engineering, Reykjavik University, Reykjavik, Iceland

Solange Assunção Quintella Curso de pós-graduação em Química, Universidade Federal do Ceará, Fortaleza, Brazil

Mona Azarbayjani Integrated Design Labs, Energy Performance Laboratory, School of Architecture, College of Arts + Architecture, University of North Carolina at Charlotte, Charlotte, NC, USA

F Azizpour Solar Energy Research Institute, Universiti Kebangsaan Malaysia, Bangi, Selangor, Malaysia

L-V. Bénet Socotec Industries, Montigny-le-Bretonneux, France

Viorel Badescu Faculty of Mechanical Engineering and Mechatronics, Thermodynamics Department, Polytechnic University of Bucharest, Bucharest, Romania

Manoel Barbosa Dantas Curso de Licenciatura em Química, Instituto Federal da Paraíba—Campus Sousa, Fortaleza, Brazil

Jorge Barrios Corpa School of Architecture, University of Malaga, Málaga, Spain

Syeda Qamar Batool Rawalpindi, Pakistan

Olivier Baume Resource Centre for Environmental Technologies—Public Research Centre Henri Tudor, Esch-sur-Alzette, Luxembourg

Peter Bayer Department of Earth Sciences, ETH Zurich, Zurich, Switzerland

Markus Beck Wilhelm-Schickard-Institute for Computer Science (WSI), University of Tübingen, Tübingen, Germany

H. Bencheikh Laboratoire de génie civil, Université Amar Telidji, Laghouat, Algeria

Jörg Bendfeld Department of Sustainable Energy Concepts, University of Paderborn, Paderborn, Germany

Hyun Beom-Soo Division of Naval Architecture and Ocean System Engineering, Korea Maritime and Ocean University, Busan, Korea

Julian D. Booker Faculty of Engineering, University of Bristol, Bristol, UK

Matthew Bouchet School of Engineering, Robert Gordon University, Aberdeen, UK

Yassin Bouyraaman Department of Sustainable Energy Concepts, University of Paderborn, Paderborn, Germany

Christian Braun Resource Centre for Environmental Technologies—Public Research Centre Henri Tudor, Esch-sur-Alzette, Luxembourg

Dan J L Brett Electrochemical Innovation Lab, Department of Chemical Engineering, University College London (UCL), London, UK

Philipp Breymann Department of Sustainable Energy Concepts, University of Paderborn, Paderborn, Germany

Klaus Broekel Institute of Engineering Design/CAD, University of Rostock, Rostock, Germany

B. Butkutė Institute of Agriculture, Lithuanian Research Centre for Agriculture and Forestry, Kėdainiai district, Lithuania

Juan de la Casa Higueras School of Architecture, University of Malaga, Málaga, Spain

Valentina Cecchi Integrated Design Labs, Energy Performance Laboratory, School of Architecture, College of Arts + Architecture, University of North Carolina at Charlotte, Charlotte, NC, USA

J. Cesevičienė Institute of Agriculture, Lithuanian Research Centre for Agriculture and Forestry, Kėdainiai district, Lithuania

Priyanka Chawla Department of Chemistry, C.M.P. Degree College, University of Allahabad, Allahabad, India

Kai Chen School of Architecture, Harbin Institute of Technology, Harbin, China

Diego Chulde National Institute of Energy Efficiency and Renewable Energy (INER), Quito, Ecuador

Dania González Couret Faculty of Architecture, Instituto Superior Poiltécnico José Antonio Acheverría, Havana, Cuba

Cristiana Croitoru Building Services Department, Technical University of Civil Engineering in Bucharest, Bucharest, Romania

Ruxandra Crutescu Faculty of Architecture, Spiru Haret University, Bucharest, Romania

Spiteri Staines Cyril University of Malta, Msida, MSD, Malta

G. Dabkevičienė Institute of Agriculture, Lithuanian Research Centre for Agriculture and Forestry, Akademija, Kėdainiai, Lithuania

Maitha M. Bin Dalmouk Department of Architectural Engineering, UAE University, Al-Ain, United Arab Emirates

Brynhildur Davidsdottir School of Engineering and Natural Sciences, Environment and Natural Resources, University of Iceland, Reykjavik, Iceland

Jazaer Dawody Volvo Group Trucks Technology, Advanced Engineering & Research, Gothenburg, Sweden

Zahir Dehouche Brunel University, Uxbridge, UK

Y.G. Dessouky Electrical and Control Engineering Department, Arab Academy for Science and Technology & Maritime Transport, Alexandria, Egypt

P. K. Dey Aston Business School, Aston University, Birmingham, UK

Vishaal Dhamotharan Department of Engineering Design, IIT Madras, Chennai, Tamilnadu, India

Richard M. Dinsdale Sustainable Environment Research Centre (SERC), Faculty of Computing Engineering and Science, University of South Wales, Pontypridd, Mid-Glamorgan, UK

K. Dogan School of Science and Technology, International Hellenic University, Thessaloniki, Greece

Xu Dong School of Architecture, Harbin Institute of Technology, Harbin, China

Nenad Doroslovac Faculty of Mining and Geology, Department of Hydrogeology, University of Belgrade, Belgrade, Serbia

David Thomas Duarte Arruda Curso de pós-graduação em Química, Universidade Federal do Ceará, Fortaleza, Brazil

A. El-Deberky Department of Architecture, Minya University, Minya, Egypt

Eriyatno Graduate School of Business Management, Bogor Agricultural University, Bogor, Indonesia

Sebastián Espinoza National Institute of Energy Efficiency and Renewable Energy (INER), Quito, Ecuador

Robert N. Farrugia Institute for Sustainable Energy and Department of Mechanical Engineering, University of Malta, Msida, MSD, Malta

A. M. Foley School of Mechanical and Aerospace Engineering, Queen's University Belfast, Belfast, UK

Matt Folley SPACE, Queen's University Belfast, Belfast, Northern Ireland

Eric S Fraga Electrochemical Innovation Lab, Department of Chemical Engineering, University College London (UCL), London, UK

Centre for Process Systems Engineering, Department of Chemical Engineering, University College London (UCL), London, UK

Carwyn Frost School of Engineering, Cardiff University, Cardiff, UK

Junichiro Fukutomi Institute of Science and Technology, The University of Tokushima, Tokushima-city, Japan

Ben Futrell Integrated Design Labs, Energy Performance Laboratory, School of Architecture, College of Arts + Architecture, University of North Carolina at Charlotte, Charlotte, NC, USA

Paola Gallo Department of Architecture DIDA, University of Florence, Florence, Italy

Alberto García Marín School of Architecture, University of Malaga, Málaga, Spain

M. Garoum Laboratoire d'Energétique, Matériaux et Environnement (LEME), Université Mohammed V Rabat-Agdal, EST de Salé, Salé Medina, Maroc

S. K. Ghosh Mechanical Engineering Department, Centre for Quality Management System, Jadavpur University, Kolkata, India

Moa Ziethèn Granlund Royal Institute of Technology, Stockholm, Sweden

Liam Griggs Faculty of Science, Engineering and Computing, Kingston University, London, UK

E. Gruzdevienė Upytė Experimental Station, Lithuanian Research Centre for Agriculture and Forestry, Panevėžys district, Lithuania

Nabil M Guirguis Housing and Building Research Center, Giza, Egypt

Fredrik Gunnarsson Chalmers University of Technology, Gothenburg, Sweden

Alan J. Guwy Sustainable Environment Research Centre (SERC), Faculty of Computing Engineering and Science, University of South Wales, Pontypridd, Mid-Glamorgan, UK

Kristian Gvozdenovic Royal HaskoningDHV, Rotterdam, The Netherlands

Faculty of the Built Environment, University of Technology Eindhoven, Eindhoven, The Netherlands

Hanna Härelind Chalmers University of Technology, Gothenburg, Sweden

Olle Högblom Chalmers University of Technology, Gothenburg, Sweden

Reinaldo Haas Department of Physics, Federal University of Santa Catarina, Florianópolis, Brazil

Reinhard Haas Vienna University of Technology, Vienna, Austria

Jana Hadler Institute of Engineering Design/CAD, University of Rostock, Rostock, Germany

M.S. Hamad Electrical and Control Engineering Department, Arab Academy for Science and Technology & Maritime Transport, Alexandria, Egypt

Mohamed Hamdy Department of Energy Technology, Aalto University, Espoo, Finland

Building Physics and System Unit, Eindhoven University of Technology, Eindhoven, The Netherlands

George Bassili Hanna Housing & Building National Research Center, Cairo, Egypt

Rolf Hansen LM Wind Power, Kolding, Denmark

Ala Hasan VTT Technical Research Centre of Finland, Espoo, Finland

Mahmoud A Hassan Housing and Building Research Center, Giza, Egypt

Rickard Heijl Chalmers University of Technology, Gothenburg, Sweden

Ebtisam K. Heikal Egyptian Petroleum Research Institute, Nasr City, Cairo, Egypt

P. Higgins School of Mechanical and Aerospace Engineering, Queen's University Belfast, Belfast, UK

Lennart Holmgren Termo-Gen AB, Lärbro, Sweden

Angus Hood School of Engineering, Robert Gordon University, Aberdeen, UK

Mamdud Hossain School of Engineering, Robert Gordon University, Aberdeen, UK

Iva Hrgovan Department of Wind Energy, Technical University of Denmark, Kongens Lyngby, Denmark

Jin Hwan Ko Korea institute of Ocean Science and Technology, Ansan, South Korea

Chulsang Hwang Smart Distribution Research Center, Korea Electrotechnology Research Institute, Changwon, Republic of Korea

R. A. Ibrahim Electrical and Control Engineering Department, Arab Academy for Science and Technology & Maritime Transport, Alexandria, Egypt

Kim In-Chul Department of Mechanical Engineering, Graduate School, Korea Maritime and Ocean University, Busan, Korea

Sheikh Zahidul Islam School of Engineering, Robert Gordon University, Aberdeen, UK

Muhammad Azzam Ismail Department of Architecture, University of Malaya, Kuala Lumpur, Malaysia

Piyush Jadhav Department of Engineering Design, IIT Madras, Chennai, Tamilnadu, India

Serm Janjai Department of Physics, Solar Energy Research Laboratory, Silpakorn University, Nakhon Pathom, Thailand

Z. Jankauskienė Upytė Experimental Station, Lithuanian Research Centre for Agriculture and Forestry, Panevėžys district, Lithuania

Jin-Hong Jeon Smart Distribution Research Center, Korea Electrotechnology Research Institute, Changwon, Republic of Korea

Hong Jin School of Architecture, Harbin Institute of Technology, Harbin, China

Jian Kang School of Architecture, Harbin Institute of Technology, Harbin, China

School of Architecture, University of Sheffield, Sheffield, UK

M.C. Katafygiotou Department of Environmental Science and Technology, Cyprus University of Technology, Limassol, Cyprus

Amandeep Kaur Sustainable Environment Research Centre (SERC), Faculty of Computing Engineering and Science, University of South Wales, Pontypridd, Mid-Glamorgan, UK

Nila I Keumala Department of Architecture, Faculty of Built Environment, University of Malaya, Kuala Lumpur, Malaysia

Fouad Khaldi Energetics Applied Physics Laboratory, Department of Material Physics, Faculty of Sciences, Hadj Lakhdar University, Batna, Algeria

Salah A. Khalil Egyptian Petroleum Research Institute, Nasr City, Cairo, Egypt

Eung-Sang Kim Smart Distribution Research Center, Korea Electrotechnology Research Institute, Changwon, Republic of Korea

Gyeong-Hun Kim Smart Distribution Research Center, Korea Electrotechnology Research Institute, Changwon, Republic of Korea

Jihoon Kim Korea institute of Ocean Science and Technology, Ansan, South Korea

Jong-Yul Kim Smart Distribution Research Center, Korea Electrotechnology Research Institute, Changwon, Republic of Korea

Seul Ki Kim Smart Distribution Research Center, Korea Electrotechnology Research Institute, Changwon, Korea

Maria Kolokotroni Brunel University, Uxbridge, UK

T. Koutra Department of Environmental Science and Technology, Cyprus University of Technology, Limassol, Cyprus

Andrew Kozyn School of Engineering, University of Guelph, Guelph, ON, Canada

Stefan Krauter Department of Sustainable Energy Concepts, University of Paderborn, Paderborn, Germany

Kanzumba Kusakana Department of Electrical, Electronic and Computer Engineering, Central University of Technology, Free State, Bloemfontein, South Africa

David Arroyo López-Carro E.T.S. Ingenieros Industriales, Universidad de Valladolid, Valladolid, Spain

Jesús López-Villada National Institute of Energy Efficiency and Renewable Energy (INER), Quito, Ecuador

N. Laaroussi Laboratoire d'Energétique, Matériaux et Environnement (LEME), Université Mohammed V Rabat-Agdal, EST de Salé, Salé Medina, Maroc

F. Lacroux AREVA T&D, Paris La défense cedex, France

Per-Olof Larsson Höganäs AB, Höganäs, Sweden

Jonathan Leaver Department of Civil Engineering, Unitec Institute of Technology, Auckland, New Zealand

Kwang-Soo Lee Korea institute of Ocean Science and Technology, Ansan, South Korea

Evita H. Legowo Swiss German University, Serpong, Indonesia

N. Lemežienė Institute of Agriculture, Lithuanian Research Centre for Agriculture and Forestry, Akademija, Kėdainiai, Lithuania

Ulrich Leopold Resource Centre for Environmental Technologies—Public Research Centre Henri Tudor, Esch-sur-Alzette, Luxembourg

Baizhan Li Key Laboratory of Eco-environments in Three Gorges Reservoir Region, Ministry of Education, Chongqing University, Chongqing, China

National Centre for International Research of Low-Carbon and Green Buildings, Chongqing, China

Ming Li School of Architecture, Harbin Institute of Technology, Harbin, China

Xin-xin Li School of Architecture, Harbin Institute of Technology, Harbin, China

Ž. Liatukas Institute of Agriculture, Lithuanian Research Centre for Agriculture and Forestry, Akademija, Kėdainiai, Lithuania

I. Liaudanskienė Institute of Agriculture, Lithuanian Research Centre for Agriculture and Forestry, Kėdainiai district, Lithuania

C. H Lim Solar Energy Research Institute, Universiti Kebangsaan Malaysia, Bangi, Selangor, Malaysia

Célio Loureiro Cavalcante Curso de pós-graduação em Química, Universidade Federal do Ceará, Fortaleza, Brazil

William David Lubitz School of Engineering, University of Guelph, Guelph, ON, Canada

Mulè Stagno Luciano Institute of Sustainable Energy, University of Malta, Msida, MSD, Malta

Wim Maassen Royal HaskoningDHV, Rotterdam, The Netherlands

Faculty of the Built Environment, University of Technology Eindhoven, Eindhoven, The Netherlands

Jesper Madsen LM Wind Power, Kolding, Denmark

Ch. Maragkos Soft Energy Applications and Environmental Protection Laboratory, Mechanical Engineering Department, Technological Educational Institute of Piraeus, Athens, Greece

M. Markides Department of Environmental Science and Technology, Cyprus University of Technology, Limassol, Cyprus

G. Martinopoulos School of Science and Technology, International Hellenic University, Thessaloniki, Greece

Allan Mason-Jones School of Engineering, Cardiff University, Cardiff, UK

Alessio Mastrucci Resource Centre for Environmental Technologies—Public Research Centre Henri Tudor, Esch-sur-Alzette, Luxembourg

Dipartimento di Ingegneria Civile, Edile e Architettura—Università Politecnica delle Marche, Ancona, Italy

S Mat Solar Energy Research Institute, Universiti Kebangsaan Malaysia, Bangi, Selangor, Malaysia

Ranjana Meena Department of Engineering Design, IIT Madras, Chennai, Tamilnadu, India

Dejan Milenic Faculty of Mining and Geology, Department of Hydrogeology, University of Belgrade, Belgrade, Serbia

Hossein Mirzaii Kingston University, London, UK

Faculty of Science, Engineering and Computing, Kingston University, London, UK

S Moghimi Solar Energy Research Institute, Universiti Kebangsaan Malaysia, Bangi, Selangor, Malaysia

Tathilene Bezerra Mota Gomes Arruda Curso de pós-graduação em Química, Universidade Federal do Ceará, Fortaleza, Brazil

K. Moustris Soft Energy Applications and Environmental Protection Laboratory, Mechanical Engineering Department, Technological Educational Institute of Piraeus, Athens, Greece

Helmut F. O. Mueller Green Building R&D, Duesseldorf, Germany

Navid Nahi Department of Architecture, East Azarbaijan Science and Research Branch, Islamic Azad University, Tabriz, Iran

Lee Nak-Joong Department of Mechanical Engineering, Graduate School, Korea Maritime and Ocean University, Busan, Korea

Ricardo A. Narváez C. National Institute of Energy Efficiency and Renewable Energy (INER), Quito, Ecuador

Centre for Renewable Energy Systems Technology (CREST), Loughborough University, Loughborough, UK

Ilinca Nastase Building Services Department, Technical University of Civil Engineering in Bucharest, Bucharest, Romania

Kanokwan Ngaosuwan Chemical Engineering Division, Engineering Faculty, Rajamangala University of Technology Krungthep, Bangkok, Thailand

J. D. Nixon Faculty of Science, Engineering and Computing, Kingston University, London, UK

E. Norkevičienė Institute of Agriculture, Lithuanian Research Centre for Agriculture and Forestry, Akademija, Kėdainiai, Lithuania

Bubele Papy Numbi Department of Electrical, Electronic and Computer Engineering, University of Pretoria, Pretoria, South Africa

Daphne O'Doherty School of Engineering, Cardiff University, Cardiff, UK

Tim O'Doherty School of Engineering, Cardiff University, Cardiff, UK

Patrick R. Osborne Lee Evans Partnership, London, UK

Anders Palmqvist Chalmers University of Technology, Gothenburg, Sweden

Matti Palonen Department of Energy Technology, Aalto University, Espoo, Finland

Michael de Paly Wilhelm-Schickard-Institute for Computer Science (WSI), University of Tübingen, Tübingen, Germany

Nikos Papamanolis School of Architecture, Technical University of Crete, Chania, Greece

Jin-Soon Park Korea institute of Ocean Science and Technology, Ansan, South Korea

Joachim Pasel Forschungszentrum Jülich GmbH, Institute of Energy and Climate Research, IEK-3: Electrochemical Process Engineering, Jülich, Germany

Júlio C. Passos Department of Mechanical Engineering, Federal University of Santa Catarina, Florianópolis, Brazil

Mechanical Engineering Department, Federal University of Santa Catarina, Florianópolis, Brazil

Ralf Peters Forschungszentrum Jülich GmbH, Institute of Energy and Climate Research, IEK-3: Electrochemical Process Engineering, Jülich, Germany

Lars J. Pettersson Royal Institute of Technology, Stockholm, Sweden

Poorang A.E. Piroozfar Division of Built Environment and Civil Engineering, School of Environment and Technology, University of Brighton, Brighton, UK

Francesco Pomponi Division of Built Environment and Civil Engineering, School of Environment and Technology, University of Brighton, Brighton, UK

Nágila Maria Pontes Silva Ricardo Curso de pós-graduação em Química, Universidade Federal do Ceará, Fortaleza, Brazil

Ruth Potopsingh University of Technology, Kingston, Jamaica

K. Arul Prakash Department of Applied Mechanics, IIT Madras, Chennai, Tamilnadu, India

Giuliano C. Premier Sustainable Environment Research Centre (SERC), Faculty of Computing Engineering and Science, University of South Wales, Pontypridd, Mid-Glamorgan, UK

Stephen Pretlove School of Architecture & Landscape, Kingston University London, Grange Road, UK

Tuyen Quang Le Korea institute of Ocean Science and Technology, Ansan, South Korea

Valeria Ramírez National Institute of Energy Efficiency and Renewable Energy (INER), Quito, Ecuador

Palaniappan Ramu Department of Engineering Design, IIT Madras, Chennai, Tamilnadu, India

A. Rizk Department of Architectural Engineering, Tanta University, Tanta, Egypt

Rosa Romano Department of Architecture DIDA, University of Florence, Florence, Italy

Krisztian Ronaszegi Electrochemical Innovation Lab, Department of Chemical Engineering, University College London (UCL), London, UK

N. Saboohi Department of Environmental Science and Technology, Cyprus University of Technology, Limassol, Cyprus

S. Sagi School of Engineering and Applied Science, Aston University, Birmingham, UK

Yoshiaki Sakagami Department of Health and Service, Federal Institute of Santa Catarina, Florianópolis, Brazil

Department of Mechanical Engineering, Federal University of Santa Catarina, Florianópolis, Brazil

Yoshiaki Sakagami Health and Science Department, Federal Institution of Santa Catarina, Florianópolis, Brazil

Tawfeeq Wasmi M. Salih Department of Mechanical Engineering, Al-Mustansiriyah University, Baghdad, Iraq

E Salleh Solar Energy Research Institute, Universiti Kebangsaan Malaysia, Bangi, Selangor, Malaysia

Remzi Can Samsun Forschungszentrum Jülich GmbH, Institute of Energy and Climate Research, IEK-3: Electrochemical Process Engineering, Jülich, Germany

Tonio Sant Institute for Sustainable Energy and Department of Mechanical Engineering, University of Malta, Msida, MSD, Malta

Pedro A. A. Santos Department of Mechanical Engineering, Federal University of Santa Catarina, Florianópolis, Brazil

Mechanical Engineering Department, Federal University of Santa Catarina, Florianópolis, Brazil

Aliyah N. Z Sanusi Department of Architecture, Kulliyyah of Architecture and Environmental Design, International Islamic University Malaysia, Kuala Lumpur, Malaysia

Francesco Sasso Green Building R&D, Duesseldorf, Germany

Francesca Scalisi Department of Architettura, University of Palermo, Palermo, Italy

J. A. Scott Aston Business School, Aston University, Birmingham, UK

D.K. Serghides Department of Environmental Science and Technology, Cyprus University of Technology, Limassol, Cyprus

Despina K. Serghides Department of Environmental Science and Technology, Cyprus University of Technology, Limassol, Cyprus

Ehsan Shafiei School of Engineering and Natural Sciences, Environment and Natural Resources, University of Iceland, Reykjavik, Iceland

Li Shao Division Institute of Energy and Sustainable Development, University of Reading, Reading, UK

Teng Shao School of Architecture, Harbin Institute of Technology, Harbin, China

Wen Zhong Shen Department of Wind Energy, Technical University of Denmark, Kongens Lyngby, Denmark

Toru Shigemitsu Institute of Science and Technology, The University of Tokushima, Tokushima-city, Japan

Maryam Singery Department of Architecture, Tabriz Branch, Islamic Azad University, Tabriz, Iran

Hermanto Siregar Graduate School of Business Management, Bogor Agricultural University, Bogor, Indonesia

Shaun Slater School of Engineering, Robert Gordon University, Aberdeen, UK

K Sopian Solar Energy Research Institute, Universiti Kebangsaan Malaysia, Bangi, Selangor, Malaysia

Cesare Sposito Department of Architettura, University of Palermo, Palermo, Italy

Bernard H. Stark Faculty of Engineering, University of Bristol, Bristol, UK

M. Stathopoulos Soft Energy Applications and Environmental Protection Laboratory, Mechanical Engineering Department, Technological Educational Institute of Piraeus, Athens, Greece

Francesca Stazi Dipartimento di Ingegneria Civile, Edile e Architettura—Università Politecnica delle Marche, Ancona, Italy

Hlynur Stefansson School of Science and Engineering, Reykjavik University, Reykjavik, Iceland

Detlef Stolten Forschungszentrum Jülich GmbH, Institute of Energy and Climate Research, IEK-3: Electrochemical Process Engineering, Jülich, Germany

RWTH Aachen University, Aachen, Germany

Kazuyoshi Sumi Department of Marine Engineering, Marine Technical College, Hyogo, Japan

Erwin Susanto Sadirsan Bogor Agricultural University, Bogor, Indonesia

Donald Swift-Hook Kingston University, Woking, Surrey, UK

S. N. N. Syed Yahya Department of Architecture, University of Malaya, Kuala Lumpur, Malaysia

Chihiro Tanaka Graduate School of Advanced Technology and Science, The University of Tokushima, Tokushima-city, Japan

Sarah Tatum School of Engineering, Cardiff University, Cardiff, UK

Frederico F. Taves Tractebel Energia S.A. (GDF Suez), Florianópolis, Brazil

Javier Terrados Cepeda School of Architecture, University of Malaga, Málaga, Spain

Sant Tonio University of Malta, Msida, MSD, Malta

Evangelia Topriska Brunel University, Uxbridge, UK

Mridula Tripathi Department of Chemistry, C.M.P. Degree College, University of Allahabad, Allahabad, India

G. Tzanes Soft Energy Applications and Environmental Protection Laboratory, Mechanical Engineering Department, Technological Educational Institute of Piraeus, Athens, Greece

Ioana Udrea Faculty of Mechanical Engineering and Mechatronics, Thermodynamics Department, Polytechnic University of Bucharest, Bucharest, Romania

Herman Jacobus Vermaak Department of Electrical, Electronic and Computer Engineering, Central University of Technology, Free State, Bloemfontein, South Africa

Ana Vranjes Faculty of Mining and Geology, Department of Hydrogeology, University of Belgrade, Belgrade, Serbia

Rungrat Wattan Department of Physics, Solar Energy Research Laboratory, Silpakorn University, Nakhon Pathom, Thailand

Trevor Whittaker SPACE, Queen's University Belfast, Belfast, Northern Ireland

B.W. Williams Electronics and Electrical Engineering Department, Strathclyde University, Glasgow, Scotland, UK

Samuel J. Williamson Faculty of Engineering, University of Bristol, Bristol, UK

Earle Wilson University of Technology, Kingston, Jamaica

D. Wright Aston Business School, Aston University, Birmingham, UK

Jun Yao School of Engineering, University of Lincoln, Lincoln, UK

Yufeng Yao Department of Engineering Design and Mathematics, University of the West of England, Bristol, UK

Lee Young-Ho Division of Mechanical and Energy System Engineering, Korea Maritime and Ocean University, Busan, Korea

Kamil M. Yousif Dept of Environmental Sciences, Zakho University, Zakho, Iraq

Wei Yu Key Laboratory of Eco-environments in Three Gorges Reservoir Region, Ministry of Education, Chongqing University, Chongqing, China

National Centre for International Research of Low-Carbon and Green Buildings, Chongqing, China

D. Zafirakis Soft Energy Applications and Environmental Protection Laboratory, Mechanical Engineering Department, Technological Educational Institute of Piraeus, Athens, Greece

15. Russel R, Davis M (2009) Evaluating a federal feed-in tariff for the United Stated—lessons from Germany. Tufts University
16. Saxena JP, Sushil, Vrat P (1992) Hierarchy and classification of program plan elements using interpretive structural modelling. Syst Pract 5(6):651–670
17. Shahabadkar P, Hebbal SS, Prasahant S (2012) Deployment of interpretive structural modeling methodology in supply chain management—an overview. Int J Ind Eng Prod Res 22(3):195–205
18. Shen L, Liu L, Yao Z, Liu G, Lucas M (2010) Development potentials and policy options of biomass in China. Environ Manage 46:539–554

Chapter 2
Features of Carbon Stock in the Biomass of Industrial Hemp and Stinging Nettle

B. Butkutė, I. Liaudanskienė, Z. Jankauskienė, E. Gruzdevienė, J. Cesevičienė and K. Amalevičiūtė

Abstract Recently, researchers have drawn their attention to industrial hemp (*Canabis sativa* L.) and stinging nettle (*Urtica dioica* L.), as feedstocks, potentially having a wide nonfood application. The aim of the present work was to compare dry matter (DM) and carbon (C) yields as well as C concentration in the above-ground biomass, stems and shives of the mentioned crops. In this chapter, extra attention has been paid to the C accumulation in stems and shives, since stems are a more environmentally friendly resource for solid biofuel compared to the whole above-ground part of the plant, and shives are an agricultural waste.

Field experiments with industrial hemp (eight varieties) and stinging nettle (one wild nettle and two treatments of fibre nettle clone) were carried out during 2010–2012. Dew retting and water retting were used to extract the fibre. C concentration in the samples of hemp and nettle was determined by wet oxidation with dichromate.

DM yield of the above-ground biomass of hemp amounted to an average of 10607 kg ha^{-1}, of stems 9063 kg ha^{-1} with high C concentrations of 555 and 568 g kg^{-1} DM, respectively. DM yield of the nettle declined along with a harvest year and ranged from 11604 kg ha^{-1} (2010) to 5596 kg ha^{-1} (2012) averaging

B. Butkutė (✉) · I. Liaudanskienė · J. Cesevičienė · K. Amalevičiūtė
Institute of Agriculture, Lithuanian Research Centre for Agriculture and Forestry,
Instituto al. 1, Akademija, Kėdainiai district, LT-58344, Lithuania
e-mail: brone@lzi.lt

I. Liaudanskienė
e-mail: inga@lzi.lt

J. Cesevičienė
e-mail: jurgita@lzi.lt

K. Amalevičiūtė
e-mail: kamaleviciute@gmail.com

Z. Jankauskienė · E. Gruzdevienė
Upytė Experimental Station, Lithuanian Research Centre for Agriculture and Forestry,
Linininkų 3, Upytė, Panevėžys district, LT-38294, Lithuania
e-mail: soja@upyte.lzi.lt

E. Gruzdevienė
e-mail: upyte@upyte.lzi.lt

A. Sayigh (ed.), *Renewable Energy in the Service of Mankind Vol I*,
DOI 10.1007/978-3-319-17777-9_2

7589 kg ha^{-1} per trial. DM yield of wild nettle was more than twice as low as that of fibre nettle clone (on average 3945 kg ha^{-1} vs 9411 kg ha^{-1}).

C stock in stems of hemp and nettle amounted to an average of 5149 and 3719 kg ha^{-1}, respectively. DM yield was a weighted factor for C yield.

Shives, which are the woody residue left over from the processing of hemp and nettle straw appeared very rich in C the concentration of which in hemp shives varied in the range of 564–602 g kg^{-1} DM and in nettle shives 543–596 g kg^{-1} DM. The retting method (R) significantly ($P<0.01$) affected the C concentration in nettle shives.

The high heating value (HHV) of biomass, stems and shives of hemp and nettle was determined, and the theoretical accumulation of CO_2 in biomass per ha was calculated.

Results of this study showed that the hemp and fibre clones of stinging nettle could be promising candidates for bioenergy production. The CO_2 content fixed into the biomass of the studied crops might contribute towards the reduction of climate warming.

Keywords Bioenergy plant · Shives · Carbon · Dry matter yield (DMY) · *Canabis sativa* L · *Urtica dioica* L

2.1 Introduction

Today, it is clear that the development of Europe's renewable energy resources is a crucial element in the battle against climate change [1]. The fight against climate change and the depletion of fossil fuels has forced humanity to "decarbonising" the economy [2]. Burning fossil fuels uses "old" biomass and converts it into "new" CO_2, which contributes to the "greenhouse" effect and depletes a non-renewable resource [3]. The need for CO_2 management, in particular capture and storage, is currently an important technological, economical and global political issue and will continue to be so until alternative energy sources and energy carriers diminish the need for fossil fuels [4]. A plant used for biomass energy grows by removing CO_2 from the air through photosynthesis [5]. Carbon (C) accumulated in biomass through CO_2 fixation can be easily converted into usable energy [2]. Biofuel from energy plants is considered at least as "C neutral": CO_2 that is released in burning is returned to the biomass from the atmosphere during photosynthesis and returned for a cycle of new growth [3, 5].

Currently, researchers are focusing attention on hemp (*Canabis sativa* L.) and stinging nettle (*Urtica dioica* L.), as high yielding multipurpose feedstocks. Growing well in Central Europe, both hemp and stinging nettle are promising candidates for nonfood market production: textile, paper, building industries as well as bioenergy [6–11]. Hemp is an annual monoecious or dioecious plant and it is considered as being a crop that requires no pesticides, since it outcompetes weeds, uses little fertiliser or irrigation in temperate areas and potentially causes little land use change since it can be grown on marginal land (though it will require more inputs if not grown on cropland) [12]. Industrial hemp (variety Futura 75) grown in Northern Europe, yielded an average of 14.4 Mg DM ha^{-1} and exhibited satisfactory net energy yields per hectare [9, 13]. The stinging nettle is a common dioecious herbaceous perennial plant that grows on ruderal sites, in gardens, at the edges of forests and in wooded areas of the riverine floodplains [6]. Nettle is practically not grown commercially in Europe [14]. The existing areas cover experimental plots. Several new nettle clones have been selected and characterised by high fibre content and strong tillering [15]. The trials with the clones of fibre nettle have been conducted in Austria, Germany and Italy [8, 16, 17]. The annual dry matter yield (DMY) of nettle in boreal growing conditions ranged from 6–10 Mg ha^{-1} [18] and DMY of stalks of five clones fluctuated from 2.3–9.7 Mg ha^{-1} [16]. Clones of fibre nettle were successfully tested as bioenergy crop, mainly for the purpose of anaerobic digestion [18, 19].

Shives are the woody residue left over after the processing of stalks (hemp, nettle, flax or other fibrous crop) on target for fibre extraction. This agricultural waste could be used as a renewable source for composites, combustion or other forms of bioenergy [20,21].

Since C containing compounds from the biomass generate a principal amount of biomass energy, we devoted our attention to the features of C stock in the above-ground biomass of hemp and stinging nettle and its fractions—stems and shives.

2.2 Material and Methods

A field experiment was carried out in the Central Lowland of Lithuania (55°39′11″N 24°13′59″E) at the Upytė Research Station of the Lithuanian Research Centre for Agriculture and Forestry on an *Eutri-Endohypogleyic Cambisol* (*CMg-n-w-eu*) [22]. The study involving eight monoecious varieties of industrial hemp and three treatments of stinging nettle (one wild nettle and two treatments of fibre nettle clone) was carried out during 2010–2012 in three replications. The hemp variety (V) "USO 31" (of Ukrainian origin) is known as a very early variety. Polish varieties "Beniko" and "Bialobrzeskie" are considered as medium-early in the country of their origin. The other five hemp varieties are French in origin and differ in earliness: "Fedora 17" is early-maturing, "Felina 32" and "Santhica 27" are medium-late maturing, "Epsilon 68" is late-maturing and "Futura 75" is a very late-maturing variety. The seed rate of hemp was 50 kg ha^{-1}, interrow spacing was 10 cm. Hemp was sown at the beginning of May, and harvested when the first mature seeds appeared (in September or October, depending on the year and variety ripening). In all experimental years, the V "USO 31" was harvested 0.5–1 month earlier than the other varieties: on September 9 in 2010, September 13 in 2011 and September 19 in 2012. The rest of the hemp varieties were harvested on October 4, September 22–23 and October 15–16, respectively.

The current study on stinging nettle was a follow-up of the experiment where nettle has been cultivated since 2008. The wild stinging nettle was planted at a density of 60×60 cm and clones of fibre nettle—at two densities: 60×60 and 60×100 cm. Stinging nettle was cut at the end of May—beginning of June for testing biomass of young plants as food ingredient. After regrowth, stinging nettle was harvested when seeds were mature in the lower part of the inflorescence: at the end of August in 2010, on September 12 and 4 in 2011 and 2012, leaving stubble of up to 15 cm height. The biomass harvested at this time was studied as a feedstock for industry (textile and bioenergy).

Since we investigated stinging nettle and hemp as multifunctional plants, the dew retting and water retting were used to extract fibre. Shives obtained after retting were evaluated as a feedstock for solid biofuel.

The growing seasons were abundant in precipitation, whose amount was distributed unevenly over the period, but generally the weather conditions were favourable for hemp and stinging nettle growing.

C concentration in the samples of hemp and nettle was determined by a spectrophotometric procedure after wet oxidation of plant material with dichromate [23, 24]. In this chapter, extra attention has been paid to the C accumulation in stems and shives, since stems are a more environmentally friendly resource for solid biofuel compared to the whole above-ground plant part, and shives are an agricultural waste. Samples of hemp ant nettle stems were analysed for C every year. C in shives was established in the samples of 2010 and 2011 harvest years and in the samples of above-ground mass of 2010 harvest year. C concentration in above-ground biomass of 2011, 2012 harvests was calculated from C concentration in stems using coefficients 0.9774 for hemp and 0.9664 for nettle. These coefficients were obtained from the data of 2010 harvest as a ratio of C in the respective samples of plant above-ground biomass and stems. Samples for the total nitrogen (N) concentration were analysed by the Kjeldahl method. The ash content was determined by the method consistent with LST EN 14775 [25] with mass incineration at (550 ± 10) °C. Klason lignin was analysed using the procedure NREL LAP 003 [26]. Gross calorific value (GCV) or high heating value (HHV) was measured using an IKA bomb calorimeter (C 200, Germany) following the LST EN 14918 [27]. Around 1 g of biomass was pelletised and introduced in the bomb, which was charged with pure oxygen (>99.99 %) to a pressure of 3.0 ± 0.2 MPa. The bomb had previously been calibrated with benzoic acid.

Two- and three-way ANOVA [28] with a three-replication design was performed on the data to determine the significance of the following factors: hemp varieties/nettle treatment V/T: (V/T; for above ground biomass, stems and shives), harvest year Yr: (Yr; for above ground biomass, stems and shives), retting method (R; for shives) on dry matter (DM) yield, C concentration and yield.

2.3 Results and Discussion

According to the data of two-way ANOVA, the DMY both of hemp above-ground biomass (further biomass) and stems, as well as C output in biomass and stems significantly ($P<0.01$ or $P<0.05$) depended on V and did not depend on the growing year (Yr; Table 2.1). Conversely, the effect of the Yr was significant ($P<0.01$), and statistically insignificant varietal impact on C concentration (C) in the biomass and stems was revealed. The effect of interaction of the V and Yr on DMY, C yield (CY) as well C in biomass was negligible. Factors' interaction was significant for C in stems (at $P<0.05$) only.

Mean biomass DMY per trial amounted to 10607 kg ha^{-1} and stems DMY to 9065 kg ha^{-1}. DMY both of stems and biomass of early-maturing varieties USO 31 and Fedora 17 were significantly lower than mean per trial. One of the most productive varieties was the latest-maturing variety Futura 75. Polish hemp varieties Beniko and Bialobrzeskie were also high yielding though they are ranked as medium early-maturing. The same patterns were obtained regarding CY accumulated in biomass and stems. As C concentration variation with variety was inappreciable DMY was a weighted factor for calculated CY. With regard to the Yr, which is an entireness of environmental conditions of the hemp growth period, lower than average per trial DMY and CY were obtained in 2010 and 2012, in contrast to 2011 where it was higher yielding, though differences from the mean were statistically insignificant. It was observed that hemp biomass, and particularly stems, contained high C concentration (on average 552 and 568 g kg^{-1} DM). The highest (at $P<0.01$) C was found in biomass and stems (558 and 573 g kg^{-1} DM) of the 2012 yield, when the lowest average DMY was recorded and vice versa. Some differences in temperature and precipitation distribution during the growing seasons of hemp and plant biomass maturity at harvesting could have an impact on C concentration: The majority of hemp varieties were harvested 2–3 weeks earlier in 2011 than in 2010 and 2012.

All the parameters of the stinging nettle presented in Table 2.2 were significantly ($P<0.01$) impacted by Yr. Nettle treatment (T; at $P<0.01$) and $T \times Yr$ interaction (at $P<0.05$ mostly) were weighty factors for DMY both of biomass and stems as well as for the respective CY. However, C in both biomass and stems was T and $T \times Yr$ independent parameter. Biomass DMY of wild nettle was more than twice as low as that of the clone of fibre nettle (on average 3945 kg ha^{-1} vs 9194 and 9629 kg ha^{-1}). Such large distinction between clones of fibre nettle and wild ecotype was observed in each year of the study and not only in DMY of biomass, but in DMY of stems as well as in CY from biomass and stems. Differences in both DM and CY between treatments of fibre nettle grown in plots of diverse density were inappreciable with a trend of higher values in the sparser plots (60×100 cm). Biomass and stems of nettle plants had high C (528 and 546 g kg^{-1} DM, respectively) and CY (4125 and 3719 kg ha^{-1}); although these values were lower in the respective fractions of hemp. C stock of clones of fibre nettle amounted to 4882–5389 kg ha^{-1} in biomass and to 4579–4810 kg ha^{-1} in stems. DMY of the nettle declined along with a harvest year and ranged from 11,604 kg ha^{-1} (2010)–5596 kg ha^{-1} (2012)

Table 2.1 Summary of two-way ANOVA for dry matter yields (DMY), C concentration (C) and C yield (CY) in above-ground biomass and stems of industrial hemp

Variable	Biomass DMY kg ha^{-1}	Stems DMY kg ha^{-1}	C in biomass g kg^{-1} DM	C in stems g kg^{-1} DM	CY with biomass kg ha^{-1}	CY with stems kg ha^{-1}
	Significance of the factor					
Variety (V)	*	**	Ns	ns	*	**
Growing year (Yr)	ns	ns	**	**	ns	ns
V × Yr	ns	ns	Ns	*	ns	ns
Mean value per trial	10607	9065	552	568	5866	5149
Variety	*Average value for variety*					
Beniko	11848*	10471**	551	566	6533*	5926**
Bialobrzeskie	11435	9953	555	570	6356	5673
Epsilon 68	9994	8534	548	563	5483	4803
Fedora 17	9272*	7777	554	569	5148*	4430*
Felina 32	10498	8956	554	569	5816	5089
Futura 75	11627	10193*	555	570	6457	5809*
Santhica 27	10877	9085	550	565	5993	5135
USO 31	9303*	7548**	552	573	5142*	4329**
P_{05}	*1110*	*950*	*5.93*	*6.12*	*623*	*549*
P_{01}	*1482*	*1268*	*7.92*	*8.17*	*832*	*733*
Growing year	*Average value for year*					
2010	10479	8991	553	570	5814	5122
2011	11164	9491	546**	561**	6100	5328
2012	10178	8712	558**	573**	5683	4998
P_{05}	*593*	*508*	*3.17.*	*3.27*	*333*	*293*
P_{01}	*792*	*678*	*4.23*	*4.37*	*444*	*392*

ns not statistically significant, significance of differences evaluated from mean
* significant at $P<0.05$ ** significant at $P<0.01$

Table 2.2 Summary of two-way ANOVA for dry matter yield (DMY), C concentration (C) and C yield (CY) in above-ground biomass and stems of stinging nettle

Variable	Biomass DMY kg ha^{-1}	Stems DMY kg ha^{-1}	C in biomass g kg^{-1} DM	C in stems g kg^{-1} DM	CY in biomass kg ha^{-1}	CY in stems kg ha^{-1}
	Significance of the factor					
Treatment (T)	**	**	ns	ns	**	**
Growing year (Yr)	**	**	**	**	**	**
T × Yr	*	*	ns	ns	*	**
Mean value per trial	7589	6715	528	546	4125	3719
Nettle treatment	*Average value for treatment*					
Wild nettle	3945**	3194**	524	542	2103**	1767**
Fibre nettle, 60 × 60 cm	9194**	8305**	526	545	4882**	4579**
Fibre nettle, 60 × 100 cm	9629**	8646**	533	552	5389**	4810**
P_{05}	*982*	*898*	*8.09*	*8.37*	*473*	*487*
P_{01}	*1353*	*1237*	*11.14*	*11.54*	*652*	*671*
Growing year	*Average value for year*					
2010	11604**	10979**	550**	569**	6584**	6234**
2011	5567**	4633**	528	546	2945**	2537**
2012	5597**	4533**	505**	523*	2845**	2385**
P_{05}	*982*	*898*	*8.09*	*8.37*	*473*	*487*
P_{01}	*1353*	*1237*	*11.14*	*11.54*	*652*	*671*

ns not statistically significant, significance of differences evaluated from mean
* significant at $P < 0.05$ ** significant at $P < 0.01$

averaging 7589 kg ha^{-1} per trial and 9412 kg ha^{-1} per clones of fibre nettle (Table 2.2). As reported by Vogl and Hartl [6] as well as Harwood and Edom [29] stinging nettle can be cultivated for 4 and more years. It is likely that in our study a sharp decrease in DMY and other related parameters in 2011 occurred due to the fact that after harvesting in 2010 before winter (November 15) the aftermath was cut. On the other hand, yielding potential of plants could fall with the harvesting year: 2011 was a third harvesting year of nettle.

Although, according to the most discussed features, stinging nettle conceded to industrial hemp, our results showed nevertheless that the annual C production per above-ground biomass and stems was distinctly higher for clones of fibre stinging nettle (5135 and 4695 C ha^{-1} yr^{-1} on average) than that of the mature forests [30]. Gower et al. [30] reported that the above-ground net primary C production of the mature forests in Canada ranged from 3490–3520 kg C ha^{-1} yr^{-1} for aspen stands to 1170–1220 kg C ha^{-1} yr^{-1} for jack pine stands. Consequently C stocks in the biomass and stems of wild nettle (2103 and 1767 C ha^{-1} yr^{-1}) were lower than annual C accumulated per ha for aspen stands; however exceeded those for jack pine stands. Our results also showed that C production per above-ground biomass and stems of hemp (5866 and 5149 kg C ha^{-1} yr^{-1} on average) noticeably exceeded that of mature forests. To accumulate these C quantities in above-ground biomasses, plants of hemp and clones of fibre nettle had utilised from atmosphere on average 21,509 and 18,828 kg CO_2 ha^{-1} yr^{-1}, respectively, (CO_2 content needed for stubble and roots biomass not included). So theoretical calculation shows, that fixation of CO_2 into biomasses of industrial hemp and clone of fibre nettle might contribute towards reducing its accumulation in the atmosphere. Our results support the conclusion of Finnan and Styles [31] that hemp could considerably reduce greenhouse gas emission. The same is true for stinging nettle. This chapter discusses the C stock in a part of the annual yield of nettle above-ground biomass and stems only, that is, in biomass accumulated during the short summer period of approximately 3 months. Therefore in order to calculate the total annual C stock, one should sum the above described C yield with C stock in biomass, cut at the end of May–beginning of June and aftermath for stinging nettle as food ingredient, as well in stubbles and roots. It is noteworthy that, first, nettle is a perennial crop with root biomass exceeding that of annuals [32] and second, nettle fields could be expected to remain productive for several years with low labour costs which will positively influence the economic viability of nettle biomass production [29].

Three-way analysis of variance (ANOVA), applied to reveal the significance of factors for shives percentage from stems, shives DMY, C and CY in shives was carried out according to the following scheme: A factor R, B factor V or nettle treatment (T), C factor harvest year (Yr; Table 2.3). ANOVA was used for hemp and stinging nettle separately. The R was significant at $P<0.01$ for shives output from stems and C concentration in nettle shives only. Yr and V or nettle treatment were factors that impacted output, DMY, CY of shives both of hemp and stinging nettle. All the tested factors and Yr interaction with both R and T were significant ($P<0.01$) for nettle shives output. Average per trial output of shives from nettle stems (60.5%) was markedly higher than that from hemp stems (45.7%). Water-retted stems provided lower percentage of shives (45.7 and 57.0%, respectively for hemp and nettle), than dew-retted stems (46.2 and 64.0%, respectively). Output of shives had a direct impact on DMY and CY in shives: DMY and CY for nettle shives were higher than for hemp shives, likewise DMY and CY for dew-retted shives were higher than for water-retted shives. The contribution of DMY of stems in combination with shives output for DMY and C stock in shives it cannot be disregarded.

As for C stock, it should be noted that C accumulated in shives of clones of fibre nettle was equal to that of the mature forests for aspen stands, CY in shives of wild nettle was close to annual C production per jack pine stands and CY in shives of hemp varieties took up an intermediate position (Table 2.3) and [30]. So, for C stock that was found in the discussed agricultural waste, the target plants consumed large quantities of atmospheric CO_2: 9068 and 9739 kg ha^{-1} on an average for hemp and stinging nettle shives, respectively.

Average HHV, concentration of components, which are important to characterise hemp and nettle biomass, stems and shives as a source for solid biofuel are presented in Table 2.4.

Table 2.3 Summary of three-way ANOVA for shives output from stems and DMY, C concentration (C) and C yield (CY) of hemp and nettle shives

Variable	Shives output % from stems		Shives DMY kg ha^{-1}		C in shives g kg^{-1}		CY in shives kg ha^{-1}	
	Hemp	Nettle	Hemp	Nettle	Hemp	Nettle	Hemp	Nettle
Factor	*Significance of the factor*							
Retting method (R)	ns	**	ns	ns	ns	**	ns	ns
Hemp variety (V)/Nettle treatment (T)	**	**	**	**	ns	ns	**	**
Growing year (Yr)	**	**	**	**	ns	**	**	**
R × V/T	ns	ns	ns	ns	*	ns	ns	ns
R × Yr	ns	**	ns	ns	ns	**	ns	ns
V/T x Yr	**	**	**	**	ns	ns	**	**
R × V/T × Yr	ns	ns	ns	ns	*	ns	ns	ns
Mean value per trial	45.7	60.5	4222	4595	586	579	2473	2656
Retting method	*Average value for retting method*							
Dew retting	46.2	64.0**	4245	4830	588	566**	2494	2741
Water retting	45.2	57.0**	4198	4360	584	591**	2452	2571
P_{05}	*0.960*	*1.11*	*158*	*353*	*2.37*	*5.02*	*92.6*	*212*
P_{01}	*1.28*	*1.51*	*210*	*479*	*3.15*	*6.82*	*123*	*289*
V/Nettle treatment	*Average value for* V*/nettle treatment*							
Beniko	37.5**		3861		584		2250	
Bialobrzeskie	44.9		4494		587		2634	
Epsilon 68	47.9		4259		585		2486	
Fedora 17	44.9		3247**		585		1898*	
Felina 32	46.8		4670		584		2720*	
Futura 75	44.4		4655		583		2715	
Santhica 27	45.1		4266		593*		2535	
USO 31	53.9**		4323		589		2546	
P_{05}	*2.54*		*418*		*6.27*		*245*	

Table 2.3 (continued)

Variable	Shives output % from stems		Shives DMY kg ha^{-1}		C in shives g kg^{-1}		CY in shives kg ha^{-1}	
	Hemp	Nettle	Hemp	Nettle	Hemp	Nettle	Hemp	Nettle
P_{01}	*3.38*		*555*		*8.33*		*326*	
Wild nettle		62.8**		2056**		579		1200**
Fibre nettle, 60 × 60 cm		57.4**		5429**		579		3162**
Fibre nettle, 60 × 100 cm		61.3		6299**		578		3606**
P_{05}		*1.57*		*499*		*7.10*		*300*
P_{01}		*2.14*		*678*		*9.65*		*408*
Growing year	*Average value for* V/*nettle treatment*							
2010	35.8**	56.6**	3208**	6245**	588	586**	1886**	3657**
2011	55.5**	64.4**	5235**	2945**	585	571**	3060**	1654**
P_{05}	*0.960*	*1.11*	*158*	*353*	*2.37*	*5.02*	*92.6*	*212*
P_{01}	*1.28*	*1.51*	*210*	*479*	*3.15*	*6.82*	*123*	*289*

ns not statistically significant, significance of differences evaluated from mean
* significant at $P < 0.05$ ** significant at $P < 0.01$

Table 2.4 Averaged data for solid fuel quality-related components in biomass and it fractions of hemp and stinging nettle

Biomass fraction	Hemp				Stinging nettle			
	HHV	Ash	Lignin	N	HHV	Ash	Lignin	N
	MJ kg−1 DM	g kg−1 DM			MJ kg−1 DM	g kg−1 DM		
Whole above ground plant part	18.1	67.1	175	5.70	18.2	99.8	165	15.6
Stems	18.5	37.1	177	4.59	18.3	74.3	171	9.70
Shives, dew retting	19.5	21.2	207	3.09	18.7	53.2	208	6.68
Shives, water retting	19.5	12.8	204	3.13	19.1	20.5	217	5.06

A problem that herbaceous energy crops pose during combustion is the ash content: high ash amounts can cause slagging. High concentration of N in combusting biomass can promote greenhouse gas NO_x emissions, and lignin is associated with HHV [33]. All fractions of hemp and nettle showed good results of HHV: from 8.1–18.2 MJ kg^{-1} DM of whole above-ground plant part biomass to 19.1–19.5 MJ kg^{-1} DM of shives. Ash, N contents in both hemp and nettle declined, whereas HHV and lignin increased in the following order: above-ground plant part—stems—shives. From the data presented in Table 2.4, one can see that combustion properties (lower ash and N contents and higher HHV) of hemp fractions seem to be more valuable for solid biofuel than those of nettle. Therefore, our results corroborate an argument of Finnan and Styles [31] that hemp is a sustainable annual crop for climate and energy policy.

Figure 2.1 shows the linear relationships between averaged for fractions HHV and C as well as lignin concentration values. Both parameters of biomass quality positively at $P<0.01$ correlated with the HHV. Demirbaş [33] showed also that the HHV of lignocellulosic fuels is highly correlated with lignin content.

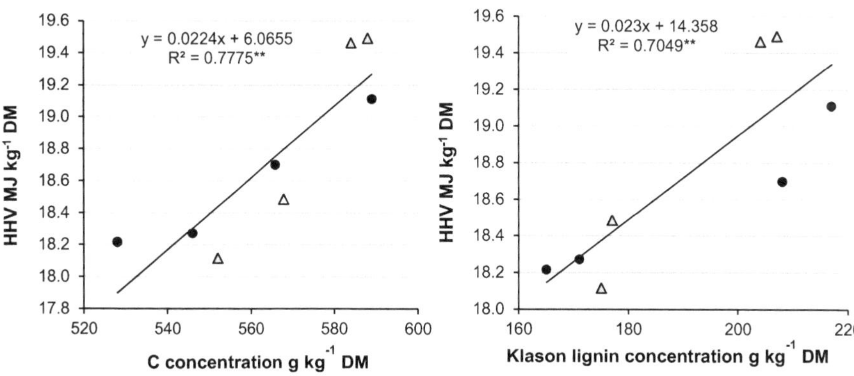

Fig. 2.1 The impact of C and Klason lignin concentration on high heating value (HHV) of biomass. Data averaged for fractions over all hemp varieties/nettle treatments, growing year (Yr) and replications, but separately for hemp (▲) and stinging nettle (•) as well as for water- and dew-retted shives

2.4 Conclusions

Hemp and clones of fibre stinging nettle could be promising candidates for bioenergy production. Annual C production per above-ground biomass and stems was distinctly higher for hemp (5866 and 5149 kg C ha^{-1} yr^{-1} on an average) and clones of fibre stinging nettle (5135 and 4695 kg C ha^{-1} yr^{-1} on an average) compared with mature forests. According to HHV, C, lignin concentration and other solid biofuel-related parameters shives were revealed to be the most valuable fraction of both crops—hemp and stinging nettle. When comparing the two crops, hemp fractions showed better properties for solid biofuel purpose than nettle. The CO_2 content fixed into biomass of studied crops might contribute towards reduction of climate warming.

Acknowledgements The article presents research findings obtained through the long-term programme "Biopotential and Quality of Plants for Multifunctional Use" implemented by Lithuanian Research Centre for Agriculture and Forestry.

References

1. Communication from the Commission to the Council and the European Parliament (2009) The renewable energy progress report: commission report in accordance with Article 3 of Directive 2001/77/EC, Article 4(2) of Directive 2003/30/EC and on the implementation of the EU Biomass Action Plan, COM (2005)628. COM;192 final. http://eur-lex.europa.eu/LexUriServ/LexUriServ.do?uri=COM:2009:0192:FIN:EN:PDF. Accessed 15 June 2014
2. Wyman CE (1994) Alternative fuels from biomass and their impact on carbon dioxide accumulation. Appl Biochem Biotechnol 45/46:897–915
3. McKendry P (2002) Energy production from biomass (part 1): overview of biomass. Bioresource Technol 83(1):37–46
4. Skjånes K, Lindblad P, Muller J (2007) BioCO_2? A multidisciplinary, biological approach using solar energy to capture CO_2 while producing H_2 and high value products. Biomol Eng 24(4):405–413
5. Field CB, Campbell JE, Lobell DB (2008) Biomass energy: the scale of the potential resource. Trends Ecol Evol 23(2):65–72
6. Vogl CR, Hartl A (2003) Production and processing of organically grown fiber nettle (*Urtica dioica* L.) and its potential use in the natural textile industry: a review. Am J Alternative Agr 18(3):119–128
7. Karus M, Vogt D (2004) European hemp industry: cultivation, processing and product lines. Euphytica 140:7–12
8. Bacci L, Baronti S, Predieri S, Di Virgilio N (2009) Fiber yield and quality of fiber nettle (*Urtica dioica* L.) cultivated in Italy. Ind Crop Prod 29(2–3):480–484
9. Prade T, Svensson SE, Andersson A, Mattsson JE (2011) Biomass and energy yield of industrial hemp grown for biogas and solid fuel. Biomass Bioenerg 35(7):3040–3049
10. Kolarikova M, Havrland B, Ivanova T (2013) Energy balance of hemp (*Cannabis sativa* L.) grown for energy purposes. Agric Trop Subtrop 46(1):10–15
11. Jankauskienė Z, Gruzdevienė E (2013) Physical parameters of dew retted and water retted hemp (*Cannabis sativa* L.) fibres. Zemdirb Agric 100(1):71–80
12. Glew DW (2012) Validating low carbon bio-renewable alternatives to petrochemicals: exploring end of life impacts. PhD thesis, University of York http://etheses.whiterose.ac.uk/4641/1/David%20Glew%20PhD%20Thesis%20for%20print%20rev2.pdf. Accessed 15 June 2014

13. Prade T, Svensson SE, Mattsson JE (2012) Energy balances for biogas and solid biofuel production from industrial hemp. Biomass Bioenerg 40:36–52
14. Heller K, Baraniecki P, Talarzyck M, Christou M, Alexopoulou E, Monti A et al (2012) WP1. Non-food crops. In: non-food crops-to-industry schemes in EU27 http://www.crops2industry.eu/images/pdf/pdf/D1.2_INFMP_final.pdf. Accessed 15 June 2014
15. Zimniewska M, Wladyka-Przybylak M, Mankowski J (2011) Cellulosic bast fibers, their structure and properties suitable for composite applications. In: Kalia S, Kaith BS, Kaur I (eds) Cellulose fibers: bio- and nano-polymer composites—green chemistry and technology. Springer, Germany, p 97–119
16. Hartl A, Vogl CR (2002) Dry matter and fiber yields, and the fiber characteristics of five nettle clones (*Urtica dioica* L.) organically grown in Austria for potential textile use. Am J Altern Agr 17(4):195–200
17. Fischer H, Werwein E, Graupner N (2012) Nettle fibre (*Urtica dioica* L.) reinforced poly (lactic acid): a first approach. J Compos Mater 46(24):3077–3087
18. Lehtomäki A, Viinikainen TA, Rintala JA (2008) Screening boreal energy crops and crop residues for methane biofuel production. Biomass Bioenerg 32(6):541–550
19. Braun R, Weiland P, Wellinger A (2008) Biogas from energy crop digestion. In: IEA bioenergy task 37 p 1–20. http://www.iea-biogas.net/files/daten-redaktion/download/energycrop_def_Low_Res.pdf. Accessed 21 July 2014
20. Mankowski J, Kolodziej J (2008) Increasing heat of combustion of briquettes made of hemp shives. In: Proceedings of international conference on flax and other bast plants p 344–352
21. Balčiūnas G, Vėjelis S, Vaitkus S, Kairytė A (2013) Physical properties and structure of composite made by using hemp hurds and different binding materials. Procedia Eng 57:159–166
22. FAO-UNESCO (1997) Soil map of the world: revised legend with corrections and updates. Technical Paper 20. ISRIC Wageningen, Netherlands
23. Bremner JM, Jenkinson DS (1960) Determination of organic carbon in soil. I. Oxidation by dichromate of organic matter in soil and plant materials. J Soil Sci 11(2):394–402
24. Slepetiene A, Butkute B (2003) Use of a multichannel photometer (Multiskan MS) for determination of humic materials in soil after their dichromate oxidation. Anal Bioanal Chem 375:1260–1264
25. LST EN 14775 (2010) Solid biofuels—determination of ash content. Lithuanian Standards Board, Vilnius
26. Templeton D, Ehrman T (1995) Determination of acid-insoluble lignin in biomass. In: Standard Biomass Analytical Methods of National Renewable Energy Laboratory. LAP-003:1-13 http://infohouse.p2ric.org/ref/40/39182.pdf. Accessed 21 July 2014
27. LST CEN/TS 14918 (2006) Solid biofuels—method for the determination of calorific value. Lithuanian Standards Board, Vilnius
28. Tarakanovas P, Raudonius S (2003) The statistical analysis of agronomic research data using the software programs *ANOVA*, *STAT*, *SPLIT-PLOT* from package *SELEKCIJA* and *IRRI-STAT*. Lithuanian University of Agriculture, Kaunas (in Lithuanian)
29. Harwood J, Edom G (2012) Nettle fibre: its prospects, uses and problems in historical perspective. Text Hist 43(1):107–119
30. Gower ST, Vogel JG, Norman JM, Kucharik CJ, Steele SJ, Stow TK (1997) Carbon distribution and above-ground net primary production in aspen, jack pine, and black spruce stands in Saskatchewan and Manitoba, Canada. J Geophys Res 102(D24):29029–29041
31. Finnan J, Styles D (2013) Hemp: a more sustainable annual energy crop for climate and energy policy. Energ Policy 58:152–162
32. Hakala K, Keskitalo M, Eriksson C, Pitkänen T (2008) Nutrient uptake and biomass accumulation for eleven different field crops. Agric Food Sci 18(3–4):366–387
33. Demirbaş A (2003) Relationships between heating value and lignin, fixed carbon, and volatile material contents of shells from biomass products. Energ Source 25(7):629–635

Chapter 3
Obtaining Long-Chain Esters with Lubricant Properties from Sesame Biomass *(Sesamum indicum)*

Tathilene Bezerra Mota Gomes Arruda, Francisco Eduardo Arruda Rodrigues, Manoel Barbosa Dantas, Solange Assunção Quintella, David Thomas Duarte Arruda, Célio Loureiro Cavalcante Jr and Nágila Maria Pontes Silva Ricardo

Abstract Biolubricants are defined like long-chain esters that have lubricant properties and can be obtained from animal or vegetal resources. In this work, the obtainment of the biolubricant by the transesterification of the sesame methyl and ethyl biodiesel in the presence of polyalcohol (trimethylolpropane—TMP) is reported. To conduct the experiment, a distillation-fractioned system was adapted. The reaction conditions were: 3/1 biodiesel/TMP molar ratio, 2 wt% of sodium methoxide catalyst, and reaction temperature and time 110 °C and 6 h respectively. Physicochemical properties like acid value, iodine value, kinematic viscosity at 40 °C, and density at 20 °C were determined. The products were characterized by infrared spectroscopy and ^{1}H and ^{13}C nuclear magnetic resonance (NMR). The thermogravimetric analysis (TGA) of the bioproducts was realized and the main thermal events were evaluated.

Keywords Biolubricant · Sesame · *Sesamum indicum* · Trimethylolpropane (TMP)

3.1 Introduction

In the current context of the global scenario, the energy resources are an important and basic factor for the economic development of any nation. Sectors such as industries, transport, and domestic needs of the people are directly impacted by it. We still live in a world where the fossil fuels, such as petroleum, natural gas, and

T. B. M. G. Arruda (✉) · S. A. Quintella · D. T. D. Arruda · C. L. Cavalcante · N. M. P. S. Ricardo
Curso de pós-graduação em Química, Universidade Federal do Ceará, Fortaleza, Brazil
e-mail: tathilene@gmail.com

F. E. A. Rodrigues · M. B. Dantas
Curso de Licenciatura em Química, Instituto Federal da Paraíba—Campus Sousa, Fortaleza, Brazil

A. Sayigh (ed.), *Renewable Energy in the Service of Mankind Vol I*,
DOI 10.1007/978-3-319-17777-9_3

coal, are the main resources of energy. It needs millions of years to be renewed and is therefore considered nonrenewable and with this the search for the new energy resources becomes of vital importance. The use of biomass as an alternative energy resource can be mentioned. Biomass is any organic material that can be converted to any energy form. The biomass has several advantages compared with fossil resources, such as it is renewable, has low cost of obtaining, allows the reuse of low-cost waste, provides more stability to carbon cycle, and is less polluting. Esters obtained by transesterification of methyl or ethyl esters in the presence of polyalcohols are denominated as impeded esters. These biolubricants have higher thermal and oxidative stability compared to other types of biolubricants. This stability is attributed to the absence of the secondary hydrogen in the β position and the presence of the central quaternary carbon [1, 2, 5]. The trimethylolpropane (TMP) has characteristics that make it a good choice for production of the environmentally acceptable biolubricants, such as moderate market price and low melting point in comparison with other polyalcohols [3].

The sesame *(Sesamum indicum)* is a plant originated in the Orient. Its cultivation has adapted well to the harsh climate of the Brazilian Northeast. It belongs to family of pedaliaceae and its seeds provide a yield of up to 50 % in oil. Its seeds are rich in antioxidant substances, such as the sesamin and the sesamolin, and types of lignans [6]. These substances provide a good thermal and oxidative stability to biodiesel obtained by the transesterification of the sesame oil. For example, the biodiesel production by homogeneous basic catalysis of the sesame oil obtained by chemical extraction with hexane is reported and the biodiesel showed physicochemical properties in accordance with American Society for Testing and Materials (ASTM) and European Standards (ENs) [4].

3.2 Material and Methods

3.2.1 Material

The seeds were cultivated in São Gonçalo's irrigated area, localized in Sousa city, Paraíba, Brazilian Northeast. The sesame oil was obtained by mechanical extraction of raw seeds in a filter press. The other additional resources were: methyl alcohol P.A. from Synth, potassium hydroxide powder from Synth (85 %), anhydrous sodium sulfate from Vetec, trimethylolpropane from Sigma-Aldrich (98 %), sodium methoxide P.A. from Merck.

3.2.2 Analytical Methods

The composition of the sesame oil was determined from the gas chromatography-mass spectrometry (GC-MS). A gas chromatograph linked to mass spectrometer

(Shimadzu QP-5050) was utilized. Thermogravimetric (TG) curves were obtained using a simultaneous Thermogravimetry/Diferencial Thermogravimetry (TG/DTG) analyzer (Q50 V20, TA Instruments) in synthetic air, in the 30–600 °C temperature range at a heating rate of 10 °C/min. ^{1}H and ^{13}C NMR data were recorded using a Bruker® DRX-500 spectrometer operating at 500 MHz of frequency and using tetrahydrofuran (THF) as solvent. Infrared data were recorded using a FT-IR Espectrum 1000 Perkin Elmer® spectrometer, with a liquid film in between KBr cells. Kinematic viscosity measurements were determined at 40 °C using Schott AVS 350 viscometer with Schott CT52 temperature controller. All viscosity measurements were made in quintuplicate and the average value was reported.

3.2.3 Experimental Procedure

3.2.3.1 Synthesis of Sesame Biodiesel by Methyl/Ethyl Route

Crude oil of sesame, denominated T0, was transesterified using sodium hydroxide as catalyst in the presence of methyl and ethyl alcohol. T0 was kept under stirring for 1 h for methyl route and 2 h for ethyl route at ambient temperature. The conditions of reaction for methyl route were: 20 wt% of methyl alcohol and 1 wt% of catalyst. For ethyl route, the conditions were: 20 wt% of ethyl alcohol and 1 wt% of catalyst. The biodiesel obtained from methyl and ethyl route were denominated T1 and T2 respectively.

3.2.3.2 Synthesis of Biolubricants

The transesterification of T1 and T2 in the presence of polyalcohol (biodiesel/TMP molar ratio 3:1) was conducted using an adapted distillation system. It utilized 2 wt% of sodium methoxide as a catalyst. The system was kept under stirring at 110 °C and constant pressure of − 1 mm bar for 6 h. The biolubricant obtained from T1 and T2 were denominated T3 and T4 respectively (Fig. 3.1).

3.3 Results and Discussion

3.3.1 Bioproducts Characterization

Acid value and iodine value were obtained by titration. Kinematic viscosity at 40 °C and density at 20 °C were also determined. The results are compiled in Table 3.1.

The presence of –C=O stretching at 1745 cm^{-1} in case of T1, 1738 cm^{-1} in case of T2, 1740 cm^{-1} in case of T3, and 1730 cm^{-1} in case of T4 as observed in the infrared (IR) spectra (Fig. 3.2), confirmed the ester functionalities in all the compounds.

Fig. 3.1 System utilized for the biolubricant production

Table 3.1 Physicochemical properties of the biocompounds

	T0	T1	T2	T3	T4
Acid value (mgKOH/g)	0.97	0.43	0.46	0.455	0.422
Iodine value (mg I_2/g)	45.01	38.86	66.6	23.69	86.67
Kinematic viscosity at 40 °C (mm^2/s)	34.58	7.39	6.81	10.32	14.72
Density at 20 °C (g/cm^3)	0.9155	0.8862	0.8987	0.9430	0.9030

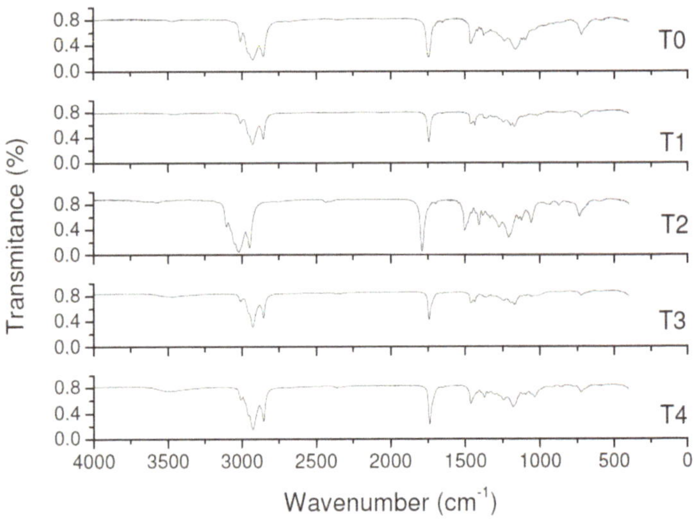

Fig. 3.2 IR spectra for the biocompounds

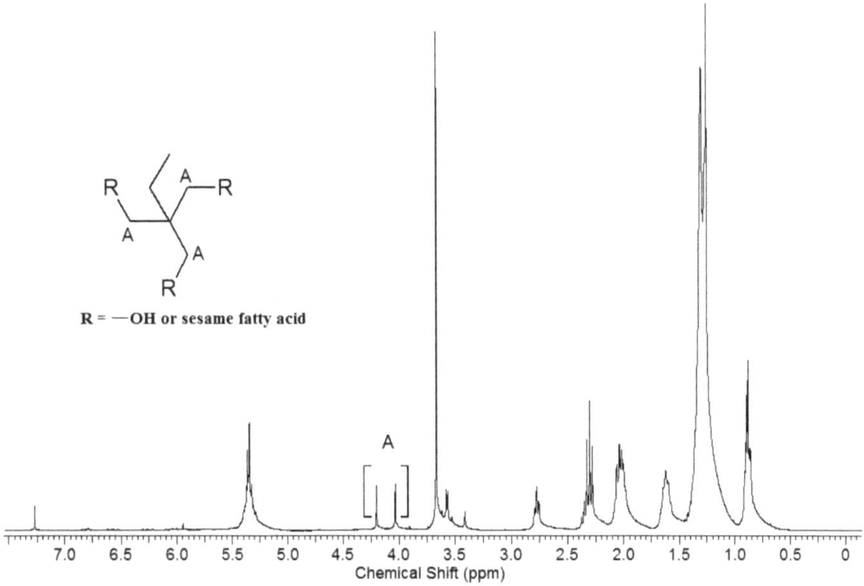

Fig. 3.3 ^{1}H Nuclear Magnetic Resonance (NMR) for T3

It was also observed –C=C– stretching around 721 and 724 cm^{-1}. The peaks were at 1100 cm^{-1} in case of T1, 1096 cm^{-1} in case of T2, 1062 cm^{-1} in case of T3, and 1070 cm^{-1} in case of T4 were attributed to ester –C–O–C– stretching.

The obtaining of the long-chain esters was confirmed by ^{1}H and ^{13}C NMR spectra analysis. The presence of peaks at δ4.03 and δ4.20 ppm for T3 and peaks at δ4.06 and δ4.21 ppm for T4 observed in ^{1}H NMR (Figs. 3.3 and 3.4) spectra confirmed the substitution of the polyalcohol in different degrees. In ^{13}C NMR (Figs. 3.5 and 3.6), signals at δ7.39 ppm for T3 and δ7.48 ppm for T4 corresponding to carbon B and signals at δ22.8 ppm for T3 and δ22.5 ppm for T4 corresponding to carbon A of the generic structure of the long-chain ester was observed. It was also observed that signals at δ174.37 ppm for T3 and δ174.03 ppm for T4 corresponding to carbonyl confirmed the obtaining of long-chain ester.

The TG curves for T3 and T4 are shown in Fig. 3.7. It was observed that the decomposition of more than 50 % of the sample in the first step corresponds to ester pyrolysis. It was also observed that the initial temperature of degradation for the samples was higher than 100 °C excluding the possibility of water presence. The thermal events and the maxima temperatures observed in TG curves are listed in Table 3.2.

3.4 Conclusions

By the transesterification of methyl and ethyl biodiesel of sesame crude oil with polyalcohol was obtained long-chain esters with biolubricant potential. It utilized a distillated fractioned system and the results obtained with this system were

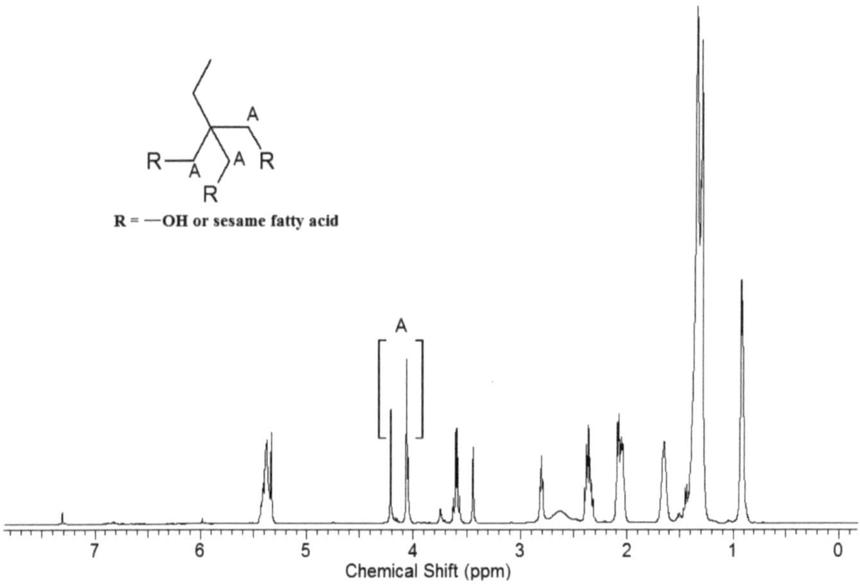

Fig. 3.4 ^{1}H NMR of T4

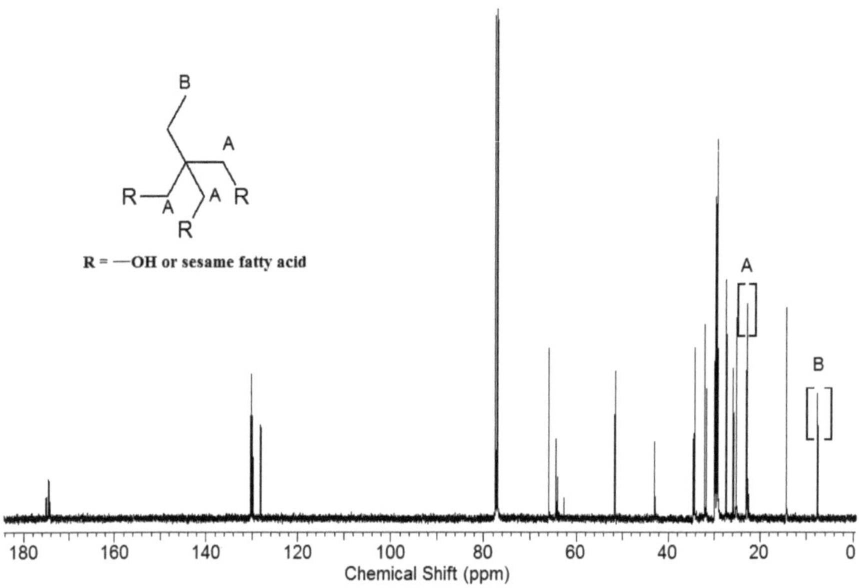

Fig. 3.5 ^{13}C NMR for T3

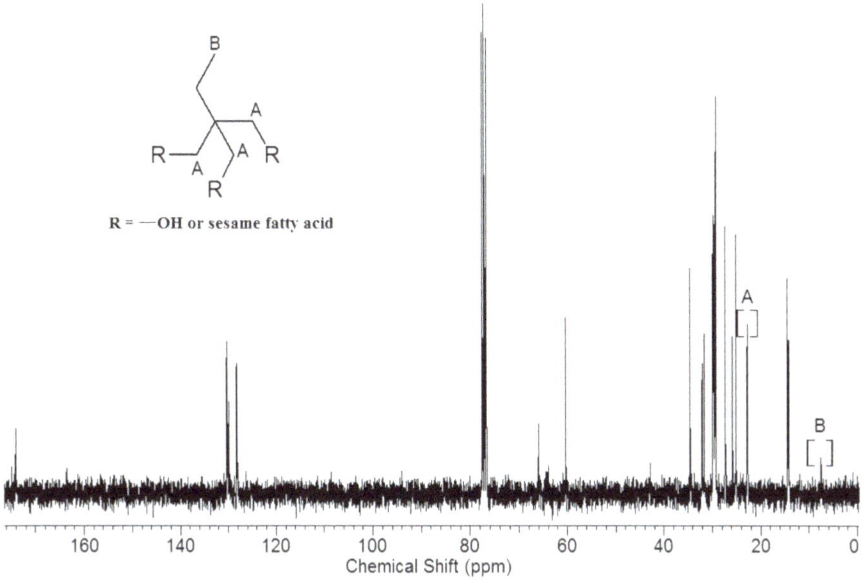

Fig. 3.6 ^{13}C NMR for T4

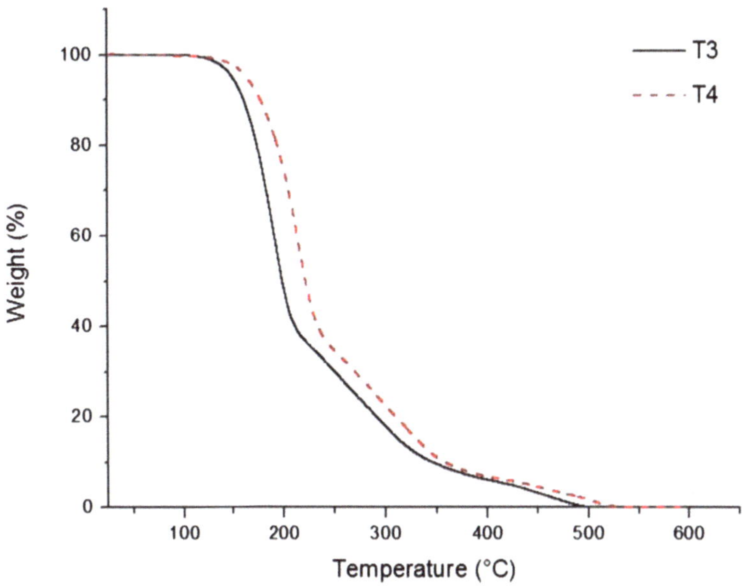

Fig. 3.7 TG for T3 and T4 at 10 °C/min in oxidative atmosphere

Table 3.2 Thermic events observed in TG to 10 °C/min

	T3		T4	
Event	Temperature (°C)	Loss weight (%)	Temperature (°C)	Loss weight (%)
1st	125.1–212.28	61.4	137.39–234.34	60.9
2nd	212.28–346.61	29.3	234.34–358.35	30.1
3th	346.61–480.59	9.3	358.35–509.95	9.0

satisfactory. The bioproducts were characterized by IR and ^{1}H and ^{13}C NMR analysis and physicochemical properties like iodine and acid value, viscosity at 40 °C, and density at 20 °C were determined. The TGA was made and the bioproduct obtained from ethyl biodiesel showed a slightly better stability than the bioproduct obtained from methyl biodiesel. To summarize, the long-chain ester obtained from sesame crude oil can be exploited for lubricant formulations in several areas.

References

1. Ebnesajjad S, Morga RA (2012) Applications of fluorinated additives for lubricants. Fluoropolym Addit 1:69–106
2. Padmaja KV, Rao BVSH, Reddy RK, Bhaskar PS, Singh AK, Prasad RBN (2012) 10-undecenoic acid-based polyol esters as potential lubricant base stocks. Ind Crop Prod 35:237–240
3. Resul MFMG, Ghazi TIM, Idris A (2012) Kinetic study of jatropha biolubricant from transesterification of jatropha curcas oil with trimethylolpropane: effects of temperature. Ind Crop Prod 38:87–92
4. Saydut A, Duz MZ, Kaya C, Kafadar AB, Hamanci C (2008) Transesterified sesame (*Sesamum indicum* L.) seed oil as a biodiesel fuel. Bioresour Technol 99:6656–6660
5. Sripada PK, Sharma RV, Dalai AK (2013) Comparative study of tribological properties of trimethylolpropane-based biolubricants derived from methyl oleate and canola biodiesel. Ind Crop 50:95–103
6. Wu Y, Li W, Zhang M, Wang X (2013) Improvement of oxidative stability of trimethylolpropane trioleate lubricant. Thermochim Acta 569:112–118

Chapter 4
Biodiesel from Jatropha Oil

Ebtisam K. Heikal, Salah A. Khalil and Ismaeil K. Abdou

Abstract In this chapter, we provide an overview of the currently available information on the different process steps of the production of biodiesel from *Jatropha curcas* L. (JCL). Based on the collection of data and information, the best available practice, the shortcomings and the potential environmental risks, and benefits are discussed for each production step. The chapter concludes with a call for general precaution and application of science.

Keywords Jatropha · Biodiesel · Greenhouse gas emissions · Wasteland reclamation

4.1 Introduction

With the ever-increasing demand for energy, coupled with the dwindling supplies of fossil fuels, the world has to look for other prospective sources of energy which could complement standing depletable fossil fuel sources and/or eventually replace them [1]. Many research organizations around the world have been studying and developing various technologies for the use of new and renewable energy sources [2].

Research activities around the world have come to the conclusion that biomass is presently the only available renewable option to produce transport fuels which are carbon neutral, and it can be stored and transported in large quantities using available petroleum infrastructure [2–4].

Biomass is already used to produce liquid transportation fuels namely, bioethanol and biodiesel from human farm feed products [2]. Yet, the trend nowadays, is to develop production technologies for both products, from agricultural wastes and nonedible seed oils [4].

Under Egypt's land and water supply availabilities, only nonedible oils such as Jatropha, jojoba, and castor oils can be considered for biodiesel production. Their plants can be grown on a large scale on non-cropped marginal land and wastelands.

E. K. Heikal (✉) · S. A. Khalil · I. K. Abdou
Egyptian Petroleum Research Institute, Nasr City, Cairo, Egypt
e-mail: ebtisam.heikal@yahoo.com

A. Sayigh (ed.), *Renewable Energy in the Service of Mankind Vol I*,
DOI 10.1007/978-3-319-17777-9_4

In this chapter, we present a state-of-the-art literature review of the whole Jatropha biodiesel production process. This collection of data and information enables us to discuss the actual best practices for the production of Jatropha biodiesel.

4.2 JCL Chemistry and Analysis

High-energy density liquid components, which can be used to make liquid fuels, are produced in plants as triglycerides or terpenes [1–4]. Triglycerides, as fats and oils, are found in the plant and animal kingdom and consist of water-insoluble, hydrophobic substance, and are made of 1 mol of glycerol and 3 mol of fatty acids [5–12].

Typically, 1 % of the vegetable oils (VOs) are made up of unsaponifiable compounds (carotenoids, phosphor lipids, tocophenols or tocotrienols, and oxydation products) [12].

The composition and characteristics of the crude JCL oil are shown in Table 4.1.

We can note from Table 4.1 that the values of free fatty acids (FFAs), unsaponifiables, acid number, and carbon residue figures show a very wide range, a fact which indicates that the oil quality is dependent on the interaction of environment and genetics. These wide ranges should be taken into consideration with regard to further processing of the oil.

It is important to point out that pure VOs cannot be used directly in diesel engines because of high viscosity, low volatility, and engine problems including coking on the injectors, carbon deposits, oil ring sticking, and thickening of the lubri-

Table 4.1 JCL oil composition and its characteristics [12]

	Range	Mean	S.D.	*n*
Specific gravity (g cm^{-3})	0.860–0.933	0.914	0.018	13
Calorific value (MJ kg^{-1})	37.83–42.05	39.63	1.52	9
Pour point (°C)	−3			2
Cloud point (°C)	2			1
Flash point (°C)	210–240	235	11	7
Cetane value	38.0–51.0	46.3	6.2	4
Saponification number (mgg^{-1})	102.9–209.0	182.8	34.3	8
Viscosity at 30 °C (cSt)	37.00–54.80	46.82	7.24	7
Free fatty acids % (kg kg^{-1}*100)	0.18–3.40	2.18	1.46	4
Unsaponifiable % (kg kg^{-1}*100)	0.79–3.80	2.03	1.57	5
Iodine number (mg iodine g^{-1})	92–112	101	7	8
Acid number (mg KOH g^{-1})	0.92–6.16	3.71	2.17	4
Monoglycerides % (kg kg^{-1}*100)	nd–1.7			1
Diglycerides % (kg kg^{-1}*100)	2.50–2.70			2
Triglycerides % (kg kg^{-1}*100)	88.20–97.30			2
Carbon residue % (kg kg^{-1}*100)	0.07–0.64	0.38	0.29	3
Sulfur content % (kg kg^{-1}*100)	0–0.13			2

S.D standard deviation, *n* number of observation used, *nd* no data cating oils [4, 12]. Yet, they can be used as base for liquid engine fuels in various ways, e.g., blends with other components, microemulsification, transesterification (TE), and hydrotreating [4].

4.3 Biodiesel Production Technologies

4.3.1 Homogeneous Catalysts for the Transesterification of Vegetable Oil [8–15]:

Transesterification Chemistry TE (also known as alcoholysis) of VOs is the reaction of the oil constituents (triglycerides) with excess of methyl or ethyl alcohol, in a three-step reaction as follows:

$$\mathbf{TG + M \rightleftarrows DG + E}$$
$$\mathbf{DG + M \rightleftarrows MG + E}$$
$$\mathbf{MG + M \rightleftarrows G + E}$$

Where TG, DG, and MG are respectively the tri-, di-, and mono glycosides, and M, G, and E indicate methanol, glycerol, and the mixture of methyl esters which form biodiesel.

All the reaction steps are reversible, hence the need to use excess of alcohol, and a catalyst mostly a base, e.g., sodium hydroxide (Na OH), Potassium hydroxide (KOH), and their alkoxides.

It has been demonstrated that the methanol–oil molar ratio, catalyst concentration, and reaction temperature are the significant parameters affecting the yield of fatty acid methyl ester (FAME).

Alkali-catalyzed TE proceeds ≈4000 times faster than that catalyzed by the same amount of an acidic catalyst.

Methanol and ethanol are the main alcohols used for the VOs TE. Methanol is commercially available in anhydrous form, and the catalyst NaOH dissolves quickly in methanol.

KOH is used as catalyst when ethyl alcohol is used for TE due to its higher solubility in ethanol.

However, producing fatty acid ethyl ester (FAEE) is of high interest because it yields an entirely agriculture-based fuel, besides the energy content and cetane number are higher.

Optimum molar ratio of alcohol/oil is 6/1 in the case of methanol, and 12/1 for ethanol.

For TE of JCL, catalyst requirement is ≈1 wt% of the oil, and the reaction is conducted close to the boiling point of methanol (60–70 °C).

The products of the reaction are in two phases: a glycerol-rich phase and a methyl ester-rich phase. These two phases are physically separated, and treated to produce the American Society for Testing and Materials (ASTM) standard biodiesel, and pure glycerin.

The optimum combination for reducing the FFAs' content in jatropha oil, from 14% to less than 1%, was found to be 1.43% H_2SO_4 acid as catalyst, 0.28 V/V methanol–oil ratio, and 88 min reaction time at a temperature of 60 °C as compared to a 5/1 molar ratio of methanol–oil, and 24 min reaction time at a temperature of 60 °C for producing biodiesel, using 0.55% W/V KOH as an alkaline catalyst.

4.3.2 Heterogeneous Catalysts for the TE of VOs

Despite industrial applicability, homogeneous catalysts have their limitations: the catalyst dissolves fully in the glycerin layer, and partially in the FAME layer, which makes the product separation arduous. As a result, biodiesel should be cleaned through a slow, tedious, and environmentally unfriendly water washing process to remove excess catalyst. Moreover, catalyst-contaminated glycerin is becoming a disposal problem. Another negative aspect is that the catalysts are nonreusable [17–19].

These problems have initiated research work and development of heterogeneous catalysts which can be easily removed from the product and recycled [16–19].

Yet, current heterogeneous catalysts pose their own set of problems, namely, they are not as active as homogeneous catalysts, and they require higher reaction temperatures (200–250 °C) and pressures [13].

Furthermore, it should be taken into consideration that the presence of FFAs in the feed will strongly poison solid base catalysts. In case of acidic catalysts, strong deactivation occurred when the catalyst was reused [10].

Although most of the reported work in the field of TE using heterogeneous catalysts deals with research and development studies, yet, it is worth mentioning that a 160,000 T/y commercial plant, using the Hester tip-H technology developed by the Institute France de Petrol (IFP), has started production since 2006. The catalyst is a mixed oxide of zinc and aluminium, the operating temperature is 200–250 °C at a pressure of 50 atm [18].

4.3.3 Enzymatic TE of VOs [20]

Lipase enzymatic catalysts can catalyze esterification and TE reactions. The advantages of lipase catalysts are their ability to catalyze both TE and E of FFAs in one step, production of glycerol side-stream with minimal water content and little or no inorganic material, and recyclability.

It is interesting to note that Jatropha seeds are reported to contain lipase activity which could also catalyze TE reactions. However, enzymatic catalysts have high costs, and deactivate due to feed impurities.

4.3.4 Non-catalytic Super Critical VOs Conversion to FAME [21]

With the aim of developing a novel methanolysis process for VOs without using any catalyst, it has been demonstrated that preheating to a temperature of 350 °C, at 100–250 atm and treatment for 240 s in supercritical (SC) methanol are sufficient to convert VOs to FAMEs with a higher yield than that obtained by alkali catalysts. Since SC methanol has a hydrophobic nature with a lower dielectric constant, non-polar TGs can be well solvated with SC methanol to form a single phase oil/methanol mixture. FFAs contained in the VOs can also be converted efficiently to FAME in SC methanol leading to an increase in the total yield of FAMEs.

The purification of the products after TE reaction is much simple, and more environmentally friendly compared with the alkali-catalyzed method. However, the supercritical method requires higher temperatures and pressures, and large amounts of methanol.

Reaction with SC methanol has the following advantages:

I. TGs and FFAs are reacted with equivalent rate.
II. The homogenous phase eliminates diffusive problems.
III. The process tolerates great percentages of water in the feedstock.
IV. The catalyst removal step is eliminated.

4.4 JCL FAME Evaluation and Performance

a) Characteristics and Composition [12] Various specifications for FAMEs such as ASTM-D6751 and EN 14214 are presented in Table 4.2 together with JCL FAME characteristics. It is clear that JCL FAMEs comply with these specifications.

Yet, there are some technical problems with biodiesel which have persisted to the present time, namely, oxidation stability, poor low-temperature properties, and a slight increase in nitrogen oxides (NOx) exhaust emissions [21]. Ethyl and isopropyl esters have improved low-temperature properties without comprising cetane number or oxidation stability [21].

b) Performance More than 100 years ago, Dr. Rudolf Diesel invented the original diesel engine and designed it to run on a host of fuels including heavy mineral oil, and VOs. His first experiments were catastrophic failures, but in time he showed his engine at the World Exhibition in Paris in 1900 and it was running on 100% peanut oil.

Using biodiesel in a conventional diesel engine substantially reduces emissions of unburned hydrocarbons, carbon monoxide, sulfates, polycyclic aromatic hydrocarbons, nitrated polycyclic aromatic hydrocarbons, and particulate matter. Neat biodiesel reduces carbon dioxide (CO_2) emissions by more than 75% over petroleum diesel [21].

Table 4.2 JCL (m) ethyl ester composition and characteristics with the corresponding values of the European (EN 14214:2003), German (DIN V 51606), and the *USA Standards (ASTM D 6751)* [12]

	JME				JEE $n=1$	EN 14214:203	DIN V 51606	ASTM D6751
	Range	Mean	S.D	n				
Density (gcm^{-3})	0.864–0.880	0.875	0.007	6	0.89	0.86–0.90	0.87–0.90	
Calorific value (MJ kg^{-1})	38.45–41.00	39.65	1.28	3	190			
Flash point (°C)	186	186	11	4	59	min 120	min 110	min 130
Cctane value	50.0–56.1	52.3	2.3	5	5.54	min 51	min 49	min 47
Saponification number (mg g^{-1})	202.6			1				
Viscosity at 30 °C (cSt)	4.84–5.65	5.11	0.47	3		3.5–5.0	3.5–5.0	1.9–6.0
Iodine number (mg iodine g^{-1})	93–106			2		Max 120	Max 115	Max 115
Acid number (mg KOH g^{-1})	0.06–0.5	0.27	0.22	3	0.08	Max 0.5	Max 0.5	Max 0.5
Monoglycerides %(kg kg^{-1}*100)	00.24			1	0.55	Max 0.8	Max 0.8	
Diglycerides %(kg kg^{-1}*100)	0.07			1	0.19	Max 0.2	Max 0.4	
Triglycerides %(kg kg^{-1}*100)	nd			0	nd	Max 0.02	Max 0.4	
Carbon residue %(kg kg^{-1}*100)	0.02–0.50	0.18	0.27	3		Max 0.3	Max 0.3	Max 0.5
Sulfur content %(kg kg^{-1}*100)	0.0036			1		Max 0.01	Max 0.01	Max 0.015[a]
Sulfur ash %(kg kg^{-1}*100)	0.005–0.010	0.013	0.002	4		Max 0.2	Max 0.03	Max 0.02
(M)ethyl ester content %(kg kg^{1}*100)	99.6			1	99.3	min 96.5	Max 0.3	
Methanol %(kg kg^{-1}*100)	0.06–0.09			2	0.05	Max 0.2	Max 0.3	
Water %(kg kg^{-1}*100)	0.07–0.01			1	0.16	Max 0.5	Max 0.02	Max 0.5
Free glycerol %(kg kg^{-1}*100)	0.015–0.030			2	nd	Max 0.02	Max 0.25	Max 0.02
Total glycerol %(kg kg^{-1}*100)	0.088–0.100			2	0.17	Max 0.25		Max 0.24

S.D standard deviation, *n* number of observations used, *nd* no data

[a] Maximum 0.015 % for S 15 Grade and maximum 0.05 % for S 500 Grade

The use of biodiesel decreases the solid carbon fraction of particulate matter since the oxygen in biodiesel enables more complete combustion of CO_2, and reduces the sulphate fraction. Emission of NO_x increases with the increase in concentration of biodiesel in the fuel. Some additives have shown promise in modifying the increase [22].

4.5 Conclusions

I- JCL Oil Chemistry Some of the properties of JCL oils produced from different species are more or less the same, while others are widely variable. The FFAs' content of the oil is one of the variable properties which should be given paramount importance since it decides the FAME production scheme and the economics of the process.

II- JCL FAME Production At present, the most widely used technology for FAME production is the homogeneous alkali-catalyzed process. Yet, between the technologies under development and those which have entered commercial application, two new technologies, namely, the SC non-catalytic TE and the newly developed heterogeneous solid catalyst process capture interest.

Yet, the SC TE process has been in commercial application, mainly in Germany, long before the introduction of the heterogeneous solid catalyst technology.

References

1. Kalay Asiri P et al (1996) Survey of oils for use as diesel fuels. J Am Oil Chem Soc 73(4):47–474
2. Gübtz GM et al (1999) Exploitation of the tropical oil seed plant *Jatropha Curcas* L. Bioresourc Technol 67:73–82
3. Barnwal BK, Sharma MP (2005) Prospects of biodiesel production from vegetable oils in India. Renew Sustain Energy Rev 9:363–378
4. Om Tapanes NC et al (2008) Transestrification of *Jatropha curcas* oil glycerides: thearetical and experimental studies of biodiesel reaction. Fuel 87:2286–2295
5. Connemann et al (1994) Process for the continuous production of lower alkyl esters of higher fatty acids. U S A Patent 11(5):354, 878
6. Pilar Dorado M et al (2004) Optimization of alkali- catalyzed Transesterification of *Brassica Carinata* oil for biodisel production. Energy Fuels 18:77–83
7. Pramanik T, Tripathi S (2005) Biodiesel: clean fuel of the future hydrocarbon processing 84(2):49–54
8. Abderrahim Bouaid et al (2005) Pilot plant studies of biodisel production using *Brassica Carinata* as row material. Catal Today 106:193–196
9. Tiwari Alok Kumar et al (2007) Biodiesel production from Jatropha oil with high free fatty acid: an optimized process. Biomass Bioenerg 31:569–575
10. Singh Alok Kumar et al (2007) Base-catalyzed fast transesterification of soybean oil using ultrasonication. Energy Fuels 21:1161–1164
11. Franceschini Gaia, Macchietto Sandro (2007) Validation of model for biodiesel production through model-based experiment design. Ind Eng Res 46:220–232

12. Achten WMJ et al (2008) Jatropha biodiesel production and use. Biomass Bioenerg 32:1063–1084
13. Saldar SK et al (in press) Studies on the comparison of performance and emission characteristics of diesel engine using three degummed non-edible vegetable oils. Biomass Bioenerg
14. Stiefel Scott, Dassori Gustavo (2009) Simulation of biodiesel production through transesterification of vegetable oils. Ind Eng Chem Res 48:1068–1071
15. Kim HJ et al (2004) Transesterification of vegetable oil to biodiesel using heterogeneous base catalyst. Catal Today 93:93–95, 315–320
16. Furuta S et al (2004) Biodiesel fuel production with solid superacid catalysis in fixed bed reactor under atmospheric pressure. Catal Commun 5:721–723
17. Loters Edgar et al (2005) Synthesis of biodiesel via acid catalysis. Ind Eng Chem Res 44(14):5353–5363
18. Bokade Vijay V, Yadav Ganapati D (2009) Transesterification of edible on nonedible vegetable oils with alcohols over heteropolyacids supported on acid-treated clay. Ind Eng Chem Res 48:9408–9415
19. José M et al (1999) Preparation and properties of biodiesesl from *Cynara cardunculus* L. oil. Ind Eng Chem Res 38(8):2927–2931
20. Fukuda Hideki et al (2001) Biodiesel fuel production by transesterifcation of oils. J Biosci Bioeng 92(5):405–416
21. Thomas D et al (2000) Effect of biodiesel, biodiesel blends, and a synthetic diesel on emissions from light heavy-duty diesel vehicles. Environ Sci Technol 34(3):349–355
22. Janet M et al (2002) Lubricity- enhancing properties of soy oil when used as blending stock for middle distillate fuels. Ind Eng Chem Res 41(5):1386–1388

Chapter 5
Solid Acid Catalyst Derived from Coffee Residue for Biodiesel Production

Kanokwan Ngaosuwan

Abstract Biodiesel is a free fatty acid methyl ester (FAME) produced from transesterification of oil and short-chain alcohol. Nowadays, the cost of conventional biodiesel production might not be competitive with petro-diesel due to the cost of refined oil as a main feedstock. Many challenging researches have proposed method or technology to efficiently produce biodiesel. One of the most important knowledge is the synthesis of the suitable catalyst for biodiesel production. Solid acid catalyst is a promising catalyst to produce biodiesel from low-cost feedstocks since it can catalyze simultaneously esterification of free fatty acid (FFA) and transesterification of triglyceride. To improve the biodiesel CO_2 cycle, the waste material, coffee residue, was selected as a supported catalyst. It provides the appropriate textural properties such as high surface area with mesoporous structure and hydrophobic properties. The sulfonation with concentrated H_2SO_4 was used for additional acidic functional group which exhibits a strong protonic acid site density. Therefore, this research aims to synthesize sulfonated activated carbon derived from coffee residue (SCAC) to catalyze esterification of caprylic acid as a model of FFA. The sulfonation temperature varied from 140 to 200 °C as named SCAC-140, SCAC-160, SCAC-180, and SCAC-200 catalysts.

After sulfonation process, the X-ray diffraction (XRD) patterns of all catalysts were similar to coffee residue activated carbon (CAC) as a support. A few, broad peak at $2\theta = 15–30°$ and 30–50°, were related to amorphous carbon and graphene sheet at the plane C (101), respectively. Fourier transform infrared spectroscopy (FT-IR) of CAC at 3400 cm^{-1} was absent which confirmed the hydrophobic property of this support. FT-IR spectra of SCAC catalysts illustrated the acid function as 3400, 1700, 1600, 1300, 1000, and 1100 cm^{-1} , which is related to O–H stretching mode of COOH and phenolic, C=O of COOH, C=C of polyaromatic, O=S=O symmetric SO_2 stretching, O=S=O asymmetric SO_2 stretching, respectively. SCAC-180 catalyst exhibited the highest initial rate due to the highest total acid site density which included sulfonated, carboxylic, and phenolic group. At 2 h, SCAC-200 showed the highest caprylic acid conversion while SCAC-140 provided the lowest caprylic acid conversion. It was probably due to the textural and structural properties of these

K. Ngaosuwan (✉)
Chemical Engineering Division, Engineering Faculty,
Rajamangala University of Technology Krungthep, Bangkok, Thailand
e-mail: kanokwan.n@rmut.ac.th

A. Sayigh (ed.), *Renewable Energy in the Service of Mankind Vol I*,
DOI 10.1007/978-3-319-17777-9_5

synthesized catalysts. All of synthesized carbon catalysts demonstrated the higher catalytic activity than that of Amberlyst-15. However, the solid acid catalyst still provided a lower catalytic activity as compared to homogeneous H_2SO_4 catalyst.

Keywords Biodiesel · Fatty acid methyl ester (FAME) · Esterification · Sulfonated activated carbon catalyst

5.1 Introduction

Biodiesel is the renewable energy which is a potential fuel to replace fossil diesel [1]. This can be used directly in any diesel engine without the engine modification and is made from renewable sources such as edible oils (vegetable oils and animal fat), nonedible oils (castor oil and rubber seed oil), or waste cooking oils. Trans-esterification of oils or fats with a low-molecular weight of alcohol using homogeneous alkaline catalyst is a conventional process to produce biodiesel. However, this method required the pretreatment and neutralization steps which is a drawback. Utilization of virgin oil as a raw material for biodiesel production is uneconomical which attributes to over 70% of the total biodiesel production cost [2]. Therefore, the waste oils or nonedible oils are becoming one of the highest potential to produce biodiesel economically. The yield of biodiesel from these oils could be reduced due to the presence of water and high free fatty acid (FFA) value. Heterogeneous acid catalyst is a promising choice and available to serve as means of reducing biodiesel production costs and has several advantages including of (1) readily usable for low-cost feedstocks without the need of pretreatment, (2) simultaneous esterification and transesterification, (3) removing the washing step of biodiesel production, (4) easy to separate as well as minimized product contamination, (5) reusable and can be regenerated, (6) minimization of vessel corrosion problem, and (7) flexible to use in continuous process [3].

It has been revealed that the heterogeneous acid catalyst for biodiesel production included Brönsted acid catalyst [4–7], Lewis acid catalyst [8], and mixed Brönsted and Lewis acid catalyst [9–11]. Sani et al. [3] summarized the ideal solid acid catalyst for biodiesel production required: (1) hydrophobic surface on solid acid catalysts to promote the selectivity, (2) interconnected system of large pores to minimize diffusion problems, and (3) moderate to high concentration of strong acidic sites to obtain the appreciable reaction rates in the mild condition.

Sulfonated carbon-based catalyst has been addressed in terms of low-cost renewable "green catalyst" due to the fact that it could be prepared either from biomass or from household waste generation [12]. Coffee residue has been selected as a carbon source to produce an activated carbon as a catalyst support via carbonization under the CO_2 atmosphere. Activated carbon derived from coffee residue (CAC) is a material that has all the required characteristics to be used as a catalyst support and additionally it has unique properties like its thermal stability, mesoporous structure with high specific surface area, and hydrophobic surface property [13]. Sulfonic group

(–SO_3H) has been functionalized onto CAC via sulfonation of concentrated sulfuric acid to prepare the strong acid catalyst with the constraint of ideal solid acid catalyst. Toda et al. [14] reported functional of sulfonated group on the amorphous carbon material with a small polycyclic aromatic carbon sheets as a promising solid acid catalyst for replacement of sulfuric acid catalyst. The preparation method of sulfonated carbon catalyst needs further improvements in biodiesel production including higher activity and lower amounts of catalyst requirement, stability, and reusability. These catalysts are susceptible to poisoning in polar media because of their surface organic nature. Therefore, the current strategy should be explored to avoid or eliminate this limitation by the use of enhance surface hydrophobization [3].

This research aims to explore the synthesis of sulfonated activated carbon derived from coffee residue (SCAC) for biodiesel production. A numerous research involving sulfonated carbon catalyst reported the difference of sulfonation temperature which strongly depended on type of carbon source [15–17]. Therefore, the effect of sulfonation temperature was evaluated to obtain the optimum condition for the synthesis of SCAC catalyst. Esterification of caprylic acid was investigated to test the catalytic performance.

5.2 Experimental

5.2.1 Materials

Coffee residue has been provided from Moccona Company (Thailand). $ZnCl_2$, concentrated sulfuric acid (98%), and hydrochloric acid (37%) were purchased from Fluka as analytical grade. Reactant and standard caprylic acid (HCp), methyl carpylate (MeCp), methyl valerate, and methanol for esterification were purchased from Sigma-Aldrich Company.

5.2.2 Sulfonated Activated Carbon Preparation

Activated carbon derived from coffee residue was prepared using $ZnCl_2$ as activated chemical via carbonization at 600 °C for 4 h under CO_2 atmosphere [13]. Sulfonation of concentrated sulfuric acid was used to introduce sulfonic group on the CAC support with ratio of CAC to H_2SO_4 as equal to 1 g: 20 mL. The sulfonation temperature varied from 140 to 200 °C so called SCAC-140, SCAC-160, SCAC-180, and SCAC-200.

5.2.3 Characterizations

The powder X-ray diffraction (XRD) patterns were collected on Bruker, D8 Advance with a Cu-Kαα radiation ($\lambda = 0.15406$ nm) over a 2θ range of 10–80°. Attentuated total reflectance Fourier transform infrared spectroscopy (ATR-FT-IR) (Thermo Electron Corporation, Nicolet 6700) was used to characterize the functional group of the catalysts. The specific surface area of the catalysts was investigated using the Brunauer–Emmett–Teller (BET) method. Ion exchange back titration using HCl and NaOH aqueous solution was performed to determine acid site densities of the catalysts.

5.2.4 Reaction Tests

5.2.4.1 Esterification of HCp

Liquid phase esterification of HCp with methanol was performed using 50 mL three-necked round bottom flask equipped with a reflux condenser at 60 °C for 2 h. The catalyst loading based on the weight of HCp was 5 %wt. and molar ratio of HCp to methanol was 1:3. When the reaction mixture reached desired temperature, the catalyst was introduced and reaction time was started. Gas chromatograph (HP 5890 series II plus) with flame ionization detector (FID) and DB-wax column were used to analyze the resulting reaction mixture.

5.3 Results and Discussion

5.3.1 Structural Characterization of the Synthesized Carbon Material

Figure 5.1 indicates that both CAC before and after sulfonation at different sulfonation temperature possessed low crystallinity of XRD patterns. The characteristic of amorphous carbon composed of the random orientation of aromatic sheets is observed from a weak and broad diffraction peak at 2θ of 10–30° at plane C(002). Whereas, a definitive plane C(101) of diffraction peak at 2θ of 30–50° elucidated an axis of the graphite structure. This also indicated that all SCAC catalysts are composed of a larger graphene sheet than that of CAC support after sulfonation process. XRD pattern of CAC support also shows the additional peaks associated with crystalline of γ-Fe_2O_3 [18]. This might be due to the iron impurity from stainless steel tube during the carbonization process. However, the iron residue was removed after the sulfonation process as seen in Fig. 5.1.

FT-IR pattern of CAC support and SCAC catalyst at different sulfonation temperature in the wave length of 500–4000 cm^{-1} are shown in Fig. 5.2. The carbonization process using $ZnCl_2$ as activated chemical has been successfully prepared CAC support with hydrophobic property. (Fig. 5.2) No band at 3400 cm^{-1} of CAC support could be assigned to the hydroxyl group (O–H) corresponding to the report of Bonnamuuayvitaya et al. [13]. After sulfonation process with concentrated H_2SO_4

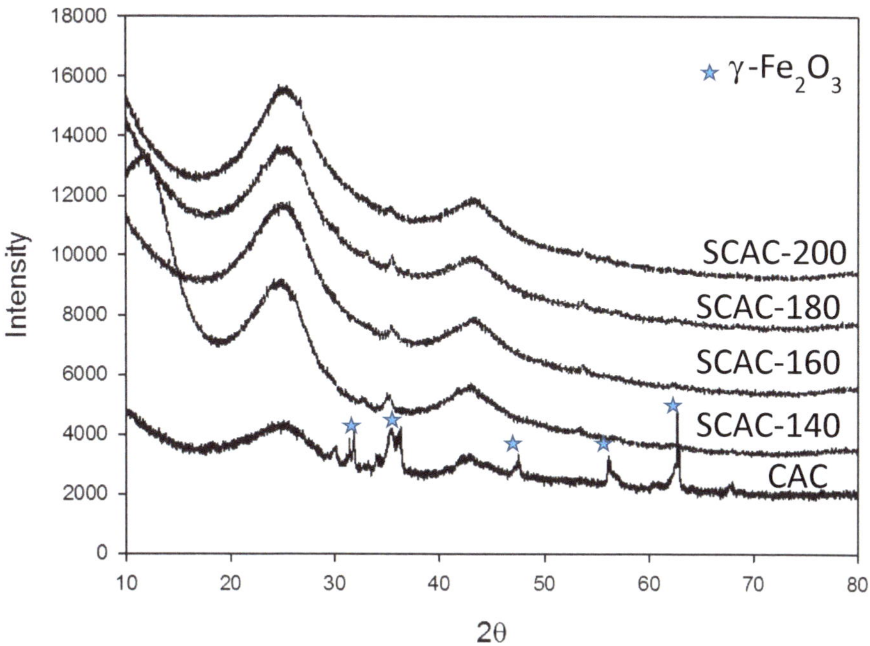

Fig. 5.1 XRD pattern of activated carbon derived from coffee residue *(CAC)* and sulfonated activated carbon derived from coffee residue *(SCAC)* at different sulfonation temperature

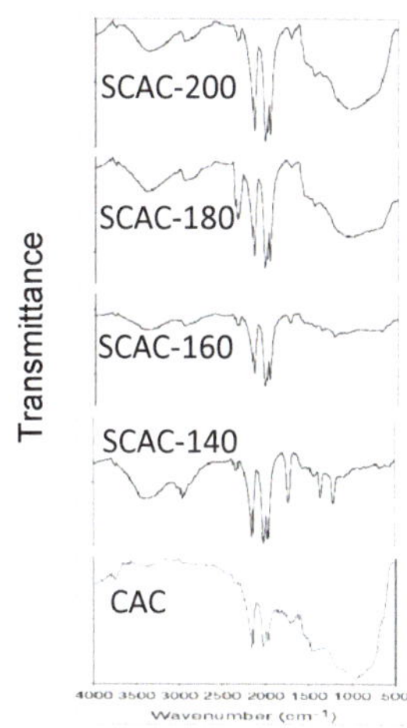

Fig. 5.2 FT-IR spectra of activated carbon derived from coffee residue *(CAC)* and sulfonated activated carbon derived from coffee residue *(SCAC)* at different sulfonation temperature

at different sulfonation temperature, the broad band at 1000 and 1100 cm^{-1} were observed which confirmed that SCAC catalysts composed of SO_3H group with respect to O=S=O of symmetric SO_2 stretching and O=S=O of asymmetric SO_2 stretching, especially in the FT-IR pattern of SCAC-140 catalyst [19]. In addition, the broad band at around 3400 $cm^{-1}$is a stretching mode of O–H of –COOH carboxylic and

Table 5.1 Texture properties of activated carbon derived from coffee residue *(CAC)* and sulfonated activated carbon derived from coffee residue *(SCAC)* at different sulfonation temperature[a]

Samples	Specific surface area (m^2/g)	Average pore size diameter (nm)	Pore volume (cm^3/g)	SO_3H acid site concentration (mmol/g)[a]	Total acid site concentration (mmol/g)[b]
CAC	956	3.4	0.8315	Na	0.003
SCAC-140	1218	3.35	1.022	0.72	0.946
SCAC-160	1065	3.79	1.010	0.66	0.931
SCAC-180	992	3.64	0.9035	0.57	1.011
SCAC-200	1091	3.51	0.9584	0.45	0.993
Amberlyst-15	45			4.30[c]	4.70[c]
H_2SO_4				20.4[d]	20.4[d]

CAC activated carbon derived from coffee residue, *SCAC* sulfonated activated carbon derived from coffee residue

[a]Based on EDS results

[b]Based on Ion-exchange titration results with the experimental error ± 0.01

[c]Data from ref [18]

[d]Data from ref [22]

phenolic OH group which is the obvious difference from CAC support. The existing band at 1700 cm^{-1} refers to C=O of –COOH while the band 1600 and 1300 cm^{-1} attributes to C=C of polyaromatic ring of hydrocarbon [19]. The remarkable bands at 2000 and 2300 cm^{-1} of CAC support and SCAC catalysts are may be due to CO_2 adsorption on the surface under atmospheric operation.

The sulfonation process using concentrated H_2SO_4 reagent tends to increase specific surface area and pore volume with similar average pore size and pore structure. Moreover, Kastner et al. [15] reported the significantly increased specific surface area due to activation and oxidation of the carbon during sulfonation process with H_2SO_4. Ion-exchange titration technique was used to measure the total acid density while the SO_3H acid site density was estimated from the sulfur content assuming all S atoms were in the –SO_3H using energy dispersive spectrometry (EDS) analysis results. CAC support has no SO_3H acid site density and small amount of total acid site due to weak acid group (e.g., –COOH) as indicated by the formation of FT-IR band at 1700 cm^{-1}. Functionalization of SO_3H group on the CAC support is hence to increase total acid site density in terms of strong Brönsted acid site. The total acid site and specific surface area of SCAC catalysts were comparable to that of sulfonated porous carbon catalyst as prepared from the carbonization of wood powder using activated chemical $ZnCl_2$ [20]. The sulfonation temperature is strongly related effect on the SO_3H acid site density of SCAC catalysts. Increase in sulfonation temperature results in decrease of SO_3H acid site density due to the condensation playing a dominating role with the further increasing of sulfonation temperature [21]. However, the other functional groups including COOH of carboxylic and OH of phenolic acid group also presented an increase in the total acid site density for all SCAC catalysts (Table 5.1).

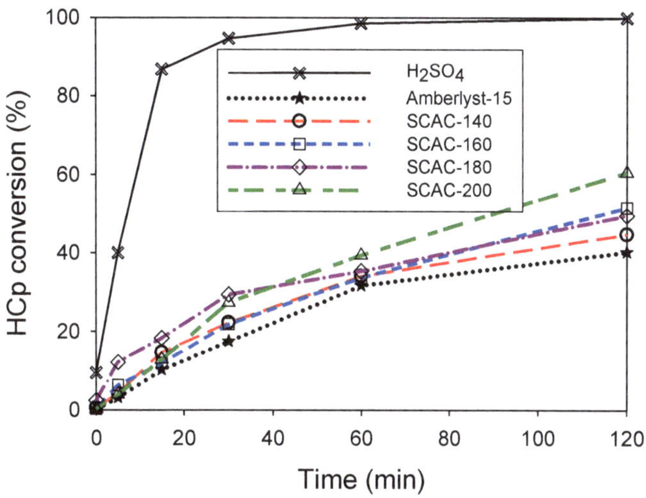

Fig. 5.3 HCp conversion along reaction time of sulfonated activated carbon derived from coffee residue *(SCAC)* at different sulfonation temperature

5.3.2 Esterification of HCp

Esterification of caprylic acid (HCp) was used to verify the catalytic activities of SCAC catalysts as shown in Fig. 5.3. Blank test of esterification without any catalyst and CAC support were also performed to confirm the catalytic activity under the similar condition. It was indicated that no formation of methyl caprylate resulted either of esterification in the absence of catalyst or using CAC support as a catalyst (data not shown). All SCAC catalysts were demonstrated to be catalytically active for esterification of HCp with methanol. HCp conversion using SCAC-140 catalyst at 2 h is lowest while using SCAC-200 achieved the highest HCp conversion at 2 h. This indicated that the sulfonation temperature has a dominating effect on the catalytic activities of carbon-based catalyst [20]. SCAC-140 catalyst with the highest SO_3H acid site density reached the lowest conversion of HCp at 2 h. Kastner et al. also reported the sulfonated wood bases activated carbon catalyst (SO_3H acid site of 0.2 ± 0.01 mmol/g) had significantly higher esterification activities than that of sulfonated peanut hull char catalyst (SO_3H acid site of 0.61 ± 0.03 mmol/g) [15]. This is probably due to its strong Brönsted acid site was also easier to catalyst deactivation by poisoning with water as an intermediate product of esterification. It was noted that all SCAC catalysts exhibit a remarkably higher HCp conversion than that of Amberlyst-15 resin as a typical catalyst for esterification. Zong et al. [23] and Lou et al. [17] also reported that sulfonated carbon catalysts performed higher catalytic activities than that of the three typical solid acid catalysts (sulphated zirconia, Amberlyst-15, and niobic acid). However, HCp conversions of all SCAC catalysts were lower than that of H_2SO_4 as a homogeneous catalyst.

5.4 Conclusions

A coffee residue derived sulfonated carbon-based catalyst, which can be synthesized through sulfonation of incompletely carbonized coffee residue, is a class of economic, eco-friendly, and promising catalyst. This consists of polyaromatic hydrocarbons and a small proportion of aromatic and alphatic hydrocarbons which comprise SO_3H, COOH, and phenolic functional groups. The sulfonation temperature is strongly affected by the catalytic activity due to their textural and structural properties. The optimum condition for synthesis of the coffee residue derived sulfonated carbon catalyst is the carbonization temperature of 600 °C for 6 h and sulfonation of 200 °C for 18 h with high surface area and total acid site density.

Acknowledgements The author would like to thank Thailand Research Fund (TRF) and Rajamangala University of Technology Krungthep, and the Commission on Higher Education for supporting by grant fund under the MRG program with ID of MRG5080040. The author also would like to thank the Center of Excellence on Catalysis and Catalytic Reaction Engineering and Moccona company (Thailand) for providing analysis facility and coffee residue, respectively.

References

1. Coronado CR, de Carvalho JA, Silveira JL (2009) Biodiesel CO_2 emissions: a comparison with the main fuels in the Brazilian market. Fuel Process Technol 90:204–211
2. Wu X, Leung DYC (2011) Optimization of biodiesel production from camelina oil using orthogonal experiment. Appl Energy 88:3615–3624
3. Sani YM, Daud WMAW, Aziz ARA (2014) Activity of solid acid catalysts for biodiesel production: a critical review. Appl Catal a-Gen 470:140–161
4. Silva VWD, Laier LO, da Silva MJ (2010) Novel H3PW12O40: catalysed esterification reactions of fatty acids at room temperature for biodiesel production. Catal Lett 135:207–211
5. Baig A, Ng FTT (2010) A single-step solid acid-catalyzed process for the production of biodiesel from high free fatty acid feedstocks. Energy Fuel 24:4712–4720
6. Suwannakarn K, Lotero E, Ngaosuwan K, Goodwin JG (2009) Simultaneous free fatty acid esterification and triglyceride transesterification using a solid acid catalyst with in situ removal of water and unreacted methanol. Ind Eng Chem Res 48:2810–2818
7. Ngaosuwan K, Mo XH, Goodwin JG, Praserthdam P (2010) Reaction kinetics and mechanisms for hydrolysis and transesterification of triglycerides on tungstated zirconia. Top Catal 53:783–794
8. Chen SY, Mochizuki T, Abe Y, Toba M, Yoshimura Y (2014) Ti-incorporated SBA-15 mesoporous silica as an efficient and robust Lewis solid acid catalyst for the production of high-quality biodiesel fuels. Appl Catal B-Environ 148:344–356
9. Jitputti J, Kitiyanan B, Rangsunvigit P, Bunyakiat K, Attanatho L, Jenvanitpanjakul P (2006) Transesterification of crude palm kernel oil and crude coconut oil by different solid catalysts. Chem Eng J 116:61–66
10. Li Y, Zhang XD, Sun L, Xu M, Zhou WG, Liang XH (2010) Solid superacid catalyzed fatty acid methyl esters production from acid oil. Appl Energy 87:2369–2373
11. Lam MK, Lee KT, Mohamed AR (2009) Sulfated tin oxide as solid superacid catalyst for transesterification of waste cooking oil: an optimization study. Appl Catal B-Environ 93:134–139

12. Konwar LJ, Boro J, Deka D (2014) Review on latest developments in biodiesel production using carbon-based catalysts. Renew Sust Energy Rev 29:546–564
13. Boonamnuayvitaya V, Chaiya C, Tanthapanichakoon W (2004) The preparation and characterization of activated carbon from coffee residue. J Chem Eng Jpn 37:1504–1512
14. Toda M, Takagaki A, Okamura M, Kondo JN, Hayashi S, Domen K, Hara M (2005) Green chemistry—biodiesel made with sugar catalyst. Nature 438:178–178
15. Kastner JR, Miller J, Geller DP, Locklin J, Keith LH, Johnson T (2012) Catalytic esterification of fatty acids using solid acid catalysts generated from biochar and activated carbon. Catal Today 190:122–132
16. Shu Q, Zhang Q, Xu GH, Nawaz Z, Wang DZ, Wang JF (2009) Synthesis of biodiesel from cottonseed oil and methanol using a carbon-based solid acid catalyst. Fuel Process Technol 90:1002–1008
17. Lou WY, Guo Q, Chen WJ, Zong MH, Wu H, Smith TJ (2012) A highly active bagasse-derived solid acid catalyst with properties suitable for production of biodiesel. Chemsuschem 5:1533–1541
18. Liu F, Sun J, Sun Q, Zhu L, Wang L, Meng X, Qi C, Xiao F-S (2012) High-temperature synthesis of magnetically active and SO3H-functionalized ordered mesoporous carbon with good catalytic performance. Catal Today 186:115–120
19. Mar WW, Somsook E (2012) Sulfonic-functionalized carbon catalyst for esterification of high free fatty acid. Procedia Eng 32:212–218
20. Kitano M, Arai K, Kodama A, Kousaka T, Nakajima K, Hayashi S, Hara M (2009) Preparation of a sulfonated porous carbon catalyst with high specific surface area. Catal Lett 131:242–249
21. Fu P, Hu S, Xiang J, Sun L, Su S, Wang J (2012) Evaluation of the porous structure development of chars from pyrolysis of rice straw: effects of pyrolysis temperature and heating rate. J Anal Appl Pyrolysis 98:177–183
22. Okamura M, Takagaki A, Toda M, Kondo JN, Domen K, Tatsumi T, Hara M, Hayashi S (2006) Acid-catalyzed reactions on flexible polycyclic aromatic carbon in amorphous carbon. Chem Mater 18:3039–3045
23. Zong MH, Duan ZQ, Lou WY, Smith TJ, Wu H (2007) Preparation of a sugar catalyst and its use for highly efficient production of biodiesel. Green Chem 9:434–437

Chapter 6
Evaluation of Energy Demand and Air Emissions by Using the Long-range Energy Alternatives Planning (LEAP) Model in Transport Sector of Punjab, Pakistan

Sheikh Saeed Ahmad and Syeda Qamar Batool

Abstract Air pollution is an emerging environmental issue in major cities of Pakistan. One of the major causes of air pollution is urban transportation. The purpose of this chapter is to provide a critical overview of air pollution due to transport emissions and evaluation of energy demand for this sector as well. Long-range Energy Alternatives Planning (LEAP) model was used to present existing scenario and future projections of energy capacity in transport sector of Punjab province of Pakistan. Diesel- and gasoline-based vehicular emissions and their growth rates in the next 30 years were anticipated according to current situation. Solar energy and biodiesel along with other alternative scenarios were proposed to avoid air emissions and shortfall of diesel fuel for passenger transport in near future. The results indicated that the number of vehicles and vehicular emissions were increasing with increasing population size leading to high energy consumption. The growth rate of vehicles resulted as 15.6 % with population growth rate of 1.82 % per annum, the highest growth rate of motorcycles was 10 % followed by 8.2 % of cars. The alternative scenarios given in this research will not only reduce pollutants in the air but also provide best replaceable cost-effective fuel for demand fulfillment.

Keywords Air pollution · Transport emissions · Air emissions · Vehicles

6.1 Introduction

The continuous and rapid growth in population, urbanization, industrialization, and transportation in Pakistan in recent years has caused tremendous damage to the environment. Air pollution has been consistently linked with substantial burdens of ill-health in developed and developing countries. The rapid growth in motor vehicle and other energy-intensive activities in Pakistan is important not only because of their locally harmful air pollution effects, but also because of their regional and

S. S. Ahmad (✉) · S. Q. Batool
Rawalpindi, Pakistan
e-mail: drsaeed@fjwu.edu.pk

A. Sayigh (ed.), *Renewable Energy in the Service of Mankind Vol I*,
DOI 10.1007/978-3-319-17777-9_6

global impacts. However, very little work has been done so far to assess the environmental implications of urban transport in developing countries. Lack of suitable tools and unavailability of reliable data are loopholes for environmental assessment in developing countries [7]. Punjab is the largest province of Pakistan by population. It is also the hub for all sectors including economy, transport, education, health, industry, etc. Major cities of Pakistan are also located in Punjab province. Due to high population rate, rapid urbanization, and increasing transportation, Punjab province is most vulnerable in attaining sustainability [1]. The high activity level in any city also leads to escalating vehicular numerical value. In low-income countries, transport sector is the major contributor towards energy consumption and emissions [3]. There are several factors responsible for ignorance of environmental implications due to urban transportation, such as lack of suitable tools, unavailability of reliable data for analysis, etc. The main purpose of the chapter is to identify the gap between the demand and supply of diesel and gasoline-based fuel in transportation sector of Punjab along with the evaluation of vehicular air emissions to avoid bad effects on public health.

6.2 Materials and Methods

The Long-range Energy Alternatives Planning (LEAP) model was used to analyze energy demand and forecast its fuel-based emissions from baseline year 2010 to 2040 by suggesting alternative strategies for the transport sector in Punjab province of Pakistan. LEAP is an energy planning system developed by the Stockholm Environment Institute, Boston (SEI-B) [6]. The LEAP model requires data for the selected base year and any of the future years. The present study was divided into two parts to evaluate energy demand and increasing emissions; first part focused the type of vehicles—both light or heavy duty—and their ownership—either public or private. This part depicted the number of vehicles of each type from 2010 to 2040. The second part analyzed the energy demand and forecasted the emissions.

6.2.1 Travel Demand

Only the road transport was considered due to its high proportion in total consumed energy in the transport sector. In the LEAP model, total passenger travel demand was expressed in terms of passenger-km. Therefore, total travel demand in the sector level was estimated as follows:

$$\text{Travel demand}(t) = \Sigma \text{Vi}(t) \times \text{VKTi}(t) \times \text{vehicle occupancy rate.} \tag{6.1}$$

Table 6.1 Critical parameters for estimation of energy in 2010

	Registered vehicles	% share in total vehicles	VKT (Avg. annual distance)	Load factor	Fuel efficiency km/L
Light-duty vehicles					
Cars, jeeps, and wagons	1,156,353	15.4	30,066	2.6	0.09
Motorcycles	5,253,715	70	9000	1.6	0.025
Rickshaws	175,921	2.35	39,000	1.8	0.081
Heavy-duty vehicles					
Bus	72,103	0.96	60,000	50	0.28
Pickups	102,681	1.37	56,000	15	0.05
Tractors	652,284	8.7		1.8	0.5
Others	6041	0.080	8000	18	0.1

6.2.2 *Fuel Efficiency*

Fuel efficiency is the average fuel consumption of a vehicle per year in km/L. The average annual distance, registered vehicles, vehicle occupancy, percentage share of vehicles, and fuel efficiency are shown in Table 6.1.

6.2.3 *Energy Demand*

Therefore, the total energy consumption of a vehicle can be calculated by the following equation (Table 6.2):

$$\text{Energy demand} = \Sigma Vi(t) \times VKTi(t) \times Fi(t). \tag{6.2}$$

Table 6.2 Emission factors for each type of vehicle

	NO_x	SO_2	PM_{10}
Light-duty vehicles	g/km	g/km	g/km
Private			
Cars, jeeps, and wagons	2.81	0.66	0.48
Diesel gasoline	3.81	0.02	0.06
Motorcycles gasoline	0.10	0.0	0.06
Public			
Rickshaws gasoline	0.54	0.01	0.18
Heavy-duty vehicles			
Bus diesel	5.40	4.78	2.50
Tractors diesel	233.34	5.58	1.25
Pickups diesel	1.56	0.67	0.23
Gasoline	1.15	1.01	0.03
Others gasoline	3.89	0.01	0.12

6.2.4 *Scenario Analysis*

One of the specialties of LEAP model is that it provides future projections in different scenarios. Business as usual (BAU) is indigenous scenario of LEAP while other scenarios are also proposed according to the feasible strategies. Different scenarios for the present research were plotted which are explained as follows:

- BAU

For this scenario, 2010 was selected as baseline year and 2040 was taken as end year for this research. BAU simulates data till the supposed end year without any changes.

- Use of biodiesel

The potential of biodiesel was considered 10% for transport sector. Policy-makers are focusing on biodiesel as an area of interest to fill the gap of shortfall of other fuels, such as diesel, CNG, gasoline, etc.

- Use of solar energy

Solar energy is a renewable source of energy and best alternative for any energy source. Solar energy use in vehicles can reduce the vehicular emission rate.

- Reduced population

Another scenario proposed in the present study was population reduction. Increasing number of individuals also demand increased transportation. It was assumed that 7% reduction in population can ameliorate the current situation.

- Replace private with public transport

Growth rate of motorcycles was highest in Punjab and it demands high supply of gasoline and diesel which results in more air pollutants. Shift from private to public transport can balance the demand and supply of vehicular fuel along with comparative cleaner atmosphere. Shift from private to public was estimated up to 8%.

6.3 Results and Discussions

The results obtained from the LEAP model can assist in formulating future energy budgets for fuel and a sustainable environment.The selected base year was 2010, and it was extended up to 2040 to study future increase in vehicle number consequently increasing the emissions of NO_x, SO_2, and PM_{10}. Figure 6.1 shows an increase in vehicle number from baseline year to end year. It was 7.4 million in 2010 which would increase up to 500 million with growth rate of 15.6% in 2040 (Figs. 6.2, 6.3).

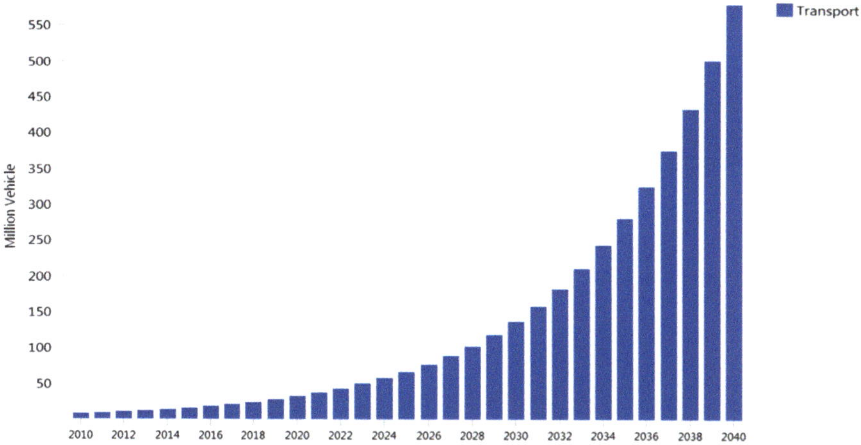

Fig. 6.1 Increased demand of transportation from 2010 to 2040

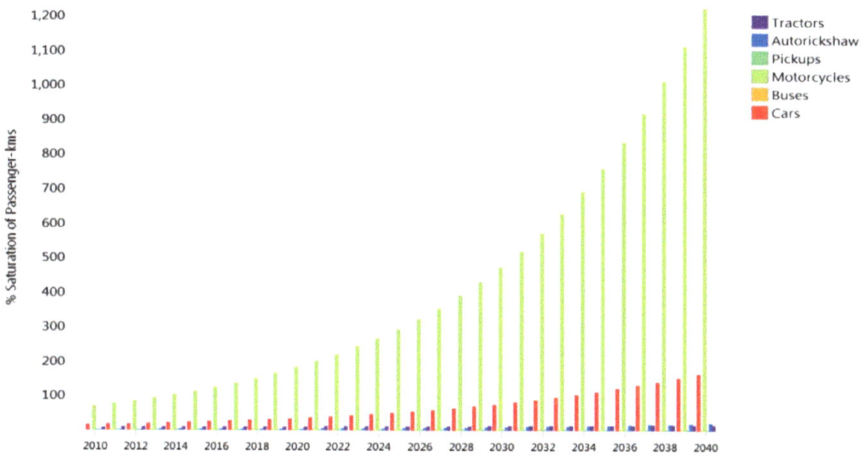

Fig. 6.2 Percentage increase in each type of vehicle till 2040

6.3.1 Comparison with Baseline Scenario

Comparison of baseline year with other alternatives is shown in Fig. 6.4. In baseline scenario, emissions are 26,000 tonnes while they are reduced to only 16,000 with biodiesel and by replacing private with public transport. However, these emissions are further cut down to 10,000 tonnes with solar energy and population reduction. The growth rate of vehicles has shown that motorcycles are increasing at highest rate of 20 % which need to be ultimately changed by improving the public transport.

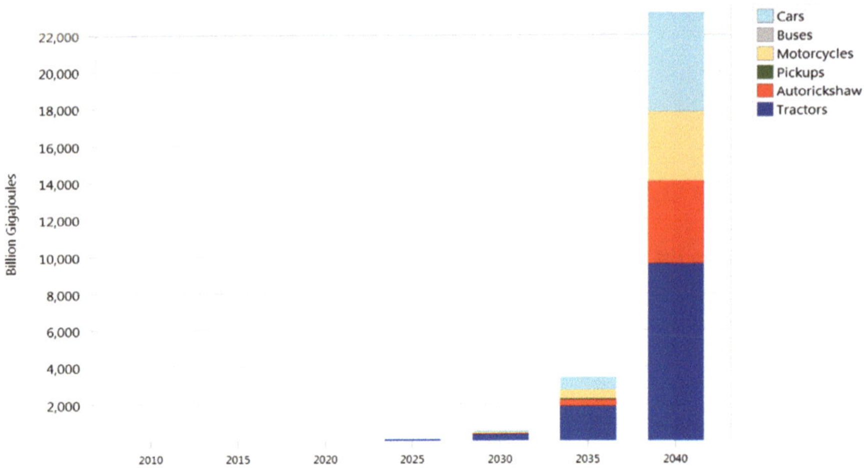

Fig. 6.3 Energy demand in baseline year and end year for each vehicle

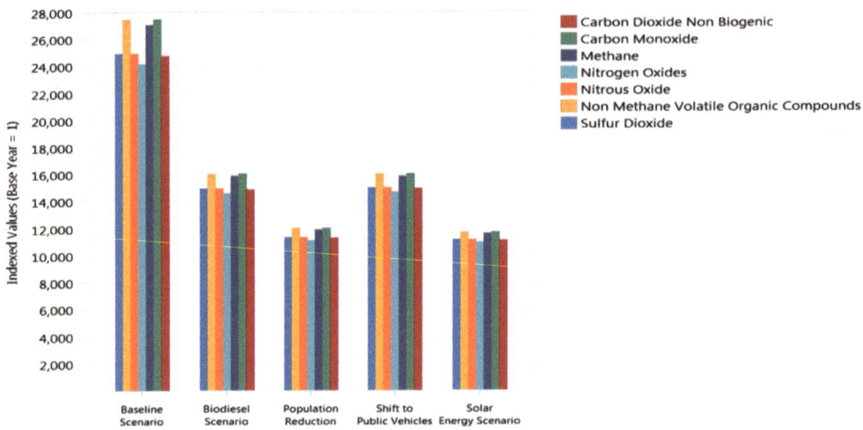

Fig. 6.4 Comparison of baseline year with alternative scenarios

6.4 Discussion

Over the past 10 years, road traffic has grown significantly faster than any other aspect in Punjab province of Pakistan. Nearly 60 % of the population of Pakistan lives in Punjab province. Population growth rate is an important factor contributing to increased vehicle ownership [1]. Due to rapid rate of motorization, urban transport is gradually increasing the state of air pollution in many Asian cities [2]. Over the past three decades, motor vehicle numbers have doubled every 10 years or so in many Asian countries, as against a 2–5 % annual growth rate in Canada, the USA, the UK, and Japan. It is estimated that by 2040 low-income countries could have as many vehicles as North America and Western Europe [5]. The number of motor vehicles will jump from 7.4 million in 2010 to 550 million in 2030, an increase of over 100% due to a tremendous increase in population. In present study of energy demand and emission analysis for the transport sector, 2010 was taken as base year and 2040 was taken as end year. The emission analysis of three pollutants, i.e., NO_x, SO_2, and PM_{10} was done. In BAU scenario, the registered vehicles, travel demand, energy consumption, and emission level increased. The present study showed that among all scenarios reduction in population and solar energy proved to be better. Population reduction is the best alternative because it will ultimately reduce transport growth rate. Best management of urban transport system needs collaboration of government and public measures. Planning policies and financial sources should consider the alternative fuels for better efficiency and cleaner air conditions.

6.5 Conclusion

The study has shown that the LEAP system combined with apposite data can provide a valuable tool for assessing future urban transport options in developing countries. Steps for promotion of solar vehicles and use of biodiesel instead of gasoline and diesel engines can also go a long way in reducing air pollution due to automobiles in Punjab and consequently in all over Pakistan. BAU has shown a rampant increase in energy demand and the resultant emissions. The alternative scenarios have shown that the increase in energy consumption and pollutant levels can be limited through measures such as encouraging a shift to buses, reducing the population, and substitution of biodiesel and solar energy for diesel and gasoline vehicles. A combination of scenarios given in this study can ameliorate the situation effectively as using only one alternative would be insufficient to control the transport situation in Punjab.

References

1. Bureau of statistics (2011) Punjab Development Statistics. Lahore, Government of Punjab
2. Faiz A, Weaver C, Sinha K, Walsh M, Carbajo J (1992) Air pollution from motor vehicles: issues and options for developing countries. The World Bank, Washington, DC
3. Gru¨ bler A (1994) The transportation sector: growing demand and emissions. In Krishnan R (ed) Growing numbers and dwindling resources. Tata Energy Research Institute, New Delhi, pp 44–57
4. Imran M (2009) Public transport in Pakistan. J Public Transp 12(2):55–83
5. Jraiw K (1989) Urban road transport in Asia's developing countries: safety and efficiency strategy. Transp Res Rec 2003
6. (2008) Long range energy alternatives planning system. SEId Stockholm Environment Institute, Tellus Institute, Boston
7. Shabbir R, Ahmed SS (2010) Monitoring urban transport air pollution and energy demand in Rawalpindi and Islamabad using LEAP model. J Energy 35(2010): 2323–2332

Chapter 7
Issues and Challenges of Implementing Waste-to-Energy Practices in India

J. D. Nixon, D. Wright, P. K. Dey, J. A. Scott, S. Sagi and S. K. Ghosh

Abstract India's rapid development and urbanisation have led to major issues and challenges with the provision of adequate waste and energy management services. Conversion of waste-to-energy (WtE) through incineration is now an established method in developed countries for reducing waste mass and volume whilst simultaneously providing heat and power. However, the uptake of WtE technologies in developing countries has not been so successful. This study aims to identify the main issues and challenges with improving municipal solid waste (MSW) management and implementing WtE technologies in India. Specifically, we provide recommendations on actions to be taken to improve waste management and promote WtE technologies. In order to achieve this, we survey and conduct a workshop in India with over 50 participants comprising 26 government body members, 20 industry practitioners and 6 academics involved in waste management in India. Data are gathered on the social, political, technical, financial and environmental barriers to WtE and MSW management. We find that over 35% of government body members perceive the major problem to be poor waste segregation. In comparison, industries feel that there is a lack of government support in the form of incentives and thought the ongoing challenge remains with finding a technology that can handle the characteristics of domestic waste in India. Interestingly, there seems to be minimal opposition to WtE from the public and environmentalist, which are some of the main issues in developed countries. Recommendations

J. D. Nixon (✉)
Faculty of Science, Engineering and Computing, Kingston University,
Roehampton Vale campus, London, SW15 3DW, UK
e-mail: j.nixon@kingston.ac.uk

D. Wright · P. K. Dey · J. A. Scott
Aston Business School, Aston University, Aston Triangle,
Birmingham, B4 7ET, UK

S. Sagi
School of Engineering and Applied Science, Aston University,
Aston Triangle, Birmingham, B4 7ET, UK

S. K. Ghosh
Mechanical Engineering Department, Centre for Quality Management System,
Jadavpur University, Kolkata 700032, India

A. Sayigh (ed.), *Renewable Energy in the Service of Mankind Vol I*,
DOI 10.1007/978-3-319-17777-9_7

arising from this study are that key improvements are required in education at institutional levels, policies and regulations on the disposal and handling of MSW and financial support from central government to make plants economically viable. There is also a greater need for an increase in collaboration and communication among central and local governments, industries and communities. Further research will be conducted to gather public opinions regarding MSW management and WtE projects in India.

Keywords Waste-to-energy (WtE) · Municipal solid waste (MSW) · Energy policy · Waste management

7.1 Introduction

India faces numerous problems regarding municipal solid waste (MSW) and energy due to rapid urbanisation and population growth. Despite the introduction of the MSW Management and Handling Rules in 2000, the majority of India's cities still do not—or are unable to—comply with regulations. The result is extremely low waste collection rates, minimal waste separation and waste primarily being disposed of through dumping and unsanitary landfill [1]. Issues regarding energy in India include many areas remaining unelectrified and regular blackouts—evidenced by a nationwide blackout in 2012 that left 700 million people without power. Thus, there is a pressing need for energy recovery from MSW solutions that will provide energy services, reduce waste mass and volume and alleviate health hazards from pollution.

India generates around 55–70 million tonnes of MSW per annum and this value is rapidly increasing every year. Approximately 70 % of MSW is collected and predominantly disposed of via landfill, where 90 % of landfill sites are considered to be unsanitary. This is a major concern as leachate can contaminate ground water. The composition of MSW in India is highly variable, particularly between urban and rural locations. Typically, it is characterised by a high percentage of organics (40 %) and industrial waste (40 %), such as stone and ash. The content of recyclables in MSW, including paper, plastics, glass and metals, is low at around 1 % [2].

Over the next few years, substantial investments will be made in India in alternative energy from waste facilities. However, there are many barriers to successfully deploying waste-to-energy (WtE) facilities regarding polices, technology selection, waste management strategies and logistical supply chain issues. It is therefore necessary to identify the main issues and challenges that have prevented or will prevent the successful deployment and operation of viable WtE facilities. This study sets out to address this by surveying a range of key stakeholders across the WtE supply chain in India, and thus provide valuable data to direct future research and development efforts. The methodology that has been adopted to achieve this is now outlined.

7.2 Methodology

The study takes a holistic approach to identify the issues and challenges to implementing WtE initiatives in India. Data have been gathered from stakeholders across the entire WtE supply chain during a facilitated workshop held at the 4th International Conference on Solid Waste Management, Hyderabad, 2014. From among several hundred delegates, more than 60 people attended the workshop session and participants included WtE providers, operators, researchers and urban local body (ULB) officers. A survey was distributed to the participants to capture individual stakeholder opinions regarding issues and challenges, their cause and effects and possible solutions. The results are evaluated to see the occurrence and types of issues raised. Responses are categorised into supply side (logistical), conversion issues (technical), economical, social and political. The study concludes by providing recommendations for actions to be taken to address the main issues and causes that have been identified.

7.3 Workshop

The workshop was held in Hyderabad, India and comprised participants from across the whole of India involved in waste management; specifically, 26 local government officers from ULBs, 20 industry practitioners and 6 academics. Each participant was asked to provide answers to the following questions:

Q1. What are the major issues and challenges regarding WtE in India?

Q2. What is the cause and effect of each of these challenges?

Q3. What do you think are the solutions to these challenges?

Tables 7.1–7.3 collates the participant responses from (i) government local bodies, (ii) industry and (iii) academia. Responses are categorised as logistical, technical, economical, social and political. A record of how many times the same issue was identified is recorded in parentheses.

After completing the surveys, a discussion session was held to capture more detailed qualitative information on the data provided. A summary of the discussed logistical, technical, economical, social and political issues is now provided.

7.3.1 Logistical

Logistical issues relate to poor compliance, communication and collaboration across the MSW to energy supply chain in India. Tendering and litigation issues cause delays; however, this is not uncommon for WtE projects. Facilities struggle with obtaining a reliable supply of waste and find its composition to be highly variable. One participant reported that truck collections are contaminated with industrial waste to increase weight, and thus also income received for its delivery.

Table 7.1 Issues regarding implementing WtE strategies in India as identified by local government bodies

	Issue	Cause	Effect	Solutions
Logistical	Inadequate collections (2)	Lack of funds	Low collection efficiency and unreliable deliveries	Government support, timed appointments, better segregation to improve transport efficiency
	Segregation (1)	Inefficient collections and contamination at intermediate dumpsites	Poor waste quality	Use segregation facilities and improve intermediate dumpsites
	Litigation issues (1)	Legal procedures	Increases cost and time	Removal of litigation matters
Technical	Viable technologies (5)	Currently technologies are not affordable or suitable	Heavy reliance on landfill	Grants and subsidies, build infrastructure and better characterisation of waste
	Lack of disposal sites (2)	Low land availability in many ULBs	Open dumping	Use technologies with compact footprint
	High moisture content of waste (2)	Segregation, no pretreatment, poor handling, high level of agricultural waste	Low calorific value and can be more difficult to process	Improvement to technologies
Economical	Inadequate funding (2)	Limited funds from central government	Lack of investment	Policy improvements and more funds from central government
	Cost of WtE (2)	Not currently affordable to ULBs		Grants needed
Social	Environmental issues from pollution control board (1)	Land lease agreement issue, environmental clearance certificate	Delay in project planning	Faster permitting so projects can start sooner
	Source segregation at homes and industries (9)	Lack of knowledge	Difficult to process due to composition (wet wastes and industrial wastes)	Increase public awareness and engagement, introduce penalty, fines, etc.
	Public awareness of waste being a problem (6)	Poor education	Perceive waste to be the government's problem	Better training, education system and literacy
	Contamination during transport (1)	Mixed with industrial waste	Difficult to process	Introduce penalty, fines, etc.
	Prevention of waste (2)		Minimise material usage and waste generation	Policies, strict laws and community engagement

Table 7.1 (continued)

	Issue	Cause	Effect	Solutions
Political	Inconsistent political campaigns (1)	Political motivation	Project delays	Improve political awareness and better engagement with society
	Political interference (1)	Lack of knowledge in ULBs	Not in my back yard policies	Only higher authorities can take up issues

ULB urban local body, *WtE* waste to energy

Table 7.2 Issues regarding implementing WtE strategies in India as identified by industry

	Issue	Cause	Effect	Solutions
Logistical	Unreliable schedule and delays to bidding process (1)		Foreign companies hesitant to invest in India	Monitoring service and incentives for private sector
Technical	Quality of waste (3)	No segregation	Difficult to process by incineration	Segregation facilities, improve source segregation and rules for segregation
	Lack of proven technology for MSW to energy in India (1)	Lack of incentives	Lack of investment	Increase R&D
	Lack of support and understanding by ULBs (2)	No expertise in ULBs for planning and implementation WtE projects		Deploy MSW expert officers in ULBs
	Lack of data on waste quality and quantity (2)		Poor availability of data for service providers	Universities need to study Indian waste quantity and quality
Economical	Lack of feed-in tariffs and incentives (1)	Government not in favour of WtE projects	No business case	More WtE plants set up by local governments
	No clear PPP model and difficult to deal with ULBs and government (3)	ULBs do not share risk	Lack of investment and causes project delays and increases project cost	Relax policies and PPP model to be defined for each state. ULBs need to share risk
	Cost of WtE plants (2)			Introduce government initiatives and tipping fee
	Limited finances for ULBs (1)		ULBs fail to support private partnerships	Strengthen local bodies financially and encourage PPP
Social	Source segregation (5)	Lack of awareness, no access to disposal or recycling facilities	Loss of valuable resources, high moisture contents and low calorific value	Research alternative methods for utilising MSW and separation technologies, introduce segregation policies
	Open dumping (2)	Lack of public awareness	Health hazards	Increase public awareness that waste is a resource
	Public attitude that waste is the state's responsibility (1)	Poor education	Dumping and segregation	Educate society that waste is a resource
Political	Lack of policies and regulation (2)	Lack of expertise in governments	Makes project feasibility and risk mitigation difficult	Clear policies from central and state governments
	Lack of credible monitoring body (1)		Emissions can be cheated making WtE plants untrustworthy	Elimination of corruption and introduce government monitoring service

MSW municipal solid waste, *ULB* urban local body, *WtE* waste to energy, *PPP* public private partnership

Table 7.3 Issues regarding implementing WtE strategies in India as identified by Indian academics

	Issue	Cause	Effect	Solutions
Logistical	Lack of supply chain linkages between waste, segregation, recycling and WtE		Ineffective waste management and WtE facilities	Development of integrated waste management strategies
Technical	Lack of advanced WtE technologies (2)	Lack of skilled expertise and funding for R&D		Increase funding by government and agencies
	High moisture content (1)	Source segregation	Difficult to process with conventional incineration	Better technology to process unsegregated waste
	Poorly planned and implement plants (1)	Lack of human resources		Accelerate capacity building programmes
Economical	WtE not economical in comparison to other power generating technologies (1)	Lack of incentives	Lack of investment	Introduce tariffs and subsidies
	Costly equipment (1)	High project cost		Seek cost reductions
Social	Lack of knowledge and awareness of general public (4)	Lack of education	Poor segregation	Improve education and community and media engagement
Political	Lack of policies (2)		No investments due to lack of incentives, heavy reliance on landfill	Introduce short, medium and long term incentives, and fines for non-supply of waste

WtE waste to energy

Whilst waste is often separated at source into dry and wet wastes, the MSW becomes contaminated at intermediate and final dumpsite locations due to poor facilities and management.

7.3.2 Technical

WtE technologies are not established in India and historically installed plants have not been successful. The workshop panel attributed this to the unreliable quantity and quality of waste and lack of expertise in local government. They believe that bespoke technologies need to be developed that can process India's variable MSW. Several government body members felt that land limitations are a major challenge for the effective disposal of MSW.

7.3.3 Economical

WtE projects are not economically feasible in India as there is no standard gate or tipping fee—like in parts of Europe—or incentives such as feed-in tariffs. This has resulted in minimal investment being made in the WtE sector. The industry participants felt that local bodies are reluctant to form public private partnerships and share risk in projects and guarantee a consistent and reliable supply of waste. The local governments remarked that central government needs to increase the money available to local bodies to spend on collections and developing facilities.

7.3.4 Social

The greatest number of issues raised by the workshop panel related to (i) poor public awareness of how to handle waste and (ii) lack of source segregation leading to vast amounts of open dumping. This is considered to be a result of poor education and the perception that waste is the government's problem. However, one attendee argued that facilities are not provided to the public to segregate, recycle and dispose of their waste, and collections are unreliable and inadequate, which suggests a deeper underlying issue in the supply chain.

7.3.5 Political

Despite the introduction of the MSW Management and Handling Rules in 2000, India has a lack of policies and regulations relating to energy recovery from waste. Political interference and corruption are a concern for investors as they delay projects

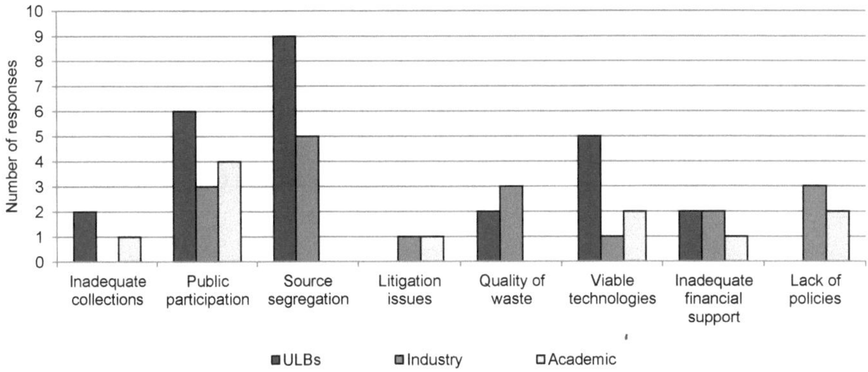

Fig. 7.1 Top issues identified by ULBs, industry and academics

and make risk mitigation difficult. A lack of monitoring of WtE facilities has also resulted in plants falsifying emission data and breaching safe operating conditions.

7.4 Results and Discussion

Many issues have been raised several times in the outputs from this study. Figure 7.1 shows the number of survey responses on a particular issue from local bodies, industries and academics. The three main issues preventing the successful deployment of WtE technologies in India are indicated to be the lack of source segregation, public participation and viable technologies. Around 25 % of the workshop participants perceived poor public engagement and source segregation to be the biggest challenge for WtE in India. With MSW in India characterised by high moisture and industrial and commercial waste content, eight attendants thought that new viable technologies need to be developed that can handle India's variable waste. Additional preprocessing of waste, increasing the settling time of MSW in bunkers and improved management of storage bunkers would dry and reduce moisture content and improve waste composition. Integrating segregation facilities and WtE processing plants is one way to help address some of these cultural and society challenges. From among the few WtE plants operating in India (Shalivahana MSW Green Energy Ltd, Rokem Power Plant and Ecopolis), most of them are using segregation facilities to try and improve the MSW quality [3, 4]. However, it is also increasing the cost of the MSW feedstock by approximately £ 30/t. Nevertheless, separated recyclables and compost can be sold for additional revenue.

Lack of polices was the second biggest concern for industry participants and the absence of policies and supportive incentives, such as a gate fee or feed-in tariffs, for WtE projects makes them economically unviable in India. The low level of financial support provided to local bodies from central government also results in a lack of collections and public private partnerships as local bodies are unwilling to share risk with private companies. Given that approximately half the capital cost of a WtE facility is for processing and cleaning flue gases, research and development needs to focus on seeking cost reductions in this area. Many of the ULBs in India are now struggling to find space to locate new landfill sites. This should drive the uptake of WtE facilities as they have a far more compact footprint in comparison to landfill sites. An example of where this has already started to occur is Pune Municipal Corporation, which has several WtE initiatives in place.

7.5 Conclusion

There is clearly a pressing need in India to make WtE projects more viable and this study has identified some of the main issues and barriers that have prevented their greater deployment in India. Poor waste segregation has repeatedly been identified as a critical challenge for WtE in India from a range of stakeholders. A number of recommendations to improve segregation so that MSW can be processed more efficiently in India arise from this study: introduce contracts for waste collections to improve reliability and reduce contamination; control and regulate intermediate dumpsites; increase public awareness and participation; provide additional disposal facilities; integrate segregation and waste processing facilities following best practices that have been developed in western countries. However, there are many other issues that also need to be addressed. The economics of WtE must be improved by introducing tax exemptions, tipping fees, feed in tariffs and soft loans to make projects feasible and increase investment. Polices need updating to guide, regulate and increase public private partnerships. Staff training at local bodies on waste management and consulting industrial experts also needs to be increased and the lack of reliable emissions monitoring is a particular concern.

This study provides a unique presentation of the current issues and challenges facing WtE operations in India from the perspective of a large number of government body members and industries operating from across the whole of India. Thus, we believe that these findings will be of particular value to policy makers, local governments and central government working on waste management and WtE in India and elsewhere.

References

1. Wolfe JM, Mahadevia D (2008) Solid waste management in Indian cities: status and emerging practices. Concept Publishing Company, New Delhi
2. Unnikrishnan S, Singh A (2010) Energy recovery in solid waste management through CDM in India and other countries. Res Conserv Recycl 54:630–640
3. Shalivahana Green Energy, MSW Projects in Operation (2011) http://www.shalivahanagroup.com/home.php?cat=msw-projects&stage=3. Accessed 12th June 2014
4. Timarpur Okhla Waste Management Company Pvt Ltd (2011) http://towmcl.com/content.aspx?MKey=33. Accessed 12th June 2014

Chapter 8
The Variation of Ash and Inorganic Elements Concentrations in the Biomass of Lithuania-Grown Switchgrass (*Panicum Virgatum* L.)

B. Butkutė, J. Cesevičienė, N. Lemežienė, E. Norkevičienė, G. Dabkevičienė and Ž. Liatukas

Abstract Switchgrass (SG) is a perennial grass of C4 type of photosynthesis and has been identified as a potential biomass crop in North America. In Europe, this species has been studied in the Atlantic maritime climate zone and in southern countries. In Lithuania, SG is a novel nonfood crop, whose investigation as a multiuse bioenergy plant, particularly as a feedstock for solid biofuel, has just been started. In this study, our attention was focused on the accumulation of biomass components that have an impact on air pollution and corrosion of combustion equipment, as well as slagging. We quantified the contents of ash and major elements of whole aboveground biomass, stems, leaves, and panicles of SG accessions, promising for growing under Lithuanian conditions. The trials were conducted during 2010–2014 at the Lithuanian Research Centre for Agriculture and Forestry on an *Endocalcari-Epihypogleyic Cambisol (CMg-p-w-can)*. Chemical composition of the biomass was assayed by reference methods.

SG biomass, cut at seed maturity had N, ash, K, Cl, Si, and S concentrations averaging 7.94, 56.5, 12.4, 2.79, 4.26, and 0.92 o kg^{-1} DM respectively. Biomass of SG plants from a collection planted in 2010 had less ash, Mg, and N in both harvesting years than SG from the collection set up in 2011. Biomass of C3 energy plant reed canary grass (RCG) contained higher ash, K, N, S, and Cl concentrations (66.8, 19.1, 14.3, 1.41, and 7.02 g kg^{-1} DM respectively) compared to those of SG. Genotypic variability in the inorganic composition occurred within SG genotype, but the differences among years were greater than those among accessions. With a delay in harvesting, the greatest reduction occurred in K and Cl concentrations: from September to February K concentration in the biomass of whole aboveground plant part (WP) changed from 12.3 to 3.25 g kg^{-1} DM and that of Cl altered from 2.26 to 0.590 g kg^{-1} DM, i.e., decreased by 74 %. Ash, N, and Mg concentration in the biomass declined by 41, 52, and 47 % compared with the biomass cut at seed

B. Butkutė (✉) · J. Cesevičienė · N. Lemežienė · E. Norkevičienė · G. Dabkevičienė · Ž. Liatukas
Institute of Agriculture, Lithuanian Research Centre for Agriculture and Forestry, Instituto al. 1, LT-58344 Akademija, Kėdainiai, Lithuania
e-mail: brone@lzi.lt

A. Sayigh (ed.), *Renewable Energy in the Service of Mankind Vol I*,
DOI 10.1007/978-3-319-17777-9_8

maturity. Concentrations of ash, Ca, Mg, Si, N, and S in the biomass of stems of all harvesting dates were the lowest compared with the biomass of leaves and panicles. Concentration of K and Cl was practically equal in all plant parts at the respective harvesting time. For all harvesting dates, N and S concentrations in panicles were the highest.

Summarising our data presented in the current and previous papers we conclude that SG could be a useful perennial crop with a high potential of dry matter (DM) yield and adequate quality for renewable energy purposes and suitable for cultivation under conditions of nemoral environmental zone of Lithuania. By quality, overwintered biomass is the most appropriate for biofuel production compared to that of harvested in the autumn.

Keywords *Panicum virgatum* · Biomass · Solid biofuel · Mineral composition

8.1 Introduction

On 23 April 2009, the European Parliament and the Council adopted Directive 2009/28/EC on the promotion of the use of energy from renewable sources and endorsed mandatory targets to be achieved by 2020 for a 20 % overall share of renewable energy in the EU [1]. The choice of renewable energy source in a country or region is determined mainly by their geographical, topographical, and hydrological conditions. The renewable energy sources in Lithuania used for heating are biomass, geothermal, and solar energy. Biomass accounts for the major part of the three resources available. According to Smeets et al. [2], Lithuania has the highest availability of surplus arable land that will be potentially available for energy crop production of all countries in the EU-25 by the year 2030. Directive 2009/28/EC [1] emphasizes, that the Member States may encourage the use for biofuel of nonfood feedstocks which gives additional benefits, including the benefits of diversification. It is required to develop existing biomass resources and mobilise new biomass resources.

Searching for novel energy crops, attention was paid to C4 herbaceous rhizomatous perennials plants. They have many positive attributes as potential and actual biomass energy crops on agricultural land [3]. One of them is SG (*Panicum virgatum* L.). Although SG is a grass of C4 photosynthetic pathway, some accessions appeared to be suitable for growing in the cool and temperate regions of the world [4]. SG is native to North America where it occurs naturally from 55°N latitude to deep into Mexico [5]. An interest in the species has increased worldwide (especially in the USA): SG as an energy crop is used for solid biofuel, lignocellulosic bioethanol, and biogas production. It has become one of the potential new energy crops in the European countries with Mediterranean climate in the South and oceanic climate in the Western Europe, i.e., in the regions that lie between the latitudes of 40°09′N (Greece) and 52°38′N (Netherlands) [5]. Lithuania is located further north—between 53°54 and 56°27′N latitudes. A SG germplasm collection was obtained from the Plant Genetic Resources Conservation Unit (PGRCU) of the Agricultural Research Service (ARS), USA opened up an opportunity to investigate this species under Lithuanian pedoclimatic conditions and the first results showed that the species may expand the diversity of renewable energy resources in the nemoral zone of Europe [6, 7]. Based on the economic situation of 2004, calculations of Smeets et al. [2] for SG cultivation in Lithuania revealed that the costs of production (including storage and transportation across 100 km) per oven dry ton of biomass would amount to 43–64 € or per GJ^{-1} higher heating value to 2.4–3.6 €.

A problem that SG and other herbaceous energy crops pose during combustion is the amount of ash in the biomass material [8]. In direct combustion systems, the mineral concentration can cause corrosion, slagging, and fouling of boilers [9], [10], [11]. Possible also are problems regarding NOx, SOx, and HCl emissions with combustion of herbaceous biomass. [12]. A successful development of a bioenergy industry depends on identification of cultivars with high yield potential and acceptable biofuel quality [13]. However, not only the species/genotype [13], [14], [15] is responsible for biomass quality but location [5], [9], [14], harvesting time [7], [9] [16], [17], [18], herbage age/harvest year [5], [7], [17], and fraction of plant aboveground biomass [7], [17], [19], [20] are also other very important factors affecting chemical composition of herbage. In this study, our attention was focused on the accumulation of biomass components that have an impact on air pollution and corrosion of combustion equipment, as well as slagging; and the challenge was to identify the extent and source of variation of their content in the biomass of Lithuania-grown switchgrass.

8.2 Material and Methods

Plant Material and Field Trial Conditions The collection trials were carried out during 2010–2014 in the Central Lowland of Lithuania at the Lithuanian Research Centre for Agriculture and Forestry (LRCAF) (55°23′49″N 23°51′40″E). Experimental germplasm collections of SG were set up using the seedlings grown in a greenhouse. Seeds were obtained from the (PGRCU) of the United States Department of Agriculture (USDA) and ARS. Three germplasm collections were established: in 2010, 2011, and 2012. The collections included cultivars, breeding lines, and wild ecotypes. The majority of the accessions were northern upland ecotypes from North Dakota. In the collection, each accession consisted of 28 plants spaced 0.5 m apart with seven plants per row, two rows per replication, and two replications per treatment. For comparison of SG chemical composition with that of traditional in temperate climate C3 perennial energy grass, reed canary grass (RCG) variety "Chiefton"" was used. RCG was planted according to the same design as SG. The biomass for chemical composition evaluation was taken in 2012 and 2013 harvesting years. Biomass of the collection set up in 2010 harvested in 2012 and 2013 was qualified as a biomass of the 2nd and 3rd year of production, herbage of collection planted in 2011—as a biomass of the 1st and 2nd year of production respectively. And finally SG collection which was planted in 2012 was approached as a stand of the first year of production in 2013 harvesting year. Grass as feedstock for solid biofuel was cut once a year at seed maturity (on 13 September), senescent plants after frost killing (on 24 October and 07 November), and owerwintered plants (on 21 February). Separation of the sampled herbage into three botanical fractions (leaves: blades+sheaths, stems, and panicles) was additionaly performed.

The soil of the experimental site is *(CMg-p-w-can)* with the following characteristics of the plough layer (0–25 cm): pH ranging from 6.3 to 7.3, humus content 1.4–3.6 %, available phosphorus (P_2O_5) 143–265 mg kg^{-1}, available potassium (K_2O) 109 and 209 mg kg^{-1}. Fertilizer at a rate of $N_{60}P_{60}K_{60}$ was applied preplanting once and N_{60} was applied each spring of the following years. Weeds were controlled by herbicides every year.

According to the environmental stratification of Europe, Lithuania is assigned to nemoral zone with the cool temperate climate and quite a short growing season of 190–195 days [21]. The weather conditions during the experimental period differed rather markedly between the seasons. The first season 2010/2011 was characterized by a cold winter, wet summer, and autumn; the second season 2011/2012—by a mild winter, early spring, and wet summer; the third season 2012/2013—by a cold and long lasting winter, very late, dry and warm spring, and changeable summer with heat waves. Autumn–winter period of 2013/2014 can be defined as unusually warm with temperatures prevailing positive; the average monthly temperature in January and February was more than 5 °C higher than long-term average.

Sample Preparation and Chemical Analyses. Fresh samples, chopped into particles of 3–5 cm, were fixed at 105 °C for 15 min, dried at 65 ± 5°C, and ground in a cyclonic mill with a 1 mm sieve. Prior to analysis, the samples prepared in the following way were assessed for dry matter (DM) according to the standard method NREL/TP-510-42621 [22]. The ash content (on DM basis) was determined by the standard method consistent with EN 14775:2009 [23]. Determination of silicon (Si) content was performed spectrophotometrically on a Cary 50 UV-Vis (Varian Inc., USA; 660 nm) according to the method reported by Kraska and Breitenbeck [24] and Dai et al. [25]. Sulphur (S) and nitrogen (N) in the samples were assessed by dry combustion on a CNS analyser (Elementar, Vario EL, Germany). Potasium, calcium, and magnesium (K, Ca, and Mg respectively) concentration in the wet digested samples were quantified by flame atomic absorption spectroscopy (FAAS). Parameters of the AAnalyst 200 instrument (Perkin Elmer, USA) were chosen in accordance with the manufacturer's instructions. Chloride (Cl) was extracted with water solution of 0.085 mol L^{-1} $Ca(NO_3)_2$ from a dried material. Cl content was determined by Liu [26] and Gaines et al. [27] spectrophotometrically after Cl substitutes the thiocyanate ions in mercury thiocyanate and the released thiocyanate forms a red iron-thiocyanate complex. Absorption was measured at the wavelength of 460 nm.

8.3 Results and Discussion

The concentrations and composition of inorganic elements of biomass are the crucial factors determining whether or not a feedstock can be used for combustion effectively and eco-friendly. Table 8.1 shows the averaged data of ash and major elements concentration in biomass of SG and RCG samples harvested at seed maturity stage for two harvesting years and 3 years of herbage production.

Comparison of ash concentration and biomass mineral composition of RCG with that of SG, proved SG to be a more valuable energy crop for combustion with lower ash, K, N, S, and Cl concentration. RCG biomass had lower Ca concentration than SG. Mg and Si values fitted in the range of respective concentration in SG biomass. Although our research results confirmed that SG biomass, cut at seed maturity had higher N, ash, K, Cl, and Si concentration (on average 7.94, 56.5, 12.4, 2.79, and 4.26 g kg^{-1} DM respectively) compared with hardwoods and softwoods composition as reported by Telmo et al. [28], McKendry [29] as well as wood of short rotation forest (willow and hybrid poplar) [30], S concentration (0.92 g kg^{-1} DM) was nearly equal to that found in wood of willow and hybrid poplar. Concerning N, S, and ash, the concentration of these components in SG biomass was noticeably lower than that in coal and lignite [30], [31]. The impact of both harvest year and plant age (year of production) on inorganic composition of SG plant biomass was unreliable. As for plant age, consistent patterns were found for ash, Mg, and N only in both harvest years as well as for the majority of tested components in 2012 harvest year. Averaged data revealed that biomass of SG plants from a collection planted in 2010 had less ash, Mg, and N in both study years: 53.4 ± 4.30 g ash kg^{-1} DM, 3.61 ± 0.31 g Mg kg^{-1} DM, and 6.57 ± 0.77 g N kg^{-1} DM in 2012 as well as 43.8 ± 3.31 30 g ash kg^{-1} DM, 2.36 ± 0.27 g Mg kg^{-1} DM, and 4.61 ± 0.91 g N kg^{-1} DM in 2013. SG biomass cut in 2012 from accessions of the collection planted in 2010 (2nd year of production) contained less K, Si, S, and Cl; however, more Ca than SG biomass from collection planted in 2011 (1st year of production). However, it must be recognized that the differences in chemical composition of biomass from different collections may be related to the chemical and physical soil properties where collections were set up. High values of standard deviation (SD) and wide range of variation in constituent's concentration within each collection, especially within that planted in 2011 is an indication of genotypic variability, which in addition to the variability in agrobiological traits described [6] would be feasible to select populations for the desirable combination of chemical and agrobiological properties through a SG development program under nemoral zone conditions.

According to the data of two-way analysis of variance (ANOVA), the differences in mineral composition of SG biomass between years were considerably larger than the variability associated with the genotype (Table 8.2).

In particular, a small variation in the investigated components among accessions were observed in the collection planted in 2010—only biomass of two ecotypes

Table 8.1 Effects of year and SG plant age on the concentration of ash and inorganic elements (g kg^{-1} DM) in the biomass of SG harvested once per year at seed maturity in comparison with RCG chemical composition. Data represent values of a number of SG accessions in collections: 2010/2nd −18; 2011/1st i10; 2010/3rd i9; 2011/2nd 7 and 2012/1stomass of SG harvested once per year at seed maturity in comparison with, 1st, 2nd, and 3rd year of production

Constituent		2012 harvest year			2013 harvest year				Mean
		2010/2nd	2011/1st	Average	2010/3rd	2011/2nd	2012/1st	Average	
Ash	Average	53.4	63.4	57.0	43.8	74.4	54.9	56.0	56.5
	Range	47.0–60.0	52.8–70.3	47.0-70.3	38.9–50.5	56.5–96.9	45.7–61.1	38.9–96.9	
	SD	4.3	5.1	12.4	3.3	14.1	4.0	12.8	
	RCG	63.1	63.3	63.2	68.9	74.8	67.6	70.4	66.8
K	Average	13.9	16.1	14.7	10.5	8.66	11.2	10.5	12.4
	Range	12.7–15.5	14.5–17.8	12.7–17.8	9.25–12.3	7.65–9.65	9.50–12.6	7.65–12.6	
	SD	0.9	1.0	3.1	1.04	0.79	0.96	1.36	
	RCG	22.1	16.0	19.1	18.4	19.6	19.4	19.1	19.1
Ca	Average	5.01	4.69	4.89	8.84	13.23	5.02	7.81	6.47
	Range	4.13–5.91	3.84–4.99	3.84–5.91	7.75–10.4	9.60–16.1	4.30–6.50	4.30–16.1	
	SD	0.57	0.38	1.05	0.92	2.25	0.56	3.49	
	RCG	4.50	3.37	3.94	4.22	3.43	3.71	3.79	3.86
Mg	Average	3.61	4.47	3.92	2.36	3.89	2.81	2.91	3.37
	Range	3.04–4.09	3.96–4.97	3.04–4.97	2.00-2.80	2.65–5.45	2.00-3.30	2.00-5.45	
	SD	0.31	0.29	0.89	0.27	1.13	0.34	0.79	
	RCG	5.46	4.45	4.96	2.75	2.81	2.14	2.56	3.76
Si	Average	4.67	5.09	4.82	3.54	6.37	2.85	3.78	4.26
	Range	4.00-5.27	4.07–5.46	4.00-5.46	3.20–3.90	5.70–7.95	2.30–3.70	2.30–7.95	
	SD	0.38	0.30	0.98	0.26	0.75	0.35	1.46	
	RCG	5.04	4.69	4.86	4.13	4.33	3.41	3.95	4.41
N	Average	6.57	13.5	9.05	4.61	6.20	8.59	7.00	7.94
	Range	5.16–7.91	11.0–16.0	5.16-16.0	3.48–6.50	4.59–8.38	5.86–10.68	3.48–10.68	
	SD	0.77	1.62	3.88	0.91	1.47	1.18	2.10	
	RCG	15.71	8.82	12.27	15.80	16.60	16.50	16.30	14.3

Table 8.1 (continued)

Constituent		2012 harvest year			2013 harvest year				Mean
		2010/2nd	2011/1st	Average	2010/3rd	2011/2nd	2012/1st	Average	
S	Average	0.67	1.02	0.79	0.95	0.68	1.28	1.06	0.94
	Range	0.46–0.85	0.84–1.18	0.46–1.18	0.79–1.21	0.51–0.84	1.04–1.47	0.51–1.47	
	SD	0.08	0.12	0.24	0.12	0.13	0.12	0.27	
	RCG	2.13	1.26	1.70	0.82	1.01	1.54	1.12	1.41
Cl	Average	2.37	2.77	2.52	2.19	1.83	4.13	3.11	2.79
	Range	1.84–3.09	1.84–3.80	1.84–3.80	1.97–2.66	1.50–2.11	2.93–6.15	1.50–6.15	
	SD	0.35	0.59	0.48	0.25	0.20	0.83	1.23	
	RCG	8.60	4.89	6.75	7.79	7.53	6.56	7.29	7.02

SD standard deviation, *RCG* reed canary grass

Table 8.2 Variability of ash and mineral concentration in biomass, harvested at seed maturity among SG accessions and between growing years

Variable	Ash	K	Ca	Mg	Si	N	S	Cl
	G kg^{-1} DM							
	Collection, planted in 2010							
	Significance of the factor							
Accession (A)	ns	ns	*	ns	ns	ns	ns	ns
Harvest year (Yr)	**	**	**	**	**	**	**	ns
A x Yr	ns	*	**	ns	ns	ns	ns	ns
Mean value per collection	48.3	12.1	7.04	2.90	4.07	5.44	0.80	2.26
Accession, state of origin	Average value for accession							
WE PI 642295, ND	51.3	11.4	7.19	3.26	4.46	5.31	0.78	2.12
WE PI 642296, ND	45.4	11.6	7.17	2.94	3.69	5.46	0.82	2.21
WE PI 642300, ND	46.3	11.5	6.20**	2.82	3.70	5.36	0.81	2.13
WE PI 642306, ND	53.5	12.4	7.53*	2.98	4.08	6.05	0.76	2.42
WE PI 642309, ND	49.6	12.4	7.14	3.05	4.36	5.75	0.74	2.02
Cv. Summer PI 642191, SD	43.8	12.4	7.15	2.52	3.98	4.87	0.84	2.50
WE PI 642194, ND	48.4	12.8	6.89	2.73	4.21	5.27	0.83	2.44
P_{05}	*5.71*	*0.82*	*0.461*	*0.405*	*0.432*	*1.09*	*0.098*	*0.308*
P_{01}	*7.96*	*1.15*	*0.642*	*0.564*	*0.603*	*1.52*	*0.137*	*0.429*

Table 8.2 (continued)

Variable	Ash	K	Ca	Mg	Si	N	S	Cl
Harvest year	Average value for year							
2012	52.6**	13.4**	4.93**	3.45**	4.53**	6.42**	0.69**	2.31
2013	44.1**	10.7**	9.15**	2.35**	3.61**	4.46**	0.90**	2.21
P_{05}	*2.33*	*0.34*	*0.190*	*0.170*	*0.180*	*0.450*	*0.040*	*0.130*
P_{01}	*3.25*	*0.47*	*0.260*	*0.230*	*0.250*	*0.620*	*0.060*	*0.180*
Variable	Collection, planted in 2011							
	Significance of the factor							
Accession (A)	**	ns	**	**	*	ns	ns	ns
Harvest year (Yr)	**	**	**	**	**	**	**	**
A x Yr	*	*	**	*	ns	*	*	ns
Mean value per collection	68.7	12.5	8.91	4.21	5.73	10.06	0.86	2.26
Accession, state of origine	Average value for accession							
WE PI 642198, ND	54.6**	11.6	7.76**	3.53**	5.26	8.68	0.72**	1.67**
WE PI 642199, ND	65.3	12.7	7.04**	3.55**	5.68	10.3	0.91	2.43
WE PI 642200, ND	63.5	12.5	8.65	3.85	5.66	9.71	0.88	2.45
WE PI 642204, ND	68.5	12.8	9.14	3.94	5.37	9.97	0.84	2.63
WE PI 642208, ND	70.0	13.0	9.14	4.33	5.74	10.6	0.96	2.17
Cv. Dacotah PI 537588, ND	79.6**	12.1	10.1**	5.04**	6.58**	9.70	0.82	2.02

Table 8.2 (continued)

Variable	Ash	K	Ca	Mg	Si	N	S	Cl
Cv. Nebraska 28 PI 477003, NE	79.5**	12.6	10.5**	5.21**	5.86	11.5	0.93	2.44
P_{05}	*6.60*	*0.950*	*0.682*	*0.444*	*0.492*	*1.50*	*0.097*	*0.439*
P_{01}	*9.20*	*1.33*	*0.951*	*0.619*	*0.686*	*2.10*	*0.135*	*0.612*
Harvest year	Average value for year							
2012	63.1**	16.2**	4.58**	4.52**	5.10**	13.9**	1.05**	2.68**
2013	74.4**	8.7**	13.2**	3.89**	6.37**	6.20**	0.68**	1.83**
P_{05}	*2.69*	*0.39*	*0.280*	*0.180*	*0.200*	*0.610*	*0.040*	*0.180*
P_{01}	*3.76*	*0.54*	*0.390*	*0.250*	*0.280*	*0.860*	*0.060*	*0.250*

ns not statistically significant, significance of differences evaluated from mean, *WE* wild ecotype, *ND* North Dakota, *SD* South Dakota, *NE* Nebraska
*, ** significant at $P<0.05$ and $P<0.01$

originated from North Dakota PI642300 and PI 642306 statistically significantly differed in Ca concentration from the mean value. Higher genotypic variability was established in the collection planted in 2011. Biomass of wild ecotype PI 642198 was characterized by the lowest concentration of all components, and that of ash, Ca, Mg, S, and Cl was significantly (at $P<0.01$) lower than respective mean per collection, while the biomass of cvs. Dacotah and Nebraska 28 was distinguished significantly by (at $P<0.01$) higher contents of ash, Ca, and Mg. Some researchers found relationship between several mineral components and latitude of genotype origin [5], [14]. Since almost all of our investigated SG accessions were from North Dakota, we could not clearly link mineral composition with latitude of origin. All biomass components, except Cl in collection planted in 2010, were significantly (at $P<0.01$) impacted by harvest year. Biomass harvested in 2013 contained on average less K, Mg, N, and Cl, but more Ca. Year effect on ash, Si, and S was collection specific. In general, our observations agree with the findings of Lemus et al. [13], that the alterations in chemical composition with the harvest year were substantially greater than differences among accessions. El-Nashaar et al. [14] also noticed a strong location effect on mineral concentrations in SG populations; and that could be related to site-specific management practices and other environmental influences.

Our results presented in Tables 8.1 and 8.2 demonstrate mineral composition of SG biomass harvested at seed maturity in order to ascertain genotypic, plant age effect on the investigated parametres. Biomass quality at this plant development stage did not show excellent properties required for feedstocks of sustainable solid biofuel. Management or other means of reducing the amount of ash, K, and Cl in biomass are exceptionally important for improving combustion quality. One of them is delaying biomass harvesting. With plant senescence in autumn, many minerals from annual plant tissues (leaves, stems, and panicles) are translocated to the perennial tissues including the crowns and roots [32]. While examining data on the issue of biomass quality change in response to delay in harvesting is presented in Table 8.3, one can see conspicuously decreasing trends in the majority of the tested components in the whole aboveground part (WP). Similarly, Adler et al. [9] reported K and Cl decreased most of all the components studied: from September to February K concentration in WP biomass changed from 12.3 to 3.25 g kg^{-1} DM, i.e., reduced by 74 %, and that of Cl declined by 74 % also and altered from 2.26 to 0.590 g kg^{-1} DM. In general, both K and Cl are located in the cell sap as components of soluble inorganic salts, therefore their translocation in plant during senescence and leaching during winter period is easier compared with other components. The lowering of Cl and K in biomass is particularly important, as both K and Cl are related to corrosion processes in the furnace, lessening the melting point of ash, causing the problems of slagging and fouling.

Elements which are more or less involved in the structures of macromolecular compounds (enzymes, proteins, nucleic acids, and other organic molecules) are fewer mobile and therefore decrease in concentration of such elements is notably less. N and Mg concentration in biomass of overwintered plants was lower by 52 and 47 % respectively. Delay of harvesting was a less effective means for Si and Ca concentration reduction and practically did not impact S concentration in WP. As to S, our results support the statement of Ogden et al. [8], Adler et al. [9], Kludze et al. [32] that S had no relationship with the harvest time. In consonance with Obernberger et al. [10], emission-related problems are to be expected at S concentrations above 0.2 or 2 g kg^{-1} in dry biomass. In our experiment, S concentration did not exceed this concentration either when SG was cut in September or later or in separate plant fractions (Table 8.3). Declining of elements, such as Ca, Mg, and K, as well as Si when harvest was delayed led to 41 % lower ash content in biomass of overwintered plants. Numerous researchers suggest that leaves resulted in a significant deterioration of biofuel quality when compared with stems [5], [20], [32] and flower heads [20]. In general, results of our study corroborate this statement. Concentration of ash, alkaline earth metals such as Ca and Mg, Si, N, and S in stem biomass of all harvesting dates was lowest compared with leaf and panicle biomass. Concentration of the most mobile elements such as K and Cl was practically equal in all plant parts at the respective harvesting time with some small exceptions. At all harvesting dates, N and S concentrations in panicles were the highest. This can be explained by the fact that both S and N are associated with protein, which is abundant in the mature seeds. And in fact a calculated correlation showed a close positive linear relationship between N and S with $r = 0.907$, $P < 0.01$. Higher heating value (HHV) was higher in WP, stems, and leaves of overwitered plants, unlike in panicles. Biomass of leaves showed the lowest and that of panicles the highest HHV among plant parts. It should also be noted that the proportion of stems in the biomass during the period from September to February markedly increased and the share of panicles declined by more than 2.5-fold. However, changes in the ratio of plant morphological parts of the hand-cut collection (Table 8.3) may differ from that of the industrial environment. As a result, it is possible to significantly change both the composition and quantity of ash in SG through management. Although the results prove a preference for late-winter harvest of SG used for direct combustion, attention should be drawn to the fact that long-term snowfalls, which could often occur in countries of nemoral climate zones, can lead to considerable decrease of biomass yield. Table 8.3 shows that the mineral composition of biomass altered unevenly from one harvesting date to another. The last harvesting in autumn was performed within the first 10 days of November. Further research is needed to determine more precisely the optimal harvesting time which would harmonize quality and yield of biomass under conditions of nemoral environmental zone.

8.4 Conclusions

Our research results confirmed that SG biomass cut at seed maturity stage had lower N, ash, K, Cl, and S concentrations compared with mineral composition of RCG biomass. The impact of plant age (year of production) on inorganic composition of SG biomass was component- and collection-dependent.

Table 8.3 Alteration in ash, mineral concentration, and calorific values of whole aboveground plant part, leaves, stems, and panicles of SG, cut between 13 September 2013 and 21 February 2014

Biomass fraction	Time of harvesting	n[a]	Constituent, g kg^{-1} DM								Plant fraction % WP	HHV[b] MJ kg^{-1} DM
			Ash	K	Ca	Mg	Si	N	S	Cl		
WP	13 Sep	14	59.3	12.3	11.2	3.12	4.99	5.33	0.794	2.02	–	18.5
	24 Oct	4	55.4	8.78	8.78	2.31	3.62	6.85	0.886	1.85	–	18.4
	07 Nov	2	47.3	5.10	8.33	1.90	4.14	5.93	1.01	1.27	–	n.a.
	21 Feb	4	34.4	3.25	6.03	1.88	3.16	3.72	0.784	0.59	–	18.7
Stems	13 Sep	14	24.6	12.1	8.71	1.26	1.74	2.88	0.906	2.03	41.6	18.6
	24 Oct	4	35.6	6.93	6.93	1.18	1.16	4.78	0.810	1.79	39.8	18.7
	07 Nov	2	36.2	3.23	6.47	1.29	2.03	3.90	0.797	1.49	45.2	n.a.
	21 Feb	4	15.6	2.78	5.24	1.36	1.20	2.34	0.593	0.61	60.6	19.2
Leaves	13 Sep	14	77.9	11.1	15.3	4.79	6.31	7.23	0.754	2.25	35.2	18.6
	24 Oct	4	77.3	10.5	10.5	3.05	4.64	13.6	1.43	2.00	32.6	18.4
	07 Nov	2	85.5	2.93	6.55	2.59	6.12	8.54	1.20	1.39	33.4	n.a.
	21 Feb	4	61.4	3.00	7.37	2.10	5.42	5.72	1.10	0.78	30.7	18.5
Panicles	13 Sep	14	56.3	10.6	10.3	2.43	5.60	11.3	1.24	1.79	23.2	19.5
	24 Oct	4	55.9	10.2	10.2	2.43	4.55	14.6	1.55	1.21	27.5	18.9
	07 Nov	2	56.6	5.10	7.03	2.30	6.19	16.3	1.45	0.85	21.4	n.a.
	21 Feb	4	47.4	2.93	6.64	1.90	4.69	9.12	1.26	0.92	8.71	19.0

WP whole aboveground plant part, *n* number of samples in group, *n.a.* not analysed

[a] Plants at seed maturity were taken from collections planted in 2010 and 2011, at later harvesting from collection planted in 2012

[b] HHV determined in the composite samples of respective fraction and harvesting date

Genotypic variability in the concentration of biomass components that have an impact on air pollution and corrosion of combustion equipment as well as slagging occurred within SG genotype, but the differences among years were greater than the differences among accessions.

During winter, the concentration of K and Cl in SG biomass reduced by 74 %, and N and Mg by 52 and 47 % respectively compared with biomass cut at seed maturity. The reduction was caused by a marked increase in stems proportion. Delay of harvesting was a less effective means for Si, Ca concentration reduction and practically did not impact S concentration. The concentration of ash,alkaline earth metals Ca and Mg, Si, N, and S in stem biomass of all harvesting dates was lowest compared with leaf and panicle biomass. The concentration of the most mobile elements K and Cl was practically equal in all plant parts at the respective harvesting date with some small exceptions. Biomass of leaves displayed the lowest and that of panicles the highest HHV among plant parts.

In summary, delay of harvesting would be the preferred option for SG to be used as a feedstock for direct combustion. Further research is needed to determine more precisely the optimal harvesting time which would harmonize quality and yield of biomass under conditions of nemoral environmental zone.

Acknowledgements This research was funded by a grant (No. MIP-073/2012) from the Research Council of Lithuania.

References

1. European Commission (2009) Directive 2009/28/EC of the European Parliament and of the Council of 23 April 2009 on the promotion of the use of energy from renewable sources and amending and subsequently repealing Directives 2001/77/EC and 2003/30/EC. OJEU 52 L 140/16 – 62. http://eur-lex.europa.eu/legal-content/EN/TXT/PDF/?uri=CELEX:32009L0028&from=EN. Accessed 21 July 2014
2. Smeets EM, Lewandowski IM, Faaij AP (2009) The economical and environmental performance of miscanthus and switchgrass production and supply chains in a European setting. Renew Sust Energ Rev 13(6):1230–1245
3. Heaton E, Voigt T, Long SP (2004) A quantitative review comparing the yields of two candidate C_4 perennial biomass crops in relation to nitrogen, temperature and water. Biomass Bioenerg 27(1):21–30
4. Robbins MP, Evans G, Valentine J, Donnison IS, Alison GG (2012) New opportunities for the exploitation of energy crops by thermochemical conversion in Northern Europe and the UK. Prog Energ Combust Sci 38:138–155
5. Elbersen HW, Christian DG, Yates NE, Bassam N E, Sauerbeck G, Alexopoulou E Switchgrass nutrient composition. In: FAIR 5-CT97-3701. Switchgrass (*Panicum virgatum* L.) as an alternative energy crop in Europe. Initiation of a productivity network. Final Report for the period from 01 to 04– 1998 to 30-09-2001 2001 www.switchgrass.nl
6. Lemežienė N, Norkevičienė E, Liatukas Ž, Dabkevičienė G, Cesevičienė J, Butkutė B (2015) Switchgrass from North Dakota – an adaptable and promising energy crop for northern regions of Europe. Acta Agr Scand B-S P 65(2):118-24
7. Butkutė B, Lemežienė N, Cesevičienė J, Liaudanskienė I (2013a) Dry matter yield and carbon partitioning in the above-ground part of switchgrass (*Panicum virgatum* L.) germplasm. Agr Food 1(1):87–97

8. Ogden CA, Ileleji KE, Johnson KD, Wang Q (2010) In-field direct combustion fuel property changes of switchgrass harvested from summer to fall. Fuel Process Technol 91(3):266–71
9. Adler PR, Sanderson MA, Boateng AA, Weimer PJ, Jung HJG (2006) Biomass yield and biofuel quality of switchgrass harvested in fall or spring. Agron J 98(6):1518–25
10. Obernberger I, Brunner T, Bärnthaler G (2006) Chemical properties of solid biofuels—significance and impact. Biomass Bioenerg 30(11):973–82
11. Szemmelveisz K, Szücs I, Palotas AB, Winkler L, Eddings EG (2009) Examination of the combustion conditions of herbaceous biomass. Fuel Process Technol 90(6):839–47
12. Sommersacher P, Brunner T, Obernberger I (2012) Fuel Indexes: A novel method for the evaluation of relevant combustion properties of new biomass fuels. Energ Fuel 26(1):380–390
13. Lemus R, Brummer EC, Moore KJ, Molstad NE, Burras CL, Barker MF (2002) Biomass yield and quality of 20 switchgrass populations in southern Iowa, USA. Biomass Bioenerg 23(6):433–442
14. El-Nashaar HM, Banowetz GM, Griffith SM, Casler MD, Vogel KP (2009) Genotypic variability in mineral composition of switchgrass. Bioresource Technol 100(5):1809–1814
15. Kanapeckas J, Lemežienė N, Butkutė B, Stukonis V (2011) Evaluation of tall fescue (*Festuca arundinacea* Schreb.) varieties and wild ecotypes as feedstock for biogas production. Zemdirbyste–Agriculture 98(2):149–156
16. Christian DG, Riche AB, Yates NE (2002) The yield and composition of switchgrass and coastal panic grass grown as a biofuel in Southern England. Bioresource Technol 83(2):115–24
17. Butkutė B, Lemežienė N, Cesevičienė J, Liatukas Ž, Dabkevičienė G (2013b) Carbohydrate and lignin partitioning in switchgrass *Panicum virgatum* biomass as a bioenergy feedstock. Zemdirbyste–Agriculture 100(3):251–260
18. Lindsey K, Johnson A, Kim P, Jackson S, Labbé N (2013) Monitoring switchgrass composition to optimize harvesting periods for bioenergy and value-added products. Biomass Bioenerg 56:29–37
19. Banowetz GM, Griffith SM, Steiner JJ, El-Nashaar HM (2009) Mineral accumulation by perennial grasses in a high-rainfall environment. Energ Fuel 23(2):984–988
20. Monti A, Virgilio N D, Venturi G (2008) Mineral composition and ash content of six major energy crops. Biomass Bioenerg 32:216–223
21. Metzger MJ, Shkaruba AD, Jongman RHG, Bunce RGH (2012) Descriptions of the European Environmental Zones and Strata. Alterra Report 2281. Alterra, part of Wageningen UR, Wageningen,p. 154 http://content.alterra.wur.nl/Webdocs/PDFFiles/Alterrarapporten/AlterraRapport2281.pdf. Accessed 21 July 2014
22. Sluiter A, Hames B, Hyman D, Payne C, Ruiz R, Scarlata C et al (2008) Determination of total solids in biomass and total dissolved solids in liquid process samples laboratory. Analytical procedure (LAP). NREL/TP-510–42621. National Renewable Energy Laboratory, Golden,
23. European Committee for Standardization. CEN EN 14775: (2009). Solid biofuels—determination of ash content. European Committee for Standardization, rue de Stassart 36, B-1050 Brussels, Belgium
24. Kraska JE, Breitenbeck GA (2010) Simple, robust method for quantifying silicon in plant tissue. Commun Soil Sci Plant Anal 41:2075–2085
25. Dai WM, Zhang KQ, Duan BW, Sun CX, Zheng KL, Cai R, Zhuang JY (2005) Rapid determination of silicon content in rice. Rice Sci 12(2):145–147
26. Liu L (1998) Determination of chloride in plant tissue. In: Kalra YP (ed) Handbook of reference methods for plant analysis. CRC Press, Boca Raton, pp 111–113
27. Gaines TP, Parker MB, Gascho GJ (1984) Automated determination of chlorides in soil and plant tissue by sodium nitrate. Agron J 76: 371–374
28. Telmo C, Lousada J, Moreira N (2010) Proximate analysis, backwards stepwise regression between gross calorific value, ultimate and chemical analysis of wood. Bioresource Technol 101(11):3808–3815

29. McKendry P (2002) Energy production from biomass (part 1): overview of biomass. Bioresour Technol 83(1):37–46
30. Jenkins BM, Baxter LL, Miles TR Jr, Miles TR (1998) Combustion properties of biomass. Fuel Process Technol 54(1):17–46
31. Vassilev S, Vassileva C, Baxter D, Andersen L (2010) Relationships between chemical and mineral composition of coal and their potential applications as genetic indicators. Part 2. Mineral classes, groups and species. Geol Balcanica 39(3):43–67
32. Kludze H, Deen B, Dutta A (2011) Report on literature review of agronomic practices for energy crop production under Ontario conditions. University of Guelph. http://www.ofa.on.ca/uploads/userfiles/files/u%20of%20g%20ofa%20project-final%20report%20july%2004–2011%20(1).pdf. Accessed 19 July 2014

Chapter 9
Microwave Pyrolysis Process Potential of Waste Jatropha Curcas Seed Cake

Ricardo A. Narváez C., Valeria Ramírez, Diego Chulde, Sebastián Espinoza and Jesús López-Villada

Abstract This chapter aims to express the potential of waste *Jatropha curcas* seed cake (JCSC) as an interesting option for energy generation. The results obtained by microwave pyrolysis (MWP) process, undoubtedly, indicate that the remaining organic liquid substances contained in JCSC, the solid residue obtained from the process, and the gaseous fraction are still a potential energy source that can be explored. The exploitation is particularly interesting in cases where an extensive production of this type of biomass waste exists and its transportation implies extensive paths and additional costs.

"*Jatropha Curcas* for Galapagos Islands" project is a current initiative aimed to reduce the requirements of fossil fuels for Galapagos Islands, which is framed under the "Zero fossil fuels for Galapagos Islands" national objective. The project is focused on gathering *J. curcas* fruit from living fences along the rural areas of the Province Manabí (Ecuador) to extract oil in small-scale facilities. The extracted oil is refined and transported to Galapagos Islands where it is used as biofuel for internal combustion engine generators to provide electricity to the islands. Although the by-products of extraction are currently not being exploited as an energy source, this possibility is seen as a remarkable choice for final disposition and energy generation improvement. In 2012, especially in this particular case, seed cake production was estimated to be 213 t/year as a result of its exponential growth since 2009.

Conventional pyrolysis presents some inconvenient situations, such as resistance to heat transfer, heat loss, and lack of fast heating. Taking into account that short operation/retention times are preferred in cases with growing demand, and low energy-intensive extraction technologies maximize the energy output of the primary energy source, MWP appears as a suitable option to be applied.

The available JCSC was assayed in a bench-scale MWP batch process that allows tar and chars formation and their separation for analysis. During the process,

R. A. Narváez C. (✉) · V. Ramírez · D. Chulde · S. Espinoza · J. López-Villada
National Institute of Energy Efficiency and Renewable Energy (INER), 6 de Diciembre Av. N33–32 & Bossano St., 170102 Quito, Ecuador
e-mail: ricardo.narvaez@iner.gob.ec; r.a.narvaez-cueva@lboro.ac.uk; ranarvaezcueva@gmail.com

R. A. Narváez C.
Centre for Renewable Energy Systems Technology (CREST), Loughborough University, Holywell Park, Loughborough LE11 3TU, UK

A. Sayigh (ed.), *Renewable Energy in the Service of Mankind Vol I*,
DOI 10.1007/978-3-319-17777-9_9

MWP operating conditions were programmed to reach 550 °C in 30 min from room temperature and maintain this condition for an additional 10 min. Also, JCSC was previously assayed to establish the content of volatile organic compounds with the purpose of estimating the organic liquid substances yield in case of a potential extraction. Liquid products (bio-oil) were analyzed through gas chromatography coupled with a mass-spectrometer detector (GC-MS) in order to determine the most representative compounds in a qualitative-basis. Their functional groups profile was obtained through a Fourier-Transform Infrared Spectroscopy (FTIR) assay. Bulk properties, such as density and calorific value, were obtained by applying existing standard methods suitable for diesel fuel. Analysis results were used to estimate its behavior as a liquid fuel by comparing the results with diesel-fuel quality standards and characteristics. In addition, calorific value, proximal and ultimate biomass analyses were developed for solid products in order to assess the suitability of a potential use as solid fuel.

The results show that the maximum content of potential liquid fuel substances is around 75 %, which leads to a conclusion that liquid fuel extraction by thermal methods is a suitable option to be evaluated regarding an augmentation of liquid fuel yield.

Keywords Microwave pyrolysis · Jatropha curcas · Seed cake · Ecuador

9.1 Introduction

Thermal technologies, such as pyrolysis, appear as suitable options for agricultural waste exploitation since this type of material can be considered as energy source [1] and/or raw material for certain bioproducts [2] Besides this affirmation, the particular case of bio-oil production has been recognized as an interesting option for final disposal of waste biomass considering the environmental and economic point of view [3, 4].

In the case of the bioenergy project *"Jatropha Curcas for Galapagos Islands,"* which is being held in Ecuador, 213 MT/year of waste biomass were generated according to the Ministry of Electricity and Renewable Energy, the developer of the project [5]. This initiative is one of the implemented actions of the *"Zero fossil fuels for Galapagos Islands"* national objective. Part of it consists in gathering *Jatropha curcas* fruits produced in living fences along the Province of Manabí, located in the Ecuadorian land territory. Once the raw material is collected, its seeds are fed into a screw-type oil extraction unit to obtain vegetable oil. After this stage is completed, the oil is transported to the Galapagos Islands to be blended with fossil diesel that will feed two 75 kWe internal combustion engine generators located in Floreana Island (Galapagos Islands, Province of Galapagos, Ecuador).

The project developer reports a yield of 30 % w/w in oil extraction. Hence, most of the collected biomass becomes waste material after the extraction process. The amount of biomass residue currently generated and the suitability of its use in other applications related to energy and chemical industry can be considered as a valid option for extra by-products benefit of the extraction process.

Despite the fact that *J. curcas* was initially considered as an energy solution with a short-term potential, further results have shown several limitations for a profitable large-scale implementation [6, 7]. High fertilizer requirements and other competitive uses like soap production are among the reasons found to support this statement [8].

Bio-oil production appears as a suitable parallel project that can be linked to any initiative where organic solid residues are generated [6, 7]. These type of projects can reach feasibility through their waste valuation, without neglecting their energy orientation, even if it may not become the main source of income [6, 7].

Bio-oil produced through liquid extraction processes such as pyrolysis, appears as a remarkable choice for organic residues exploitation since carbon volatiles content shows the aptness of waste as raw material [9, 10, 11]. Considering bio-oil yield is related to the organic liquid presence in biomass, carbon volatiles content can be considered as an initial indicator of suitability of this technology. Among the available options for biomass pyrolysis, microwave pyrolysis (MWP) bio-oil may present certain advantages for the case previously described. For instance, using microwave-induced heat, instead of partial combustion of biomass certainly would increase the liquid fraction yield since this case would not require the heat supply provided by the oxidation of part of the liquid products. Furthermore, avoiding the consumption of part of the tar generated during the pyrolysis process would increase the bio-oil yield. Therefore, the amount of liquid bio-products to be obtained would be boosted as well. Also, cases where chemical products present a visible interest without neglecting energy recovery after its separation from bio-oil might consider MWP as an appropriate path.

Other of the important aspects of the bio-oil production concerns the quality of bio-oil. The solid carbonic residue in bio-oil that has been reported in literature is likely to affect bio-oil performance as biofuel for internal combustion engines [12, 13]. In cases where thermal pyrolysis is applied and energy is self-supplied by partial combustion of biomass, it is expected that certain carbon particles produced during the oxidation reactions remain in the liquid product. Microwave-induced heat is expected to not present this situation since the combustion phenomenon is avoided. Even if a complete refining process of bio-oil after MWP includes a stage for separating chemical products, the remaining part aimed to be used as fuel may present a more appropriate quality regarding its solid content.

9.2 Materials and Methods

Jatropha curcas seed cake (JCSC) was previously characterized in order to obtain its composition and bulk properties. The analytical techniques that were applied are described in Table 9.1.

Table 9.1 Bulk properties and analytical techniques applied to the *Jatropha curcas* seed cake (JCSC)

Content element/bulk property	Applied standard technique	Piece of equipment used
Proximate analysis		
Humidity content, %wt	BS EN 14774-3:2009	–
Carbon volatiles content, %wt	BS EN 15148:2009	–
Fixed carbon content (by difference), %wt	–	–
Ashes content, %wt	BS EN 14775:2009	–
Ultimate analysis		
C, %wt	BS EN 15104:2011	Perkin Elmer® 2400
H, %wt		
N, %wt		
S, %wt		
O (by difference), %wt	–	–
Bulk density, kg/m^3	BS EN 15103:2009	–
Gross calorific value, MJ/kg	ASTM D5865—13	IKA® C2000

BS EN British Standard European Norm, *ASTM* American Society for Testing and Materials

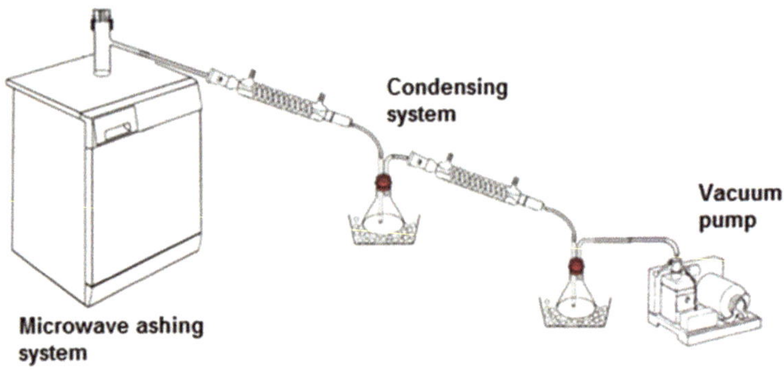

Fig. 9.1 Bench-scale equipment for *Jatropha curcas* seed cake (JCSC) bio-oil production

In order to perform the MWP, a laboratory microwave oven was coupled to a laboratory glassware assembly aimed for condensing liquid products as a part of the experiment. The equipment configuration is shown in Fig. 9.1. This arrangement was intended to assess the feasibility of MWP in bench-scale applied to JCSC and to obtain bio-oil produced by microwave-induced heat. The equipment consists of a Milestone PYRO microwave ashing system, two Graham condensers, two water traps, and a vacuum pump.

Operating conditions for laboratory microwave-oven were established to reach 550 °C in 30 min and maintain those conditions for 10 additional minutes [14]. Microwave ashing system was operated at 1100 W. Liquid products (bio-oil) were produced during this process and were condensed to analyze their properties. The solubility was analyzed by assaying several blends with polar solvents such as methanol and water, neutral polarity solvents such as chloroform, and nonpolar solvents such as n-hexane and cyclohexane.

The solvents that presented high miscibility with bio-oil were selected to be used as carrier substances during the characterization assays. Quantitative and qualitative techniques were applied to the liquid fraction to determine its potential as source for chemicals and liquid fuels.

Bio-oil was characterized by using the analytical techniques described in Table 9.2.

Finally, the remaining solid considered as biochar was also gathered to be characterized with the purpose of being used as solid fuel. The standard assays previously selected for the JCSC were also applied to this solid fraction.

Table 9.2 Analytical techniques applied to the liquid fraction of microwave pyrolysis (MWP) applied to *Jatropha curcas* seed cake (JCSC)

Quantitative analysis		
Content element/bulk property	Applied standard technique	Piece of equipment used
Ultimate analysis		
C, %wt	BS EN 15104:2011	Perkin Elmer® 2400
H, %wt		
N, %wt		
S, %wt		
O (by difference), %wt	–	–
Density, kg/m^3	BS EN 15103:2009	–
Gross calorific value, MJ/kg	ASTM D5865—13	IKA® C2000
Carbonic residue, %wt	ASTM D-189	–
Qualitative analysis		
Technique name	Purpose of application	Equipment used
Fourier Transform Infrared Spectroscopy Analysis (FTIR)	Determination of the most significant functional groups found in the bio-oil sample. Results are required to estimate the potential as fuel of chemical substances source	Jasco® 4200
Gas spectrometry coupled to mass spectrometry (GC-MS)	Determination of the most likely presence of chemical species in the bio-oil sample. It is required to recognize the suitability of isolating any particular chemical specie with a significant market price. Feasibility determination requires further analysis to quantify the actual content of chemical species in the sample	Agilent® 7890B

BS EN British Standard European Norm
ASTM American Society for Testing and Materials

9.3 Results

Characterization results of JCSC are presented in Table 9.3.

Results for liquid product fraction (bio-oil) obtained from MWP are presented in Tables 9.4, 9.5 and 9.6 respectively.

Table 9.3 *Jatropha curcas* seed cake (JCSC) characterization results

Content element/bulk property	Reported value
Proximate analysis	
Humidity content, %wt	10.48
Carbon volatiles content, %wt	75.10
Fixed carbon content, %wt	11.23
Ashes content, %wt	3.19
Ultimate analysis	
C, %wt	52.12
H, %wt	6.91
N, %wt	5.01
S, %wt	0.70
O (by difference), %wt	32.07
Bulk density, kg/m^3	1152
Gross calorific value, MJ/kg	18.62

Table 9.4 Characterization results of the liquid fraction of microwave pyrolysis (MWP) applied to *Jatropha curcas* seed cake (JCSC)

Content element/bulk property	Reported value
Ultimate analysis	
C, %wt	4.61
H, %wt	11.95
N, %wt	3.26
S, %wt	1.19
O (by difference), %wt	78.99
Density, kg/m^3	1140
Gross calorific value, MJ/kg	3.37

Table 9.5 Functional groups in bio-oil found through Fourier-Transform Infrared Spectroscopy (FTIR) analysis

Bond type	Wave number range, cm^{-1}
-C=C-	1690–1640
-N-H	1570–1515
-CH_2	2857–2849
-C-H	1392–1388
-C-H	2936–2916
-C-H	2863–2843
-OH	3400–3200
-C=N	1700–1600
-C-N-	1420–1400
-NO_2	1570–1520

Table 9.6 Chemical compounds with larger presence probability in bio-oil according to GC-MS analysis

Chemical compound	Presence probability, %
Octasulfur	98.40
Amylene dichloride	87.80
Succinimide	85.40
Acetylene Tetrachloride	75.40
Phenol	72.40

Table 9.7 Biochar characterization results

Content element/bulk property	Reported value
Proximate analysis	
Humidity content, %wt	3.69
Carbon volatiles content, %wt	12.77
Fixed carbon content, %wt	47.97
Ashes content, %wt	35.57
Ultimate analysis	
C, %wt	40.64
H, %wt	1.19
N, %wt	2.87
S, %wt	0.04
O (by difference), %wt	19.69
Bulk density, kg/m^3	1.09
Gross calorific value, MJ/kg	28.30

Table 9.8 Carbonic residue of *Jatropha curcas* seed cake (JCSC) bio-oil and other comparable fossil fuels

Carbonic residue, %wt	Reported value
JCSC Bio-oil carbonic residue	3.46
Crude oil (24.5 °API) carbonic residue	9.50
Fuel oil no. 6 (11.0 °API) [15]	16.00

Solid fraction characterization was developed in order to assay its potential as solid fuel. Results are presented in Table 9.7.

The trend of bio-oil to form carbonic residues during combustion was also analyzed through the Conradson carbon assay. Figures for some other fuels were obtained as part of the assessment in order to compare its particular characteristic (Table 9.8).

9.4 Discussion and Conclusion

Initial dissolution test showed that MWP liquid products are miscible with polar solvents, partially soluble in neutral-polarity solvents, and slightly soluble in non-polar solvents. Characterization results of the liquid fraction of MWP applied to JCSC show an acceptable potential of obtaining a remarkable yield of bio-oil according to the proximate analysis figures. In addition, if the carbon volatiles content is related to the presence of chemical compounds that are attractive for the market, it can be inferred that this particular liquid product constitutes an interesting source of fuels and chemical products due to its organic liquid content.

Even though further analyses are certainly required, it can be affirmed that this particular bio-oil is suitable to be fed into a separation process to obtain bioproducts and biofuels.

Regarding the results of the qualitative analysis applied to the bio-oil, it can be stated that it contains substances may be used as liquid biofuel. Considering that the liquid fraction shows indicatives of containing hydrocarbon and alcoholic substances, according to the FTIR analysis results, the possibility of using it as a potential source of liquid fuels can be supported.

The presence of functional groups such as single carbon–hydrogen bond and double carbon bond indicates that bio-oil might contain alkanes and also other hydrocarbons that remain in liquid state at regular conditions.

Regarding other types of liquid biofuels, JCSC bio-oil can be recognized as a source of alcohols as well. This particular application proposal can also be supported in the presence of oxygen–hydrogen bonds since the presence of this particular functional group may denote the presence of this chemical family. This fact leads to consider this bio-oil as a source of liquid fuel with suitability of being used in Otto-cycle engines [16].

It is necessary to notice that fractioning by distillation should be included as part of the further analysis that are required in a previous phase of an effective exploitation of this material due to the variety of chemical species found during this assessment.

This fact indicates the liquid material that was obtained could require some additional separation process such as distillation in order to isolate potential liquid fuels suitable for internal combustion engines. This affirmation is made since the results obtained for density, gross calorific value, and ultimate analysis (hydrogen and oxygen contents) results lead to consider that the sample may contain an amount of water formed during pyrolysis. Taking into account that MWP was the extraction method for this particular case and proximate analysis results of JCSC also show that water is present in the pyrolysis oil, humidity separation should also be considered during designing stages of separation processes [17, 18, 19].

Among other considerations, its trend to generate low proportions of carbonic residue during combustion can be mentioned as well. Although its Conradson carbon figure seems larger than Ecuadorian standards for light fuels (diesel fuel must have 0.15 % according to [20]), it should be compared with regular fossil oil which presents figures above 9.5 %wt according to the current assessment.

On the other hand, qualitative analysis also indicates a substantial probability that MWP liquid product contains several chemicals that can constitute an important source of income for the *"Jatropha Curcas for Galapagos Islands"* bioenergy project once isolated. Phenol and alcoholic functional group were identified by applying the GC-MS and the FTIR analysis techniques, respectively. The presence of phenol indicates that bio-oil has a lignin-derived fraction [2]. Therefore, it is possible to consider that a fraction of the bio-oil may be commercialized with a competitive price as raw material for phenol–formaldehyde resins [21].

Nitrogen content was also reported as one of the former elements in JCSC and its bio-oil. If this fact is related to the nitrogenated functional groups that were identified by the FTIR analytical technique, it can be affirmed that MWP liquid fraction can also contain a fraction suitable for producing fertilizers. Authors such as Bridgwater et al. (1999) [22] affirm that bio-oil based fertilizers are not only suitable but also potentially important for carbon sequestration in soils. Even if it is a field that is currently under development, technological applications already reported could exploit this option [23].

Regarding the solid fraction, it can be mentioned that it is appropriate for being used as energy source according to its composition and bulk properties. In addition, it is also necessary to mention that this particular biochar could be assayed as material for charcoal pellet in order to increase its density and its energy performance. It is also worth mentioning that chemical compounds that do not condense might be a potential energy source as well.

Considering that the amount of waste biomass reported for the project developer, the bio-oil characteristics, and properties that were obtained during the development of the current assessment, it can be affirmed that bio-oil production and exploitation certainly would increase the amount of liquid fuel generated in comparison to the current vegetable oil yield. Besides this proposal, it is also possible to include other bioproducts in addition to liquid fuels as part of a strategy to diversify the sources of income of the *"Jatropha Curcas for Galapagos Islands"* initiative.

It is necessary to develop further composition analysis owing to identity punctual compounds that can be separated and commercialized as chemical products and to design a proper separation stage. Assays such as Hempel distillation [24] should be applied in order to consider distillation or analog physical separation methods. This assay can deliver a first approach of composition and possible characteristics of light fractions of bio-oil; despite it is currently superseded. This particular case is suitable for this test since chemical composition and base of liquid products is not fully determined; hence more precise composition techniques cannot be applied as the first option due to the uncertainty about its solubility and chemical affinity.

Acknowledges Authors would like to thank the Ministry of Electricity and Renewable Energy of Ecuador (MEER) who executes the *"Jatropha Curcas for Galapagos Islands"* project through the National Direction of Biofuels & Cogeneration. In addition, authors would like to acknowledge the support of the Interamerican Institute of Cooperation for Agriculture (IICA) staff located in Ecuador and the agroenergy research group of the National Institute of Agricultural Research of Ecuador (INIAP).

References

1. Demirbas A (2001) Conversion of biomass to a pyrolytic oil for blending gasoline as an alternative fuel in internal combustion engines. Energy Sour 23(6):553–562
2. Czernik S, Bridgwater AV (2004) Overview of applications of biomass fast pyrolysis oil. Energy Fuels 18(2):590–598
3. Anex RP, Aden A, Kazi FK, Fortman J, Swanson RM, Wright M, et al (2010) Techno-economic comparison of biomass-to-transportation fuels via pyrolysis, gasification, and biochemical pathways. Fuel 89:S29–S35

4. Fan J, Kalnes TN, Alward M, Klinger J, Sadehvandi A, Shonnard DR (2011) Life cycle assessment of electricity generation using fast pyrolysis bio-oil. Renew Energy 36(2):632–641
5. Proyecto: "Biocombustibles Galápagos—Sustitución de Combustibles Fósiles por Biocombustibles en la Generación Eléctrica en las Islas Galápagos con aceite de Piñón (Jatropha curcas) procedente de la Provincia de Manabí"—INFORME FINAL: MAYO 6/2011-NOVIEMBRE 6/2012 (2012), Ministry of Electricity and Renewable Energy of Ecuador
6. Mohan D, Pittman CU, Steele PH (2006) Pyrolysis of wood/biomass for bio-oil: a critical review. Energy Fuels 20(3):848–889
7. Ng KS, Sadhukhan J (2011) Process integration and economic analysis of bio-oil platform for the production of methanol and combined heat and power. Biomass Bioenergy 35(3):1153–1169
8. Openshaw K (2000) A review of Jatropha curcas: an oil plant of unfulfilled promise. Biomass Bioenergy 19(1):1–15
9. Gerçel HF (2002) The production and evaluation of bio-oils from the pyrolysis of sunflower-oil cake. Biomass Bioenergy 23(4):307–314
10. Yorgun S, Şensöz S, Koçkar ÖM (2001) Flash pyrolysis of sunflower oil cake for production of liquid fuels. J Anal Appl Pyrolysis 60(1):1–12
11. Ucar S, Ozkan AR (2008) Characterization of products from the pyrolysis of rapeseed oil cake. Bioresour Technol 99(18):8771–8776
12. Özçimen D, Karaosmanoğlu F (2004) Production and characterization of bio-oil and biochar from rapeseed cake. Renew Energy 29(5):779–787
13. Volli V, Singh RK (2012) Production of bio-oil from de-oiled cakes by thermal pyrolysis. Fuel 96:579–585
14. Butler E, Devlin G, Meier D, McDonnell K (2011) A review of recent laboratory research and commercial developments in fast pyrolysis and upgrading. Renew Sustain Energy Rev 15(8):4171–4186
15. Standard NTE INEN 1983:02– Productos derivados del petróleo. Fuel Oil. Requisitos (2012), National Institute of Normalization of Ecuador (INEN)
16. Agarwal AK (2007) Biofuels (alcohols and biodiesel) applications as fuels for internal combustion engines. Prog Energy Combust Sci 33(3):233–271
17. Wang S, Gu Y, Liu Q, Yao Y, Guo Z, Luo Z, Cen K (2009) Separation of bio-oil by molecular distillation. Fuel Process Technol 90(5):738–745
18. Guo X, Wang S, Guo Z, Liu Q, Luo Z, Cen K (2010) Pyrolysis characteristics of bio-oil fractions separated by molecular distillation. Appl Energy 87(9):2892–2898
19. Junming X, Jianchun J, Yunjuan S, Yanju L (2008) Bio-oil upgrading by means of ethyl ester production in reactive distillation to remove water and to improve storage and fuel characteristics. Biomass Bioenergy 32(11):1056–1061
20. Standard NTE INEN 1489–7:2012–Productos derivados del petróleo. Diesel. Requisitos (2011), National Institute of Normalization of Ecuador (INEN)
21. Mann MK, Spath PL, Kadam K (1996, June) Technical and economic analysis of renewables-based hydrogen production. In Proceedings of the 11th World Hydrogen Energy Conference, Stuttgart, Germany (pp 23–28)
22. Bridgwater AV, Meier D, Radlein D (1999) An overview of fast pyrolysis of biomass. Organ Geochem 30(12):1479–1493
23. Radlein D, Piskorz J, Majerski P (1997) Method of producing slow-release nitrogenous organic fertilizer from biomass. US Patent 5,676, 727, 1997 and European Patent Application 0716056
24. UOP Standard UOP77–85, "*Crude Oil Evaluation by Hempel Distillation*" (1985) ASTM International, West Conshohocken. www.astm.org

Part II
Fuel Cells

Chapter 10
Fuel Processing of Low-Sulfur Diesel for Fuel Cell Systems

Joachim Pasel, Remzi Can Samsun, Ralf Peters and Detlef Stolten

Abstract Fuel processing is the umbrella term for a variety of chemical processes aiming at the provision of a hydrogen-rich gas mixture to be fed into the anode of a fuel cell. Among them, the most important ones are desulfurization, reforming, water-gas-shift reaction, preferential oxidation, and catalytic combustion. At Juelich, all these processes are under investigation. This contribution concentrates on autothermal reforming (ATR) of different synthetic and fossil fuels with Juelich's reactor ATR 13. ATR 13 was developed within the European project Fuel Cell Based Power Generation (FCGEN). Here, special emphasis is on using diesel fuel SD10. SD10 is a conventional low-sulfur diesel, currently used for truck engine testing. ATR 13's power class of 3–5 kWe is particularly interesting for fuel cell systems with polymer electrolyte fuel cells (PEFCs) working as auxiliary power units (APUs) for onboard power supply in trucks. During autothermal reforming, liquid fuels such as diesel are converted together with air and steam into a hydrogen-rich gas by means of catalysis. The design and construction of ATR 13 was particularly optimized with respect to its power density, since ATR 13 will be assembled in an allocated and limited space at the bottom of a conventional truck. As a result, a high value of 3.6 kW/l was reached. Intensive experimental evaluation of ATR 13's operational behavior was done and will be reported in this chapter.

Keywords Fuel processing · Diesel fuel · Fuel cells · Truck applications

J. Pasel (✉) · R. C. Samsun · R. Peters · D. Stolten
Institute of Energy and Climate Research, IEK-3: Electrochemical Process Engineering, Forschungszentrum Jülich GmbH, 52425 Jülich, Germany
e-mail: j.pasel@fz-juelich.de

D. Stolten
RWTH Aachen University, Aachen, Germany

A. Sayigh (ed.), *Renewable Energy in the Service of Mankind Vol I*,
DOI 10.1007/978-3-319-17777-9_10

10.1 Introduction

The core component of any fuel cell system operating on liquid fuels such as kerosene or diesel fuel—apart from the fuel cell itself—is the reformer. There are many different routes for reforming and of these partial oxidation (POX), heated steam reforming (HSR), and autothermal reforming (ATR) are the most widespread and mature. At the Institute of Energy and Climate Research in Jülich, autothermal reforming was chosen because it is has been experimentally proven to be the most dynamic, robust, and simple variant. Against the background of the technical application of fuel cell systems with the autothermal reforming of liquid fuels, for example, in aircraft or trucks, designing the reactor for autothermal reforming is challenging with respect to four major issues: (i) high power density and mass specific power, (ii) sufficient load change capability (e.g., from 20 to 120 %), (iii) outstanding long-term stability, and (iv) high quality of the product gas (reformate). In Pasel et al.'s work [1], the long-term stability of Jülich's reactors for autothermal reforming covering 10,000 h on stream is described in detail. The present chapter focuses on the fourth challenge, which is especially important if the reforming process is combined with a polymer electrolyte fuel cell (PEFC) operating at only 80 °C. High quality of the reformate means that it must not contain any impurities such as ethene, propene, or benzene, but only the main products of autothermal reforming, i.e., H_2, CO, CO_2, and CH_4, which are formed according to the following four reactions:

$$C_nH_m + n\,H_2O \leftrightarrow n\,CO + (m/2 + n)\,H_2 \quad \text{endothermic (1)}$$

$$C_nH_m + n/2\,O_2 \leftrightarrow n\,CO + m/2\,H_2 \quad \text{exothermic (2)}$$

$$CO + H_2O \leftrightarrow CO_2 + H_2 \quad \text{exothermic (3)}$$

$$CO + 3\,H_2 \leftrightarrow CH_4 + H_2O \quad \text{exothermic (4).}$$

In this respect, a new reformer generation was developed within the European FCGEN project (Fuel Cell Based On-Board Power Generation). The design and construction of the FCGEN autothermal reformer (ATR) was particularly optimized in terms of its power density since the reformer itself and the whole fuel cell system will be assembled in an allocated and limited space at the bottom of a conventional truck. As a result, a high value of 3.6 kW/l for the power density of the FCGEN autothermal reformer was reached.

10.2 Fundamental Reformer Design

Figure 10.1 shows the basic design of Jülich's FCGEN autothermal reformer. During start-up, a stream of hot air is fed into the fuel evaporation chamber to heat up the reformer until the dew point of water is exceeded. Then, using a steam flow with

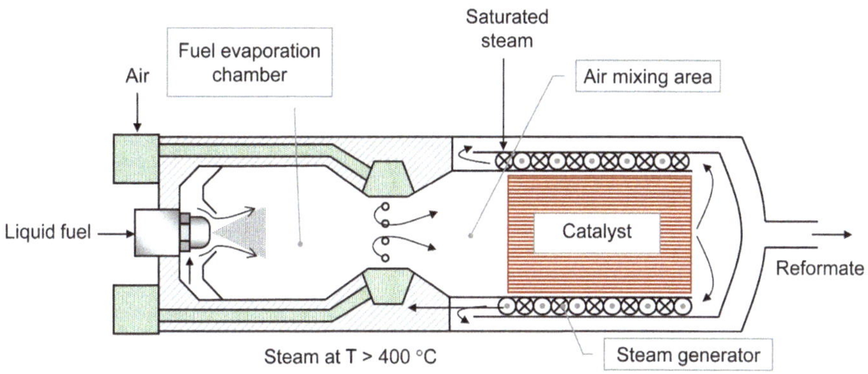

Fig. 10.1 Fundamental design of Jülich's FCGEN reactor for autothermal reforming of liquid fuels

a maximum temperature of 600 °C the reformer is further heated. The next phase begins when the temperature of the ATR catalyst reaches a value of at least 350 °C. A fuel feed can then be injected into the reformer. Within a few seconds, the catalytic partial oxidation of the hydrocarbons on the catalyst ignites. At steady state, a mixture of superheated steam coming from the catalytic burner of the fuel cell system and cold water, resulting in saturated steam, flows through the internal heat exchanger of the FCGEN autothermal reformer, in which it is superheated to more than 400 °C when it enters the fuel evaporation chamber. Simultaneously, fuel is injected via a nozzle at ambient temperature and is evaporated in the fuel evaporation chamber via the enthalpy flow of the superheated steam. Air is added and mixed with steam and evaporated fuel. This mixture is then fed into the catalyst where the autothermal reforming takes place. The heat of formation from the reforming process is transferred to the saturated steam via the heat exchanger in the FCGEN ATR.

10.3 Experimental Results, Phase 1

In phase 1, the experimental evaluation of the FCGEN ATR was done using the operating setup given below:

- Temperature of steam from catalytic burner: approx. 230 °C
- Temperature of air to the air mixing area: approx. 100 °C
- Reformer load: 50–100 %
- Oxygen-to-carbon molar ratio: 0.47
- Steam-to-carbon molar ratio: 1.70
- Fuels: SD10 diesel and NExBTL diesel
- Catalyst: Johnson Matthey HiFuel ATR catalyst
- Commercial nozzle with 80–100 μm bore diameter
  - Resulting nozzle pressure: 3–20 bar

Table 10.1 Chemical and physical properties of the fuels applied for autothermal reforming with the FCGEN ATR

Fuel	Supplier	Final boiling point, °C	Aromatics, wt.%/S fraction, ppm	Average molecular formula
NExBTL diesel	NESTE Oil Finland	321	<0.1/<1	$C_{17}H_{36}$
SD10 diesel	Prem Sweden	360	18.8/3	$C_{14}H_{26}$

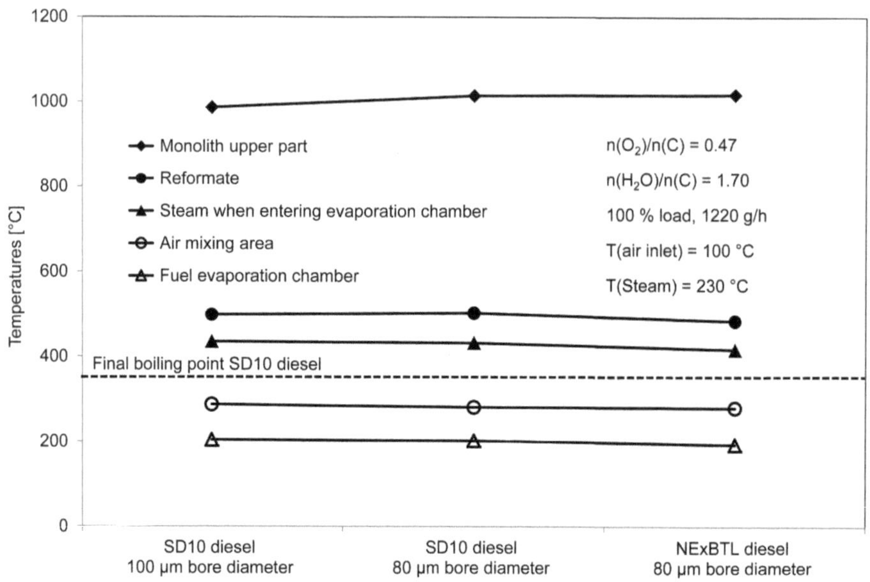

Fig. 10.2 Temperatures at different positions using the fuels from Table 10.1, $n(O_2)/n(C)=0.47$, $n(H_2O)/n(C)=1.70$, 100% reformer load, T(Air)=100 °C, T(Steam)=230 °C

Table 10.1 gives an overview of the fuels to be used in phase 1 and summarizes their physical and chemical properties. NExBTL diesel and SD10 diesel differed from each other in their boiling ranges and their mass fractions of aromatics and sulfur. NExBTL diesel was synthesized via the hydrotreated-vegetable-oil (HVO) process using plant oils. Therefore, it had a sulfur mass proportion of less than 1 ppm and fractions of aromatics far lower than 1 vol.%. SD10 diesel, however, was conventionally made from crude oil in a refinery where it underwent a desulfurization step (hydrocracking) also leading to sulfur mass fractions lower than 10 ppm. SD10 diesel contained mass proportions of aromatics of approx. 18.8 vol.%,

Figure 10.2 shows relevant temperatures inside the FCGEN autothermal reformer during operation with both fuels from Table 10.1 at 100% reformer load. It can be seen that the reaction temperatures in the upper part of the monolith were in

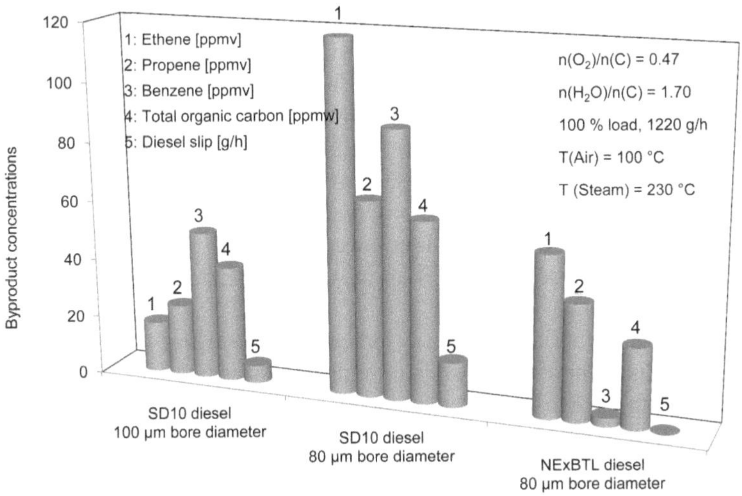

Fig. 10.3 Concentrations of by-products in the product gas using the fuels from Table 10.1, $n(O_2)/n(C)=0.47$, $n(H_2O)/n(C)=1.70$, 100% reformer load, T(Air) = 100 °C, T(Steam) = 230 °C

the range of 1000 °C. These high temperatures were due to the exothermic partial oxidation of the hydrocarbon molecules in both fuels. The reformate temperature was about 500 °C. The temperature of steam when it enters the fuel evaporation chamber was around 435 °C at maximum. As explained above, the enthalpy flow of the superheated steam is used in the evaporation chamber to vaporize the fuel. However, the temperatures established in the fuel evaporation chamber and in the air mixing area were only in the range of 200 and 280 °C, respectively. These values were far too low to ensure complete evaporation of both fuels from Table 10.1, whose final boiling points were 320 and 360 °C, respectively. From these findings it can be concluded that there might have been some larger fuel drops among the droplets in the diesel spray which was injected into the evaporation chamber via the nozzle. Those larger drops might not have been fully evaporated. They might have impinged on the monolith and could have undergone pyrolysis on the catalyst, thus possibly damaging active sites of the catalyst. Additionally, it was found that the nozzles applied in the experimental phase 1 for fuel injection were not manufactured with sufficient accuracy and that they displayed poor spray characteristics especially at part load.

The conversion of the carbon contained in liquid fuels was defined as being complete when no molecules besides H_2, CO, CO_2, or CH_4 were detected in the product streams of the reforming process [2]. During experimental phase 1, the concentrations of H_2, CO, CO_2, and CH_4 showed typical values for autothermal reforming with both fuels from Table 10.1. Those of H_2 varied between 34.5 and 36.5 vol.%, those of CO and CO_2 were almost constant at approx. 12 vol.%, while those of CH_4 amounted to 0.3–0.6 vol.%. However, with respect to the amounts of undesired by-products in the gas phase of the FCGEN autothermal reformer during phase 1, which are shown in Fig. 10.3, the consequences of the low temperatures in the fuel evaporation chamber and the air mixing area (cf. Fig. 10.2) and of the poor quality of the nozzles for fuel injection become obvious.

Considerable amounts of by-products such as ethene, propene, and benzene were detected in the gas phase when using SD10 diesel and NExBTL diesel. Additionally, the unconverted condensed water stream leaving the reactor contained traces of organic carbon in the range of 30–60 ppmw. In the case of the experiments with SD10 diesel, even a diesel slip was observed. The use of a nozzle with a smaller bore diameter (80 μm) did not result in lower amounts of by-products. On the contrary, they even increased, which is an indication of catalyst deactivation during operation with SD10.

These experimental findings clearly indicate that the operating setup as described above during the experimental evaluation phase 1 was not suitable for producing a reformate with the FCGEN autothermal reformer meeting all the quality requirements of a PEFC.

10.4 Experimental Results, Phase 2

Five main approaches to further reformer development were mainly monitored in phase 2 aiming at improved performance of the FCGEN ATR. In doing so, the requirements and potential of the FCGEN system had to be kept in mind.

- Temperature of air to the mixing area: approx. 380 °C
- Steam-to-carbon molar ratio: 1.90
- Fuel: only SD10 diesel
- Catalyst: fresh sample of Johnson Matthey HiFuel ATR catalyst
- Newly manufactured nozzles with 80 μm bore diameter
  - Resulting nozzle pressure: 10–40 bar

The curves for the differential pressures between the inlet and outlet of the newly manufactured nozzles as a function of the mass stream to be injected showed very good alignment. This proves the excellent reproducibility and thus the high quality of the manufacturing processes applied for the new nozzles. Additionally, all nozzles showed a suitable range for the differential pressure between approx. 10 and 40 bar at part load and full load of the FCGEN ATR, respectively. The experiments in Jülich's nozzle test rig showed that for all nozzles good diesel spray characteristics were already found at a differential pressure of 10 bar. At 40 bar, they were much better. It can be concluded from these findings that all newly manufactured nozzles with a bore diameter of 80 μm had significantly better quality and provided a much more suitable range for the differential pressure than the nozzles in phase 1. They were, therefore, used during the autothermal reforming experiments in phase 2.

It was found in the experiments in phase 2 that the most efficient way to increase the temperature of the steam when it enters the fuel evaporation chamber and to raise the temperature in the fuel evaporation chamber and in the air mixing area was to increase the temperature of the air from 100 °C as in phase 1 to approx. 380 °C.

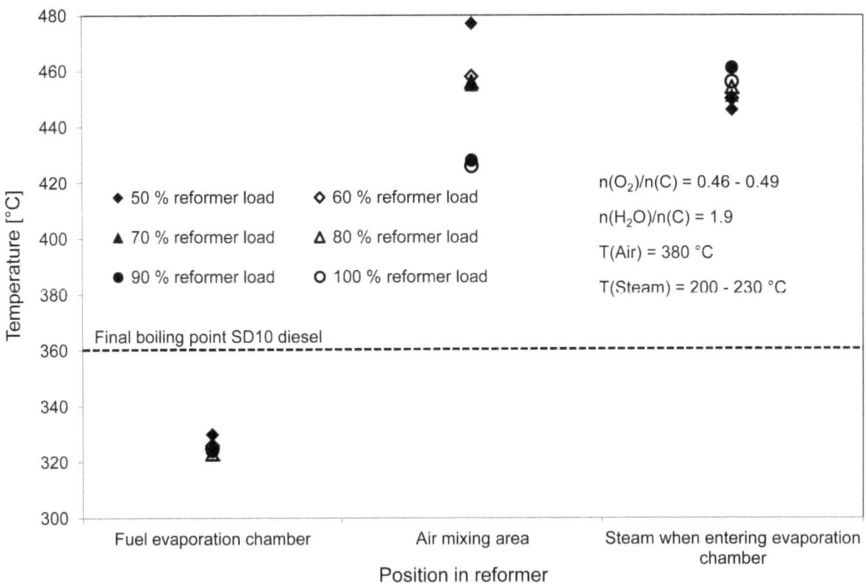

Fig. 10.4 Temperatures at different positions in FCGEN ATR during experiments at various reformer load points using SD10 diesel, T(Air) = 380 °C, $n(H_2O)/n(C) = 1.9$

Other routes, for example, increasing the steam temperature or feeding air through the internal heat exchanger of the FCGEN ATR were not efficient and/or not practical from the system perspective.

In that respect, Fig. 10.4 displays relevant temperatures at different positions in the FCGEN ATR during experiments at various reformer load points between 50 and 100% with an inlet air temperature of 380 °C. The temperature of the steam when it enters the internal heat exchanger of the FCGEN ATR varied between approx. 200 and 230 °C depending on the reformer load.

Figure 10.4 shows that by raising the temperature of the air from 100 °C (phase 1) to 380 °C, it was possible to increase the steam temperature when it enters the fuel evaporation chamber from approx. 435 °C in phase 1 to about 450–460 °C in phase 2. Fortunately, a very drastic increase in the temperature in the ATR air mixing area from 280 °C, as observed in phase 1, to 430–480 °C was found during the experiments in phase 2. Obviously, a homogeneous pre-reaction between the hydrocarbon molecules in SD10 diesel and oxygen was induced in the air mixing area due to the increased temperature of air. This means that under these reaction conditions the temperatures in the air mixing area were significantly higher than those at the end of the boiling range of SD10 diesel fuel (approx. 360 °C) and, therefore, high enough to ensure complete evaporation of the fuel droplets. The temperature of the reformate leaving the reactor amounted to approx. 535–540 °C, which is close to the allowable maximum of 550 °C at this position in the reformer.

During experimental phase 2, the concentrations of H_2, CO, CO_2, and CH_4 again showed typical values for the autothermal reforming of diesel fuels such as SD10,

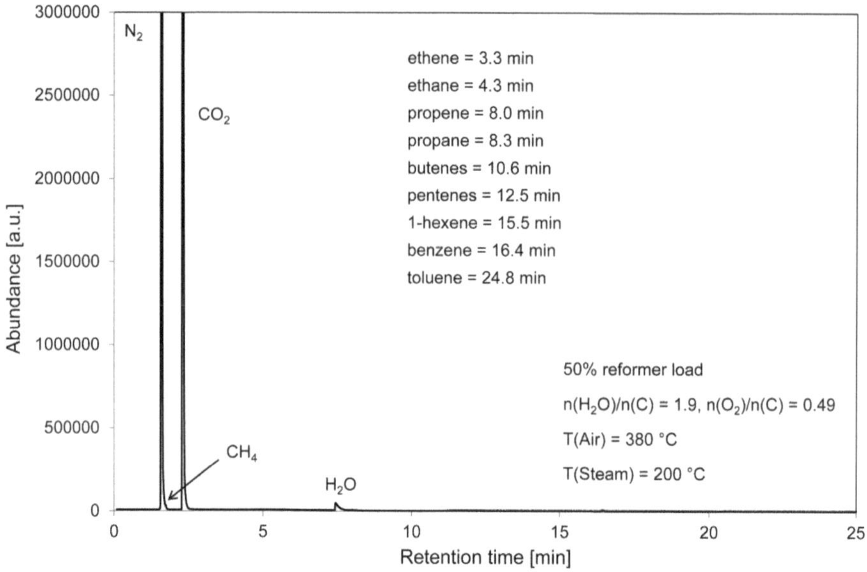

Fig. 10.5 Gas chromatogram of the dry product of the FCGEN ATR using SD10 diesel fuel, $n(O_2)/n(C) = 0.49$, $n(H_2O)/n(C) = 1.90$, T(air inlet) = 380 °C

which contain considerable mass fractions of aromatic substances. Depending on the load point of the reformer, the concentrations of H_2 varied between 34.5 and 36.0 vol. %, those of CO and CO_2 ranged between 10.5 and 12.0 vol. %, while the values for the concentration of CH_4 amounted to less than 0.1 vol. %.

Figure 10.5 shows the gas chromatogram of the dry reformate observed during an experiment at 50 % reformer load using SD10 diesel fuel at an oxygen-to-carbon molar ratio of 0.49, a steam-to-carbon molar ratio of 1.9, and a temperature of the air entering the reformer of 380 °C. On the left-hand side of the chromatogram, at short retention times two large peaks can be seen, whose intensities far exceeded the upper limit of the scale of the *y*-axis. They can be attributed to nitrogen and carbon dioxide, respectively. At the bottom of the peak for carbon dioxide, a very small shoulder was found belonging to the adsorption of methane on the GC column. At a retention time of 7.4 min, a small peak for residual water in the dried reformate was found. However, peaks for the undesired by-products of the reaction, whose retention times are displayed in Fig. 10.5, were not found in most cases or they were so small that their corresponding integration areas were at or under the detection limit of the analytical instrument. Figure 10.5 serves as an example of the other load points from 60 to 100 %. These chromatograms were very similar. Based on these results the carbon conversions at the different load points were calculated to be more than 99.95 %.

10.5 Summary and Conclusion

In phase 1, it was found experimentally that the commercial nozzles for fuel injection with bore diameters of 80 and 100 μm involved significant manufacturing inaccuracies leading to nonsymmetrical spraying of the fuel droplets to be injected into the fuel evaporation chamber. New nozzles with excellent reproducibility and high quality were therefore designed and constructed. All nozzles showed suitable ranges for the pressure drop between approx. 10 and 40 bar (part load and full load of the FCGEN ATR) and very good diesel spray characteristics. A precise production technology for the nozzles was successfully implemented.

In each of the experiments in phase 2 with increased temperature of the air entering the reformer and a newly manufactured nozzle, the reaction conditions in the fuel evaporation chamber and the air mixing area were much more suitable to ensure complete evaporation of SD10 diesel when it enters the monolithic catalyst. As a consequence, conversion of SD10 diesel was more than 99.95 % with concentrations of short-chain hydrocarbons and benzene at or under the detection limit.

Based on the results presented for phase 2, Jülich's FCGEN autothermal reformer is able to deliver a PEFC-quality reformate flow containing H_2, CO, CO_2, N_2, CH_4, and Ar with concentrations to the thermodynamic equilibrium at the temperature of the outlet of the reformer catalyst. This reformate can be processed in any of the downstream components of the FCGEN fuel cell system including the PEFC itself. This conclusion holds for reformer loads between 50 and 100 %.

Acknowledgement The research leading to these results has received funding from the European Union's Seventh Framework Programme (FP7/2007–2013) for the Fuel Cells and Hydrogen Joint Technology Initiative under grant agreement no. [277844] 10.

References

1. Pasel J, Samsun RC, Peters R, Thiele B, Stolten D (2014) Long-term stability at fuel processing of diesel and kerosene. Int J Hydrog Energy. doi:10.1016/j.ijhydene.2014.03.148
2. Pasel J, Samsun RC, Peters R, Stolten D (2013) Fuel processing of diesel and kerosene for auxiliary power unit applications. Energy Fuels 27:4386–4394

Chapter 11
Basic Study on the Application of the Fuel Cell System Operated by Kerosene to Vessel

Kazuyoshi Sumi

Abstract The CO_2 emissions from ocean-going vessels in 2007 increased by 86 % to about 900 million in comparison with 1990, and it was 3 % of total CO_2 emissions. If measures are not taken in reducing CO_2 emissions from ocean-going vessels, it is assumed, by 2050, to increase six times against 1990.

Therefore, from January 1, 2013,the International Maritime Organization (IMO) started to regulate the CO_2 emissions from the ocean-going vessels. Japanese shipping and marine machinery companies had developed the system that reduced the CO_2 emissions from the ocean-going vessels. With regard to the regulation of the NO_x emissions from the ocean-going vessels, IMO had the phased regulation of it, and IMO will do NO_x Tier III regulations in 2021.

With regard to the application of renewable energy to vessels, some vessels with photovoltaic panels were operated by Japanese shipping companies. Therefore, the author suggested the application of the fuel cell system that did not release air pollution gases to vessels.

Keywords IMO · PEFC · Protection of marine environment · Air pollution gases

11.1 Introduction

For protecting the maritime environment, the International Maritime Organization (IMO) carried out the regulation of the CO_2 emissions and the NO_x emissions from the ocean-going vessels; some systems for reducing them have been designed and realized. Therefore, the author carried out simulation of the operation about the application of the fuel cell system that did not release air pollution gases to vessel. The vessel that had carried out the simulation of the operation was the training vessel of the Marine Technical College, the Kaigi Maru.

K. Sumi (✉)
Department of Marine Engineering, Marine Technical College,
12–24, Nishikura-cho, Ashiya, Hyogo, Japan
e-mail: k-sumi@mail.mtc.ac.jp

A. Sayigh (ed.), *Renewable Energy in the Service of Mankind Vol I*,
DOI 10.1007/978-3-319-17777-9_11

First, the author examined the type of fuel cell and its fuel for the simulation of the operation. Second, the author designed the electric power system combined with the fuel cell based on the electric power system and the consumed electric energy of the vessel. Based on the actual value of electric power that was measured by the author and the specifications of each electric apparatus, load power model was made.

Finally, the author carried out the simulation of the operation of the electric power system combined with the polymer electrolyte fuel cell (PEFC) based on the navigation schedule of the vessel from April 1, 2011, to March 31, 2012. From the result of the simulation of the operation, the author calculated the reduction volume of fuel consumption, NO_x emissions by the application of the fuel cell instead of the diesel generator of the vessel, and the author considered some problems that the system was operated.

11.2 Consideration of Electric Power System Applying Fuel Cell

11.2.1 Outline of the Kaigi Maru

The principal specification of the Kaigi Maru is shown in Table 11.1.

The Kaigi Maru had a length of 34.0 m, breadth of 6.8 m, and gross tonnage of 157 t. She had one main propulsion engine with an output of 500 kW and navigated Osaka bay and Seto inland sea in Japan at a speed of about 11 knot (about 22 km/h). The ship had one main diesel generator with an output of 150 kVA and one shaft generator with the same output as electric sources. Normally, the main diesel generator was providing electric power. As emergency source, she had the lead–acid battery with a capacity of 300 Ah.

Table 11.1 The principal specification of the Kaigi Maru

Hull	
Length	34.0 m
Breads	6.8 m
Gross tonnage	157 t
Service speed	11 knot (about 22 km/h)
Main propulsion engine *Diesel engine: 1 set*	
Output	500 kW
Main diesel generator: 1 set	
Output	150 kVA
Voltage	450 V
Frequency	60 Hz
Shaft generator: 1 set	
Output	150 kVA

11.2.2 Consideration of the Type of Fuel Cell

From the difference of operation temperature, the fuel cells were categorized as high-temperature operation type and low-temperature operation type. The author selected the fuel cell of low-temperature operation type taking the operation and maintenance of the fuel cell on board into consideration. Among the fuel cell of low-temperature operation type, the author adopted PEFC that had capacity up to 100 kW and was widely used for houses.

11.2.3 Consideration of the Fuel of the PEFC

The fuel of the PEFC was natural gas and methanol that consisted of hydrocarbons. Natural gas and methanol were reformed before being supplied to the fuel cell and hydrogen of the fuel was only sent to it. Considering storage and handling on board, the author concluded that liquid fuel was more useful than gas fuel and selected low-sulfur kerosene (sulfur content: 0.001 % (m/m); average composition: $C_{12}H_{26}$) that was standardized in Japan in 2009 as the fuel of the fuel cell [1].

11.2.4 Component of Electric Power System Combined the PEFC

The electric power system combined with the fuel cell that was considered for the simulation is shown in Fig. 11.1. The kerosene that was stored in the fuel tank was sent to the vaporizer by the fuel supply pump, and it was vaporized by the heat of the electric heater. After the kerosene that was vaporized was desulfurized by desulfurizer that was heated by the electric heater, it was sent to the reforming unit. When the load power was changing, the revolution speed of the fuel supply pump was adjusted, and the volume of fuel that was sent to the reforming unit was controlled.

The fuel gas that was sent to the reforming unit was mixed with the steam supplied by the boiler of the vessel, and it was heated by the burner, then it was divided into hydrogen (H_2) and carbon monoxide (CO) by the reformer. The density of CO in the fuel gas was decreased by CO shift converter and CO remover, and the fuel gas was sent to the anode of the stack of the PEFC.

The electric power was generated by the reaction of the hydrogen of the fuel gas and the oxygen of air that was sent to the cathode of the PEFC. The electric power from the stack of the PEFC was changed from direct current (DC) power to alternating current (AC) power by the inverter and was stepped up by the transformer, and then it was supplied to the load. The conversion efficiency of the inverter was 0.96 and that of the transformer was 0.98.

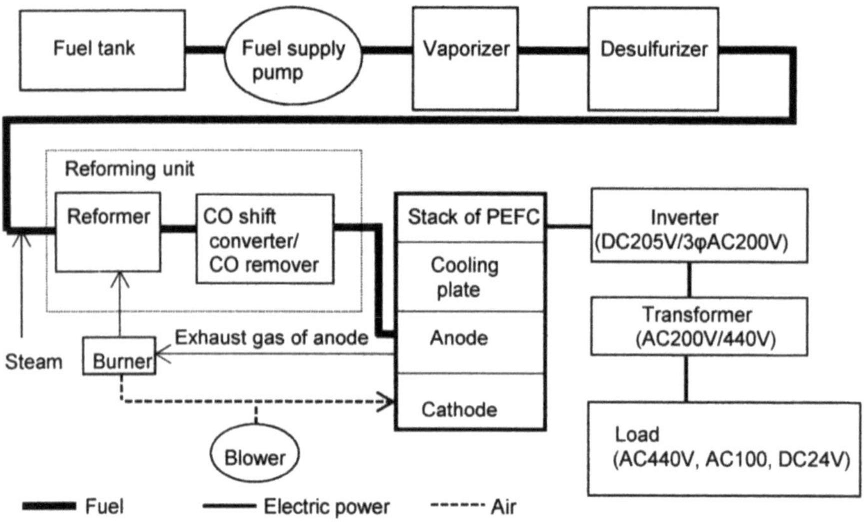

Fig. 11.1 The electric power system combined with the fuel cell. *PEFC* polymer electrolyte fuel cell

The blower sent air to the cathode of the stack and the burner of the reforming unit. The exhaust gas that contained the residual hydrogen from the anode of the stack of the PEFC was sent to the burner. The cooling plate of the stack of the PEFC was cooled by pure water because its temperature increased from 80 to 100 °C.

11.3 Simulation Method of the Operation of the Electric Power System Combined with the PEFC

11.3.1 Load Power Model

Based on the actual value of electric power that was measured by the author and the specifications of each electric apparatus, the load power model was made. Here, the load power of the auxiliary apparatus (the fuel supply pump, the blower, and so on), as shown in Fig. 11.1 for the PEFC, was estimated 3 kW.

The load power model when the vessel was navigating is shown in Fig. 11.2. The maximum electric energy of 1 h when the vessel was navigating was 39.0 kWh, the minimum was 24.0 kWh, and the average was 31.5 kWh.

The load power model when the vessel was at anchor is shown in Fig. 11.3. The maximum electric energy of 1 h when the vessel was at anchor was 29.0 kWh, the minimum was 18.0 kWh, and the average was 20.3 kWh.

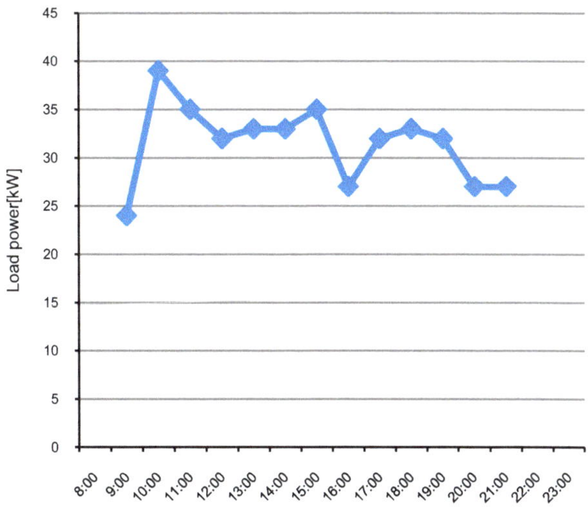

Fig. 11.2 The load power model when the vessel was navigating

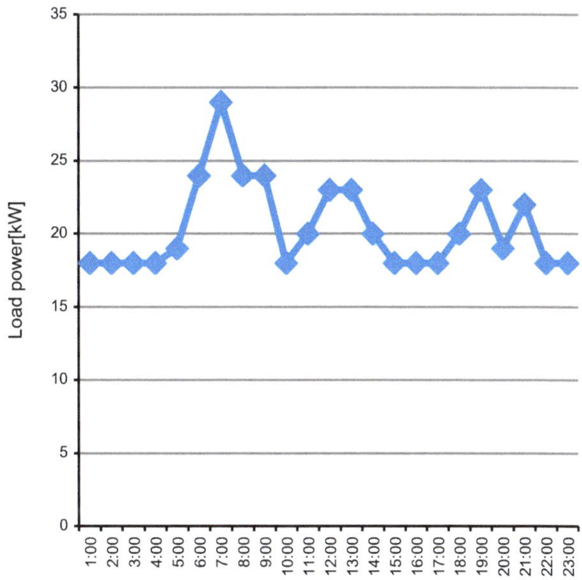

Fig. 11.3 The load power model when the vessel was at anchor

11.3.2 Volume of Fuel and Air Supplied to the PEFC

The volume of the fuel supplied to the reformer was calculated as follows [2]. Here, the structure of stack of the PEFC was that many unit cells were layered. First, the power output of the stack of the PEFC was calculated by Eq. (11.1); here, the net power output (the power output of the transformer) was P_{FC}:

$$P_S = \frac{P_{FC}}{\eta_{INV} \cdot \eta_T}, \tag{11.1}$$

where P_S is the power output of the stack of the PEFC (kW), P_{FC} is the net power output (kW), η_{INV} is the conversion efficiency of the inverter—it was 0.96, and η_T is the conversion efficiency of the transformer—it was 0.98.

Second, the voltage of the unit cell V_C was calculated as follows. The stack current I_S was calculated by Eq. (11.2); here, the stack voltage V_S was 205 V:

$$I_s = \frac{P_S \times 1000}{V_S}, \tag{11.2}$$

where I_S is the stack current (A), P_S is the power output of the stack (kW), and V_S is the stack voltage (V).

From the stack current I_S and the effective electrode area of the unit cell A_C, the current density i_D was calculated by the equation:

$$i_D = \frac{I_S \times 1000}{A_C}, \tag{11.3}$$

where i_D is the current density (mA/cm^2), I_S is the stack current (A), and A_C is the effective electrode area of the unit cell (cm^2)—it was 225 cm^2.

The voltage of the unit cell V_C was evaluated using the *I–V* characteristics that expressed the relation between the current density i_D and the voltage of the unit cell V_C. The *I–V* characteristics mentioned in Ref. [1] are shown in Fig. 11.4.

The number of unit cells that were stacked N_C was calculated by the equation:

$$N_C = \frac{V_S}{V_C}, \tag{11.4}$$

where N_C is the number of unit cells that were stacked, V_C is the voltage of the unit cell (V), and V_S is the stack voltage *(V)*.

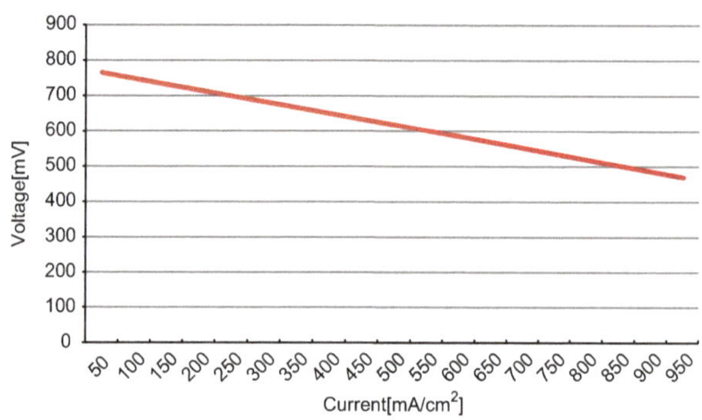

Fig. 11.4 The *I–V* characteristics

From the abovementioned equations, the volume of the hydrogen supplied to the stack of the PEFC F_H was calculated by the equation:

$$F_H = \frac{I_S \cdot N_C}{2391.5}, \tag{11.5}$$

where F_H is the volume of the hydrogen supplied to the stack of the PEFC (Nm^3/h), I_S is the stack current (A), and N_C is the number of unit cells that were stacked.

The volume of the fuel supplied to the reformer F_F was calculated by the equation:

$$F_F = \frac{Q_H}{U_F \cdot U_R \times 37}, \tag{11.6}$$

where F_F is the volume of the fuel supplied to the reformer (Nm^3/h), F_H is the volume of the hydrogen supplied to the stack of the PEFC (Nm^3/h), U_F is the fuel utilization rate—it was 0.80, and U_R is the reforming efficiency—it was 0.95.

The volume of the air supplied to the stack of the PEFC F_A was calculated by the equation; here, the volume rate that oxygen was contained in air was 20.8 %:

$$F_A = \frac{Q_H}{U_O \times 0.208 \times 2}, \tag{11.7}$$

where F_A is the volume of the air supplied to the stack of the PEFC (Nm^3/h), F_H is the volume of the hydrogen supplied to the stack of the PEFC (Nm^3/h), and U_O is the oxygen utilization rate—it was 0.50.

11.4 Result of the Simulation of the Operation

Based on the load power model, the load electric energy when the vessel was at anchor after navigating (The number of the day was 74 days of the simulation of the operation period, called navigation day), and the vessel was at anchor all day (The number of the day was 292 days of the simulation of the operation period, called anchor day), was obtained. The average load electric energy of 1 h of the navigation day was 23.5 kWh and that of the anchor day was 20.3 kWh. Then, the volume of the fuel supplied to the reformer was calculated by Eqs. (11.1–11.6) based on the load power.

The fuel oil consumption and the NO_x emissions in case where not the PEFC but the diesel generator supplied the electricity were calculated as follows. The fuel oil consumption was calculated by the characteristics between the output and the fuel oil consumption that was made from the test record of the diesel engine of the generator. The characteristics between the output and the fuel oil consumption of the diesel engine of the generator are shown in Fig. 11.5.

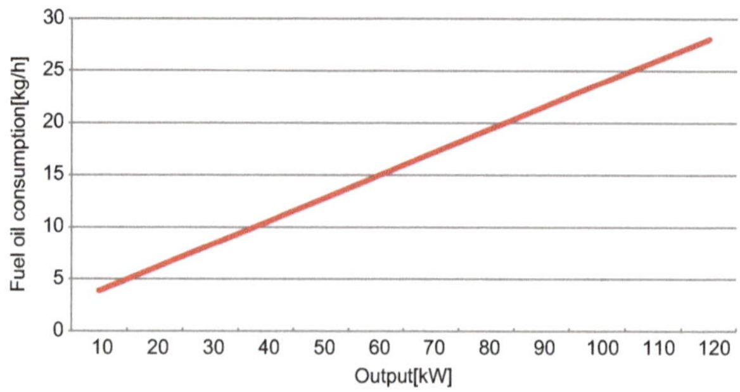

Fig. 11.5 The characteristics between the output and the fuel oil consumption of the diesel engine of the generator

The NO_x emissions were calculated considering the load factor of the diesel engine, as given below [3].

$$N_{NO}\{2.11-1.92(1-z)\}P_D\times10^{-3}\times46/22.4, \tag{11.8}$$

where N_{NO} is the NO_x emissions (kg/h), z is the load factor of the diesel engine, and P_D is the output of the diesel engine of the generator (kW).

Here, the output of the diesel engine of the generator was calculated by the equation:

$$P_D=\frac{P_G}{\eta_G}, \tag{11.9}$$

where P_D is the output of the diesel engine of the generator (kW), P_G is the output of the generator (kW), and η_G is the generating efficiency of the generator, it was 0.80.

The volume of the fuel supplied to the reformer and the fuel oil consumption of the diesel engine of the generator are shown in Table 11.2. The NO_x emissions in case where the diesel generator supplied the electricity are shown in Table 11.3. From Table 11.2, the fuel oil consumption for supplying the electricity proved to be reduced by using the PEFC. The daily average reduction rate of the navigation day was 36.3 %. The daily average reduction rate of the anchor day was 40.0 %. From Table 11.3, NO_x emissions proved to be reduced by 300.3 kg for 1 year.

Finally, the author considered the validity of the calculating process of the volume when the fuel was supplied to the PEFC by evaluating the generating efficiency of the PEFC. The volume of the fuel supplied to the PEFC for the load power was calculated by Eqs. (11.1)–(11.6), then the calorific value of the fuel was calculated by the equation:

$$Q_F=F_{FL}\cdot L_{HV}, \tag{11.10}$$

Table 11.2 The volume of the fuel supplied to the reformer and the fuel oil consumption of the diesel engine of the generator

		The volume of the fuel supplied to the reformer (kg)	The fuel oil consumption of the diesel engine of the generator (kg)
Navigation day	Average value of 1 day	112.0	175.9
	Average value of 1 h	4.7	7.3
Anchor day	Average value of 1 day	93.1	155.2
	Average value of 1 h	3.9	6.5

Table 11.3 The NO_x emissions in case the diesel generator supplied the electricity

	NO_x emissions (kg)
Total value of navigation day	76.2
Total value of anchor day	224.1
Total value	300.3

where, Q_F is the calorific value of the fuel (kcal/h), F_{FL} is the volume of the fuel supplied to the PEFC (l/h), and L_{HV} is the low calorific value of the fuel (kcal/l), it was 8329 kcal/l.

Then, the generating efficiency of the PEFC was calculated by the equation:

$$\eta_{FC} = \frac{P_{FL} \cdot 1000}{Q_F \cdot 1.163}, \tag{11.11}$$

where η_{FC} is the generating efficiency of the PEFC, P_{FL} is the load power (kW), and Q_F is the calorific value of the fuel (kcal/h).

The generating efficiency of the PEFC is shown in Fig. 11.6. The maximum generating efficiency of PEFC was 43.4 %, and its minimum generating efficiency

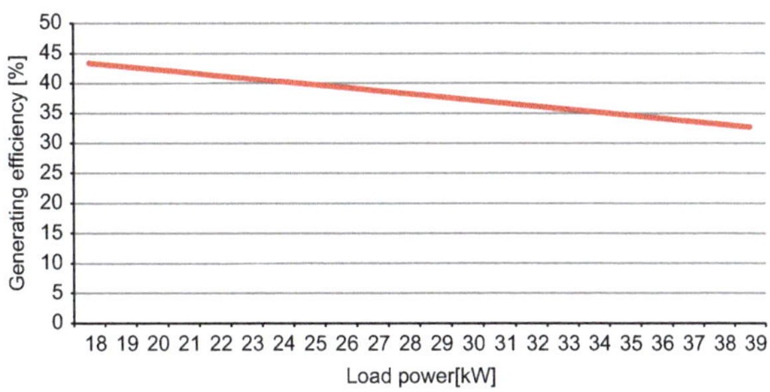

Fig. 11.6 The generating efficiency of the PEFC

was 32.7 %. From the reference [4], the generating efficiency of the PEFC was from 33 to 44 %, and the calculating process of the volume of the fuel that was supplied to the PEFC was proved to be valid.

11.5 Conclusion

The author carried out the simulation of the operation with regard to the application of the fuel cell system that was operated by kerosene on the training vessel of Marine Technical College, the Kaigi Maru. From the result of the simulation of the operation, the fuel oil consumption for supplying the electricity proved to be reduced by using the PEFC. The daily average reduction rate of the navigation day was 36.3 %. The daily average reduction rate of the anchor day was 40.0 %. Then, NO_x emissions proved to be reduced by 300.3 kg for 1 year. The author considered the following problems in the application of the fuel cell system to the vessel:

1. The structure and the pipe arrangement of the space in the engine room where the vaporizer, the desulfurizer, the reforming unit, and the stack of PEFC were installed (these was shown in Fig. 11.1)
2. The safety control system when the hydrogen was leaking from the pipe, and the engine room had fire

In future, the author will examine the abovementioned problems and continue to discuss the application of the fuel cell system to the vessel.

References

1. http://www.jisc.go.jp/app/pager?id=244570 accessed 29 Jan 2014.
2. Uematsu H (2000) Fuel cell generation system and heat calculation. Ohmsha Tokyo
3. Nishikawa E, Takarada T (1994) Estimation of amount of air pollutants discharge from ships sailing in Osaka Bay. Mech Eng Sci J 29(6):412–421
4. https://www.jema-net.or.jp/Japanese/res/fuel/about.html accessed 7 March 2014.

Chapter 12
System Modelling for Hybrid Solar Hydrogen Generation and Solar Heating Configurations for Domestic Application

Krisztian Ronaszegi, Dan J L Brett and Eric S Fraga

Abstract Hydrogen generation has the potential to deliver an environmentally friendly, low-cost, and renewable energy source. One promising generation method is solar water splitting via a photoelectrochemical (PEC) reaction as an alternative to a combined photovoltaic-electrolyser system. Although PEC technology shows potential, the efficiency of this technology is currently limited by thermodynamics and technical issues in implementation. The development of novel materials is one route for improvements in PEC system efficiencies. In particular, with multiple band-gap electrodes, the thermodynamic efficiency, and so the overall generated hydrogen quantity, can be increased.

In the case of applications where there are heating requirements beyond the need to generate hydrogen, there are further options for extracting energy from the solar resource. Longer wavelength radiation not used by the PEC system may be available for use. Just as it is possible to have a photovoltaic–thermal (PV/T) hybrid system which generates both electricity and heat, a PEC unit may also be combined with a solar thermal unit as a hybrid PEC/T system. This combined heat and power (CHP) system will deliver heat directly and also both heat and power through the use of the hydrogen as a fuel in, for instance, a fuel cell.

Despite the promise of PEC technology, there is little research in modelling and system simulation and especially for hybrid systems. Systems' modelling is a prerequisite for optimal design, especially for the design and exploration of novel

E. S. Fraga (✉) · K. Ronaszegi · D. J. L. Brett
Electrochemical Innovation Lab, Department of Chemical Engineering, University College London (UCL), London, UK
e-mail: e.fraga@ucl.ac.uk

E. S. Fraga
Centre for Process Systems Engineering, Department of Chemical Engineering, University College London (UCL), London, UK

D. J. L. Brett
e-mail: d.brett@ucl.ac.uk

K. Ronaszegi
e-mail: krisztian.ronaszegi.10@ucl.ac.uk

A. Sayigh (ed.), *Renewable Energy in the Service of Mankind Vol I*,
DOI 10.1007/978-3-319-17777-9_12

configurations. A system model of a dwelling, with varying heat and power demands, together with a hybrid PEC/T system for meeting these demands, has been developed and implemented in Matrix Laboratory (MATLAB). The full system integrates a PEC unit for hydrogen generation, a solar thermal unit, a proton exchange membrane (PEM) fuel cell, a hydrogen storage tank, and a buffer tank for heat storage. The model has been evaluated through a case study consisting of a typical three-person household in the UK. The aim of the case study is to investigate present and near-future capabilities of renewable energy supply and CO_2 emission reduction subject to the UK building energy regulations. Results show that single band-gap photo-electrode materials are not able to meet the energy demands of the household adequately if the demand includes power and both space and hot water heating. However, with novel multiple band-gap electrodes, in a hybrid CHP system, the system efficiency can be significantly increased, and we demonstrate the potential to help meet the comprehensive demands of a typical household through the development of novel materials for PEC reactions.

Keywords Hydrogen · Photoelectrochemical · Combined heat and power · Domestic energy

12.1 Introduction

A key challenge with many forms of renewable energy is the intermittent nature of generation. Efficient energy storage provides one mechanism for alleviating the impact of this intermittency. One of the most promising storage mechanisms is in the form of hydrogen, a good environmentally neutral energy carrier. Hydrogen can be generated from water using energy from renewable energy sources. Hydrogen is typically generated by direct electrolysis, but alternative technologies exist including the use of a photoelectrochemical (PEC) process.

PEC solar water splitting method is a developing technology, and the current research focus is on electrode materials development to increase the hydrogen generation. More hydrogen can be generated with multiple band-gap electrodes, but these are expensive and currently not efficient enough.

When the eventual use of the hydrogen is considered, especially in a country such as the UK, we find that a significant proportion is used to produce heat, as opposed to electricity. In practice, we observe that a PEC device will heat up due to the solar energy not used by the electrochemical process while generating hydrogen. This heat could be extracted and used directly. Combining the thermal capability with the electrochemical process could lead to more attractive overall energy efficiencies. Such hybrid methods have already been considered with PV cell applications where heat reduction through its use for meeting heating demands proves to be essential due to the inverse efficiency effect [1–4].

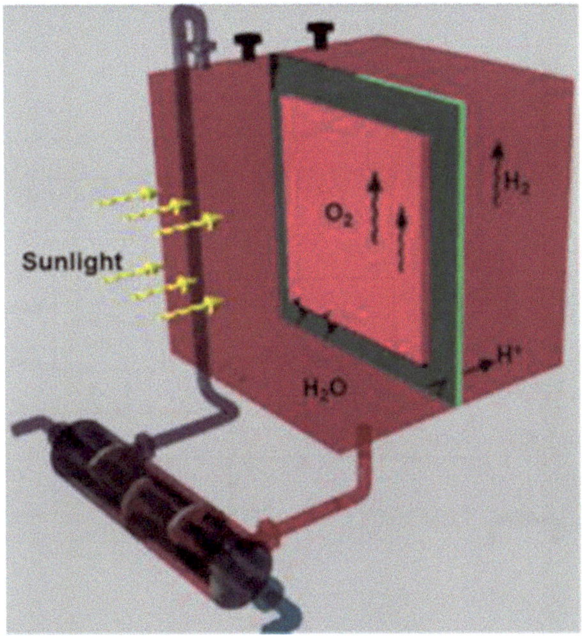

Fig. 12.1 A hybrid photoelectrochemical thermal solar water splitting and heat collector unit. The solar radiation both heats the water and is used to split the water into oxygen and hydrogen. The heated water is recirculated to send heat to the rest of the system. The hydrogen is subsequently used to generate electricity and heat

In this chapter, we explore the potential of a hybrid PEC/T unit, shown in Fig. 12.1, through the use of a full-system simulation with application to the domestic sector. The domestic sector is a key sector for energy savings [5]. The system includes a CHP unit, using hydrogen as a fuel, for the generation of both electricity and heat. A proton exchange membrane fuel cell (PEMFC) has been considered for this part of the system.

12.2 The Energy System

Figure 12.2 shows the full energy system for a dwelling. It includes three sources of energy: solar energy, the electricity grid, and the natural gas grid. Both power and heating demands are modelled. Connecting the energy inputs to the demands is a system comprising the novel hybrid PEC/T unit for generating hydrogen and the direct delivery, of heat, a PEMFC for power generation, as well as heat delivery and an auxiliary multi-fuel burner which can combust both hydrogen and natural gas. Ancillary units include an inverter, a compressor for hydrogen storage, and storage tanks for both hydrogen and heat. Two-heat storage tanks are used, one for direct space heating and the other for hot water.

Fuel cells offer a good option for micro-combined heat and power for residential dwellings. A PEM fuel cell was chosen for the system because of its good dynamic responsiveness to load, even though its heat to power ratio is low. The low heat

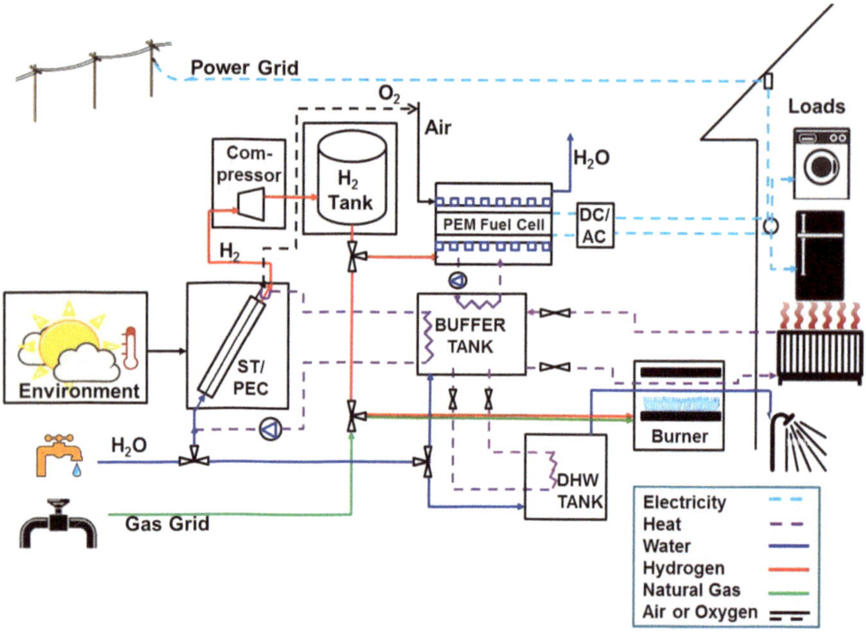

Fig. 12.2 Schematic of a potential solar PEC/T hybrid hydrogen domestic micro-CHP-based energy system. *PEM* proton exchange membrane, *AC* alternating current, *DC* direct current, *PEC* photoelectrochemical cell

generation will be partially ameliorated by the heat captured by the solar thermal collector unit. The fuel cell is operated at a constant voltage, 0.7 V and at a constant operating temperature, 80 °C.

The fuel cell uses hydrogen. When the rate of hydrogen generation is insufficient to meet the electricity or heat demands, the unmet demands are satisfied by the grid energy sources. Natural gas is used in an auxiliary burner for both space and water heating. If there is a surplus of heat but not electricity, the surplus heat is stored in a buffer tank. If the opposite situation exists, the surplus electricity is not stored, although it could be sold back to the national grid. Finally, if there is more hydrogen generated than can be used for both power and heating, the surplus is stored in a storage tank under pressure.

The buffer tank accumulates heat from the fuel cell and from the solar collector. It transfers the heat to the room via radiators and to the hot water tank via an internal heat exchanger. Water goes to the user from the hot water tank which is heated further by the burner if the temperature is less than the required 60 °C. The burner also covers the deficit for space heating.

A detailed dynamic model of this system has been developed and implemented in (MATLAB). The models for the key components of this system are described in the next section.

12.3 The Models

Existing models for PEC units estimate efficiency from tabulated data although some models exist for photocurrent prediction. However, there is no complete model suitable for system design. In particular, the existing models do not capture the dynamic nature of a PEC unit when incorporated into an energy system, taking into account, for instance, variations in solar radiation, the temperature of the water in the PEC unit itself [6–9]. A model should take into account the differences due to the location of the unit and hourly, daily, and seasonal variations in solar radiation.

We have developed a model based on thermodynamic principles. The model also incorporates semi-empirical elements derived by fitting data from the literature. Since the data are for laboratory-scale devices, generated by different methods even when the photoelectrodes and electrolytes are the same, the model has an element of uncertainty. Nevertheless, it is more rigorous than a *black-box* model and represents the dynamic nature.

The model for thermodynamic limits for solar conversion to chemical energy was created by Ross and Hsiao [10]. In this model, the power is equal to the quantity of photoproduct multiplied by the free-energy difference between the initial state of the material and the product state. The power and the derived efficiency are given by

$$\eta_{\mathrm{td}} = \frac{J_{\mathrm{S}}\mu\left(1-\phi_{\mathrm{loss}}\right)}{\int_{0}^{\infty}\phi(\lambda)d\lambda},$$

where the numerator is the power for chemical product generation, for example, hydrogen, and the denominator is the total solar energy. In the above equation, η_{td} is the thermodynamic efficiency, J_{S} is the flux of absorbed photon, $\propto$ is the chemical potential generated by the system as free-energy difference between states, ϕ_{loss} is the quantum yield that is lost via radiative mechanism from the excited state, but without leading to a photoproduct, and so the $(1-\phi_{\mathrm{loss}})$ term represents the fraction of the absorbed photons which turn to photoproduct, and $\phi(\lambda)$ is the spectral irradiance.

Ross and Hsiao assumed that the absorber is ideal which means that the absorption cross section is a step function at the band-gap wavelength (λ_{g}). Furthermore, there was no consideration about the non-radiative loss such as electron-hole recombination. In our model, the absorbed photon flux has been modified by adding the quantum efficiency which takes consideration of the non-radiative loss, as suggested by Bolton [11]:

$$J_{\mathrm{S}} = \int_{0}^{\lambda_{\mathrm{g}}}\eta_{\mathrm{IPCE}}(\lambda)N_{\mathrm{S}}(\lambda)d\lambda,$$

where $\eta_{\mathrm{IPCE}}(\lambda)$ is the incident photon to current efficiency (or quantum efficiency), and N_{S} is the incident solar photon flux.

Further parameters, such as optical efficiency and chemical efficiency, are required to obtain the solar to hydrogen efficiency. The optical efficiency (η_0) describes the effect of light reflecting back or being absorbed by the electrolyte before it reaches the photoelectrode. The chemical efficiency (η_{chem}) is the kinetic loss of the charge transfer at the surface reaction. According to Bolton, it is assumed that 0.8 eV must be lost in each photochemical step. Thus, the final efficiency of the solar thermal hybrid unit, η_{sth} is given by

$$\eta_{sth} = \eta_{td}\eta_0\eta_{chem}.$$

The hydrogen generation rate is therefore:

$$m_{H_2} = \frac{A_{PEC}\eta_{sth}G}{\Delta H_C^\circ},$$

where A_{PEC} is the area of the PEC device, G is the hourly solar intensity, and ΔH_C° is the combustion enthalpy of hydrogen.

The thermal hybrid concept considers a simple solar collector where the electrolyte flows across the PEC device, transporting the accumulated heat to a heat buffer tank. The model is an energy balance for a solar collector [12]:

$$\frac{d}{dt}T_{out} = \frac{A\eta G}{V\rho c_p} - \frac{U_{coll}A}{V\rho c_p}\left(\frac{T_{in}+T_{out}}{2} - T_{amb}\right) + \frac{v}{V}\left(T_{in} - T_{out}\right)$$

where T are temperatures for the collection inlet and outlet, A is net surface area of the collector, η is the collector efficiency, G is the hourly solar radiation, V is the volume of the PEC/T hybrid device, ρ and C_p are the density and heat capacity of the electrolyte, U is the overall heat loss transfer coefficient, and v is the flow rate in the collector loop. The left term in this equation describes the rate of heat accumulation in the collector implicitly by the outlet temperature of the electrolyte. On the right hand side, the first term is the inlet energy from the solar radiation, the second term is the heat lost by convection and radiation between the collector and the environment, and the last term is the absorbed heat by the electrolyte which flows to the buffer storage.

The collector efficiency was modelled with linear approximation of the temperature difference between the inlet temperature and the ambient temperature [13]:

$$\eta = \eta_0 - b_0\frac{T_{coll,in} - T_{amb}}{G},$$

where η_0 is the optical efficiency of the collector, the maximum efficiency assuming no heat loss, and b_0 is the first-order heat loss coefficient.

Combining these two units requires modelling the allocation of the solar spectrum to each part. Figure 12.3 shows a typical daily solar spectrum with two photoelectrodes band-gap wavelengths indicated [14]. The area for a photoelectrode compared to the total area is much higher than the solar to hydrogen efficiency.

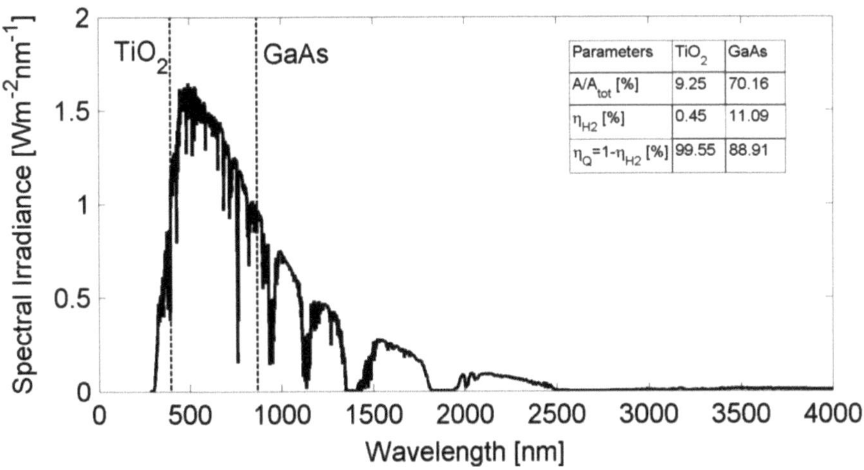

Fig. 12.3 Solar spectrum with two electrode parameters and information about the fractions which turn to hydrogen generation and the rest which turns to heat accumulation

Thus, in the model, first the PEC efficiency was calculated with given electrode properties. Since there is no available PEC device to calculate all the reflection–transmittance term as function of wavelength, the known optical efficiency from the solar collector was used at the PEC efficiency calculation as well. After calculating the PEC efficiency, the rest was used for the solar collector efficiency.

The temperatures in the thermal units (the buffer tank and the storage tank) are each modelled by a simple energy balance and a heat exchanger:

$$V_{\mathrm{unit}}\rho c_{\mathrm{p}}\frac{d}{dt}T_{\mathrm{unit}}=\sum Q_{\mathrm{in}}-\sum Q_{\mathrm{out}},$$

where the inlet and outlet heat rates are:

$$Q_{\mathrm{in/out}}=mc_{\mathrm{p}}\left(T_{\mathrm{in/out}}-T_{\mathrm{return}}\right),$$

and where the return temperature is calculated with the heat exchanger efficiency [12]:

$$T_{\mathrm{return}}=\left(T_{\mathrm{unit1}}-T_{\mathrm{unit2}}\right)e^{-\frac{UA}{\dot{m}c_{\mathrm{p}}}}+T_{\mathrm{unit2}}$$

12.4 Case Study

A "typical" three-person household in London was simulated. The total area covered by hybrid PEC/T cells is 18 m^2 and a 1-kW PEM fuel cell was considered. Demands were 2540 kWh year^{-1} for electricity, 9610 kWh year^{-1} for space heating,

Table 12.1 Summary of hydrogen production (m_{H_2}), electricity required from the grid (E_g), natural gas required from the grid (m_{CH_4}), and the resulting savings obtained in grid usage for different electrode materials in a hybrid PEC/T unit

Material	m_{H_2} (kg)	E_g (kWh)	Saving (%)	m_{CH_4} (kg)	Saving (%)	CO_2 reduction (kg)
None	0	2540	0	1034	0	0
TiO_2	4	2540	0	975	6	146
CdS	26	2522	1	932	10	262
$GaInP_2$	54	2476	3	900	13	366
GaAs	88	2408	5	875	15	463

and 4344 kWh year^{-1} for hot water. The heat demands can be met with 1034 kg of natural gas from the grid. The simulation of the system was carried out in MATLAB using the *rkf45* ODE solver. The aim was to quantify potential fuel reductions and resulting reductions in CO_2 emissions.

A range of photoelectrode materials have been considered in a single band-gap PEC cell: TiO_2 (400 nm), CdS (517 nm), $GaInP_2$ (689 nm) and GaAs (870 nm). The electrodes represent both recent and possible future hydrogen generation technologies with different efficiencies. It should be noted that hydrogen generation efficiency trades-off with thermal efficiency.

The simulations, using hourly solar spectral data [15] and hourly demand profiles are summarised in Table 12.1. The model, as used in the case study, has a number of design parameters which have been set, parameters such as the flow rate of electrolyte, the cell area, and the size of the fuel cell. These would typically be given values which optimise the performance of the complete system. For instance, the larger the fuel cell, the greater the amount of waste heat produced, heat which can be used to meet heating demands. However, fuel cells are intended to operate continuously so the electricity produced may be more than that required. The model presented could provide the basis for an optimisation-based design procedure, and this is the scope for further work in this area. The results in the table are based on non-optimised values for these parameters, however, so these results should be taken as a guide for what is possible. Better performance is likely to be achievable.

12.5 Conclusion

A system model of the energy system for a domestic dwelling has been presented. The energy system is centred on a hybrid PEC/T unit which produces hydrogen and simultaneously heats up the electrolyte (water in our case). The hydrogen is then used in a PEMFC to generate electricity, with waste heat also made available to help meet the dwelling's heating demands. A number of materials for the PEC electrode have been considered. These differ in the range of the solar spectrum they can make use of. The results show that, with a reasonable coverage of 18 m^2 of roof space, significant savings in both electricity and natural gas requirements from national grids are possible.

References

1. Dubey S, Sandhu GS, Tiwari GN (2009) Analytical expression for electrical efficiency of PV/T hybrid air collector. Appl Energy 86:697–705
2. Joshi AS, Dincer I, Reddy BV (2009) Performance analysis of photovoltaic systems: a review. Renew Sustain Energy Rev 13:1884–1897
3. da Silva RM, Fernandes JLM (2010) Hybrid photovoltaic/thermal (PV/T) solar systems simulation with Simulink/Matlab. Sol Energy 84:1985–1996
4. Kumar R, Rosen MA (2011) A critical review of photovoltaic-thermal solar collectors for air heating. Appl Energy 88:3603–3614
5. DECC (2011) 2050 Pathways analysis. Part 1, 2
6. Conibeer GJ, Richards BS (2007) A comparison of PV/electrolyser and photoelectrolytic technologies for use in solar to hydrogen energy storage systems. Int J Hydrogen Energy 32:2703–2711
7. Wingens J, Krost G, Ostermann D, Damm U, Hess J (2008) Application of photo-electrochemical hydrogen production for autonomous solar based electricity supply. Third international conference on electric utility deregulation and restructuring and power technologies, 2008 DRPT. pp. 2470–2475
8. James BD, Baum GN, Perez J, Baum KN, Square OV (2009). Technoeconomic analysis of photoelectrochemical (PEC) hydrogen production. DOE report
9. Pinaud BA, Benck JD, Seitz LC, Forman AJ, Chen Z, Deutsch TG et al (2013) Technical and economic feasibility of centralized facilities for solar hydrogen production via photocatalysis and photoelectrochemistry. Energy Environ Sci 6:1983–2002
10. Ross RT, Hsiao T-L (1977) Limits on the yield of photochemical solar energy conversion. J Appl Phys 48:4783–4785
11. Bolton JR (1978) Solar fuels. Science 202:705–711
12. Buzás J, Farkas I, Biró A, Németh R (1998) Modelling and simulation aspects of a solar hot water system. Math Comput Simul 48:33–46
13. Kalogirou SA (2004) Solar thermal collectors and applications. Prog Energy Combust Sci 30:231–295
14. International A (2003) ASTM Standard C33, 2003—specification for concrete aggregates. West Conshohocken, PA
15. Gueymard CA (2001) Parameterized transmittance model for direct beam and circumsolar spectral irradiance. Sol Energy 71:325–346

Chapter 13
An Integrated System for Energy-efficient Exhaust Aftertreatment for Heavy-duty Vehicles

Jazaer Dawody, Lennart Andersson, Lars J. Pettersson, Moa Zieth̀en Granlund, Hanna Härelind, Fredrik Gunnarsson, Anders Palmqvist, Rickard Heijl, Ronnie Andersson, Olle Högblom, Lennart Holmgren, Per-Olof Larsson and Fredrik Andreasson

Abstract This chapter presents a unique system approach applied in a joint academic–industrial research programme, E4 Mistra, to attain the goals of high energy efficiency and low emissions in an exhaust aftertreatment system for heavy-duty vehicles. The high energy efficiency is achieved by heat recuperation, onboard hydrogen production for NO_x reduction, and by finding new solutions for making the aftertreatment system active at low exhaust temperatures. To reach low particulate emissions, a mechanical filter using a sintered metal powder is developed and coated with catalytic material to improve the soot oxidation efficiency. Low NO_x emissions are achieved by an efficient NO_x reduction catalyst.

The integrated E4 Mistra system comprises four technological advances: thermoelectric (TE) materials for heat recuperation, catalytic reduction of NO_x over innovative catalyst substrates using either the onboard diesel or biodiesel, H_2 from a high-efficiency fuel reformer, and particulate filtration over a porous metal filter. The TE materials are used in a TE generator (TEG) which converts thermal energy into electricity. The TEG is used to recuperate heat from the exhaust-gas recirculation (EGR) circuit of heavy-duty trucks and is expected to generate ~1 kW electric

J. Dawody (✉) · L. Andersson
Volvo Group Trucks Technology, Advanced Engineering & Research, Sven Hultins gata 9D, 412 88 Gothenburg, Sweden
e-mail: jazaer.dawody@volvo.com

L. J. Pettersson · M. Z. Granlund
Royal Institute of Technology, 100 44 Stockholm, Sweden

H. Härelind · F. Gunnarsson · A. Palmqvist · R. Heijl · R. Andersson · O. Högblom
Chalmers University of Technology, 412 96 Gothenburg, Sweden

L. Holmgren
Termo-Gen AB, Hangvar, Olarve 609, 624 54 Lärbro, Sweden

P.-O. Larsson
Höganäs AB, Bruksgatan 35, 263 83 Höganäs, Sweden

F. Andreasson
Alfa Laval AB, Rudeboksvägen 1, 226 55 Lund, Sweden

A. Sayigh (ed.), *Renewable Energy in the Service of Mankind Vol I*,
DOI 10.1007/978-3-319-17777-9_13

power from 20 kW heat in the exhaust gas. The TEG is integrated in a plate heat exchanger (HEX) designed particularly for this application. Apart from the knowledge and experiences in TEG and heat exchange technologies, a thorough fluid dynamics and TE analysis are performed in this project to understand the governing processes and optimize the system accordingly.

The components of the E4 Mistra system are explained in the chapter in addition to test results, which show the system's capacity for H_2 production, NO_x conversion, particulate matter filtration and soot oxidation, and finally electric power generation via heat recuperation from the exhaust gas using the developed TEG–HEX system.

Keywords Energy efficiency · Emission control · Thermoelectric generators · Hydrogen production

13.1 Background

The transport industry plays an important role in the prosperity of communities and their economics. Vehicles are used in building roads, installing water pipes, transporting people and goods, and facilitating trade and travel. However, the transport sector accounts for a high proportion of greenhouse gas (GHG) emissions as well as hazardous emissions such as nitrogen oxides (NO_x), unburned hydrocarbons, and particulate matters (PM) which have adverse effects on the climate [1, 2]. To protect the environment from the harmful effects of these emissions, stringent legislations have been introduced in order to regulate the amount of harmful exhaust-gas emissions from vehicles. For Europe, the prevailing legislation, EURO VI, requires NO_x reduction to 0.4 g/kWh and PM reduction to 0.01 g/kWh [3]. The corresponding legislation in the USA (US10) puts higher requirement on NO_x (0.26 g NO_x/kWh) but lower on PM (0.013 g PM/kWh) [3]. The abatement of harmful emissions from vehicle exhaust gases requires advanced technologies to secure high conversion of pollutants in highly durable and compact systems. Many exhaust aftertreatment technologies have been developed and implemented in vehicles to meet the stringent emission regulation targets. However, exhaust aftertreatment systems still need further optimization to reach the durability, size, cost requirements, and make them more energy efficient.

To reduce the CO_2 emission from vehicles, there is no simple solution. The consumption of fuel used for commercial transportation may be reduced by several methods. The transport efficiency may be increased by allowing longer or heavier vehicles or by improving the transport logistics. The vehicle's fuel consumption may decrease by lowering the air drag or rolling resistance or by incorporating hybrid technology. The impact of these improvements is typically larger than the improvements that may be achieved in the diesel engine. Typically, less than 40 % of the energy content of diesel consumed in a long-haul truck is used for actual work, while more than 50 % is lost as heat either in the exhaust gases or in the cooling system. Recovering part of the energy that is lost in the exhaust gases and using this energy for propelling the vehicle can have a large impact on the overall fuel consumption [2]. The option, which is used in this research programme, is to use a TE device for recovering heat from the exhausts. A temperature difference in a heat exchanger (HEX) is transferred to a thermoelectric generator (TEG) and directly converted to an electric current, and the electric power may be used for charging the batteries in a conventional or hybrid drive train. All efforts to use the energy in the exhaust must be balanced by the requirements of the exhaust aftertreatment system. The exhaust temperature for many common types of commercial vehicles is close to the lower limit where the catalyst is active, and a careful integration of waste heat recovery technologies is therefore required to improve the catalytic activity.

13.2 The Structure of the E4 Mistra Programme

To tackle the described situation, E4 Mistra uses a system approach based on recent developments in lean deNO_x (technology for nitrogen oxides reduction) catalysis, fuel reforming catalysis, TE materials, and HEX and particle filter technology. The intertwined structure of the E4 Mistra programme is illustrated in Fig. 13.1. The programme has the goal to reach very low emission limits (0.1 g NO_x/kWh and 0.005 g PM/kWh) while decreasing the overall CO_2 emissions of the vehicle by using efficient heat recuperation and renewable fuels.

The exhaust-gas recirculation (EGR) unit in the system allows the engine to run at fairly low NO_x emissions from the engine, compared to an engine without EGR.

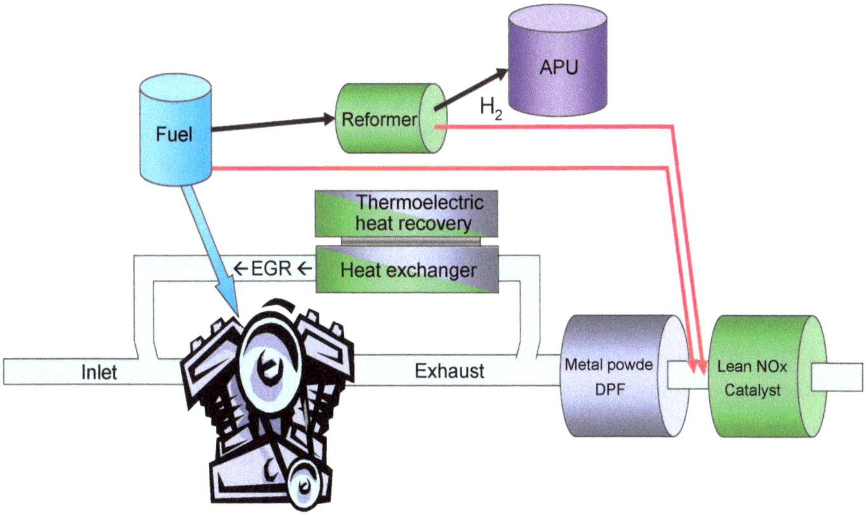

Fig. 13.1 A schematic view of the E4 Mistra system layout. *EGR* exhaust-gas recirculation, *DPF* diesel particulate filter, APU auxiliary power unit

The TE generator (TEG) is preferentially located in the EGR circuit where the TEG may be incorporated within the required HEX, and where the temperature level is higher than that after the catalytic system has been used.

13.3 Results and Discussion

13.3.1 The Strategy for NO_x and PM Control in E4 Mistra

To reach the programme's goal of developing energy-efficient new aftertreatment systems for combustion engines, hydrogen-assisted lean NO_x reduction technology was used to reduce the NO_x engine-out emissions with biofuels as reducing agents; while, for PM filtration, a sintered metal filter stack was designed to eliminate pressure drop problems, thereby providing higher energy efficiency compared to conventional diesel particulate filters (DPFs). The development and test results for the deNO_x and PM filter systems are explained in the next section.

13.3.1.1 Hydrogen-assisted Hydrocarbon Lean NO_x Reduction

The hydrogen-assisted hydrocarbon lean NO_x reduction (HC-SCR) system using hydrogen-rich gas produced in a separate fuel reformer has been evaluated, focussing on the effect of reformate addition [4]. The HC-SCR catalyst (4 wt.% Ag supported on γ-Al_2O_3, doped with 100 ppm(w) Pt [5, 6] and the fuel reforming catalyst (3 wt.% Rh supported on high surface area CeO_2–ZrO_2) [7] have been thoroughly evaluated and characterized separately prior to the evaluation of the combined system. The combined evaluation ranges from bench-scale reactor (BSR) experiments, with simulated exhaust gases, to pilot-scale reactor (PSR) experiments using real engine exhaust gases. The fuel used for the fuel reformer and as reducing agent for the HC-SCR catalyst was commercial biodiesel NExt generation Biomass To Liquid fuel (NExBTL). The BSR experiments were performed at four hydrogen concentrations (0, 1000, 1500, and 3250 ppm H_2), with C to N ratio of 6. The promotional effect from adding hydrogen is clearly seen (Fig. 13.2, top panel); however, the effect is not proportional to the H_2 concentration. The improvement in NO_x reduction is most prominent at low temperatures and decreases as the temperature increases and is negligible for temperatures above 400 °C. When hydrogen is added, the NO_x conversion increases from 13 to 68 % with an addition of 3250 ppm H_2, and the catalyst shows a distinct change in activity with increasing H_2 concentration. The unusually high low-temperature activity, which was observed during the BSR experiments (~70 % below 300 °C with 3250 ppm H_2 from reformate), is likely owing to hydrocarbon species in the hydrogen-rich gas from the fuel reformer. In the PSR experiments, the HC-SCR catalyst was exposed to close to real operating conditions, with exhaust gases generated by a gen-set engine (DX 6000 TE XL C, Yanmar). The HC-SCR catalyst was evaluated using both bottled hydrogen (~1400,

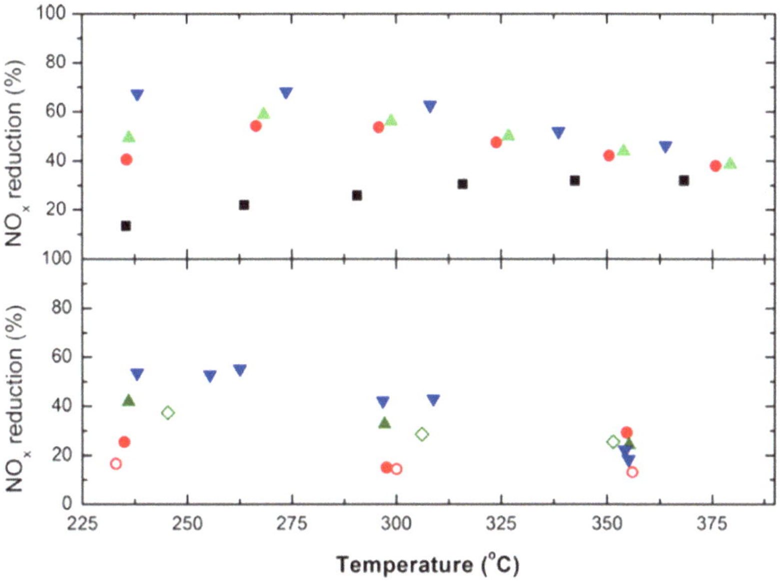

Fig. 13.2 NO_x reduction results for a 4 wt.% Ag catalyst supported on γ-Al_2O_3, doped with 100 ppm(w) Pt. *Top panel* Bench-scale reactor experiments where hydrogen concentrations are denoted as 0 ppm (■), 1000 ppm (●), 1500 ppm (▲) and 3250 ppm (▼). Inlet gas composition: 200 ppm NO, 10 vol.% O_2, 5 vol.% H_2O, and $N_{2,balance}$ with NExBTL as reducing agent. The C to N molar ratio was 6 and GHSV=33,200 h^{-1}. *Bottom panel:* Pilot-scale reactor setup with bottled hydrogen *(solid markers)* and hydrogen from a fuel reformer *(open markers)*. Bottled hydrogen concentrations are denoted as ~1400 ppm (●), ~2800 ppm (▲), and ~7000 ppm (▼). Hydrogen was supplied from the fuel reformer at a low concentration (~1200 ppm,○) and a high concentration (~2000 ppm,◇). Inlet gas composition: ~420 ppm NO_x, ~10 vol.% O_2, ~5 vol.% H_2O, $N_{2,balance}$, and NExBTL as reducing agent with a C to N molar ratio of ~4 and GHSV ~ 22,300 h^{-1}.*GHSV* gas hourly space velocity

~2800, and ~7000 ppm; Fig. 13.2, bottom panel, filled symbols) and hydrogen from the fuel reformer (~1200 and ~2000 ppm; Fig. 13.2, bottom panel, open symbols). The NO_x reduction was calculated by analyzing the total NO_x concentration, upstream and downstream of the HC-SCR catalyst, with a Fourier transform infra-red (FTIR) gas analyzer. The C to N ratio was approximately 4. In the PSR experiments, an NO_x reduction of ~60 % was observed with 7000 ppm H_2 (bottle) at 275–300 °C. In addition, ~40 % NO_x reduction was achieved with 2000 ppm H_2 originating from the fuel reformer compared to 2800 ppm H_2 needed from bottle to reach the same NO_x reduction.

The results from the two different reactor scales show similar trends. The NO_x reduction activity is significantly increased with the addition of hydrogen at low temperatures and, as the temperature is increased, the promotional effect is less pronounced (Fig. 13.2). It is clearly illustrated in the integrated system studies that it is vital to evaluate the HC-SCR catalysts in combination with a fuel reformer at realistic operating conditions, owing to the effects of the hydrogen-rich gas from the fuel reformer. It further illustrates that adding hydrogen-rich gas from a fuel reformer results in a higher NO_x reduction compared to equivalent amount of hydrogen from bottle. This effect is observed in both bench-scale and pilot-scale experiments. The increase in NO_x reduction activity is likely owing to the reactive HC species in the hydrogen-rich gas in combination with the hydrogen. This work gives a unique approach to an improved understanding of the total hydrogen-assisted lean NO_x reduction system. It covers the development of catalytic materials evaluated in controlled bench-scale experiments, followed by integrated bench-scale experiments of the lean NO_x and fuel reformer to a final evaluation of the integrated system in an engine bench reactor setup. Therefore, this study proves a viable concept for lean NO_x reduction together with onboard hydrogen production.

As seen from the Fig. 13.2, the 6 wt.% Ag–alumina sample shows promising NO_x reduction performance also at relatively low temperatures, which has not been shown previously for biodiesel. Moreover, the results (not shown in this chapter) indicate that the NO_x reduction performance is sustained also after hydrothermal ageing up to 650 °C. With biodiesel as reductant, the samples are more sensitive to ageing. However, the fresh samples show promising NO_x reduction performance, which has not been shown previously for biodiesel.

13.3.1.2 Metal Powder Filter

For medium and heavy-duty truck operation, it is estimated that a B20 fuel (20 % biodiesel in regular diesel) approximately doubles the ash amount compared to engine operation using 100 % regular diesel [8]. Using pure biodiesel would lead to a sixfold increase in ash, resulting in an unacceptably high service interval for the closed channel-based ceramic DPFs mainly used today. A DPF design, with the potential to accumulate more ash, has been developed by Höganäs AB and Volvo AB in the E4 Mistra project. It is a cross-flow filter design, made from sintered porous metal powder sheets in between corrugated solid stainless sheets twisted 90° every second layer, with an ashtray to collect noncombustible inorganic residue. The filter has been coated with a Pt/Al_2O_3 washcoat (~0.25 g Pt/l substrate). Compared with an uncoated filter, the coated filter has a lower soot accumulation speed, giving less pressure drop, and lower soot regeneration temperature (275–375 °C compared to ~500 °C for an uncoated filter). Filtration efficiencies above 99.95 % have been obtained. The gas hourly space velocity (GHSV) has been ~22,000 h^{-1}, which is corresponding to a heavy-duty (HD) truck operating around half load with a 24 l DPF. With this high filtration efficiency, the E4 Mistra goal with <0.005 g PM/kWh is easily obtained with raw emissions from the engine containing 0.1 g PM/kWh. Similar filtration results are obtained with regular diesel and biodiesel (rapeseed methyl ester; RME). The filter has been tested in an accelerated ash storage test at Volvo AB corresponding to about 700,000 km of operation. Roughly, 45 % of the ash ended up in the ashtray and the other 45 % in the filter, and then mainly on the outlet towards the ashtray (Fig. 13.3). The main part of this ash in the filter was removed easily by shaking the filter carefully and

Fig. 13.3 The filter stack after the accelerated ash loading test corresponding to 700,000 km of operation. *(left)* the ashtray, *(middle)* the end facing the ashtray, *(right)* one of the ends facing the exhaust-gas outlet *(clean)*

blowing with compressed air. Less than 10 % of the ash was strongly trapped in the filter and could not be removed. During the operation of a truck it is likely that a large amount of the easily removed ash will fall into the ashtray due to vibrations and variations in the exhaust-gas flow, conditions which were not obtained in the accelerated steady-state ash storage test. The conclusion is that most likely up to 90 % of the ash can be removed from the filter and stored in the ashtray. This will result in lower pressure drop and fuel penalty compared to if the ash could not be removed continuously. In the existing ceramic DPF solutions, 100 % of the ash is accumulated. This shows that the cross-flow metal-powder-based DPF developed in the E4 Mistra project offers exciting filtration possibilities in future biodiesel transport solutions.

13.3.2 Thermoelectric Waste Heat Recovery

13.3.2.1 Design and Evaluation of Thermoelectric Waste Heat Recovery Systems

Development of TE power generation methods for heat recuperation in the exhaust-gas recirculation system (EGR) has been the focus of this work. This includes prototyping and evaluation in engine bench, TEG device modelling, and TE material development. When designing TEGs for waste heat recovery in automotive exhaust-gas applications, there are several challenges to meet. Among others, a high heat transfer rate from the gas bulk to the TE material is required, while maintaining low pressure drop [9]. Furthermore, the TEG unit must be stable with very low long-term performance degradation rate, that is, the TEG must withstand frequent thermal cycling, and modules should not be overheated at high engine load.

Two different TEG designs that utilize an increased heat transfer area have been developed in this work: (i) heat pipe design and (ii) plate HEX design with integrated fins. The heat pipes are sealed pipes containing a mixture of gas and liquid in which the heat transport, from the exhaust gas to the TE modules, is based on evaporation and condensation. This HEX design, shown in Fig. 13.4a, has the advantage that the heat pipes have a maximum heat cut-off of approximately 500 W

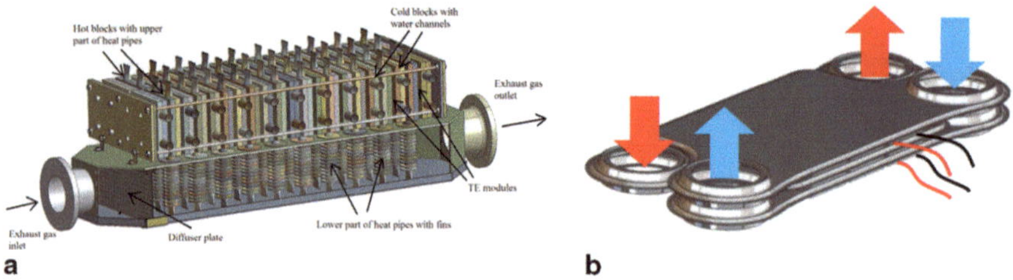

Fig. 13.4 **a** Heat pipe prototype, **b** principal heat exchanger design

per pipe, that is, they act as thermal fuse. Consequently, the TE modules will not be overheated, and there will be a more even temperature distribution inside the TEG.

The TEG heat pipe prototype was tested in the engine laboratory at Volvo AB using different gas-flow pattern, open flow, and with a diffuser plate at the entrance of the TEG, shown in Fig. 13.4a. The maximum TEG power was measured to 239 W using the diffusor plate. The plate HEX design, shown in Fig. 13.4b, is characterized by its compactness and flexibility in design regarding the number of plates. A brazed cassette solution with separated hot and cold cassettes to withstand thermal fatigue was evaluated in this programme. This design concept has several benefits, among others, resistance to high temperatures and resistance to corrosion by use of brazed stainless steel plates, no bypass of heat flow by separated gas and water channels, flat surfaces with integrated fins for improved heat transfer and structural integrity, no exposure of exhaust gas or cooling media to the TE modules, accessibility of cables for easier connection, external frame to enable pressure on sealings, and contact pressure on TE modules. The test was done in a one-cylinder rig, which represents the full flow through an EGR circuit. The results from the full-size HEX, containing 25 cassettes and 216 modules, gave an electric power of 415 W. Thus, these tests confirmed that TE modules can be integrated in a plate HEX.

In order to understand and predict the performance of TEGs, a multiphysics simulation model was developed. The model is based on coupled computational fluid dynamics (CFD) and TE simulations. In the simulations, the thermoelectric phenomenon is described by two differential equations; one for the voltage:

$$\nabla V = -S\nabla T - J/\sigma\,,$$

and one for the thermal energy:

$$-\nabla(\kappa\nabla T) = mc_{\mathrm{p}}\frac{\partial T}{\partial t} + J^2/\sigma + \nabla STJ$$

For solving these equations, a geometry, temperature-dependent material data (Seebeck coefficient S, thermal conductivity κ, and electric conductivity σ) and boundary conditions are needed [10]. The model was further used to predict the performance of the heat pipe prototype. It is concluded that the perforated diffuser plate in front of the heat pipes is forming impinging jets, thereby increasing the heat transfer onto the first row of heat pipes, Fig. 13.5. This plate has gradually increasing size of the holes towards the side, which explains the increased heat transfer on the outermost heat pipes.

A comparison between measurement and simulation results is shown in Table 13.1. The agreement is excellent in terms of pressure drop over the system and surface temperatures on the modules.

The power generation was calculated for modules working at optimal conditions,that is, the modules were simulated individually, working with an optimal external load. This was done in order to reduce the complexity of the model. In the hardware test, the TEG–EGR system was not working at optimal conditions, since the modules were connected electrically in groups in series. This was done in order to achieve useful voltage output from the system and to avoid complete power failure if one module was damaged. A consequence of this action is that optimal conditions can never be reached in real TEG systems. However, the extent to which nonideal conditions affect the TEG performance can be quantified by the use of simulations, thereby allowing assessment of robustness and performance of various designs.

13.3.2.2 Development of Thermoelectric Materials

The aim of the material development was to develop new, efficient TE materials that can be used in devices to convert waste heat to electricity in a temperature range suitable for combustion engines. TE devices require two types of TE materials for their function, a p-doped and an n-doped semiconductor. A high-performing TE

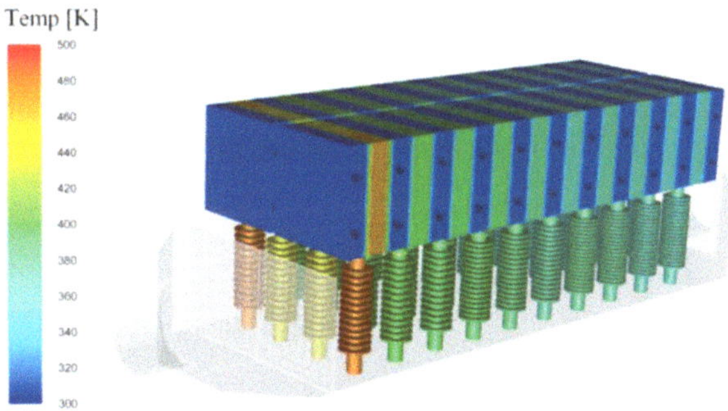

Fig. 13.5 Simulated surface temperatures on heat pipes and modules

Table 13.1 Comparison between experiments and simulations for the heat pipe prototype

	Pressure drop (kPa)	$T_{h, inlet}$(°C)	$T_{h, outlet}$(°C)	Power output(W)
Measurement	12.8	165	83	137
Simulation	13.1	158	79	165

material has a high figure of merit, $ZT = S^2\sigma T/\kappa$, at the desired temperature T, where S is the Seebeck coefficient, σ is the electrical conductivity and κ is the thermal conductivity, all of which are temperature dependent. Thermal conductivity consists of lattice vibrations and heat conduction through electronic conduction, which means that to improve ZT one should develop materials with crystal structures that have glass-like lattice conductivity contributions to κ to enable a maintained high σ. For this, we have followed the phonon–glass–electron–crystal (PGEC) concept [11] in which host–guest structures such as the germanium-based clathrates shown in Fig. 13.6a are exploited to efficiently reduce the lattice thermal conductivity of the material while maintaining a high electronic conduction. We have evaluated additional means to further improve the performance of the clathrates, for example, by preparing quaternary compositions with atomic substitution on the framework sites or by introduction of nanoinclusions of hetero compounds [12]. These studies assess the possibility to further improve ZT by reducing thermal conductivity via introduced scattering centres for different phonon wavelengths and by improving S via the energy-filtering concept. Figure 13.6b shows the improvement in ZT achieved by introduction of titanium dioxide (TiO_2) nanoparticles in a matrix of the $Ba_8Ga_{16}Ge_{30}$ clathrate[13]. Magnesium silicide-based materials have also been evaluated for this concept [14, 6] and undoped Mg_2Si shows the most significant improvement (almost threefold) in ZT by introducing 1 vol% TiO_2 nanoparticles in the material, Fig. 13.6c. For both series of materials, an optimum amount of nanoinclusions was found at which ZT was higher than that of the reference materials without nanoinclusions. In both series, the optimum temperature was shifted to lower values compared to the reference materials.

13.4 Conclusions

This chapter summarizes the work performed within a Swedish national programme to develop an energy-efficient low-emission exhaust aftertreatment system suitable for heavy-duty applications. The system energy efficiency is improved by heat recuperation using several generations of TE generators developed within the project

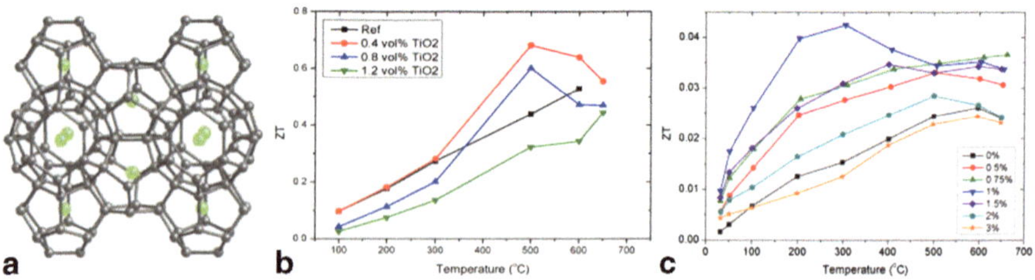

Fig. 13.6 **a** Structure of the $Ba_8Ga_{16}Ge_{30}$ clathrate. Temperature-dependence of ZT of (**b**) a series of $Ba_8Ga_{16}Ge_{30}$ clathrates with varying amounts of TiO_2 inclusions and **c** a series of Mg_2Si materials with varying amounts of TiO_2 inclusions

from novel TE materials. The best performance has been achieved by a TEG integrated in a plate HEX where ~0.4 kW electrical energy was recuperated from a 20 kW thermal energy, which is only 40 % of the project target. For NO_x reduction Ag/Al_2O_3 catalysts with different Ag loadings were synthesized and tested under various conditions using biofuels as reducing agents and hydrogen gas generated from biofuels over a reformer catalyst developed in the project. For PM abatement, sintered metal powder particle filter stacks were developed and coated with oxidation catalyst to improve the soot oxidation efficiency. Test results showed that the DPF was able to reduce the PM far below the project targets, while NO_x conversion reached at best 70 % of the defined target for NO_x reduction.

Acknowledgements The authors of this chapter are grateful for the financial support from Mistra, the Swedish Foundation for Strategic Environmental Research, and the Swedish Energy Agency.

References

1. U. S. Department of Energy (2011) Quadrennial technology review, U. S. department of energy [Online]. http://energy.gov/sites/prod/files/QTR_report.pdf
2. Knittel CR (2012) Reducing petroleum consumption from transportation. J Econ Perspect 26:93–118
3. Dieselnet. http://www.dieselnet.com/standards. http://www.dieselnet.com/standards/us/fe_hd.php. Accessed 15 Aug 2014.
4. Fredrik G, Granlund ZM, Mattias E, Jazaer D, Pettersson JL, Hanna H (2015) Combining HC-SCR over Ag/Al_2O_3 and hydrogen generation over Rh/CeO_2-ZrO_2 using biofuels: an integrated system approach for real applications. Appl Catal B: Environ 162:583–592
5. Gunnarsson F et al (2014) Improved low-temperature activity of silver-alumina for lean NO_X reduction- Effect of Ag loading and low-level Pt doping. Appl Catal B 152–153:218
6. Gunnarsson F et al (2013) Influence of ageing, silver loading and type of reducing agent on the lean NO_x reduction over Ag-Al_2O_3 catalysts. Topics Catal 56:416
7. Granlund ZM, Oliver G, Peter P, Pettersson JL (2014) Comparison between a micro reactor with multiple air inlets and a monolith reactor for oxidative steam reforming of diesel. Int J of Hydrog Energy 39:18037–18045
8. Williams AEA (2011) J Fuels Lubr 4:110
9. Högblom O et al (2012) CFD modeling of thermoelectric generators in automotive EGR-coolers. AIP Conf Ser 1449
10. Högblom O et al (2014) Analysis of thermoelectric generator performance by Use of simulations and experiments. J Electron Mater 43:2247–2254
11. Slack G (1995) CRC handbook of thermoelectrics. In: DM Rowe (ed). CRC Press, Boca Raton
12. Ma Y et al (2013) Composite thermoelectric materials with embedded nanoparticles. J Mater Sci 48:2767–2778
13. Heijl R et al (2012) Thermoelectric properties of Ba8Ga16Ge30 with TiO_2 nanoinclusions. J Appl Phys 112:044313
14. Cederkrantz D et al (2012) Enhanced thermoelectric properties of Mg_2Si by addition of TiO_2 nanoparticles. J Appl Phys 111:023701

Chapter 14
Cost-effectiveness and Potential of Greenhouse Gas Mitigation through the Support of Renewable Transport Fuels in Iceland

Ehsan Shafiei, Brynhildur Davidsdottir, Jonathan Leaver, Hlynur Stefansson and Eyjolfur Ingi Asgeirsson

Abstract The system dynamics model of Iceland's energy systems (UniSyD_IS) is used to explore the potential transition paths towards renewable transport fuels with implications for greenhouse gas (GHG) emissions and mitigation costs. The study focuses on Iceland's potential fuel pathways including renewable electricity, hydrogen from electrolysis, biogas from municipal wastes, bioethanol from lignocellulosic biomass, and biodiesel from oil seeds and waste oils.

The vehicle fleet is divided into light- and heavy-duty vehicles, and each fleet consists of different alternative fuel vehicles. The model allocates the forecasted fleet growth among different vehicle types based on consumers' preferences towards vehicle attributes and social network influences.

Oil price, carbon tax, renewable fuel supply–push, and government incentives are selected as the fundamental factors for scenario analysis. The results show that the transitions to renewable transport fuels seem to be feasible economically, initially, through biogas and then through uptake of hydrogen and electric vehicles. The cost-effectiveness analysis in UniSyD_IS indicates that the initial momentum

E. Shafiei (✉)
School of Engineering and Natural Sciences, University of Iceland,
Oddi Building, Sturlugata 3, 101 Reykjavik, Iceland
e-mail: ehsan@hi.is

B. Davidsdottir
School of Engineering and Natural Sciences, Environment and Natural Resources,
University of Iceland, Oddi Building, Sturlugata 3, 101 Reykjavik, Iceland
e-mail: bdavids@hi.is

J. Leaver
Department of Civil Engineering, Unitec Institute of Technology, Auckland, New Zealand
e-mail: jleaver@unitec.ac.nz

H. Stefansson · E. I. Asgeirsson
School of Science and Engineering, Reykjavik University, Reykjavik, Iceland
e-mail: hlynurst@ru.is

E. I. Asgeirsson
e-mail: eyjo@ru.is

A. Sayigh (ed.), *Renewable Energy in the Service of Mankind Vol I*,
DOI 10.1007/978-3-319-17777-9_14

of alternative fuels will not only mitigate GHG emissions but also could provide net benefits from an overall energy system and consumer perspective.

Keywords Alternative fuel vehicles · Renewable fuels · GHG mitigation · Cost-effectiveness

14.1 Introduction

Iceland is a leading country in developing and utilizing renewable energy resources to generate heat and electricity. Almost all of the generated electricity comes from hydro- and geothermal sources. However, Iceland has remained entirely dependent upon gasoline and diesel imports to sustain its transportation needs. Due to the growth of vehicles per capita and travel activities, greenhouse gas (GHG) emissions from road transport have been increasing rapidly during the past decade. Iceland's long-term vision includes deep reductions of net GHG emissions by 50–75% by 2050, from the 1990 level [1]. However, to achieve these deep reductions in GHG emissions in the transport sector, a transition to alternative fuel vehicles (AFVs) will be required.

To examine the transition process towards a renewable transport, the system dynamics model of Iceland's energy systems (UniSyD_IS) is applied. UniSyD_IS is founded on the UniSyD_NZ model of New Zealand's energy economy, created by the Unitec Institute of Technology, New Zealand, in cooperation with Stanford University, USA. UniSyD_NZ has been employed to assess the impacts of AFVs on New Zealand's energy economy [2–4].

In this chapter, we focus on the renewable energy resources and apply UniSyD_IS model to evaluate the possible transition paths towards a renewable future transport in Iceland. The impact of climate mitigation policies to support the transition to renewable fuels in the transport sector is evaluated. Cost-effective policies to achieve deep cuts in GHG emissions are also identified.

In Sect. 14.2, model structure and methodology are briefly presented. Section 14.3 describes the main assumptions and data for analysis of transition to renewable transport fuels. Scenarios are defined in Sect. 14.4, simulation results are presented in Sect. 14.5, and Sect. 14.6 concludes the chapter.

14.2 Methodology

14.2.1 Model Structure

The UniSyD_IS model focusses on the energy supply sector with endogenous analysis of road transport energy demand. UniSyD_IS is a detailed resource and technology-specific model in which equilibrium interactions act across six key markets: electricity, hydrogen, biogas, bioethanol, biodiesel, and vehicle fleets. Gasoline and diesel supply sectors are exogenous. The detailed structure of the

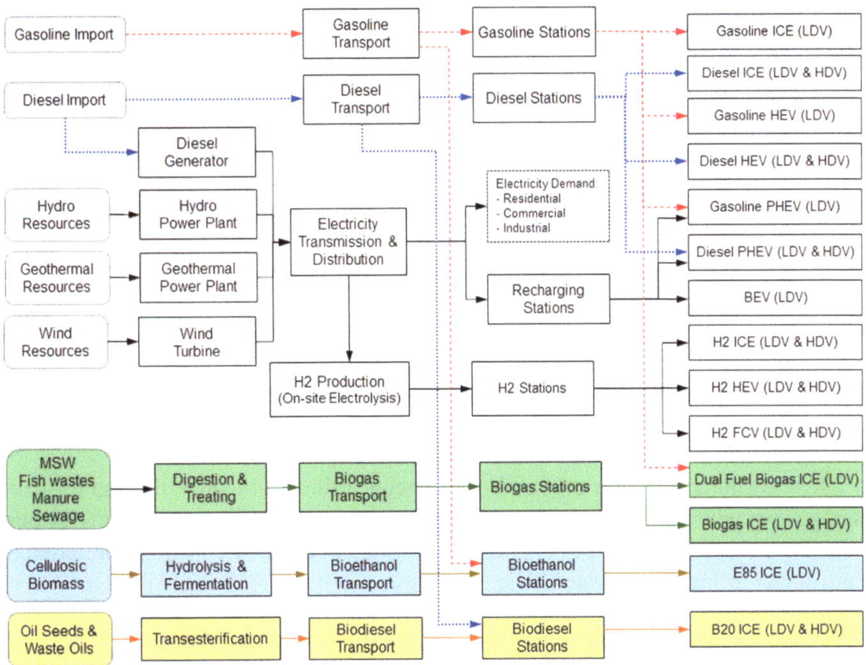

Fig. 14.1 Fuel supply pathways and the corresponding powertrain technologies. *LDV* light-duty vehicle, *HDV* heavy-duty vehicle, *ICE* internal combustion engine, *HEV* hybrid electric vehicle, *PHEV* plug-in hybrid electric vehicle, *BEV* battery electric vehicle

well-to-wheel (WtW) pathways modelled in UniSyD_IS is illustrated in Fig. 14.1. It shows Iceland's energy supply system, which includes conventional and alternative fuel supply pathways and the corresponding vehicle powertrains. Besides the imported petroleum products, the entire domestic fuel supply system begins from renewable resources (hydro, geothermal, wind, and biomass) and ends up to the different end users of transport sector. The transport fleet is divided into light(LDV)- and heavy-duty vehicle (HDV) fleets with the upper weight limit for LDVs being 3.5 tonnes.

14.2.2 Evolution of Vehicle Fleets

The total vehicle fleet is expected to continue growing based on historical rates until the vehicles per capita indicators approach a saturation level. Once the fleet growths of the LDV and HDV fleets are determined, the model allocates these fleets among vehicle types. The allocation is accomplished using a vehicle choice algorithm to forecast the market shares of new adopted vehicles. Consumers purchase the vehicles based on both their own preferences and the behaviour of other consumers.

The multinomial logit (MNL) model, according to Eq. (14.1), gives the probability $P_{k,t}$ that the consumers purchase vehicle type k at time t, on the basis of their preferences towards the vehicles' attributes [5]:

$$P_{k,t} = \exp\ (U_{k,t}) / \sum_k \exp\ (U_{k,t}), \tag{14.1}$$

where consumer utility U_{kt} relies on a set of monetary and nonmonetary attributes. The attributes are included in the consumer's utility function as in Eq. (14.2):

$$U_{k,t} = \beta_1 C_{k,t} + \beta_2 F_{k,t} + \beta_3 M_{k,t} + \beta_4 G_{k,t} + \beta_5 B_{k,t} + \frac{\beta_6}{R_{k,t}} + \beta_7 \exp(-\beta_8 A_{k,t}), \tag{14.2}$$

where $C_{k,t}$ is the purchase price for vehicle k at time t (\$), $F_{k,t}$ fuel cost (\$/km), $M_{k,t}$ annual maintenance cost (\$/year), $G_{k,t}$ GHG emissions (gCO_2eq/km), $B_{k,t}$ battery replacement cost (\$), $R_{k,t}$ vehicle range (km), $A_{k,t}$ fuel availability (relative to gasoline), and β_i consumer preferences towards vehicle attributes. Among these attributes, GHG emissions, fuel cost, battery cost, and fuel availability are calculated endogenously.

It is assumed that the probability of social influences equals the share of the consumers who have already adopted a particular vehicle. Total probability of purchasing vehicle k at time t ($\mathrm{TP}_{k,t}$) is the weighted average of consumer preferences and social influences probabilities:

$$\mathrm{TP}_{k,t} = \lambda_t \times \mathrm{SN}_{k,t} + (1-\lambda_t) \times P_{k,t} \quad , \quad 0 \le \lambda_t \le 1, \tag{14.3}$$

where SN and P are probabilities for social influences and individual preferences, respectively. The weighting coefficient λ reflects the importance and strength of social influences for consumers.

14.3 Data and Key Assumptions

14.3.1 Electricity Supply Sector

Hydropower, geothermal, and wind are the main potential electricity generation technologies in Iceland. The cost of electricity produced from existing hydro plants is assumed to be 3.7 c/kWh [6]. Geothermal plants are assumed to emit 50 g/kWh of carbon dioxide [7]. Existing geothermal plants provide electricity at an aggregate cost of 3.6 c/kWh [6]. The future costs of hydro and geothermal capacities are modelled using supply curves in which unit generation cost is expected to increase with cumulative installed capacities. The average generation cost of 9.82 c/kWh is estimated for wind turbines [6]. A cost reduction of 1.8 % per year is applied to wind generation [4]. The costs of transmission and distribution are simplified to 1.05 and 3.5 c/kWh tariffs, respectively [8] , and the total grid loss is about 5.6 % [9].

14.3.2 Hydrogen Sector

Hydrogen can be produced by alkaline electrolysis technology. The forecourt electrolysis is modelled with the capacity of 1500 kg of hydrogen per day that can support about 1600 vehicles. The unit capital and operation and maintenance (O&M) cost of generation is US$ 1.1 per kg. The cost of compression, storage, and dispensing is US$ 2.47/kg [10]. A simple linear reduction is assumed for the costs of hydrogen production and delivery. Electrolysis electricity use will be reduced from 53.4 kWh/kg in 2015 to 44.7 kWh/kg in 2025 [10].

14.3.3 Biofuel Generation

Bioethanol in Iceland can be produced from cellulosic residuals. The lignocellulosic resource in Iceland consists of cereals, hemp, waste hay, paper, timber, and garden wastes. Total estimated resource potential used in the model is 0.78 pJ per year [11]. A constant supply curve is assumed for the lignocellulosic resource cost with US$ 0.4 per l produced bioethanol [12]. Five plant sizes of 2.5, 5, 10, 20, and 40 million l per year are assumed for bioethanol production. A simple power series trend line governs the capital cost of new bioethanol plant construction. The yearly operational and maintenance costs vary with bioethanol plant sizes [13]. A cost reduction of 0.8 % per year is applied to the capital and O&M costs during the study period [14].

Biodiesel in Iceland can be harvested from oil seeds, fish wastes, animal fat waste, and waste vegetable oils. Total resource potential for biodiesel production is 0.4 pJ per year [11]. UniSyD_IS uses an economy of scale supply curve for the total unit cost of biodiesel. The possible plant sizes are similar to bioethanol.

Potential resources for biogas production in Iceland are municipal solid wastes (MSW), fish wastes, manure, and sewage. Biogas resource potential is 1.72 pJ per year [11]. The average cost of raw material, including collection and transport, is US$ 0.33/m^3 produced biogas. The investment and O&M costs are approximately US$ 0.3 and US$ 0.28/m^3 for a production plant serving 30 kt of waste per year. This assumed plant can produce about 3.3 Mm^3 biogas per year [15]. New biogas plant sizes are considered to be of any of the 2.5, 5, 10, 20, and 40 Mm^3/year. Based on data in [16], a scaling factor exponent of −0.6 is used to adjust the unit cost of various plant sizes.

14.3.4 Vehicle Fleet

The average annual growth rates of LDVs and HDVs during the past decade were 4.7 and 4.1 %, respectively [17]. The saturation level used for vehicle-per-capita indicator is 0.85 for LDVs (0.8 for passenger LDVs) and 0.043 for HDVs. The starting values of 12,000 and 28,000 km per year are assumed for LDVs and HDVs, respectively [17, 18]. We use a fuel cost elasticity of −0.33 [19] to adjust the annual travel demand with respect to changes in fuel cost per km.

Table 14.1 shows the purchase price, annual vehicle price reduction, and maintenance costs of different alternatives for LDVs. Using the rough expert estimates, the price and cost data used for LDVs are scaled up by a factor of 10 to provide the average estimations for HDV data [20, 21].

It is assumed that the fuel economy of new vehicles follows a linear path from the current values to a high bound over time. The range of each vehicle in the model is increased corresponding to the fuel economy improvements.

14.4 Scenario Definition

Based on the changes in oil price, fuel availability, carbon tax, and government subsidies, eight scenarios are defined as explained in Table 14.2. Government incentives are the subsidies directed to battery electric vehicle (BEV) and fuel cell vehicle (FCV) as the promising zero emission vehicles (ZEVs).

To examine the effects of biodiesel, bioethanol, hydrogen, and electricity infrastructure, we define three cases of demand–pull with no initial infrastructure and supply–push with both low and high initial momentum. Table 14.3 defines the meaning of the supply–push cases used in the scenarios. Since a biogas plant is operating in Iceland, further initial supply–push is not needed for this fuel.

14.5 Simulation Results

Figure 14.2 shows the evolution of vehicle fleets for four selected scenarios. In the REF scenario, HEV_gasoline, HEV_diesel, and biogas vehicles are the attractive alternatives for LDVs due to the lower fuel cost per kilometre. For HDVs, the market share of diesel internal combustion engine (ICE) is gradually reduced and replaced by the hybrid option. In the REF+LP case, electric vehicles (EVs), including plug-in hybrid electric vehicles (PHEVs) and battery electric vehicles (BEVs), along with H2 vehicles (ICE_H2, HEV_H2 and FCV) take a share of less than 20 % in 2050 for both LDV and HDV fleets.

In the HC scenario, the market introduction of EVs and hydrogen (H2) vehicles are observed after 2040 when oil price and carbon tax are significantly increased. In the high cost+high push (HC+HP) scenario, the share of EVs rises quickly after 2030 due to the lower renewable electricity cost and enhanced infrastructure availability. The introduction of hydrogen vehicles starts from 2030 to 2035 and results in a market share by 2050 of 10 and 24 % for LDV and HDV fleets, respectively. Biogas keeps its contribution to the fleet mix and bioethanol and biodiesel account for maxima of only 1.7 and 1 %, respectively of the LDV fleet by 2050. The share of biodiesel in HDV fleet in 2050 would be 2.4 %. Zero-emission fuels, including

Table 14.1 Annual maintenance cost of vehicles, purchase price, and annual purchase price reduction

Powertrain (fuel type)	ICE (Gasoline)	ICE (E85)	ICE (Diesel)	ICE (B20)	ICE (Biogas)	ICE (Biogas dual)	ICE (H2)	HEV (Gasoline)	HEV (Diesel)	HEV (H2)	PHEV (Gasoline)	PHEV (Diesel)	BEV (Electricity)	FCV (H2)
Maintenance Cost ($/year) [21]	742	742	742	742	802	840	1176	660	660	1046	566	566	412	693
Purchase Price (k$) [22–26]	24.2	24.2	25.7	25.7	26.4	27.2	39.8	30.7	32.7	46.0	30.3[a]	32.3[a]	28.9[a]	117.7
Purchase Price Reduction (%/year) [23, 24, 26, 27]	–	–	–	–	0.06	0.06	0.5	0.236	0.236	0.736	0.47	0.47	0.43	2.9

ICE internal combustion engine, *HEV* hybrid electric vehicle, *PHEV* plug-in hybrid electric vehicle, *BEV* battery electric vehicle, *FCV* fuel cell vehicle

[a] The prices exclude the battery cost. Assuming the range values of 200 and 60 km for BEV and PHEV, the initial battery costs of US$ 19k and US$ 8.7k are estimated, respectively. The costs are reduced to US$ 6.2k and US$ 2.8k by 2050 (based on [28])

Table 14.2 Scenarios based on the changes in oil price, fuel availability, carbon tax, and government subsidies

Scenarios	Oil price	Carbon tax	Fuel availability	Government incentives
Reference (REF)	Constant (100$/bbl)	Constant	Demand–pull	–
	0 % per year	US$ 25/tonne	no initial momentum	
Reference + Low supply–push (REF + LP)	Constant (100$/bbl)	Constant	Supply–push	–
	0 % per year	US$ 25/tonne	low initial momentum	
Reference + High supply–push (REF + HP)	Constant (100$/bbl)	Constant	Supply–push	–
	0 % per year	US$ 25 $/tonne	high initial momentum	
High costs (HC)	Increasing	High and Increasing	Demand–pull	–
	2 % per year	100 $/tonne + 2 % pa	no initial momentum	
High costs + Low supply–push (HC + LP)	Increasing	High and Increasing	Supply–push	–
	2 % per year	100 $/tonne + 2 % pa	low initial momentum	
High Costs + High Supply–Push (HC + HP)	Increasing	High and Increasing	Supply–push	–
	2 % per year	100 $/tonne + 2 % pa	high initial momentum	
High costs + High supply–Push + 10 %Subsidy (HC + HP + LS)	Increasing	High and Increasing	Supply–push	ZEVs Subsidy
	2 % per year	100 $/tonne + 2 % pa	high initial momentum	10 % of vehicle price
High Costs + High Supply–Push + 20 %Subsidy (HC + HP + HS)	Increasing	High and Increasing	Supply–push	ZEVs Subsidy
	2 % per year	100 $/tonne + 2 % pa	high initial momentum	20 % of vehicle price

Table 14.3 Assumptions on initial supply–push

Alternative fuels	Low initial momentum	High initial momentum
Biodiesel	One small plant size with capacity of 2.5 Ml/year	One medium plant size with capacity of 5 Ml/year
Bioethanol	One small plant size with capacity of 2.5 Ml/year	One medium plant size with capacity of 5 Ml/year
Hydrogen	One forecourt electrolyzer unit with capacity of 1500 kg/day	Two forecourt electrolyzer units each with capacity of 1500 kg/day
Electricity	1 % as initial fuel availability index and increasing with growth of EVs	2 % as initial fuel availability index and increasing with growth of EVs

electricity and hydrogen, account for 39 % of total energy use by 2050. This share can reach to 53 % by including low-emission biofuels.

To explore the impact of renewable supply–push policies, the differences in annual and cumulative GHG emissions between the main scenario variants are shown in Fig. 14.3 (left). Renewable fuel supply–push policies result in an increasing trend

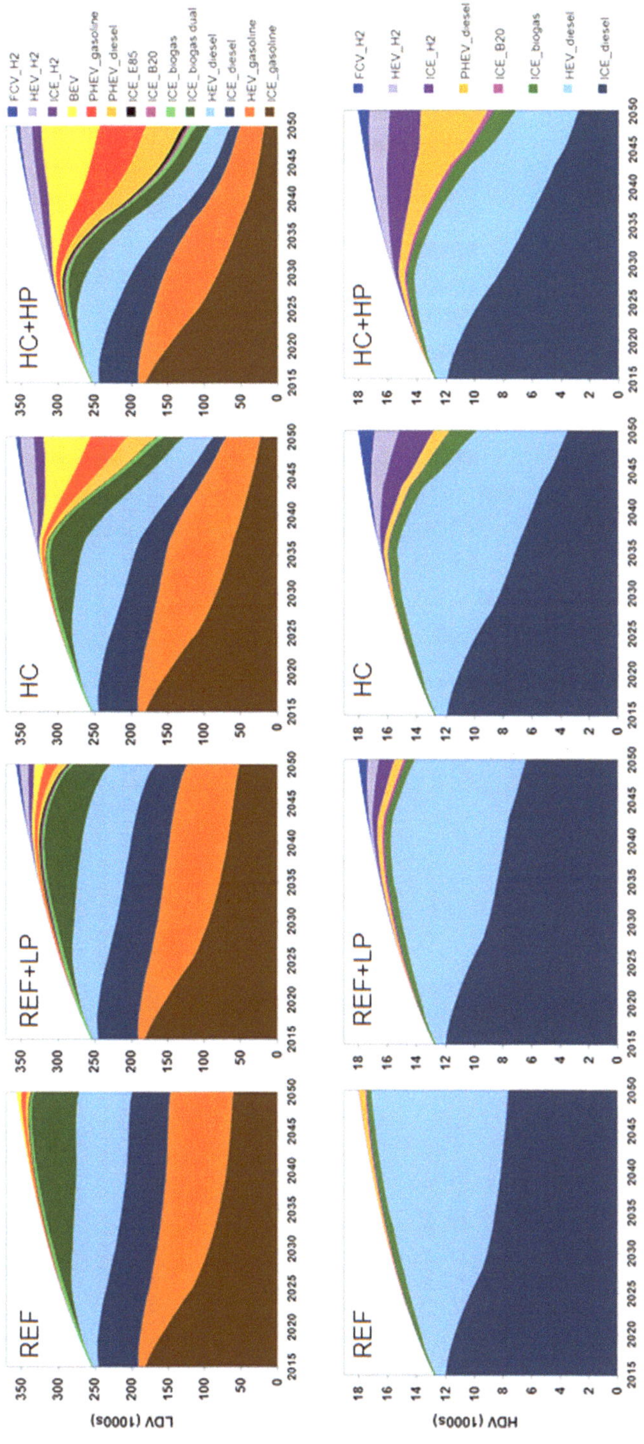

Fig. 14.2 Evolution of Iceland's LDV (*top*) and HDV (*bottom*) fleets in different scenarios. *FCV* fuel cell vehicle, *HEV* hybrid electric vehicle *ICE* internal combustion engine, *BEV* battery electric vehicle, *PHEV* plug-in heavy electric vehicle

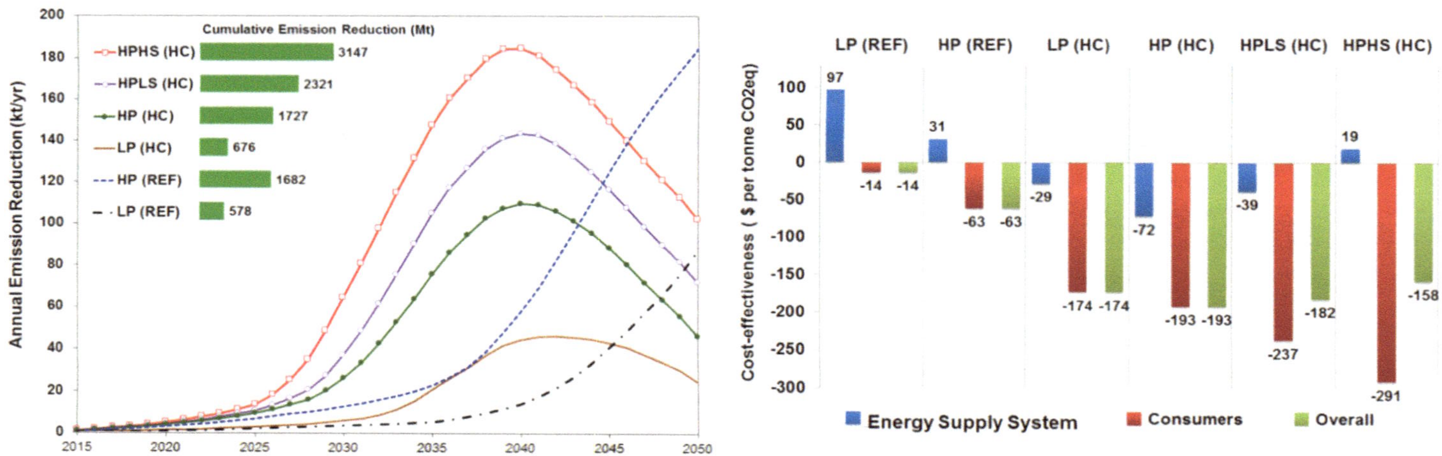

Fig. 14.3 Potential of GHG mitigation (*left*) and cost-effectiveness of mitigation measures from different perspectives (*right*)—LP (*REF*): impact of low supply–push in REF scenario; HP (*REF*): impact of high supply–push in REF scenario; LP(*HC*): impact of low supply–push in HC scenario; HP (HC): impact of high supply-push in HC scenario; HPLS (*HC*): impact of high supply-push with 10 % subsidy in HC scenario; HPHS (*HC*): impact of high supply–push with 20 % subsidy in HC scenario. *HC* high cost, *REF* reference

for GHG mitigation over time. Supply–push policies in the high-cost condition provide greater mitigation potential. The maximum annual mitigation is reached by 2040 and, thereafter, the higher levels of fuel costs reduce the mitigation potential through supply–push strategies.

Figure 14.3 (right) compares the cost-effectiveness of supply–push and government incentives towards renewable fuels. The cost-effectiveness of each mitigation strategy is expressed as the increased cumulative discounted costs divided by the resulted cumulative avoided emissions. The costs can be explored from two perspectives—energy system or consumers. Energy system costs include gasoline and diesel imports, supply costs of renewable fuels (i.e. biofuels, hydrogen, and electricity), and the entire WtW GHG emission cost. In addition, government incentives through ZEV subsidies can be included with the above cost items.

Consumer costs in the transportation sector include the cost of conventional and alternative fuels as well as the capital and operating costs of vehicles. Since the fuel cost is composed of all of the energy system costs excluding the subsidies, consumers' vehicle ownership costs reflect the overall cost of fuel chains and transport system. To estimate the overall benefits (energy system and consumers) in the cases with government incentives, total vehicle subsidies are subtracted from the net consumer benefits.

Comparing the total present value of costs (assuming the discount rate of 7 %) and cumulative emissions of REF + LP scenario against the REF scenario reveals the mitigation cost of US$ 97 per tonne CO_2eq from the energy system perspective. The greater supply–push in the REF scenario enables significant reductions in emissions with the cost of US$ 31 per tonne CO_2eq. In the HC conditions, any supply–push policy leads to a negative mitigation cost, which means that emission reductions can provide net benefits for the energy system. However, the higher subsidy rate of 20 % for ZEVs raises the government expenses and, hence, the mitigation cost rises to US$ 19 per tonne CO2eq.

The cost-effectiveness of supply–push policies will be always negative from consumer and overall perspectives. Higher government subsidy along with strong supply–push would be the most cost-effective scenario from the consumer perspective with the net benefit of US$ 291 per tonne. High supply–push in HC condition could be the most cost-effective at reducing GHG emissions from energy system and overall perspectives with the net benefits of US$ 72 and US$ 193/ per tonne, respectively.

14.6 Conclusions

The comparative analysis of scenarios shows that until 2030–2035, the majority of alternative vehicles in Iceland will be hybrid (diesel and gasoline) and biogas vehicles. In the long run, the study suggests EVs as a winner among AFVs, as biofuel vehicles suffer from the limited resource potential and the higher costs of small-capacity biofuel-generating plants in Iceland, while a hydrogen pathway needs longer time to be well established. H2 vehicles would be expected to play a significant role after 2040 with maximum H2 demand requiring 19 forecourt electrolyzer units with capacity of 1500 kg/day by 2050. Maximum transport electricity demand requires about 1 TWh of additional electricity generation out of total generation of 60 TWh by 2050 (through each of wind, hydro, and geothermal), around 46% of which will be needed for hydrogen production.

The cost-effectiveness analysis indicates that the initial momentum of renewable fuels for mitigation of GHG emissions can provide overall long-term benefits through consumers' fuel cost saving and reduced fuel import costs. Government incentives through ZEV subsidies can achieve a deeper reduction in GHG emissions. A higher subsidy rate, which causes a greater mitigation, is more cost-effective from a consumer perspective and less cost-effective from an overall perspective.

Infrastructure development costs for refuelling/charging stations have not been fully explored in the current version of UniSyD_IS model. It implies that the full picture of total costs has not been captured in the analysis and, hence, future work will be required to capture the development and cost of refuelling and recharging station.

Acknowledgments The preparation of this chapter has been supported by the Norden Top-level Research Initiative sub-programme "Effect Studies and Adaptation to Climate Change" through the Nordic Centre of Excellence for Strategic Adaptation Research (NORD-STAR).

References

1. NORDEN and IEA (2013) Nordic energy technology perspectives: pathways to a carbon neutral energy future. Paris: IEA
2. Leaver J, Gillingham K, Leaver L (2009) Assessment of primary impacts of a hydrogen economy in New Zealand using UniSyD. Int J Hydrogen Energy 34(7):2855–2865
3. Leaver J, Gillingham K (2010) Economic impact of the integration of alternative vehicle technologies into the New Zealand Vehicle fleet. J Cleaner Prod 18(9):908–916
4. Leaver J, Gillingham K, Baglino A (2012) System dynamics modeling of pathways to a hydrogen economy in New Zealand. Unitec ePress, Auckland
5. Train K (2009) Discrete choice methods with simulation. Cambridge University Press, Cambridge
6. Johannsson S (2011) IceScot: submarine power cable from Iceland to Scotland. Annad veldi ehf, Reykjavik. http://www.veldi.is/Frett/13858. Accessed 2013
7. NEA (2012) http://www.nea.is/the-national-energy-authority/energy-statistics/gas-emissions-from-geothermal/. Accessed 2013
8. Orkuveita Reykjavíkur (2012) http://www.or.is/Fyrirtaeki/Rafmagn/Verdskra. Accessed 2012
9. Statice, "Energy Statistics" (2012) http://www.statice.is/Statistics/Manufacturing-and-energy/Energy. Accessed 2013
10. NREL, "DOE H2A Production Analysis" (2012) http://www.hydrogen.energy.gov/h2a_production.html. Accessed 2013
11. Sundberg M, Guðmundsson J, Guðmundsson M (2010) Biofuel production in Iceland: survey of potential raw materials and yields to 2030. MANNVIT Engineering, Reykjavik
12. IEA-ETSAP and IRENA (2012) Technology policy brief: production of liquid biofuels. http://www.iea-etsap.org. Accessed 2012

13. Gnansounou E, Dauriat A (2010) Techno-economic analysis of lignocellulosic ethanol: a review. Bioresource Technol 101:4980–4991
14. IEA (2011) Technology roadmap: biofuels for transport. International Energy Agency (IEA), Paris
15. Danielsson DI (2011) Bio-methane production expansion in Iceland, a cost-benefit analysis. Master Thesis, Aarhus University
16. Murphy J, McCarthy K (2005) The optimal production of biogas for use as a transport fuel in Ireland. Renewable Energy 30:2111–2127
17. Shafiei E, Davidsdottir B, Leaver J, Stefansson H, Asgeirsson EI (2014) Potential impact of transition to a low-carbon transport system in Iceland. Energy Policy 69:127–142
18. Road Traffic Statistics (2013) Iceland road traffic statistics. http://ww2.us.is/node/52. Accessed 2013
19. Dahl CA (2012) Measuring global gasoline and diesel price and income elasticities. Energy Policy 41:2–13
20. Schmid E, Knopf B, Bauer N (2012) REMIND-D: a hybrid energy-economy model of Germany. Potsdam Institute for Climate Impact Research, Potsdam
21. IEA & ETSAP (2011) Energy demand technologies. http://iea-etsap.org/Energy_Technologies/Energy_Demand.asp. Accessed 2013
22. Van Vliet O, Kruithof T, Turkenburg W, Faaij A (2010) Techno-economic comparison of series hybrid, plug-in hybrid, fuel cell and regular cars. J Power Sources 195:6570–6585
23. Kloess M, Muller A (2011) Simulating the impact of policy, energy prices and technological progress on the passenger car fleet in Austria- a model based analysis 2010–2050. Energy Policy 39:5045–5062
24. Thiel C, Perujo A, Mercier A (2010) Cost and CO_2 aspects of future vehicle options in Europe under new energy policy scenarios. Energy Policy 38:7142–7151
25. JRC-IE (2011) Well-to-wheels analysis of future automotive fuels and powertrains in the European context, tank-to-wheels report, version 2c (March). European Commission Joint Research Centre, Luxembourg
26. Offer GJ, Howey D, Contestabile M, Clague R, Brandon NP (2010) Comparative analysis of battery electric, hydrogen fuel cell and hybrid vehicles in a future sustainable road transport system. Energy Policy 38:24–29
27. Thomas C (2009) Fuel cell and battery electric vehicles compared. Int J Hydrog Energy 34:6005–6020
28. Gerssen-Gondelach SJ, Faaij AP (2012) Performance of batteries for electric vehicles on short and tonger term. J Power Sources 212:111–129

Chapter 15
Parametric Study of Polymer Electrolyte Membrane Fuel Cell Performance Using CFD Modelling

Angus Hood, Shaun Slater, Matthew Bouchet, Sheikh Zahidul Islam and Mamdud Hossain

Abstract A comprehensive three-dimensional, isothermal, steady-state, straight-channel proton exchange membrane (PEM) fuel cell model was developed to investigate the transport limitations of fresh reactants at high current densities. The model is created based on the existing models in literature to predict the reactant's transport limitations at higher current densities using three-dimensional framework. A user-defined function (UDF) code was developed considering source terms for porous zones, effective diffusivity models for species transport inside cells and electrochemistry algorithm to predict cell voltage at an average current density. Water transport through membrane was implemented considering electroosmotic drag and back diffusion inside PEM fuel cell. Simulation-predicted cell performances for different average current densities were validated with experimental results, and the effect of design parameters on cell performance is obtained using parametric studies. Parametric studies were performed to determine the best possible operating and geometrical design parameters of PEM fuel cell.

Keywords PEM fuel cell · Single phase · Porosity · Permeability · Parametric study

S. Z. Islam (✉) · A. Hood · S. Slater · M. Bouchet · M. Hossain
School of Engineering, Robert Gordon University, Aberdeen, UK
e-mail: s.z.islam1@rgu.ac.uk

A.Hood
e-mail: a.hood2@rgu.ac.uk

S. Slater
e-mail: s.slater@rgu.ac.uk

M. Bouchet
e-mail: m.bouchet@rgu.ac.uk

M. Hossain
e-mail: m.hossain@rgu.ac.uk

A. Sayigh (ed.), *Renewable Energy in the Service of Mankind Vol I*,
DOI 10.1007/978-3-319-17777-9_15

15.1 Introduction

Fuel cells are one of the clean sources of energy that can make real contribution in the reduction of CO_2 emission. Among many fuel cells (such as alkaline, direct methanol, solid oxide, molten carbonate), proton exchange membrane (PEM) fuel cell is the most versatile that can be used in both residential and transport sectors. PEM fuel cell has shown and demonstrated that potential to be a suitable alternative power source because of its simple design, load following capabilities, efficiency, feasibility and quick start-up. Although PEM fuel cells received much attention at fundamental research and technology development levels, the commercialization of PEM fuel cells is still in initial stages because of its relatively high cost of fuel cell prototypes, durability, hydrogen storage and hydrogen fuel infrastructure, and insufficient lifetime for stationary power generation [5, 12].

One of the major solutions to reduce cost is to have cheaper catalyst layers with reduced Pt loading. In addition, different membrane thicknesses play a key role to the cost of PEM fuel cell. In order to develop a cost-competitive PEM fuel cell system, researchers' are performing material research to innovate new design and material for catalyst layer and membrane. An alternative approach to reduce cost and improve durability is to have design and operation optimization through fundamental understandings. PEM fuel cell experimental researches only address the global parameters such as polarization characteristics at different current densities by examining a voltage–current curve. However, detailed understanding of species transport inside the PEM fuel cell is not possible due to the size constraint. The operation of PEM fuel cell not only involves electrochemical reactions but also current distribution, fluid mechanics, species transport, water and thermal management. It is difficult to understand these complex electrochemical reactions and the transport limitations of fresh reactants using experiments, while modelling of PEM fuel cell provides better understanding of these electrochemical reactions and transport phenomena.

Water content in polymer membrane plays an important role in PEM fuel cell performance. The proton conductivity of the membrane depends on the water content of membrane. However, too much water can cause flooding inside the cell. Therefore, it is necessary to investigate the performance of PEM fuel cell at higher current densities and higher humidity.

Lum and McGuirk [14] formulated a steady-state, isothermal, three-dimensional, single-phase complete PEM fuel cell model. The model was validated both globally and locally using experimental data. A deviation in the simulated results from experimental results was observed at high current densities and high humidity condition. Berning and Djilali [6] performed a parametric study to investigate the effect of operational parameters (e.g., pressure, temperature) and geometrical and material parameters (e.g., gas diffusion electrode thickness, porosity, and channel width and land area ratio) on PEM fuel cell performance. According to their study, it was observed that increasing temperature and pressure significantly improves the cell performance. The predicted cell performance reproduced the similar trends of experimental results [19, 11]. Stoichiometric flow ratio, gas diffusion porosity and thickness have a significant influence on limiting current density of PEM fuel cell as it controls the concentration of reactants to catalyst layer.

Wang et al. [20] conducted a parametric study based on experiments and compared model predicted results with an experimental data. Increasing fuel cell temperature and pressure showed an increasing trend in the cell performance. However, considerable effect of anode humidification temperature was observed at low current density. Similar trends were observed in the modelling results of Al-Baghdadi and Al-Janabi [2, 3] and Yuan et al. [21]. Shimpalee and Zee [18] investigated different serpentine flow-field configurations to minimize the water flooding. According to their observation, narrow channel with wider land area configuration gives higher cell performance for stationary conditions. Therefore, flow-field geometrical configurations and shoulder width play a vital role on PEM fuel cell performance at high current density [1].

Lum and McGuirk [14] found that decreasing shoulder width increases the cell potential significantly. Effect of decreasing permeability of porous regions on cell performance was investigated, and a limiting permeability value (10^{-9} m^2) was identified from their study. Dawes et al. [8] further investigate the effect of gas diffusion layer (GDL) permeability on cell performance and found the limiting value of 5×10^{-11} m^2. Rismanchi and Akbari [17] showed that increase in gas diffusion parameters increases the cell performance, while no marked effect was observed by changing catalyst diffusion parameters.

The objective of this study is to develop a complete three-dimensional single-phase model to investigate transport limitations and its effect on fuel cell performance. The model was used to investigate the cell performance at different current densities. Parametric studies were performed to identify most favourable operating conditions and design parameters for the present model. The single-phase model is used to investigate the cell performance at high current densities and fully humidified inlet conditions. Reactant flow along the channel is observed, and possible transport limitation zones were identified.

15.2 Numerical Model Development

The diversity and complexity of electrochemical reactions and transport phenomena occurring at disparate length and scale require a systematic framework and computational fluid dynamics (CFD) code to solve the comprehensive mathematical model. The computational fuel cell dynamics involve the following steps:

1. Physicochemical model development;
2. advanced numerical algorithms to solve the model
3. material characterization and appropriate boundary conditions and
4. finally, the model validation.

The focus of this research is to develop a complete three-dimensional steady-state PEM fuel cell model based on electrochemistry, species transport considerations,

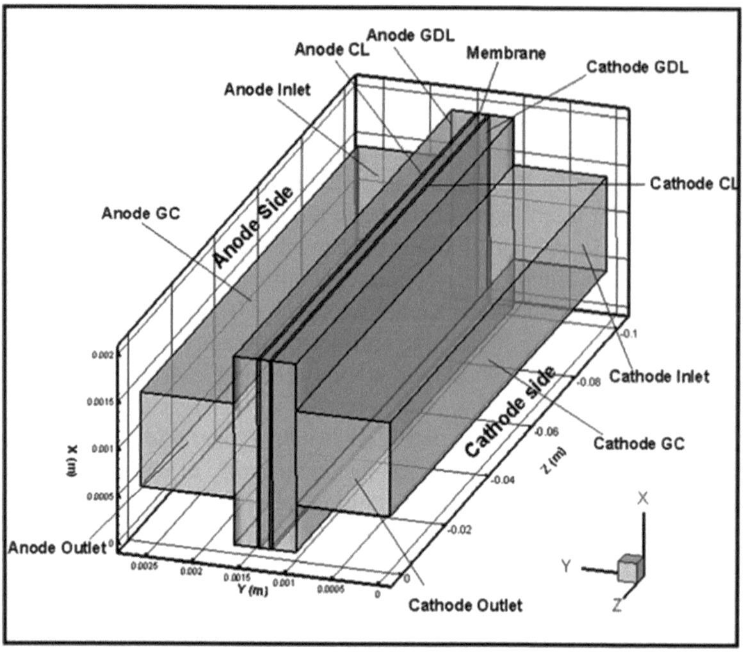

Fig. 15.1 Schematic diagram of a three-dimensional PEM fuel cell model. *CL* catalyst layer, *GC* gas channel, *GDL* gas diffusion layer

and appropriate CFD code to solve the model. Figure 15.1 shows a three-dimensional straight channel numerical model which consists of anode and cathode gas channels transporting the reactants (oxygen and hydrogen) and the product (water vapour) mixed in air to and from the porous GDL. The electrochemical reactions occur at the catalyst layers, and the polymer electrolyte membrane is sandwiched between the catalyst layers. Protons and water are transported through the membrane. Detailed model development, governing and auxiliary equations were presented in Hossain et al. [9].

15.3 Numerical Procedure

The governing equations and auxiliary equations are solved to investigate the complex electrochemical process and transport phenomena. The governing equations were solved along with the auxiliary equations using finite volume method. The finite volume method is based on the SIMPLE (semi-implicit method for pressure-linked equation) algorithm. Appropriate source terms are applied to the catalyst layers using user-defined functions (UDFs). Source terms, diffusivity model and electrochemistry algorithm were written in C++UDFs which were interpreted by the CFD solver [4]. The relative error is considered less than 10^{-6} for convergence. The electrochemistry algorithms and polarization overpotential were worked out at specified cell current density, and the cell voltage was computed for that specified cell current density. The fuel cell geometry is similar to the computational work of Min [16], Mazumder and Cole [15] and Liu et al. [13]. Physical dimensions of the computational domain as well as relevant fuel cell parameters are given in Table 15.1 [7, 10].

Table 15.1 Physical parameters and boundary conditions used for the simulations

Gas channel length	$L=100$ mm
Gas channel width	$W=1$ mm
Gas channel height	$H_{ch}=1$ mm
Diffusion layer height	$H_d=0.254$ mm
Catalyst layer height	$H_{ct}=0.0287$ mm
Land area width	$W_l=1$ mm
Membrane thickness	$t_m=0.23$ mm
Permeability	$K=1.76\times10^{-11}$ m^2
Operating pressure	$P=101325$ Pa
Operating temperature	$T=323$ K
Dry mass of membrane	$M_{m, dry}=1.1$ kg mol^{-1}
Dry density of membrane	$\rho_{m, dry}=2000$ kg m^{-3}
Fuel/air stoichiometric ratio	$\xi_a / \xi_c = 5/5$
Electron number of anode reaction	$n_a=4$
Electron number of cathode reaction	$n_c=2$
Relative humidity of inlet fuel	$RH_a=100\%$
Relative humidity of inlet air	$RH_c=0\%$
H_2 diffusion coefficient	$D_{h, ref}=0.915\times10^{-4}$m^2 s^{-1}
O_2 diffusion coefficient	$D_{O, ref}=0.22\times10^{-4}$m^2 s^{-1}
Water vapour diffusion coefficient	$D_{w, ref}=0.256\times10^{-4}$m^2 s^{-1}
Anode exchange current density	$i_{a, ref}=2.0\times10^{8}$A m^{-3}
Cathode exchange current density	$i_{c, ref}=160$ A m^{-3}
Hydrogen reference concentration	$C_{h, ref}=56.4$ mol m^{-3}
Oxygen reference concentration	$C_{o, ref}=3.39$ mol m^{-3}
Anode transfer coefficient	$\alpha_a=0.5$
Cathode transfer coefficient	$\alpha_c=0.5$

15.4 Results and Discussion

The predicted fuel cell polarization curve of the single-phase model is shown in Fig. 15.2. As the model is similar to the geometry of Mazumder and Cole [15] and Liu et al. [13], the predicted cell voltage is compared with the predicted results from these models. Mazumder and Cole [15] and Liu et al. [13] compared their model predicted results with experimental results by Ticianelli et al. [19] to establish the accuracy of the simulated results. A good agreement with experimental results was observed at lower current densities for the present single-phase model. While with increasing current densities, deviation from experimental results was observed for single-phase model. This variation occurs due to single-phase model cannot illuminate the liquid water effect on PEM fuel cell performance at higher current densities. This agrees with the work of Mazumder and Cole [15], Liu et al. [13] and Islam [10].

Parametric studies are conducted to investigate the effect of different parameters on PEM fuel cell performance. Based on the parametric study, most favourable PEM fuel cell design can be proposed for improved performance. The effect of stoichiometric flow ratio is summarized in Fig. 15.3. An increase in stoichiometric flow ratio results in increasing the gas mixture velocities in both anode and cathode

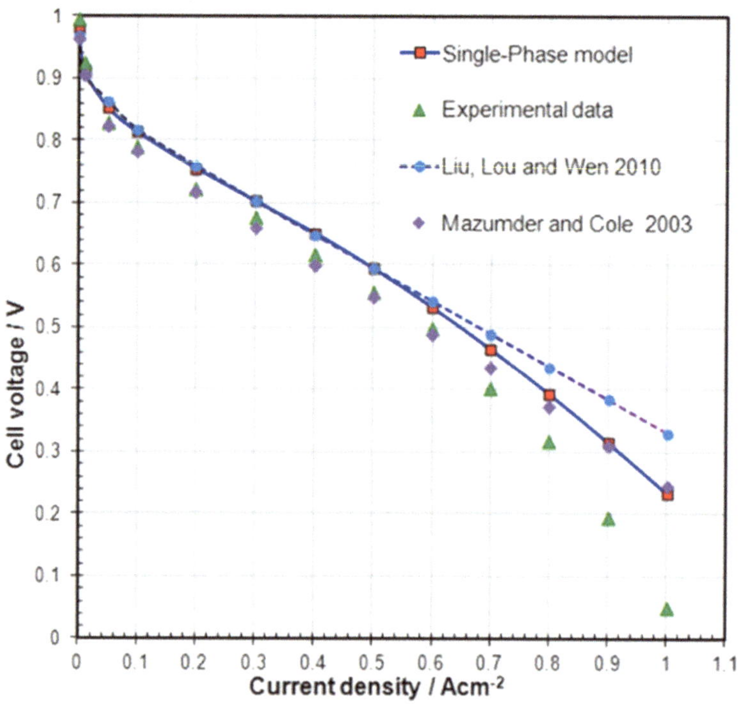

Fig. 15.2 Comparison between the single-phase model results with different model results and experiments [19]

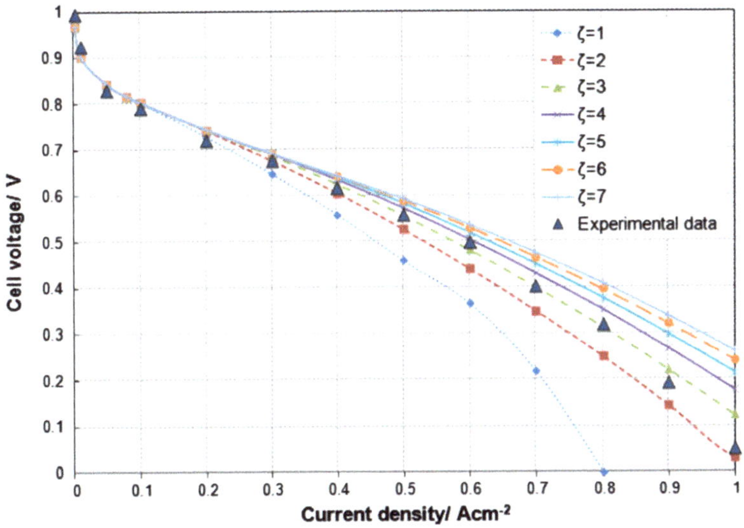

Fig. 15.3 Effect of stoichiometric flow ratio on cell performance at different average current densities

sides, while all other parameters remain constant. Simulation studies were carried out for a range of stoichiometric ratio, and significant influence on cell performance was observed. It is shown in Fig. 15.3 that cell performance is improved for higher values of stoichiometric ratio from 1.0 to 7.0. The model-predicted results show an excellent agreement with experimental results for value equal to 3.0. However, the deviation at higher current density still exists. Therefore, this value can be used for better cell performance prediction at lower and intermediate current densities. A higher stoichiometric ratio means higher gas mixtures velocity. Therefore, more oxygen reaches at cathode catalyst layer and improves the reaction rate. In addition, excessive water productions at higher current densities cause flooding inside cell. Higher gas mixture velocities help to improve water disposal rate from cathode. At low stoichiometric ratio, cell voltage dropped rapidly due to low amount of oxygen in catalyst layer at high current density. A large gain was observed for changing the value from 1.0 to 3.0 and increasing trend for increasing stoichiometric ratio. However, small increase in cell voltage was observed for a stoichiometric ratio above 5.0. In addition, higher gas mixture velocities cause excess water removal from anode side and dehydration state of membrane. Therefore, an optimum value of stoichiometric ratio is between 3.0 and 5.0 for maximizing cell performance at various operating conditions [10].

Simulations were performed for different values of GDL permeability, and the results are summarized in Fig. 15.4. The cell performance decreases for decreasing GDL permeability up to a certain range. From Fig. 15.4, it is clear that no change in cell voltage is observed for a permeability value lower than 10^{-11} m^2. At lower current densities, the change in GDL permeability has no effect on cell performance. However, GDL permeability influences the cell at higher current densi-

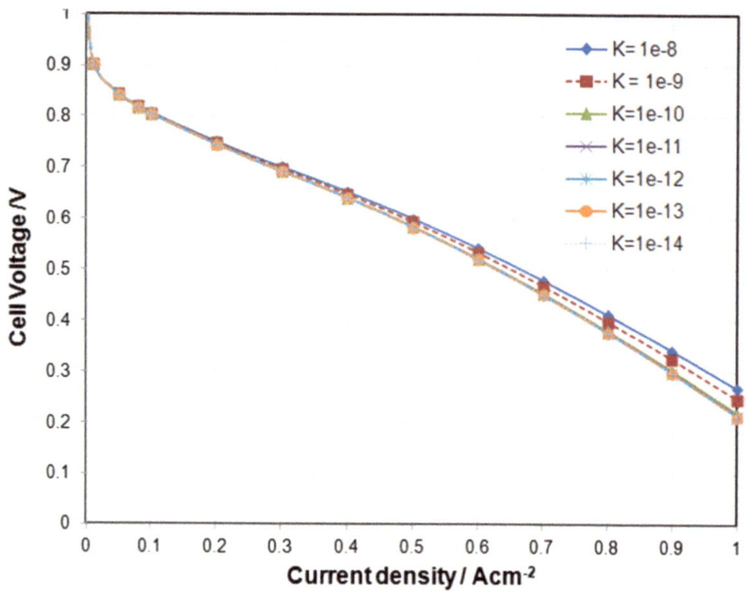

Fig. 15.4 Effect of GDL permeability on cell performance at different average current densities

ties. Lower permeability exhibits higher resistance to flow through porous medium, hence limits the disposal of water. Simulations were carried out for different values of Catalyst Layer (CL) permeability, and the results are summarized in Fig. 15.5. It is clear that no significant change in cell voltage is observed for the range of

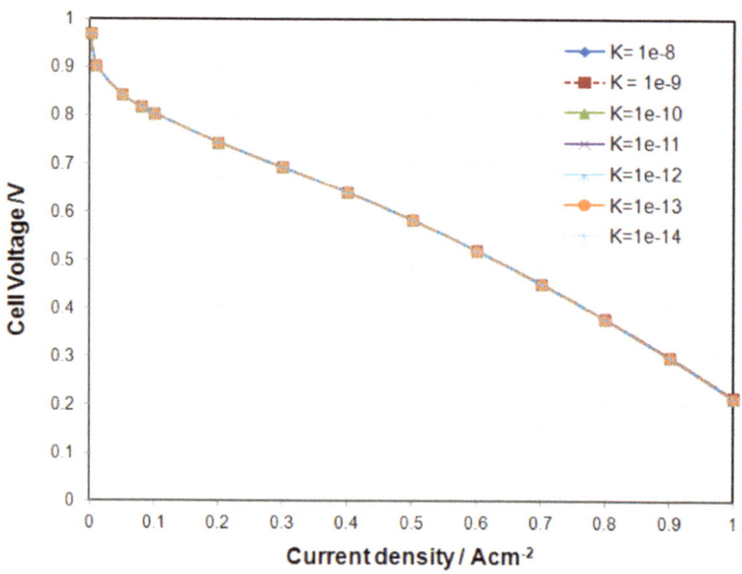

Fig. 15.5 Effect of CL permeability on cell performance at different average current densities

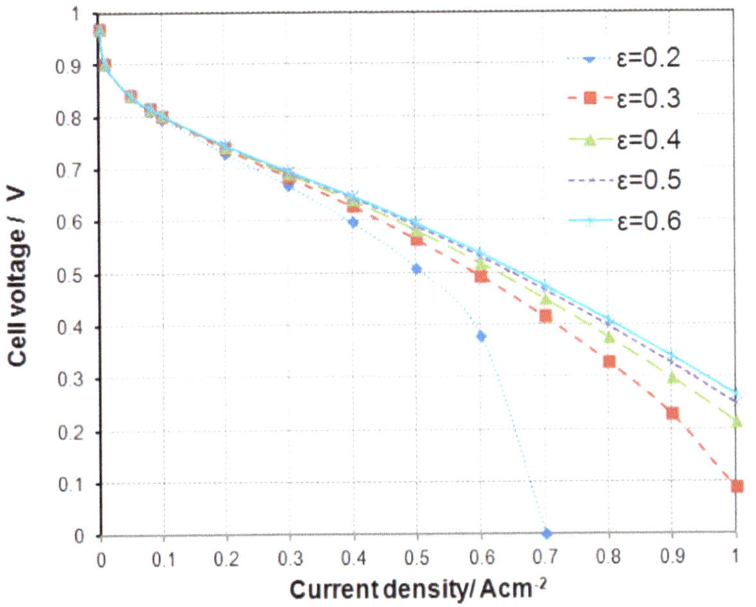

Fig. 15.6 Effect of GDL porosity on cell performance at different average current densities

permeability values. This is due to the same value used in the model for GDL permeability (10^{-10} m^2). The reactant receives similar restriction in the GDL before coming to the catalyst layer, and oxygen concentration remains unchanged before reaching CL. Simulations were performed for different values of GDL porosity, and the results are summarized in Fig. 15.6. It is observed that porosity has a strong influence on cell performance as the transport of reactant is primarily diffusive in porous medium. Higher values of porosity mean higher void fraction available for the reactants in the porous medium with lower restriction to flow. Higher concentrations of reactants lead to higher limiting current density for the fuel cell. Fig. 15.6 shows that cell performance increases with increasing GDL porosity. No significant influence was observed for a value higher than 0.5. In addition, rapid drop in cell voltage is observed for lower GDL porosity. Oxygen mass fraction contours along the length are plotted. Starvation for reactants was observed at GDL and CL for lower GDL porosity. A reduction in land area width increases the oxygen transport to catalyst layer. Simulations are performed for a constant channel area width of 1 mm and land area width values between 0.8 and 1.2 mm. From Fig. 15.7, it is observed that decreasing the land area width has no effect at lower current densities. However, it has a strong influence at higher current densities. The cell voltage increases with decreasing land area width. This is due to more oxygen concentration in the CL and evenly distribution under land area. At high current density, increase in land area width shows starvation of oxygen under the land area. Small land area facilitates more oxygen to reach CL layer for reaction. Membrane thickness affects the resistance of proton transport across the membrane. The ohmic overpotential occurs due to resistance to proton transfer across the membrane from anode CL to

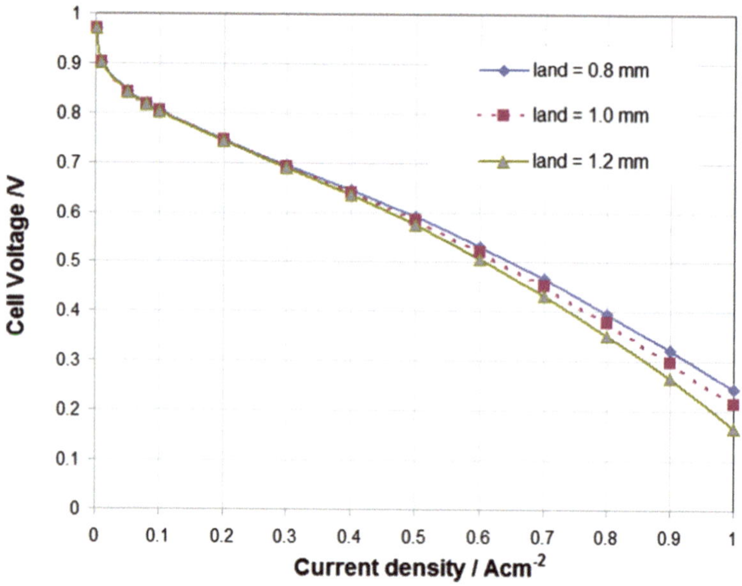

Fig. 15.7 Effect of land area on cell performance at different average current densities

cathode CL. Lower values in membrane thickness suggest lower path need to be travelled and hence lower potential loss. Reducing membrane thickness leads to higher cell voltage and shown in Fig. 15.8. In addition, the thickness of membrane has a strong influence on net water transfer across membrane. Lower membrane thickness facilitates more water to be transferred from anode side to cathode side. However, dehydration in anode side is observed at higher rate of net water transfer and therefore increases the ohmic overpotential loss [10].

The multifactor design of experiment (DOE) technique was utilized in the simulations. The reason the multifactor technique was used for the simulations was to establish the most suitable operating parameters for the fuel cell, taking two parameters as the factors for each simulation and plotting the results. The factors that were used stoichiometric ratio (ratio of between 3 and 7), diffusivity (values between 0.3 and 0.7) and cathode relative humidity (temperatures between 40 and 80 °C) are very important parameters to the successful operation of a fuel cell. The response output of each simulation allowed for the cell voltage to be plotted against each factor. Since 2 parameters were weighed against each other in each study, three-dimensional plots of the results could be created to show which parameters allow most optimal use.

It can be see that at the lowest cathode GDL, porosity of 0.3 results in zero cell performance regardless of stoichiometric ratio at high current density. This implies that the small pore space within the GDL is insufficient to allow both inward

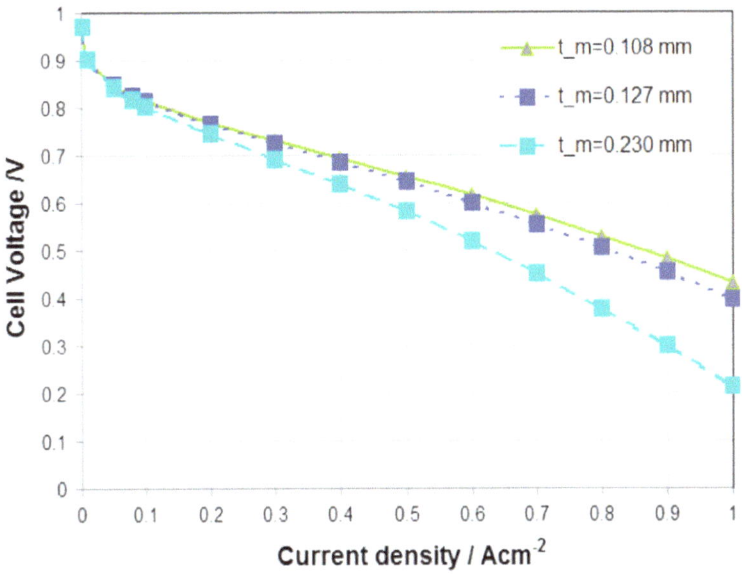

Fig. 15.8 Effect of membrane thickness on cell performance at different average current densities

diffusion of oxygen and outward diffusion of water. Increasing values of the stoichiometric ratio and cathode GDL porosity show an increased cell output. The operating range for optimum performance in this investigation was identified between 0.5 and 0.7 cathode GDL porosity, and stoichiometric ratios were between 6 and 7. It can be seen that the optimum operating conditions were when the porosity was at its highest (0.7), and the temperature was at its lowest (40 °C) which provided an output cell voltage of 0.432 V. The least optimum conditions were when high temperatures were coupled with a low porosity as can be depicted by the locations on the plot at 0 V. The contours can be seen to increase significantly as the porosity is increased from 0.3 to 0.4 before the incline slows down when the porosity increases up to 0.7. The curve of the temperature variations is less severe when temperatures are increased showing that in these simulations, the dominant parameter was the porosity. From Fig. 15.9, an area of optimum performance can be established by examining the plot and defining an area where the output results do not differ from each other significantly. For the DOE procedure 2, the optimum range of parameters was defined to be when porosity was between 0.5 and 0.7 and when the cathode temperature was between 40 °C and 60 °C. This can be seen in Fig. 15.9 by the flatter area of the graph in the light blue contour.

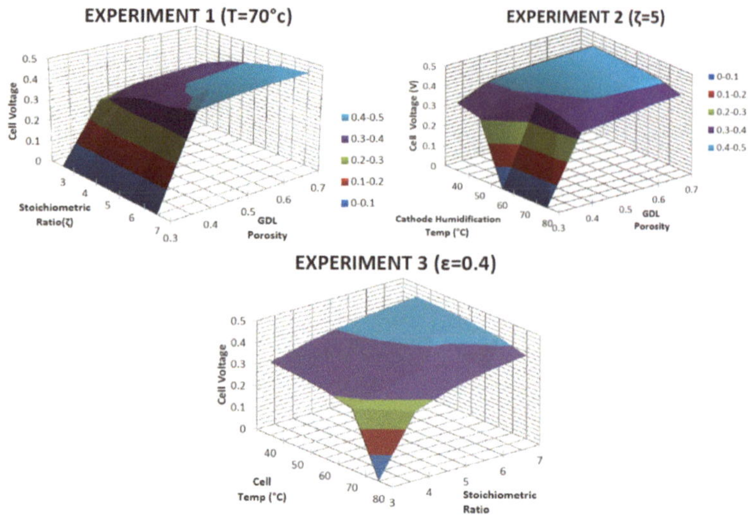

Fig. 15.9 Response surface plot at $I_{av}=0.7$ A cm^{-2}. *GDL* gas diffusion layer

15.5 Conclusion

Overall, the parametric studies conducted in this investigation provided strong evidence for the variation of cell performance under various conditions and can be used to identify the most effective operating conditions for a particular application. Increasing stoichiometric flow ratio from 2.0 to 5.0 shows improved fuel cell performance at higher current densities. GDL permeability has no effect on cell performance for a value lower than 10^{-11} m^2. GDL porosity is one of the major design parameters which have significant influence on limiting current density, hence on cell performance. A GDL porosity value between 0.4 and 0.6 shows improved cell performance over the entire operating range. Decreasing land area width for a fixed channel width shows improved performance. Low membrane thickness provides higher cell performance, and approximately 50 % reduction in membrane thickness results approximately 100 % improvement in cell performance at high current density of 1.0 A cm^{-2} [10]. Overall, this study serves to confirm the difficulties in commercializing fuel cell technology, with cell performance varying under many different operating conditions due to the complex nature of cell water management. In order to develop more robust understanding of cell performance under varying conditions, with the aim of optimizing cell performance, it was necessary to develop a methodology for analysing multiple operating conditions at a given time. The DOE investigation conducted in this investigation went some way towards satisfying this. However, this method itself does have some limitations, most notably the vast amount of simulations required for sufficiently comprehensive results as well as the difficulty in plotting these graphically.

Reference

1. Ahmed DH, Sung HJ (2006) Effects of channel geometrical configuration and shoulder width on PEMFC performance at high current density. J Power Sources 162:327–339
2. Al-Baghdadi MARS, Al-Janabi HAKS (2007a) Parametric and optimization study of a PEM fuel cell performance using three-dimensional computational fluid dynamics model. Renew Energy 32(7):1077–1101
3. Al-Baghdadi MARS, Al-Janabi HAKS (2007b) Modeling optimizes PEM fuel cell performance using three-dimensional multi-phase computational fluid dynamics model. Energy Convers Manag 48:3102–3119
4. Ansys Fluent 12.0 UDF Manual, 2010
5. Barbir F (2007). PEM Fuel cells. In: L Sammes (ed). Fuel cell technology: reaching towards commercialization. Springer, Germany, pp 27–51
6. Berning T, Djilali N (2003) Three-dimensional computational analysis of transport phenomena in a PEM fuel cell—a parametric study. J Power Sources 124:440–452
7. Hossain M, Islam S. Z. and Pollard P (2013) Investigation of species transport in a gas diffusion layer of a polymer electrolyte membrane fuel cell through two-phase modelling. Renewable Energy, Elsevier 51:404–418. doi:10.1016/j.renene.2012.10.008
8. Dawes JE, Hanspal NS, Family OA, Turan A (2009) Three-dimensional CFD modelling of PEM fuel cells: an investigation into the effects of water flooding. Chem Eng Sci 64:2781–2794
9. Hossain M, Islam SZ, Pollard P (2012) Numerical study of the effect of effective diffusivity coefficient and permeability of gas diffusion layer on fuel cell performance. IMECHE J Power Energy 226(7):907–921
10. Islam SZ (2012) Computational fluid dynamics modelling of PEM fuel cells to investigate transport limitations, PhD thesis, Robert Gordon University, UK
11. Kim J, Lee SM, Srinivasan S, Chamberlin CE (1995) Modelling of proton exchange membrane fuel cell performance with an empirical equation. J Electrochem Soc 142(8):2670–2674
12. Larminie J, Dicks A (2003) Fuel cell systems explained. 2nd edn. Wiley, England, pp 75–79
13. Liu X, Lou G, Wen Z (2010) Three-dimensional two-phase flow model of proton exchange membrane fuel cell with parallel gas distributors. J Power Sources 195:2764–2773
14. Lum KW, Mcguirk JJ (2005) Three-dimensional model of a complete polymer electrolyte membrane fuel cell-model formulation, validation and parametric studies. J Power Sources 143:103–124
15. Mazumder S, Cole JV (2003) Rigorous 3-D mathematical modelling of PEM fuel cells I. Model prediction without liquid water transport. J Electrochem Soc 150(11):A1503–A1509
16. Min CH (2010) A novel three-dimensional, two-phase and non-isothermal numerical model for proton exchange membrane fuel cell. J Power Sources 195:1880–1887
17. Rismanchi B, Akbari MH (2008) Performance prediction of proton exchange membrane fuel cells using three-dimensional model. Int J Hydrogen Energy 33:439–448
18. Shimpalee S, Zee JWV (2007) Numerical studies on rib & channel dimension of flow-field on PEMFC performance. Int J Hydrogen Energy 32:842–856
19. Ticianelli EA, Derouin CR, Srinivasan S (1988) Localization of platinum in low catalyst loading electrodes to attain high power densities in SPE fuel cells. J Electroanal Chem 251:275–295
20. Wang L, Husar A, Zhiu T, Liu H (2003) A parametric study of PEM fuel cell performances. Int J Hydrogen Energy 28:1263–1272
21. Yuan W, Tang Y, Pan M, Li Z, Tang B (2010) Model prediction of effects of operating parameters on proton exchange membrane fuel cell performance. Renew Energy 35:656–666

Chapter 16
Proton Modified Pt Zeolite Fuel Cell Electrocatalysts

Jun Yao, Yufeng Yao and Hossein Mirzaii

Abstract NaY zeolite has been selected as a suitable material to host 1.5 wt.% platinum (Pt) loading on zeolite using ion-exchange methods of $Pt(NH_3)_4(NO_3)_2$ without excess NH_4NO_3 nitrate and $Pt(NH_3)_4(NO_3)_2$ with excess NH_4NO_3 nitrate. The structure/reactivity relationship of Pt nanoparticle was experimentally studied via Nafion@-bound electrodes to investigate the interaction nature of Pt with zeolite and electron transfer. By using extended X-ray adsorption fine-structure (EXAFS) technique, Pt particle size was predicted as 0.7–1.5 nm. It was found Pt oxides could be electrochemically reduced via a hydrogen 'spillover' phenomenon. A highly dispersed small Pt particle distribution can be achieved with excessive H^+ ions on zeolite acidic sites.

Keywords Pt/Y zeolite electrocatalyst · Nanoparticle · CV · EXAFS · Fuel cell

16.1 Introduction

Platinum (Pt) is an important catalyst of polymer electrolyte membrane fuel cell (PEMFC), because of its ultra-high catalytic properties that can largely promote high energy density. Unfortunately, Pt is an extremely expensive metal material, and its performance degrades very rapidly that will hinder its wider applications in fuel cell. Hence, the creation of high Pt active surface with less metal content is one of important but challenging themes in fuel cell electrocatalyst research.

J. Yao (✉)
School of Engineering, University of Lincoln, Brayford Pool, Lincoln LN6 7TS, UK
e-mail: jyao@lincoln.ac.uk

Y. Yao
Department of Engineering Design and Mathematics, University of the West of England, Coldharbour Lane, Bristol BS16 1QY, UK
e-mail: yufeng.yao@uwe.ac.uk

H. Mirzaii
Kingston University, Roehampton Vale, Friars Avenue, London SW15 3DW, UK
e-mail: H.Mirzaii@kingston.ac.uk

A. Sayigh (ed.), *Renewable Energy in the Service of Mankind Vol I*,
DOI 10.1007/978-3-319-17777-9_16

The use of carbon nanotube or zeolite to facilitate the adsorption of small catalyst particles was used as a solution for particle cluster size control problem [1]. The excessive hydrogen-ion-exchanged zeolite has rich protons at zeolite acidic centre to form hydroxyl groups with oxygen, leading to the interaction of metal precursors with zeolite and thus decrease the sintering propensity of the metal [2]. Nevertheless, the surface oxygen groups, which form the anchoring sites of metallic precursors and metals, are well known in determining the catalyst supporting material properties [3]. Koningsberger et al. [4] has investigated the influence of oxygen atom charge transfer on zeolite using the extended X-ray adsorption fine structure (EXAFS). The increasing of electronegativity nearby the support oxygen atoms was found to increase zeolite alkalinity. The interaction between Pt and oxygen would move Pt electrons more nearer to oxygen. A charge transfer can be subsequently occurred from or to Pt metal particles. This has particularly accounted for the PtHY zeolite catalyst, as H^+ ions exchanged into zeolite structure can result in a significant increase of the O, Si and Al binding energy.

A highly dispersed Pt/zeolite catalyst using O_2 calcination and H_2 reduction process was reported by Sachtler and Gallezot [5]. Pt was found favourable to locate in supercages at 300 °C with a particle size of 6–13 Å. The X-ray photoelectron spectroscopy (XPS) study has detected Pt particle migration through zeolite supercages to sodalite cages at calcination temperature of 450 °C [6]. Pt constrained within zeolite supercages was varied from single atom to a particle in an order of 1–2 nm [7], forming a cluster of 10–25 atoms [8]. The Pt–Pt binding energy was found to increase with Pt electron deficiency. An average of approximately 20–30 atoms with 11.3 Å per particle was predicted at a reduction temperature of 300 °C and was increased to 16 Å at 500 °C in H_2 [9]. Yokoyama et al. [10] also revealed that the Pt–Pt bonding distance of Pt cluster encaged in zeolite supercages was shorter than the bulk Pt metal. This suggested that the charge transfer from Pt clusters to the substrate was stronger in the Pt/Y zeolite catalyst. The contraction of Pt–Pt bond distance on zeolite was mainly due to the increase of Pt–Pt bonding energy caused by electron deficiency.

Although zeolite is a well-known direct current (DC) insulator [11], its capability as a solution-like ionic conduction linkage commonly used within electrochemical environment indicates that the electrochemical reaction on Al–Si zeolite is able to occur in a charge-balanced ionic environment [12]. The Pt metal particles are able to be completely hydrated and dehydrated due to a great capacity of water absorption on zeolite [12]. The role of high water content in electrochemical reaction is considered to facilitate the spillover of hydrogen [13]. In fact, the water can act as a 'vehicle' to transport ions through zeolite free volume space by hitching a ride on water. This phenomenon is called 'spillover' [14], mainly involving hydrogen- and oxygen-containing species, and has been observed on catalyst such as Pt/C [15], Pt/SiO_2 [16], Pt/Al_2O_3 [17] and Pt/zeolite [18] in gas phase. Therefore, it is important to investigate H^+ spillover effect in zeolite-supported Pt Nafion@-bound Pt zeolite electrode system [7].

This study contributes to the understanding of electron transfer at the interface between Pt/zeolite electrode and the solution, and the Pt particle size associated electrocatalytic efficiency in electrochemical reaction. The structure of the Pt particle will be investigated using the EXAFS, and the correspondent Pt electrocatalytic performance will be determined via cyclic voltammetry (CV) measurement.

16.2 Experimental

The 1.5 wt.% Pt loading was made by ion exchange of Pt^{2+} ions with Na^{+} on Y zeolite using either $Pt(NH_3)_4(NO_3)_2$ or $Pt(NH_3)_4(NO_3)_2/NH_4NO_3$ solution. The ion exchange procedure was followed by methods developed by Boyan [8] and Lalchan [19] at natural condition. The Pt microstructure synthesis was then followed the procedure developed by Gallezot and co-workers [6]. The samples were calcined at 350 °C in O_2 and later reduced in H_2 at 400 °C.

The Nafion@-bound electrode was made by spreading a thin paste of 1.5 wt.% Pt/Y zeolite or Pt/HY zeolite and carbon powder (untreated XC-72R carbon powder) mixture with Nafion@ solution (5 wt.% solution in lower alcohols from Aldrich) to carbon paper (E-TEK TGHP-90). A disc of 2.5 cm^2 diameter of electrode was cut for CV measurement and 1.3 cm^2 diameter for EXAFS study via Hg/Hg_2SO_4 mercury/mercurous sulphate (MMS) reference electrode. The electrode was placed in 2.5 and 1 mol cm^{-3} sulphuric acid (H_2SO_4) solutions for laboratory CV and in situ EXAFS measurements, respectively, using a computer-controlled potentiostat (Auto-Lab PGSTAT20) at a constant scan rate (dE/dt) of 1 mV/s [20]. The oxidation/reduction of the electro-active species of interest was detected between −0.65 and 0.5 V, in which no any decomposition occurred [21]. The EXAFS measurements were performed using the synchrotron radiation source (SRS) at Daresbury Laboratory, the UK. The wiggler beam line was operated at 2 GeV and 100 mA. The high-order harmonics were removed using a double-crystal Si220 monochromator. The Pt L_{III} absorption edge was located by 50 % detuning of a harmonic beam with a Pt foil used as a reference sample for EXAFS data collection.

16.3 Results and Discussion

16.3.1 Zeolite BET surface areas measurement

Zeolite crystalline structure was characterised by Brunauer–Emmett–Teller (BET) surface area measurement. Figure 16.1 depicts the zeolite BET surface area measurements for electrocatalysts, such as (a) 1.5 wt.% Pt loading on zeolite calcined at 350 °C and reduced at 400 °C (i.e. 15PtANCR4) and (b) 1.5 wt.% Pt loading on zeolite made with excess NH_4NO_4 ion exchange calcined at 350 and 400 °C (i.e. 15PtANXCR4), respectively. The plain zeolite and carbon XC-72R powders were used as references, and their surface areas were measured at 632 and 200 m^2 g^{-1}

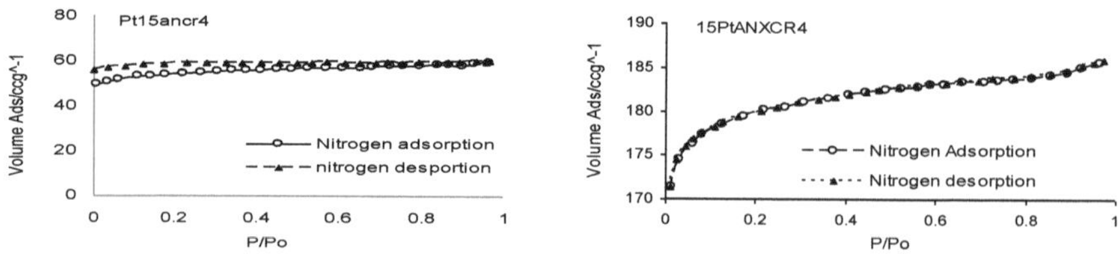

Fig. 16.1 Volumetric uptake of nitrogen at 77 K with absorption in open dotted line and desorption in filled dotted line. **a.** Sample15PtANCR4; **b.** Sample 15PtANXCR4

Table 16.1 BET surface area measurement for plain zeolite, untreated carbon XC-72R, 1.5 wt.% Pt/Yzeolite (i.e. 15PtANCR4) and 1.5 wt.% Pt/HY zeolite (i.e. 15PtANXCR4) electrocatalysts

Acronym name	Plain zeolite catalyst	Carbon XC-72R	15PtANCR4	15PtANXCR4
Surface area (m^2/g)	623.2±0.1%	200±0.1%	168±0.1%	539±1.0%

(see Table 16.1). The discrepancy of sample 15PtANXCR4 zeolite surface area is very small compared to that of plain zeolite. A well-preserved zeolite crystalline structure was determined for sample 15PtANXCR4 made by excess NH_4NO_4 ion exchange method. The Pt zeolite electrocatalyst with no excess NH_4NO_4 ion change (i.e. 15PtANCR4) has shown a significant reduction of zeolite surface area which was measured at 168 $m^2\ g^{-1}$, indicating the blockage of zeolite opening pores in a small channel.

16.3.2 The CV Measurement

The CV measurement was performed in an electrochemical cell to determine the electroactivity of species using Nafion@-bound electrode made by a mixture of Pt/zeolite electrocatalyst and carbon powder. The response of electrochemical reaction was determined by monitoring current change via potential. The CV measurement of 1.5 wt.% Pt loading on Pt/Y zeolite carbon | Nafion@ electrodes (i.e. 15PtANCR4 and 15PtANXCR4) and 1.5 wt.% Pt/C | Nafion@ electrode (i.e. 15Pt/XC-72R) made from a commercially available 40 wt.% Pt/XC-72R carbon catalyst mixed with the extra XC-72R carbon is depicted in Fig. 16.2 to investigate hydrogen adsorption/desorption activity on Pt at −0.65 V in 2.5 mol dm^{-3} H_2SO_4 solution. Figure 16.2a depicts the CV of 1.5 wt.% Pt/CX-72R electrode consisting of three distinguished regions, a hydrogen adsorption/desorption region (peak A) starting at potential region of −0.32 to −0.65 V, that is, oxide region and the double-layer region. The peak B is responsible for a hydrogen evolution, and the peak C is linked with the weakly adsorbed H^+ ions forming covalent bond on Pt. The H reoxidation (peak D) is observed on the hydrogen reoxidation (HRO) region of −0.65 to −0.36 V, and the peak E is associated with weakly bound H desorption. The current increase between

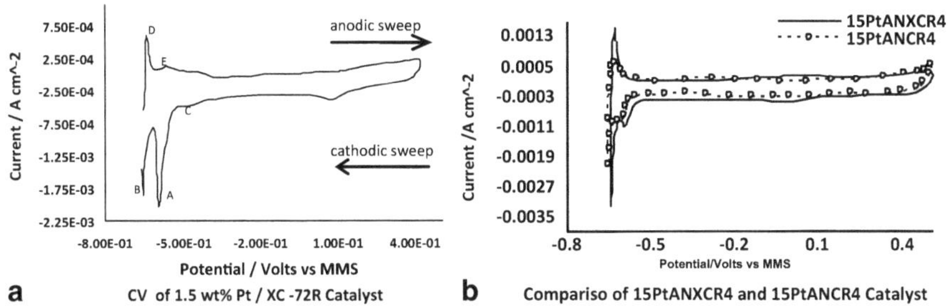

Fig. 16.2 Steady-state CV curves are measured at a scan rate of 1 mV/s in 2.5 mol dm^{-3} H_2SO_4 solution. **a.** Sample 15Pt/XC-72R, **b.** comparison of CV for samples 15PtANCR4 and 15PtANXCR4

0.12 and 0.5 V indicates the formation of Pt oxide species correspondent to an oxide stripping region of 0.5–0.02 V. A charge separation is determined in a double-layer region from 0.02 to 0.36 V and −0.36 to 0.12 V in a reverse sweep manner. Figure 16.2b shows a consistent electrochemical oxidation/reduction features for sample 15PtANCR4 and 15PtANXCR4. A strong hydrogen adsorption peak has been observed at −0.6 and −0.56 V for both electrodes. No weakly bound hydrogen adsorption peak can be determined, as both weakly and strongly bound hydrogen may adsorb similar amount of energy, resulting in difficulties to distinguish them unless a much slow sweep rate less than 1 mV/s used. A hydrogen desorption peak was detected at −0.56 V, followed by hydrogen evolution at −0.65 V. The large hydrogen coverage is observed for 15PtANXCR4. The electrochemical responses coincide to previous findings reported by Konigsberg and his co-workers [22]. The excess NH_4NO_3 on zeolite can also result in small but well-dispersed Pt clusters inside the zeolite cage.

The presence of electron/ion transfer at the interface of Nafion@-bound Pt/zeolite electrode and electrolyte solution can be demonstrated by CV method. Gallezot et al. [22, 23] predicted Pt particles more likely remained in zeolite supercage at a reduction temperature of 400 °C, where electrochemical species were easy to access on zeolite Pt active site. Zeolite is a type of DC electronic insulator material that cannot be used as direct electrode materials. However, zeolite is capable of solution-like ionic conduction which can act as an electron bank to contribute electrons to or garner electron from a reactant [24]. This commonly links with electrochemical environment for reaction to occur in a charge-balanced ionic environment.

The hydrogen 'spillover' pathway on Pt/C catalyst was investigated by Srinivas and Rao [25]. They found the adsorption/desorption can occur on carbon-based Pt active sites. Later, study by McBreen [26] has further revealed the current generation on Pt active sites, remotely from conducting material surface with fast surface mobility of adsorbed species on the electrode. Therefore, the surface active sites of Pt particles that are not in direct contact with solid conductive polymer electrolytes might be also involved in an interfacial process through the pathway of surface conductance. Two possible pathways either the surface mobility of adsorbed species or hydrogen ad-atoms/H^+ ion may spill over through zeolite framework or on electrode surface [13] on which hydrogen ions (H^+) were able to form OH group with carbon/zeolite acidic surface oxides. The hydrogen spillover on Pt and zeolite acidic sites has also been studied in gas phase [18], in which an anti-bonding state of Pt–H interaction relating to the electronic properties of the catalytic active sites has been observed [4]. This finding can also be used to explain the hydrogen spillover between Pt electrode and solution in electrochemical environment. In addition, H^+ spillover may occur during the surface diffusion process of the absorbed species [18]. Water is an important spice to facilitate hydrogen migrating between zeolite and Pt. The electron transfer may take place between electrode and solution via free species, such as H^+ and H_3O^+. Ions are transported through free species by hitching a ride on water during their migration through zeolite channels and between the interface of electrodes and solution.

16.3.3 In situ EXAFS Study for 1.5 wt.% Pt/Y Zeolite Electrocatalysts

Figure 16.3 illustrates the EXAFS data fitting for electrocatalysts 15PtANCR4 (with no excess nitrate ion exchange) and 15PtANXCR4 (with excess nitrate ion exchange). The correspondent fitting results are presented in Table 16.2. The in situ EXAFS data are collected at Pt L_{III} edge at room temperature and potential of − 0.65 V in 1.0 mol dm^{-3} H_2SO_4 solution. For electrocatalyst 15PtANCR4, the curve fitting (in dotted lines) matches the general form of the raw data (in dashed lines) very well, and the data fitting is extended up to three shells with shell 1 and shell 2 predicted for Pt and shell 3 for oxygen, respectively. The Pt coordination numbers in shell 1 and shell 2 are approximately 7.45 and 2.05. The introducing of third Pt–O shell has a significant improvement of data fitting quality. The first shell Pt cluster is predicted with Pt–Pt distance of 2.77 Å, implicating the preserving of the bulk Pt characteristics. The Pt–O bond may attribute to the direct link of Pt associated with oxygen on zeolite Lewis acidic site. The increase of Pt electron deficiency associated to either zeolite acidity, or the decrease of Pt–Pt bonding energy is leading to a slight increase of Pt–Pt bond (2.77 Å), compared to the standard metallic Pt bond of 2.75 Å. The curve fitting for 15PtANXCR4 is extended up to two shells. The Pt atom is fully reduced as no oxygen atom is predicted at neighbouring Pt shells. The total Pt atoms are fitted up to two shells at 7.86, indicating the Pt particle size is small to excess nitrate salt ion exchanged electrocatalyst 15PtANXCR4, compared to sample with no excessive H^+ presence (i.e. 15PtANCR4) with an average of 9.50 atoms. The Pt particle size is estimated at 1.5 and 0.8 nm for two samples 15PtANCR4 and 15PtANXCR4, respectively.

a. 15PtANCR4, test data and Fourier transformation graph
b. 15PtANXCR4, test data and Fourier transformation graph

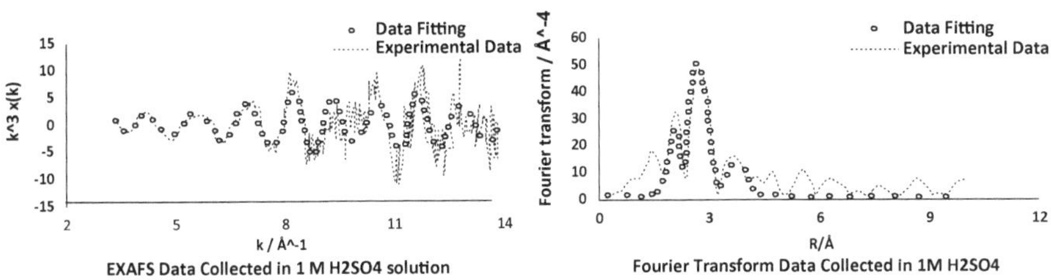

a 15PtANCR4, test data and Fourier transformation graph

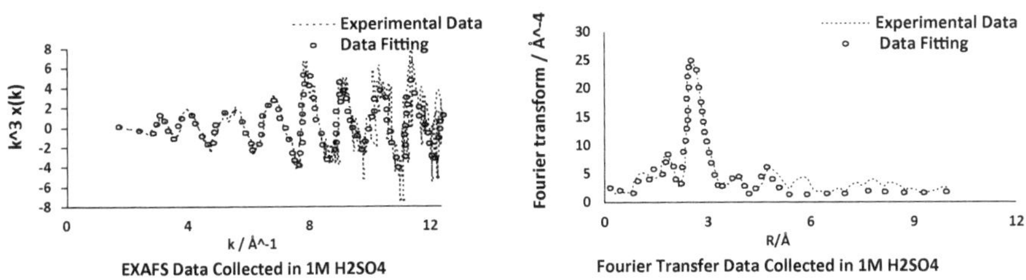

b 15PtANXCR4, test data and Fourier transformation graph

Fig. 16.3 The in situ EXAFS data fitting for 1.5 wt.% Pt Y zeolite catalysts with phase correction. The experimental data and the fitting results are presented in dashed line and dotted line, respectively. Data are collected at −0.65 V versus MMS reference electrode in 1.0 mol dm^{-3} H_2SO_4 solution under potential deposition. The EXAFS data were collected at Pt L_{III} edge at room temperature. **a.** Electrocatalyst 15PtANCR4, **b.** Electrocatalyst 15PtANXCR4. *EXAFS* extended X-ray adsorption fine structure, *MMS* mercury/mercurous sulphate

Table 16.2 The data fitting results for 1.5 wt.% Pt/Y zeolite for electrocatalysts 15PtANCR4 and 15PtANXCR4 at −0.65 V potential in 1.0 mol dm^{-3} H_2SO_4 solution

	Atom coordination	Shell 1—Pt	Shell 2—Pt	Shell 3—O
15PtANCR4	N	7.45	2.05	0.75
	R (Å)	2.77	3.85	2.19
15PtANXCR4	N	6.38	1.48	No
	R (Å)	2.77	3.90	No

N = coordination number, R = coordination distance

16.3.4 Pt Active Surface Area Measurement and Pt Distribution on Zeolite

The Pt active surface area under hydrogen peak is measured at 103.57 m^2/g for 15PtANXCR4 and 51.29 m^2/g for 15PtANCR4, correspondent to 0.77 and 1.52 wt.% Pt loading on zeolite as seen in Table 16.3 and Fig. 16.4, respectively. The results are consistent to Pt surface atom distributions. The total Pt atoms per square centimetre correspondent to Pt loading for sample 15PtANXCR4 have shown 50% reduction, compared to sample 15PtANCR4. The Pt surface atom dispersion is 29.44 and 14.66% for two catalyst samples 15PtANXCR4 and 15PtANCR4, respectively.

Table 16.3 Pt active surface area (m^2/g)

Sample	15PtANCR4	15PtANXCR4	15Pt/XC-72R	Dispersion N_S/N_T (%)
Pt active surface area under H (area/m^2/g)	51.29	103.57	0.133	14.66
Wt.% Pt	1.52	0.77	1.49	29.44

Fig. 16.4 Comparison of Pt active surface area distributions of two test samples

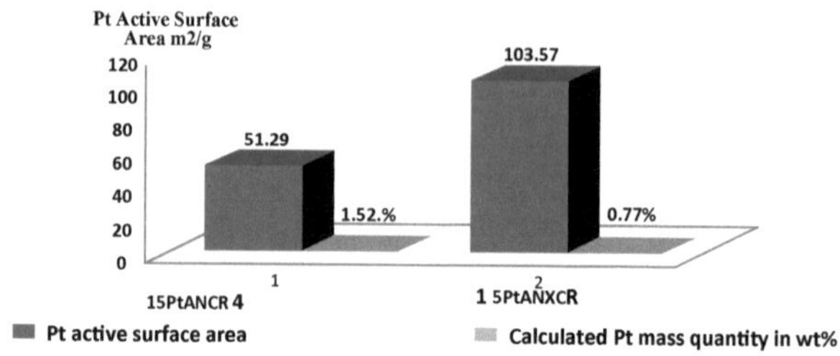

This study has shown that a highly dispersed Pt particle distribution on zeolite can be achieved with the presence of excessive H^+ ions on zeolite. The zeolite acidity may affect Pt catalytic activity, as the formation of metal—proton adduct could results in Pt electron change and hydrogen spillover phenomenon. The spillover H^+ ions can also be transmitted at the surfaces of the zeolite support to increase the surface conductivity of the support [27].

16.4 Conclusion

This study has contributed to the understanding of electron transfer between zeolite and Pt metal particle. The relations of Pt electrocatalytic efficiency correspondent to Pt particle size, Pt distribution and electron transfer pathway have been investigated by electrochemical control and the EXAFS. The cyclic voltammogram feature of hydrogen oxidation/reduction peak for Nafion@ electrode made with excess or no excess nitrate (NH_4NH_3) ion exchange (i.e. 15PtANXCR4 or 15PtANCR4) is found consistent to that of 1.5 wt.% Pt/XC-72R. The electrochemically active surface area (EASA) measurement indicates significantly high Pt distributions for catalyst 15PtANXCR4 than that of catalyst 15PtANCR4. The presence of excess H^+ can facilitates electrochemical reaction on Pt active sites, and H^+ is also able to improve metal–proton adduct formation and increase electrode surface conductivity by spillover H^+ on zeolite. Water is important for hydrogen migration via free species, such as H^+ and H_3O^+, by hitching a ride on water. The Pt particle cluster is predicted at 0.8 nm for catalyst 15PtANXCR4 and 1.5 nm for catalyst 15PtANCR4, correspondent to Pt distributions of 29.44 and 14.66 % with Pt loading of 0.77 and 1.54 wt.%, respectively. Conclusively, the 1.5 wt.% Pt loading on zeolite made by excess nitrate (i.e. 15PtANXCR4) has shown a great advantage to provide high surface area with efficient mass transport capacity for electro-active species to and from electrode surface in comparison of that without excess nitrate ion-exchanged electrocatalysts (i.e. 15PtANCR4).

References

1. Yang HN, Lee JY, Na Y, Yi SC, Kim WJ (2012) Effect of functionalization for carbon molecular sieve (CMS) synthesized using zeolite template on the incorporation of Pt nanoparticle and performance of the electrodes in PEMFC. Microporous Mesoporous Mater 152:148–56
2. Rodriguez-Reinoso F (1998) The role of carbon materials in heterogeneous catalysis, carbon 36-3:159–175
3. Auer E, Freund A, Pietsch J, Tacke T (1998) Carbons as supports for industrial precious metal catalysts. Appl Catal A 173:259–271
4. Koningsberger DC, de Graaf J, Mojet BL, Ramaker DE, Miller JT(2000) The metal -support interaction in Pt/Y zeolite: evidence for a shift in energy of metal d-valence orbitals by Pt -H shape resonance and atomic XAFS spectroscopy. Appl Catal A 191:205–220
5. Sachtler WMH, Zhang Z(1993) Catalysis of zeolite supported transition metals. Adv in Catal 39:129–220
6. Gallezot P, Alarcon-Diaz A, Dalmon JA, Renouprez AJ, Imeuk B (1975) Location and dispersion of platinum in PtY zeolite. J Catal 39:334–349
7. Rolison DR, Heys EA, Rudzinski WE (1989) Electrode-modified zeolite: electrode microstructures contained in and on a heterogenerous catalyst. J Phys Chem 93:5524–5531
8. Boyanov BI, Morriso TI (1996) Supported and temperature effects in platinum clusters. 1. Apatial structure. J Phys Chem 100(40):16310–16317
9. Pandya KI, Heald SM, Hriljac JA, Petrakis L, Fraissard J (1996) Bimetallic copper-platinum particles supported in Y zeolite: structural characterization by EXAFS. J Phys Chem100 (12):5070–5077
10. Yakoyama T, Kosugi N, Asakura K, Iwasawa Y, Kuroda H (1986) Temperature dependence of the Pt L3 - edge EXAFS of platinum clusters supported on NaY - zeolite. J Phys Colloque December 47:C8: 273–276
11. Breck DW (1974) Zeolite molecular sieves: structure, chemistry and use. Wiley, New York
12. Rolison DR (1990) Zeolite-Modified Electrodes and Electrode-Modified Zeolites. J Chem Rev 90(5):867–878
13. Liu WJ, Wu BL, Cha CS (1999) Surface diffusion and the spillover of H-adatoms and oxygen-containing surface species on the surface of carbon black and Pt/C porous electrode. J Electroanal Chem 476(2):101–108
14. Sermon PA, Bond GC(1974) Hydrogenspillover. Catal Rev 8(1):211–239
15. Bagotzky VS, Kanevsky LS, Palanker VS (1973) Adsorptive and catalytic properties of platinum microcrystals deposited on inert supports. Electrochim Acta 18(7):473–483
16. Candau JFC, Conner WC (1987) The spillover of hydrogen onto silica: IV. The use of scanning FTIR to follow spillover from a point source. J Catal 106(2):378–385
17. Cavanagh RR, Yates JT Jr (1981) Hydrogen spillover on alumina—A study by infrared spectroscopy. J Catal 68(1):22–26
18. Zhang A, Nakamura I, Fujimoto K (1997) A New Probe Reaction for Studying the Hydrogen Spillover Phenomenon. J Catalysis 168(2):328–333
19. Persaud L, Bard AJ, Campion A, Fox MA, Mallouk TE, Webber SE, White JM (1987) A new method for depositing platinum exclusively on the internal surface of zeolite L. Inorg Chem 26(22):3825–3827
20. Gerischer H, Tobias CW (1978) Advances in electrochemsitry and electrochemical engineering, vol. 11. Willey, New York

21. Brett C, Brett AMO (1998) Electroanalysis. Oxford University Press, Oxford
22. Zhang Z, Wong TT, Sachtler WMH (1991) The effect of Ca2+ and Mg2+ ions on the formation of electron-deficient palladium-proton adducts in zeolite Y. J Catal 128(1):13–22
23. Gallezot P(1979) The state and catalytic properties of platinum and palladium in Faujasite - type zeolites. Catal Rev Sci Eng 20(1):121–154
24. Larsen G, Haller G (1989) Metal-support effects in Pt/L-zeolite catalysts. Catal Lett 3(1):103–110
25. Srinivas S, Rao P(1994) Direct observation of hydrogen spillover on carbon-supported platinum and its influence on the hydrogenation of benzene. J Catal 148(2):470–477
26. McBreen J (1985) Voltammetric studies of electrodes in contact with ionomeric membranes. J Electrochem Soc 132(5):1112–1116
27. Herrmann JM (1984) Electronic effects in strong metal-support interactions on titania deposited metal catalysts. J Catal 89(2):404–412

Chapter 17
Improved Dynamic Response and Range in Microbial Fuel Cell-Based Volatile Fatty Acid Sensor by Using Poised Potential

Amandeep Kaur, Richard M. Dinsdale, Alan J. Guwy and Giuliano C. Premier

Abstract Low-cost measurement of the concentration of key volatile fatty acids (VFAs) would be very useful to improve the operation of a number of important bio-processes, through monitoring and control. In situ microbial fuel cells (MFC)-based VFA sensors could replace the current generation of relatively complicated and expensive techniques for online VFA analysis. Cyclic voltammetry (CV) was used to determine the oxidation peak potentials for three VFA species (acetic, butyric and propionic). The dynamic behaviours and static sensitivities were recorded for 10-ml single chamber cubic MFC-based VFA sensors, separately acclimated on acetate, butyrate and propionate, while the anode was poised at a potential corresponding to the oxidation peak from CVs. The current responses at the fixed working electrode potentials were analysed for each VFA sensor in response to different concentration (0–250 mg/L) of the corresponding and other VFA species. When corresponding VFAs were supplied to the sensors, the range of the sensor was increased from 40 to 220 mg/L with no response to other VFA species in acetate- and propionate-enriched reactors.

Keywords Microbial fuel cell · Volatile fatty acid · Sensor · Poised potential · Offline implementation

17.1 Introduction

Microbial fuel cell (MFC) technology when first discovered was known almost exclusively for electricity generation, but later, its applications to biosensing have attracted the attention of many researchers, primarily for mainly toxicity analyses [1, 2] and biochemical oxygen demand (BOD; [3, 4]. A few reports have also been published in relation to analysing specific substrates such as lactate, for example, the work of Kim et al. [5] and volatile fatty acid (VFA) by Kaur et al. [6].

G. C. Premier (✉) · A. Kaur · R. M. Dinsdale · A. J. Guwy
Sustainable Environment Research Centre (SERC), Faculty of Computing Engineering and Science, University of South Wales, Pontypridd, Mid-Glamorgan CF37 1DL, UK
e-mail: gcpremier@glam.ac.uk

A. Sayigh (ed.), *Renewable Energy in the Service of Mankind Vol I*,
DOI 10.1007/978-3-319-17777-9_17

In previous studies, an MFC-based VFA sensor using cyclic voltammetry (CV) as one of the methods was introduced by Kaur et al. [6]. The methods were improved later by using immobilization of microbial community with natural and conductive polymers [7]. In CV-based sensor, the measured parameter is oxidation potential of an electrochemical reaction. The working principle relies on the fact that when a voltage is applied to an electrode in solution, a current flow occurs because of an electrochemical reaction. The voltage at which these reactions occur is indicative of a particular reaction and/or particular species [8]. However, for a sensor to meet the expected criteria of a robust deployable VFA sensing, the range and selectivity required improvement.

In an MFC, the biocatalyst present in the anodic chamber influences several aspects of the overall performance. Anode potential is one of the important factors controlling the synergistic interactions within and between microbial populations and current flow as it can influence the electron-liberating capacity of the biocatalyst [9, 10]. Setting the anode potential to a fixed voltage using a potentiostat in bioelectrochemical system (BES) can exert effects on the system performance, bacterial growth and/or composition of consortia [9, 11]. The positive effects of setting anode potential have already been presented in relation to pure cultures with ranges 0.095–0.595 V (vs. Standard Hydrogen Electrode (SHE), *Shewanella putrefaciens*) [12], −0.16–0 V (vs. SHE, *Geobacter sulfurreducens*) [13] and 0.27–0.32 V (vs. SHE, *Pseudomonas putida* F1; [14]. While the use of set potentials for efficient performance at bioanodes has received significant attention in the context of improving start-up time [9, 15, 16], and power generation and bioelectrochemical behaviour [14, 17], only a limited number of reports have been published regarding the application of set potentials to improve the degradation of recalcitrant organics, and likewise, the importance of set potentials in measuring different substrates has not been published so far for MFCs-based sensors. This study presents results that indicate an improved range for the sensor, achieved by using specific poised potentials on the basis of oxidation peaks obtained from CVs, in relation to a set of specific VFA-enriched MFC sensors. Offline testing of samples collected from a scaled-up MFC running on sucrose and an electrodialysis (ED) cell with sucrose and grass separately as feed was also undertaken to investigate if the industrial implementation of the sensor could be plausible.

17.2 Material and Methods

17.2.1 MFC Sensor Set-Up

Three cubic MFCs were constructed from Perspex material with external dimensions (50 × 50 mm) to have a single cylindrical chamber with internal diameter of 25 mm. The thickness of the front and outer cover was 6 mm, and the middle chamber was 14.20-mm thick. The three individual parts were joined together using aluminium screws to hold the sandwiched assembly, including the membrane electrode assemble (MEA) of anode, cation exchange membrane (CMI-7000, Membrane International Inc. NJ, USA) and cathode. For anode material, a plain carbon cloth (BASF Fuel Cell Inc., NJ, USA) was used, with an exposed surface area of 4.9 cm^2. The cathode was prepared on plain carbon cloth (CCP40, Fuel Cell Earth LLC, MA, USA) as described by Cheng et al. [18]. The system was sealed to make it watertight, giving a total liquid reactor volume in the anode chamber of 10 ml. These MFCs were enriched with 0.8 ml of sludge using 20 mM of acetate, propionate and butyrate as electron donor and 9.2 ml of media with composition as described previously by Kim et al. [19].

17.2.2 Calculations

Current (I) and power ($P = IV$) were calculated as described in the previous report [20]. The current (Cs^{-1}) was calculated according to $I = V/R$, where V is the measured voltage and R is the resistance. The percentage error was calculated for comparing the MFC-measured responses to the gas chromatograph (GC) responses obtained from various samples as follows:

$$\text{Error} = (C_{\text{approx}}) - (C_{\text{exact}}), \tag{17.1}$$

$$\text{Error}\ \% = C_{\text{approx}} - C_{\text{exact}} / C_{\text{exact}} \times 100, \tag{17.2}$$

where C_{approx} is the concentration from MFC sensor array and C_{exact} is the concentration from GC.

17.2.3 Electrochemical and Chemical Analysis

The poised potential and CV experiments were carried out to characterize the oxidation–reduction reaction of VFAs on the anode biofilm by using PC 4/750 potentiostat, Gamry. A conventional three-electrode system was employed with the anode as the working electrode, the cathode (2.5 cm^2, 0.35 mg/cm^2; 10% Pt; E-Tek, NJ) as the counter electrode and an Ag/AgCl reference electrode as shown in Fig. 17.1.

The potentials were shifted from −750 to +750 mV at a scan rate of 50 mV/s while monitoring the current response. The voltage and current across the load were monitored using a virtual instrument-based data logging system (National Instruments, LabVIEW™) connected to an analogue input on the input/output (I/O) card, respectively. Acetate, butyrate and propionate were measured using a Perkin Elmer headspace gas chromatograph (model number HS40XL, Perkins Elmer, Waltham, Massachusetts, USA) in conjunction with a flame ionization detector (FID) and a Nukol free fatty acid phase (FFAP) column (30 m × 0.32 mm; Supleco) running at 190 °C and 14 psi with nitrogen as a carrier gas according to the method by Cruwys et al. [21].

Fig. 17.1 Experimental set-up during cyclic voltammetry and anode poised potential with Ag/AgCl as reference electrode saturated in KCl. *PEM* proton exchange membrane

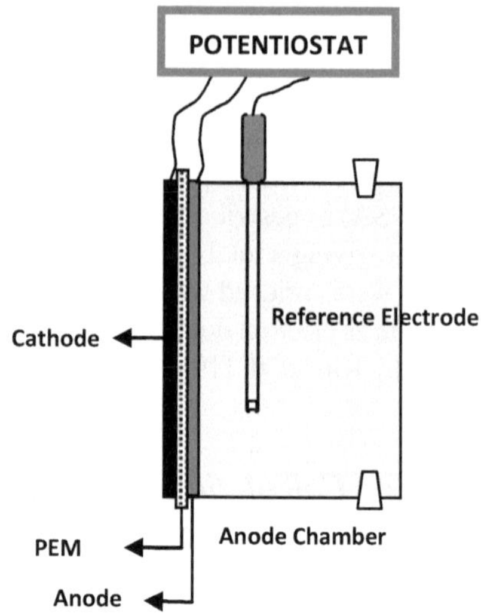

17.3 Results and Discussion

17.3.1 Anode Posed Potential Investigation

CV tests were conducted in previous study by Kaur et al. [6] to demonstrate that current generation was coupled to VFA oxidation and to examine if the concentration of the substrate can be detected from the current produced. It was concluded that CV can be used to measure and speciate VFAs in MFCs for a limited range of concentrations. The characteristic shapes of the cyclic voltammogram obtained were very distinctive for each VFA-enriched MFC. The peak potential for acetate- and propionate-enriched reactors (AMFC and PMFC, respectively) was consistent (i.e. −0.260 and −0.220 V) with ± 30 mV of peak shift. In contrast to AMFC and PMFC, butyrate-enriched reactor (BMFC) showed a shift of peak current potential ranging between ± 0.100 V. This potential shift is believed to be because of the degradation of butyrate to short-chain fatty acids as the BMFC sensor gives a current response to all three VFAs. The potential for BMFC was then selected on the basis of peak potential measured just before the calibration of the sensor, and the potential was −0.100 V. The selection of potentials was based on the experimental results, as theoretically, the redox potential for all the VFAs lies between −0.290 and −0.280 mV. Figure 17.2 shows the different oxidation peak potentials in AMFC, PMFC and BMFC, that is, −260, −220 and −100 mV, respectively.

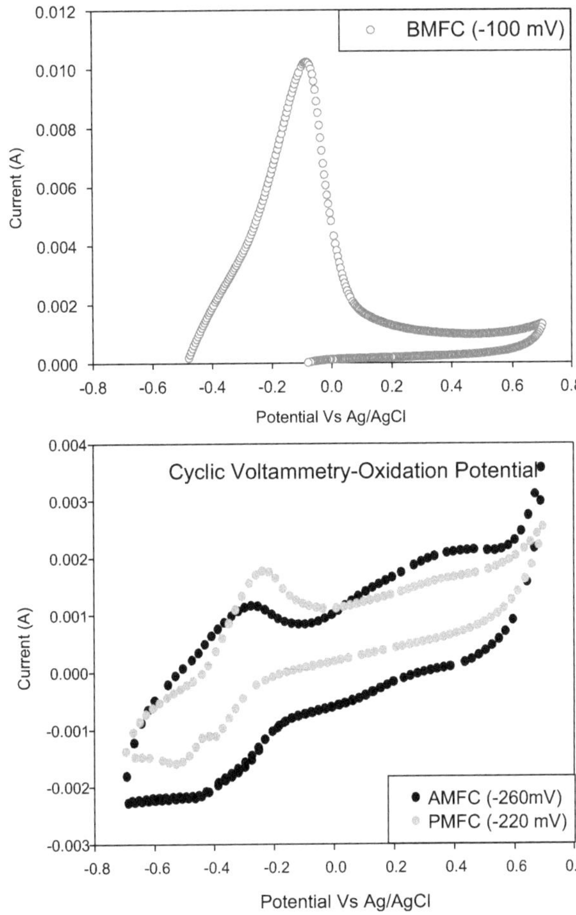

Fig. 17.2 Oxidation potential selection for individual VFA by using cyclic voltammetry. *AMFC* acetate microbial fuel cells, *PMFC* propionate MFC, *BMFC* butyrate microbial fuel cells

17.3.2 Calibration with Standard and Mixed VFA

After enrichment and oxidation peak potential selection, the constant voltage/current was maintained, operating the MFCs in fed-batch mode with 5 mM substrate concentration. The calibration of individual VFA-enriched sensors was done by using standard concentrations of respective VFA ranging from 0 to 200 mg/L. Before calibration, reactors were shifted to the anaerobic chamber to keep the anode environment anaerobic and also to maintain the temperature at 30 °C. The anode chamber was rinsed with phosphate buffer and then refilled with media without any substrate in it. The current was stabilized for 1 h at preselected and respective poised potential for the individual VFAs. This stable current was taken as a baseline current. Then, concentration of the respective substrate was increased stepwise and measured for 10 min to achieve the stable current. Before the anolyte was refreshed in each measurement, the anode chamber was washed using deionized (DI) water

Table 17.1 The R^2 and range achieved from calibration of individual VFA-enriched MFC using specific poised potential

MFC	VFA	R^2	Range (mg/L)	Equation of straight-line fit	Static sensitivity
AMFC	Acetate	0.9833	20–160	$y=0.0009x+0.1731$	0.1731
PMFC	Propionate	0.9833	20–140	$y=0.0012x+0.5098$	0.5098
BMFC	Total VFA	0.9955	30–250	$y=0.0019x+0.0164$	0.0164

VFA volatile fatty acid, *MFC* microbial fuel cell, *AMFC* acetate MFC, *PMFC* propionate MFC, *BMFC* butyrate MFC

to remove the organic matter that may have remained on the chamber wall. The current responses at the fixed working electrode potentials were analysed for each VFA sensor in response to different concentration (0–200 mg/L) of the corresponding and other VFA species at the appropriate poised potentials. When corresponding VFAs were supplied to the sensors, the range of the sensor was increased from 40 to 200 mg/L with no response to other VFA species in the case of AMFC and PMFC. This implies that the specific poise potentials also retained the selectivity as previously observed by Kaur et al. [6] and a further improved applicability of the sensor. The range achieved from individual VFA-enriched MFC and R^2 values is summarizedin Table 17.1 with the equation of linear fit and respective static sensitivities also included.

17.3.3 Measurement of Samples in MFC Sensor

A study of the application of biosensors to the real world is necessary before moving toward the ultimate ambition for any such device's commercialization. Keeping in mind the application of the sensor presented in this study, we considered the samples from a tubular MFC and an ED cell running on complex substrates. A total of ten samples were collected from tubular MFC and ED cell and were analysed for their VFA content through GC. The maximum and minimum concentrations of mixed VFA in all samples are summarized in Table 17.2. In order to test the ability of the MFC sensor array to measure mixed VFAs contained in samples, the specific potential was poised for the individual MFC using miniaturized Ag/AgCl saturated in KCl as reference. The sensor responses were checked and compared against the GC results. The results obtained from individual MFC sensor device on addition of these samples are discussed in the following sections:

17.3.3.1 AMFC and PMFC Sensor Response

Most of the samples obtained from the sucrose-fed tubular MFC and ED cell were within the range of AMFC and PMFC and were not diluted further for measurement. All samples were measured twice, and the mean value of current was plotted on a calibration curve, analysed at the same time and under the same conditions, to derive a value of equivalent target compound concentrations. Table 17.2 summarizes the

Table 17.2 Comparison between GC and MFC responses

Sample number	Source	Ac	Pr	i+n Bu	TVFA (1:2) dilution	AMFC	%Error	PMFC	%Error	BMFC	%Error
Gas chromatography response (mg/L) MFC responses (mg/L)											
1	MFC S1	40	0	180	220 (110)	32	20 %	< 10	–	140	−21.4
2	MFC S2	56	24	200	270 (135)	44	21.4	< 20	–	150	−10
3	MFC S3	62	20	240	332 (166)	54	12.9	< 20	–	80	51.8
4	MFC S4	84	20	158	262 (131)	68	19.04	< 20	–	100	23.6
5	MFC S5	100	20	152	272 (136)	75	25	< 20	–	100	26.5
6	ED S1	62	32	242	336 (168)	88	−29	24	25 %	98	41.6
7	ED S2	0	0	72	72	<20	–	–	–	104	−30.7
8	ED S3	0	0	36	36	<20	–	–	–	62	−48.4
9	ED S4	148	68	76	282 (141)	134	9.45	40	41.7 %	172	−18
10	ED S5	72	72	144	288 (144)	98	−26.5	58	19.4 %	189	−28.8

GC gas chromatography, *MFC* microbial fuel cell, *Ac* acetate, *Pr* propionate, *Bu* butyrate, *TVFA* total volatile fatty acid, *AMFC* acetate MFC, *PMFC* propionate MFC, *BMFC* butyrate MFC typical responses obtained from AMFC with respect to each sample. The maximum and minimum concentrations of acetate in the samples were 148 and 0 mg/L, respectively. For zero concentration, the response was < 20 mg/L on the calibration curve, which indicates a low amount of acetate in the sample, possibly due to the degradation of the butyric acid in the samples. Both concentrations gave very close correlations, with < 10 % error when compared with GC response, even in the presence of other VFAs. This shows that AMFC could be a promising candidate for measuring acetate concentrations in low VFA concentration samples. The propionate sensor was then tested for measuring propionate concentrations within the samples and also the effect of other VFAs in the samples. In contrast to AMFC, the concentration for propionate in all samples was very low, that is, < 70 mg/L with a maximum in the sample from ED S10, and the minimum was 0 mg/L. When measured in c-PMFC, these samples showed responses < 50 mg/L. Table 17.2 shows the responses obtained from MFC- and ED-derived samples in PMFC. Although the concentrations were not measured with high accuracy compared to the GC, no higher current was generated even in the presence of high butyric acid concentrations. This shows that the propionate sensor was specific to its corresponding VFA and might be inhibited by other unknown factors, resulting in a lower response. It is important to analyse the system performance with a larger number of samples with higher concentrations of propionate to assess how accurate the quantification of the target compound is.

17.3.3.2 BMFC Sensor Response

The samples were diluted in the ratio of (1:2) wherever required, with the same buffer solution that was used during the calibration. As BMFC responded to all VFAs, it was calibrated for both butyrate and total VFA (TVFA; Ac + Pr + Bu) concentrations with increasing and equal proportions of each VFA. Where AMFC and PMFC showed comparable results to GC measurements, it was found that in BMFC, it was difficult to determine the accurate concentration of the VFAs. The current responses obtained in BMFC were higher than expected in almost all samples. The current responses obtained indicate a low accuracy and repeatability of the measurement outputs from BMFC. In Table 17.2, the highlighted negative values of %error show the greater responses and high % error with respect to GC responses. As the butyric acid-enriched sensor always gives a higher response to acetate and propionate concentration as opposed to the butyric concentration, the higher current responses were in fact expected. The synergistic effect of combined VFAs in samples affected the BMFC more than AMFCs and PMFCs. The reliability of the sensor response decreases when butyric concentration in samples is ≥ 100 mg/L. In sample MFC S3 and ED S1, the concentration of the butyric acid was > 100 mg/L, which resulted in the randomly decreased current output with high %error.

This decrease might be due to the maximum detection limit of the BMFC for butyric acid, beyond which the concentration inhibits the microbial respiration. It is expected that anaerobic processes in, for example, still waters, sewers and primary waste waters, industrial wastewaters and other diluted environments may usefully be monitored for VFAs at low concentrations. Also, the application of such sensors at very low concentration in, for example, BES such as microbial electrolysis or desalination, or MFCs able to reduce concentrations to low levels of VFAs, could be important. The control of these devices or indeed aerobic or membrane effluent polishing systems could all benefit from the measurement of VFAs in or close to the ranges presented. In the present state of the sensor, it is recommended that the calibration should be done frequently to extract the more useful data. The calibration methods can be improved according to the implementation of the sensor system and the electrochemical method used for the analysis. Further detailed analysis of the bacterial communities involved should be considered for better understanding of the sensor signals. This would also help in understanding the detailed mechanisms behind the electron transfer between different bacterial communities and the electrode on the addition of individual VFA. Appropriate data processing in order to disaggregate coupled information should be considered in order to deliver VFA species-specific measurements from mixed samples. Given existing advances in biological sciences, and coupled with advances in various other scientific and engineering disciplines, it is plausible that existing VFA measurement methods can be replaced by MFC-based VFA sensor.

17.4 Conclusions

The results obtained showed a behaviour which might support interpretation from the developed array of VFA biosensors, when applied to environmental applications. The MFCs were able to generate significant responses, although complexity in analysing the samples may present interferences which defy simple analysis. The range of the sensor was improved from 40 to 200 mg/L. Where the acetate and propionate were poised 10 ml, MFCs sensors gave comparable results; the butyrate sensor needs further calibration, and deeper study is required to understand the complex chain of mechanisms. More analyses of the VFA biosensor are recommended, submitting it to different environmental samples. Convincing data have been produced which demonstrate a degree of robustness, selectively good measurement accuracy (in AMFC and PMFC) when compared with GC analysis. It is proposed with some confidence that the presented study on VFA sensing will contribute to understanding the place of MFC technology in the sensing of VFAs.

Acknowledgements This research was funded by the RCUK Energy Programme, SUPERGEN Biological Fuel Cell Consortium project (EP/H019480/1). The Energy Programme is an RCUK cross-council initiative led by EPSRC and contributed by ESRC, NERC, BBSRC and STFC.

References

1. Patil S, Harnisch F, Schroder U (2010) Toxicity response of electroactive microbial biofilms—a decisive feature for potential biosensor and power source applications. ChemPhysChem 11(13):2834–2837
2. Stein NE, Hamelers HMV, van Straten G, Keesman KJ (2012) On-line detection of toxic components using a microbial fuel cell-based biosensor. J Process Control 22(9):1755–1761

3. Chang IS, Jang JK, Gil GC, Kim M, Kim HJ, Cho BW, Kim BH (2004) Continuous determination of biochemical oxygen demand using microbial fuel cell type biosensor. Biosens Bioelectron 19(6):607–613
4. Di Lorenzo M, Curtis TP, Head IM, Velasquez-Orta SB, Scott K (2009) A single chamber packed bed microbial fuel cell biosensor for measuring organic content of wastewater. Water Sci Technol 60(11):2879–2887
5. Kim HJ, Moon Sik Hyun, In Seop Chang, Kim BH (1999) A microbial fuel cell type lactate biosensor using a metal-reducing bacterium, Shewanella putrefacians. J Microb Bio 9(3):365–367
6. Kaur A, Kim JR, Michie I, Dinsdale RM, Guwy AJ, Premier GC (2013) Microbial fuel cell type biosensor for specific volatile fatty acids using acclimated bacterial communities. Biosens Bioelectron 47(0):50–55
7. Kaur A, Ibrahim S, Pickett CJ, Michie IS, Dinsdale RM, Guwy AJ, Premier GC (2014) Anode modification to improve the performance of a microbial fuel cell volatile fatty acid biosensor. Sens Actuators B Chem 201(0):266–273
8. D'Souza SF (2001) Microbial biosensors. Biosens Bioelectron 16(6):337–353
9. Aelterman P, Freguia S, Keller J, Verstraete W, Rabaey K (2008) The anode potential regulates bacterial activity in microbial fuel cells. Appl Microbiol Biotechnol 78(3):409–418
10. Schroder U (2007) Anodic electron transfer mechanisms in microbial fuel cells and their energy efficiency. Phys Chem Chem Phys 9(21):2619–2629
11. Kumar A, Siggins A, Katuri K, Mahony T, O'Flaherty V, Lens P, Leech D (2013) Catalytic response of microbial biofilms grown under fixed anode potentials depends on electrochemical cell configuration. Chem Eng J 230(0):532–536
12. Carmona-Martinez AA, Harnisch F, Kuhlicke U, Neu TR, Schroder U (2013) Electron transfer and biofilm formation of *Shewanella putrefaciens* as function of anode potential. Bioelectrochemistry 93:23–29
13. Wei J, Liang P, Cao X, Huang X (2010) A new insight into potential regulation on growth and power generation of *Geobacter sulfurreducens* in microbial fuel cells based on energy viewpoint. Environ Sci Technol 44(8):3187–3191
14. Friman H, Schechter A, Nitzan Y, Cahan R (2012) Effect of external voltage on *Pseudomonas putida* F1 in a bio electrochemical cell using toluene as sole carbon and energy source. Microbiology 158(2):414–423
15. Chou T-Y, Whiteley CG, Lee D-J (2014) Anodic potential on dual-chambered microbial fuel cell with sulphate reducing bacteria biofilm. Int J Hydrogen Energy 39(33):19225–19231
16. Huang L, Chai X, Chen G, Logan BE (2011) Effect of set potential on hexavalent chromium reduction and electricity generation from biocathode microbial fuel cells. Environ Sci Technol 45(11):5025–5031
17. Wagner RC, Call DF, Logan BE (2010) Optimal set anode potentials vary in bioelectrochemical systems. Environ Sci Technol 44(16):6036–6041
18. Cheng S, Liu H, Logan BE (2006) Increased performance of single-chamber microbial fuel cells using an improved cathode structure. Electrochem Commun 8:489–494
19. Kim JR, Premier GC, Hawkes FR, Rodriguez J, Dinsdale RM, Guwy AJ (2010) Modular tubular microbial fuel cells for energy recovery during sucrose wastewater treatment at low organic loading rate. Bioresour Technol 101(4):1190–1198
20. Min B, Kim J, Oh S, Regan JM, Logan BE (2005) Electricity generation from swine wastewater using microbial fuel cells. Water Res 39(20):4961–4968
21. Cruwys JA, Dinsdale RM, Hawkes FR, Hawkes DL (2002) Development of a static headspace gas chromatographic procedure for the routine analysis of volatile fatty acids in wastewaters. J Chromatogr A 945(1–2):195–209

Chapter 18
The Application of Solar-Powered Polymer Electrolyte Membrane (PEM) Electrolysers for the Sustainable Production of Hydrogen Gas as Fuel for Domestic Cooking

Evangelia Topriska, Maria Kolokotroni, Zahir Dehouche, Ruth Potopsingh, Earle Wilson

Abstract Hydrogen generation systems coupled with renewable energy systems such as photovoltaic panels can have a role as sustainable energy generation systems for independent applications that can also have a significant part in tackling energy poverty. This chapter examines a system that consists of a polymer electrolyte membrane (PEM) electrolyser powered by photovoltaic panels with the purpose to use the produced hydrogen as cooking fuel for domestic applications. It describes a small-scale experimental study of a PEM electrolyser powered by operational data from a photovoltaic panel system installed in Brunel University, UK, and the University of Technology, Kingston, Jamaica. The results show that maximum efficiency of the electrolyser was 62.8 % for a day with constantly high irradiance and 47 % for a day with volatile irradiance.

Keywords Hydrogen · Electrolysis · Photovoltaic · Cooking in developing countries

18.1 Introduction

Over 2.5 billion people depend on firewood, charcoal, agricultural and animal waste to satisfy their energy needs for cooking. The use of these fuels can have a negative impact on health. Statistics show that death related to indoor air pollution from the use of solid fuels is rated fourth after malnutrition, human immunodeficiency virus infection and acquired immune deficiency syndrome (HIV/AIDS) and lack of clean water, and is responsible for more fatalities than malaria and almost same as tuberculosis [1]. Subsequently, many developing countries have turned to the

E. Topriska (✉) · M. Kolokotroni · Z. Dehouche
Brunel University, Uxbridge, UK
e-mail: Evangelia.Topriska@brunel.ac.uk

R. Potopsingh · E. Wilson
University of Technology, Kingston, Jamaica

A. Sayigh (ed.), *Renewable Energy in the Service of Mankind Vol I*,
DOI 10.1007/978-3-319-17777-9_18

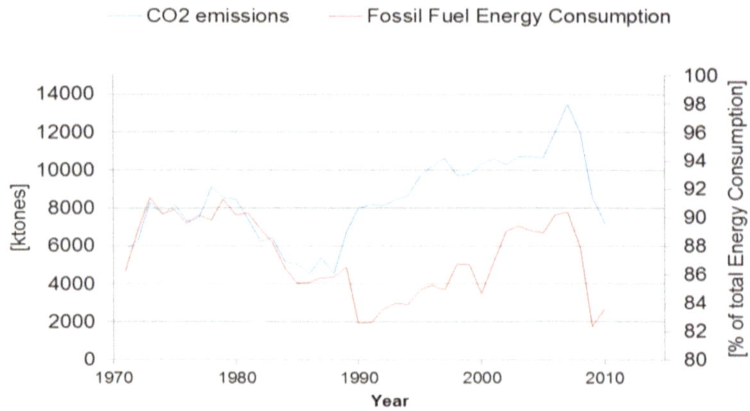

Fig. 18.1 CO_2 emissions in relation to fossil fuel consumption in Jamaica. [5]

importation of petroleum-based fuels as their main energy source for cooking. The affordability and volatility of oil prices has proven to have a negative impact on the economic and social development of these countries.

This chapter presents work focussed on Jamaica. Jamaica is a Caribbean State of the Great Antilles with a population of 2.8 million. It is a country well endowed with natural resources, mainly solar and hydro, but nevertheless has a high dependency (over 90 %) on imported petroleum fuels for meeting its energy needs [2]. In particular, cooking is one of the activities highly affected by the instability of the energy supply, as its main source is liquefied petroleum gas (LPG). The import of oil represents more than 30 % of the value of all imports and more than 100 % of export earnings [3]. This situation, in combination with the low technology used in the oil refinery stations and by end users results in high cost of energy generation and distribution.

Eighty-six percent of the Jamaican households use LPG for cooking, with this percentage being further distributed into 92.8 % for urban households and 78.6 % for rural households [3]. In 2011, the country spent US$ 2.2 billion on the importation of crude oil and petroleum products. Households that cannot afford LPG, use firewood and charcoal; their use is related to high CO_2 emissions, respiratory problems and deforestation [3]. Charcoal is produced from cutting and burning of trees. This is a non-managed process, as the trees cut by charcoal burners are not replanted, resulting in high rates of deforestation. In the past 20 years, the forest area in Jamaica has steadily declined by almost 0.1 % per year, leading to a 2.5 % reduction which corresponds to a total of over 250 km^2 forest area [4]. The wood is usually converted to charcoal in outdated, inefficient charcoal furnaces, which visibly pollute the atmosphere, with smoke, particulates, CO_2 and other gases. Figure 18.1 shows that there is a trend between CO_2 emissions and fossil fuel consumption.

This condition intensifies further in periods of economic crisis and in periods of high volatilities in oil prices. Deforestation is of particular concern in Jamaica because of the frequency of extreme climate-related events like droughts, flooding, tropical storms and hurricanes that the Caribbean region has experienced in recent times. These events cause environmental impacts such as loss of soil in agricultural areas, silting of the oceans and loss of biodiversity.

The energy security of the country is susceptible to the global oil prices and availability. In the past Jamaica has suffered greatly from periods of oil crisis, and there have always been efforts to become independent from oil imports. Within this context, the Government of Jamaica engaged into the national energy policy 2009–2030, which seeks to create an efficient, diversified and environmentally sustainable energy sector [6].

Within this framework, this research project promotes sustainability in an important subdivision of the energy sector, which is fuel for cooking and in which renewable energy use is usually ignored. The main objective of this research is to investigate the application of a novel stand-alone solar-powered polymer electrolyte membrane (PEM) electrolyser plant to produce hydrogen gas at a scale suitable for storage and distribution to multiple households for cooking. Producing electrolytic hydrogen using solar energy is a sustainable process to provide an energy system with very little environmental cost [7]. The outcomes of the project implementation would be: increase the use of renewable energy resources, reduce respiratory problems and deforestation caused by the cutting of trees and charcoal burning, contribute to the reduction of the energy import bill for Jamaica, make more resources available to satisfy other basic socioeconomic needs as well as de-risk energy dependency by diversifying the energy source in Jamaica and any other countries that adopt the technology.

18.2 Methodology

18.2.1 Photovoltaic-Powered Electrolyser System Description

The technical challenge of this research project is to optimise the efficiency of a stand-alone solar powered PEM electrolyser plant powered by photovoltaic (PV) panels to produce hydrogen gas at a scale suitable for distribution to households. The hydrogen is produced at regulated pressure suitable for storage without the need of a compressor. This is important in terms of energy savings in the hydrogen storage process as the energy required for the compression process is avoided. Moreover, hydrogen storage for distribution for domestic usage requires pressure at approximately 10 bar and less, which is within the pressure range of commercial PEM electrolysers. Apart from this, PEM electrolysers present a variety of advantages that set them highly suitable for use in renewable energy systems: (i) they consist of a proton exchange polymeric membrane for the separation of gases and therefore no toxicity is included as in the case of aqueous potassium hydroxide (KOH) of alkaline electrolysis, (ii) they prevent very efficiently mixing of gases and thus offer a safe hydrogen generation process, (iii) they offer fast system response to variable power supply [8].

Storage of hydrogen presents a challenge because of its low volumetric energy density (8491 MJ/m^3). In comparison, gasoline has a volumetric energy density of 31,150 MJ/m^3. On a weight basis however this is opposite and hydrogen has much higher energy density than gasoline (140.4 MJ/kg contrasted with 48.6 MJ/kg) [9]. Therefore, to achieve high gravimetric capacity in a hydrogen storage system, high pressure or high volume is required. The most common and traditional hydrogen storage methods are high-pressure storage of hydrogen gas and liquid storage. Liquid storage has the drawback of a refrigeration unit that is necessary to maintain hydrogen at cryogenic state, thus adding extra energy costs and complicating the system. High-pressure storage is restricted by the cylinders weight and the possibility of leaks and therefore safety issues. Metal hydrides present higher energy density (6.5 H atoms/cm^3 for MgH_2) than compressed hydrogen gas (0.99 H atoms/cm^3) and liquid hydrogen (4.2 H atoms/cm^3) [10]. In this context, the storage of hydrogen in metal hydride form presents a low-pressure safe and promising solution for stationary applications, such as domestic use, and will be examined within this project in the future. Another low-pressure storage option is the cascading compressed gas storage, where hydrogen can be stored in 13 bar tanks and is regulated down to 3 bar and stored in low-pressure cylinders suitable for household distribution. The low-pressure hydrogen will then be fed to a regular gas stove that is specially adapted for hydrogen use.

18.2.2 Experimental Setup Overview

The proposed system consists of two parts: (a) photovoltaic modules as a primary electricity source of a PEM electrolyser which produces hydrogen gas at the required pressure for storage and (b) storage and burner for cooking. This chapter focusses on the first part of the system. Two experimental rigs have been constructed (one in Jamaica and one in the UK) each including a PEM electrolyser, powered by a 1.5-kWp and 1.765-kWp photovoltaic installation respectively for the two locations. The electrolyser used in this study consists of four-cell stack unit and it is rated at a maximum hydrogen production of 600 cc/min at current of 30 A DC. It is a compact commercial unit, and apart from the electrolyser stack it includes some supplementary features that provide the necessary management of the produced gases and water. The water used is deionised with a conductivity of less than 1 μS/cm, is inserted into a water tank inside the electrolyser and is circulated by a small pump. It flows through the stack where it is separated into hydrogen and oxygen. Oxygen is vented to the ambient environment and hydrogen passes through a drying and purification module that removes the excess water. Hydrogen gas exits through a back-pressure control valve at pressures within the range of 3–13.8 bar. This internal system is powered and controlled by a DC power supply unit that provides the necessary power to each component.

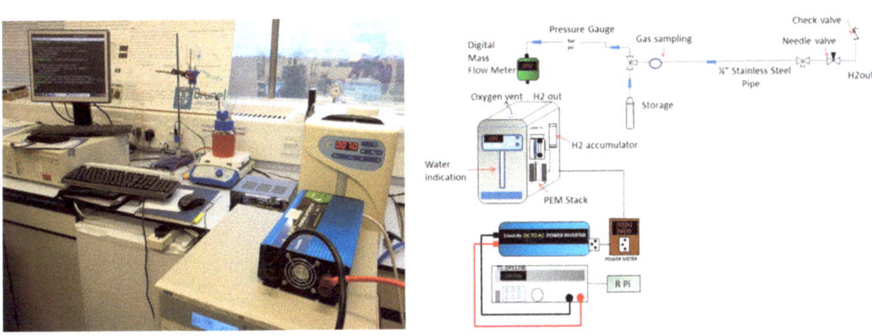

Fig. 18.2 Experimental setup in the UK laboratory

18.2.2.1 UK Experimental Rig and Experimental Process

The experiments have a dual focus: (i) to evaluate the electrical performance of the whole electrolyser unit as well as the electrochemical behaviour of the stack itself and (ii) to assess the operation of the electrolyser and the balance of plant when it is coupled to a PV system. The amount of hydrogen produced under different irradiance levels and therefore power output of PV panels, as well as under the distinctive operational conditions of the electrolyser, is analysed and the system performance evaluated.

The electrolyser is rated at 1200 W, 110–240 V, and 6 A single-phase AC supply. For the purpose of the experiments, it is powered by a 1500-W, 240-V AC output and 24 V DC input, pure sine output inverter. The inverter is in turn connected to a DC power supply unit, rated at 1200 W, 60 V and 50 A maximum, within a power envelope. The DC power supply unit is controlled by digital signals sent from a Raspberry Pi computer through a 12-bit digital to analogue converter. This setup (Fig. 18.2) was constructed to provide the flexibility to simulate the operation of PV panels and to test for different sets of irradiances in the laboratory. The inverter has a measured efficiency varying between the values of 86–93 %, depending on the fraction of power that goes through and the AC power that is used.

Irradiance data from Brunel University's weather station and power data from a 1.765-kW photovoltaic array installed by Brunel Institute of Power Systems (BIPS) of Brunel University has been used. The array consists of four Sanyo HIT 210 W modules of 16.8 % efficiency and five Sharp monoSi 185 W modules of 14.1 % efficiency. The characteristics of the modules are presented in Table 18.1:

Data selection for the conduction of the experiments is a critical part of the process. After taking into consideration 3 years of recorded daily irradiance levels, from 2010 to 2013, one specific week was selected. This was the first week of June 2013. June is the month with the highest irradiance in the UK and the selected set of days offer many hours of sunshine and irradiance levels that correspond to the highest PV power in the year. As the application of the system is in Jamaica, the tests should emulate conditions that correspond to high solar irradiance. Additionally, the selected data offer days with clear sky and therefore high irradiance throughout the day as well as days with unstable cloud coverage, where the output of the PV panels

Table 18.1 Electrical and thermal characteristics of the PV modules of the experimental PV emulator

PV type	Sanyo heterojunction with intrinsic thin layer (HIP)-210NHE1	Sharp MonoSi NUS5E3E/NU185E1
Nominal power output (P_{MP})	210	185
V_{MP} (V)	41.3	24
I_{MP} (A)	5.09	7.71
V_{OC} (V)	50.9	30.2
I_{SC} (A)	5.57	8.54
Temperature coefficient of power (%/°C)	−0.30	−0.485

PMP: Power at Maximum Power Point
VOC: Open Circuit Voltage
ISC: Short Circuit Current

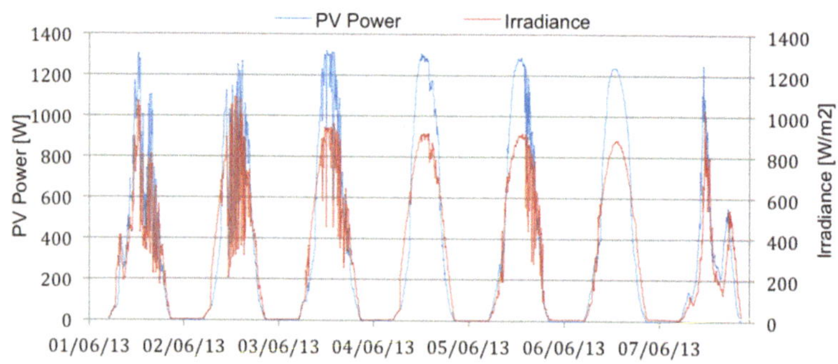

Fig. 18.3 Power output data of the PV array for the first week of June 2013.

varies significantly. Figure 18.3 shows the irradiance and power output values of the PV array for the first week of June 2013 that is used and analysed as input to the DC power supply unit.

For the tests, the DC power supply is set at 24 V constant to match the inverter specifications and the current changes accordingly so that the unit operates as the PV array. This is controlled by a Raspberry Pi computer which works on a Linux operational system. Python programming is used and a code is developed that serves the purpose of sending signals to the DC power supply unit, so that the whole operation can be automated and simulate real conditions.

The data that is sent from the Raspberry Pi correspond to current values in Amperes. The current values derive from analysis on the power data of the PV panels based on the fact that for the experiments a 24-V pure sine output inverter was used. The process is described by Eq. (18.1):

$$I[A] = \frac{P[W]}{\eta \times V[V]}, \tag{18.1}$$

where,

P	inverter power output (W)
η	efficiency of the inverter
$V = 24$ V	the voltage of the inverter
I	current (A)

Therefore, for each of the PV power values there is a current value derived from Eq. (18.1) that is used as an input in the developed code.

The experimental setup consists of various measuring instruments for recording the power supply and hydrogen production. A 1000-ml/min digital mass flow meter is used for the recording of the hydrogen flow rate and the rates are recorded through a 4–20-mA analogue signal sent to the computer. An AC power analyser and a current recorder are used to analyse the energy use in the electrolyser. The electrolyser is supplied through the inverter, the operation is monitored and the recorded data is: (a) electrolyser power consumption, (b) electrolyser response to the intermittent and variable power supply and (c) hydrogen production so that a production pattern in relation to irradiance is created.

18.2.2.2 Jamaica Experimental Rig and Experimental Process

Experiments are also being conducted in the University of Technology, Jamaica, Kingston and they focus on the storage and cooking aspect of the system. The same commercial electrolyser is used and it is connected to a 1.5-kWp PV system. The electrolyser is powered through a battery and is connected to a small hydrogen storage cylinder, through a cascading storage system, which is in turn connected to a modified hydrogen gas test stove.

The cascading storage system receives the produced hydrogen from the electrolyser at elevated pressure and stores it in a high-pressure cylinder. This high-pressure cylinder is connected to a low-pressure cylinder through a pressure reducing valve, where gas is stored at pressure levels acceptable for domestic use (up to 13 bar). The low-pressure cylinder supplies the modified hydrogen stove with gas and testing includes evaluation of the charging times of the low-pressure cylinder and storage of hydrogen gas at 13 bar, the gas supply process to the modified stove and the use of hydrogen as a cooking gas.

18.3 Results

The experiments give a solid understanding of the electrical behaviour of a commercial PEM electrolyser to enhance subsequent modelling and integration of these electrolysers in renewable energy systems applications. The electrolyser is a compact commercial unit which can produce hydrogen gas at user set pressures (within the range of 3–13.8 bar). Under different pressure set points the flow rates are also

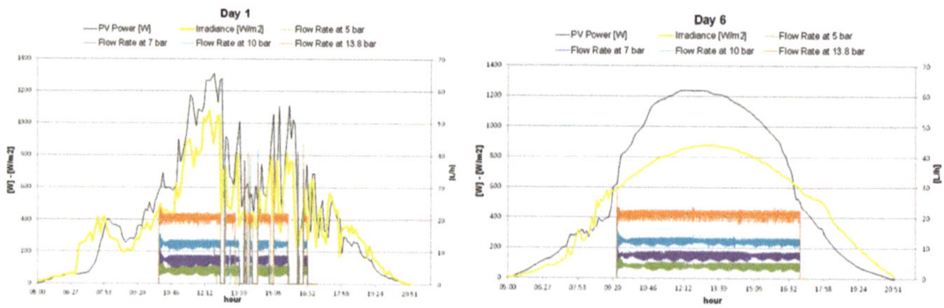

Fig. 18.4 Hydrogen production rate in L/h at different pressure set points during day 1 that presents unstable irradiance levels and during day 6 that has consistently high irradiance levels.

different. Depending on the desired pressure of the produced hydrogen gas the flow rate is regulated. At higher pressures, the flow rate is higher so that the necessary pressure is achieved at the exit of hydrogen gas. The controls of the unit regulate the DC current to the stack and thus the volumetric hydrogen production changes accordingly, in agreement with Faraday's first law of electrolysis [11], as can be seen in Eq. (18.2):

$$V_{H_2} = \int_0^t \frac{\eta_F * n_c * R * I * T}{F * p * n} dt, \tag{18.2}$$

where, I is the current to the electrolyser (A),n_c is the number of cells, η_F is the Faraday efficiency of the electrolyser,T is the temperature (K),p is the atmospheric pressure,R is the ideal gas constant,n is the number of electrons which is 2 andF is a Faraday constant (96,485 C/mol).

Additionally, the amount of generated hydrogen varies according to the irradiance levels of each day. In days with a clear sky and constant high irradiances the production rate is higher for the same current to the stack. The energy consumption and efficiency of the electrolyser stack is measured and evaluated under different PV panels operating conditions. Figure 18.4, shows the hydrogen production trend at four selected pressure points and for 2 days of the week selected for the experiments. Day 1 is a day with highly variable irradiance due to cloud coverage and day 6 is a clear day. On day 1 the maximum irradiance is 1076 W/m^2 at 12.25 with a PV output of 1309 W. It should be noted that the greatest variation in irradiance and thus PV power output during this day is at 12.55 that irradiance drops by 619 W/m^2 and PV power by 742 W. Day 6 has a maximum of 883 W/m^2 and peak PV power 1230 W. As the electrolyser is a commercial unit comprised of supplementary components that manage its operation, it requires a minimum energy supply. This depends mainly on the heater function that is variable, and when it is on it adds 500 W to the system. This in combination with the stack that requires at maximum 300 W reaches a total 800 W demand. The heater works during the start-up of the electrolyser and at certain points during operation that are dependent on the amount of generated hydrogen. At higher pressures, the stack generates hydrogen at higher

Table 18.2 Experimental results on the energy use, hydrogen production and energy efficiency at different operation conditions

		5 bar	7 bar	10 bar	13.8 bar
Day 1	Specific energy consumption (kWh/m^3)	11.2	8.4	7.71	7.4
	Produced hydrogen (L)	22.12	35.66	59.64	97.2
	Consumed energy (kWh)	0.248	0.300	0.459	0.721
	Energy efficiency of the stack (%)	31	41.5	45.24	47
Day 6	Specific energy consumption (kWh/m^3)	9.41	6.9	6.05	5.55
	Produced hydrogen (L)	27.97	48.53	84.32	146
	Consumed energy (kWh)	0.263	0.335	0.510	0.810
	Energy efficiency of the stack (%)	37	50	57.7	62.8

flow rates and the heater operation is more frequent. Thus, at points where the PV panels do not provide the necessary power and the heater goes on, the electrolyser stops producing.

As can be observed in Table 18.2, the results of the experimental measurements indicate that the stack operates more efficiently at higher pressure set points and the operation at maximum pressure, which is 13.8 bar, is the most efficient in terms of energy consumption and hydrogen production. This is because the flow rates and the total amount of generated hydrogen increase for higher pressure set points. Operation during the day with clear sky (day 6) is more efficient than the variable irradiance day (day 1). This is because the stack uses more energy after every stop and restart to reach a steady state operation, than during constant operation. On day 1, the electrolyser produced 97.2 l, used 0.721 kWh and has a specific energy use of 7.4 kWh/m^3. On day 6, it produced 146 l, used 0.810 kWh and has a specific energy use of 5.55 kWh/m^3.

It is important to examine the case of the Jamaican household cooking demand in combination with the different system operations. The daily cooking demand of an average household in Jamaica is 1.98 kWh, and taking a stove efficiency of 50 %, this is satisfied by 0.10 kg of hydrogen. Figure 18.5 shows the energy required by the stack, to produce this amount of hydrogen during a clear day (day 6) and a day with variable irradiance (day 1) at different pressure set points. It can be seen that operation at the highest producing pressure is the most energy efficient. It can also be adequately combined with the suggested storage methods of metal hydride and cascading system, as the pressure of 13.8 bar fits their pressure range requirements.

18.4 Conclusions and Future Work

This chapter examines the application of a solar-powered PEM electrolyser system to produce hydrogen at a large scale as a cooking fuel for domestic applications focussing on Jamaica. A performance analysis of a small-scale experimental setup of the system is presented and results indicate that the balance of plant is important

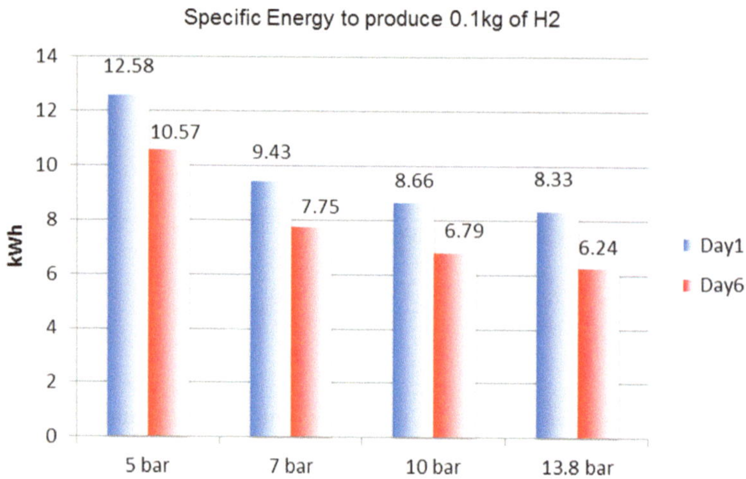

Fig. 18.5 Specific energy consumption with a reference production of 0.1 kg hydrogen gas

as the electrolyser performs more efficiently at highest hydrogen production pressures and when it is not subject to continuous stops and start-ups.

The results will be used for validation and calibration of a mathematical model (within TRNSYS) to simulate the operation of the complete system for application to different geographical regions.

Acknowledgements This project is partly funded by ACP Caribbean and Pacific Research Programme for Sustainable Development of the European Union (EuropeAid/130381/D/ACT/ACP).

References

1. OECD/IEA (2006) Energy for cooking in developing countries. OECD/IEAParis Cedex 15, France
2. Ministry of Energy and Mining (2009) Oil import statistics. Ministry of Energy and Mining, Kingston
3. The Planning Institute and the Statistical Institute of Jamaica (2007) Residential consumer end use survey. The Planning Institute and the Statistical Institute of Jamaica, Kingston
4. Owen EB, Camirand R (2000) Forest cover and deforestation in Jamaica-an analysis of forest cover estimates over time. Forestry Department, Kingston
5. The World Bank (2014) Jamaica development indicators, Latin America and Carribean (developing only). http://data.worldbank.org/country/jamaica?display=default. Accessed 1 May 2014
6. Ministry of Energy and Mining (2010) National renewable energy policy, 2009–2030. Ministry of Energy and Mining, Kingston
7. Nowotny J, Veziroglou N (2011) Impact of hydrogen on the environment. Int J Hydrogen Energy 36:13218–13224
8. Carmo M, Fritz D, Mergel J, Stolten D (2012) A comprehensive review on PEM water electrolysis. Int J Hydrogen Energy 38:4901–4934

9. Gupta RB (2009) Hydrogen fuel: production, transport and storage. Taylor & Francis Group, LLC, Boca Raton
10. Sakintuna B, Lamari-Darkrim F, Hirscher M (2007) Metal hydride materials for solid hydrogen storage: a review. Int J Hydrogen Energy 32:1121–1140
11. Zini G, Tartarini P (2012) Solar hydrogen energy systems. Springer, Italy

Chapter 19
Surface Modification and Optimization of Semiconductor ns-TiO_2–WO_3 Admixed Photoelectrode in Regard to Solar Hydrogen Production

Mridula Tripathi and Priyanka Chawla

Abstract To minimize the harmful effects of commercial energies like fossil fuel and nuclear energy on our environment, research is going on to find clean and renewable sources of energy. The main efforts of researchers nowadays is to harness solar energy for the production of clean hydrogen fuels by a photoelectrochemical (PEC) cell which represents a very attractive but challenging alternative. Some strategies have been developed to improve PEC performances of the photoelectrode materials, including doping for enhancing visible light absorption in the wide bandgap semiconductor or promoting charge transport in the narrow bandgap semiconductor, respectively. This chapter deals with the investigation on the optimization of ns-WO_3–TiO_2 admixed/Ti with respect to optimum photoelectrode area for semiconductor septum (SC-SEP) PEC solar cell. The motivation of the present work was to prepare an electrode having high-effective surface area and hence better quantum yield and improved PEC activity. Several attempts have been made to bring spectral response of TiO_2 into visible or near visible region. It is known that the spectral response of the TiO_2 films can be improved through admixing with appropriate oxides. The surface morphology, structural, and PEC characterization of the bare TiO_2 as well as the TiO_2 overlaid with WO_3 thin film admixtures have been investigated in relation to hydrogen production through SC-SEP PEC solar cell. The PEC response of ns-WO_3–TiO_2 photo electrodes for four different electrode areas has been measured to explore the effect of electrode area on the output power in a chemical fuel (i.e., H_2) produced by SC-SEP PEC cell. This was done for determining the electrode area for optimum electrical output and hydrogen production. The PEC cell having ns-WO_3–TiO_2 admixed/Ti photoanode of several geometric areas like 0.5, 1.0, 1.5, 2.0, and 2.5 cm^2 were fabricated and characterized. It has been found that the photoanode area corresponding to optimum electrical output and hydrogen production rate corresponds to 1.0 cm^2. The ns-WO_3–TiO_2 exhibited a high photocurrent and

M. Tripathi (✉) · P. Chawla
Department of Chemistry, C.M.P. Degree College, University of Allahabad, Allahabad 211002, India
e-mail: mtcmpau@gmail.com

A. Sayigh (ed.), *Renewable Energy in the Service of Mankind Vol I*,
DOI 10.1007/978-3-319-17777-9_19

photovoltage of 15.6 mA cm^{-2}, 960 mV, respectively. The ns-WO_3–TiO_2 electrode exhibited a higher hydrogen gas evolution rate of 13.8 l h^{-1} m^{-2}.

Keywords Photoelectrochemical cell · Hydrogen · ns-WO_3–TiO_2 · Semiconductor

19.1 Introduction

The depleting trend of the present source of commercial energies, and current concerns about pollution leading to environmental degradation, has made it imperative to accelerate research and development in the area of renewable energy [1]. The main efforts of researchers nowadays is to harness solar energy for the production of clean hydrogen fuels by a photoelectrochemical (PEC) cell which represents a very attractive but challenging alternative because of convenience in fabrication and in situ storage facilities [2].

Photocatalytic water splitting is a chemical reaction for producing hydrogen by using two major renewable energy resources, namely, water and solar energy. As the feed stocks for the reaction, water is clean, inexpensive, and available in a virtually inexhaustible reserve, whereas solar energy is also infinitely available, nonpolluting and appropriate for the endothermic water-splitting reaction. Thus, the utilization of solar energy for the generation of hydrogen from water has been considered as an ultimate solution to solve the crisis of energy shortage and environmental degradation [3]. The following is the dissociation of the water molecule to yield hydrogen and oxygen: this simple process has gathered a big interest from an energetic point of view because it holds the promise of obtaining a clean fuel, H_2, from a cheap resource of water [4, 5]. The use of a photocatalyst makes the process feasible with photons within solar spectrum since the discovery of the PEC performance for water splitting on TiO_2 electrode by Fujishima and Honda [6]. To date, the above water splitting can be photocatalyzed by many inorganic semiconductors such as titanium dioxide (TiO_2), which was discovered in 1971 by Fujishima and Honda [7].

Among the various types of widely investigated semiconductor materials, titanium dioxide (TiO_2) has been considered the most active photocatalyst due to its low cost, chemical stability, and comparatively high photocatalytic efficiency [8, 9].Though TiO_2 is found to be a suitable semiconductor electrode for PEC cells, it suffers from the major drawback of high bandgap of 3 eV (410 nm) to significantly absorb the solar energy. Several attempts have been made to bring the spectral response of TiO_2 into visible and near visible region. It is known that the spectral response of the TiO_2 films can be improved through admixing with appropriate oxides or by deposition of noble metal nanoparticles on TiO_2 [10–13].

The motivation of the present work was to obtain improved TiO_2 photoelectrode by admixing WO_3 with the nanostructured prepared TiO_2 photoelectrode; the surface morphology, structural, and PEC characterization of the TiO_2 overlaid with WO_3 admixtures have been investigated in relation to hydrogen production through semiconductor septum photo-electro-chemical (SC-SEP PEC) solar cell. WO_3 is an interesting candidate because it has wide range of applications such as gas sensor, photochromic, and electrochromic [14, 15]. Determination of effective photoelectrode area that will lead to high electrical and hydrogen production output is one of the important aspects which we have also considered in the present research work. This would help in making modular PEC cell configuration for hydrogen productions. For the investigation of PEC solar cells with modular configuration leading to "hydrogen production reactor," estimates of optimum photoelectrode geometric areas form an essential ingredient.

19.2 Experimental

The WO_3–TiO_2 oxide was prepared in the form of a solgel by the hydrolysis process. For this preparation, first $Ti[OCH(CH_3)_2]_4$ was added slowly to propanol drop by drop. Deionized water was slowly added under vigorous stirring conditions for the duration of 10 min. Then concentrated HNO_3 solution was added under magnetic stirring. Then calculated amount of Na_2WO_4 solution was added dropwise into above mixture under continuous stirring. The mixture was stirred at room temperature for 6 h and the resulting gel of WO_3–TiO_2 oxides was obtained.

Titanium sheets were used as a conducting substrate. Titanium sheets (~0.33 mm thickness) of different areas like 0.5, 1.0, 1.5, 2.0, and 2.5 cm^2 were taken and mechanically polished with emery papers of various grades. These were ultrasonically cleaned in acetone for 15 min. The nearly uniform films were deposited by placing ultrasonically cleaned substrate over vacuum interlocked spinner holder. One drop of sol was spread over the substrate rotating at 3000 rpm. The films so obtained were dried in an air oven for 15 min at 80 °C and then fired at 450 °C for 30 min. This process was repeated for about four to five times to increase film thickness. Finally, samples were annealed in an argon atmosphere at 550 °C for 4 h. Electrode thus obtained (having ~6 μm, ns-TiO_2–WO_3 layer over Ti) were then subjected to PEC characterization.

The rectangular characterization of ns-TiO_2–WO_3 electrode materials was carried out through X-ray diffraction (XRD) by employing a Phillips PW 1710 diffractometer equipped with a graphite monochromator.

The microstructural characteristics of the thus synthesized mixed oxide ns-TiO_2/WO_3 electrodes were also performed by employing scanning electron microscope (SEM; GEOL JXA_8100 EPMA).

To explore the spectral response, the absorption spectra of ns-TiO_2/WO_3 on indium–tin-oxide (ITO) glass substrate were carried out through double beam spectrophotometer (Systronics 2201).

Rectangular SC-SEP PEC cell made of Lucite had a quartz window for illumination. The cells were divided into two compartments by a Lucite separator which had a hole at its centre on which the ns-TiO_2–WO_3 admixed/Ti septum electrode of different areas was glued with araldite (Fig. 19.1). The configuration of present

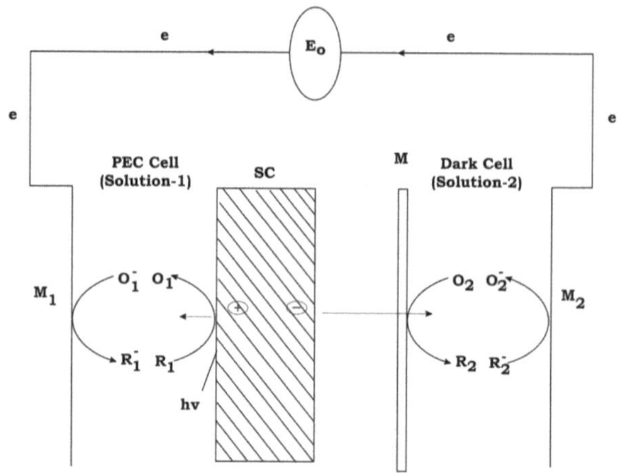

Fig. 19.1 Cell diagram of semiconductor septum (SC-SEP) photoelectrochemical (PEC) solar cells

SC-SEP PEC cell consisting of two chambers connected through ns-TiO_2–WO_3/Ti septum electrode is:

$$SCE / 1M\ NaOH / ns - TiO_2 / WO_3 admixed / Ti / H_2SO_4 + K_2SO_4 / Pt_{CE}, Pt_{WE},$$

CE counter electrode
SCE saturated calomel electrode
WE working electrode

All electrochemical measurements of electrodes of different areas were carried out by using a Princeton Applied Research (PAR) model 173 potentiostat/galvanostat PAR 175 universal programmer coupled with a Houston 2000 *x*–*y*/*t* recorder. A 1000-W xenon-mercury lamp was used as the illumination source for ns-TiO_2–WO_3 photoelectrodes. The intensity of incident radiation was adjusted and fixed at 450 mW/cm^2. The hydrogen production kinetics of the SC-SEP PEC cell has been monitored by collecting hydrogen in the dark compartment over platinum (Pt) electrode using inverted burette under illumination.

19.3 Results and Discussions

Figure 19.2 shows XRD patterns (CuKα radiation) of the ns-TiO_2–WO_3 materials. Analysis of XRD patterns revealed that on alloying ns-TiO_2 with WO_3 did not lead to formation of any new compound material. Prepared photoelectrode materials were found to be polycrystalline.

Figure 19.3 shows SEM images of the ns-TiO_2–WO_3 nanomaterial and the observed pattern suggest nanostructured characteristics, reveals a fine-grained structure and shows a nanocrystalline like matrix. As clear from SEM, the island

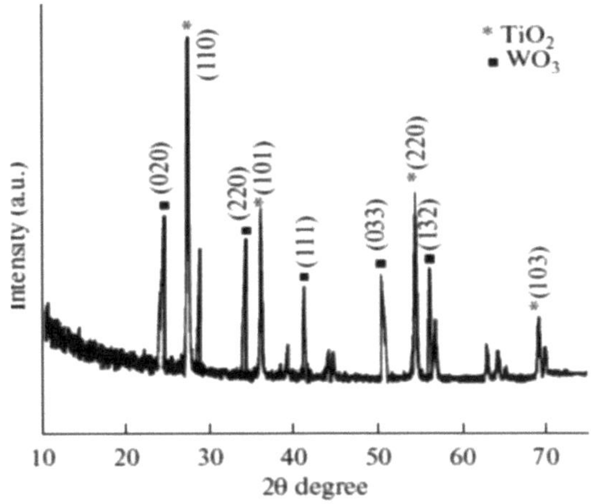

Fig. 19.2 XRD pattern of ns-TiO_2–WO_3 materials

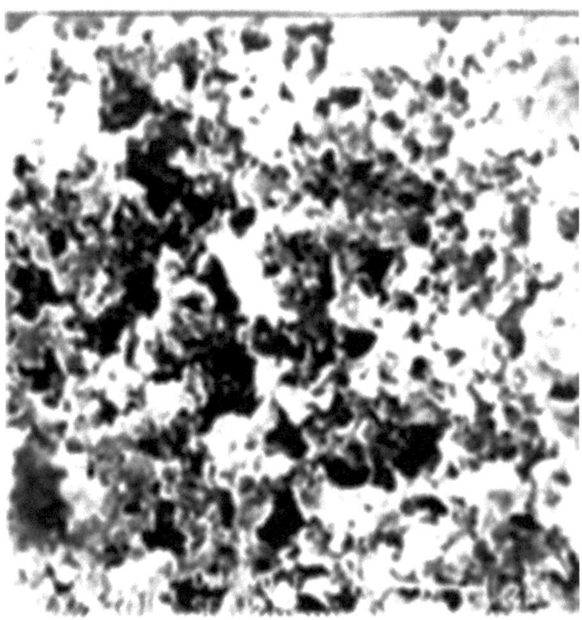

Fig. 19.3 SEM image of ns-TiO_2–WO_3 materials

of WO_3 from over the surface of TiO_2 is thought to catalyze the anode reaction probably with added advantage of colloidal photochemical system. The enhanced photoactivity added by increased gas evolution kinetics can also be thought to be due to surface modification with thin WO3 islands with high conductivity and wide-absorption range (from ~200 to 500 nm).

The average grain size of the photoelectrode was found to be ~20–40 nm for ns-TiO_2–WO_3.

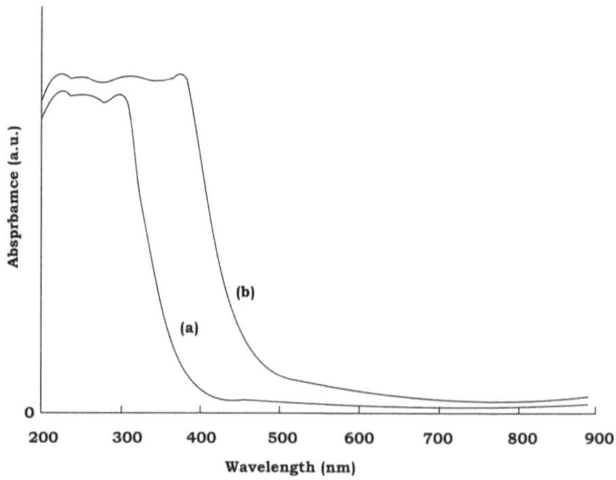

Fig. 19.4 UV diagram of ns-TiO_2–WO_3

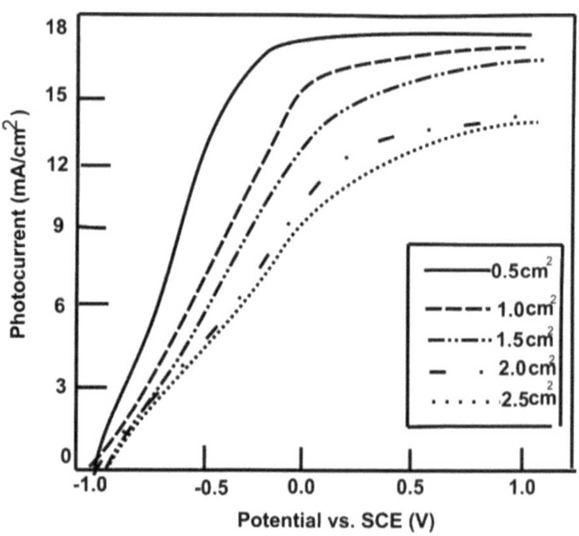

Fig. 19.5 Current–potential curves for ns-TiO_2–WO_3 electrodes for five different photoanode areas. *SCE* saturated calomel electrode

Figure 19.4 shows the UV–Vis absorption curves of the ns-TiO_2, and ns-TiO_2–WO_3 nanomaterials. There is a clear enhancement in absorption ranges caused by addition of WO_3.

The I–V of SC-SEP PC cell (SCE/1M NaOH/ns-TiO_2/WO_3 admixed/Ti/H_2SO_4+K_2SO_4/Pt_{CE}, Pt_{WE}) which was measured for different geometric areas of photoelectrodes (Fig. 19.5) have been drawn. Figure 19.5 represents variation in photocurrent as a function of electrode potential versus saturated calomel electrode (SCE) for five different photoelectrode areas and the photocurrent values for different photoelectrodes have been extracted and plotted against electrode of different areas (Fig. 19.6). It has been observed that the photocurrent density decrease rapidly on increasing the electrode area and then saturates. Such a behavior of photocurrent

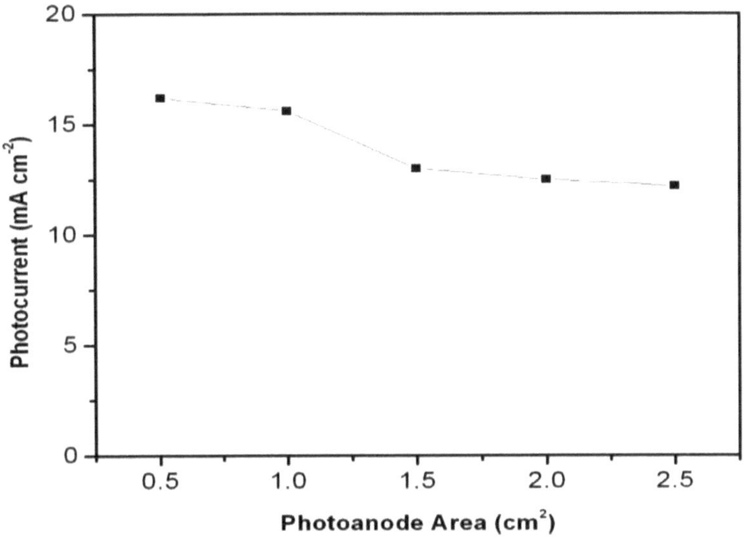

Fig. 19.6 Variation of photocurrent density with photoanode area

is known to be due to increase in surface state, originating mostly from grain boundaries/surface defects, and acting as recombination centre for a charge carrier.

In regard to hydrogen production, this is due to combining the photoproduced electrons with H^+ (from acidic solution) or electron may reduce H_2O producing $H_2+OH-(2H_2O+2e-H_2+2OH^-)$, the hydrogen gas may be liberated in the dark compartment. The SC-SEP PEC solar cell can drive energetically demanding overall electrochemical process, for example, the decomposition of water to hydrogen and oxygen.

The hydrogen production studies were carried out based on said SC-SEP PC solar cells with different photoelectrode areas (0.5, 1.0, 1.5, 2.0, 2.5 cm^2). Figure 19.6 shows the variation of hydrogen production rate with photoanode area and it can be noted that the highest value of photocurrent density and hydrogen production rate corresponds to the smallest electrode area and starts decreasing for increasing electrode area.

Figure 19.7 shows the variation of hydrogen production rate with photoanode area, and it can be noted that the highest value of photocurrent density and hydrogen production rate corresponds to the smallest electrode area and starts decreasing for increasing electrode area. The observed decrease in HPR is slow beyond a specific area, viz. 1.5 cm^2, in the present case. Since the modular PEC cells would require adequate photoelectrode area to reduce the complexity for better hydrogen production rate leading to adequate hydrogen production and at the same time workable adequate photoelectrode area to reduce the complexity. It would be appropriate to choose the electrode area that is in between the smallest and the one beyond which, decrease in photocurrent density and hydrogen production rate nearly saturates. For

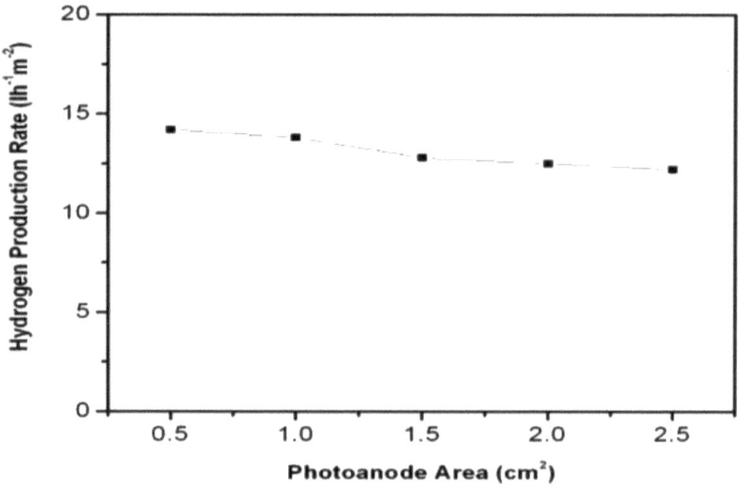

Fig. 19.7 Variation of hydrogen production rate with photoanode area

the present case, the appropriate photoelectrode area for modular cell designs for hydrogen production can be taken as 1.0 cm^2.

As clear from SEM the island of WO_3 formed over the surface of TiO_2 are thought to catalyze the anode reaction probably with added advantage of colloidal photochemical system. The enhanced photoactivity added by increased gas evolution kinetics can also be thought to be due to surface modification with thin WO_3 islands with high conductivity and wide-absorption range (from ~200 to 500 nm). It may, therefore, be taken that ns-TiO_2 overlaid with surface clusters of WO_3 represents a new improved photoelectrode for hydrogen production through SC-SEP PEC cell.

19.4 Conclusion

In conclusion, it can therefore be said that the ns-TiO_2–WO_3 represent improved version of photoelectrode and it enhances the photo production of hydrogen. The PEC response of ns-TiO_2–WO_3 photoelectrode for five different electrode areas has been measured to explore the effect of electrode area on the output power in a chemical fuel (i.e., H_2) producing SC-SEP PEC cell. Based on the present studies, it has been found that the optimum ns-TiO_2–WO_3 photoelectrode area for adequate photoconversion efficiency and hydrogen production rate corresponds to 1.0 cm^2.

Acknowledgment We are thankful to Board of Research in Nuclear Science, Bombay, India (2012/34/33/BRNS) for the financial support.

References

1. Primo A, Marino T, Corma A, Molinari R, García H (2011) Efficient visible-light photocatalytic water splitting by minute amounts of gold supported on nanoparticulate CeO_2 obtained by a biopolymer templating method. J Am Chem Soc 133:6930–6933
2. Puangpetch T, Sreethawong T, Yoshikawa S, Chavadej S (2009) Hydrogen production from photocatalytic water splitting over mesoporous- assembled $SrTiO_3$ nanocrystal-based photocatalyst. J Mol Catalysis A 312:97–106
3. Osterloh FE (2008) Inorganic materials as catalysts for photochemical splitting of water. Chem Mater 20:35–54
4. Serpone N, Lawless D, Terzian R (1992) Solar fuels: Status and perspectives. Solar Energy 49:221–234
5. Hernández-Alonso MD, Fresno F, Suárez S, Coronado Development of alternative photocatalysts to TiO_2: Challenges and Opportunities. JM. Energy Environ Sci 2:1231–1257
6. Fujishima A, Honda K (1972) Electrochemical photolysis of water at a semiconductor electrode. Nature 238:37–38
7. Fujishima A, Honda KB (1971) Electrochemical evidence for the mechanism of the primary stage of photosynthesis. Chem Soc Jpn 44:1148–1150
8. Hoffmann MR, Martin ST, Choi W, Bahnemann DW (1995) Environmental applications of semiconductor photocatalysis. Chem Rev 95:69–96
9. Wang L, Wang WZ (2012) Graphene/Semiconductor Nanocomposites: Preparation and Application for photocatalysis. Int J Hydrogen Energy 37:3041–3047
10. Tripathi M, Pandey K, Kumar SD (2007) Surface modification of semiconductor photoelectrode for improved solar cell performance. Solar Energy Solar Mater 91:1663–1668
11. Karn KR, Misra M, Srivastava ON (2000) Semiconductor- septum photoelectrochemical cell for solar hydrogen production. Int J Hydrogen Energy 25:407–413
12. Wang X, Caruso RA (2011) Enhancing photocatalytic activity of titania materials by using porous structures and the addition of gold nanoparticle. J Mater Chem 21:20–28
13. Silva CG, Carabineiro SAC, Lima MJ, Drazic G, Figueiredo JL, Faria JL (2014) Titanium dioxide-Based Photocatalyst for the conversion of water pollutants. Chemical Physics Research Journal 21:23–35
14. Ni M, Leung MKH, Leung DYC, Sumathy K (2007) A review and recent development in photocatalytic water splitting using TiO_2 for hydrogen production. Renewable Sustainable Energy Rev 11:401–425
15. Zach M, Hagglund C, Chakarov D, Kasemo B (2006) Curr Opin Solid State Mater Sci 10:132–143

Chapter 20
Hydrogen for Mobility: An Assessment from Economic, Energetic, and Ecological Point of View

Amela Ajanovic and Reinhard Haas

Abstract In this chapter, the role of hydrogen as storage for surplus electricity from renewable energy sources (RES) and as a fuel for transport is discussed. The different energy supply chains based on the use of electricity from RES in passenger cars are analyzed. These analyses are conducted mainly from economic point of view but also energetic and ecological aspects are shortly addressed.

Keywords Hydrogen · Methane · Costs · Emissions · Efficiency

20.1 Introduction

The global energy system faces currently two major challenges. On the one hand, it is important to have sufficient and secure energy supply. However, at the same time, it is important to reduce energy-related greenhouse gas (GHG) emissions and to move towards more sustainability by increasing the use of renewable energy sources (RES). The key goals of the EU climate and energy policy are mitigation of global warming, improvement of air quality, and reduction of energy consumption. The requirements of strategies for the transformation of current energy systems towards more sustainable ones include among other measures further technical developments of the concepts of converting renewable power into easy storable energy carriers such as hydrogen and methane as well as implementation of the corresponding supporting policies.

Due to the supporting policy measures in the EU the share of RES in total energy supply is continuously increasing. In the past years, especially high photovoltaic (PV) and wind penetration in some regions has been noticed. However, the use of intermittent RES for electricity production leads to the need for increased

A. Ajanovic (✉) · R. Haas
Vienna University of Technology, Gusshausstr. 25-27/370-3, Vienna, Austria
e-mail: ajanovic@eeg.tuwien.ac.at

R. Haas
e-mail: haas@eeg.tuwien.ac.at

A. Sayigh (ed.), *Renewable Energy in the Service of Mankind Vol I*,
DOI 10.1007/978-3-319-17777-9_20

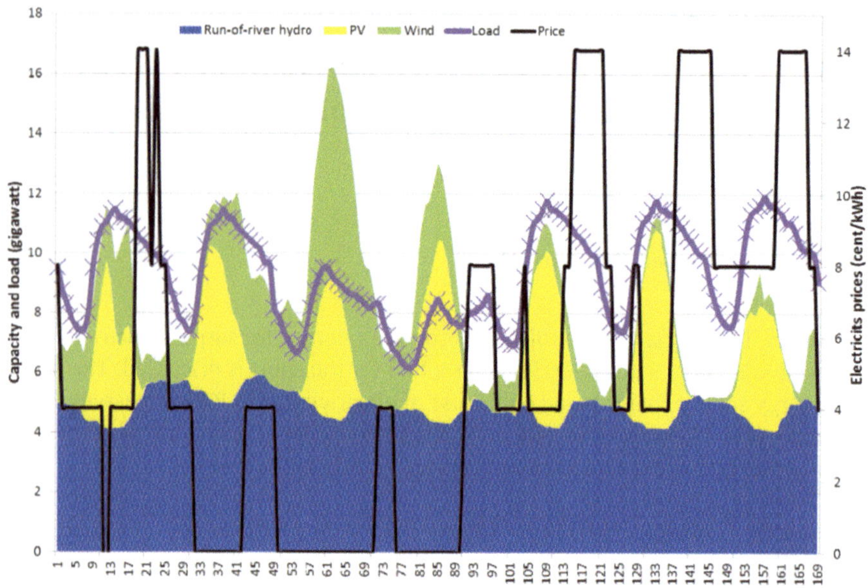

Fig. 20.1 Example: Development of variable renewables from wind, PV, and run-of-river hydro plants over a week in summer on an hourly base in comparison to demand and resulting electricity market prices with total costs charged for conventional capacities [1]

balancing activities between supply and demand to meet the so-called residual load, see Fig. 20.1.

The important issue is how these changes in the residual load will impact the price of electricity on the wholesale spot markets. The effects on electricity prices are shown in Fig. 20.1 where a hypothetical scenario of high levels of electricity generation from wind, PV, and run-of-river hydro plants over a week in summer is depicted using synthetic hourly data for an average year in Austria. The graph shows significant volatilities in electricity market prices with total costs charged for conventional capacities fluctuating rapidly from 0 to 14 cents/kWh (black solid line). In practice, of course, prices may not only go to zero but even become negative [1].

Hence, in peak-demand periods for residual load the price of electricity is significantly higher than in periods where excess electricity is available, see Fig. 20.1. With the increasing use of PV and wind for electricity generation, increasing amounts of cheap or even free surplus energy also is sometimes available. This surplus can be stored and used later to cover demand at times with undercoverage of variable electricity from RES (RES-E) [2].

One opportunity to store electricity from variable energy sources (such as hydro, wind, and PV) is via hydrogen. For example, excess electricity can be converted into hydrogen by electrolysis. Hydrogen can be stored much easier than electricity, and it can be re-electrified if required in times of lack of RES-E generation. However, the total round trip efficiency of such a process is very low, in the best case about 40 % [3]. Moreover, given the low overall levels of electricity market prices expected in the next years [1] and competing technologies such as hydro pump storages or power to heat which are cheaper, the prospects for such energy chains are very low.

However, there is another opportunity for hydrogen use—in the transport sector. Since the transport sector, especially road transport, is one of the largest emitter of GHG emissions, it is very important to increase use of RES for mobility. According to the 2020 climate and energy package, by 2020 share of energy consumption produced from RES in the EU should increase for 20%. However, the options for the use of RES in transport are much more limited than in the heating or electricity sector. Currently, the mostly used alternative fuels are biofuels which have been increasingly criticized in recent years [4, 5]. In the past years, huge effort has been put on an increasing penetration of electric vehicles. Unfortunately, there are still major technical and economic barriers, such as limited driving range and high investment costs, curtailing the broader use of battery electric vehicles (BEV) or cars. Hydrogen-powered fuel cell vehicles (FCV) are another possible zero-emission automotive technology for the long term. However, BEV and FCV are really environmentally friendly only in combination with RES.

In this chapter, the role of hydrogen as storage for surplus electricity from RES and as a fuel for transport is discussed. Former version of this chapter has been presented at the World Renewable Energy Congress 2014 and the BIEE Academic Conference 2014. We compare three different energy supply chains based on the RES-E: (i) direct use of electricity in BEVs, (ii) the use of RES-E for hydrogen production, which is finally used in FCV, and (iii) the methane (CH_4) production via methanization of hydrogen, which is fed into the natural gas grid and used in vehicles powered by compressed gas (CGV). These analyses are conducted mainly from economic point of view but also energetic and ecological aspects are shortly addressed. The electricity used in these chains comes only from the variable energy sources such as hydro, wind, and PV.

20.2 Hydrogen as a Storage and Energy Carrier

Hydrogen is the simplest, lightest, and most abundant element in the universe, but it does not exist on the earth in the elemental form in quantities associated with energy use. Huge advantage of hydrogen is that it can be produced from different energy sources—fossil energy, nuclear energy as well as renewable energy. Hydrogen, which is considered as one of the cleanest energy carriers, has a great potential to reduce GHG emissions, global warming, and to contribute to energy diversity and supply security. However, the largest environmental benefit can be reached in the case that hydrogen is produced from RES. With the increasing production of electricity from wind and PV, the production of hydrogen from RES is now more and more interesting.

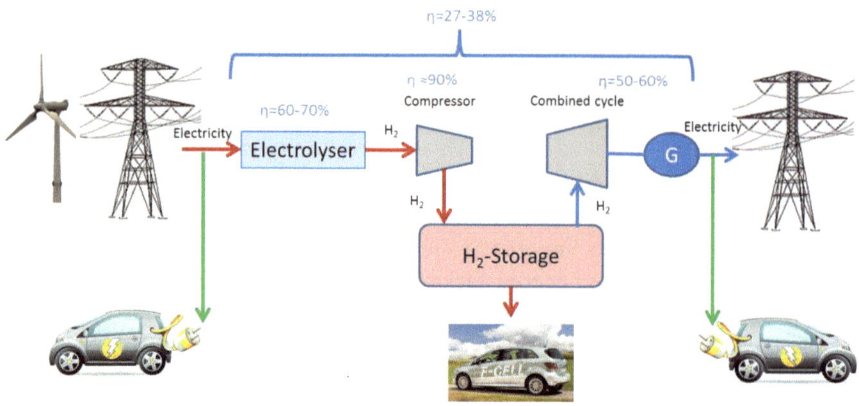

Fig. 20.2 Energy supply chains: Storage and/or use of RES for mobility

The largest amount of hydrogen today is produced by steam reforming of natural gas, since this process is very well developed and cost-effective. Hydrogen production via electrolysis of water is also well established, but significantly more expensive technology. Currently, electrolysis is mostly limited to applications requiring very pure hydrogen. It is a very clean process, as long as the necessary electricity for electrolysis comes from clean, RES. The costs of electricity used are the major cost factor for electrolysis [6].

Which type of RES is used for electrolysis has a huge impact on hydrogen costs. These costs could be significantly reduced if only surplus electricity from variable RES is used for hydrogen production, see Fig. 20.1. As mentioned above, hydrogen can be used as storage for surplus electricity. How renewable energy can be stored and/or used in transport sector is shown in Fig. 20.2.

As illustrated in Fig. 20.2, electricity from RES can be directly used in electric vehicles. However, the operating range of BEV is still relatively low, about 100–150 km. This is a major reason for the low acceptance of these vehicles. For the future, a more interesting type of electric vehicles is the fuel cell car. It is powered by hydrogen and already now, in the immature stage of this technology, they have operating range between 200 and 690 km [7].

In addition to the hydrogen use as a fuel for mobility, it can be also used as storage for electricity. A typical hydrogen storage system consists of electrolyzer, compressor, a hydrogen storage tank and a fuel cell, gas motors, gas turbines, or combined cycles of gas and steam turbines for electricity generation from hydrogen [8, 2]. Hydrogen can be stored as a gas under pressure, a liquid at very low temperature, adsorbed on metal hydrides, or chemical bonded in complex hydrides. However, for stationary applications, gaseous storage under high pressure is the most popular choice [2].

One of the main problems of energy carriers are the losses occurring in the whole energy supply chain. The energy losses depend on the efficiency of the conversion process at the each transformation stage as well as efficiency of the finally used technology. The efficiency of the electrolysis is usually in the range from 60 to 70 %. Efficiency of compressing and storing hydrogen is about 90%. For example, total efficiency of the energy chain shown in Fig. 20.2 is between 27 and 38 %. The energy losses due to converting hydrogen into electricity are between 40 and 50 % [8]. These additional losses can be avoided if hydrogen is used directly, for example, in the transport sector.

20.3 Hydrogen in Transport

In the transport sector, in opposite to other sectors, GHG emissions have been continuously increasing. One of the major challenges for the EU is to find policies and measures with which this trend can be stopped. One possibility is increasing use of RES in the transport sector in combination with zero-emission vehicles, such as BEV and FCV. These vehicles have no emissions at the point of use, but total emissions in the well-to-wheel (WTW) energy supply chain are very dependent on type of primary energy used for electricity and hydrogen production, see Fig. 20.3. Gasoline used in conventional internal combustion engine (ICE) vehicles is here taken only for comparison.

Hydrogen costs are very dependent on primary energy used too. Energy costs per 100 km driven of different types of vehicles and fuels are shown in Fig. 20.4. Beside energy costs, different taxes (excise tax, value add tax (VAT), and electricity fee) are also illustrated.

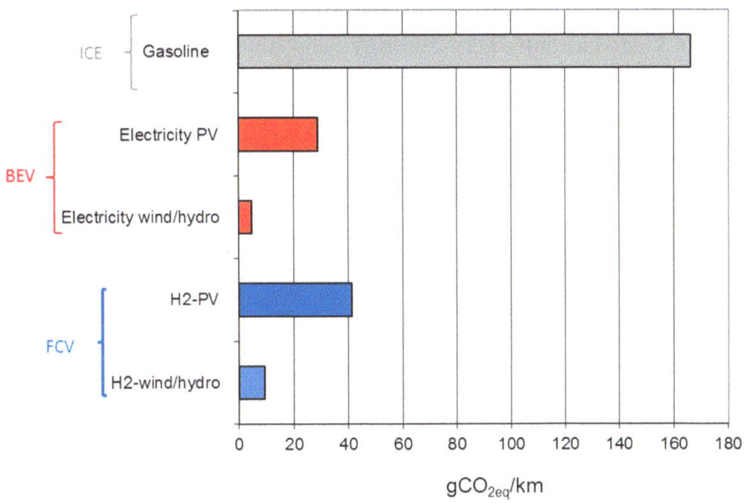

Fig. 20.3 WTW CO_2 emissions in 2010. *ICE* internal combustion engine, *BEV* battery electric vehicles, *FCV* fuel cell vehicles [9]

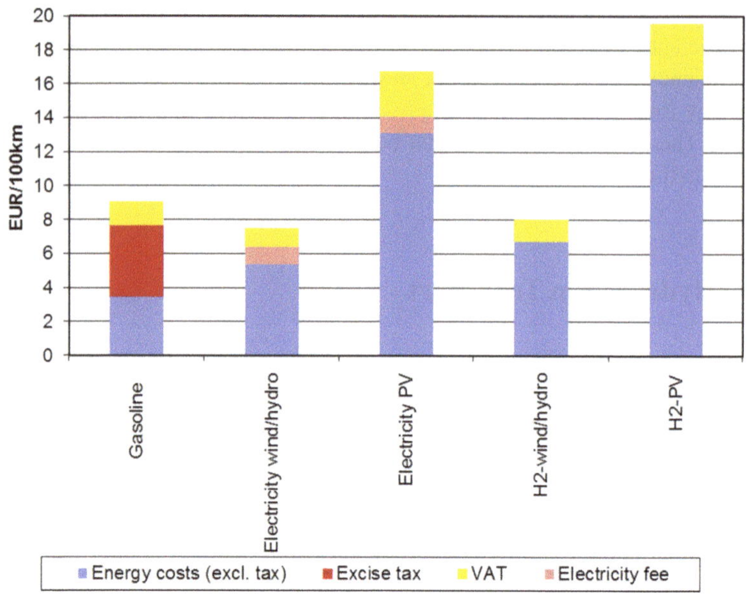

Fig. 20.4 Energy costs of mobility per 100 km based on average of EU-15 countries in 2010. *PV* photovoltaic

To reduce relatively high costs related to the hydrogen use in the transport sector, two different energy supply chains are considered. First one is shown in Fig. 20.2 and includes hydrogen production from surplus electricity only via electrolysis, and direct hydrogen use in FCV (without re-electrification). In this case, full-load hours, which are dependent on availability of surplus electricity from wind and PV, have a crucial impact on the total hydrogen costs.

Since FCV are currently very expensive and necessary infrastructure for their use is missing, we have here considered also indirect use of hydrogen via methanization, as shown in Fig. 20.5. It is assumed that hydrogen produced by electrolysis

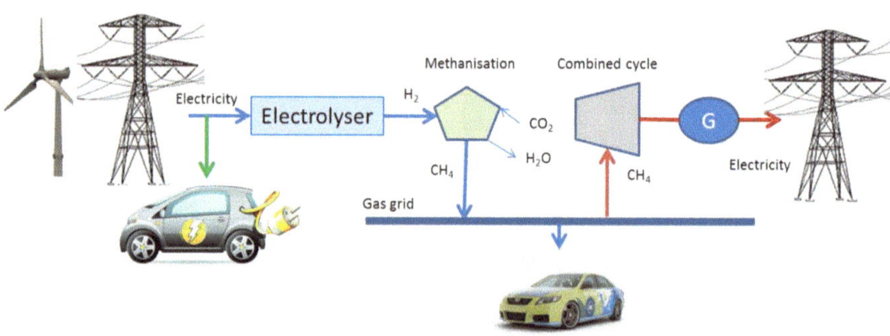

Fig. 20.5 Energy supply chains: Methane for mobility

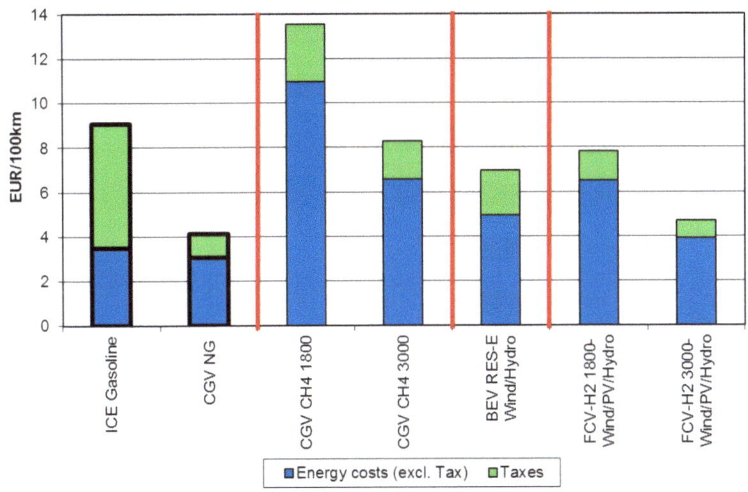

Fig. 20.6 Energy costs of mobility per 100 km in 2010 depending on full-load hours of the electrolysis for hydrogen production. *CGV NG* compressed natural gas vehicles, *CGV* CH_4 vehicles powered by methane, *BEV RES-E* battery electric vehicle powered by electricity from renewable energy resources, *FCV*-H_2 hydrogen-powered fuel cell vehicles

will be converted to methane (CH_4) which can be easily stored in gas grid and used in CGV. Methane is also an option to store electricity as chemical energy. It can be stored in pressure tanks, underground, or it can be directly fed in the gas grid, as shown in Fig. 20.5. The main advantage of this approach is that the already existing gas grid can be used as storage. Up to a certain concentration also pure hydrogen can be stored in the gas grid. However, in the case of hydrogen methanization, this additional step reduces total efficiency significantly, 20–35%.

The advantage of such an energy supply chain is a significant reduction of total transport costs due to much lower costs of CGV in comparison to BEV and especially to FCV. Figure 20.6 shows energy costs per 100 km for hydrogen and methane depending on full-load hours (1800 and 3000 h). It is obvious that the number of full-load hours have significant impact on energy costs. Compressed natural gas vehicles (CGV NG) are taken as a reference for vehicles powered by methane (CGV CH_4).

However, the impact of the energy cost on total travel cost is relatively low, especially in the case of FCV and BEV, see Fig. 20.7.

For the future use of hydrogen in the transport sector, a remarkable reduction of capital costs is the key issue. From economic point of view, methane use in CGV is already now competitive with conventional gasoline cars. Although the use of compressed natural gas as a transport fuel is mature technology, it is widely used only in some parts of the world (e.g., Pakistan, Argentina, Iran, and Brazil). In the rest of the world, the availability of fueling stations is limited. Due to that as well as the fact that CGV have lower driving range than gasoline cars, the largest part of CGV are bi-fuel.

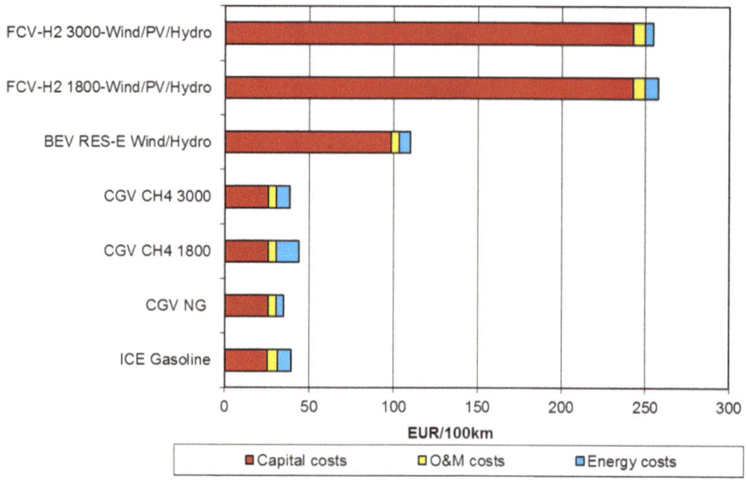

Fig. 20.7 Total specific costs per 100 km in 2010 depending on full-load hours of the electrolysis for hydrogen production

20.4 Conclusions

With increasing electricity generation from variable RES the need for storing electricity increases. At the same time, the need for environmentally friendly technologies in the transport sector increases too. In this context, mostly discussed automotive technologies are BEV, FCV, and CGV, as well as corresponding energy carriers, electricity, hydrogen, and methane.

Every considered energy carrier and automotive technology has some advantages and disadvantages. In opposite to BEV and CGV, which are mature technologies already available in the market, FCV are a long-term option mostly due to high costs and missing infrastructure.

Hydrogen as well as methane could be used as a fuel for mobility but also as energy storage for volatile RES. However, very long energy supply chains are always connected with energy losses at the every conversion step. Energy chains with storage of RES via hydrogen and methane and re-electrification are relatively inefficient. A better efficiency could be reached by the direct use of hydrogen and methane in the transport sector without re-electrification. Due to very good energy efficiency of BEV and FCV, even in the case of the low efficiency in a well-to-tank (WTT) energy supply chains, total WTW efficiency could be competitive with conventional fuels and vehicles.

Since FCV are zero-emissions vehicles (with longer driving ranges than BEV) political support for the market penetration of these vehicles can be expected. However, it is likely that the economic competiveness of these vehicles will be reached only beyond 2030.

References

1. Haas R, Auer H (2013) On the future design of electricity markets. Capacity payments or smart solutions? Proceedings of the 8th SDEWES-conference
2. European Comission (2012) DG ENER Working Paper. The future role and challenges of Energy Storage
3. Energy Storage Association (2014) Hydrogen energy storage. http://energystorage.org/energy-storage/technologies/hydrogen-energy-storage. Accessed 1 Dec 2014.
4. Ajanovic A (2010) Biofuels versus food production: does biofuels production increase food prices? Energy 36(4):2070–2076. doi:10.1016/j.energy.2010.05.019
5. Ajanovic A, Haas R (2010) Economic challenges for the future relevance of biofuels in transport in EU countries. Energy 35(8):3340–3348, 35
6. Ajanovic A (2008) On the economics of hydrogen from renewable energy sources as an alternative fuel in transport sector in Austria. Int J Hydrog Energy 33(16):4223–4234
7. Mazloom K, Gomes C (2012) Hydrogen as an energy carrier: prospects and challenges. Renew Sustain Energy Rev 16:3024–3033
8. Felberbauer KP, Kloess M, Jungmeier G, Haas R, Könighofer K, Prüggler W, Pucker J, Rezania R, Beermann M, Wenzel A (2012) Energiespeicher für erneuerbare Energie als Schlüssel-Technologie für zukünftige Energiesysteme. FFG-Projekt Nr. 821935
9. Ajanovic A, Haas R, Beermann M, Jungmeier G, Zeiss C (2012) "ALTETRA"—perspectives for alternative energy carriers in Austria up to 2050. Final report

Part III
Geothermal Energy

Chapter 21
Hydrogeothermal Potential of the Belgrade City Area, the Capital of Serbia First Assessment

Dejan Milenic, Ana Vranjes and Nenad Doroslovac

Abstract The first assessment of hydrogeothermal potential of the Belgrade city area was carried out up to the depth of 300 m. The criterion for defining the in-depth cut of the research was based on financial viability of resource exploitation, and it is related to the method of drilling exploitation boreholes. The temperatures of the groundwater resources up to 300 m deep, range from 12 to 25 °C, so their use for the sake of the building air-conditioning is determined by the application of heat pumps. According to the assessed quantities of hydrogeothermal resources (kg/s), the temperature difference which can be achieved on the temperature exchanger (°C) and, finally, the specific heat capacity of the water (KJ/kg/°C), the total hydrogeothermal potential of the Belgrade city area is approximately 1200 MW. Half of the total heating requirements of the urban residential and business funds in the Belgrade city area is met with the district heating system, and the other half is met with the electric energy, up to 20 %; gas, 20 %; and hard mineral fossil fuels (coal, wood), 10 %. The total installed capacity of the heating plants is about 2500 MW, and it can be concluded that hydrogeothermal resources could meet about 50 % of the city's heating needs. According to the above-mentioned data, hydrogeothermal resources are a strategically significant resource for the further development of the Belgrade city area.

Keywords Hydrogeothermal energy · Energy potential · Heat pumps · Sustainable management

21.1 Introduction

In comparison to the use of geothermal resources of high and medium enthalpy, the use of geothermal resources of low enthalpy, that is, subgeothermal resources (temperatures up to 30 °C), is just beginning to develop. A few countries in the world actually know what subgeothermal potential they have at their disposal, unlike the

A. Vranjes (✉) · D. Milenic · N. Doroslovac
Faculty of Mining and Geology, Department of Hydrogeology, University of Belgrade, Belgrade, Serbia
e-mail: vranjes_ana@yahoo.ie

A. Sayigh (ed.), *Renewable Energy in the Service of Mankind Vol I*,
DOI 10.1007/978-3-319-17777-9_21

potential of fossil fuels whose figures are very well known. The reason for this is the fact that the groundwaters with temperatures up to 30 °C were a little time ago studied from the aspect of water supply, bottling, balneology, etc. but not from the aspect of possible buildings' heating and cooling. With the technology development, first of all in the domain of heat pumps, and the introduction of buildings energy efficiency measures, subgeothermal resources are becoming an important segment of a total energy balance of each and every country. This is also supported by the fact that since 2005 in the territory of Europe more than 4.5 million heat pumps have been installed and their energy production equals 34.9 TWh and the annual savings are about 8.13 Mt of greenhouses gases [1]. According to the statistical forecast in the European territory, the installed capacities of heat pumps can reach about 35.6 GW_{th} by 2020 and this equals reduction of final energy consumption by 34.4 Mt [2]. Only in the German territory the number of installed geothermal heat pumps on the annual level will reach 65,000 units by 2017 [3].

Hydrogeothermal energy can also be used in the heating and cooling of buildings of different size and purpose, starting with small family houses, over big blocks of buildings, hotels, business facilities, educational facilities, warehouses, etc. Likewise, hydrogeothermal energy can be used by individual consumers and by systems of remote heating. A good instance is the use of subgeothermal resources in the system of remote heating in the city of Milan, where the heat pumps system water–water was integrated in the system of two gas power stations for the production of both heat and electric energy. The heat pumps use the groundwater formed in the alluvial deposits with the temperature up to 15 °C. The combined system was deployed in 2010 and during its 2-year-long exploitation it has proved extremely reliable and efficient. Positive results of the heat pumps' work have caused spreading of distribution network toward consumers and reaching the maximum heat energy production of 216 TJ/year which equals the primary energy savings of about 72 TJ on annual basis, that is, savings of about 3200 Mg/year CO_2 [4].

Potential barriers which can slow down and threaten the development of hydrogeothermal use are numerous. In the foreground, there are financial and technological problems and the greatest stumbling stone is actually the lack of or insufficient knowledge of hydrogeothermal terrain features. Thus, the bases for a successful development of increasing subgeothermal resources' share in the total energy balance make up the bases for estimating the potential and that is exactly the first step toward any further activity. Quantification of subgeothermal resources and their evaluation, energetically speaking, are the bases for any strategy which envisages a sustainable management in order not to endanger the resources and simultaneously use their potential as much as possible.

A similar situation regarding subgeothermal resources is found in the territory of Serbia. The first evaluation of subgeothermal resources potential in the area of Belgrade city was performed in 2012 after a 2-year-long research. The research was multidisciplinary. Apart from hydrogeologist whose task was to evaluate subgeothermal potential of the Serbian territory, the significance was placed on the possibility of using these resources, so the experts in engineering and architecture were also research team members.

21.2 Methodology

21.2.1 Geographical Position of Research Area

The Republic of Serbia is located in southeastern Europe (Fig. 21.1). The capital city of Serbia is Belgrade, with the population of 1,650,000 people. The city area occupies the surface of 770 km^2. The city area belongs to the moderate–continental climate. Mean annual rainfall is about 680 mm, whereas average annual air temperature values are about 12 °C, relative air humidity is about 70 %, and atmospheric pressure is about 1001 Mbar.

In the past few years, there has been a trend of people's migration to urban areas, which has resulted in the spread of the Belgrade urban area. An intensive urbanization process along with the disappearance of "free" green areas in comparison to asphalt and concrete areas, on the one hand, leads to the change of conditions and quality of the environment and on the other, to the increase in energy needs, since buildings are regarded as one of the greatest energy consumers. In order to meet the energy needs (heating and cooling, lighting), the construction sector "spends" about 40 % of the world energy, 16 % of fresh water, and 25 % of wood timber, and in this process 70 % of the total emission of sulfur dioxide is released into the atmosphere and 50 % of the total carbon dioxide emission [5].

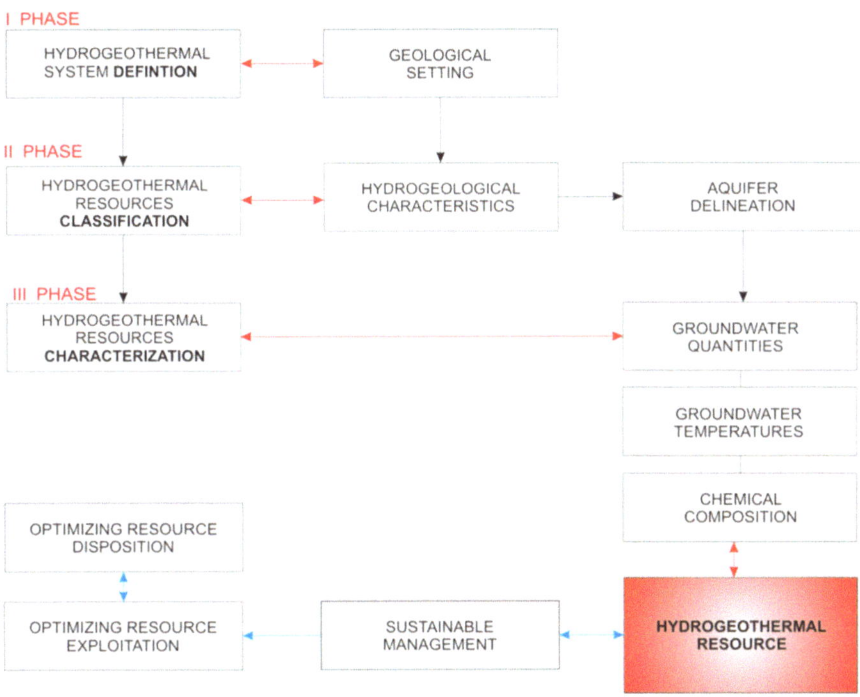

Fig. 21.1 Flowchart of carried out research methodology

In conformity with the city's sustainable development, which is based on the increase of energy efficiency in the buildings as well as on the decrease of a negative impact on the environment and increase of renewable energy resources' share, the first assessment of the area's hydrogeothermal potential has been made. Simultaneously, with the assessment of hydrogeothermal potential, what was also analyzed is the possibility and ways of applying hydrogeothermal resources according to the energy needs of the city of Belgrade. A very important aspect of research was related to the development of mechanisms of sustainable management of hydrogeothermal resources, that is, protection of both resources and the environment as a whole.

21.3 Applied Research Methodology

Complex geological and tectonic features of the area, along with a range of hydrogeological specific qualities and different conditions of the resources exploitation (urban and rural parts of the city area), have given rise to a new methodological approach to researching hydrogeothermal energy. The investigation were carried out in several phases, and each phase was conditioned by the results of the previous one (Fig. 21.2).

In phase one of the research, geological structure of the terrain was observed and it is characterized by an expressed complexity, and this was contributed to by different conditions of sedimentation and paleorelief features of sedimentation basin as well as active tectonics. The oldest rocks in the Belgrade area belong to Paleozoic and the youngest ones to the Quaternary period. Starting with Paleozoic, in the terrain there are Mesozoic and Cenozoic rocks' sections, and here hiatus are spot-

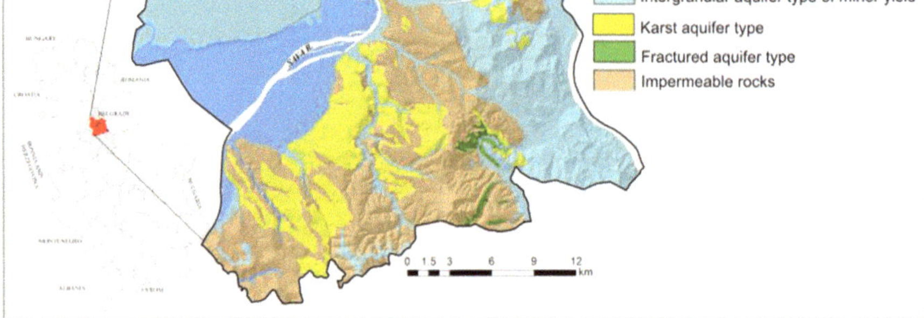

Fig. 21.2 Simplified hydrogeological map of the area of Belgrade

ted in the sedimentation. The oldest members of the Mesozoic era in the inner city area are the Jurassic rocks. The complete development of cretaceous period can be followed in the city area. Flysch sediments of both early and late cretaceous age represent the widest spread sediments in the Belgrade territory. Apart from flysch sediments during the cretaceous period, the facet of drift and subdrift limestone of urgon development was also formed. In the course of tertiary, the facet of limestone of Tortonian and Sarmatian age was formed and from the hydrogeological aspect it has an important role in the formation of groundwater. Apart from these limestone sediments, an important role in the formation of groundwater resources is played by sand–gravel sediments of alluvial deposits of the two big rivers: the Sava and the Danube.

After a detailed analysis of geological structure of the city area, numbers and types of hydrogeothermal systems in the city area have been defined. The definition of hydrogeothermal systems has been carried out according to the lithostratigraphic features of the rocks where the aquifer was previously formed (Table 21.1).

In phase two, in each of the hydrogeothermal systems the classification of hydrogeothermal resources was carried out and it previously required the analysis of the terrain's hydrogeological features.

From the aspect of groundwater formation in the Belgrade area, the most significant aquifer types are intergranular and karst. The intergranular aquifer type of higher yield was formed within sand–gravel deposits from *Corbicula fluminalis* (alluvial deposits of the Sava and Danube rivers), whereas intergranular aquifer type of lower yield was formed in the remaining sand–gravel deposits of Quaternary age and neogene sand deposits. The average thickness of alluvial deposits is about 14 m, whereas the average thickness of sand layers of the Neogene complex ranges from 24 to 30 m. Average yields of wells within sand–gravel deposits of Quaternary age range from 7 to 11 l/s, whereas within Neogene sand deposits yields range from 3 to 6 l/s. Karst aquifer type in the city area was formed within the limestone of urgon age and Torton–Sarmatian limestone whose distribution is significantly wider in comparison to urgon limestone. The average thickness of Torton–Sarmatian limestone in central parts of the city is about 35 m and the wells' yield is about 6 l/s.

Apart from the intergranular and karst aquifer types in the city area, there are also fractured aquifers within serpentinite of Jurassic age in the outermost southeastern terrain parts and parts of the terrain without any hydrogeological function, that is, insignificant aquifers.

Figure 21.2 is showing hydrogeological map of the Belgrade territory.

Table 21.1 Hydrogeothermal systems in the area of Belgrade

Hydrogeothermal system			Identification
1.	Within Quaternary sediments		$_{HGTS}Q$
2.	Within pre-Quaternary sediments		$_{HGTS}T$
	Within tertiary sediments	Within cretaceous sediments	

HGTS hydrogeothermal systems

Table 21.2 Classification of hydrogeothermal resources

Hydrogeothermal system		Lithology	Identification labels
1.	$_{HGTS}Q$	Sands and gravel from *Corbicula fluminalis*	A_1
2.		Quaternary sands	A_2
3.	$_{HGTS}T$	Pliocene sands	B_1
4.		Sarmatian and Tortonian sands	B_2
5.		Tortonian and Sarmatian limestones	B_3
6.		Cretaceous limestones	B_4

HGTS hydrogeothermal systems

After a detailed analysis of hydrogeological features of Belgrade, there have been registered six types of hydrogeothermal resources with the corresponding identification labels (Table 21.2).

In phase three, for each type of hydrogeothermal resources there has been carried out the characterization including the following: an aquifer's geometry (aquifers distribution, bedrock, and overlying bed), zones of aquifer recharging and discharging, basic hydrodynamics characteristics, average several-years temperatures, and basic chemical composition of the groundwater.

21.4 Results

According to the assessed quantities of hydrogeothermal resources (l/s), then the temperature difference which can be achieved on the temperature exchanger (°C) and, finally, the specific heat capacity of the water (KJ/kg/°C), total hydrogeothermal potential of the Belgrade city area is approximately 1200 MW_t (Table 21.3).

Table 21.3 Estimated heat capacity of hydrogeothermal resources in the Belgrade area

Hydrogeothermal system		Hydrogeothermal resource	Quantities	Groundwater temperature	Temperature difference	Potential heat capacity
			Q (l/s)	$T_{av.}$ (°C)	ΔT (°C)	MW_t
1.	$_{HGTS}Q$	A_1	13,775	12.4	7.4	430
2.		A_2	5700	15.0	10.0	240
3.	$_{HGTS}T$	B_1	11,500	14.0	9.0	435
4.		B_2	375	20.0	15.0	24
5.		B_3	900	18.5	13.5	51
6.		B_4	–	–	–	–
Total						1180

HGTS hydrogeothermal systems

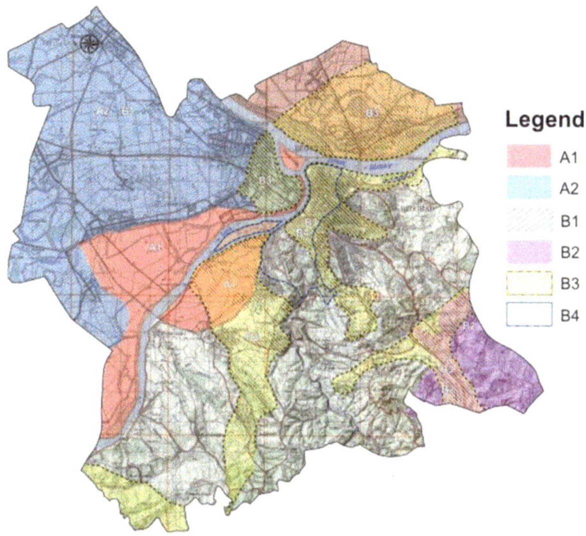

Fig. 21.3 Map of hydrogeothermal potential of the Belgrade city area

When calculating heat capacity, the adopted groundwater temperature value (ΔT) which can be realized on the heat exchanger is up to 5 °C.

From Table 21.3, one may see that the highest hydrogeothermal potential is found in the terrain parts built of sands of Pliocene age, then in the terrain parts built of the alluvial deposits of the Sava and Danube rivers of Quaternary age, whereas the lowest value is found in the sands of Torton–Sarmatian age. Due to an insufficient number of input data, hydrogeothermal resources formed in the limestone of cretaceous age could not be energetically evaluated.

The direct outcome of the performed research is a map of hydrogeothermal potential of the area, which is included into main basis for making detailed regulation plans for the city (Fig. 21.3).

21.5 Discussion

The first assessment of hydrogeothermal potential of the Belgrade city area was carried out up to the depth of 300 m. The criterion for defining the in-depth cut of the research was based on financial viability of resource exploitation, and it is related to the method of drilling exploitation boreholes. The temperatures of the groundwater resources up to 300 m deep, range from 12 to 25 °C, so their use for the sake of the building air-conditioning is determined by the application of heat pumps.

The results of a 2-year-long research pointed to a very significant hydrogeothermal potential of the Belgrade city area. The highest hydrogeothermal potential is found in the terrains built of Pliocene sands (435 MWt), and that is almost as much as thermal power of hydrogeothermal resources formed within alluvial deposits from *C. fluminalis*. Then, comesthe terrains made of the sands of Quaternary age (240 MWt), then terrains built of Torton–Sarmatian limestone (51 MWt), whereas Torton–Sarmatian sands have the lowest geothermal potential (24 MWt). When it comes to the entire territory, the estimated thermal power of hydrogeothermal resources is about 1200 MWt.

Comparing the obtained results with the current energy needs of the city, it is concluded that hydrogeothermal resources could meet about 50 % of the city heating needs. Namely, half of the total heating requirements of the urban residential and business funds in the Belgrade city area is met with the district heating system, whose total installed capacity equals 2500 MWt and the other half is met with the electric energy, up to 20 %; gas, 20 %; and hard mineral fossil fuels (coal, wood), 10 %. According to the above-mentioned data, hydrogeothermal resources are a strategically significant resource for the further development of the Belgrade city area.

Along with the research of the terrain's hydrogeothermal potential, the background for sustainable management of these resources was formed, namely, with the aim of sustainable management of hydrogeothermal resources in the city area, which is based on optimizing resources exploitation; a systematization and organization of the data were carried out by means of geographic information system (GIS) technology. Two main categories of data were formed. The first category includes the data related to a hydrogeothermal resource, its quantity, temperature, and quality, and the other category consists of the data related to the city development plans (urban basis) and the legal regulations for the sphere of exploitation and application of hydrogeothermal resources.

References

1. EHPA and Delta Energy & Environment (2013) Unleashing the opportunity, European best practice in building successful heat pump markets,
2. EHPA (2014) Heat pumps and EU targets. http://www.ehpa.org/technology/heat-pumps-and-eu-targets
3. Platt M et al (2010) Analyse des deutschen Wärmepumpenmarktes–Bestandsaufnahme und Trends, GeothermieZentrum Bochum, Hochschule Bochum, Bochum University of Applied Sciences
4. Spadoni L et al (2013) Geothermal heat pump for A2A district heating service in Milan (Italy). Proceedings of European Geothermal Congress, 3–7 June, Pisa, Italy
5. Santamouris M (2007) Energy in buildings and citizenship. Proceedings of 38 international congress on heating, cooling and air conditioning. 5–7 December, Belgrade, pp 13–26, Belgrade

Chapter 22
Automatic Optimization of Multiple Borehole Heat Exchanger Fields

Peter Bayer, Markus Beck and Michael de Paly

Abstract Borehole heat exchangers (BHE) are often applied in multiple BHE fields. In current planning practice, interaction between adjacent BHEs is rarely considered, and all BHEs are operated in the same mode. This means, potential adverse effects from superimposed cold or heat plumes, which simultaneously evolve around individual neighboring BHEs, are neglected. The long-term heat extraction over decades, however, may lead to a significant local cooling, especially in the interior of the field. As a consequence, the performance of the complete ground source heat pump (GSHP) system is attenuated, and ground temperatures below regulation thresholds may develop. In our work, we employ mathematical optimization techniques to strategically operate and arrange BHEs in such fields. Linear programming and an evolutionary algorithm are applied in combination with analytical equations to solve realistic problems. The presented methodology is flexible and robust, and it can be applied to various conditions. The two scenarios studied in this chapter represent conditions with negligible and significant groundwater flow. We inspect a field with 36 BHEs, which has a seasonally variable heating energy demand. It is demonstrated, by taking the maximum temperature decline in the ground as objective, that the BHE field performance can be improved by both case-specific ideal arrangement and time-dependently regulated individual BHE operation. It is found that instead of standard lattice arrangements, optimized geometries are favorable, with BHEs concentrated along the fringe of a field. Apparently, this enhances lateral conductive heat provision into the field. Groundwater flow means additional energy provision by advection towards the field.

Keywords BHE · Optimization · Modeling · Groundwater

P. Bayer (✉)
Department of Earth Sciences, ETH Zurich, Sonneggstrasse 5, 8092 Zurich, Switzerland
e-mail: bayer@erdw.ethz.ch

M. Beck · M. d. Paly
Wilhelm-Schickard-Institute for Computer Science (WSI), University of Tübingen, Sand 1, 72076 Tübingen, Germany

A. Sayigh (ed.), *Renewable Energy in the Service of Mankind Vol I*,
DOI 10.1007/978-3-319-17777-9_22

22.1 Introduction

The combined use of multiple borehole heat exchangers (BHEs) in large-scale low-enthalpy geothermal applications is a common practice, for example, to supply district heating systems or big office buildings. Multiple simultaneously operating BHEs can access the ground over a larger area than single BHEs, and it is not required to drill deep boreholes, which may be costly and/or face regulative barriers [1, 2, 3]. A common approach is to arrange the BHEs in a lattice, line(s), or L-shape with sufficient distance between them (e.g., Refs. [4, 5, 6]). For example, Signorelli et al. [7] recommend a distance of 7–8 m, which is considered adequate to avoid substantial interaction among neighboring installations. However, even smaller distances of around 5 m are chosen in practical applications (e.g., Ref. [8]).

BHEs are commonly operated for a long time. Mostly, an annually balanced operation with the same amount of heat extracted as injected is not achieved. When full regeneration is not possible, permanent thermal anomalies develop and grow in the ground [9, 10]. Their size mainly depends on the thermal conditions of the ground, potential hydrogeological factors, intensity of heat extraction, and operation mode. If, as is often the case, the prevailing application of a BHE field is for heating, the ground is cooled. In the long term, the thermal anomalies from individual BHEs expand and potentially overlap. Even with a distance of several meters, such interaction could be relevant, as conductive heat supply to the individual BHEs is mitigated. With unchanged heat extraction rates, further local temperature decline in the field could become significant [11, 12]. Strong cooling of the ground is not desirable, as an altered ground(water) temperature influences the seasonal performance factor of the heat pump(s), and for ecological reasons it is often not even allowed. In several countries, regulations set or recommend minimum temperature thresholds for ground and groundwater [13].

Our goal is to find a strategy that minimizes the local cooling of the ground. For a given volume of the ground, this can be achieved either by applying different heat extraction rates to the BHEs in the field or by positioning the BHEs in an optimal way. In the following, we first present a methodology, which integrates superimposed BHE line source models in an optimization framework. The objective is to achieve a balanced cooling from the ground by avoiding extreme temperatures, to improve system performance, and to promote sustainable operation this way. Then, based on our previous work [3, 11, 12, 14], we inspect the optimization potential of two scenarios with 36 BHEs.

22.2 Methodology

22.2.1 Analytical Simulation

The temperature evolution, T, in the ground during BHE operation is modeled by the infinite moving line source model [15, 16, 17]. This analytical solution of heat transport is a fast and efficient means to simulate simplified homogeneous conditions with conduction and advection. The thermal effect from heat extraction is described as temperature change ($\Delta T = T_u - T$) in comparison to undisturbed conditions with natural background temperature, T_u. For a specific heat extraction rate (or load) per unit borehole length, q, we obtain for a single BHE at time t at distances Δx and Δy:

$$\Delta T(\Delta x, \Delta y, t, q) = \frac{q}{4\pi L\sqrt{\lambda_u \lambda_v}} \exp\left(\frac{\rho_w c_w n_p v_a \Delta x}{2\lambda_u}\right) \times \int_0^b \exp\left[-\left(\frac{\Delta x^2}{\lambda_u} + \frac{\Delta y^2}{\lambda_v}\right) \times \frac{(\rho_w c_w n_p v_a)^2}{16\lambda_u \Theta} - \Theta\right] \frac{d\Theta}{\Theta}, \tag{22.1}$$

where

$$b = \frac{(\rho_w c_w n_p v_a)^2 t}{4\rho_m c_m \lambda_u}.$$

The effective thermal conductivity in the longitudinal and transverse direction is given by λ_u and λ_v. Parameters c_w and c_m are the volumetric heat capacities of water and porous matrix, ρ_w and ρ_m are the densities, n_p is the porosity, and v_a is the horizontal groundwater flow velocity. Equation (22.1) is linear with respect to q assuming that the thermal and hydraulic properties of the ground are independent from the temperature. This allows spatial superpositioning, and we obtain for n BHEs in a borehole field:

$$\Delta T_{i,j}(t, q_{k=1,\ldots,n}) = \sum_{k=1}^{n} \Delta T_k (i - x_k, j - y_k, t, q_k). \tag{22.2}$$

The total temperature change at location (i, j) is the sum of all changes ΔT_k by each BHE k at position (x_k, y_k), with an energy transfer rate q_k. Energy extraction varies during the year, typically in a seasonal mode and higher demand during cold winter periods. To describe time dependency of q, m time steps with constant q_l for time step l, with $q_0 = 0$ and $t_0 = 0$, are distinguished for each BHE. By spatial and temporal superposition, the temperature change by operation of multiple BHEs can be calculated by [3]:

$$\Delta T_{i,j}(t, q_{k=1,\ldots,n,1,\ldots m}) = \sum_{l=1}^{m}\sum_{k=1}^{n} q_{k,l}\,\omega_{k,l}^{t,i,j}(i - x_k, j - y_k), \tag{22.3}$$

with

$$\omega_{k,l}^{t,i,j}(\Delta x, \Delta y) = \frac{1}{4\pi L\sqrt{\lambda_u \lambda_v}} \exp\left(\frac{\rho_w c_w n_p v_a \Delta x}{2\lambda_u}\right) \times \left(\int_0^{b_{l-1}} \exp\left[-\left(\frac{\Delta x^2}{\lambda_u} + \frac{\Delta y^2}{\lambda_v}\right) \times \frac{(\rho_w c_w n_p v_a)^2}{16\lambda_u \Theta} - \Theta\right]\frac{\mathrm{d}\Theta}{\Theta} - \int_0^{b_l} \exp\left[-\left(\frac{\Delta x^2}{\lambda_u} + \frac{\Delta y^2}{\lambda_v}\right) \times \frac{(\rho_w c_w n_p v_a)^2}{16\lambda_u \Theta} - \Theta\right]\frac{\mathrm{d}\Theta}{\Theta}\right), \tag{22.4}$$

with

$$b_{l-1} = \frac{(\rho_w c_w n_p v_a)^2 t_{l-1}}{4\rho_m c_m \lambda_u}, \quad b_l = \frac{(\rho_w c_w n_p v_a)^2 t_l}{4\rho_m c_m \lambda_u}.$$

Coefficient $\omega_{k,l}^{t,i,j}$ is the response factor of BHE *k* on *(i, j)* at time step $l \in \{l, \ldots, m\}$, in relation to the current time step $t \in \{l, \ldots, m\}$ and $l'' \; t$. In vectorized form, Eq. (22.3) reads

$$\Delta \vec{T}_{i,j}(t, \vec{q}) = \vec{q}(\vec{\omega}^{t,i,j})^{\mathrm{T}}, \tag{22.5}$$

with

$$\vec{\omega}^{t,i,j} = (\omega_{1,l}^{t,i,j}, \ldots, \omega_{n,1}^{t,i,j}, \ldots, \omega_{n,m}^{t,i,j})$$
$$\vec{q} = (q_{1,1}, \ldots, q_{n,1}, \ldots, q_{n,m}).$$

22.2.2 Multiple BHE Scenarios

We demonstrate the combined simulation–optimization approach for a hypothetical BHE field [14]. In total, 36 BHEs are arranged on a regular grid on a 50 m × 50 m (Fig. 22.1). Accordingly, a grid distance of 10 m between neighboring BHEs is chosen, which is several meters above thresholds we find in common practice and recommendations. The depth per BHE is set fixed to 100 m. The field is operated with time-dependent loads to supply a given and annually constant seasonal energy demand over an operation time of 10 years (Fig. 22.2). For this demonstration example, only two periods are subdivided per year, but this could be easily expanded and a finer time discretization could be used (e.g., Hecht-Mendez et al. 2012).

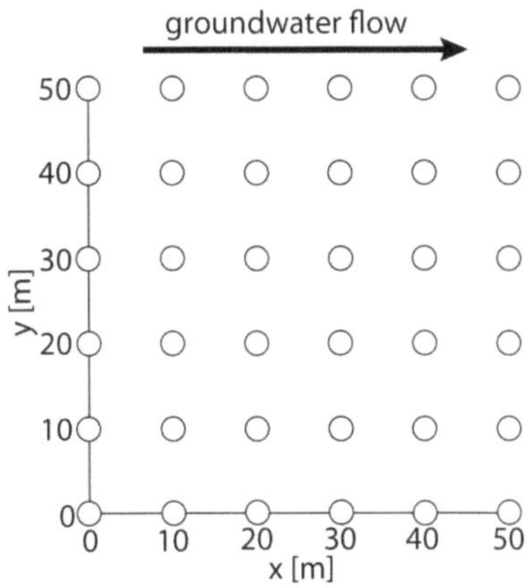

Fig. 22.1 Lattice arrangement of 36 BHEs on regular grid with a step size of 10 m

Two scenarios are distinguished, oriented at realistic conditions [14]:

1. *Conduction-dominated case:* The BHEs are operated in water-saturated ground with low groundwater flow velocity of $v_a = 2.4 \times 10^{-10}$ m/s. Such conditions reflect a silt/clay dominated ground, and accordingly, we specify the other thermal parameters as follows: volumetric heat capacity is 2.4×10^6 Jm^{-3} K^{-1}, thermal conductivity is 1.7 Wm^{-1} K^{-1}, and porosity is 46%.
2. *Advection-dominated case:* The BHEs are installed in a sand aquifer with substantial groundwater flow of $v_a = 2.0 \times 10^{-7}$ m/s. Volumetric heat capacity here is 2.6×10^6 Jm^{-3} K^{-1}, thermal conductivity is 2.4 W m^{-1} K^{-1}, and porosity is 30%.

22.2.3 *Optimization Problem Formulation and Solution Procedure*

We choose the temperature change of the ground as the optimization criterion. The purpose is to keep the induced thermal anomalies as small as possible in order to maximize the heat pump efficiency and to avoid extreme local cooling in the field. This is implemented in the optimization procedure by minimization of the strongest cooling, simulated by Eq. (22.4):

$$min\left[max\left(\Delta T_{i,j}\right)\right]. \tag{22.6}$$

The sum of all loads $q_{k,l}$ has to match the energy demand E_l for each time step l:

$$E_l = \sum_{k=1}^{n} q_{k,l} \quad l = 1,\ldots,m. \tag{22.7}$$

Four reference points at a distance of 0.5 m around each BHE k at position (x_k, y_k) are defined to measure the ground temperature changes.

There are two ways to optimize the field. Either individual loads of the BHEs are separately adjusted [3, 11] or their positions are optimized [12, 14]. Load optimization is possible by linear programming, and consequently, this is applied for minimization of the maximum temperature change during each time step l, combining Eqs. (22.5) and (22.6):

$$argmin\left(w \cdot max\left(\Delta\vec{T}_{i,j}\right)\left(t,\vec{q}\right) + \sum_{l=1}^{m} max\left(\Delta\vec{T}_{i,j}\left(l,\vec{q}\right)\right)\right) \quad \forall(i,j,t) \in S, \tag{22.8}$$

where S is the set of all spatial and temporal reference points, defined by (i, j, t); w is introduced as weighting factor to assign a high priority to the primary objective (Eq. 22.5) by $w = 100$.

A general formulation is possible by introducing virtual variables $z_0,\ldots,m$. The max-norm terms of the combined objective expressed by Eq. (22.6) can be formulated as $m+1$ linear programs, which minimize $z_0,\ldots,m$. For every z, the linear program reads

$$\begin{aligned} &min(z) \\ &\Delta\vec{T}_{i,j}\left(l,\vec{q}\right) - z\vec{e} \prec 0 \\ &-\Delta\vec{T}_{i,j}\left(l,\vec{q}\right) - z\vec{e} \prec 0, \end{aligned} \tag{22.9}$$

with $\vec{e}$ being the unity vector. With the linear relationship between $\Delta\vec{T}$ and q, respectively, between $\Delta\vec{T}_{i,j}$ and $\vec{q}$, the linear optimization problem can be stated as:

$$min\left(w \cdot z_0 + \sum_{l=1}^{m} z_l \right) \tag{22.10}$$

subject to

$$\begin{aligned} &\Delta\vec{T}_{i,j}(t,\vec{q}) - z_0 \prec 0 \\ &-\Delta\vec{T}_{i,j}(t,\vec{q}) - z_0 \prec 0 \\ &\Delta\vec{T}_{i,j}(l,\vec{q}) - z_l \prec 0 \\ &-\Delta\vec{T}_{i,j}(l,\vec{q}) - z_l \prec 0 \\ &\forall(i,j,t) \in S, \end{aligned} \tag{22.11}$$

with $l = 1,\ldots,m$ and $z_0,\ldots,m$. and q as optimization variables. The linear program is constrained by the energy demand for each time (Eq. (22.6), see Fig. 22.2). As only heating is considered, the loads are positive:

$$q_{k,l} \geq 0 \quad l = 1,\ldots,m \text{ and } k = 1,\ldots,n. \tag{22.12}$$

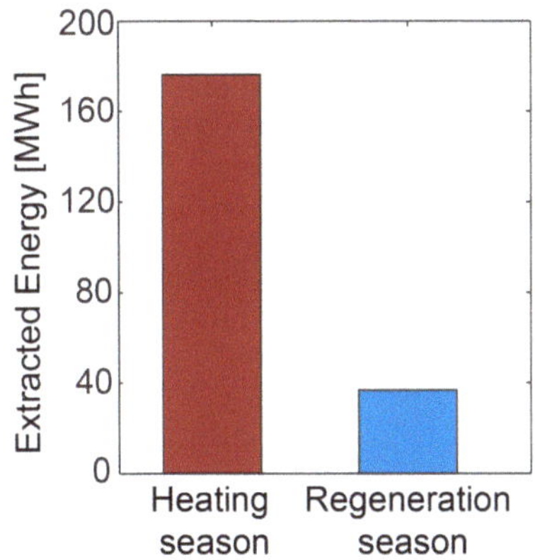

Fig. 22.2 Constantly extracted energy during winter (heating) and summer (regeneration) season for each year of operation

Aside from load optimization, another option is to adjust the positions of the BHEs. Since the temperature $\vec{T}_{i,j}$ at a position *(i, j)* is nonlinearly dependent on the position (x_k, y_k) of a borehole, a nonlinear optimization method is needed. For this objective, we selected a heuristic method, differential evolution (DE) [18]. This is an established evolutionary algorithm, and we chose standard settings with population size of 30, algorithm-specific tuning parameter $F = 0.8$, and cutoff probability of $Cr = 0.7$. This variant revealed to be favorable to alternative evolutionary global search methods [14].

22.3 Results

The results are depicted in Figs. 22.3 and 22.4. In Fig. 22.3, we show the maximum temperature changes for the two different scenarios. For scenario 1, with small groundwater flow velocity, the same equal load operation would result in a stronger maximum temperature decline. ΔT of 9.75 K is computed for this variant, whereas the relative temperature change reaches only $\Delta T = 5.6$ K for scenario 2. This reflects the more pronounced additional advective heat component, which is characteristic for higher groundwater flow velocities and which promotes seasonal regeneration.

It is demonstrated that any mathematical optimization mitigates the maximum temperature change. Applying linear programming for load optimization only means a substantial reduction to $\Delta T = 7.25$ K and 3.52 K for scenarios 1 and 2, respectively. This is a relative improvement of around 30 %. When only positions are adjusted by DE, the resulting BHE arrangements perform better than the grid shaped BHE field. However, for both cases, load optimization is favorable to position optimization. The boxplots also show that repeated applications of the heuristic algorithm generate slightly variable results. This reflects that DE is a stochastic

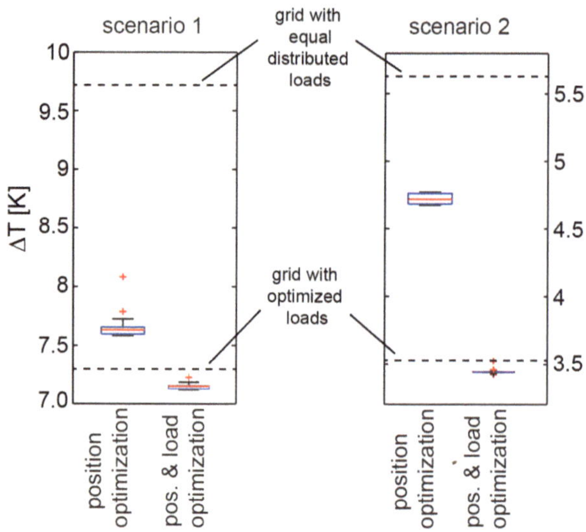

Fig. 22.3 Results for the two scenarios in comparison to default case with lattice arrangement and equal loads. Both boxplots show the resulting maximum temperature changes for exclusive position optimization, load optimization, and combined position/load optimization

search procedure and needs to be employed several times to arrive at a reliably good solution. In contrast to linear programming, the stochastic search is computationally more demanding and does not guarantee to find the global optimum. Even so, experience with this and alternative evolutionary algorithms indicates that the detected "optima" are close to the global optimum [14].

Combining position and load optimization in a sequential procedure turns out to be the best choice, which exploits all degrees of freedom. However, as shown in Fig. 22.3, the improvement potential is only minor. Apparently, by concentrating on the loads only, already potentially satisfactorily close optimal solutions are generated.

The best solutions for the two scenarios are illustrated in Figs. 22.4 and 22.5. BHEs tend to concentrate on the fringe of the given area, where lateral heat conduction is not insulated by adjacent BHEs (see also Refs. 12, [14]). As expressed in the objective function, strong local thermal anomalies are avoided and a balanced cooling within the BHE field is achieved. For the second scenario, cooling is less pronounced due to stronger advection. What is also shown in Fig. 22.5 with substantial groundwater flow, the thermal plume mainly generated during the winter moves downstream. This means advection not only does provide additional energy but also moves cooled regions away from the interior of the BHE field. This accelerates regeneration but may be critical for neighboring geothermal applications in the downstream.

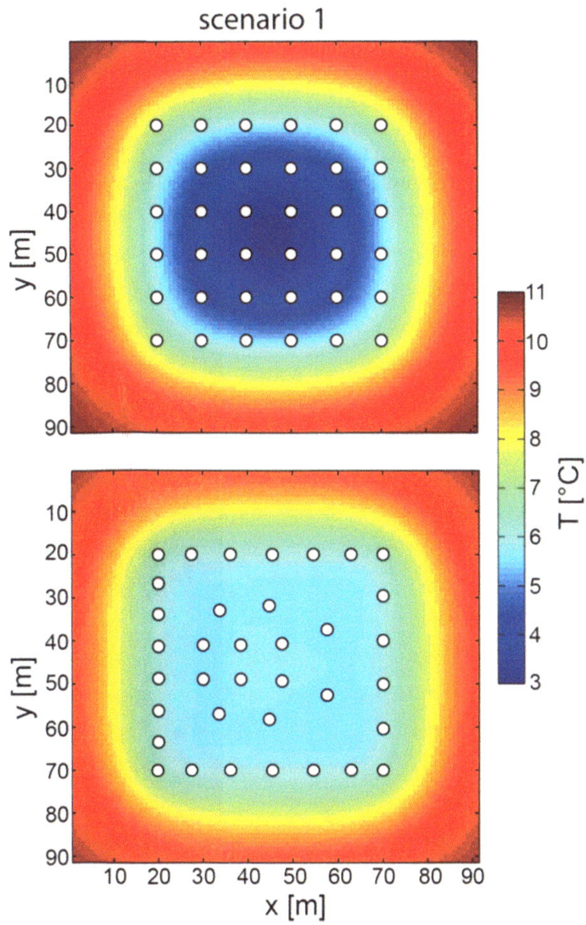

Fig. 22.4 Scenario 1: Absolute temperatures for position optimized BHE fields with groundwater flow from left to right (90 m × 90 m area). *Top* figure shows equal loads, and *bottom* one shows load optimized results

22.4 Conclusions

With the two scenarios, we demonstrated the potential of combined simulation–optimization methods to properly adjust positions and/or extraction rates of individual boreholes in multiple BHE fields. This is shown for conditions with mainly conduction and negligible groundwater flow, as well as for conditions with pronounced advection.

The selected scenarios are idealized examples but clearly show that mathematical optimization can be employed to achieve a more balanced heat extraction from the borehole field than is typical for standard application without BHE regulation. The nonregulated reference case with equal loads, however, only approximates

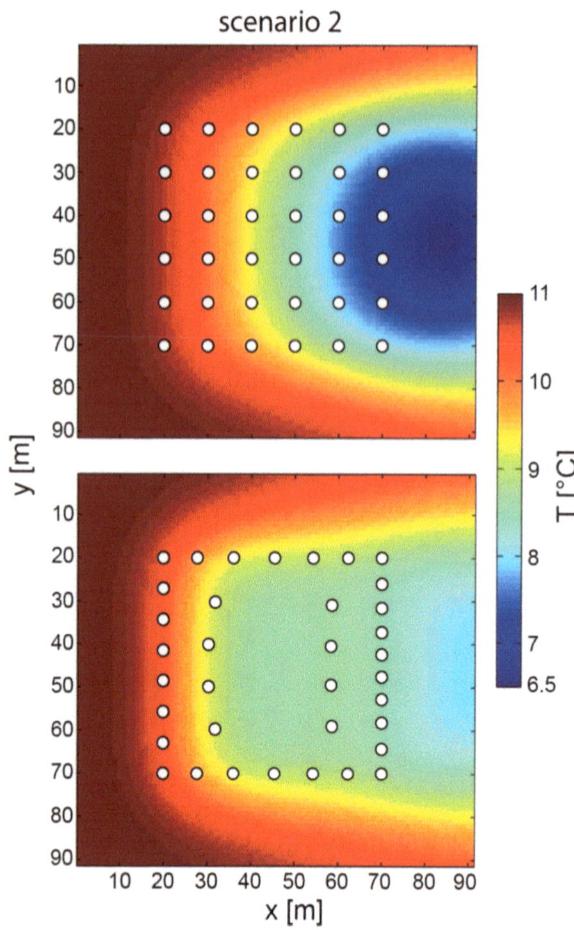

Fig. 22.5 Scenario 2: Absolute temperatures for position optimized BHE fields with groundwater flow from left to right (90 m × 90 m area). *Top* figure shows equal loads, and *bottom* one shows load optimized results

conditions in practice. For given total extraction rates, the BHE-specific extraction rates typically are not the same. In fact, by equal flow rates of the heat carrier fluids in the BHEs, heat extraction will depend on the local temperature gradient at each borehole. This can be interpreted as a self-regulating mechanism which automatically mitigates extreme local cooling. Still, further work in this direction [3, 11, 12] with numerical models shows that such equal flow conditions are commonly far from being optimal. Using load-based assessment and equal loads as proxy, thus, is a viable and computationally efficient work-around. Future work will concentrate on real-case applications and integration of combined heating and cooling modes.

References

1. Sanner B, Mands E, Sauer MK (2003) Larger geothermal heat pump plants in the central region of Germany. Geothermics 32:589–602.
2. Katsura T, Nagano K, Narita S, Takeda S, Nakamura Y, Okamoto A (2009) Calculation algorithm of the temperatures for pipe arrangement of multiple ground heat exchangers. Appl Therm Eng 29:906–919
3. de Paly M, Hecht-Mendez J, Beck M, Blum P, Zell A, Bayer P (2012) Optimization of energy extraction for closed shallow geothermal systems using linear programming. Geothermics 43:57–65
4. Fujii H, Itoi R, Fujii J, Uchida Y (2005) Optimizing the design of large-scale ground-coupled heat pump systems using groundwater and heat transport modelling. Geothermics 34: 347–364
5. Choi JC, Park J, Lee SR (2013) Numerical evaluation of the effects of groundwater flow on borehole heat exchanger arrays. Renew Energy 52:230–240
6. Koohi-Fayegh S, Rosen MA (2012) Examination of thermal interaction of multiple vertical ground heat exchangers. Appl Energy 97:962–969
7. Signorelli S, Kohl T, Rybach L (2004) Sustainability of production from borehole heat exchanger fields. Proceedings of 29th workshop on geothermal reservoir engineering, Stanford University, California, SGP-TR-175, pp 1–6
8. Fossa M, Minchio F (2013) The effect of borefield geometry and ground thermal load profile on hourly thermal response of geothermal heat pump systems. Energy 51:323–329
9. Rybach L, Eugster WJ (2010) Sustainability aspects of geothermal heat pump operation, with experience from Switzerland. Geothermics 39:365–369
10. Zanchini E, Lazzari S, Priarone A (2012) Long-term performance of large borehole heat exchanger fields with unbalanced seasonal loads and groundwater flow. Energy 38:66–77
11. Hecht-Mendez J, de Paly M, Beck M, Bayer P (2013) Optimization of energy extraction for vertical closed-loop geothermal systems considering groundwater flow. Energy Convers Manage 66:1–10
12. Beck M, Bayer P, de Paly M, Hecht-Mendez J, Zell A (2013) Geometric arrangement and operation mode adjustment in low enthalpy geothermal fields for heating. Energy 49:434–443
13. Haehnlein S, Bayer P, Blum P (2010) International legal status of the use of shallow geothermal energy. Renew Sustain Energy Rev 14(9):2611–2625
14. Beck M, de Paly M, Hecht-Méndez J, Bayer P, Zell A (2012) Evaluation of the performance of evolutionary algorithms for optimization of low-enthalpy geothermal heating plants. Proceedings of 14th international conference on genetic and evolutionary computation, Philadelphia, PA, USA, pp 1047–1054
15. Sutton MG, Nutter DW, Couvillion RJ (2003) A ground resistance for vertical borehole heat exchangers with groundwater flow. J Energy Resour—ASME 125:183–189
16. Diao N, Li Q, Fang Z (2004) Heat transfer in ground heat exchangers with groundwater advection. Int J Therm Sci 43:1203–1211
17. Molina-Giraldo N, Blum P, Zhu K, Bayer P, Fang Z (2011) A moving finite line source model to simulate borehole heat exchangers with groundwater advection. Int J Therm Sci 50:2606–2513
18. Storn R, Price K (1997) Differential evolution—a simple and efficient heuristic for global optimization over continuous spaces. J Global Optim 11:341–359

Chapter 23
Thermo-economic Study of Hybrid Thermal Solar and Geothermal Heat Pumps *System* in Algeria

Mounir Aksas, Fouad Khaldi and Rima Zouagri

Abstract In this chapter, the sizing and thermo-economic analysis of hybrid solar thermal and ground source heat pump (GSHP) systems for water and space heating and cooling was carried out. The houses selected for the proposed study are located in rural areas at three different climate regions in Algeria.

The software RETScreen was used to evaluate the technical and economical feasibility of the hybrid system. The software can be used worldwide to evaluate the energy production and savings, costs, emission reductions, financial viability, and risk for various types of clean energy technologies.

Different scenarios into the combined system (geothermal heat pump and thermal solar collectors) have been defined and analyzed using RETScreen software. A comparative analysis of these different scenarios was conducted to determine the best configurations of the hybrid system in terms of energy, economical, and environmental performances.

The result of this study showed that this hybrid system is a feasible choice for space conditioning for cooling–heating houses: Annual energy saving is about 15.21×10^6 MWh for 900,000 rural houses during the period 2010–2014. This energy corresponds to the equivalent of 42.12×10^5 oil barrels.

Keywords Geothermal heat pump · Hybrid renewable energy system · Thermal solar collector · RETScreen

23.1 Introduction

Energy consumption in buildings is a large share of the world's total end use of energy; it accounts for close to 40% of total end use of energy [1]. Energy demand in buildings includes space heating and cooling, water heating, and electrical loads

M. Aksas (✉) · F. Khaldi · R. Zouagri
Energetics Applied Physics Laboratory, Department of Material Physics, Faculty of Sciences, Hadj Lakhdar University, Batna, Algeria
e-mail: m_aksas@hotmail.com

A. Sayigh (ed.), *Renewable Energy in the Service of Mankind Vol I*,
DOI 10.1007/978-3-319-17777-9_23

[2]. However, as the majority of global energy supply is derived from fossil fuels which are not only finite but also harmful resources to the environment, renewable energy resources represent a key solution in this context. Hence, solar energy is often exploited to answer this energy demand [3, 4], and, as sunshine is intermittent, it is frequently supplied by another resource such as geothermal energy; this is made possible with solar-assisted heat pumps. This process, called GEOSOL, meets domestic hot water and heating–cooling building energy needs [5].

Solar energy can be integrated into heat pump systems in a parallel or a series arrangement, making it possible to shift from one resource to another following the availability of the solar input [6, 7].

Several softwares have been developed to integrate the different renewable energy systems either by analyzing the feasibility or by simulating in detail these energy systems. RETScreen is the software we chose to conduct this study. It is a unique tool for decision support, developed in collaboration with many industry experts, the Canadian government and academia. Free, it can be used around the world to assess the production and energy savings, the cost over, the life cycle, emission reductions, and the financial viability and risk for various types of Renewable-energy and Energy-efficient Technologies (RETs).

We conducted this study in order to assess the performance and techno-economic feasibility of a hybrid system made by pulling together a geothermal heat pump with solar thermal collectors. This assessment was set up using a simulation method of the chosen hybrid system in heating rural houses. Indeed, many of these houses in several regions in Algeria are far from the electrical grid and such renewable energy system can provide an alternative to electrical energy resource. The ultimate aim of our study is to provide applicable results for researchers and building contractors who can use them when designing and installing hybrid geothermal heat pump systems in heating-dominated buildings in Algeria.

23.2 Solar and Geothermal Potential in Algeria

The main resource of renewable energy is solar energy. With its ideal location, Algeria has the largest solar field in the Mediterranean Basin. The average duration of sunshine in Algerian territory exceeds 2000 h/year and may reach nearly 3900 h of sunshine in the Sahara desert. The daily energy received on a horizontal surface of 1 m^2 is about 5 kWh of most of the national territory and corresponds nearly to1700 kWh/m^2/year in the northern region and 2263 kWh/m^2/year in the south of the country. The total energy received is estimated at 169,400 TWh/year, or 5000 times the annual electricity consumption of the country [8]. Figure 23.1 shows the maximum solar radiation received by Algeria on the unmodified plane in July.

Regarding geothermal potential, Algeria has an important one, and currently balneology constitutes its main use. Direct applications in the thermal field are very limited [11].

The Jurassic limestones of North Africa which represent important geothermic reservoirs are at the origin of more than 200 hot springs mainly located in the

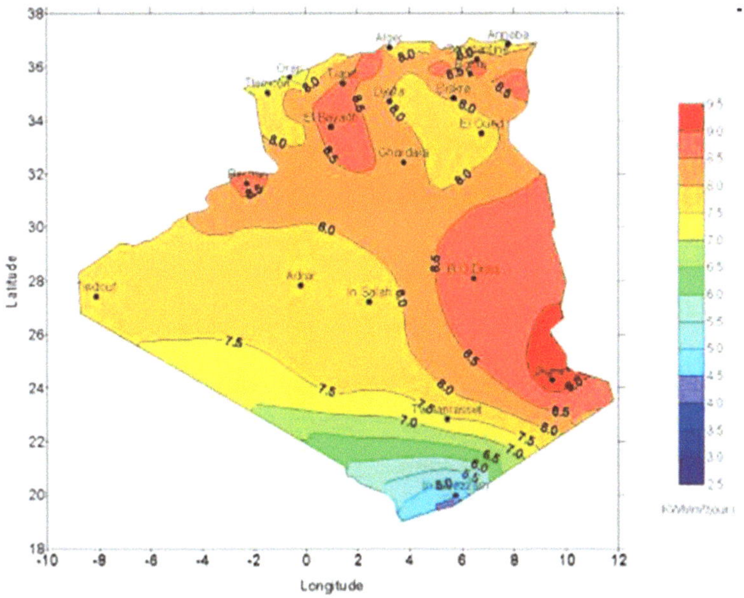

Fig. 23.1 Daily global irradiation received on normal plane in July [9]

Northeast and the Northwest regions of the country [10]. The temperature of these hot springs varies between 22 and 98 °C [11], and about one thirds temperatures are superior to 45 °C [12].

Deeper in the South, the continental rock formation constitutes a great geothermal reservoir extending over several 1000 km^2. This reservoir commonly called the "albian platform" is exploited through drilling, at a flow rate of over 4 m^3/s (Fig. 23.2). This water has an average temperature of 57 °C [10]. Broadly, the Algerian desert (Sahara) has a mean geothermal gradient of 4 °C/100 m.

23.3 Description of Hybrid Solar–Geothermal Technology

Hybrid solar–geothermal systems should gather heating, cooling, and hot water needs, while ensuring a satisfactory level of comfort and quality of service in all circumstances [5]. These systems utilize solar radiation as a thermal input to offset heat extraction by the heat pump. They include a heat collector surface (which can be installed horizontally or vertically) [13], heat transfer circuit which transfers energy derived from the sensor to the storage element (storage of energy), heat storage, and a distribution network [14]. Solar heat can be utilized either for heating of domestic hot water, heating the building, increasing the temperature in the evaporator in the heat pump, recharging the borehole, or a combination of all [15]. The

Fig. 23.2 Albian platform situation [10]

ground is the heat source for the heat pump and the compressor in the heat pump is using electricity to produce heat for the building and the domestic hot water. Solar collectors may supply heat directly to the domestic hot water system, the building heat distribution system, or support the heat pump in different ways [15].

In general, the sizing of the ground source heat pump (GSHP) will depend on whether the heat pump covers part of the load (bivalent system) or the whole load (monovalent system). In the latter, the heat pump system also includes a hot water heater and an electric booster [15, 16].

When using unglazed solar collectors, the heat can be used either to increase the temperature to the evaporator or for recharging the borehole. When using glazed solar collectors, one possible system is to only use the solar collectors for domestic hot water (Fig. 23.3a), and the other possibility is to use it also for heating the building, the evaporator or the borehole (Fig. 23.3b) [15].

23.4 Application of the Geothermal–Solar Hybrid System for a Rural House

In order to show advantages of geothermal–solar hybrid system installation in rural regions in Algeria, we chose the following case study.

23.4.1 Parameters of the Selected Regions

Algerian climate is highly diversified due to its location in the Mediterranean Sea and the changes in latitude and altitude. We can distinguish three major types of

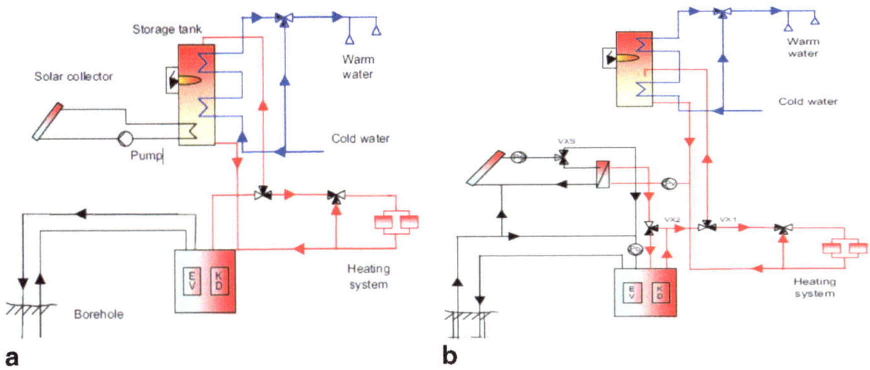

Fig. 23.3 Examples of hybrid solar–geothermal technology. **a** System with a heat pump and glazed solar collectors used for heating domestic hot water. **b** System with a heat pump and solar collectors. Solar heat may be used for (1) heating of domestic hot water, (2) heating the building, (3) heating the evaporator in the heat pump, or (4) to recharge the borehole [15]

Table 23.1 Location and climate of the selected regions in this study

The town	Latitude	Longitude	The climate
Algiers	36.8°N	3.1°E	Tell
Batna	35.8°N	6.3°E	High plains
Ghardaïa	32.4°N	3.8°E	Sahara

climate: the tell climate (a temperate humid Mediterranean climate), high plains climate (a continental-type climate), and the Sahara climate (arid and dry climate). Hence, we chose three towns in Algeria representing, by order, the three different climates: Algiers, Batna, and Ghardaïa. Table 23.1 shows the exact location and the climate of each town.

We used the available meteorological data in RETscreen software to get climatic data of the three towns. The two metrological parameters that play a very important role in solar water heater (SWH) and heat pump installations are solar radiation and soil temperature, Fig. 23.4a and b represents a comparison between metrological parameters of the three sites.

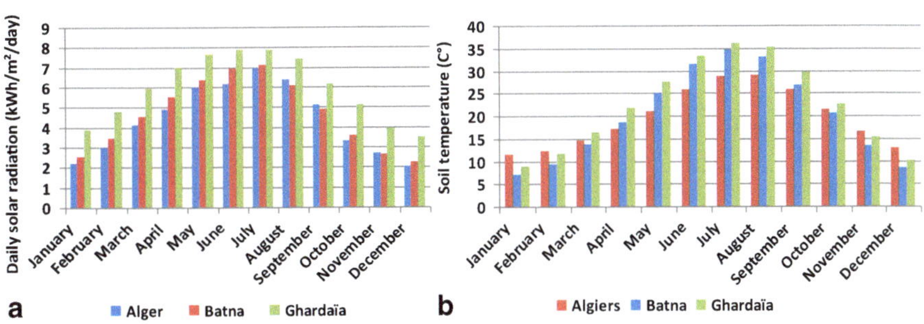

Fig. 23.4 Daily solar radiation (**a**) and soil temperature (**b**) for the three sites

23.4.2 Techno-economical Analysis of the Geothermal–Solar Hybrid System

We studied three scenarios for both SWH and heat pump installations.

23.4.2.1 Heating Water with SWH System

The base case study is an individual house with a surface of 100 m^2, located in the rural area and the number of persons occupying the house is six people. We used the house energetic model of RETScreen software and entered the selected parameters for base case scenario 1 (Table 23.2).

The first scenario is established for a rural house in Algiers. Fuels used in rural sites in Algeria for the production of hot water are diesel or butane, where the price of a liter of diesel is now US$ 0.171, and the price of a kilogram of butane is US$ 0.192. It is, therefore, important to analyze two scenarios of the SWH project

Table 23.2 Input parameters in RETscreen software (basic scenario model 1) for SWH

Parameter	Value	Parameter	Value
Occupant's number	6	Gross area of one collector (m^2)	2.4
Occupancy rate (%)	100	Aperture area of one collector (m^2)	2.2
Domestic hot water demand (L\day)	360	Number of collectors	2
Desired hot water temperature (°C)	60	Ratio of storage capacity to collector area (L\m^2)	75
Operating days per week (day)	7	Miscellaneous losses (%)	5
Tilt angle (°)	36.8	Seasonal efficiency for base case	80
Azimuth (°)	0	Seasonal efficiency for proposed case	80
Solar collectors type	Glazed	Retail price of electricity ($/kWh)	0.022
Equipment manufacturer name	Thermocad	Cold water temperature—minimum (°C)	16.1
Name of the equipment model	2100 T	Cold water temperature—maximum (°C)	20.7
Fr (tau alpha) coefficient	0.69	System capacity (KW)	3.08
Fr UL coefficient (W/m^2)/°C	4.85	Storage capacity (L)	333
Area per collector (m^2)	4.8	Heating energy delivered (MWh)	3.1

for both types of fuel. Area per collector is 4.8 m^2, the recovery rate of the load equals to 49 %, and heating energy delivered is 3.1 MWh. Other scenarios may be developed also for houses that are situated in other rural regions.

The results of the cumulative cash flows are shown in Fig. 23.5. They are produced by RETScreen for the base scenario 1, with a financial support of 45 % and considering that the fuel cost escalation rate is equal to 2 %.

The internal rate of return (IRR) before tax, which represents the actual performance of the project during its lifetime before tax, called also "return on equity invested", in the case of diesel use is 44.2 %. Simple payback and the return on investment (equity; ROI) are of 2.4 and 2.3 years, respectively. In the case of butane use, the IRR is equal to 40.1 % and simple payback and the ROI are of 2.7 and 2.6 years, respectively (see Table 23.3).

It is clear from these results that SWH will be very profitable in this case either when diesel or butane are used as fuel for heating the rural house. On the other hand, reduction in greenhouse gas (GHG) emission is as important as the latter results.

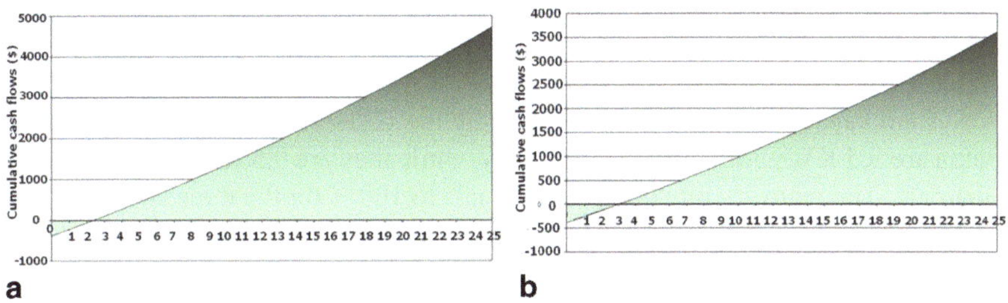

Fig. 23.5 Cumulative cash flow for SWH base case scenario when diesel *(left panel)* or butane *(right panel)* are used as fuel

Table 23.3 Results of SWH alone or GSHP–SWH hybrid systems for rural houses following fuel types and in the three studied regions

Region	Fuel type	Fuel price	Annual energy (MWh)		Annual GHG emission reduction (tCO_2)		Equivalent in oil barrels		Initial cost ($)		IRR (%)	Simple payback (year)	Equity (year)
			A	B	A	B	A	B	A	B	A	A	A
Algiers	Diesel	0.171[a]	3.1	16.1	2	2.5	4.7	5.8	688	1935	44.2	2.4	2.3
	Butane	0.192[b]			1.3	1.7	3	3.9			35	3.1	3
Batna	Diesel	0.171[a]	3.4	16.4	2.1	2.5	4.9	5.8	688	1935	47.2	2.3	2.2
	Butane	0.192[b]			1.3	1.6	3	3.6			37.3	2.9	2.8
Ghardaïa	Diesel	0.171[a]	4.2	18.2	2.1	2.6	4.9	6.1	688	1935	41.1	2.6	2.5
	Butane	0.192[b]			1.8	1.9	4.2	4.4			37.8	2.7	2.6

A, SWH system; B, GSHP–SWH hybrid system

[a] $/L

[b] $/Kg

GHG greenhouse gas, *IRR* internal rate of return

Indeed, net emissions of GHG are reduced by 1.3–2 t of CO_2 equivalent (tCO_2) which corresponds to 3–4.7 barrels of not consumed crude oil. This quantity is important and may become more significant when load recovery rates are increased and when all rural houses in Algeria will be considered.

This study was done for a recovery rate of the load equal to only 49 % and for an "area per collector" equal to 4.8 m^2. We studied other scenarios of SWH projects for other sites (Batna and Ghardaïa) and with higher values for the load recovery rate but with the same value of "area per collector." Results show that heating energy delivered for these two sites is equal to 3.4 and 4.2 MWh for a recovery rate of 50 and 70%, respectively (Table 23.3). We can notice that more the solar radiation more is the economic profitability of the SWH.

23.4.2.2 Heating and Cooling by GSHP

We used RETScreen for GSHP project and chose the energetic model for a house. Input parameters are shown in Table 23.4.

The first scenario was established for both fuel types used as a heating energy resource in rural houses at the three sites and for an inflation rate equal to 2 % (Table 23.5).

System capacity and the provided heat by the GSHP for houses in Algiers and Batna are 9.5 KW and 13 MW, respectively, while they are 9.5 KW and 14 MW for Ghardaïa. The recovery rate of the load equals to 100 % for the three sites.

The second scenario was made for an inflation rate of 9.9 %, which was the inflation rate in the beginning of 2013 [9] (Table 23.5).

The third scenario was made for an inflation rate of 9.9 % and a government subvention estimated at 45 % initial costs (Table 23.5). The government would provide a subvention for SWH but not for GSHP projects.

Table 23.4 Input parameters in RETscreen software for GSHP project

	Heating system		Cooling system	
Reference case	Heated floor area of the building (m^2)	100	Cooled floor area of the building (m^2)	100
	Fuel type	Diesel/butane	Energy source	Electricity
	Seasonal efficiency (%)	75	Seasonal cooling coefficient of performance (COP)	5
	Fuel price ($)	0.171/0.192	Retail price of electricity ($/kWh)	0.022
Proposed case	Technology	GSHP	Technology	GSHP
	Energy source	Electricity	Energy source	Electricity
	Retail price of electricity ($/kWh)	0.022	Retail price of electricity ($/kWh)	0.022
	System capacity (kW)	6	System capacity (kW)	5.5

GSHP ground source heat pump

Table 23.5 Results of GSHP system for rural houses following fuel types, in the studied regions and for the three scenarios

Region	Fuel type	Fuel price	Annual energy (MWh)[a]	Annual GHG emission reduction (tCO_2)[a]	Equivalent in oil barrels[a]	Initial cost ($)[a]	IRR (%)			Simple payback (year)			Equity (year)		
							1st	2nd	3rd	1st	2nd	3rd	1st	2nd	3rd
Algiers	A	0.171[b]	13	0.5	1.1		3.1	11	17.2	21.9	21.9	12.1	18.1	11.5	7.8
	B	0.192[c]		0.1	0.9	1247	0.1	7.8	13	32.4	32.4	17.8	24.8	14.4	10.1
Batna	A	0.171[b]	13	0.4	0.9	12	7.3	15.6	23.5	13.9	13.9	7.6	21.1	8.6	5.5
	B	0.192[c]		0.3	0.6	47	3.1	11.1	17.3	21.8	21.8	12	17.9	11.5	7.7
Ghardaïa	A	0.171[b]	14	0.5	1.2	12	4.3	12.3	18.9	19.1	19.1	10.5	16	10.6	7
	B	0.192[c]		0.1	0.2	47	1.1	8.9	14.4	28.1	28.1	15.5	22.2	13.4	9.2

A, diesel; B, butane. 1st, 2nd, and 3rd refer to the scenario number [a] Values are applicable to each one of the three scenarios [b] $/L [c] $/Kg *GHG* greenhouse gas, *IRR* internal rate of return

Table 23.6 Rural towns in Algeria

Climate	Urban towns		Rural towns	
	Number	Percentage (%)	Number	Percentage (%)
Tell	449	42	622	58
High plains	114	32	246	68
Sahara	53	33	111	67
The sum	562	36.5	979	63.5

Table 23.7 Summary of results for the case study of 900,000 rural houses

Total annual provided energy (MWh)	Total annual net reduction of GHG (tCO_2)	Total oil barrels equivalent
15.21×10^6	19.17×10^5	42.12×10^5

GHG greenhouse gas

Results of the advantages of using a SWH–GSHP hybrid system in rural houses for the three sites are summarized in Table 23.3. These results show that SWH–GSHP hybrid system for the three sites is significantly profitable and there is a considerable gain in both energy saving and GHG emission reduction. The fallout from this return will be on both the economy and the environment.

23.4.2.3 Results of SWH–GSHP Project if Applied to More Rural Houses in Algeria

Algeria contains some 1541 (rural or urban) towns dispersed in all the country (Table 23.6).

The proportion of rural housing program in the housing sector during the period 2010–2014 reached 900,000 units, equivalent to 40% of the overall program. Table 23.7 shows the advantage of applying the GEOSOL project for these 900,000 units. We can see that there is a considerable gain for both energy saving and reduction of GHG emission.

23.5 Conclusion

This work shows that the annual energy saved when applying GEOSOL project for the expected 900,000 rural houses during the period 2010–2014 is about 15.21×106 MWh for the province of Batna. This energy is equivalent to 42.12×10^5 oil barrels.

GHG emission reduction is another important parameter of renewable energy profitability, and our study shows that it reaches 19.17×10^5 tCO_2. This is equivalent to 42.12×10^5 barrels of oil not consumed, and a 2.28% reduction in greenhouse gas emissions at a national level, where GHG emissions due to final energy consumption reached 46 million tCO_2 [11]. These values become very important when we consider all rural towns in the country.

Parameters that play a critical role in the profitability of SWH–GSHP hybrid systems are fuel type and price, project site, inflation rate, and government subvention. As we can see (Table 23.3), fuel-type influenced annual GHG emission reduction, while project site influenced the annual provided energy and load recovery rates, and the other parameters influenced the economic effectiveness of the project. However, Algeria is ranked fifth worldwide in terms of lowest price of fuel and the fact that the government subsidizes fuel prices and that the initial investment of such renewable projects is very high; all this hinders the development of these projects. Thereby, it is very important that the government finances them as much as necessary. Actually, Algerian government subsidizes family housing in rural population by an amount of US$ 9000–10,000 and offers a 45 % initial cost subvention for SWH installation, this should be completed by subsidizing GSHP installation as well, and it will not cost more than US$ 90 per household.

It is noted that the GEOSOL system needs extra electricity to run solar loop pumps and GSHP, which will have an impact on project performance in terms of reducing GHG emissions.

References

1. Laustsen J (2008) Energy efficiency requirements in building codes, energy efficiency policies for new buildings. International Energy Agency (IEA) information paper. In support of the G8 Plan of Action
2. Kyoung-Ho L, Dong-Won L, Nam-Choon B, Hyeok-Min K, Chang-Jun L (2012) Preliminary determination of optimal size for renewable energy resources in buildings using RETScreen. Energy 47:83–96
3. Nelson VC, Starcher KL, Foster RE, Clark RN, Raubenheimer D (2002) Wind hybrid systems technology characterization. West Texas AM University-AEI, New Mexico State University-SWTDI, USDA-ARS Bushland, Vista University, May 2002
4. Baring-Gould EI, Flowers L, Lundsager P (2004) Worldwide status of wind-diesel applications. Pre-workshop wind-diesel 101, Anchorage, Alaska
5. Trillat-Berdal V, Souyri B, Achard G (2007) Coupling of geothermal heat pumps with thermal solar collectors. Appl Therm Eng 27:1750–1755
6. Duffie AJ, Beckman WA (2006) Solar engineering of thermal processes. Wiley, New York, pp 525–529, 534–535
7. Rapp D (1981) Solar energy. Prentice-Hall Inc., New Jersey, pp 295–300
8. Ministry of Energy and Mining (2012) Solar energy. http://www.mem-algeria.org
9. Hania A-A (2007) Algérie énergie solaire et hydrogène. Office des publications universitaire, Ben Acnoun, Algérie
10. Stambouli B, Khiat Z, Flazi S, Kitamura Y (2012) A review on the renewable energy development in Algeria: current perspective, energy scenario and sustainability issues. Renew Sustain Energy Rev 16(7):4445–4460
11. Kedaid FZ (2006) Développement de la base de données géothermique de l'Algérie par un système d'information géographique. Rev Energies Renouvel 9(4):253–258
12. CDER Algeria. /http://www.cder.dzS

13. Self SJ, Reddy BV, Rosen MA (2013) Geothermal heat pump systems: status review and comparison with other heating options. Appl Energy 101:341–348
14. Thonon B, Kohler J, Mathieu A Project promotion of efficient heat pumps for heating (ProHeatPump). Deliverable No. 21. Report on solar energy and heat pumps. EIE/06/072/S12.444283
15. Kjellsson E, Hellström G, Perers B (2009) Optimization of systems with the combination of ground-source heat pump and solar collectors in dwellings. Energy xxx:1–7
16. Dotta R, Genkinger A, Afjei T (2012) System evaluation of combined solar & heat pump systems. Energy Procedia 30:562–570

Part IV
Hydropower and Ocean Energy

Chapter 24
Study on Tandem Configuration of a Flapping Tidal Stream Generator

Jihoon Kim, Tuyen Quang Le, Jin Hwan Ko, Jin-Soon Park and Kwang-Soo Lee

Abstract So far, most tidal stream generators have been developed in three types: a horizontal axis turbine, a vertical axis turbine, and a flapping-type hydrofoil. The first two are operated by one-way rotation, but the last one is oscillated. The flapping system inspired by nature has attracted researchers due to its eco-friendly characteristics, applicability to shallow water, and nonconventional oscillating movement.

In this study, we propose a flapping tidal steam generator of a mirror-type dual configuration, and the effects of the configuration on their performance are investigated through the experimental analysis in a towing tank and numerical analysis by computational fluid dynamics simulation. Throughout the numerical and experimental studies, we found that the rear system in the dual configuration can extract power close to that of the front system with a short distance between them such as two chords and an optimum phase difference, 90°.

Keywords Flapping tidal stream generator · Tandem configuration · Computational fluid dynamics · Towing tank

24.1 Introduction

Tidal stream energy has been considered as a good candidate for an alternating energy source to reduce our dependency on fossil fuels. So far, most tidal stream generators have been developed in three types. Among them, flapping hydrofoil devices are still in a premature status while rotary-type turbines are typically used for tidal power extraction. *Stingray* designed by Engineering Business Ltd., is a famous flapping-type generator for the first ocean installation in 2002 and several systems have been designed, developed, or installed. Even though various researches have conducted the related studies in very recent years, the flapping systems still require the improvement of power-generating capability and the power transmitting

J. Hwan Ko (✉) · J. Kim · T. Quang Le · J.-S. Park · K.-S. Lee
Korea institute of Ocean Science and Technology, Ansan 426-444, South Korea
e-mail: jhko@kiost.ac

A. Sayigh (ed.), *Renewable Energy in the Service of Mankind Vol I*,
DOI 10.1007/978-3-319-17777-9_24

efficiency. Moreover, other critical characteristics such as tandem configuration, control and structural safety should be explored in order to be considered as a viable alternative to the rotary-type generators despite their advantages such as eco-friendliness and power generation in low-level water or a low flow speed.

Most previous researches of the flapping systems are focused on the performance of the single system such as optimal kinematics [1], wing design [2], fully passive system [3], and optimal pitch-axis position [4]. Some research results of dual configurations are recently exposed. Once the kinematics of a single flapper is determined, the distance and phase between two flappers become the main parameters, and the effects by their variations on the power extraction efficiency have been explored. Jones et al. [5] first developed a system with the two foils of a swing motion in the dual configuration, which are located in the same L-shape along the flow direction. The foils were separated by a distance of 9.6 chords and oscillated with a phase of 90° such that the interval for changing the direction of one was coincided with the power stroke of the other. The tandem system of Kinsey et al. [6] is similar to the design of the system of Jones et al. [5]. They used a swing arm to create the plunge motion by a seesaw-type mechanism. Thus, the phase difference between the foils was fixed at 180° and the distance was given by seven chords. They estimated a highest efficiency of 40% with nondimensional frequency between 0.11 and 0.12 after all mechanical losses . Ashraf [7] studied the flow physics associated with a tandem foil arrangement, using a Navier–Stokes solver with pitch and plunge motion of a NACA0014 foil at Re=20,000. It was shown from the simulations with sinusoidal motion that a distance of four chords was found to generate higher power at a higher efficiency when the phase difference was less than or equal to 90°, and a distance of two chords generated higher power at higher efficiency than other cases when 135° and 180° phase differences were chosen.

In our study, numerical and experimental analyses are conducted for a mirror-type dual configuration with a swing motion. Due to that the motion of the numerical simulation is fully described while that of the experiment is from flow-driven generation, the kinematics of both are a little different. Using those kinematics, the power and efficiency are obtained numerically and experimentally, and those are compared each other.

24.1.1 Materials and Methods

Since 2011, we have designed, fabricated, and explored the performance of a flapping tidal stream generator by the experiments in an open water channel and a towing tank or by numerical simulations. For the generator in a tandem configuration, the experiments in the towing tank are conducted.

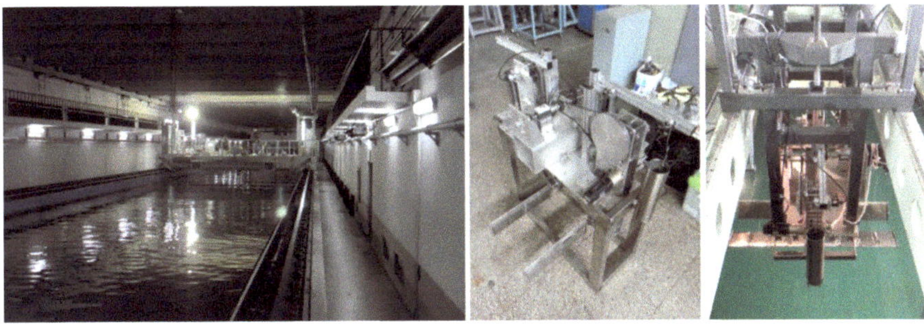

Fig. 24.1 A snapshot of the towing tank*(left)*, flap-type tidal turbine system *(middle)*, and installation on the towing tank carriage *(right)*

24.1.1.1 Experimental Apparatus

Figure 24.1 shows a panorama view of the towing tank in Seoul National University of Korea. The specification of the towing tank is a width of 8 m, a length of 100 m, and a depth of 3 m. Its maximum speed is around 5 m/s. The figure also shows the view of the experiment setup for the flapping-type turbine (FTT) system in a dual configuration. Each hydrofoil is connected with a moveable arm. To change a pitch angle for the hydrofoil, a spar gear is used. Both flappers are connected with one shaft, and then it is rotated in one direction. When the hydrofoils of the flappers move up and down, the main shaft cannot rotate smoothly due to null spot when the hydrofoil changes the moving direction. So, we use a flywheel in order to stabilize the rotational energy. At the end of the main shaft, the cylindrical type sensor (T20WN, HBM Inc.) is connected to measure the resultant torque and revolutions per minute (RPM). Figure 24.2 shows an example of the stabilized torque and RPM in case of applied torque,0.4 Nm. Due to the phase difference between the flappers of 90°, the rotational speed has two peak-to-peak values in a period. On the contrary, the torque is not much fluctuated as compared to RPM. Thus, it is indicated that the flywheel plays a good role in balancing the rotational power.

For data acquisition (DAQ), the signal processing system is required. The system contains a DAQ board (QuantumX 840A, HBM Inc.), noise-filtering amplifier (VK20A, HBM Inc.), and an IBM computer in which an in-house managing software (Catman Easy, HBM Inc.) is used. To unify the data sets, we built a total managing program. Also this program can control the mechanical parts such as on/off in servo motor. While the turbine is operated, the whole data are stored though the managing program. On the other hand, pitch and arm angle are managed by an embedded, controlling software. The algorithm of the control software drives oscillating movements and acquires the signals of the rotating encoder, which senses the flapping pitch and arm angle of the hydroplane .

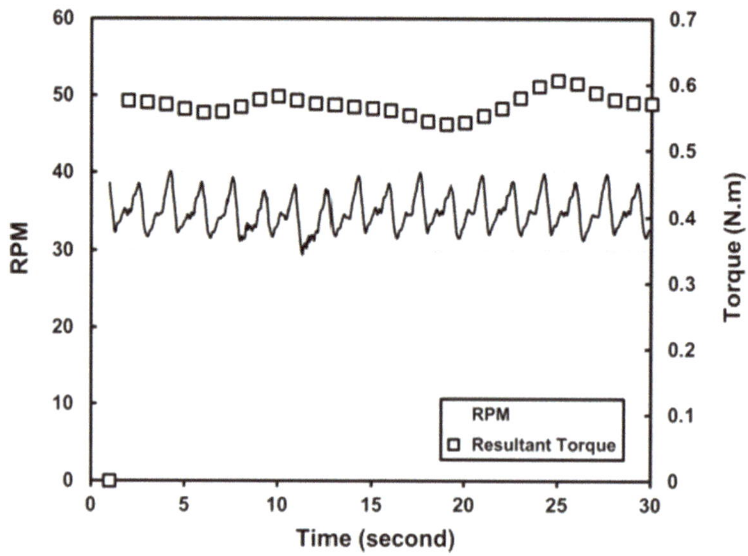

Fig. 24.2 Measurement for torque and RPM of power transmission with a flywheel

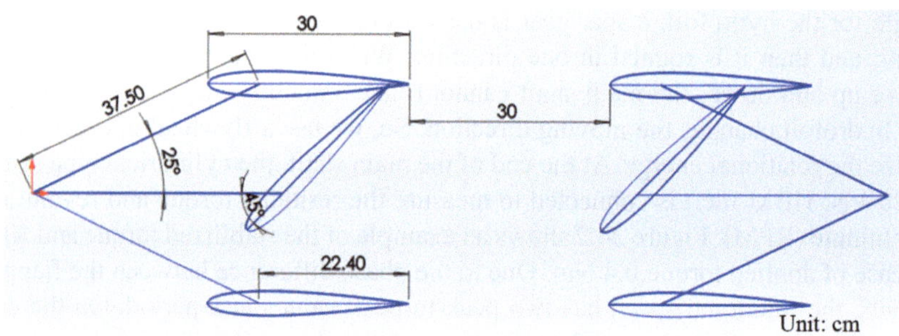

Fig. 24.3 Kinematics of flapping foils with a distance between one pitch axis and the other of 2 chords: $d=2c$

24.1.2 Numerical Setup for Computational Fluid Dynamic Simulation

The kinematics of FTT in numerical simulations is drawn based on the conditions of the conducted experiments. The kinematics of the dual configuration in numerical set-up is depicted in Fig. 24.3.

The hydrofoil is subjected to swing and pitch motion as a function of time as follows:

$$\begin{aligned} \alpha(t) &= \alpha_0 \sin(\omega t + \varphi) \\ \theta(t) &= \theta_0 \sin(\omega t) \end{aligned}, \tag{24.1}$$

where α is the flapping angle, α_0 is the amplitude of flapping angle (α_0=25°), ω is the angular velocity in radians, θ is the instantaneous pitch angle, θ_0 is the pitch amplitude (θ_0 =45°), φ is the phase ($\varphi = 90^\circ$), and t is the time. The x and y coordinate of the hydrofoil is driven by

$$\begin{aligned} x(t) &= R\cos\alpha(t) \\ y(t) &= R\sin\alpha(t) \end{aligned}, \tag{24.2}$$

where R is the length of flapping arm (R=37.5 cm). The pitching axis is located at 25 % of the chord from the leading edge. The frequency of the flapping and pitching motions is 0.3 Hz and the incoming flow velocity is 0.6 m/s. Hence, nondimensional parameters are calculated as follow: nondimensional frequency ($f^*=fc/U$) is 0.15, Strouhal number (St) is 0.1584, and Reynolds number based on the incoming flow velocity is 1.8e5.

The power extracted from the flapping hydrofoil is composed of the components of the translational and pitch motions; thus, for the two-dimensional simulation, the power-extraction efficiency is defined as the total extracted power from the flapper over the flow energy using the following definition:

$$\begin{aligned} P = P_Y + P_X + P_M &= L\dot{y} + D\dot{x} + M\dot{\theta} \\ P_{flow} &= 0.5\rho U_\infty^3 l_s \\ \eta &= P/P_{flow} \end{aligned}, \tag{24.3}$$

where L, D, M are the lift, drag, and moment, l_s is the total swept distance of the hydrofoil, U_∞ is incoming flow velocity, and ρ is the water density.

The hydrodynamic forces of the flapping foil are obtained by an in-house code parallelized multi-block structured Navier–Stokes solver. In order to conduct the simulation of the flapping tidal power extraction, a large pitch angle is mandatory; hence, a chimera mesh is utilized in the code [2].

24.1.3 Results and Discussion

24.1.3.1 Experimental Results

The operating conditions such as the incoming flow velocity have to be determined before conducting experiments. Usually, the critical parameter for the experiment is known as Reynolds number. When we consider the resultant velocity, which is combined with the incoming flow velocity and a flapping velocity, Reynolds number can be defined as

$$Re = \frac{\sqrt{U_\infty^2 + (2\pi f h_a)^2}}{\nu} c, \tag{24.4}$$

Table 24.1 Operating condition for experiments

Input arm angle	Velocity (m/s)	Frequency (Hz)	Nondimensional frequency
20°	0.6	0.2	0.1
30°		0.3	0.15
		0.4	0.2
		0.5	0.25

where f is the frequency, h_a is the amplitude of flapping motion, c is the chord length, and v is the kinematic viscosity.

Based on the definition, large size turbines will inevitably have large Reynolds numbers, but small experimental turbines will have much lower values. If Reynolds number is too small, the amplitude of flapping motion does not exhibit Reynolds number independence, namely very sensitive on the Reynolds number. Therefore, inflow velocity should be sufficient to increase Reynolds number.

Table 24.1 shows a operating range of the parameters in the kinematics. Referring to the study of [6], we select a specific frequency and arm angle to get the higher efficiency in our system. Counter-torque is imposed by 1.5 Nm for the power extraction and 90° phase difference and 2c distance are used for this experiment. For the condition of Table 24.1, the Reynolds numbers based on the resultant velocity are between 1.86e5 and 2.53e5. In the range, it is assumed that the dependence for the Reynolds number is not strong much.

In the control scheme to give a specific frequency and then adjust the arm angle, as shown in Fig. 24.4, the resultant arm angles are rapidly decreased as the frequency increases because of inevitable limitation between target frequency and arm length by the slow movement in the water. Hence, the high frequency such as 0.4 and 0.5 Hz could not reach to an arm angle of 20°.

In the operational condition, the total efficiency for the three dimensional experiment can be calculated by

$$\eta = \frac{T\omega}{0.5\rho U_{\infty}{}^{3} S l_{s}}, \tag{24.5}$$

where T is the torque, ω is the rotational speed in radian, and S is the span length of the hydrofoil.

Here, the power for the pitch control is not considered.

Table 24.2 lists the powers and efficiencies of different frequencies with different arm angles. In terms of the efficiency, the maximum case is 0.5 Hz but the swept length is smaller than that of low frequency due to the prescribed limitation. Thus, in terms of the power extraction as well as balanced power between the two foils, the suitable case is 0.3 Hz at 30° input arm angle, in which the highest power is extracted. For comparison, we chose computational fluid dynamic (CFD) simulation of 0.3 Hz at 25° arm angle because the resultant arm angle of the suitable case is lower than the input arm angle, around 21°.

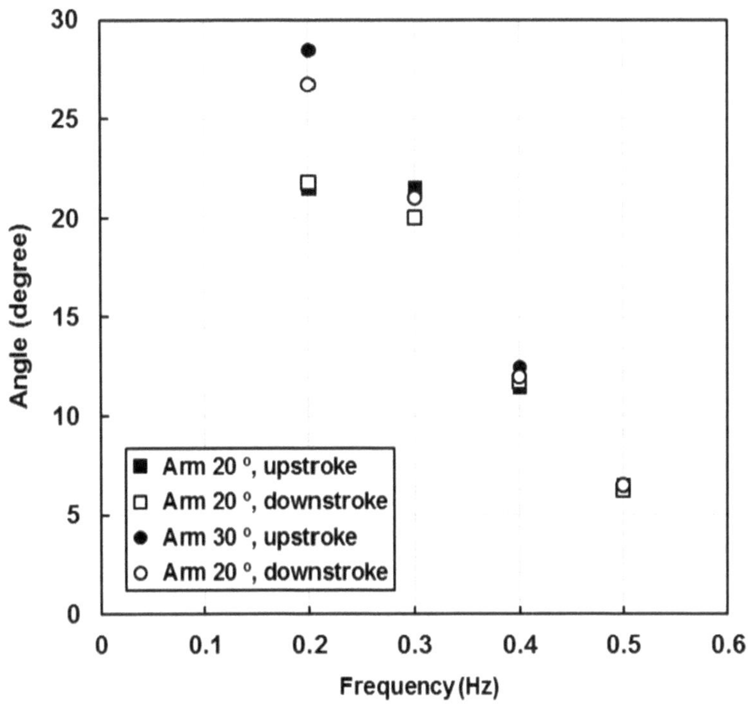

Fig. 24.4 Distribution of resultant arm angle

Table 24.2 Powers and efficiencies of different input arm angles and frequencies

Input Arm angle (degree)	Frequency (Hz)	Power (W)	Front eff.	Rear eff.	Total eff.
20	0.2	2.54	9.30	9.32	18.62
20	0.3	3.97	15.24	16.12	31.36
20	0.4	2.83	20.01	20.33	40.34
20	0.5	1.9	23.73	26.52	50.26
30	0.2	2.91	9.17	9.48	18.65
30	0.3	4.07	16.28	16.75	33.03
30	0.4	2.9	22.25	21.88	44.13
30	0.5	1.88	23.16	27.54	50.70

24.1.3.2 Numerical Results

The extracted powers and corresponding efficiencies from the flapping hydrofoils are listed in Table 24.3 in the case of single and several dual configurations by the numerical simulations with the distance between the front and rear foil varied.

Definitely, the distance between the front and rear foils as well as the phase difference between them strongly affect vortex shedding from the front foil to the rear foil, eventually changing the phase between the velocity and the corresponding force peaks. Figure 24.5 shows the lift coefficient curve of each foil in case of

Table 24.3 Power extraction and efficiency of each foil

Case	Foil	$P_H = P_X + P_Y$	P_M	$P = P_H + P_M$	η_H	η
Single system						
	Front foil	10.53	−4.22	6.31	0.308	0.184
	Rear foil	11.14	−3.68	7.46	0.325	0.217
Dual system						
d=2c	Front foil	9.61	−4.23	5.37	0.280	0.156
phase=90°	Rear foil	11.09	−2.63	8.46	0.324	0.246
d=2c	Front foil	10.27	−4.27	6.00	0.300	0.175
phase=180°	Rear foil	8.80	−2.55	6.25	0.257	0.182
d=3c	Front foil	10.04	−4.20	5.84	0.293	0.171
phase=90°	Rear foil	9.84	−2.51	7.32	0.287	0.214
d=3c	Front foil	10.61	−4.29	6.32	0.309	0.184
phase=180°	Rear foil	7.37	−4.03	3.34	0.214	0.102

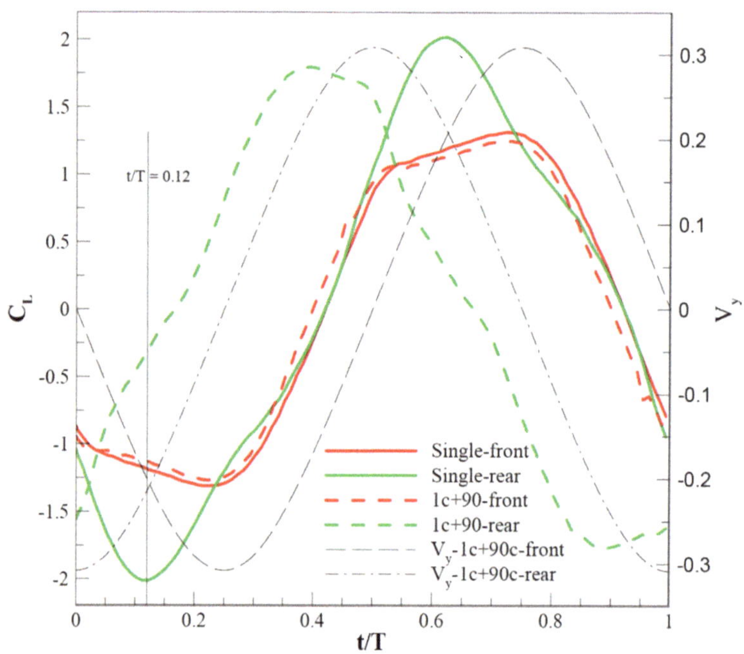

Fig. 24.5 Lift coefficient of each foil during one cycle

the single and a picked-up dual configuration ($d=2c$ and phase=90°) during one cycle. The y velocity component is also plotted in order to clearly show synchronization between their peaks. There is a slight change in lift generation by the front foil with and without the rear foil, while considerable phase shift in the rear foil is observed. Thus, the phase shift is important for final power extraction as well as the corresponding efficiency, and the dual configuration ($d=2c$ and phase=90°) is an optimum case among them in terms of this aspect.

Before the comparison between the experimental and numerical results, it is noted that the experimental operating condition, in which the flapping motion is flow driven by the pitch motion given by an actuator, is different from the numerical condition, in which pitch and flapping motions are fully prescribed. Also the efficiencies from the experiments do not contain the P_M term. Regarding the power extraction, 21 W in the simulation and 4 W in the experiment have a large gap due to the tip loss, the mechanical loss of the power transmission, smaller arm angle than input arm angle, and ideal kinematics used in the numerical simulation. Regarding the efficiency, 60 % in the simulation is almost double to 33 % in the experiment due to the similar reason.

In terms of power balance between the two foils, the efficiencies in the experiment are similar to each other while the efficiency of the rear foil is larger than that of the front foil. This difference is mainly caused by the different operating condition. Interestingly, in our mirror-type dual configuration, the efficiency of the rear foil is close to or larger than that of the front foil in all cases of the experiments as well as the simulations. Therefore, the final power in the dual configuration is expected not to be fluctuated much due to the well-balanced power between the two foils and the stabilization by the flywheel in the power transmission. Meanwhile, the considerable P_M term should be minimized for the improvement of the final performance of the flapping tidal turbine. More information and results of our study will be provided in the presentation.

24.2 Conclusion

In this study, we proposed the flapping tidal stream generator in a mirror-type dual configuration and investigated its performance numerically and experimentally. It is found throughout both approaches that the rear system along the flow direction can extract power close to or larger than that of the front system with a short distance between them such as two chords and the corresponding optimum phase difference, 90°.

The proximity power generation in the tandem configuration of the flapping-type generators will have strong advantages in competition with the rotary-type generators, especially for an array development.

Acknowledgments This work was supported by a grant from the New & Renewable Energy R&D program of the Korea Institute of Energy Technology Evaluation and Planning (KETEP) funded by the Ministry of Trade, Industry and Energy (No. 20113020070010). This work was also supported by the PLSI supercomputing resources of the Korea Institute of Science and Technology Information.

References

1. Zhu Q (2011) Optimal frequency for flow energy harvesting of a flapping foil. J Fluid Mech 675:495–517
2. Le QT, Ko JH, Byun D (2013) Morphological effect of a scallop shell on a flapping-type tidal stream generator. Bioinspiration Biomim 8(3):036009
3. Young J, Ashraf MA, Lai J (2013) Numerical simulation of fully passive flapping foil power generation. AIAA J 51(11):2727–2739
4. Zhu Q, Peng Z (2009) Mode coupling and flow energy harvesting by a flapping foil. Physic Fluids 21:033601
5. Jones KD, Lindsey K, Platzer MF (2003) An investigation of the fluid-structure interaction in an oscillating-wing micro-hydropower generator. In: Chakrabarti SK, Brebbia CA, Almoza D, Fonzalez-Palma R (eds) Fluid structure interaction II. Wessex Institute of Technology Press, Southampton, pp 73–82
6. Kinsey T, Dumas G, Lalande G, Ruel J, Méhut A, Viarouge P, Lemay J, Jean Y (2011) Prototype testing of a hydrokinetic turbine based on oscillating hydrofoils. Renew Energy 36:1710–1718
7. Ashraf MA (2010) Numerical simulation of the flow over flapping airfoils in propulsion and power extraction regimes, Doctoral thesis, The University of New South Wales

Chapter 25
An Economic Approach for the Design of Small Hydropower Converter

Jana Hadler and Klaus Broekel

Abstract The worldwide use of hydropower potential of huge dams, weirs and waterfalls is significant—these energy sources have almost been exploited. What remains are the small waters, with low flow rates and the use of low potential in remote areas for decentralised electricity supply. This small hydropower in the range of a few kilowatts in the European and global market did not attract much attention in recent years.

Based on the European project HYLOW (the EU-Project HYLOW lasted from 2008 to 2012. Partners from five European countries took part; hydropower converters for very low head differences), the following idea has been developed: to combine various technical solutions for the use of water currents of low potential under economic and environmental optimal conditions. All relevant data were analysed and finally prepared for an interactive planning tool.

Keywords Small hydropower · Low head differences · Interactive planning tool

25.1 Motivation

The floating energy converter is considered as an assembly consisting of a waterwheel between two pontoons. The wheel is installed on the pontoons which are connected via a base plate below the water level (see Fig. 25.1).

Small hydropower plants require a bundling of multiple specialists: It needs knowledge from the fields of engineering (electrical and construction), banking and accounting as well as experience in dealing with the government agencies and regulating bodies. Making predictions about the amortisation of the systems described

J. Hadler (✉) · K. Broekel
Institute of Engineering Design/CAD, University of Rostock, Rostock, Germany
e-mail: jana.hadler@uni-rostock.de

K. Broekel
e-mail: klaus.broekel@uni-rostock.de

A. Sayigh (ed.), *Renewable Energy in the Service of Mankind Vol I*,
DOI 10.1007/978-3-319-17777-9_25

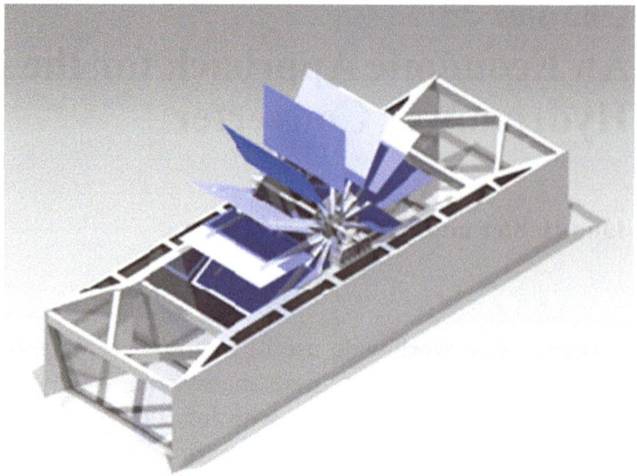

Fig. 25.1 Model of a floating energy converter (CATIA V5)

are extremely difficult because many factors are too complex to estimate in detail for long time periods.

Investments are generally calculated with the anticipated operating hours per year and the prevailing site conditions (flow rate, water data and access roads). Hence, the monetary benefit needs to be estimated.

However, regardless of the monetary benefit the qualitative benefit can be the deciding factor for the realisation of the investment. The use of energy converters for decentralised power supply in remote areas and a pollution-free power generation are unbeatable arguments for investing.

25.2 Conditions

25.2.1 Technical Conditions

The technical data based on tests with a large-scale model developed within the EU project *HYLOW*. During the project, towing and stationary tests with a large-scale model were done, and a large amount of data were collected and analysed.

The structure of the model is relatively simple owing to its field of operation—remote areas without electricity (but near running waters). The main assemblies are the pontoons with two hulls connected via a base plate (identical inlet and outlet areas) and the waterwheel with twelve straight blades. All components are made of steel. The internal space of each of the hulls will be divided into chambers, which are designed for trimming with sandbags and/or water and for carrying technical equipment and measurement. The length of the large-scale model is 7500 mm, the width 2400 mm and the height 3500 mm. The wheel diameter is 3200 and 1100 mm in width [1].

Based on this output model, four technical versions were calculated and designed with the computer-aided design (CAD) programme CATIA V5. Due to the experiences and as a result of theoretical considerations and detailed calculations, all models are in a noncritical floating behaviour.

25.2.2 *Economic Conditions*

In many countries of the world, feed-in tariffs are granted. This shortens the payback period of the energy converter, when loans are used to finance.

The amount of the feed-in tariff depends on the year of commissioning and the power output of the converter. In Germany, this tariff for the power generated is guaranteed for the next 20 years in accordance with the German renewable energies law (EEG). If the converter produces less than 500 kW, the amount of the feed-in tariff will be 12.67 ct/kW [1].

Other countries have different durations and mostly different tariffs.

Moreover, it is important to decide between two different methods of procurement—purchase and self-production. On the one hand, there are acquisition costs; on the other hand, there are personnel, direct (e.g. material and standard parts) and indirect costs (usage of electricity, equipment and machines) for producing the converter.

In both cases, the following costs are every year of the operating life considered: costs for service and maintenance, rental costs for used waters, insurances, taxes and administration.

25.3 Method

Providing the market with an economically and technically proven method for selecting suitable floating energy converters in different regions of Europe to the prevailing conditions, the collected data were implemented in a first version of a prefabricated worksheet (Excel-Visual Basic for Applications (VBA)) in Microsoft® Office.

First of all, the compilation of data was necessary; the data were put into the Excel sheet and connected by comparative, logical and mathematical operators. Figure 25.2 presents the way through the programme.

All program steps are implemented in single tabs of Excel-VBA. The following section describes the tabs in detail.

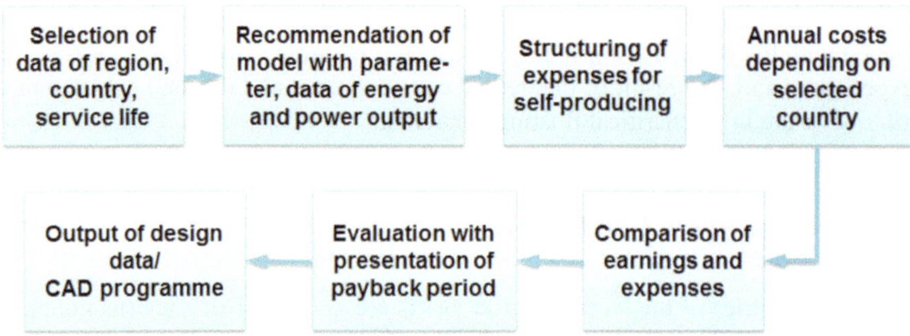

Fig. 25.2 Description of individual programme steps

25.4 Interactive Planning Tool

25.4.1 Overview of the Tabs

On the top of the page, a fixed header, displayed on all pages, is integrated. In the main area, there are the various superposed pages (in Excel-VBA called multi-pages), to be selected via the tabs. The following tabs are displayed:

- Choice
- Model
- Costs
- Annual costs
- Amortisation
- Results

For optimal results, a selection of pages in sequence is recommended; departing from this, the results are incomplete and not usable. In the following structure, function and operation of each page are described. In addition, every tab shows explanatory notes.

The first tab (see Fig. 25.3 on the left) is for the selection of the data of the region (flow velocity, water depth, channel width, direction of current), the country and the service life. On the lower right side, the results are presented due to the blockage ratio. The cross area of the water, which is calculated from the input data, is provided in intervals.

From all the inputs, the program determines a proposal which is presented on the tab "Model" on the top left (see Fig. 25.3 on the right). Directly below is a corresponding illustration of the model. The data of four model variants (see Sect. 4.2) are implemented in the program. The appropriate model is determined by the conditions and calculations which were realised by the programme. In addition, the data of energy, electricity and power output are given as well.

The third and the fourth tab in Fig. 25.4 reflect the costs. On the left side (tab "Costs") the costs of self-production are compiled. First, the manufacturing costs

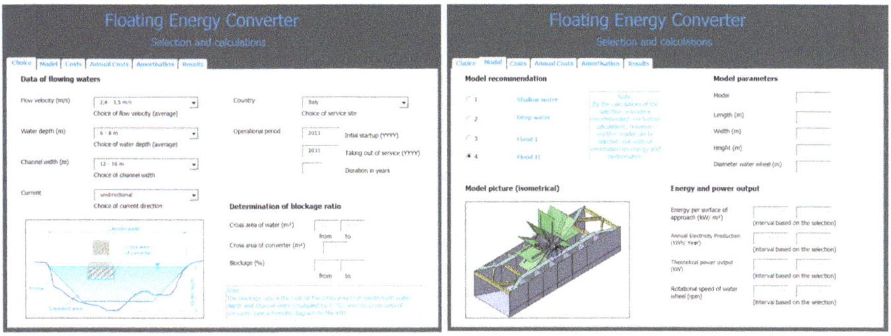

Fig. 25.3 The tabs: choice and model

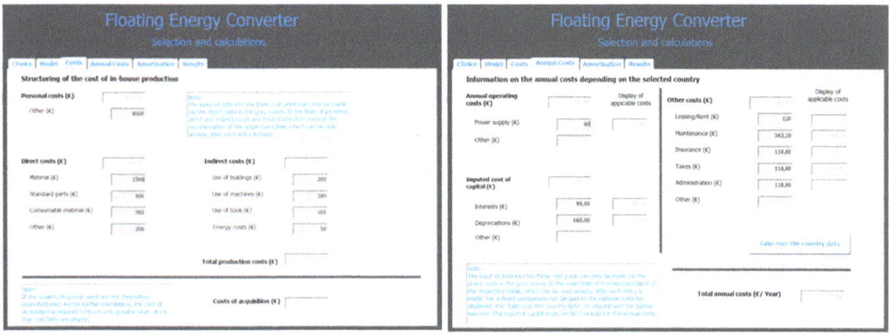

Fig. 25.4 The tabs: costs and annual costs

incurred to be determined. To the costs belong personnel, direct and indirect costs. While the direct costs include the costs of materials, standard parts and consumables, the indirect costs include the costs of building use, the use of machines and tools as well as electricity costs. In case of purchasing an energy converter, the costs of acquisition can be filled in the appropriate box.

On the right side, there is an overview to the annual costs. The calculation of annual costs gives the decision for or against the purchase and installation of the energy converter. The annual costs should be calculated within the limits of the financial resources and according to commercial rules.

To the annual costs belong operating costs, the imputed costs of capital and other costs. The latter are divided into costs for leasing or rental, the cost of maintenance, taxes, insurance and administration. Costs, which are not listed, can be summarised under "Other". The imputed costs of capital include interests and depreciations.

In Fig. 25.5, the amortisation and the results are shown. In the tab "Amortisation" (see on the left) costs are compared with benefits. On the upper left, the total annual costs are divided by the annual electricity production: With these costs per kilowatt hour it can be seen whether or not the converter will be profitable. If the costs per kilowatt hour are above the level of the feed-in tariff (see on the right), then there is no chance to make monetary profit from today's perspective. Ultimately, the balance is calculated.

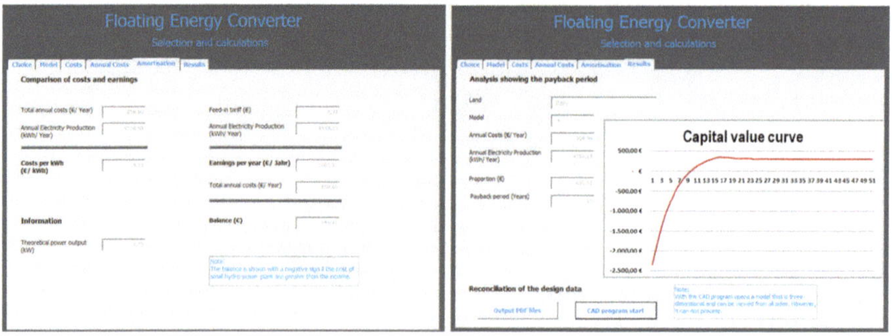

Fig. 25.5 The tabs: Amortisation and results. *CAD* computer-aided design

The tab “Results” (see on the right) gives an overview of inputs and calculations done before. On the right, the development of the capital value is presented in a diagram. Finally, the possibility of using the connected CAD programme is offered.

According to technical and cost-related calculations construction drawings and data can be queried. Thus, the design and manufacturing preparatory documents of the four basic models offered (three-dimensional (3D) models, technical drawings and cutting plans) are available in a minimum of time (by using the Portable Document Format or PDF files or by starting the CAD programme).

By transferring the existing parameters for the models of the converter to a CAD system (e.g. CATIA V5, Creo Parametric or NX5) the 3D model can be opened.

It is possible to look at the corresponding 3D model with all movement functions (scrolling, zooming and rotating). Changes in the model cannot be made. The same can also be achieved by a multi-format file viewer.

25.4.2 Four Models

From the basic model, which was developed in the EU project *HYLOW,* four variants for different applications can be derived. Because of their different fields of application, the energy converters are named as follows and shown in Fig. 25.6:

I. Shallow water
II. Deep water
III. Flood I
IV. Flood II

The energy converters I and II have a blunt stern, and the flow direction should be unidirectional. Both are suitable for water with moderate flow velocities. Converter III (an upscale of the basic model) is carried out bidirectional and suitable for the use in tidal areas. Likewise variant IV, which has wide pontoons, is designed for flow velocities up to 3 m/s. The qualitative benefit results from the allocation of suitable variant to the appropriate environment.

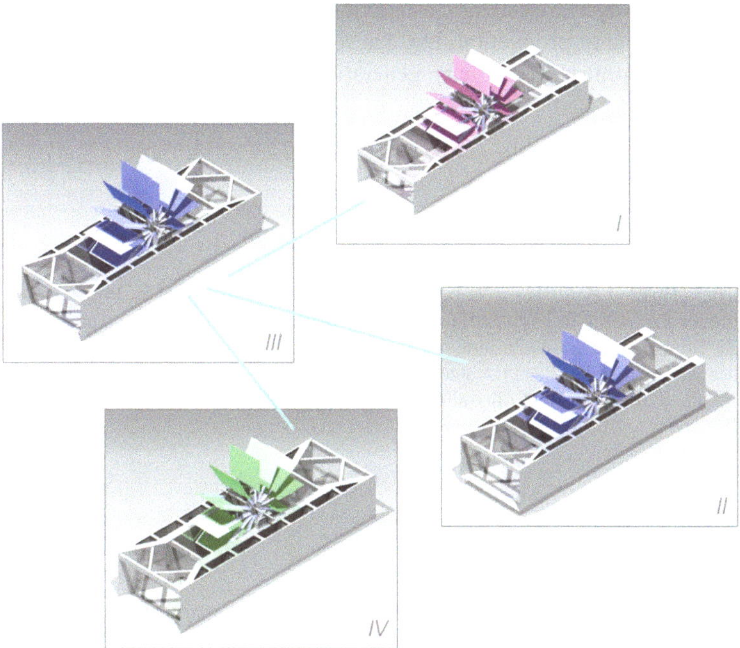

Fig. 25.6 Four variants of the energy converter [2]

25.5 Outlook

The extensive studies of interdisciplinary approaches of economical, ecological and engineering design are reflected through the use of the described interactive planning tool.

As a result of the entered data into this tool, data of the annual electricity production and the expected power output are presented. In addition to an overview of the costs, the user got information on the profitability and the payback period. Final result is the output of the CAD data of an eligible small hydropower plant.

The decision for small hydropower plants and the subsequent realisation of the installation of these requires a high effort of planning, data implementation and conscientious calculations. The interactive planning tool has been developed to minimise the effort of selecting energy converters for different regions. All necessary calculations are integrated and performed by the programme. However, a permanent maintenance of the database is necessary.

For highly industrialised regions, where electricity is available day and night, small hydropower plants, like these energy converters, in the described power output range are dispensable. In remote areas, however, where research stations, residential houses or medical service centres are not connected to any kind of power grid, such systems could play a significant role if suitable streams are available. The converters can be used in any case of power shortages.

All in all, however, free stream installations may not be a solution for large-scale renewable electricity production, but due to their simplicity they can be a viable option for decentralised electricity generation in developing countries or remote locations [1].

References

1. Hadler J, Broekel K (2011) Cost-benefit analysis for low potential hydropower converters. In: Proceedings of 34th IAHR 2011 Congress—balance and uncertainty: water in a changing world, Brisbane, Australia, 26 June–1 July 2011
2. Hadler J, Broekel K (2012) Nutzenanalyse schwimmender Energiekonverter für geringe Potentiale. Poster for 10. Colloquium Design Technology, Dresden, Germany, 14–15 June 2012

Further Readings

1. Muller GU, Batten WMJ (2010) Potential, performance limits and environmental effects of floating mills for energy generation in rivers. In: Proceedings of River Flow 2010, Braunschweig, 8–10 Sept 2010
2. Hadler J (2012) Interdisziplinärer Ansatz für die Entwicklung von Produkten zur Nutzung der Wasserkraft bei niedrigem energetischen Potential. PhD thesis, defended at Rostock University 2012

Chapter 26
Identifying Promising Wave Energy Converter Technologies

Matt Folley and Trevor Whittaker

Abstract The exploitation of wave energy is in the early stages of development. Like the development of wind energy during the first half of the twentieth century, there is currently no standard system for the commercial extraction of wave energy. This creates both an opportunity and challenge for developers and investors. An opportunity, because it is likely that significant IP remains to be developed, with its associated rewards, and a challenge, because it is not yet clear which fundamental method(s) of exploitation will become dominant in the future.

Currently, it is possible to identify at least five fundamentally different types of wave energy converters, and within each one of these categories there are typically many variations resulting in over a 100 different wave energy converter technologies being developed. Undoubtedly, the developers of each technology believe that they have a winning concept and in addition have some evidence to support this view; unfortunately, it is likely that the vast majority of these developers will be disappointed. For wave energy to reach commercial viability, it will be necessary to accurately identify promising wave energy technologies.

This chapter investigates the reasons why such a large range of technologies is currently being supported and developed. This investigation centres on identifying the particular assumptions, views and perceptions that may be held by technology stakeholders and how these factors may influence wave energy converter design to both the benefit and detriment of the technology and industry. In addition, it includes further analysis of how these factors may be related to both the hydrodynamics and economics of wave energy converters. The chapter concludes with a graphical organisation of wave energy converter technologies that illustrates their fundamental development characteristics and supports the identification of promising technologies.

Keywords Wave energy · Design ·Hydrodynamics · Economics · Risk · Investment

---

M. Folley (✉) · T. Whittaker
SPACE, Queen's University Belfast, Belfast, Northern Ireland
e-mail: m.folley@qub.ac.uk

A. Sayigh (ed.), *Renewable Energy in the Service of Mankind Vol I*,
DOI 10.1007/978-3-319-17777-9_26

26.1 Introduction

Today, in 2014, wave energy remains a commercially unexploited source of renewable energy, although studies have shown that the worldwide resource is significant [1, 2]. Moreover, the existence of wave energy is evident to anybody who has spent time at sea or on the coast and witnessed the potentially destructive power of the waves. However, unlike wind and solar energy, there has been no historical commercial development of wave energy technology. Ironically, this may be because the power in the waves made the task too challenging. It will be essential to exploit wave energy to meet the increasing requirements to generate power using renewable energy resources. Moreover, modern-day prototypes of wave energy technologies have shown that it is indeed possible to both generate power and resist the destructive power of the waves [3, 4]. Unfortunately, what has not been done is to show how this can be done economically.

A consequence of the lack of historical precedence in wave energy technology is that there is no standard archetypal technology to reference new concepts; unlike wind energy where the existence of a historical reference point and experience of extracting energy significantly simplified the assessment of new and alternative technologies.

In addition, it can be argued that the wave energy resource is significantly more complex that either the wind or solar energy resources. At any particular time, there are waves with different amplitudes, frequencies and directions that may all contain significant amounts of energy that can be extracted. This wave energy is associated with motion of the water in both horizontal and vertical directions, so that power can be extracted for the waves in either or both of these directions, with the reaction source required being obtained either from the seabed or another body that may or may not be excited by the waves. A final complication is that the characteristics of the waves change with water depth so that the offshore wave energy resource is different from the nearshore and shoreline wave energy resources.[1]

The combination of a large number of potentially viable wave energy concepts and the lack of a historical reference point or experience make it extremely difficult to identify the most promising wave energy converters (WEC). In particular, the novel requirements of wave energy exploitation mean that accurately predicting the performance of any single concept would require more time and resource than is typically available. Consequently, it is necessary to assess the different concepts based on current and limited knowledge to determine which deserve further investment.

It is proposed that the identification of promising wave energy concepts should follow a two-step process:

[1] It is not the intention of this chapter to review the range of potential options as a number of useful classifications of wave energy technology already exist [5–7], it is sufficient to recognise that a large number of fundamentally different concepts for wave energy exploitation exist and that this presents a challenge to the industry.

1. To analyse the performance of the WEC concept to guide the design process
2. To perform a risk–benefit analysis of the WEC concept to provide a measure of the investment potential

Sections 26.2 and 26.3 of this chapter describe these two steps in more detail, which are then followed by the application of the process to a fictitious wave energy converter concept and a discussion of the implications of the application of the proposed process.

26.2 Understanding the WEC Concept

Identifying whether a WEC concept is promising requires a thorough understanding of the underlying factors that will affect the specific cost of energy (pence/cents per kilowatt-hour). Fundamentally, the specific cost of energy is simply the discounted life-time cost of the WEC (construction, installation, maintenance and decommissioning) divided by the discounted life-time energy production. Unfortunately, at no stage of the WEC concept development is it possible to know the exact cost or energy production of the WEC so that assumptions/approximations need to be made to produce an estimate of the WEC's specific cost of energy. There is nothing peculiar in the need to make assumptions and approximations in developing a new concept; this is done for all technologies. However, the novel and relatively immature status of wave energy means that many of these assumptions/approximations have not been critically reviewed and are therefore often applied inappropriately.

It is sensible to base assumptions/approximations used to assess a WEC concept on similar technologies, ideally other wave energy technologies. However, this must be done with care because the relatively large number of potential WEC concepts, with different operating principles and implementations, means that there is not always a simple translation of an assumption/approximation from one WEC concept to another. Indeed, the design perspective gained from applying an inappropriate assumption/approximation can result in a misunderstanding of how a particular WEC concept operates. This may result in potentially promising concepts being prematurely dismissed.

Recognising the assumptions/approximations that have been made when assessing a WEC concept and appreciating their implications and validity is the key factor in accurately identifying the potential of a WEC concept. The following subsections contain discussion of commonly held assumptions in wave energy, including identification of when their application is appropriate and when their application may be misleading. These discussions are not intended to be either comprehensive or definitive, but provide a set of extended examples of how assumptions should be critically reviewed.

26.3 Hydrodynamics

The hydrodynamics is defined as how the WEC interacts with the waves. It is an area where the design of WECs is significantly different from other marine structures in that the objective is to maximise the interaction with the waves, whilst typically the requirement for other marine structures is to minimise the effect of the waves. Consequently, an important and distinct set of considerations exist for WECs regarding their hydrodynamics. Unfortunately, it is evident that in many cases these have not been fully appreciated. Although, in general, the hydrodynamics of a WEC can only be calculated using a complex numerical model, there are two key aspects of a WEC's hydrodynamics that, if fully understood, can help to identify the potential performance of a WEC concept. These two aspects are:

1. the source of wave force
2. the scale of the WEC

The "source of wave force" is the hydrodynamic wave forces on the WEC from which power is extracted, for example, the heave force on a heaving WEC. Typically, the wave force on a WEC has two sources: the Froude–Krylov force and the diffraction force [8]. Before proceeding, it is important to note that this assumes that wave-induced drag forces, surface tension forces, etc. are insignificant. The Froude–Krylov force is the force on a body if the undisturbed wave pressure is integrated on the body surface and typically dominates the heave force on surface-piercing WECs; it is approximately proportional to the water-plane area of the body. The diffraction force is the force on a body due to the change in water flow due to the presence of the body and typically dominates the surge forces and the heave force on a fully submerged body; it is approximately proportional to the square of the wave frequency and the volume of disturbed water [9].

The scale of the WEC refers to its size relative to the incident wavelengths. At the extremes of the scales, there is the point absorber, which is small enough relative to the wavelength that it can be approximated as a single point, and the terminator, which is sufficiently large relative to the wavelength that it can be considered as a two-dimensional body orthogonal to the direction of wave propagation. In general, all WEC concepts sit somewhere between these two extremes, with most current concepts more closely related to point absorbers. Point absorber theory has played an important role in the development of wave energy because it states that assuming the hydrodynamics are linear and the motion unconstrained then the maximum power capture is independent of the WEC size and proportional to the wavelength [10].

Using the understanding of hydrodynamics developed above, it is possible to critically review some of the common assumptions in WEC concept assessment. A common implicit assumption is that WEC concepts that extract wave energy in surge (or pitch) are simply (surface-piercing) heaving devices translated through 90°. This is not a reasonable assumption as the two cases have different principle sources of wave force; the heave force is typically dominated by the Froude–Krylov force and the surge force is dominated by the diffraction force. Therefore, they should be considered, at least hydrodynamically, as completely different concepts.

In particular, it has been found that the capture factor of a surging WEC increases (at least initially) with WEC size [11], whilst that for a (surface-piercing) heaving WEC it typically decreases [12]. This relates to another common assumption, which is that the power capture of a WEC can simply increase with size as may be expected for a wind turbine. However, for an axisymmetric heaving WEC, the point absorber effect means that there is a ceiling to the power capture irrespective of how large the WEC becomes, and there is also a limiting capture factor for surging devices as the hydrodynamics become equivalent to that of a two-dimensional body.

26.4 Dynamics

Together with the WEC hydrodynamics, the WEC dynamics also need to be fully understood for effective WEC design. The maximum power transfer theorem states that the power capture is maximised when the WEC velocity is in-phase with the incident wave force. This occurs when the natural frequency of the WEC is the same as the wave frequency and results in the system being in resonance. Furthermore, additional numerical modelling has indicated that in some circumstances the power capture can more than treble between a resonant and nonresonant WEC [13]. This has led to the common perception that the benefits to a WEC from resonance (or quasi-resonance) are so substantial that WEC concepts must be tuned, or be tuneable, to the incident wave frequencies. However, this perception neglects the importance of motion constraints [14] and the ratio of reactive to dissipative forces, both of which have a significant effect on the relevance of tuning.

The ratio of reactive to dissipative force is important for the significance of tuning because it defines the phase angle between the wave force and WEC velocity; when the WEC is tuned, the reactive force is zero, the wave force and WEC velocity are in-phase and thus the phase angle is zero. As the relative amplitude of the reactive force increases, the phase angle between the wave force and WEC velocity also increases and consequently the potential power capture decreases.

The maximum amplitude of motion will depend on the WEC size and design. Detailed examinations have shown that particularly large increases in power capture with tuning are also associated with large increases in the required amplitude of motion. If the required amplitude of motion is larger than that which is available, then the motion must be restricted, which results in a reduction in power capture and a smaller difference between the power capture of resonant and nonresonant designs.

Thus, the benefits of tuning depend on the importance of motion constraints, the ratio of reactive to dissipative forces and how they vary with frequency. In general, it is necessary to determine the significance of these factors for each WEC concept assessed. However, it is instructive to recognise that for many heaving WECs, the ratio of reactive to dissipative forces is large resulting in the potential for significant increases in power capture with tuning, whilst for many surging/pitching WECs the motion constraints are significant and the ratio of reactive to dissipative forces is small so that the increase in power capture from tuning is much smaller [15].

26.5 Techno-economics

Compared to the research into the potential power capture of WECs, the life-cycle costs of WECs have received very little attention. Undoubtedly, this is due to a range of factors including:

- Difficulty in identifying appropriate cost centres
- Complexity of associating costs with particular aspects of WEC concepts
- Lack of experience in the design of WECs (due to their novel nature and the immaturity of the industry)

However, without some estimate of WEC life-cycle costs, it is not possible to produce an estimate of the specific cost of energy and thus the relative promise of the WEC concept. Consequently, the techno-economic potential of a WEC concept is typically based on the twin pillars of parametric performance and engineering intuition/guidelines.

Although other parametric measures of performance exist, the two most commonly used parameters are capture factor and power per unit volume [7]. Both are based on an implicit assumption that the costs of the WEC concept are approximately proportional to the WEC size, that is, a characteristic dimension such as width for capture factor and WEC volume for power per unit volume. Whilst these measures seem reasonable and may be suitable for identifying particularly weak WEC concepts, on further inspection, their general application is flawed. The first flaw is that the relationship between actual costs and the WEC width or volume is unlikely to be the same for different WEC concepts and so it cannot be used sensibly to differentiate between different concepts. A WEC with a higher capture factor is not necessarily more viable. The second flaw is that the maximisation of a particular measure of performance, such as capture factor, will not necessarily result in a minimisation of the cost of power. Indeed, the relationship between the parametric performance and cost is weaker than often perceived. This is often apparent to the WEC concept developers and can result in confused WEC designs where the concept appears to be more promising with a higher parametric measure of performance, but the actual cost of energy will be higher. Consequently, the cost function should reflect the actual expected cost rather than a parametric measure of performance. This may appear to be less objective than using a standard performance parameter since the cost function is likely to be different for each WEC concept, but in actuality, is likely to be more accurate. Undue confidence in a simple measure such as capture factor or power per unit volume can hide the cost uncertainty. Having an accurate calculation of an inaccurate measure does not result in a more accurate assessment.

In the absence of extensive experience or analysis engineering, intuition/guidelines play a very important role in the design process, but it is important to separate the guidelines that are associated with the engineering aspects of the design and those that are associated with the risks to the project. An example of the confusion that can occur without making this distinction is in the treatment of complexity, where it is commonly stated that a WEC must be simple to be reliable. The implication of this statement is that anything that is not simple is unreliable. However, there are many cases of complex designs that are highly reliable, with the modern-day motor car a prime example—whose reliability has increased with complexity. In fact, what the statement is really saying is that the WEC concept risk is lower for a simple system because the development effort and the uncertainty in WEC concept performance are reduced. It is important to separate the factor of development risk from the WEC concept design since they have different impacts on how WEC concepts may be developed; this is discussed in the following section.

26.6 Risk–Benefit Analysis of the WEC Concept

Risk–benefit analysis is the comparison of the risk of a situation to its related benefits. In the case of identifying promising WEC concepts, it is the risk that the WEC concept will not be commercially viable compared to the potential economic yield of its successful exploitation. For WECs this is the product of the expected number of WECs deployed and the expected profit generated by each WEC once it has been fully developed less than the cost of development. Converting the risk into a probability of success, it is possible to calculate an expected value of the WEC concept, which is simply the product of the probability of success and the economic yield if the WEC concept is successful. For example, a WEC concept with a potential economic yield of £ 1000 million and a probability of success of 10 % has an expected return of £ 100 million, which is the same as another WEC concept with a 20 % probability of a yield of £ 500 million or a WEC concept with a guaranteed yield of £ 100 million. Thus, it is possible to draw lines of constant expected return on a risk–benefit graph that all have the same expected return and are shown as dotted lines in Fig. 26.1. However, it has been found that in general when people make decisions they are loss adverse, so that as the probability of success reduces the perceived value of the expected return also reduces, which results in the constant value contours shown as solid lines in Fig. 26.1 [16]. A particularly important contour on this graph is the zero-value contour (shown in bold), which is the threshold below which a concept is considered to be not worth developing. It is clear that as we become more risk adverse, fewer concepts become worth developing and also that this affects highly risky concepts more than less risky ones as the increase in required return on success is greater.

Before looking at a more detailed application of the risk–benefit graph, it is instructive to look at how it can be used to understand how different technologies may be supported legitimately. To achieve this, three well-known WECs are considered; Pelamis [17]—an articulated attenuator, Oyster [4]—a bottom-hinged flap and LIMPET [18]—a shoreline oscillating water column (OWC). All of the concepts have received funding but are very different in both their risks and benefits. Pelamis, as conceived with full complex control, may be considered as the concept with the highest potential economic yield as it can be widely deployed in relatively deep water; however, it may also be considered the most risky because of the re-

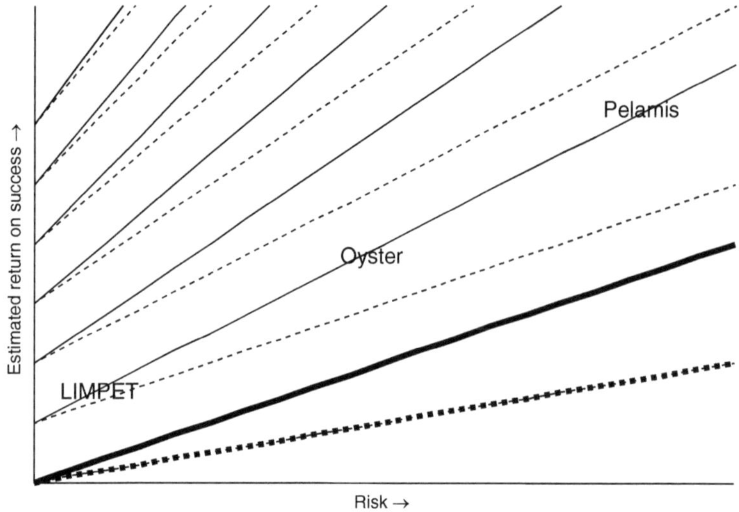

Fig. 26.1 Risk–benefit graph

quirement to develop the complex control technology. Oyster does not have as high a potential economic yield because it is deployed in shallow water so there are less potential sites; however, it may be considered less risky because it is expected that the technology requires less development as it uses a simple power-takeoff (PTO) system. Finally, LIMPET has the lowest potential economic yield because of the relatively limited number of suitable shoreline sites, but the technology is relatively mature and so the development risk is relatively small. Thus, it is possible that all three technologies have a similar perceived value and could explain why all three have received support for development. In reality, the perceived values will depend on both how their risks and benefits have been analysed together with the level of risk aversion of the investors/analysts. The more risk-adverse investors/analysts will value LIMPET more and the less risk-adverse investors/analysts will value Pelamis more.

To illustrate the use of the risk–benefit graph, consider a fictitious WEC concept called NewWEC. It is assumed that the power production and costs of NewWEC have been analysed, together with the estimated risks to the project and the design is placed at A on the risk–benefit graph of Fig. 26.2. The developers of NewWEC can be pleased because this implies the concept has value. Of course, this does not mean that NewWEC will receive support since there may be other WEC concepts on higher value contours that have currently monopolised the finite development resources. In this case, although NewWEC is promising, it is relatively unpromising because of the higher value of the other concepts. However, for this illustration, let us assume that NewWEC has attracted some investment and development on the concept continues.

As development on NewWEC progresses, its location on the risk–benefit graph will change. If the development works go as planned, then the concept will move

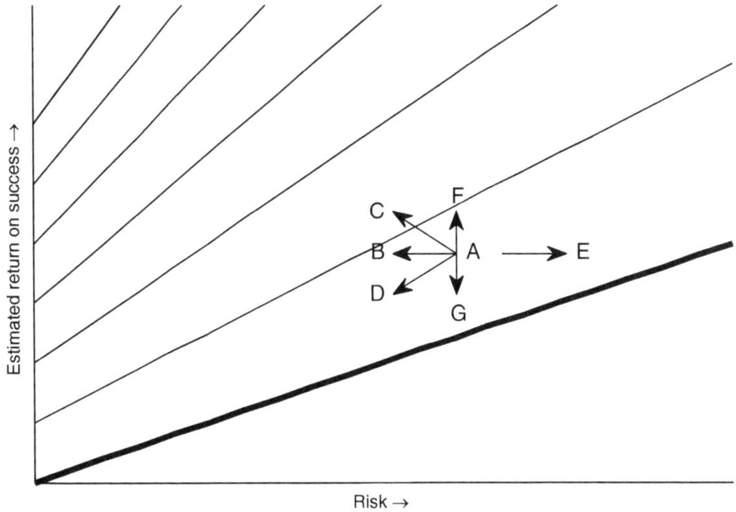

Fig. 26.2 Potential development paths for the fictitious NewWEC

to B on Fig. 26.2, as the expected yield has not changed, but the risk is reduced due to the improved understanding of the concept. Thus, the value of NewWEC has increased and further development is justified. Of course, the expected yield may also change as a result of the work, with a pleasing increase in expected yield at location C or a disappointing reduction in expected yield at location D. Unfortunately, in this final case, it would be necessary to reassess whether the concept has a positive value and deserves further development. A final possible change is that the works uncover more uncertainty than originally estimated and the risk actually increases and the concept moves to location E. Sadly, this circumstance is not as uncommon as one would hope and is especially common where the concept comes from relatively inexperienced or overconfident developers. Clearly, in these circumstances, the value of the concept must be reassessed.

In addition to any changes in the value of NewWEC due to efforts from the NewWEC developers, changes in the market may also cause movements of NewWEC on the risk–benefit graph. For example, a simple increase or decrease in the price paid for the energy would cause a movement of NewWEC to locations F and G, respectively, as shown in Fig. 26.2. Furthermore, there may also be movements due to development of other technologies. It is possible that another developer will demonstrate a technology that makes NewWEC less attractive and so reduces the expected yield from NewWEC. However, NewWEC may become more attractive when another developer demonstrates the successful use of a particular technology that is used by NewWEC and that had previously contributed to the concept risk. This clearly illustrates the benefit of generic technological developments such as advanced hydraulic actuators, linear generators or particular marine operations since these can result in reductions of risk that benefit the whole industry.

26.7 Discussion

A two-stage process for the identification of promising WEC concepts has been presented. The first stage is to identify the fundamental potential of the concept with the assumption that the technology can be designed, manufactured and operated as anticipated. It is important at this stage to understand the hydrodynamics, dynamics and techno-economics of the concept, with particular emphasis on the assumptions used to produce the estimated cost of energy. The second stage is to determine the risk that the concept will not achieve its estimated performance, which will depend on the assumptions that have been made in the analysis of the concept and also the required degree of technological development required for its implementation. This will inform the relative position on the graph and provide us with an approach to identify the most promising WECs.

It can also be seen from the analysis of risk that there is significant benefit to the industry where there is sharing of knowledge and development of common themes. For example, an improvement in the understanding of mooring loads and how they can be calculated appropriately for WECs can benefit all concepts dependent on this technology. Whilst there is currently some effort in identifying and tackling common challenges, much of the current effort is still focused on individual concepts with work undertaken by individual developers. Unfortunately, the consequence of this approach is that the full benefit of any breakthrough made by the developer is not fully exploited and the new knowledge may disappear with the developer if the concept is not successful. Clearly, developers wish to maintain their commercial advantage, but the consequence of not sharing knowledge is that development of the industry is slowed, which puts all developers at risk including developers that may have particular knowledge. This chapter does not claim to determine how this sharing of knowledge could be achieved, but it does highlight the impact that it may have on investment and, thus, provides additional evidence for the support of sharing knowledge.

It should be recognised that the process presented is intended to provide insight into the process and not fully define the activities that are required for identification of promising WEC concepts or how the risk and concept potential should be calculated. Enumeration of the risk is complex and necessarily depends on the perspective of those assessing the risk. This will vary with the relative wealth of the assessors, their risk aversion and potentially other factors specific to the particular circumstances. For example, a large company, with significant wealth, is more likely to assess the risk as lower than a smaller company, whose existence would be at risk if the project were not a success. Similarly, assessment of the concept potential is not simple since it should not only consider the potential for the concept as perceived but also the effect of partial success as well as potential spin-off technologies. Clearly, partial success can provide a reduction in the risk as well as the potential for spin-off technologies can help to make a project viable when it would otherwise be unattractive. However, these additional subtleties do not reduce the correctness of the analysis but simply mean that the analysis may be more complex than it may at first appear.

In conclusion, wave energy is a highly promising potential source of renewable energy. The methods described in this chapter allow the most promising WEC concepts to be identified from the wide range of WEC concepts currently being developed enabling resources to be used most effectively for their development.

Acknowledgments This chapter is based on research funded by the UK Centre for Marine Energy Research (CMER); however, the views expressed in the chapter are the authors' own and do not necessarily reflect those of UK-CMER.

References

1. Cruz JMBP (2008) Ocean wave energy—current status and future prospects. Springer, Berlin
2. Gunn K, Stock-Williams C (2012) Quantifying the global wave power resource. Renew Energy 44(0):296–304
3. Yemm RW, Collee M, Gibson C, Almond J, Taylor C, Henderson RM (2003) Pelamis wec—the full scale prototype programme. In: 5th European wave energy conference. Cork, Ireland
4. Oyster (2014) www.aquamarinepower.com. Accessed 30 May 1999
5. Thorpe T (1999) A brief review of wave energy. The UK department of trade and industry, Report number ETSU-R120
6. Falnes J (2007) A review of wave-energy extraction. Mar Struct 20(4):185–201
7. Babarit A, Hals J, Muliawan MJ, Kurniawan A, Moan T, Krokstad J (2012) Numerical benchmarking study of a selection of wave energy converters. Renew Energy 41:44–63
8. Newman JN (1977) Marine hydrodynamics. MIT Press, Cambridge
9. Simon MJ, Hulme A (1985) The radiation and scattering of long water waves. In: IUTAM Symposium - Hydrodynamics of ocean wave-energy utilisation. Lisbon, Portugal
10. Evans DV (1976) A theory for wave-power absorption by oscillating bodies. J Fluid Mech 77(1):1–25
11. Whittaker T, Folley M (2012) Nearshore oscillating wave surge converters and the development of oyster. Philos Trans R Soc A 370:345–364
12. Budal K, Falnes J (1982) Wave power conversion by point absorbers. pp 59–67
13. Clément AH, Babarit A, Gilloteaux J, Josset C, Duclos G (2005) The searev wave energy converter. In: 6th European wave and tidal energy conference. Glasgow, UK
14. Evans DV (1981) Maximum wave-power absorption under motion constraints. App Ocean Res 3(4):200–203
15. Henry A, Doherty K, Cameron L, Whittaker T, Doherty R (2010) Advances in the design of the oyster wave energy converter. In: Marine and offshore renewable energy conference. London, UK
16. Kahneman D (2011) Thinking, fast and slow: farrar, straus & giroux, New York
17. Pelamis Wave Power (2014) www.pelamiswave.com
18. Ellen H, Boake C, Whittaker T (2003) Operational experience with the limpet full-scale, grid connected, Shoreline Owc Device. In: 5th European wave energy conference. Cork, Ireland

Chapter 27
Optimal Operation Control of Hydrokinetic-based Hybrid Systems

Kanzumba Kusakana, Herman Jacobus Vermaak and Bubele Papy Numbi

Abstract This chapter develops a mathematical programming model to optimize the operation of a hybrid energy system consisting of a hydrokinetic, photovoltaic, wind system, a battery bank, and diesel generator. The optimization approach is aimed at minimizing the cost function subject to the availability of renewable resources, total load energy requirements, as well as the diesel generator and the battery operational constraints. Furthermore, the mathematical models of all other components of hybrid, the proposed system, and the optimization control algorithm is also developed. The main purpose of the control algorithm proposed here is to minimize the use of the diesel generator in the electricity generation process, while maximizing the use of the hydrokinetic system and other available renewable energy sources. For simulation purpose, hourly water velocity, solar irradiation, wind data, and load demand data have been collected and used as an input data. The economic analysis has resulted in the calculation of optimized daily operation cost of the proposed hybrid system in summer and winter conditions. The obtained results represent also a helpful tool for energy planners and justify the consideration of hydrokinetic-based hybrid energy systems more seriously.

Keywords Kinetic energy · Micro-hydropower generation · Hydrokinetic systems · Hybrid energy system

K. Kusakana (✉) · H. J. Vermaak
Department of Electrical, Electronic and Computer Engineering, Central University of Technology, Free State, Bloemfontein, South Africa
e-mail: kkusakana@cut.ac.za

H. J. Vermaak
e-mail: hvermaak@cut.ac.za

B. P. Numbi
Department of Electrical, Electronic and Computer Engineering, University of Pretoria, Pretoria, South Africa

A. Sayigh (ed.), *Renewable Energy in the Service of Mankind Vol I*,
DOI 10.1007/978-3-319-17777-9_27

27.1 Introduction

For a sustainable and clean electricity production in isolated rural areas, renewable energies appear to be the most suitable and usable supply options [1]. Apart from all being renewable and sustainable, each of the renewable energy sources has its specific characteristics and advantages that make it well suited for specific applications and locations [2].

The solar photovoltaic (PV) and wind turbines are well established and are currently the mostly used renewable energy sources for electricity generation in small-scale rural applications. However, for areas where adequate water resource is available, micro-hydro is the best supply option compared to other renewable resources in terms of cost of energy produced [3]. Different from traditional hydropower technology, hydrokinetic (HKT) systems generate electricity by extracting kinetic energy of flowing water instead of potential energy of falling water. This makes HKT far less site specific and more competitive compared to traditional micro-hydropower [4].

It has to be noted that apart from being capital-cost intensive, the other main disadvantages of the HKT system as well as of other renewable energy technologies are their resource-dependent output powers and their strong reliance on weather and climatic conditions [5]. Therefore, they cannot continuously match the fluctuating load energy requirements each and every time. Standalone diesel generators (DG), on the other hand, have very low initial capital costs and can generate electricity on demand, but their operation and maintenance costs are very high, especially when they run at partial loads [6]. In order for the renewable sources to reliably respond to the load energy requirements, they can be combined in a hybrid energy system with backup DG and energy storage systems [7]. The most important feature of such hybrid system is to generate energy at any time by optimally using each available energy sources [6]. The fact that the renewable resources available at a given site are a function of the season of the year implies that the fraction of the energy provided to the load is not constant. This means that for hybrid systems comprising HKT, DG, and other renewable sources in their architecture, the renewable energy fraction and the energy storage capacity are projected to have a significant impact on the DG fuel consumption, depending on the interaction between the daily variation of renewable resources and the ones of load demand. A high renewable output power will be able to respond to the load requirements and/or charge the battery; this will result in less fuel consumption by the DG.

Several research works have been conducted on the subject of optimal operation control of hybrid renewable energy systems with DG [8–10]. Currently, few publications are available on standalone HKT systems; however, it is very rare to find research dealing with HKT systems combined with other renewable sources and DG. In [11], the author investigated the potential use of hydrokinetic-based hybrid systems for low cost and sustainable electrical energy supply. Different hybrid system configurations with PV, wind, and diesel generator are modeled and simulated using the Hybrid Optimization Model for Electric Renewable (HOMER) software.

The simulation results from two different case studies show that hybrid systems with HKT modules incorporated in their architectures have lower net present costs as well as lower costs of energy compared to all other supply options where the HKT modules are not included.

The hybrid system considered in this study is composed of micro-HKT, PV, wind modules with backup DG, as well as a battery storage system. Unlike the abovementioned chapters, this chapter focuses on minimizing the operation cost of micro-HKT-based hybrid system during a 24-h period. The aim of this study is also to highlight the daily fluctuation of load demand and energy supply, as well as real operational challenges in standalone electricity generation. Considering a short time horizon, the operation costs of the battery and the renewable systems are negligible, therefore only the fuel cost of the DG is considered. The micro-HKT/PV/wind/DG/battery hybrid system operation costs are nonlinear, as they depend on the component size and type, the nonlinear load variations, the nonlinear resources fluctuations, and the dispatch strategy [12].

The various optimization techniques available in literature such as probabilistic, iterative, and other classical approaches do not take into account the short-term or seasonal variation of the resources as well as the one of the load energy demand. The optimization model developed in this study takes into account the nonlinearity of the operation costs linked with the considered micro-HKT/PV/wind/DG/battery hybrid system, and this necessitates the use of nonlinear programming. Heuristic approaches such as the one used in this work are fast and more efficient than classical approaches in terms of their ability to solve complex nonlinear problems [13]. To get the best of this modern optimization approach, a detailed and accurate model describing the nonlinear hybrid system performance has been derived. Most of research works published previously have assumed a fixed load demand as well as same daily operational cost, which can be extrapolated to get any cost in the selected time horizon. Therefore, in this model, the power flow is considered taking into account more practical and realistic daily and seasonal fluctuation in the load energy demand in order to evaluate more accurately the corresponding fuel costs of the micro-HKT/PV/wind/DG/battery hybrid system.

The subsequent sections focus on the proposed hybrid system and the component models. The optimization model, which comprises the objective function, constraints, and other parameters, is presented, followed by the analysis of the simulation results, discussion, and conclusion.

27.2 Hybrid Energy System Components

The power flow of the proposed hybrid system is shown in Fig. 27.1. The load demand (P_L) is primarily met by the sum of the output power from renewable components, and the battery starts discharging within its operating limits as soon as the renewable powers do not meet the demand. If the sum of the renewable output

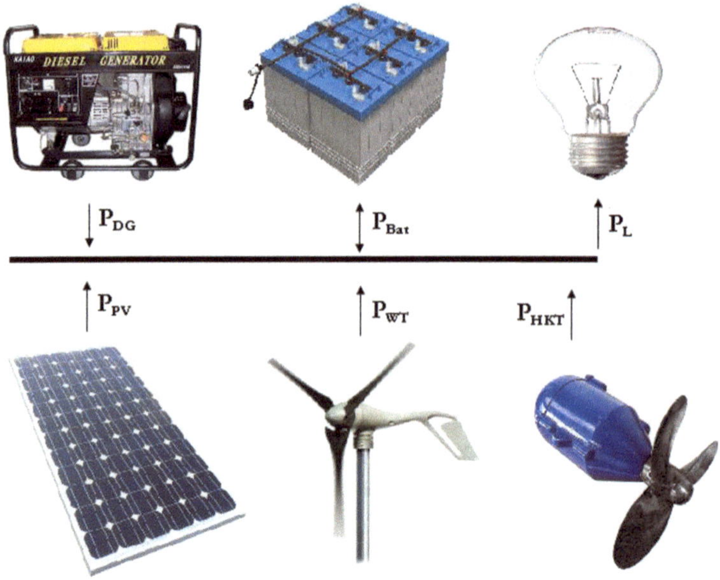

Fig. 27.1 Hybrid system layout. P_{HKT} power output of the hydroelectric system, P_{GD} power rating of the diesel generator, P_{Bat} is the power flowing from the battery system, P_L Load demand, P_{pv} output power of the solar PV system, P_{wT} the output power of the solar PV system

power is above the load energy demand, the battery is charged. The DG is used only when the power from the renewable output and the battery cannot respond to the load energy requirements. The mathematical models of the system's different components are presented in the following subsection:

27.2.1 Hydrokinetic System

HKT systems extract kinetic energy from moving water without the need for a dam, barrage, or penstock. The operation principle and the advantages of this power generation technology compared to other renewable options have been discussed in detail in reference [11].

The power output (P_{HKT}) of the HKT system within a sampling time can be expressed as follows:

$$P_{HKT} = \frac{1}{2} \times \rho_W \times A_{HKT} \times C_{p,H} \times \eta_{HKT} \times \int_{t_0}^{t} v_W{}^3(t) \times f(t) \times dt, \qquad (27.1)$$

where ρ_W is the density of water (1000 kg/m^3), $C_{p,H}$ is the coefficient of the HKT turbine performance, η_{HKT} is the combined efficiency of the HKT turbine and the generator, A_{HKT} is the turbine area (m^2), v_W is the water current velocity (m/s), and *f(t)* is the water probability density function.

27.2.2 Wind Energy System

Wind energy systems convert the kinetic energy of moving air into mechanical and then electrical energy [14]. As for the HKT system, the power output (P_{WT}) of the wind system within a sampling time can be expressed as follows:

$$P_{WT} = \frac{1}{2} \times \rho_a \times A_{WT} \times C_{p,WT} \times \eta_{WT} \times \int_{t_0}^{t} v_a^3(t) \times f(t) \times dt, \qquad (27.2)$$

Where ρ_a is the air of water (1225 kg/m^3), $C_{p,W}$ is the coefficient of the wind turbine performance, η_{WT} is the combined efficiency of the wind turbine and the generator, A_{WT} is the wind turbine area (m^2), v_a is the wind velocity (m/s), and *f(t)* is the wind probability density function.

27.2.3 Photovoltaic System

When light strikes a silicon, gallium arsenide, or cadmium sulphide cell, an electric current is generated through the PV effect [15]. The power rating of a PV panel is expressed in peak Watts (Wp) indicated at "standard test conditions" conducted at a temperature of 25 °C and irradiance of 1000 W/m^2. The output power (P_{PV}) of the solar PV system can be expressed as follows [16]:

$$P_{PV} = A_{PV} \times \eta_{PV} \times \int_{t_0}^{t} I(t) \times f(t) \times dt, \qquad (27.3)$$

where A_{PV} is the total area of the PV generator (m^2), η_{PV} is the module efficiency, I is the hourly irradiance (kWh/m^2), and *f(t)* is the radiance density function.

27.2.4 Diesel Generator

A DG is normally diesel engine coupled to generator. The DG is usually designed in such a way that it always operates between 80 and –100 % of its power rating (P_{GD}) to achieve high efficiency [17]; this condition can be used later as an operation constraint. With this operation strategy as well as operation constraint, the DG is expected to run at high load factors, which will result in a decrease of the fuel consumption and increase of the DG life span [10].

27.2.5 Battery Storage System

The output power from the renewable components, and the load demand at given time t, determine whether the battery is charging or discharging. At any time, the battery state is associated to the previous state of charge and to the energy produced and consumed situation of the system during the time t_0–t. The battery dynamic equation can be expressed as follows [18]:

$$SOC_t = SOC_{t_0} - \frac{\eta_{Bat}}{E_{nom}} \times P_{Bat(t_0)} \times (\mathrm{t} - \mathrm{t}_0) \tag{27.4}$$

where SOC is the state of charge of the battery, η_{Bat} is the battery charging or discharging efficiency, E_{nom} is the battery system nominal energy, and P_{Bat} is the power flowing from the battery system.

27.3 Optimization Model and Proposed Algorithm

27.3.1 Model Development

In the model developed in this work, the battery is charged by the renewable components only; this means that the DG is switched on only to supply the load. This configuration guarantees the optimum use of HKT, PV, and WT outputs, while no energy is wasted when the DG runs, since the total power produced matches the load demand. Therefore, the economical operation problem is to find the optimal scheduling of energy production at any given time that minimizes the DG fuel expenses, while totally responding to the load energy requirements within the system's operating limits and constraints.

27.3.1.1 Objective Function

The objective is to minimize the fuel consumption cost from the DG during the operation time. This can be expressed as:

$$\min \Delta t \times \sum_{j=1}^{N} (aP_{DG(j)}^2 + bP_{DG(j)} + c), \tag{27.5}$$

where N is the number of sampling intervals within the operation range or period of the system; a ,b, and c are the fuel cost coefficients; j is the jth sampling interval; $PDG_{(j)}$ is the output power from the DG at jth sampling interval, and Δt is the sampling time.

27.3.1.2 Constraints

The different constraints on the operation are as follows:

- Power balance:

$$P_{HKT(j)} + P_{PV(j)} + P_{WT(j)} + P_{DG(j)} + P_{Bat(j)} = P_{L(j)} \tag{27.6}$$

- Variable limits:

$$\begin{aligned} &0 \le P_{HKT(j)} \le P_{HKT(j)}^{\max} \\ &0 \le P_{PV(j)} \le P_{PV(j)}^{\max} \\ &0 \le P_{WT(j)} \le P_{WT(j)}^{\max} \\ &0 \le P_{DG(j)} \le P_{DG}^{\max} \\ &-P_{Bat}^{rated} \le P_{Bat(j)} \le P_{Bat}^{rated} \end{aligned} \tag{27.7}$$

- Battery state of charge:

$$SOC^{\min} \le SOC_{(j)} \le SOC^{\max} \tag{27.8}$$

27.3.2 Proposed Algorithm

The objective function to minimize has been modeled as a nonlinear function of the DG output power. The nonlinear optimization problem can be solved using the "fmincon" function in Matrix Laboratory (MATLAB) [19]. This function solves problems in the form:

$$\min_x f(x) \text{subject to}: \begin{cases} c(x) \le 0 \\ c_{eq}(x) = 0 \\ A.x \le b \\ A_{eq}.\text{x} = b_{eq} \\ l_b \le \text{x} \le u_b \end{cases} \tag{27.9}$$

Where x, b, b_{eq}, l_b, and u_b are vectors, A and A_{eq} are matrices, $c(x)$ and $c_{eq}(x)$ are functions that return vectors, and $f(x)$ is a function that returns a scalar, $f(x)$, $c(x)$, and $c_{eq}(x)$ can be nonlinear functions.

27.4 Case Study

27.4.1 Resources and Load Data

A rural household is selected as case studies to analyze the behavior of the hybrid system. This system has to be optimally sized such that demand is always met at any given time. The hourly renewable resources as well as load demand for a 24-h period are shown in Table 27.1. These data are used as input to the energy optimization model developed in Sect. 27.3 above.

It has to be noted that these data have been collected for a day in September where the water resource is the lowest in the year. Therefore, the considered turbine is sized in such a way to give a rated power of 1 kW at 1.4 m/s water velocity. Thus, it is assumed that the produced HKT power remains constant for water velocities above 1.5 m/s.

Table 27.1 Resources and load data

Time (h)	Summer				Winter			
	Global solar (kW/m^2)	Wind speed (m/s)	Water speed (m/s)	Load (kW)	Global solar (kW/m^2)	Wind speed (m/s)	Water speed (m/s)	Load (kW)
00:00	0.000	0.821	1.41	0.3	0.000	2.505	1.41	0.3
01:00	0.000	1.665	1.41	0.2	0.000	2.440	1.41	0.2
02:00	0.000	0.998	1.41	0.1	0.000	1.332	1.41	0.1
03:00	0.000	0.956	1.41	0.0	0.000	2.540	1.41	0.0
04:00	0.000	2.549	1.41	0.3	0.000	2.430	1.41	0.3
05:00	0.000	2.558	1.41	0.0	0.000	2.190	1.41	0.0
06:00	0.000	2.775	1.41	2.4	0.000	2.385	1.41	3.0
07:00	0.002	3.754	1.41	0.6	0.000	1.072	1.41	0.7
08:00	0.141	2.948	1.41	4.3	0.145	1.431	1.41	8.0
09:00	0.417	2.828	1.41	5.6	0.244	0.876	1.41	5.6
10:00	0.687	2.870	1.41	3.2	0.306	1.907	1.41	2.6
11:00	0.940	2.522	1.41	1.6	0.512	1.894	1.41	3.0
12:00	1.062	1.766	1.41	0.3	0.611	2.096	1.41	0.5
13:00	1.061	2.576	1.41	2.0	0.614	2.123	1.41	3.4
14:00	0.978	2.017	1.41	0.4	0.568	2.133	1.41	0.7
15:00	0.846	2.282	1.41	0.8	0.428	3.038	1.41	1.3
16:00	0.679	3.116	1.41	3.9	0.460	2.521	1.41	1.4
17:00	0.464	2.626	1.41	1.8	0.266	2.227	1.41	1.5
18:00	0.208	3.427	1.41	1.7	0.000	1.819	1.41	3.8
19:00	0.043	2.972	1.41	1.9	0.000	2.825	1.41	4.6
20:00	0.000	2.543	1.41	2.2	0.000	3.571	1.41	5.9
21:00	0.000	2.336	1.41	0.9	0.000	2.070	1.41	2.1
22:00	0.000	1.863	1.41	0.7	0.000	2.537	1.41	0.8
23:00	0.000	1.231	1.41	0.3	0.000	1.523	1.41	0.3

Table 27.2 Simulation parameters

Item	Figure
Sampling time	30 min
Battery nominal capacity	5.6 kWh
Battery maximum SOC	95 %
Battery minimum SOC	40 %
Battery charging efficiency	85 %
Battery discharging efficiency	100 %
Diesel fuel price	1.4$/l
A	0.246
B	0.0815
C	0.4333

SOC state of charge of the battery

27.4.2 Components Size and Model Parameters

The focus of the current study is principally on the optimal energy management of the hybrid system; therefore, the optimal sizing of the system components has been done using the HOMER. As a result, the combination of HKT (3 kW)/PV (1 kW)/WT (1 kW)/DG (1 kW)/13 batteries has been considered for simulation purposes of the developed optimal operation control algorithm. The battery as well as the DG parameters used in the simulation is shown in Table 27.2. The case where the DG alone is supplying the load is also considered for comparison purposes.

27.5 Simulation Results and Discussion

The objective function expressed in Eq. (27.6) means that the operation time (resulting in fuel consumption) of the DG needs to be minimized. Keeping that in mind, from the first constraint (27.7), the load has to be met first by the renewable resources as well as by the battery if the SOC is within the operating ranges; otherwise, the DG switches on to compensate the deficit in demand.

Figure 27.2 shows how the load demand as well as the maximum and optimum output power flow from the HKT, PV during the 24-h interval (winter case). Figure 27.3 shows the power balance between the load and the renewable resources; the DG output power, the battery power flow as well as the battery SOC during a 24-h period are also displayed. The following observation can be made after analyzing these two figures:

- From Fig. 27.2, it can be noticed that during the night and early morning the load demand is low; therefore, it is successfully met mainly by the HKT system; the wind and PV systems are not able to generate during these periods because of the lack of wind and solar resources. P_{BAL} (on Fig. 27.3) is the difference between the sums of the optimum renewable output power minus the load demand P_L. If

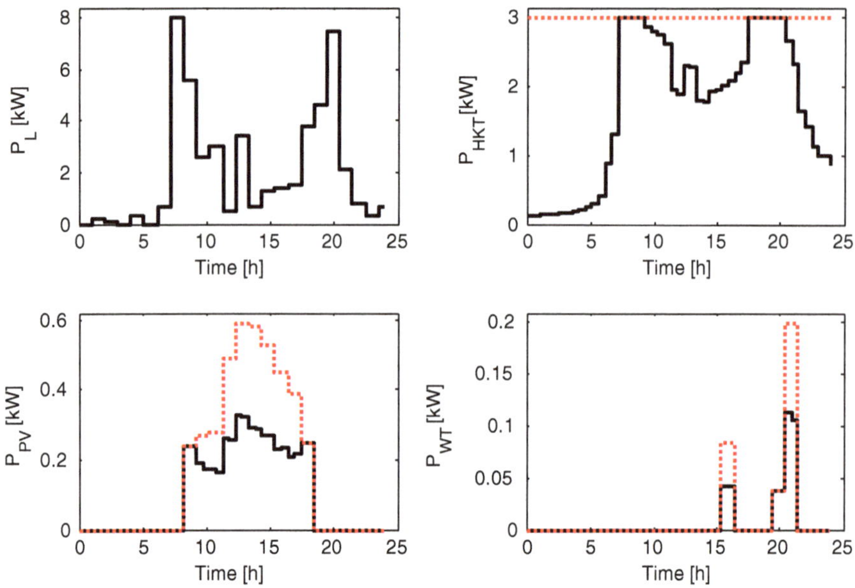

Fig. 27.2 Daily load profile and power flow from renewable systems (winter case). P_{pv} the output power of the solar PV system, P_L load demand

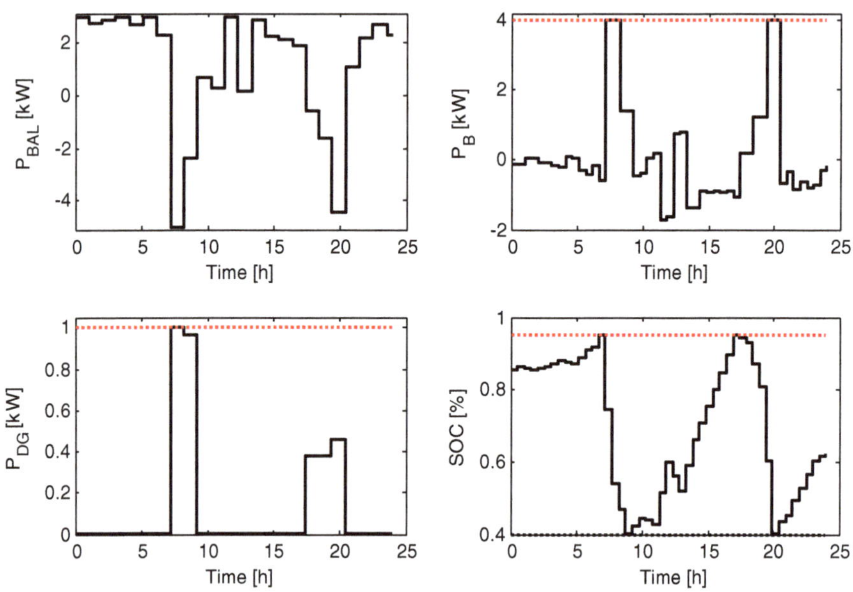

Fig. 27.3 Daily DG, battery power flow, and battery SOC (winter case). *SOC* state of charge of the battery, P_{DG} power rating of the diesel generator, P_{BAL} (on Fig. 27.3) is the difference between the sums of the optimum renewable output power minus the load demand P_L

P_L is positive, the surplus of generated power can be used to charge the battery; if P_L is negative, the shortage of energy can be brought in by using the battery (P_B) first then the DG (P_{DG}) within their operating limits. During the day, when the solar and wind resources are available, the load is met by the HKT, PV, WT, and battery bank.

- The morning peak load demand occurs between 7 a.m. and 9 p.m.; therefore, the HKT, PV, and the battery are used at their maximum output to supply the load; the DG also switches on only to balance the energy needed and then switches off as soon as there is enough power from the other power sources. The DG operating time and output power depend on the load demand, battery SOC, and the amount of power from the renewable resources. It can be seen that the DG is not used to charge the battery but only to supply the load.
- After the morning peak, the SOC of the battery is at its minimum operation limit (40 %); therefore, the renewable outputs produce more power than the load requirement. This surplus is used to charge the battery bank to an SOC of 95 % which is reached at the end of the afternoon as shown in Fig. 27.3, where the negative part of the battery power flow (P_B) represents the charging process.
- In the evening, gradually from 5 p.m., reaches the peak between 7 p.m. and 8 p.m., and then finally decreases at 9 p.m. Therefore, from 5 to 6 p.m., the HKT and PV are used at their maximum outputs in conjunction with a small contribution of the DG and the battery. After 6 p.m., the PV system can no more provide energy and the load demand is increasing; therefore, the contribution of the battery is maximal and the DG output power also increases to balance the energy needed by the load.
- A poor WT output is noticeable in the afternoon and in the evening; this power is also used to supply the load and contributes to the battery charging power requirement.

The developed model can also be used to analyze the difference in power flow during summer and winter due to different climatic conditions and load requirements which have significant effects on the diesel dispatch strategy and fuel consumption. Using the date from Sect. 27.4 above, the simulation results reveal that the DG supplies more power in winter than in summer; this is due to lower PV and WT outputs, lower SOC of the battery, as well as higher load demand in winter than in summer.

The actual daily fuel expense for the selected DG can be found by multiplying the diesel price ($/L) by the amount of fuel used (L/day). It has to be highlighted that this daily fuel expense is highly dependent on the size and type of DG (fuel cost curve and fitting parameters from the manufacturer) used in the simulation.

Table 27.3 Daily fuel cost savings

	Winter		Summer	
	Consumption (L)	Cost ($)	Consumption (L)	Cost ($)
DG only	139	195	68	95
Hybrid system	22	31	0	0
Savings	117	164	68	95

DG distributed generation

Table 27.3 shows how much operation fuel can be saved by using the hybrid system instead of the selected DG in a winter or a summer day. These results demonstrate that it is very important to take into account the variations of the load and renewable energy resources with seasons when calculating the system's daily operation cost.

27.6 Conclusion

A mathematical model to optimize the operation of a hybrid energy system consisting of a HKT, PV, wind system, a battery bank, and DG has been developed. This model aims to minimize the use of the diesel generator while maximizing the use of the available renewable energy sources. As already mentioned, this work considers the nonlinearity of the renewable resources, load demand, as well as diesel fuel consumption resulting in uniform daily operational costs. The hourly water velocity, solar irradiation, wind speed, load demand, as well as the diesel generator fuel consumption curve parameters data have been used as input data for simulation purposes. The simulation results show that by using the proposed hybrid system and taking into account the nonlinearity in daily and seasonal variations of the load demand, more accurate operation costs (fuel consumption) can be obtained.

The discrepancies in daily operations cost achieved highlight the potential of the proposed optimization model to reduce fuel consumptions for the hydrokinetic-based hybrid system compared to the diesel-only scenario. These results also demonstrate that it is very important to take into account the variations of the load and renewable energy resources with seasons when calculating the system's daily operation cost.

References

1. Kusakana K, Vermaak HJ (2013) Hybrid renewable power systems for mobile telephony base station in developing countries. Renew Energy 51:419–425
2. Hepbasli A (2008) A key review on exergetic analysis and assessment of renewable energy resources for a sustainable future. Renew Sust Energy Rev 12(3):593–661
3. Paish O (2002) Small hydropower: technology and current status. Renew Sustain Energy Rev 6(6):537–556
4. Vermaak HJ, Kusakana K, Koko SP (2014) Status of micro-hydrokinetic river technology in rural applications: a review of literature. Renew Sustain Energy Rev 29:625–633
5. Chen F, Duic N, Alves LM, Carvalho MG (2007) Renewislands—renewable energy solutions for islands. Renew Sustain Energy Rev 11(8):1888–1902
6. Kusakana K, Vermaak HJ (2013) Hybrid diesel generator—battery systems for off-grid rural applications. IEEE international conference on industrial technology (ICIT 2013), 25–28 Feb 2013, Cape Town, pp 839–844
7. Gupta A (2010) Modeling of hybrid energy system. Ph. D. Thesis, alternate hydro energy centre Indian institute of technology Roorkee. Roorkee—247 667 (India) July, 2010

8. Bernal-Agustın JL, Dufo-Lopez R (2009) Simulation and optimization of stand-alone hybrid renewable energy systems. Renew Sustain Energy Rev 13:2111–2118
9. Bajpai P, Dash V (2012) Hybrid renewable energy systems for power generation in stand-alone applications: a review. Renew Sustain Energy Rev 16:2926–2939
10. Hu Y, Solana P (2013) Optimization of a hybrid diesel-wind generation plant with operational options. Renew Energy 51:364–372
11. Kusakana K, Vermaak HJ (2013) Hydrokinetic power generation for rural electricity supply: case of South Africa. Renew Energy 55:467–473
12. Kusakana K, Vermaak HJ (2013) Hybrid diesel generator/renewable energy system performance modeling. Renewable Energy 67:97–102
13. Koutroulis E, Kolokotsa D, Potirakis A, Kalaitzakis K (2006) Methodology for optimal sizing of stand-alone photovoltaic/wind generator systems using genetic algorithms. Solar Energy 80:1072–1088
14. Ghedamsi K, Aouzellag D (2010) Improvement of the performances for wind energy conversions systems. Int J Electric Power Energy Syst 32:936–945
15. Skoplaki E, Palyvos JA (2009) On the temperature dependence of photovoltaic module electrical performance. A review of efficiency/power correlations. Solar Energy 83:614–624
16. Singh GK (2013) Solar power generation by PV (photovoltaic) technology: a review. Energy 53:1–13
17. Dufo-Lopez R, Bernal-Agustın JL (2008) Influence of mathematical models in design of PV-Diesel systems. Energ Convers Manage 49:820–831
18. Sechilariu M, Wang BC, Locment F (2014) Supervision control for optimal energy cost management in DC microgrid: design and simulation. Electric Power Energy Syst 58:140–149
19. Kumar AS, Munda JL (2014) Optimisation of voltage and frequency regulation in an isolated wind-driven six-phase self-excited induction generator. J Energy Inst. http://dx.doi.org/10.1016/j.joei.2014.03.007. Accessed 9 July 2015

Chapter 28
Experimental and Numerical Investigation of Blade Angle Variation on a Counter-Rotating Tidal Current Turbine

Lee Nak-Joong, Kim In-Chul, Hyun Beom-Soo and Lee Young-Ho

Abstract Importance of renewable energy has become paramount due to its perennial source and no adverse environmental impact. Ocean is one of the major source of renewable energy where the sun's energy is converted into various natural phenomenon. Southwestern sea in particular, in Korea, has large range of tidal currents with potential for tidal current power generation. Tidal power has great potential for future power and electricity generation because of the massive size of the oceans. The major benefit of tidal power and difference from most renewable energy sources is that it is independent of seasons and weather, that is, it is always constant which makes power generation predictable and makes tidal power a reliable energy source. Horizontal-axis-type turbine appears to be the most technologically and economically viable option for the generation of tidal power. Several studies have shown that single-rotor turbines can obtain a theoretical maximum power coefficient of 59.3%, whereas dual rotor can obtain a maximum of 64%. Hence, with the optimization of counter-rotating turbines, more power can be obtained than the single-rotor turbines.

Previous studies focus on the performance analysis of the turbine with the variation of distance between the blades. This chapter primarily concentrates on the investigation of the performance analysis and power output of a counter-rotating current turbine by the variation of blade angles by both computational fluid dynamics (CFD) simulation and experiments. Numerical simulations were performed using a commercial finite volume method solver, ANSYS CFX ver.13.0. Experiments were conducted in the water tank with a vertically circulating water channel in

L. Young-Ho (✉)
Division of Mechanical and Energy System Engineering,
Korea Maritime and Ocean University, Busan, Korea
e-mail: lyh@kmou.ac.kr

L. Nak-Joong · K. In-Chul
Department of Mechanical Engineering, Graduate School,
Korea Maritime and Ocean University, Busan 606-791, Korea

H. Beom-Soo
Division of Naval Architecture and Ocean System Engineering,
Korea Maritime and Ocean University, Busan, Korea

A. Sayigh (ed.), *Renewable Energy in the Service of Mankind Vol I*,
DOI 10.1007/978-3-319-17777-9_28

the laboratory of Korea Maritime and Ocean University (KMOU) to validate the numerical results. Several experiments were conducted with the fixed front blade angle and varying the rear blade angle and vice versa at various water flow rate. Surface streamlines, torque, total power output, power coefficient (C_p) etc., were characterized and compared for CFD and experimental cases. The results obtained find good agreement with each other.

Keywords Blade angle · Counter rotating · Current turbine · CFD · Experiment

28.1 Introduction

Renewable energy is reliable and plentiful and will potentially be very cheap once technology and infrastructure improve. Many countries are looking to increase the share of renewable energy to their total consumption these days. It includes solar, wind, geothermal, hydropower, and tidal energy plus biofuels that are grown and harvested without fossil fuels. Ocean is one of the major source of renewable energy, where the sun's energy is converted into various natural phenomenon. Southwestern sea in particular, in Korea, has tidal currents with potential for tidal current power generation. Tidal power generation is predictable and tidal power a reliable energy source because it does not rely on weather or seasons. Tidal turbines are placed in areas with high tidal movements, and are designed to capture the kinetic motion of the ebbing and surging of ocean tides to produce electricity. The surrounding marine environment is not disturbed significantly as seawater is still able to flow naturally through the tidal current turbines as it produces power from the water velocity. The horizontal-axis-type turbine appears to be the most viable option at this stage. Single-rotor turbines are calculated to have a theoretical efficiency of 59.3 %; a dual rotor turbine can obtain a maximum of 64 %. Hence, with the optimization of counter-rotating turbines, more power can be obtained than single-rotor turbines.

According to blade element momentum theory (BEMT), a three-bladed 40-W horizontal-axis-type marine current turbine was designed. This turbine is used for small-scale experimental tests to analyze its performance characteristics. The turbine is planned to be scaled up to a size to produce 10 kW within a submerged floating structure in future.

Previous experiments on tidal turbines had focused on the effect of varying the distance between the front and rear blades on the power characteristics of the turbine. This research work solely focuses on the effects of changing the front and rear blade angles on the performance of the turbine.

Along with the experimental results, the turbine setup and experimental conditions are simulated using a commercial computational fluid dynamics (CFD) software. The results from both CFD calculations and experimental investigations are compared and further discussed in this chapter.

28.2 Numerical Procedure

A 40-W horizontal-axis three-bladed tidal current turbine was designed using BEMT. Figure 28.1 shows the single airfoil of NACA-63-421 with various twist angles and chord length (left) and three-dimensional (3D) modeling of a blade (right). The turbine has the diameter (*D*) of 500 mm with designed rotational speed of 190 rpm. The rated water speed was 1 m/s.

Figure 28.2 shows the 3D model of the counter-rotating turbines and the experimental setup. Figure 28.3 (left) shows the computational grid of the turbine. The distance from the blade to the inlet, top, and outlet was 3, 5, and 7 times, respectively, the diameter of blade. Single blade was meshed with the remaining blades accounted for by periodic conditions. The nondimensional distance of the first node from the wall or the *y*-plus value was less than 5 in this calculation with approximately 6.2 million nodes in the hexahedral mesh.

Figure 28.3 (right) shows the computational domain used for the simulation. *k–ω* shear stress transport (SST) turbulence model was used, and all the calcula-

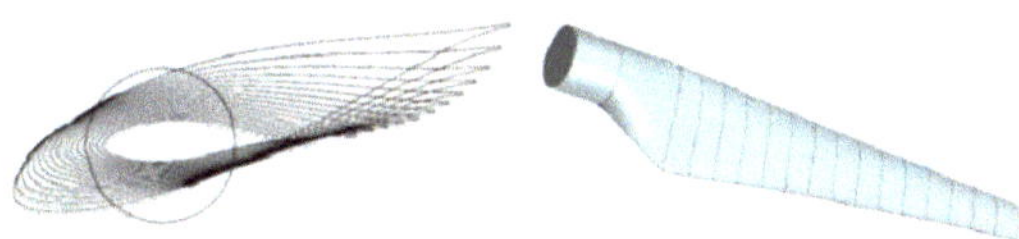

Fig. 28.1 3-D rotor blade configuration

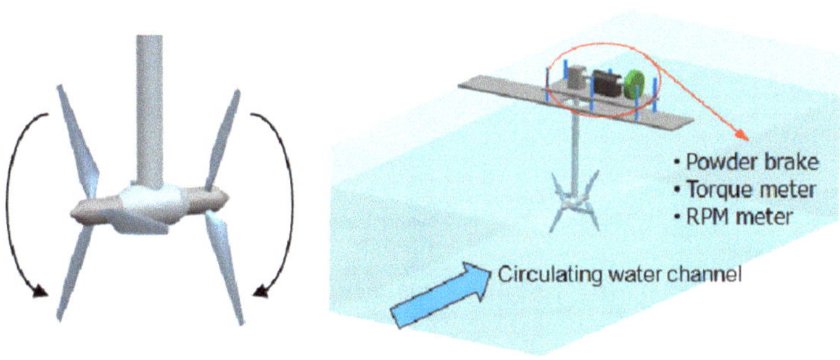

Fig. 28.2 3-D turbine configuration. *RPM* revolutions per minute

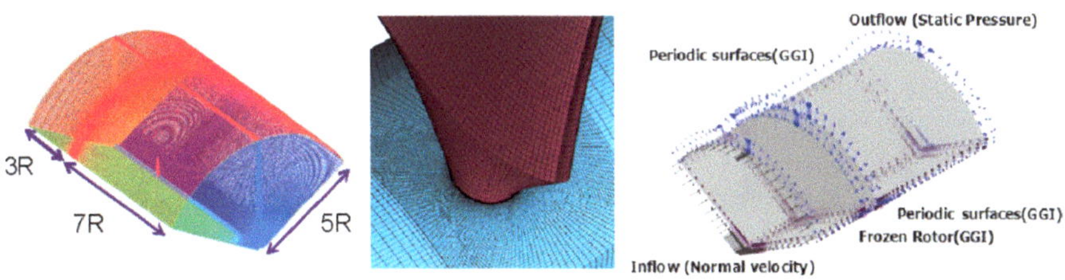

Fig. 28.3 Computational mesh and domain of rotor blade

tions were conducted under steady-state condition using a commercial finite volume method solver, ANSYS CFX ver.13.0. The water velocity was varied from 0.8 to 2.0 m/s during simulation.

28.3 Experimental Procedure

Experiments were conducted in the water tank with a vertically circulating water channel located in the laboratory of KMOU. The height, width, and length of the tank was 1200, 1800, and 4000 mm, respectively. Height of the water level used for this experiment was 900 mm. Figure 28.4 shows the water channel, the turbine, control panel for the channel, and the two fans used to control the water velocity in the water channel. Figure 28.5 shows a schematic diagram of the experimental setup.

Fig. 28.4 Circulating water channel

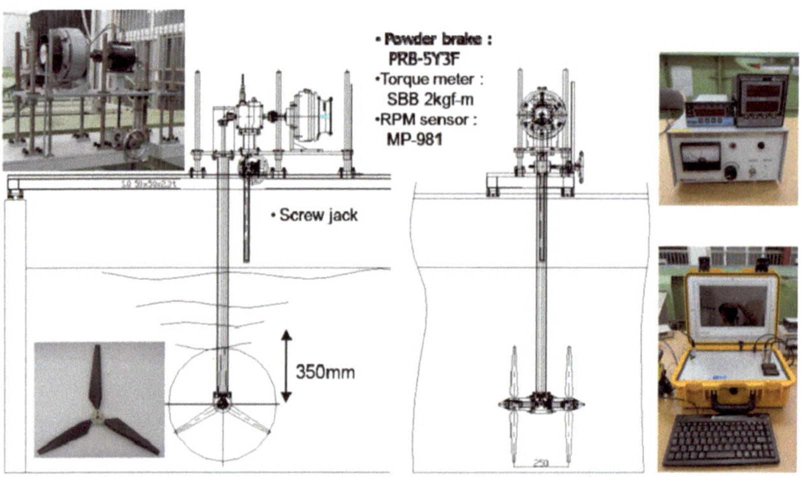

Fig. 28.5 Schematic of the experimental setup

Powder brake, torque transducer, and revolutions per minute (RPM) sensor were attached in the turbine setup for measurement, and the data were recorded in the data logger. The torque meter (model SBB) had a measurement range from 0 to 2 kgf m. The torque meter contained an RPM sensor (MP-981) with a measurement range of up to 10,000 rpm. A forced air-cooled-type powder brake (model PRB-5Y3F) was used to control the RPM of the turbine.

The blockage ratio of the system was 12 %. Therefore, a blockage correction factor was applied to the results [4, 6]. Experiments were conducted by varying the water flow from 0.6 to 1.4 m/s. Pitot tube kept at 1 m in front of the turbine was deployed to measure water velocity. The distance between the dual rotors was fixed at 0.5 D.

Aluminum was used as the material for blades, and each blade was fabricated from a single piece using a five-axis milling machine. The total number of blades manufactured was six.

28.4 Results

Figure 28.6 depicts the streamlines observed at the suction side of front and rear blades at a water velocity of 0.8–2.0 m/s in which the arrows indicate the direction of the flow. The arrangement is such that the front blade (*F*) is positioned at 0° and the rear blade (*R*) at −5°. Both blades show stable flow over the blade surface with little to no turbulent flow at lower velocities; whereas, as the flow velocity increases, the flow becomes less likely to remain stable on the surface. Flow separation can be seen on the front blade from 1.2 to 2.0 m/s. The nature of streamlines suggest that the flow at rear blade does not separate to the extent as in the front blade. The major reason behind this can be due to the reduction of velocity in the wake of the front blade. Figure 28.7 shows the suction side streamlines for front and rear blades both fixed at 0° to the flow obtained from CFD calculations. In this, the flow remains stable below the design speed of 1 m/s in a similar pattern as described earlier for

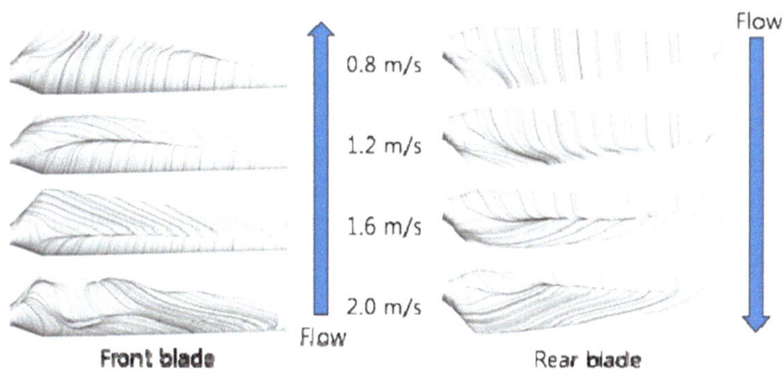

Fig. 28.6 Suction side surface streamlines ($F = 0°$, $R = -5°$)

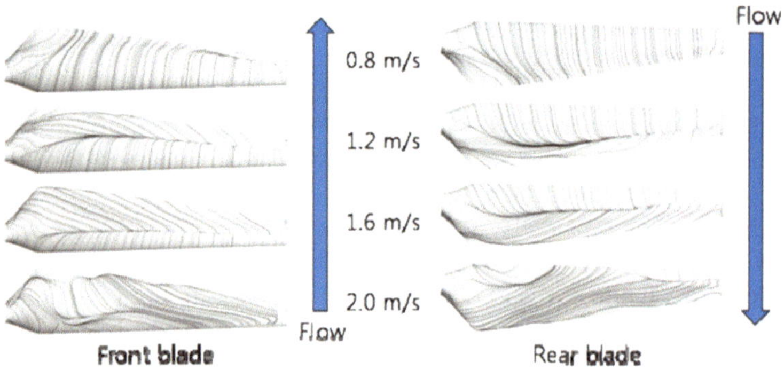

Fig. 28.7 Suction side surface streamlines ($F=0°$, $R=0°$)

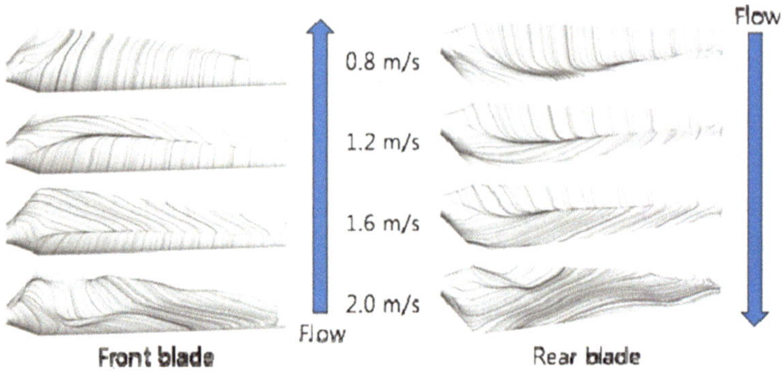

Fig. 28.8 Suction side surface streamlines ($F=0°$, $R=5°$)

Fig. 28.6. As the water velocity increases, the flow also becomes more turbulent over the surface of both blades. The flow was observed to be more turbulent on the rear blade than the previous one at 2.0 m/s. Figure 28.8 shows the streamlines from the CFD calculations when the rear blade was angled at 5° to the flow, while the front blade was fixed at 0°. The flow pattern over the front blade surface shows that it was quite similar to the previous cases (Figs. 28.6 and 28.7), whereas the rear blade shows the increase in turbulent flow occurrence for all water velocities. Figures 28.9 and 28.10 show the surface streamlines cases when the rear blade is fixed at 0°. Figure 28.9 shows the streamlines when the front blade is at −5° to the incoming flow. At water velocity of 0.8 m/s, there is relatively a lesser amount of turbulent flow on both the blades. But as the velocity increases, the flow begins to separate on the front blade much earlier than the rear. The flow on the rear blade does not fully become turbulent until 2 m/s. Figure 28.10 shows the surface flow on the blades when the front blade is angled at 5°. In this case, it can be observed that the flow on the front blade does not completely separate at higher velocities, and the streamlines show that the most of the flow stays on the blade at the lower velocities. But the rear blade streamlines are similar to the previous cases.

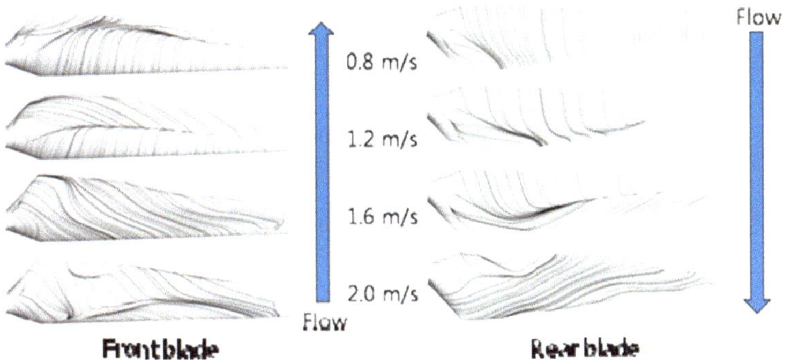

Fig. 28.9 Suction side surface streamlines ($F = -5°$, $R = 0°$)

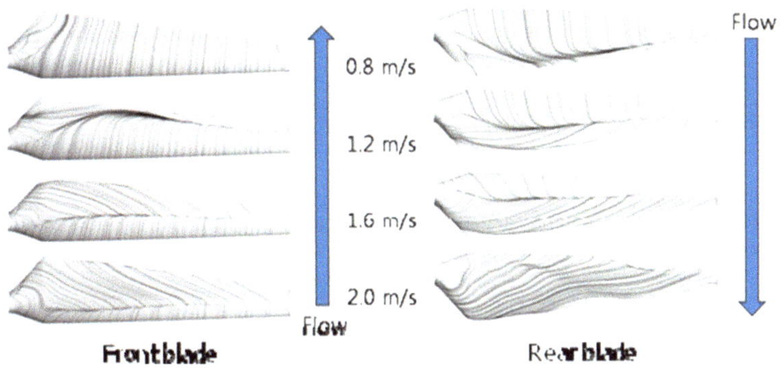

Fig. 28.10 Suction side surface streamlines ($F = 5°$, $R = 0°$)

Figure 28.11 shows the comparison between the power output developed during the experiments and obtained by CFD calculations at varied water velocities with the fixed front blade at 0° and varying rear blade. The limitation of the experimental setup was that it was unable to produce water velocities larger than 1.4 m/s. However, velocities over this speed caused a lot of problems as sudden stoppage in operation of the water channel during experimentation. The trends of the graph show that both experiment and CFD results are in good agreement over all velocities. The experimental results showed slightly lower power from 0.8 to 1.1 m/s, but slightly higher power at velocities over this range. It was observed that difference between power produced was little when the rear blade angle was changed except at 2.0 m/s. The maximum power obtained from CFD was about 183 W at 1.8 m/s when the rear blade angle was positioned at 5°.

The comparison between the power coefficients (C_p) obtained from CFD calculations and experimental result at a fixed front blade angle is presented in Fig. 28.12. The nature of the graph suggests that the power coefficient obtained from experimental data is lower than the CFD results. The difference between the power coefficients can be easily noticed in the graph as the rear angle is altered. For both

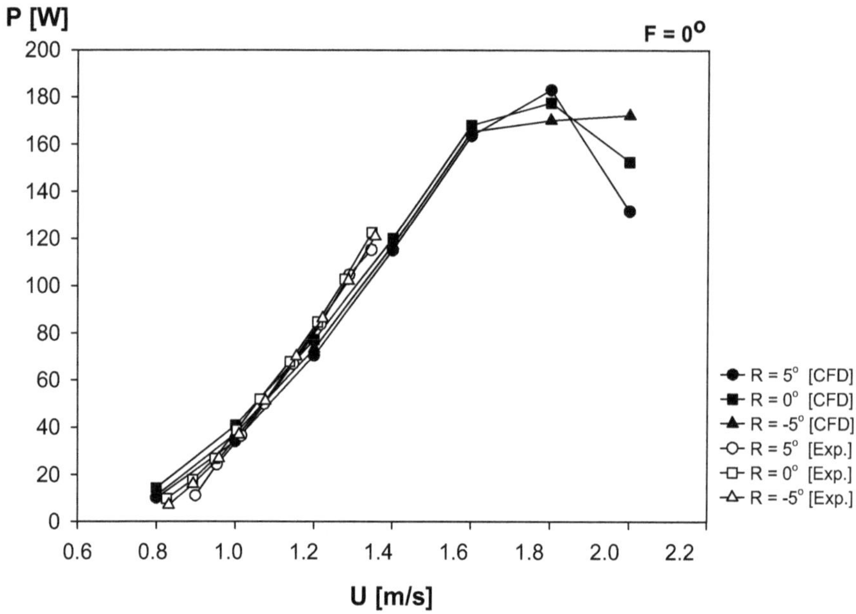

Fig. 28.11 Power output comparison between CFD and experiments (F=0°, R is varied). *CFD* computational fluid dynamics, *Exp* experimental

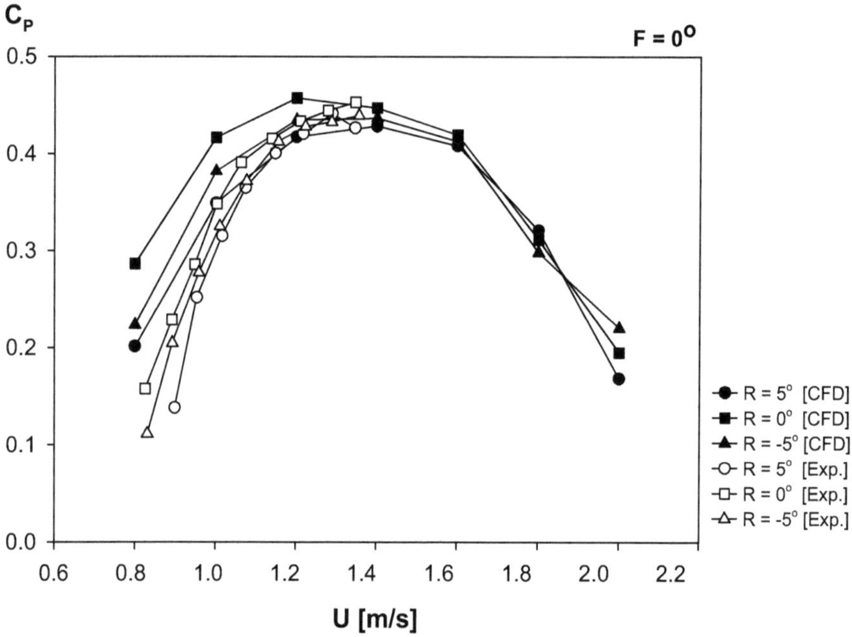

Fig. 28.12 Power coefficient (C_p) comparison between CFD and experiment (F=0°, R is varied). *CFD* computational fluid dynamics, *Exp* experimental

experiments and CFD calculations, high-power coefficients are produced as both the front and rear angles are positioned at 0° to the flow. The trend clearly shows an increase in power coefficient as the water velocities increase until a maximum is reached around 1.2–1.4 m/s and then decreases until 2.0 m/s. The highest power coefficient value of 0.457 was obtained at 1.2 m/s from CFD, whereas it was 0.453 at 1.4 m/s water velocity from experiments.

Figure 28.13 shows the power output comparison of the turbine between CFD and experimental results when the front blade angle is varied, while the rear blade angle is fixed at 0°. Each case shown in this figure shows the increase in power output as the velocity increases until around 1.4 m/s before decreasing until 2 m/s. This drop can be accounted for the loss of flow stability on the front blade shown by the streamlines. The graph shows the close agreement between the CFD and experimental results. As the experiments were conducted up to the water velocity of 1.4 m/s, only CFD data are plotted above this velocity. The power output obtained is 123.514 and 123.405 W from CFD and experiments, respectively, at 1.4 m/s when the front blade is angled at 5°, the value being almost equal. The maximum power attained is 258 W at 2 m/s water velocity for front angle at 5°.

Figure 28.14 shows the plot of power coefficient (C_p) versus the water velocity as the rear blade is fixed. The graph obtained when the front blade is angled at 5° show the higher values of power coefficient compared to other cases. For this case, the maximum C_p obtained from CFD and experiment is 0.460 at 1.4 m/s and 0.461 at 1.302 m/s water velocity, respectively. When both blade angles are fixed at 0°, maximum C_p obtained is 0.457 at 1.2 m/s for the CFD case and 0.461 at 1.302 m/s

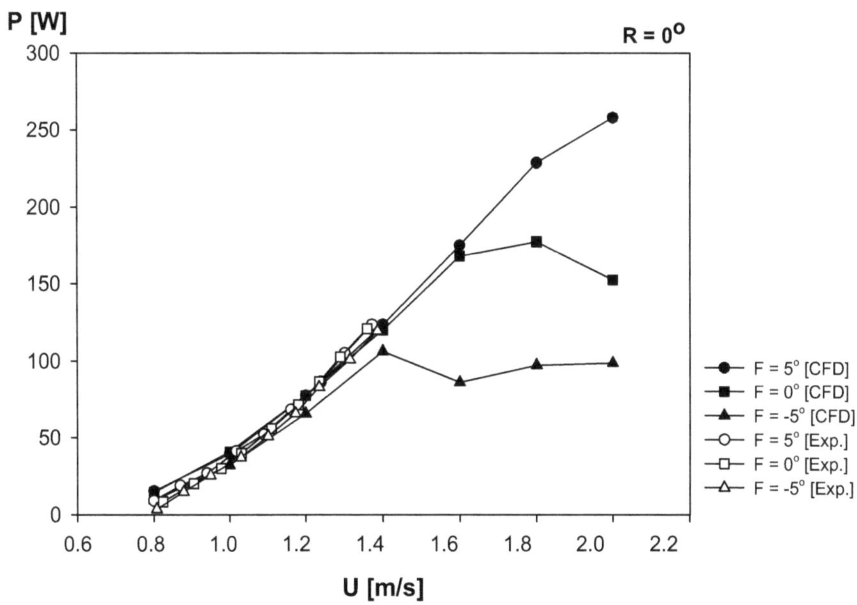

Fig. 28.13 CFD of the power characteristics when the $R=0°$, front blade (F) angle is varied

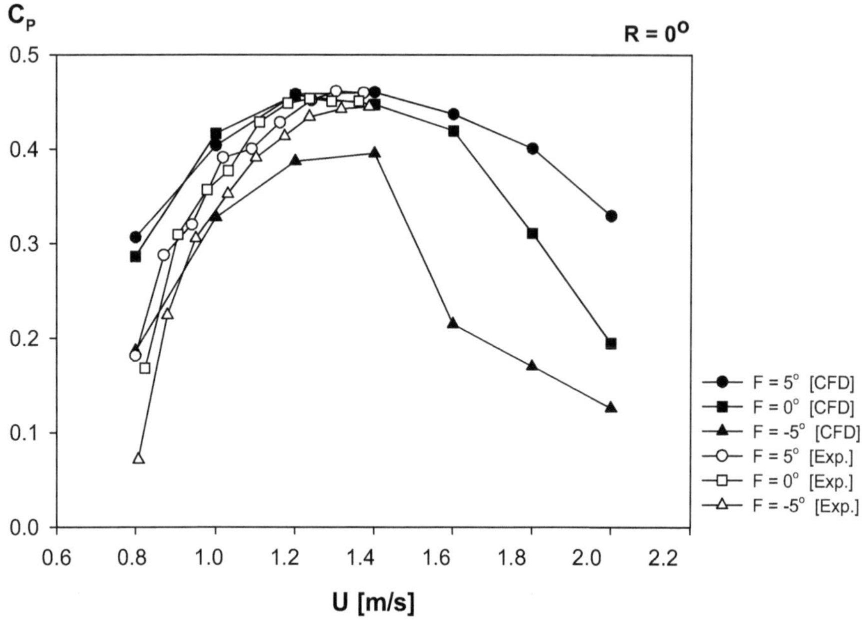

Fig. 28.14 Power coefficient (C_p) comparison between CFD and experiment (R=0°, F is varied). *CFD* computational fluid dynamics, *Exp* experimental

for the experimental case. In contrast, when the blade is angled at −5°, the values of C_p is much lower compared to other two cases in which maximum C_p obtained is 0.395 at 1.4 m/s for the CFD case and 0.445 at 1.372 m/s for experimental case. The highest power output obtained during the experiment was 123.405 W for 1.372 m/s water velocity.

Figures 28.15 and 28.16 show the torque versus the water velocity for both blades for all the CFD cases. Torque produced by the front and rear blades when the front blade angle is fixed at 0° with the rear blade angle varied is presented in Fig. 28.15, whereas Fig. 28.16 depicts the same parameters with rear blade angle fixed at 0° and the front blade angle varying. The data obtained suggest that at lower velocities, there is the occurrence of counter torque. As the velocity increases, the counter torque is removed and the value of torque increases significantly. This phenomenon of counter torque might be related with the starting water velocity and torque balance in the system.

28.5 Conclusions

To determine the effects of blade angle changes on the performance characteristics of the tidal current turbine, CFD analysis was done. Experiments were conducted in the water tank with a vertically circulating water channel at KMOU to validate

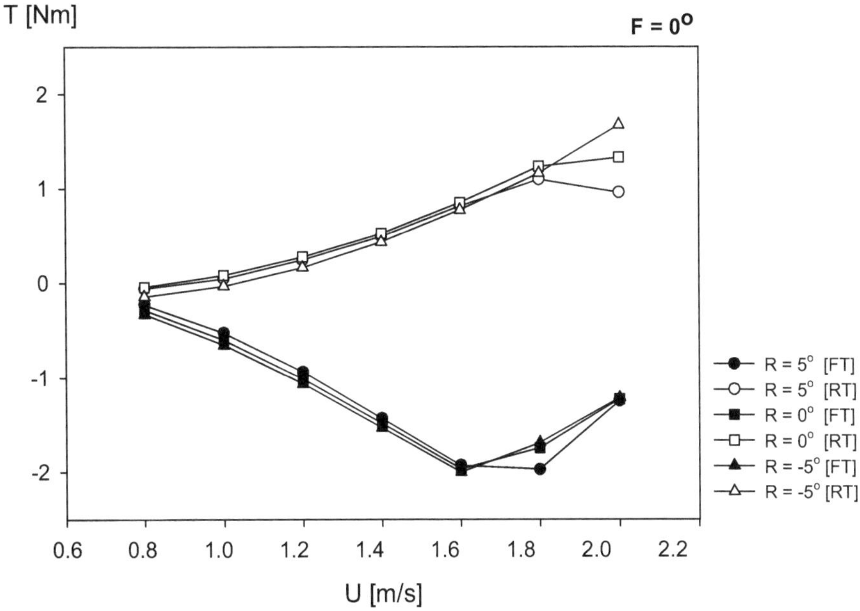

Fig. 28.15 Torque produced by the front blades and rear blades ($F=0°$, R is varied). *FT* front torque, *RT* rear torque

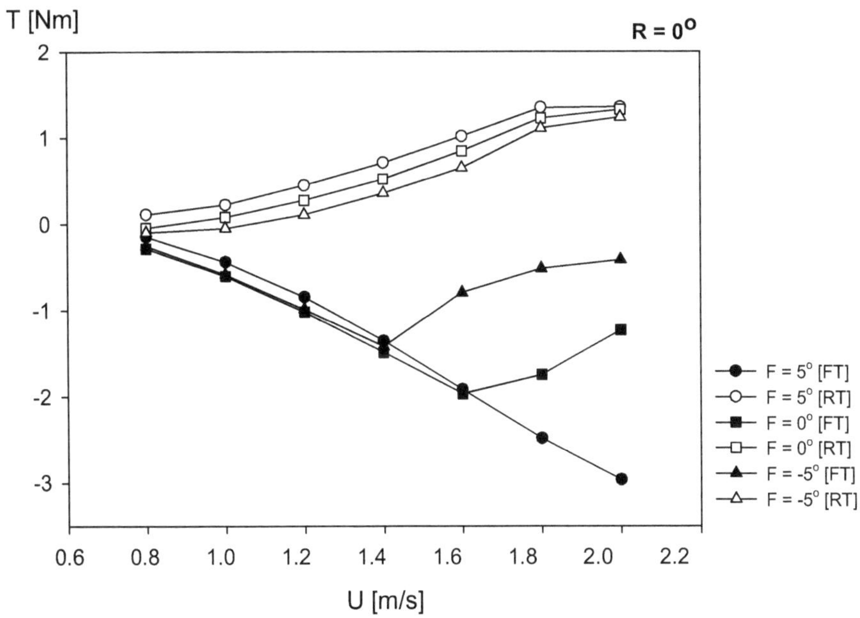

Fig. 28.16 Torque produced by the front blades and rear blades ($R=0°$, F is varied). *FT* front torque, *RT* rear torque

the numerical results. Streamlines provided the flow pattern over the front and rear blades at various water velocities when the front blade angle was fixed, while the rear blade angle was varied from −5° to 5° and vice versa. The pattern suggests that as the rear blade angle increases, more turbulent the flow becomes. The trends of the graph show that both experiment and CFD results are in good agreement over all velocities. When the front blade was fixed, experimental results showed slightly lower power from 0.8 to 1.1 m/s but slightly higher power at velocities over this range. Also, when the rear blade angle was fixed and the front blade varying, 5° front blade angle showed better flow stability on the surface. The power outputs produced and power coefficients obtained were also presented for the abovementioned cases. Power output from CFD and experiment was 115 W at 1.4 m/s with maximum C_p of 0.46 and 115.17 W with maximum C_p of 0.45, respectively, when the front blade was fixed. Similarly, when the rear blade was fixed, the power output from CFD and experiment was 123.5 W at 1.4 m/s and 123.41 W at 1.37 m/s, respectively, with maximum C_p of 0.46 for both.

Acknowledgments This work was supported by the New and Renewable Energy Core Technology Program of the Korea Institute of Energy Technology Evaluation and Planning (KETEP), granted financial resource from the Ministry of Trade, Industry and Energy, Republic of Korea (20133030000260).

References

1. Ng K-W, Lam W-H, Ng K-C (2013) 2002–2012: 10 years of research progress in horizontal-axis marine current turbines. Energies 6(3):1497–1526
2. Hyun BS, Jo CH, Kim MC, Moon CJ (2011) Technology status and prospect of tidal current power generation. Bull Soc Naval Architects Korea 48(3):12–19
3. Kim BG, Kim CJ, Choi MS (2012) A study on the performance of an 100kW class tidal current turbine. J Korean Soc Marine Environ Safety 18(2):145–152
4. Bahaj AS, Molland AF, Chaplin JR, Batten WMJ (2007) Power and thrust measurements of marine current turbines under various hydro dynamic flow conditions in a cavitation tunnel and a towing tank. J Renew Energy 32(3):407–426
5. Batten WMJ, Bahaj AS, Molland AF, Chaplin JR (2007) Experimentally validated numerical method for the hydrodynamic design of horizontal axis tidal turbines. Ocean Eng 34(7):1013–1020
6. Chen TY, Liou LR (2011) Blockage corrections in wind tunnel tests of small horizontal-axis wind turbines. Exp Therm Fluid Sci 35(3):565–569
7. Coiro DP, Maisto U, Scherillo F, Melone S, Grasso F (2006) Horizontal axis tidal current turbine: numerical and experimental investigations. Proceeding of Offshore wind and other marine renewable energies in Mediterranean and European seas, European Seminar, Rome, Italy
8. Kim BS, Yang CJ, Kim JH, Lee YH (2005) Software development for the optimum rotor design and the performance analysis of the HAWT by BEMT. Trans Korean Soc Mech Eng 5:3141–3146
9. Yang CJ Optimal rotor blade design for tidal in-stream energy. J Korean Soc Mar Environ Safety 17(1):75–82

Chapter 29
Challenge to Use Small Hydropower by Contra-rotating Small Hydro Turbine

Toru Shigemitsu, Junichiro Fukutomi and Chihiro Tanaka

Abstract Small hydropower generation is alternative energy, and there is a significant potential for a small hydro turbine. Small hydropower facilities that generate about 100–1000 kW have spread widely; however, it causes environmental destructions by a foundation construction and a set up of a draft tube. On the other hand, there are a lot of places that can generate about 100 W–1 kW (pico-hydropower) in agricultural water and a small stream. Efficiency of small hydro turbines is lower than that of large ones, and it can be used in wide flow rates range. Then, there are demands for small hydro turbines to keep high performance in wide flow rates range. Therefore, we adopted contra-rotating rotors, which could be expected to achieve high performance. In this study, a significant compact hydro turbine is named as a small hydro turbine. Final goal on this study is development of a small hydro turbine-like electrical goods, which has high portability and makes an effective use of the unused small hydropower energy resource.

In this research, we selected some places in Tokushima Prefecture in Japan, where a small hydropower could be generated, and conducted field tests of head, flow rate and water quality. Then, it was found that it was necessary to develop a small hydro turbine that could generate electricity in wide flow rates range because the small hydropower potential was different at each place. Therefore, we designed and manufactured an experimental apparatus to clarify the performance of our proposed contra-rotating small hydro turbine and investigated flow condition at the inlet of each front and rear rotor by the numerical analysis. Furthermore, the circumferential velocity distributions downstream of the rear rotor were also clarified.

T. Shigemitsu (✉) · J. Fukutomi
Institute of Science and Technology, The University of Tokushima,
Tokushima-city, Japan
e-mail: t-shige@tokushima-u.ac.jp

J. Fukutomi
e-mail: fukutomi@me.tokushima-u.ac.jp

C. Tanaka
Graduate School of Advanced Technology and Science, The University of Tokushima,
Tokushima-city, Japan
e-mail: c501332012@tokushima-u.ac.jp

A. Sayigh (ed.), *Renewable Energy in the Service of Mankind Vol I*,
DOI 10.1007/978-3-319-17777-9_29

Keywords Small hydro turbine · Contra-rotating rotors · Performance · Internal flow

29.1 Introduction

There is a strong demand to change energy resources of fossil fuels into renewable energy such as hydropower, wind power, solar energy and so on. Small hydropower generation is alternative energy, and there is a significant potential for small hydro turbine. Small hydropower facilities that generate about 100–1000 kW have spread widely; however, it causes environmental destructions by a foundation construction and a set up of a draft tube. On the other hand, there are a lot of places that can generate about 100 W–1 kW (pico-hydropower) in agricultural water and a small stream. Small hydropower installations are expected to have lower environmental impacts. Therefore, Darrieus and gyro-type turbines, which were suitable for design specifications of low head in agricultural water and a small river, were investigated, and the performance characteristics and the optimum design parameter were discussed [1, 2]. Internal flow of a spiral water turbine with wide flow passage, which had small environmental impact, was investigated [3]. Further, a small cross-flow turbine used for a small stream as an environmentally friendly pico-hydro turbine and a Savonius turbine with low cost were suggested, and performance improvement by installation of the shield plate and selecting the optimum position for it were reported [4–6]. Efficiency of small hydro turbines is lower than that of a large one, and these small hydro turbine's common problems are out of operation by foreign materials [7]. Then, there are demands for the small hydro turbines to keep high performance and wide flow passage. Therefore, we adopted contra-rotating rotors, which could be expected to achieve high performance and enable us to use low-solidity rotors with wide flow passage, in order to accomplish high performance and stable operation. In this study, a significant compact hydro turbine is named as a small hydro turbine. The final goal of this study is to develop a small hydro turbine-like electrical goods, which has high portability and makes an effective use of the unused small hydropower energy resource.

In this research, we selected some places in Tokushima Prefecture in Japan, where a small hydropower could be generated, and conducted field tests of head, flow rate and water quality. Then, it was found that it was necessary to develop a small hydro turbine that could generate electricity in wide flow rates range because change of flow rates in pico-hydropower was large. In a previous research, we investigated performance of the test model by the numerical analysis and the prospect to apply small hydro turbine for pico-hydropower was confirmed [8]. Then, we designed and manufactured an experimental apparatus to verify its performance and investigated internal flow condition by the numerical analysis results [9, 10].

In this chapter, the field test results of head, flow rate and water quality were shown. Then, the performance in wide flow rates range of our proposed contra-rotating small hydro turbine was clarified by the experiment. Furthermore, the total pressure efficiency of each rotor, the flow condition at the inlet of each front and

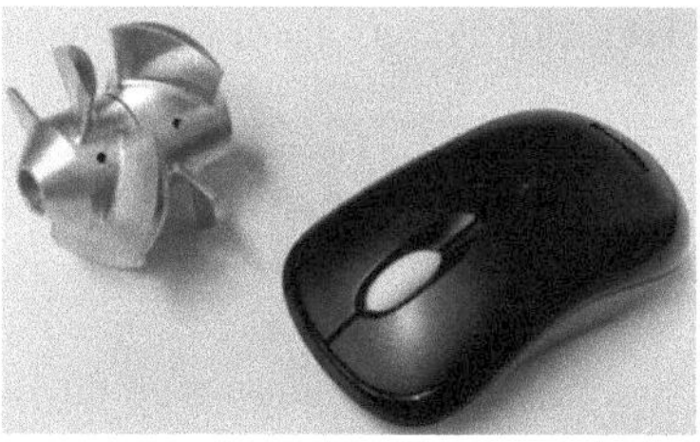

Fig. 29.1 Test hydro turbine rotors

rear rotor and the circumferential velocity distributions downstream of the rear rotor were shown by the unsteady numerical analysis results.

29.2 Experimental and Numerical Analysis Condition

29.2.1 Rotor Design Method and Design Parameter

A test turbine was supposed to be set up in a pipe of agricultural water with a diameter of about 2 in. and a small-scale water-supply system. Design flow rate and head were Q_d=4.825 l/s and H_d=2.6 m, respectively, based on power (P=10–100 W), head (H=1–4 m), flow rate (Q=2–10 l/s) assumed in a pipe of agricultural water and a small-scale water-supply system. Design heads of each front and rear rotor were the same as $H_{df}=H_{dr}$=1.3 m, and the rotational speeds of each front and rear rotor of the test turbine were $N_f=N_r$=2300 min^{-1} considering a characteristic of a small generator which could produce about 10–100 W. The rotor and primary dimensions of the contra-rotating small hydro turbine are shown in Fig. 29.1 and Table 29.1, respectively. The diameter of the rotor was 58 mm and a mouse is also shown in Fig. 29.1 to compare the size of the test rotor to the mouse. The hub tip ratios of the front and rear rotors were $D_{hf}/D_{tf}=D_{hr}/D_{tr}$=29 mm/58 mm=0.50. Each design parameters were determined by power, head, flow rate and rotational speed. In this study, blade number of each front and rear rotor should be a coprime; the front rotor Z_f=4 and rear rotor Z_r=3 to suppress the blade rows interaction of contra-rotating rotor. A guide vane was not set at the inlet of the front rotor because the test turbine was designed as compact as possible. A stagger angle of the rear rotor was determined on an assumption that the swirling flow downstream of the front rotor, which was calculated by the design efficiency of the front rotor (η_{fd}=65 %), directly went to the rear rotor. The test turbine was designed so that swirling flow did not remain downstream of the rear rotor at the design flow rate.

29.2.2 Experimental Apparatus and Method

A schematic diagram of the experimental apparatus is shown in Fig. 29.2. A diameter of a casing was D=60 mm and a tip clearance was c=1 mm because the

Table 29.1 Primary dimensions of turbine rotors

		Hub	Mid	Tip
Front rotor	Diameter (mm)	29	43.5	58
	Blade number	4		
	Blade profile	NACA6512		
	Solidity	1.4	1.07	0.85
	Stagger angle (°)	25.5	20	15.8
Rear rotor	Blade number	3		
	Blade profile	NACA6512		
	Solidity	0.86	0.71	0.59
	Stagger angle (°)	44.6	29.1	18.9
		Hub	Mid	Tip
Front rotor	Diameter (mm)	29	43.5	58
	Blade number	4		
	Blade profile	NACA6512		
	Solidity	1.4	1.07	0.85
	Stagger angle (°)	25.5	20	15.8
Rear rotor	Blade number	3		
	Blade profile	NACA6512		
	Solidity	0.86	0.71	0.59
	Stagger angle (°)	44.6	29.1	18.9

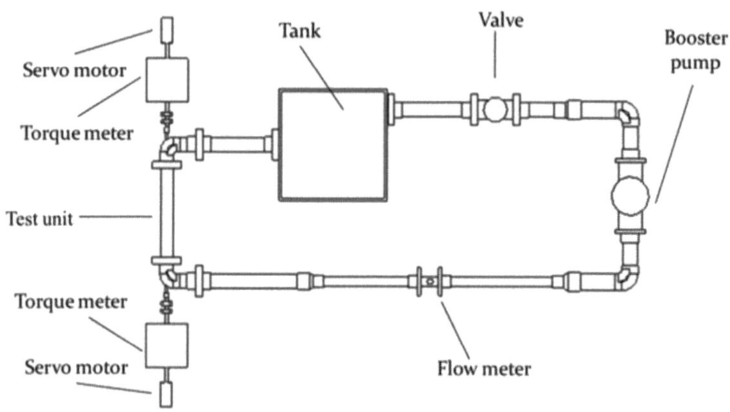

Fig. 29.2 Schematic diagram of experimental apparatus

rotor diameter was $D_r = 58$ mm. A test section was made of a transparent acrylic resin for visualization. Water was used for the experiment. For head evaluation, the static head differences on the casing wall between about 2*D* upstream of the front rotor and about 2*D* downstream of the rear rotor were measured. Each rotor was connected to respective shaft and driven by respective motor set upstream and downstream of the test section. The length of the transparent test section was 500 mm and straight pipes over 4*D* were ensured upstream and downstream of the

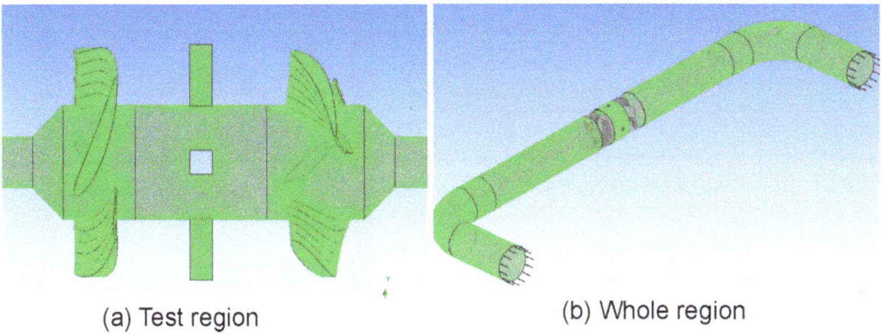

Fig. 29.3 Numerical grids. **a** Test region. **b** Whole region

test rotors to suppress the influence of the swirl flow from the 90° bend near the test section. The flow rate Q was obtained by a magnetic flow meter installed far downstream of the hydro turbine, and torques of each front and rear rotor were measured by respective torque meter. Then, shaft power was calculated by the torque and rotational speed measured by a rotational speed sensor. The shaft power was evaluated by the torque eliminating mechanical loss in this performance test. Hydraulic efficiency of the hydro turbine η was calculated as the ratio of the shaft power to the water power.

29.2.3 Numerical Analysis Method and Condition

In the numerical analysis, the commercial software ANSYS-CFX was used under the condition of 3D unsteady flow condition. Fluid was assumed to be incompressible and isothermal water and the equation of the mass flow conservation and Reynolds-averaged Navier–Stokes equation were solved by the finite volume method. The standard wall function was utilized near the wall, and the standard k–ε model was used as the turbulence model. A numerical model was the same with the test section of the experimental apparatus and included the inlet and outlet bend of the experimental apparatus. Straight pipes over 4D were ensured for the inlet and outlet pipes, which was the same with the experimental apparatus, to suppress the influence of the bend at the inlet and outlet. Each front and rear rotor's shaft was modelled in the numerical analysis and rotated in the front and rear rotor's rotational direction, respectively. The constant velocity perpendicular to the inlet boundary and the constant pressure having only axial velocity were given as the boundary condition at the inlet and outlet, respectively. The coupling between the rotating and stationary regions was accomplished by the transient rotor/stator. The time step number per one rotor rotation was 120 and the time step was $t=2.174\times10^{-4}$ s. The data of one rotor rotation were obtained after five rotor rotations in unsteady numerical analysis. The numerical grids of the test section and whole numerical domains used for the numerical analysis are shown in Fig. 29.3. The numerical domains were

Fig. 29.4 Discharge pipe of agricultural water

consisted of the inlet pipe, rotors and outlet pipe regions. The numerical grid points of each region were 495,241 for the inlet pipe region, 6,891,946 for the rotors' region and 316,811 for the outlet pipe region, respectively.

29.3 Results and Discussions

29.3.1 Field Test Results

There are so many agricultural water channels in Japan. Further, pipelines from a dam are prepared for the agricultural water in some prefectures in Japan. Then, field tests for the head, velocity and water quality in the pipe for the agricultural water were conducted. Figure 29.4 shows a picture of a discharge pipe of the agricultural water. The pipelines provide the agricultural water at each field of rice and other vegetables at some areas of Tokushima Prefecture in Japan. The flow rate of the agricultural water could be controlled by the valve installed near the discharge of the pipe. The static pressure on the pipe inner wall was measured by a pressure sensor, and velocity at an axial center of the pipe was measured by the one-hole pitot tube. The field test results of the static pressure and velocity are shown in Fig. 29.5a, b. Figures 29.5a, b show the results under the maximum flow rate condition with the valve full opening and the constant static head H=4.1 m condition with any valve opening, respectively. It was observed in Fig. 29.5a that the average velocity at the axial center of the pipe was V=4.8 m/s, which was over 2.5 times of the cross-sectional average velocity (V_d=1.7 m/s) at the design flow rate and the average static head was H=5.5 m, which was over two times of the design head (H_d=2.6 m). When, the head was kept H=4.1 m by the valve, the velocity was over 4.0 m/s. Therefore, it was found that there was enough hydraulic power potential in these pipelines for the agricultural water. The water quality was also visually checked in the field test. Algae, sand and pieces of leaves were included in the agricultural water. It was important to solve the problem related to foreign materials: adhesion of the foreign materials to the rotor, influence of the foreign material on seals, sand erosion of the rotor and so on.

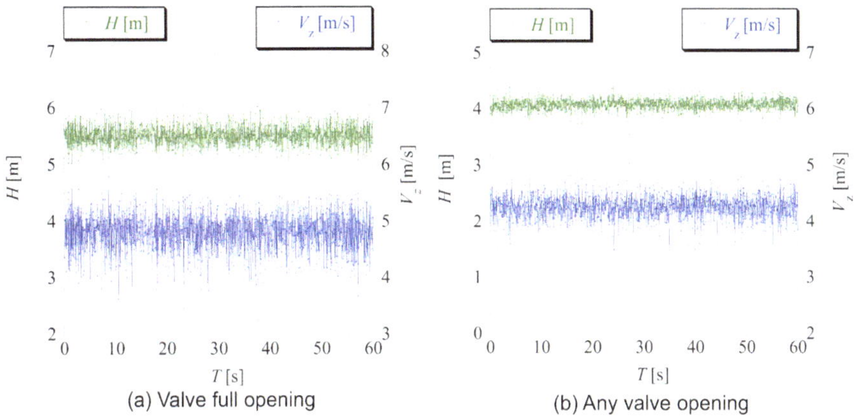

Fig. 29.5 Field test results of static head and axial velocity. **a** Valve full opening. **b** Any valve opening

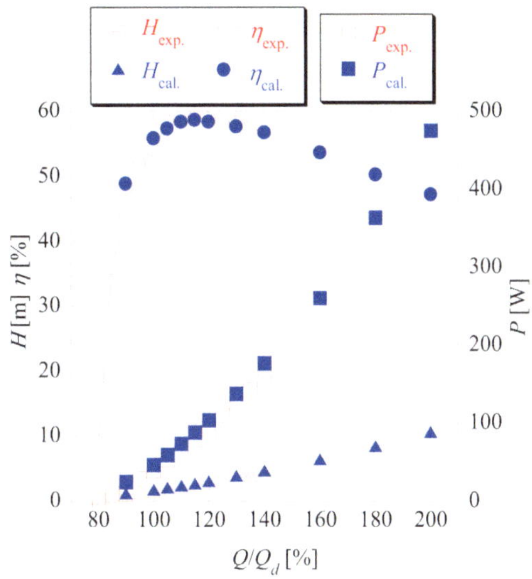

Fig. 29.6 Performance curves

29.3.2 Performance Curves

Figure 29.6 shows performance curves of the test turbine obtained by the experiment. The unsteady numerical analysis results are also shown to compare the numerical results to the experimental results. The horizontal axis is the flow rate. The first vertical axis is the turbine head and efficiency. The second vertical axis is the shaft power. The rotational speeds of each front and rear rotor were $N_f = N_r = 2300\ \mathrm{min}^{-1}$. The maximum flow rate of the experiment was 1.4 Q_d because of the limitation of measuring instruments. The turbine head and shaft power increased according to

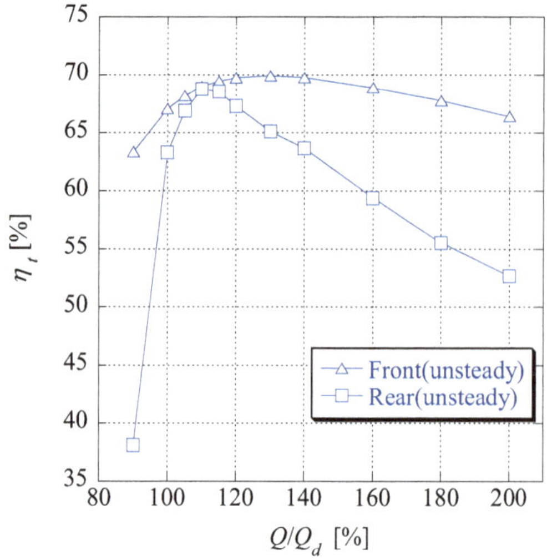

Fig. 29.7 Total pressure efficiency

the increase of the flow rate. The maximum efficiency η_{max} = 59 % was obtained at around 1. 1 Q_d–1.2 Q_d, although the contra-rotating small-sized hydro turbine was extremely compact with 60 mm casing diameter. Furthermore, efficiency more than 50 % was obtained in relatively wide flow rates range of 95–180 % of the design flow rate. On the other hand, the efficiency drastically decreased at the partial flow rates smaller than 0.9 Q_d. It was found in Fig. 29.6 that the numerical results could capture the tendency of the performance curves of the experimental results.

Figure 29.7 shows total pressure efficiency curves of each front and rear rotor obtained by the unsteady numerical analysis. The total pressure efficiency of the front rotor was calculated from the ratio of the front rotor shaft power to the input power of the front rotor. The input power of the front rotor was obtained by the multiplication of the flow rate and the mass flow averaged total pressure difference of the front rotor. Further, the total pressure efficiency of the rear rotor was also calculated by the same method of the front rotor. The total pressure efficiency of the front rotor was slightly affected by a flow rate change, however, the total pressure efficiency of the rear rotor drastically decreased in the partial flow rate. Further, high total pressure efficiency region of the rear rotor was narrower than that of the front rotor. On the other hand, the maximum total pressure efficiency of the rear rotor was about 68.8 % at 1.1 Q_d, which was almost the same with the maximum total pressure efficiency of the front rotor at 69.8 %. Then, the inlet flow condition of each front and rear rotor according to the flow rate variation were investigated by the numerical analysis results.

29.3.3 Internal Flow Condition at Inlet and Outlet of Rotors

Figures 29.8a, b show the attack angle distributions of each front and rear rotor, respectively, according to the flow rates variation. The horizontal axis of Figs. 29.8a, b

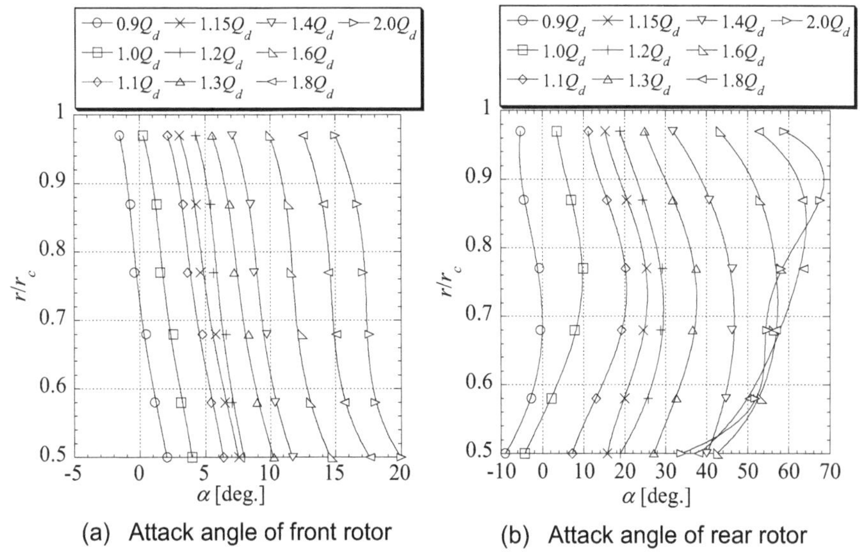

Fig. 29.8 Attack angle distributions. **a** Attack angle of front rotor. **b** Attack angle of rear rotor

are the attack angle 10 mm upstream of the front rotor and rear rotor, respectively. The vertical axis is a nondimensional radial position divided by the radius of the casing interior wall. The front rotor attack angles were around 5–15° at the flow rate (1.3 Q_d) near the maximum front rotor total pressure efficiency, and the rear rotor attack angles were also around 5–20° at the flow rate(1.1 Q_d) near the maximum rear rotor total pressure efficiency. Therefore, the attack angles around 5–20° was the suitable attack angle for this turbine rotor. On the contrary, the front rotor attack angles showed the minus value near the shroud side at the partial flow rate 0.9 Q_d. Furthermore, the rear rotor attack angles showed the minus value at all radial positions at the partial flow rate 0.9 Q_d. The minus attack angle of the rear rotor at the partial flow rate 0.9 Q_d would cause the sudden decrease of the total pressure efficiency in Fig. 29.7. On the other hand, the attack angles of the front rotor at large flow rate of 2.0 Q_d were smaller than 20°, but that of the rear rotor at the large flow rates range 1.4Q_d–2.0 Q_d were larger than 30°. The variation of the attack angle with the flow rate change for the front rotor was smaller than that of the rear rotor. The inlet flow condition of the rotor influenced the efficiency of the turbine; therefore, the variation of the total pressure efficiency by the flow rate change became large in the case of the rear rotor as shown in Fig. 29.7.

The test turbine was designed under the condition that the blade loads of the front and rear rotors were the same ($H_{df}=H_{dr}=1.3$ m), and the circumferential component of absolute velocity downstream of the rear rotor did not exist at the design flow rate. Then, the circumferential velocity distributions downstream of the rear rotor was investigated by the unsteady numerical analysis. The circumferential velocity distributions 10 mm downstream of the rear rotor in wide flow rates range are shown in Fig. 29.9. The circumferential velocity downstream of the rear rotor was

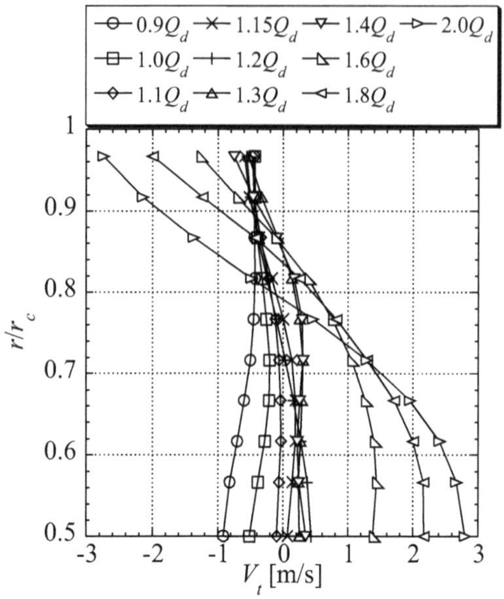

Fig. 29.9 Circumferential velocity distributions downstream of rear rotor

small from a partial rate of 0.9 Q_d to a large flow rate 1.4 Q_d. It was found that an exhaust loss was small in this flow rate range, in which a good performance was obtained.

29.4 Concluding Remarks

The field test of the head and velocity of the pipeline for the agricultural water was conducted. Then, the performance, inlet and outlet flow conditions of the contra-rotating small hydro-turbine were investigated by the unsteady numerical analysis. As a result the following conclusions were obtained:

1. The average velocity at the axial center of the pipe was $V=4.8$ m/s, which was over 2.5 times of the cross-sectional averaged velocity ($V_d=1.7$ m/s) at the design flow rate and the average static head was $H=5.5$ m, which was over two times of the design head ($H_d=2.6$ m). It was found that there was enough hydraulic power potential in the pipelines for the agricultural water.
2. The maximum efficiency $\eta_{max}=59\,\%$ was obtained around 1.1 Q_d–1.2 $Q_{d,}$ and an efficiency more than 50 % was obtained in relatively wide flow rates range of 95–180 % of the design flow rate. On the other hand, the efficiency drastically decreased at the partial flow rates smaller than 0.9 Q_d. The high total pressure efficiency region of the rear rotor was narrower than that of the front rotor.
3. The attack angles around 5–20° was the suitable attack angle for the test turbine rotor. The minus attack angle of the rear rotor at the partial flow rate 0.9 Q_d

would cause the sudden decrease of the total pressure efficiency. The variation of the attack angle with the flow rate change for the front rotor was smaller than that of the rear rotor. The circumferential velocity downstream of the rear rotor was small from the partial rate 0.9 Q_d to large flow rate 1.4 Q_d.

Acknowledgements This work was supported by JKA and its promotion funds from KEIRIN RACE and the Grant-in-Aid for Young Scientists (B) under Grant No.24760138 from Japan Society for the Promotion of Science. We would like to show our special thanks for the foundations.

References

1. Furukawa A, Okuma K, Tagaki A (1998) Basic study of low head water power utilization by using Darrieus-type turbine. Trans Japan Soc Mech Eng B (in Japanese) 64(624):2534–2540. doi:10.1299/kikaib.64.2534
2. Kanemoto T, Inagaki A, Misumi H, Kinoshita H (2004) Development of Gyro-type hydraulic turbine suitable for shallow stream (1st Report, rotor works and hydroelectric power generation). Trans JSME (in Japanese) 70(690):413–418
3. Matsui J (2010) Internal flow and performance of the spiral water turbine. Turbomachinery (in Japanese) 38(6):358–364
4. Ikeda T, Iio S, Tatsuno K (2010) Performance of nano-hydraulic turbine utilizing waterfalls. Renewable Energy 35:293–300. doi:10.1016/j.renene.2009.07.004
5. Nakajima M, Iio S, Ikeda T (2008) Performance of Savonius rotor for environmentally friendly hydraulic turbine. Journal of Fluid Science and Technology 3(3):420–429. doi:10.1016/j.renene.2009.07.004
6. Iio S, Uchiyama F, Sonoda C, Ikeda T (2009) Performance improvement of Savonius hydraulic turbine by using a shield plate. Turbomachinery (in Japanese) 37(12):743–748
7. Iio S, Oike S, Sato E, Ikeda T (2011) Failure events in the field test of environmentally friendly nano-hydraulic turbines. Turbomachinery (in Japanese) 39(3):162–168
8. Sonohata R, Fukutomi J, Shigemitsu T (2012) Study on contra-rotating small-sized axial flow hydro turbine. Open J Fluid Dyn 2(4A):318–323. doi:10.4236/ojfd.2012.24A039
9. Shigemitsu T, Fukutomi J, Tanaka C (2013) Performance and internal flow of contra-rotating small hydro turbine. Proceedings of the ASME fluids engineering division summer meeting, Incline village, Nevada, FEDSM2013-16274, pp V01BT12A002. doi:10.1115/FEDSM2013-16274
10. Shigemitsu T, Fukutomi J, Tanaka C (2013) Unsteady flow condition of contra-rotating small-sized hydro turbine. Proceedings of the 12th Asian international conference on fluid machinery, Yogyakarta, AICFM12-0019

Chapter 30
Site Implementation of a Low-Head Pico-Hydro Turgo Turbine

Samuel J. Williamson, Julian D. Booker and Bernard H. Stark

Abstract This chapter discusses how to integrate a 1.3-kW low-head Turgo turbine into existing water infrastructure. The proposed scalable system is formed by connecting a number of identical Turgo turbine units at the same site or nearby to form an off-grid network. Three types of existing water infrastructure are considered applicable for hydropower integration: irrigation canals, reservoirs and weirs. Designs for the civil works for the low-head Turgo turbine integrated into these structures are given, and the costs of construction are estimated. Integrating the Turgo unit into these existing water structures is shown, in theory, to be cost-effective, especially in developing countries, with civil works and installation costs greatly reduced.

Keywords Pico-hydro · Turgo · Site implementation · Cost

30.1 Introduction

Micro- and pico-hydropower represent a cost-effective form of generating electricity off grid, where the resource exists [1]. Many unexploited sites are at low head, as the economically viable high-head sites have already been developed [2]. Typically, propeller turbines have been used at low head, as shown in Fig. 30.1a. However, a Turgo turbine has been shown to be suitable for low-head, variable flow pico-hydro systems, with low-speed generators and inverter interfaces [3]. A particular emphasis has been to design a system that is scalable such that multiple identical units can be installed individually or in parallel or series to increase the output capacity at a site, and, in cooperation with other sites, to form a reliable off-grid network.

The viable range of this particular low-head Turgo turbine covers flow rates of 23 l/s at 1-m head to 44 l/s at 3.5-m head [4]. The extension of the application range

S. J. Williamson (✉) · J. D. Booker · B. H. Stark
Faculty of Engineering, University of Bristol, Queen's Building, University Walk, Bristol BS8 1TR, UK
e-mail: sam.williamson@bris.ac.uk

A. Sayigh (ed.), *Renewable Energy in the Service of Mankind Vol I*,
DOI 10.1007/978-3-319-17777-9_30

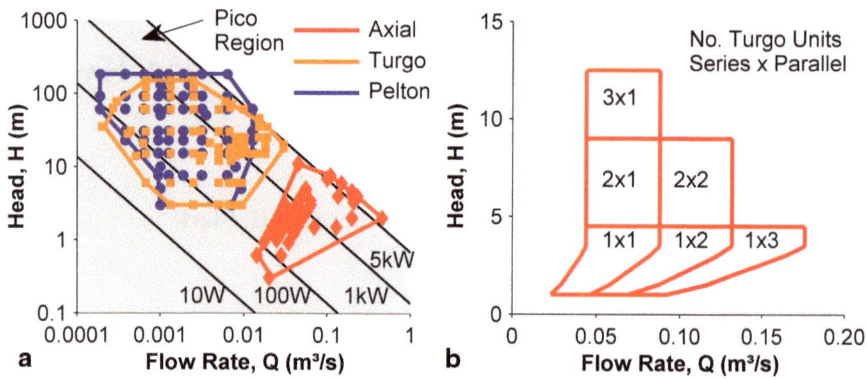

Fig. 30.1 **a** Application range in pico region with example turbines. **b** Application range of low-head Turgo turbine in series and parallel

by series and parallel operation of multiple units is illustrated in Fig. 30.1b. Series connections make use of excess head, and parallel operation scales the flow. Multiple unit systems can thus be installed at the same time or expanded as additional funds become available.

This chapter discusses the advantages of integrating this type of pico-hydro system into existing water infrastructure or natural features and proposes several designs for different types of infrastructure, and their associated cost.

30.2 The Costs of Micro- and Pico-Hydropower

The installed cost of micro- and pico-hydro systems varies depending on the location. Differences in the amount of civil works required at the intake and for the canal, the length and type of penstock, the complexity of the turbine design as well as the distance to the distribution area and number of connected households can all cause variation in the cost. Figure 30.2 illustrates the cost per kilowatt for 38 micro-hydro projects installed in Nepal and completed in 2011, data taken from [5], showing the cost per installed kilowatt varies between £ 2000/kW and £ 4000/kW, with an average of £ 2836/kW, and does not appear to have any correlation to the plant size.

One difference between the low and high cost per installed kilowatt is shown by two similar micro-hydro projects in Ilam district, eastern Nepal. The two micro-hydro schemes have been installed within a 2–3 km of each other on the Fakfok River, and are known simply as Fakfok 2 and Fakfok 3. Photos of the sites are shown in Fig. 30.3. Both systems are similar, use crossflow turbines and have comparable distribution areas. Details for each system are shown in Table 30.1.

Fakfok 2 is built on the river at the point of a waterfall, so the intake canal is very short, Fig. 30.3a, around 50 m, and the desilting only takes place in the forebay tank. In comparison, the intake structure for Fakfok 3 is much longer and more complex,

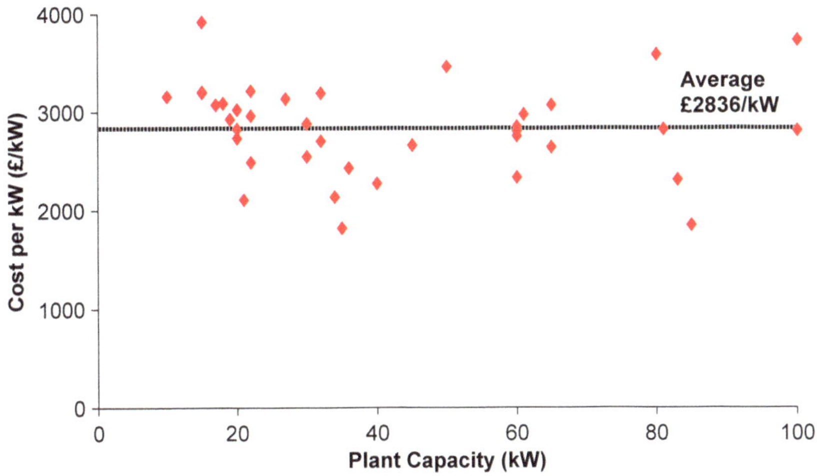

Fig. 30.2 Cost per kilowatt of installed capacity for 38 different projects in Nepal, 2011. (Data from [5])

Fig. 30.3 **a** Fakfok 2 micro-hydro scheme **b** Fakfok 3 microhydro scheme

Table 30.1 Details for the Fakfok 2 and 3 micro-hydro projects

	Fakfok 2	Fakfok 3
Turbine	Crossflow	Crossflow
Head	15.5 m	18 m
Flow rate	263 l/s	270 l/s
Output power	20 kW	28 kW
Households fed	230 hh	280 hh
Cost	£ 25,549	£ 89,285
Specific cost	£ 1277/kW	£ 3188/kW

Fig. 30.4 **a** A propeller turbine installation in Lao [6]. **b** A propeller turbine powerhouse in Nepal. **c** An intake canal, spillway and powerhouse in Nepal

as shown in Fig. 30.3b. The intake canal is over 300 m long, with a 50-m aqueduct section, supported by pillars from below and cables from the cliff above. There are two desilting tanks in addition to the forebay.

The overall cost of Fakfok 3 is over three times that of Fakfok 2, with the cost per kilowatt of electrical power generated at £ 3177/kW compared with just £ 1277/kW for Fakfok 2. These examples show that for very similar mechanical schemes with comparable distribution areas, the civil works can have a very significant impact in the overall cost of the scheme, and that integrating the civil works into existing water infrastructure or features may help to reduce the cost of the overall system.

For low-head pico-hydro systems, there are similar situations. Examples of low-head pico-hydro civil works are shown in Fig. 30.4.

Figure 30.4a shows how the structure is built into existing water features, using the fall in the river to allow a short canal, and simple structure is used to reduce the costs. However, this structure will require constant maintenance and rebuilding after flooding [6]. Figure 30.4b, c shows a stone and concrete construction for the intake canal and powerhouse. The intake canal is fed from a pre-existing irrigation canal, extended to reach the powerhouse location. The cost breakdowns for two different low-head propeller turbine systems are shown below in Fig. 30.5. Figure 30.5a itemises the average installation costs of 23 different 200 W propeller turbines in Ecuador, where the average cost of the civil works was £ 83, whilst Fig. 30.5b refers to the installation costs of a 1.1 kW propeller turbine in Nepal, where the civil works cost £ 900.

The civil works in both cases are almost 30 % of the overall cost of the project, with the installation being another 15 %; therefore, any reduction of these costs, as well as those of the turbine units, would certainly be advantageous.

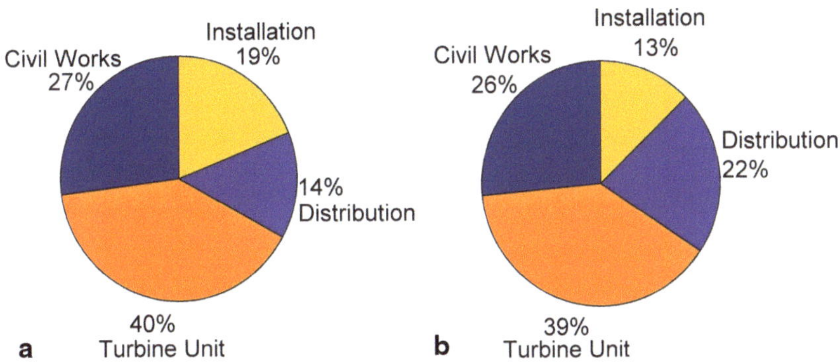

Fig. 30.5 Cost breakdown for two different pico-hydro systems: **a** Average costs for installations in low-head propeller turbines in Ecuador [7]. **b** Cost for low-head propeller system in Humla, Nepal [8]

30.3 Examples of Theoretical Low-Head Turgo Turbine Installations

In this section, three theoretical case studies are presented for the installation of low-head Turgo turbine units into three real sites: an irrigation canal, reservoir and weir. Itemised installation costs are estimated.

30.3.1 Irrigation Canal Case Study

The first evaluation considers Bhanbhane, western Nepal, where a river passes through terraced farmland with a minimum flow rate of 100 l/s. There is already an irrigation canal in place, Fig. 30.6, which feeds the paddy fields in the area, which is built to support a flow rate of around 60 l/s. At the proposed powerhouse site, there

Fig. 30.6 Irrigation canal site, in Bhanbhane, Nepal

Fig. 30.7 Low-head Turgo unit installed in irrigation canal site

is a head of around 3.5 m, which can be increased or reduced as required through repositioning of the powerhouse location.

We assume the following: The existing irrigation canal infrastructure is used, as illustrated in Fig. 30.7, although it may be required to be made deeper to accommodate the larger water flow. A concrete forebay tank is built next to the irrigation canal with a trash rack at the penstock inlet. The high-density polyethylene (HDPE) penstock is 5 m long, connecting the forebay to the turbine. The turbine is bolted onto a concrete section of the tailrace, which is dug into the field to maximise the head available. The water is then returned through an earth channel to the nearby river. A small powerhouse is built over the turbine, from concrete, corrugated steel and local stone, to protect the equipment from environmental effects. Any excess water at the inlet continues through the irrigation canal. If the water is required to irrigate the fields downstream of the turbine, then the turbine can be taken off-line, whilst the water is diverted. This design uses 1.84 m^3 of concrete, and assuming a 3:2:1 ratio of stone/aggregate to sand to cement, 2.2 t of stone, 1.5 t of sand and 0.7 t of cement is required. The stone can be locally sourced from the river, and so only sand and cement are required. The powerhouse uses a 4 m length of corrugated steel for the roofing. It is estimated that the work would take 20 person days to complete.

If the head at the forebay was greater than the rated head of the turbine unit, a second unit could be installed in series, as shown in Fig. 30.8a, or if the flow available was greater than the maximum flow for one unit, a second could be installed in parallel, Fig. 30.8b. The electrical outputs of the two generation units would be connected together, and there would be a larger supply for the consumers.

30.3.2 Reservoir

This evaluation considers an example reservoir site as illustrated in Fig. 30.9. We assume the following: The reservoir is fed through a river, with a minimum flow

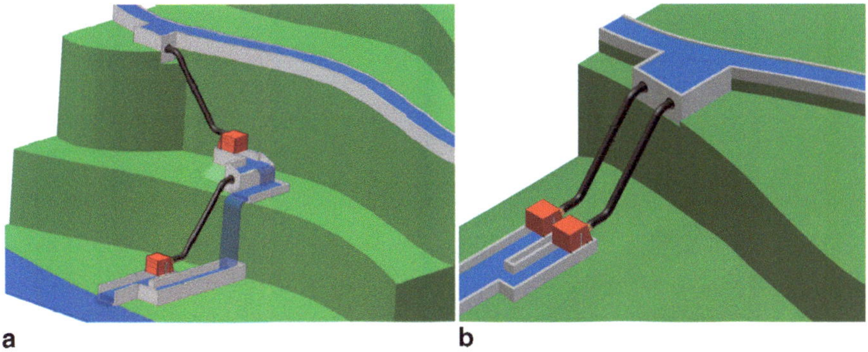

Fig. 30.8 Low-head Turgo unit installed in **a** series (excess head) or **b** parallel (excess flow)

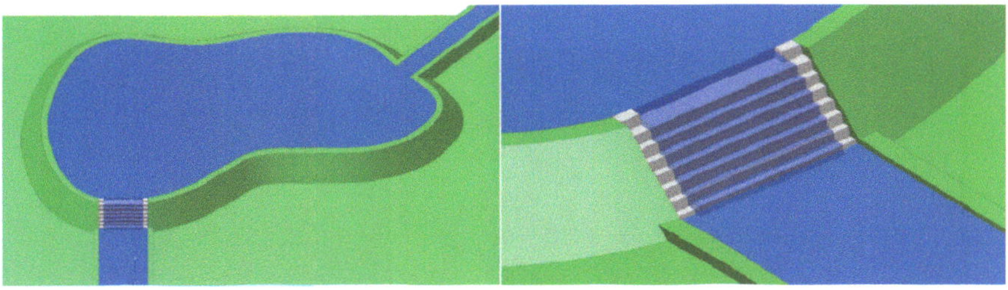

Fig. 30.9 Example reservoir site

rate of over 200 l/s. The lake has a head of 4 m, and a weir is located in one corner, with the river continuing on from there.

Figure 30.10 illustrates the reservoir installation. The forebay tank is installed next to the weir, with a 6 m HDPE penstock attached to the front. The low-head Turgo unit is located next to the outflow from the reservoir, the tailrace returning the water back into the river. This design uses 2.11 m^3 of concrete, so 2.5 t of stone, 1.7 t of sand and 0.84 t of cement is required. The powerhouse is identical to the irrigation canal design. Again, the stone is assumed to be locally sourced. The labour time for this installation is again assumed to be 20 person-days.

30.3.3 Weir

This evaluation considers a site with a sharp crested 'slat' weir, similar to that shown in Fig. 30.11a. The slat weir is constructed from a number of planks set into slots in the channel, which means they can be easily lifted in and out of the watercourse to change the weir height. We assume that the weir has a head of 2.5 m and a minimum flow rate of 150 l/s.

The design shown in Fig. 30.11b uses no concrete. Instead, the turbine is supported from the weir by a steel structure, made up from ten pieces of 3-m-long angle steel. The 1.5 m penstock is fed through a siphon intake, built into an additional slat added onto the weir. As there is little civil works, the installation time is estimated at 2 person-days.

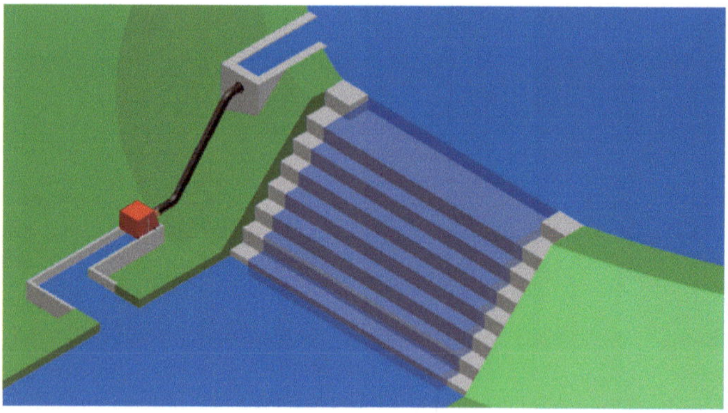

Fig. 30.10 Low-head Turgo unit installed at an example reservoir site

a b

Fig. 30.11 **a** Example of a slat weir in Gloucestershire, UK. **b** Low-head Turgo unit installed on example slat weir

30.4 Discussion

The costs of each of the designs proposed in Sect. 30.3 are assessed using the material prices shown in Table 30.2 for both the UK and Nepali installations. These costs are shown in Table 30.3, with the breakdown for each material at the different sites.

The costs in Table 30.3 show that both the reservoir and irrigation canal are very similar, and the weir installation is less, due to the large reduction in labour costs. In the UK, the labour costs make up the largest proportion of the civil works construction costs in the irrigation canal and reservoir installations, as they require a lot of manual labour to build the structures. However, with the simple design of the weir, this is reduced dramatically making it much more cost-effective. Therefore, another way to reduce costs further is to make the installation process as simple and short as possible.

Table 30.2 Material costs [9, 10],

	Nepal	UK
HDPE pipe (Penstock)	£ 13/m	£ 28/m
Cement	£ 0.10/kg	£ 0.20/kg
Sand	£ 0.005/kg	£ 0.05/kg
Gravel/aggregate	Locally sourced	£ 0.05/kg
Angle Steel (3 m length, 70 × 70 × 10 mm)	£ 10/length	£ 40/length
Corrugated steel (1 m width)	£ 1.50/m	£ 8.30/m
Labour	£ 3.65/day	£ 82.50/day

Table 30.3 Cost breakdown for different installation examples

Items	Irrigation canal		Reservoir		Weir	
	UK	Nepal	UK	Nepal	UK	Nepal
Penstock	£ 140	£ 65	£ 168	£ 78	£ 84	£ 39
Cement	£ 146	£ 73	£ 168	£ 84	N/A	N/A
Sand	£ 74	£ 7	£ 85	£ 9	N/A	N/A
Stone/gravel	£ 110	N/A	£ 127	N/A	N/A	N/A
Angle steel	£ 0	£ 0	£ 0	£ 0	£ 400	£ 100
Corrugated steel	£ 33	£ 6	£ 33	£ 6	N/A	N/A
Labour	£ 1650	£ 72	£ 1650	£ 72	£ 165	£ 7
Total	*£ 2153*	*£ 223*	*£ 2231*	*£ 249*	*£ 649*	*£ 146*

If the low-head Turgo unit did not use the existing infrastructure, such as the irrigation canal, and required a 50 m intake canal to be built, this adds another £ 379 in Nepal and £ 1580 in the UK, not including labour costs, which is another 170 % (Nepal) or 73 % (UK) on top of the original building costs. Comparing the costs in Nepal with the sites shown previously in Fig. 30.5, and normalising these costs per kW, the proposed implementations are over 45 % cheaper at £ 223/kW compared with £ 820/kW in Humla and £ 415/kW in Ecuador.

30.5 Conclusions

Civil works in micro- and pico-hydropower development can contribute up to 30 % of the overall costs of the system, and so integrating a turbine system with existing water infrastructure or features can be very advantageous. The civil works designs for a low-head Turgo system are proposed for three different sites: an irrigation canal, reservoir and weir, and the costs are estimated to be between £ 146 and £ 249 in Nepal and £ 649 and £ 2231 in the UK. This shows a saving of £ 379 (Nepal) and £ 1580 (UK) when installing the system without the infrastructure available.

Acknowledgments This research has been funded by Renishaw, and by Engineering and Physical Sciences Research Council (EPSRC) through an Impact Acceleration Award, and supported by Engineers Without Borders, UK.

References

1. ESMAP (2007) Technical and economic assessment of off-grid, mini-grid and grid electrification technologies. World Bank, New York
2. Meier T, Fischer G (Dec 2011) Assessment of the pico and micro-hydropower market in Rwanda. GVEP, Nairobi
3. Williamson SJ, Stark BH, Booker JD (2014) Low head pico hydro turbine selection using a multi-criteria analysis. Renew Energy 61:43–50
4. Williamson SJ, Stark BH, Booker JD (2013) Performance of a low-head pico-hydro Turgo turbine. Appl Energy 102:1114–1126
5. AEPC (2011) Approved microhydro projects for additional support program till July 2011. http://www.aepc.gov.np/old/images/pdf/ref-approved.pdf. Accessed 4 Aug 2014
6. Susanto J, Stamp S (2012) Local installation methods for low head pico-hydropower in the Lao PDR. Renew Energy 44:439–447
7. Taylor SD et al (2003) Stimulating the market for pico-hydro in ecuador. Department for International Development, London
8. Bhusal P et al (2007) Energy-efficient innovative lighting and energy supply solutions in developing countries. Int Rev Electr Eng 2(5):665–670
9. Panchakanya article rates. www.panchakanya.com/article-rates. Accessed 4 Aug 2014
10. Wickes UK website. www.wickes.co.uk. Accessed 4 Aug 2014

Chapter 31
Design of a Linear Electrical Machine for a Wave Generation System in the Maltese Waters

Xuereb Annalise, Spiteri Staines Cyril, Sant Tonio and Mulè Stagno Luciano

Abstract This research concerns the design of an electrical machine that shall be implemented on a wave energy converter (WEC). Analysis of the available wave data and optimisation of the WEC were carried out so as to design the optimum system that will interact efficiently with the conditions of the selected location. The electrical model was designed for wave heights and periods that yielded the maximum power throughout the year. The initial finite element simulations were undertaken on the design of a four-pole tubular permanent magnet (PM) linear synchronous machine. The performance of the system was optimised for a wave height and period of 3 m and 6 s, respectively. The capacity of the linear generator was set at 15 kVA and the dimensions of the WEC (floating point absorber) were determined accordingly through the application of a simplified model for predicting the coupled dynamic response of the absorber and the linear generator. It was necessary for the electrical machine to yield good performance even when operating under the slower and most common waves. Thus, an 8-pole and a 16-pole tubular linear PM machine were also simulated, and their performance was compared. Through this analysis, the 16-pole design resulted in the best performance, especially at low speeds. Simulations were initially compared at fixed speed but were later simulated with variable speed conditions so as to represent more accurately the wave's motion and monitor the power generator performance at variable loadings.

Keywords Wave energy converter · Linear generator · Tubular machine · ANSYS maxwell

X. Annalise (✉) · S. S. Cyril · S. Tonio
Faculty of Engineering, University of Malta, Msida, MSD 2080, Malta
e-mail: annalisexuereb@gmail.com

S. S. Cyril
e-mail: cyril.spiteri-staines@um.edu.mt

S. Tonio
e-mail: tonio.sant@um.edu.mt

M. S. Luciano
Institute of Sustainable Energy, University of Malta, Msida, MSD 2080, Malta
e-mail: luciano.mule-stagno@um.edu.mt

A. Sayigh (ed.), *Renewable Energy in the Service of Mankind Vol I*,
DOI 10.1007/978-3-319-17777-9_31

31.1 Introduction

Utilisation of renewable resources of energy is constantly becoming a necessity in the power generation industry. This led to the design of new concepts, while improving those which are already in operation. Wave energy is one important source of renewable energy for small countries like Malta which have severe spatial restrictions onshore due to a high-population density but, on the other hand, have a large accessibility to the coastline.

The design of a wave energy converter (WEC) is driven by the need to achieve efficient operation in the local conditions. Thus, analysis of the actual wave power data was an important step in the machine design optimisation process. In this chapter, wave data were gathered through instrumentation located on a floating buoy that was anchored at a selected location in Maltese waters. Analsyis of the measurement data confirmed that wave conditions are characterised by low-frequency waves with relatively small amplitudes. This indicated that a point absorber converter would be a suitable concept [1]. The study considered a point absorber converter coupled to a linear electrical generator as a power takeoff system to generate electricity from vertical oscillations imparted by wave motion.

Different linear machine configurations were investigated, the ultimate aim being that of maximising the amount of power extracted at relatively low speeds at the selected locations. A design concept involving a toothed tubular linear machine was selected. The latter was designed with a fixed armature and a moving magnetic translator. The linear generator design was optimised through simulations that were carried out in ANSYS Maxwell [2]. ANSYS Maxwell is a finite element modelling software for the design and analysis of electromagnetic fields and electromechanical devices. Further simulations were also carried out on ANSYS Simplorer [3] so as to monitor the electrical characteristics of the machine.

31.2 Analysis of Local Wave Power Data

Wave monitoring at the location chosen for implementation was an important parameter for the system's design. Data were gathered through wave-measuring instruments enclosed in a floating buoy anchored on the northwest side of Gozo (the most northern island of the Maltese Islands) close to 2 km away from land at a depth of 200 m. Due to its rounded shape, the direction of the wavefront does not influence the buoy's motion, and the buoy oscillated vertically on the wave's surface. The location selected is far away from land or any other floating structure. Thus, the effects of diffraction, reflection and radiation resulting from coastal features or other buoyant structures could be assumed to be minimal.

The wave data were collected every 30 min and spanned over a year from the beginning of October 2011 till the end of September 2012. The measurements and subsequent analysis were conducted by the International Oceanographic Institute of the University of Malta. The analysis showed that the lowest wave heights were about 10 cm and had a corresponding period of 2.26 s. As expected, an increase in wave height brings with it an increase in wave period, and the buoy even experienced an extreme 7.5 m wave height with a period of 8.9 s. The most common waves at the wave buoy station had a time period between 3 and 3.5 s and a wave height that varies between 0.2 and 0.4 m. Further calculations also showed a peak vertical velocity for the recorded waves in the range of 0.14 and 2.65 m/s. The energy content is high for waves with larger wave heights and periods, although these may have a lower frequency of occurrence they might still result in a larger quantity of energy being generated over a period of time. Thus, to examine which waves are in possession of the highest energy potential throughout the year, wave data were also analysed in terms of its probability of occurrence across the recorded period. Figure 31.1 shows that the most common waves did not result in the highest power distributed throughout the recorded year. These calculations show that the waves which yielded most energy over the recorded year have a period between 5.5 and 6 s, a wave height varying between 2.8 and 3 m and a peak vertical velocity reaching 1.57 m/s.

31.3 Dynamic Modelling of the WEC

The aim of this project was to develop the design of a linear electrical generator for a direct-drive WEC system. Due to the wide range of wave speeds at the selected locations, the linear generator being proposed was coupled to a point absorber converter. Point absorbers are floating devices that can be either semi-submerged or fully submerged. These converters are anchored to the seabed and operated by varying their displacement from the mean level of the water's surface with every crest or trough [4].

The dynamic response of a point absorber system is modelled by a simple mass–spring–damper system having a single degree of freedom. The system consists of a floating buoy coupled to a power take-off (PTO) system as shown in Fig. 31.2. The proposed design concept includes a linear generator, which is modelled by a damper and anchored to the seabed through additional retracting mechanical springs. These springs are used so as to maintain the link between the floating buoy and the translator in tension.

In the present analysis, a single floating mass was assumed to be anchored to the seabed through a PTO system. During the analysis, it was also assumed that the WEC is subjected to single-frequency waves and that the floating mass is allowed to move only in heave mode. Thus, only the vertical components of the forces acting on the system were considered. It was also assumed that the wave's amplitudes and motions are relatively small for the governing equations to work out. Using these assumptions, the response may then be also analysed in the frequency domain [5]. In practice, there will exist both vertical and horizontal components which require a more complex mechanical system model. It was also assumed that the

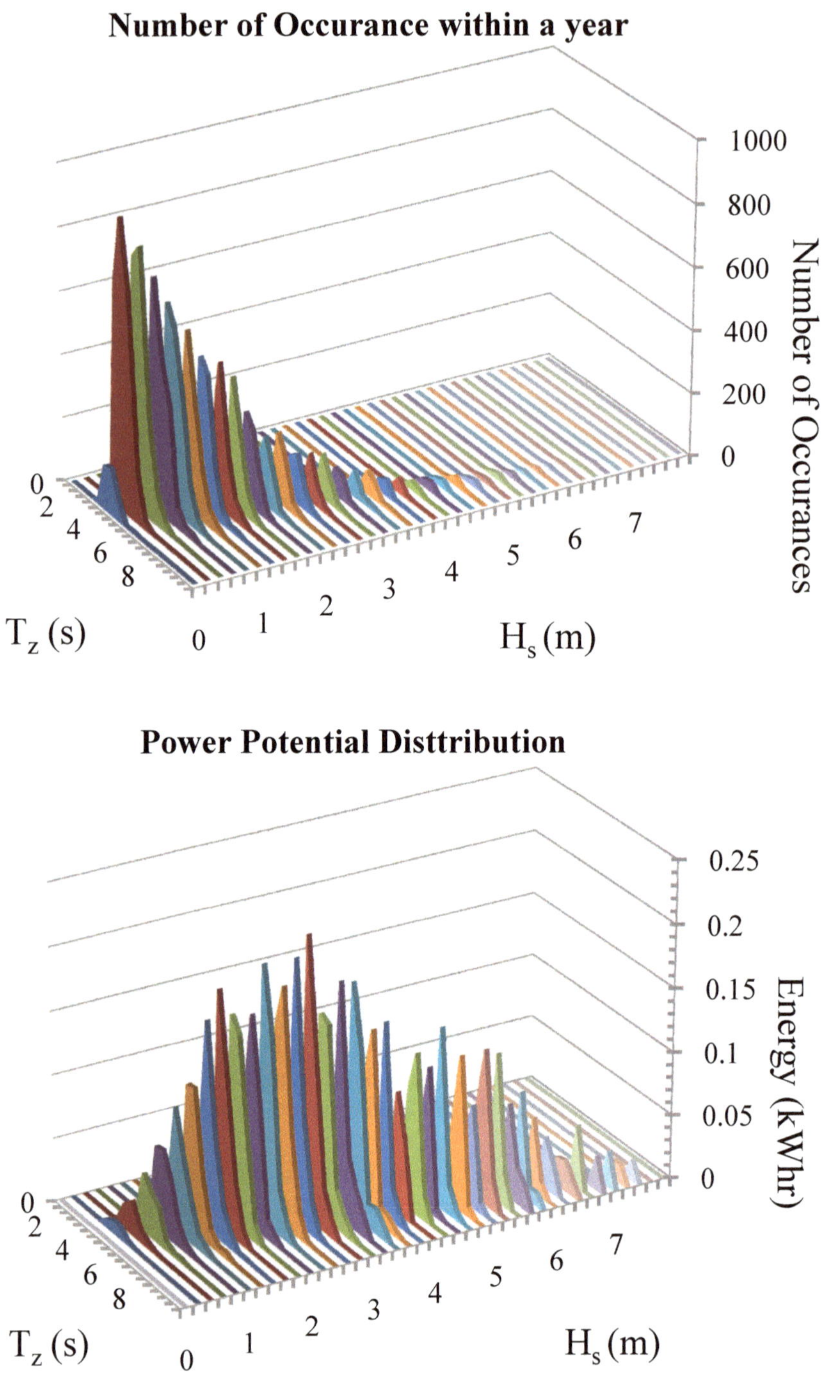

Fig. 31.1 Number of wave occurrences and power potential distributions of waves in a year for different wave height H_s(m) and zero up-crossing period T_z (s)

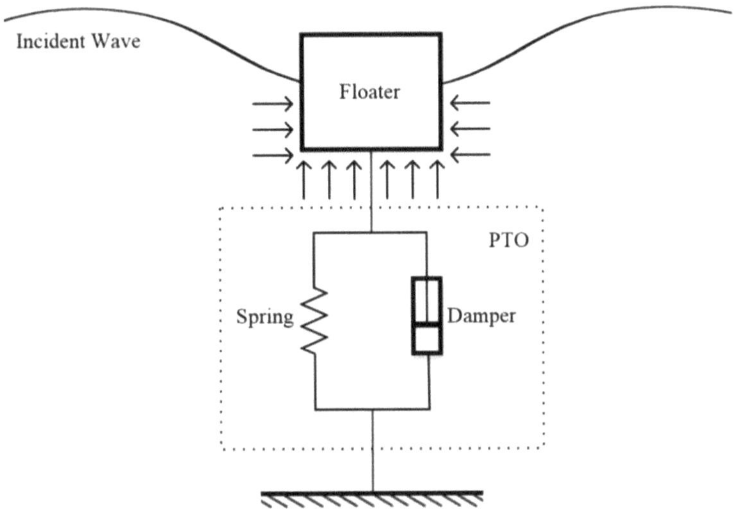

Fig. 31.2 Point absorber model with the linear generator represented by a damper

buoy will be large enough to be able to pull the translator's weight upwards and will be continuously floating at the sea water surface, unless the wave height is beyond the allowable displacement of the linear generator.

Following the above assumptions the linear generator was therefore considered to follow a vertical motion. The linear machine design shall be based on a sinusoidal wave motion with a maximum vertical velocity of 1.57 m/s for the 3-m and 6-s waves selected. Following this analysis, the final generator design shall also be tested for variable speed conditions.

The heaving force of the incident wave f_d can be represented as [5]

$$f_d = (m + m_a)\ddot{x} + (b_{hyd} + D_{pto})\dot{x} + (K_{pto} + \rho g A_0)x, \tag{31.1}$$

where m is the mass of floating buoy, m_a is the hydrodynamic coefficient of added mass to cater for the effect of the inertia of the water on the floater, b_{hyd} is the hydrodynamic damping coefficient to account for the damping of the floating buoy when energy is transferred through the radiated waves, d_{pto} is the damping coefficient of the linear damper, k_{pto} is the linear spring stiffness constant, ρ is the sea water density, g is the acceleration due to gravity, A_0 is the cross-sectional area of the floater and x, $\dot{x}$ and $\ddot{x}$ are the position, the vertical speed and vertical acceleration of the buoy.

Equation (31.1) shows that the behaviour of a heaving point absorber converter can be modelled by a second-order expression. Such systems achieve their optimum performance when operating at the resonant frequency ω_n. The natural frequency of the complete WEC is equal to:

$$\omega_n = \sqrt{\frac{\rho g A_0 + K_{pto}}{m + m_a}} \tag{31.2}$$

Equation (31.1) may be expressed in the s-domain in terms of $X_0(s)$as

$$X_o(s) = \frac{F_d}{s^2(m+m_a) + s(b_{hyd} + D_{pto}) + (K_{pto} + \rho g A_0)} \tag{31.3}$$

The energy absorbed by the PTO system is mainly due to the damping component. The stiffness of the PTO system k_{pto} does not contribute to the effective power absorbed by the electrical machine. This leads to a time-averaged power absorbed of:

$$\overline{P}_e = \frac{\Delta W}{T} = \frac{\pi D_{pto}\omega |X_0|^2}{\frac{2\pi}{\omega}} = \frac{1}{2} D_{pto}\omega^2 |X_0|^2 \tag{31.4}$$

Thus, Equation (31.4) shows that the power is dependent on motion's frequency ω and the maximum displacement X_0. Besides that, in [5], the author also shows that for maximum power absorption, apart from operating at the resonant frequency ω_n, the hydrodynamic constant b_{hyd} needs to be equal to the PTO damping D_{pto}.

The dynamic response of a WEC system was analysed in detail and optimisation of the floating buoy was carried out using a dedicated tool, WAMIT, which is a commercial software used to simulate hydrodynamic loads on offshore structures, not only WECs. This analysis was possible through the invaluable support of the research institute in the Wave Energy Centre (WavEC) Offshore Renewables in Portugal. Simulations were carried out on a 1 and 4 m draft buoys, which were expected to reach their resonance frequency at 2 rad/s (3.1 s wave) and 1.5 rad/s (4.2-s wave), respectively. The shorter buoys were able to absorb power from slower-moving waves, but a larger range of waves was absorbed by the 1-m draft buoy.

In [5], Falcão shows that for maximum power absorption, the damping of the floating system needs to be equal to the damping created by the linear machine. The latter was calculated using the machine's force and velocity parameters and a 2-m-radius buoy with a draft of 1 m was selected. It was also required to ensure that the selected buoy will float when selecting the mass required for the correct operation with the PTO system at 1.05 rad/s. Since a 2-m radius would not be able to accommodate the mass calculated, a 4-m-radius buoy was considered. Consequently, this wider buoy obtained a larger hydrodynamic damping effect than that obtained for the linear generator. Thus, the effect of different values of PTO damping of the 4-m-radius buoy was analysed through the equation of motion. It could be noted that for low values of PTO damping, the peak response is closer to 2 rad/s; however, a system with higher values of PTO damping exhibit a wider-banded wave energy absorption characteristic which allow for correct operation even with the linear generator selected. More details on the above results can be found in [6].

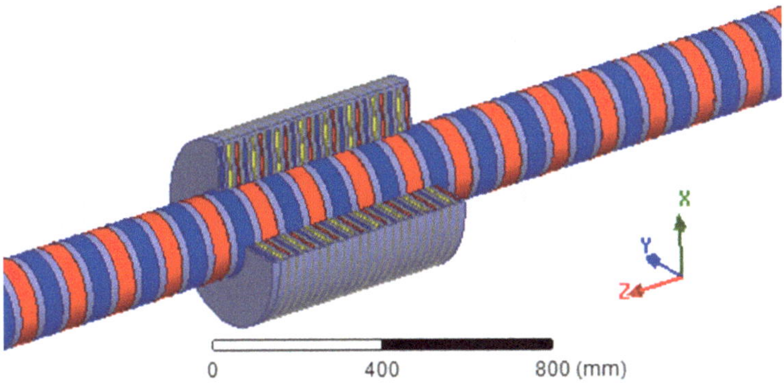

Fig. 31.3 3D view of the 16-pole tubular linear machine

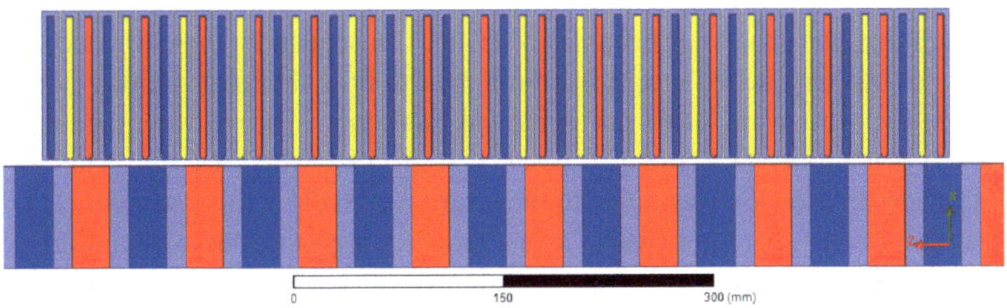

Fig. 31.4 Sectional view of the 16-pole tubular linear machine

31.4 Linear Generator Design

The tubular linear permanent magnet (PM) synchronous machine model was initially based on a design reported in [7] with a fixed armature and a moving magnetic translator. Ring magnets magnetised parallel to the direction of motion were mounted in alternating directions on an aluminium shaft to form the translator. Ring phase windings make up the stator of the tubular design and are enclosed in steel covers and separated by aluminium rings as in Figs. 31.3 and 31.4.

Simulations were carried out in ANSYS Maxwell. The model was initially simulated through the transient solver and later coupled to the electrical loading circuitry in Simplorer. In Simplorer, the model is coupled to a star-connected resistive load of 22 Ω per phase, and the machine is operated at a speed of 1.57 m/s. The simulation schematic is shown in Fig. 31.5.

Initial simulations were carried out on a four-pole configuration to optimise a 15 kVA machine at 1.57 m/s. However, the electrical outputs of this machine at the required speeds had a very low electrical frequency and a very distorted output. Thus, simulations on models with shorter-pitched translators were carried out. To keep the four-pole configuration, the width of each armature segment was reduced with a similar factor and, to maintain a similar performance, the total number of

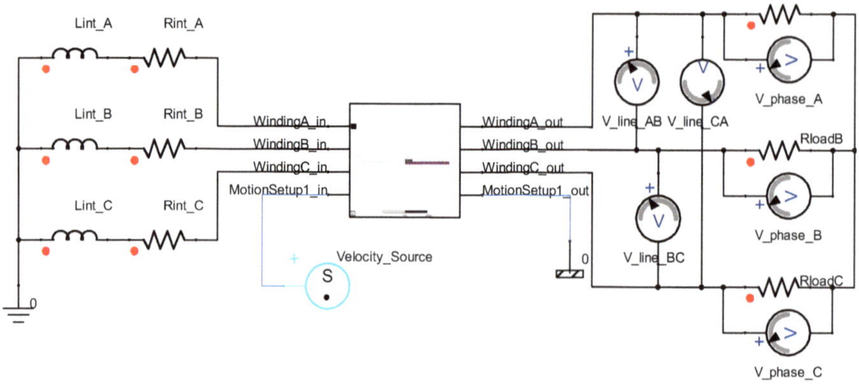

Fig. 31.5 Maxwell model block coupled to the electrical components in simplorer

winding turns per phase must remain constant. Consequently, the coils need to be thinner for shorter-pitched machines, but longer to fit in the same number of coils.

Thus, the number of turns per coils was subdivided into different stator segments so as to model an 8- and a 16-pole linear machine which are equivalent to the 4-pole design. Attention was given to the saturation levels in the stator which led to the modification of the stator teeth. Thus, the teeth are modelled with a semi-open configuration so as to allow the flux lines to flow through a wider path and reduce the saturation stresses on the teeth edges.

Simulations comparing the different configurations showed that the translator of the larger pole configuration exerted a smoother force and managed to achieve a smoother voltage with an increase in the electrical frequency for the same velocity of 1.57 m/s, as shown in Fig. 31.6.

It was also important to analyse how the linear generator would perform at variable speed conditions. Thus, a sinusoidal velocity input was prescribed to the model block in Fig. 31.5 with a peak velocity of 1.57 m/s and a period of 6 s (the period of the selected waves for which the linear generator is being designed). Under these conditions, the translator will move a peak-to-peak distance of 3 m. For a continuous output from all the coils on the stator, the length of the translator needs to be longer than the sum of the distance that the translator will move and the length of the stator. Thus, the PM machine was designed with a translator of 4 m length for a stator of 0.7 m length and an outer radius of 0.9 m. However, since there exists the possibility that the wave displacement can even exceed this length, a mechanical braking system needs to be incorporated in the WEC to limit its oscillations when the buoy is hit by large wave heights. Waves recorded with a displacement close to 3.5 m had a peak velocity of 1.83 m/s.

Energy in larger waves can also be absorbed if a longer translator is used. Slower conditions were also analysed and the speed of 0.359 m/s was selected to represent waves with a period of 3.4 s and a maximum wave height of 0.6 m. This condition was selected based on the most common type of waves recorded. Figure 31.7 shows the simulations of the three linear machines were carried out with maximum speeds

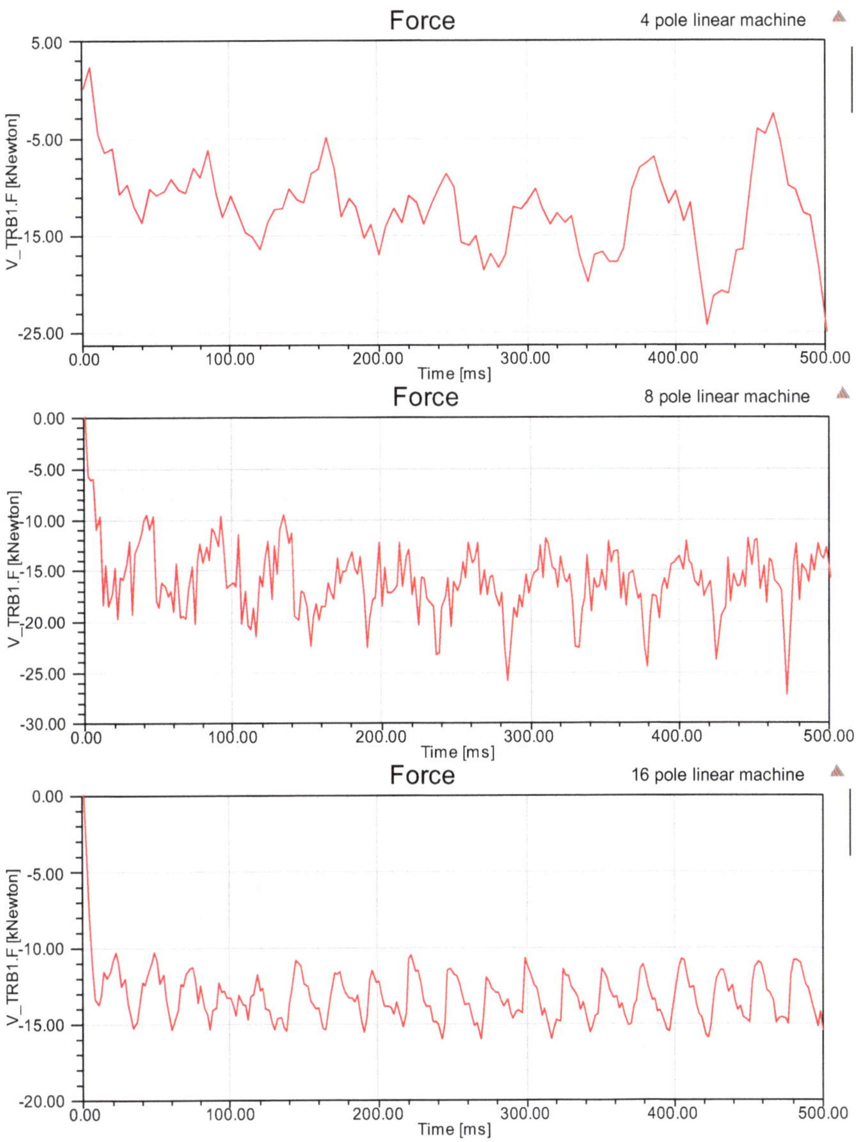

Fig. 31.6 The force exerted on the translator of the 4-, 8-, and 16-pole machines

of 1.83 m/s (the highest speed allowed by the 4-m translator system) and 0.359 m/s (speed of the most common waves) (Fig. 31.7).

The variable speed scenarios highlight the importance of having a high number of poles, especially at slow-speed conditions. Had there been less poles, the sinusoidal-shape wave would have been lost due to the ripples in the force waveform. When the input velocity was representing slower, but the most common waves, a

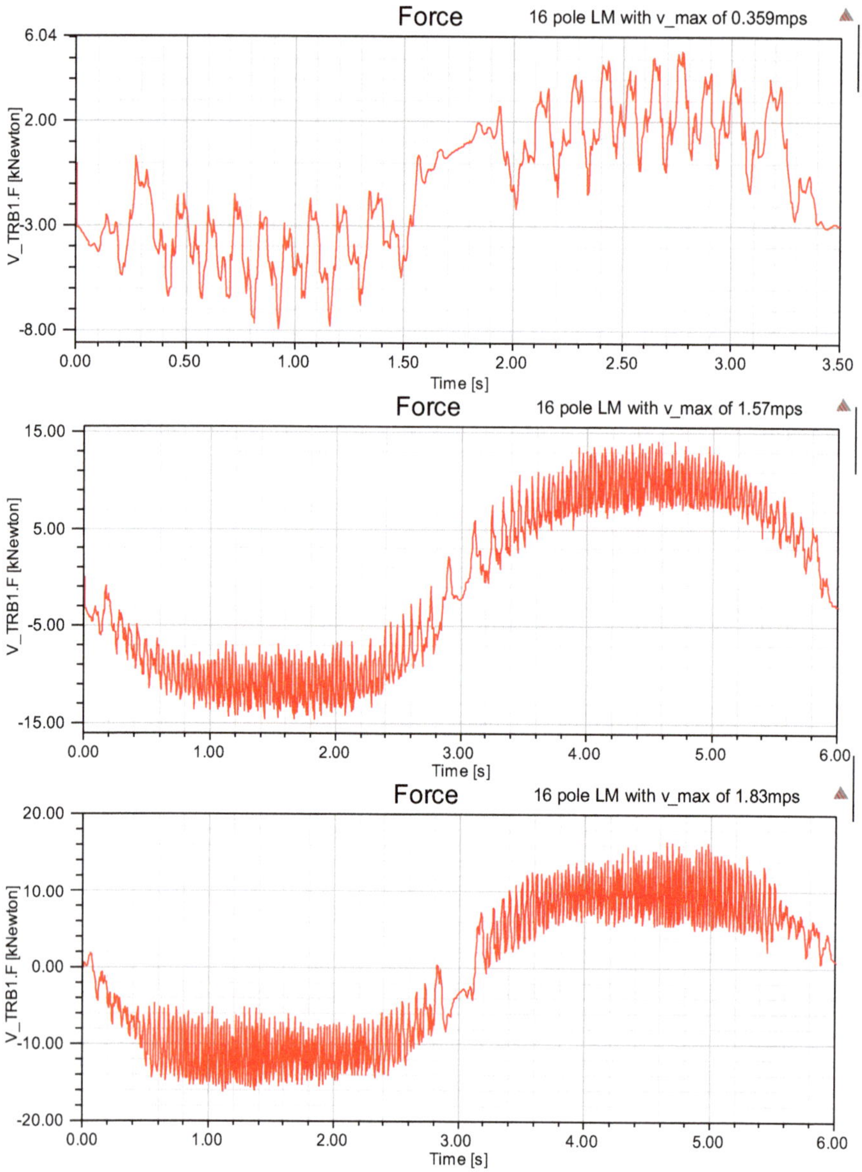

Fig. 31.7 Performance of the 16-pole machine with variable speed, peak velocities of 0.359, 1.57 and 1.83 m/s, respectively

significant difference was clearly visible between the performances of the linear machines.

The variable speed conditions have a direct effect on the generator's electrical outputs and the force on the translator. Even though the direction of the force is continuously changing, the machine is constantly in generating mode as the direction of the velocity is always opposing that of the force. Through this analysis, one could also notice that the ripple in the force is lower for machines with a larger number of poles especially at slow speeds. The voltages induced at slow speeds in the 4-pole machine will have a larger harmonic content than that induced in the 16-pole machine. The frequency of the electrical outputs are higher as there are more complete cycles in one wave cycle in the 16-pole machine. This results in a smoother current flowing through the windings and a smoother output force from the 16-pole machine. These slow speeds represent the most common waves, thus a smoother output will increase the efficiency of the linear machine design.

For comparative reasons, the double-sided flat equivalent of the 16-pole machine was also simulated. The model resulted with a lower flux density in the air gap when compared to the tubular equivalent. A lower-induced voltage was obtained and consequently a lower current for the same input velocity and resistive loading. Further details can be found in [6].

31.5 Conclusions and Further Work

Through this analysis, it was therefore concluded that the 16-pole machine would be more suitable in the given application. The selected machine resulted to have a lower force ripple on the moving translator. Furthermore, this machine design was also found to offer superior performance at slow speeds due to its higher-output frequency. The higher the frequency, the less rippled the output shall be. However, it was not possible to lower the frequency further due to constraints related to the dimensioning of the mechanical system.

The analysis presented in this chapter was mainly focused on the design of the electromechanical linear PM machine. Testing of the performance of the linear machine was carried out by operating the linear generator design together with a three-phase resistive load bank. Further testing can also be carried out on the effects of nonlinearities introduced when connecting such a machine to the rectification stage of a grid-connected inverter.

References

1. Xuereb A, Spiteri-Staines C, Mule' Stagno L, Tonio Sant, Aldo Drago (2013) Analysis of the potential of a wave energy conversion system for maltese waters. University of Malta, Malta
2. Ansys Maxwell. Ansys. http://www.ansys.com/maxwell/. Accessed 29 June 2014
3. Ansys Simplorer. Ansys. http://www.ansys.com/Products/Simulation+Technology/Systems+&+Multiphysics/Multiphysics+Enabled+Products/ANSYS+Simplorer. Accessed 29 June 2014
4. Drew B, Plummer R, Sahinkaya M (2009) A review of wave energy converter technology. IMechE 223:887

5. Falcão A (2010) Wave energy utilization: a review of the technologies. Elsevier 14(3):899–918
6. Xuereb A (2013) Design of an electrical machine for a wave energy converter. Masters in Electrical Engineering Thesis, University of Malta.
7. Piech JZ, Gieras Jacek F (2011) Linear synchronous motors: transportation and automation systems. CRC Press, USA

Chapter 32
Modelling Tidal Stream Turbines

Sarah Tatum, Carwyn Frost, Daphne O'Doherty, Allan Mason-Jones and Tim O'Doherty

Abstract The transient behaviour of the sea and the rotation of a turbine rotor can result in high asymmetric loadings, which are transmitted to the drive shaft. A turbine mounted on a circular stanchion has been used to highlight the effects of introducing more realistic boundary conditions, over a rotational cycle of the turbine. The consequences on the turbine's performance characteristics and crucial structural loading are shown. The position of the turbine relative to the support structure and its alignment to the flow direction can have significant temporal hydrodynamic and structural effects. Depending on their wavelength, waves can also have a significant effect on the overall design decisions and placement of devices. Thrust loading and bending moments applied to the drive shaft can be of the order of hundreds of kN and kNm, respectively. This leads to the need to not only size the drive shaft and bearings to account for axisymmetric thrust but also consider large asymmetric loads.

Knowledge of the flow regime can allow the designers to evaluate material selection for components (i.e. for blades, etc.) and incorporate some deformation capability of the turbine blades to increase the power output and potentially alleviate some of the stress distribution through key structural points, that is, drive shaft, bearing connectors, etc. The resulting data can then be used to estimate component life via fatigue prediction.

This chapter includes a multi-physics approach to modelling tidal energy devices and the potential for modelling to inform device condition monitoring.

S. Tatum (✉) · C. Frost · D. O'Doherty · A. Mason-Jones · T. O'Doherty
School of Engineering, Cardiff University, Queen's Buildings, The Parade, Cardiff CF24 3AA, UK
e-mail: tatumsc@cardiff.ac.uk

C. Frost
e-mail: frostc1@cardiff.ac.uk

D. O'Doherty
e-mail: odohertydm@cardiff.ac.uk

A. Mason-Jones
e-mail: mason-jonesa@cardiff.ac.uk

T. O'Doherty
e-mail: odoherty@cardiff.ac.uk

A. Sayigh (ed.), *Renewable Energy in the Service of Mankind Vol I*,
DOI 10.1007/978-3-319-17777-9_32

Keywords Tidal turbines · Computational modelling · Fluid–structural interaction

32.1 Introduction

The EU has targeted renewable energy to provide 20% of the total energy mix by 2020 [1], in comparison the UK target is 15% of the UK energy demands from renewable sources by 2020 [2]. In order to meet and sustain the targets set by 2020 and beyond, the UK and EU at large must continue to address the imbalance in the renewable energy mix. The potential for sustainable production through wave and tidal energy conversion has resulted in large investment from industry and governments. A practical and economical resource of tidal current and wave energy resource of 70 TWh/year around the UK shores have been identified, which would contribute to 20% of UK's total industrial, commerce and domestic electricity demands [3].

The long-term predictability of tides is the main advantage of tidal power over other renewable energy sources since it allows any phase change in power productions between wind and/or other tidal stream and wave sites to be balanced. The two leading techniques in energy conversion for tidal range power generation are impoundment schemes such as a barrage or tidal lagoon and tidal stream turbines (TSTs). The impoundment schemes offer large-scale solutions, as documented in the La Rance Tidal Power Plant run by EDF Energy which produces 0.54 TWh/year [4], whilst TSTs offer bespoke solutions to fit the local environment and, since they are submerged, they are less intrusive than impoundment schemes and minimise the impact on the marine and costal environment.

The introduction of TSTs into the UK energy mix can only be a positive step since the UK has some of the strongest currents in the world, with a number of areas identified as viable sites for installing tidal energy devices [3]. Ideal conditions for TSTs are a free stream velocity of 2–3 m/s and a depth of 20–30 m, at least for early-stage implementation with deeper water designs (> 40 m) introduced as the industry matures. Current developments include the DeltaStream device which is being installed in the Ramsey Sound, Pembrokeshire [5], and the array, of SeaGen S turbines, in the Skerries site off Anglesey [6]. The success of these initiatives is dependent on a thorough understanding of the hydrodynamic forces applied to the TST and experiential knowledge acquired from environmental monitoring, which, for example, has led to actively pitch controlling the blades to limit the maximum rotational speed of the rotors to 14 rpm [7]. The hydrodynamic forces cause considerable loading on the blades resulting in blade deformation. The magnitude of the deformation will be dependent on the design of the blades, their internal structure and materials used. However, the combined loading on the blades and nose cone of the rotor is transmitted to the drive shaft, and knowledge of these resultant transient forces is essential for a developer to optimise the success of their design.

The simplest scenario for determining the hydrodynamic forces is to consider the effects of a uniform tidal current. However, in reality, the current is typified by a high shear in the velocity profile through the water column, turbulence and surface waves, and these all need to be considered in determining the loads on a TST. Work on the effects of highly shearing profiles and levels of turbulence has been considered by Mason-Jones et al. (2008) [8] and Morris (2014) [9]. Wave effects are likely to be critical as they are known to penetrate the water column to a depth of approximately 50 % of their wavelength [10]. This can be further complicated by the fact that the direction of the wave does not necessarily align itself to the direction of the current as shown by Lewis et al. (2014) [11].

The aim of this work was therefore to investigate, by the means of numerical modelling, the effects of surface waves on the performance characteristics of a tidal turbine. The approach taken started with quasi-static uncoupled computational fluid dynamics (CFD) modelling of the turbine through a range of tip-speed ratios (λ) for constant current flow (the uncoupled CFD model), followed by a fully coupled two-way transient fluid–structural interaction (FSI) model at peak power for the same current flow (the FSI model). Finally, a transient uncoupled CFD model was modelled at peak power for the same current flow, with waves in the same direction as the current (the volume of fluid; VOF model). In all cases, the study focussed on the cyclic blade loading from axial thrust loads.

32.2 Fluid–Structural Interaction Modelling

Previous FSI studies of marine turbines include a loosely coupled approach, via a surface panel code to link the CFD to the finite element analysis (FEA) [12]. This work gives a valid insight to the effects of FSI and its consequences on cavitation, power attenuation and stresses on the structure. Whilst it is stated by Nicholls-Lee et al. (2011) [12] that the panel method performs well for undeviating flows, it becomes necessary to include intricacy in the hydrodynamic flow field, such as the Reynolds-averaged Navier–Stoke (RANS), detached eddy simulation (DES) and large eddy simulation (LES) solvers to accurately capture the hydrodynamics. Moreover, with knowledge of the hydrodynamics under extreme conditions, it becomes necessary to consider the strong coupling effects between the fluid and the structure.

A fully coupled two-way FSI model using a symmetry boundary and single blade was undertaken by Park et al. (2013) [13]. From the results it was suggested that although the deflection of the blade was relatively small in their study, it was still necessary to consider blade deformation due to a 1.7 % drop in maximum power extraction. Morris (2014) [9], established the performance characteristics of a tidal stream, whilst allowing for blade deformation a series of quasi-static-coupled two-way FSI models constructed with a homogenous blade material, each subjected to a uniform flow inlet boundary condition. The kinetic energy extraction and its sensitivity to structural stiffness attenuation were investigated by incrementally reducing the materials modulus and noting its effect on the rotor's performance.

Fig. 32.1 Side and front view of turbine-stanchion arrangement for all the models

32.3 Geometry Specifications

The 10-m diameter, three-bladed steel TST discussed in this chapter, utilised a Wortmann FX 63-137 profile, with a 33 ° twist from the blade root to tip. The blade pitch angle was set to the optimum of 6 °. The turbine support structure used was of a simple cylindrical design, based on the work of Mason-Jones et al. (2012) [14] and consisted of a 2.4-m diameter stanchion as shown in Fig. 32.1. Based on the work of Frost et al. (2014) [15], the turbine rotor was positioned with a clearance of ~2 m from the stanchion and was orientated to be aligned to the tidal flow.

32.4 Modelling Approach

32.4.1 The Uncoupled CFD Model

The uncoupled CFD model consisted of a rectangular domain, 350 m long × 50 m wide × 35 m deep. The turbine was located 100 m downstream from the inlet boundary such that its rotational axis was set at 25 m below the water surface. An axially aligned cylindrical domain was subtracted from the rectangular domain to form a rotational interface domain with a nonconformal interface between the sea and turbine volumes to model the rotation of the turbine. The blades were modelled using tetrahedral elements with face sizing of 10 mm for the top third and 20 mm for the bottom two thirds, leading to the root. The rotational interface domain was set with an internal body size of 200 mm and the same value was applied to the external faces of it. In total, the water domain had 3.5 million elements with 2 million concentrated about the rotational interface and turbine surfaces. The uncoupled CFD models have been previously validated using a recirculating water flume and an uncoupled CFD model [16, 17].

Inlet boundary conditions were defined as a simple uniform flow of 3.086 m/s (6 knots). The outlet was a pressure outlet, set to 0 Pa. All walls of the domain were set to free slip, apart from the seabed which was set to a ‘no slip’ wall boundary.

For this and subsequent CFD models, the RANS equations were applied to relate the Reynolds stresses to the mean velocity gradients. To close the equations, the shear-stress transport (SST) was used as the viscous model. Previous work has used the Reynolds stress model as the viscous model; however, the SST was used due to the increased model complexity and computational expense of FSI and free surface modelling. The SST model was also found to have a dissipation rate more suited to this work and had the added benefit of being computationally less expensive [9].

32.4.2 The FSI Model

The FSI model utilised the same domain definition as the uncoupled CFD model for the CFD analysis, whilst the mesh for the FEA analysis had the same face sizing on the blades and hub as the CFD model in order to limit mesh mapping errors for the transfer of data. The interface boundaries for the data transfer were set to receive the total mesh displacement from the FEA model and send the total force back from the CFD model.

The setup of the boundary conditions for both the FEA and CFD was done sequentially beginning with the FEA model which was set to match the requirements of the CFD model. That is, a rotational velocity of 2.25 rad/s ($\lambda = 3.64$), which is the angular velocity at which peak power is extracted for this turbine configuration. (To ensure that the blade tip deflection occurred gradually, allowing the mesh to 'adjust' to the new geometry, the angular velocity was ramped up to the required value). The fluid–structural interfaces were then specified, with each blade separated into two faces, front and back. The labelling of these faces is significant as they are required to set the data transfer from the CFD to the FEA and vice versa. The foot of the stanchion was fixed to the floor using a fixed support. It was assumed for this case that there is no movement of the foundation, and the entire bottom face of the stanchion would remain in contact with the seabed. Although not discussed in this chapter, it is worth noting that the FEA model also had a number of outputs including the total deformation, equivalent stress (Von-Mises), equivalent elastic strain, fatigue life and biaxiality, etc., to provide the hydrodynamic effects on the structure.

Once both the CFD and FEA models were set up, they were linked/coupled, thus enabling data transfer between the two models. Coupling was set to suit the CFD model with appropriate time steps for the total time equivalent to approximately six full rotations. The six rotations are required to ensure that the transient models have 'settled' and are producing stable solutions.

32.4.3 The VOF Model

The VOF model was based on the same definition as the uncoupled CFD model for the water domain, but an additional air domain 350 m long × 50 m wide × 15 m deep was added on top of the water domain to allow for a VOF-based analysis.

The boundary conditions defining the height of the free surface as well as inlet conditions were specified. The outlet was set as an opening, as this prevents reflection of the waves disrupting the domain and allows the sub-surface wave motion to move back and forth in an orbital fashion. A modification of the linear Airy theory was applied to account for the nonlinearity of the waves due to their high amplitude of 3 m. Nondimensionalisation of the wave height and water depth was achieved by dividing by gravitational acceleration and the square of the wave period, which was then compared to the work found in the literature to confirm the nonlinear nature of the waves in this study [18–20]. An inflation layer was applied to the top surface of the water and the bottom surface of the air domain, extending 40 layers into the mesh to allow a much more densely packed mesh in that region.

In order to specify the wave using linear theory, the wave number *k* was required. This was calculated from the dispersion relation (Eq. 32.1), which specifies that a wave of a particular frequency must have a given value of *k*, defining the wavelength (Eq. 32.2):[21].

$$\omega^2 = gk \tan h(kd), \tag{32.1}$$

$$L = \frac{2\pi}{k}, \tag{32.2}$$

$$kd \tan h(kd) = \frac{\omega^2 d}{g}, \tag{32.3}$$

where ω is angular velocity, k wave number, L wavelength, h wave height, d water depth and g gravity.

An inverse tan h function based on a set of data pairs for kd and tan $h(kd)$ was then used to determine the wave number from Eq. 32.3, given that the right-hand terms are known or can be calculated [21]. The wave height was selected to be small enough that quasi-linear theory was sufficient to describe it, but large enough that its graphical representation and effects were visibly observable.

The other constraint was that the wavelength needed to be sufficiently large for the effect of the wave to penetrate to the turbine depth. Consequently, a wavelength of 30 m was chosen, giving a wave number, $k = 0.209\ \mathrm{m}^{-1}$. Given the depth of water was 35 m, the angular velocity was calculated as 1.43 rad/s, giving a time period of 4.38 s.

The rotational speed of the turbine was once again set to 2.25 rad/s to allow direct comparison with the uncoupled CFD model. Hence, a single turbine revolution takes approximately 2.79 s, thus the period of the wave was not the same as the period of rotation of the turbine.

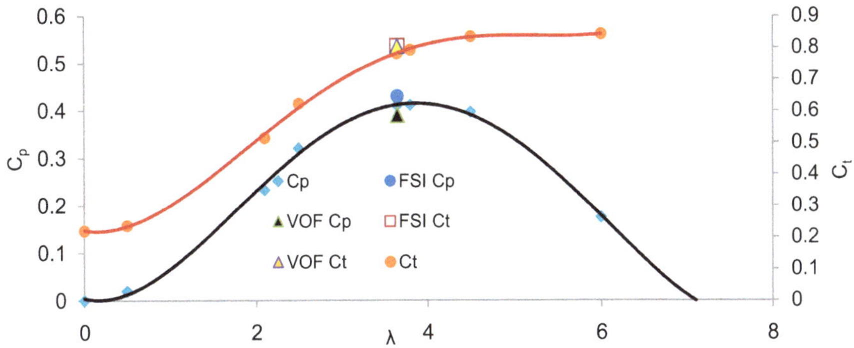

Fig. 32.2 Turbine performance characteristics (uncoupled CFD) with peak power and thrust values for the VOF and FSI models. *VOF* volume of fluid, *FSI* fluid–structural interaction

32.5 Results

32.5.1 Performance Curves

The turbine performance characteristics were determined from the thrust and power curves plotted from the results obtained from the quasi-static model used in the uncoupled CFD analysis. Figure 32.2 shows the results of both the power and thrust coefficients for the turbine with the simple support structure, which agree with previously published data that used the Reynolds stress model as the viscous model, that is, peak values of $C_p \sim 0.41$ at $\lambda = 3.64$ and $C_t \sim 0.80$ [16, 17].

Comparing the C_p and C_t values at peak power, it can be seen that when the CFD analysis is undertaken as part of the FSI model, and consequently changed to a transient model, the value of both these characteristics increases. The small increase in the power output with the FSI model ($C_p = 0.43$, i.e. 480 kW) is due to deformed shape of the blades which deflect away from the non-loaded setting [9]. There is no significant change in the thrust with this level of blade deformation. However, as shown by Morris (2014) [9], an increasing blade deflection results in a reduction in the level of thrust, which again should be expected as the projected area reduces.

Likewise, there is a similar increase in the time-averaged C_t at peak power for the VOF model, though the time-averaged C_p at peak power has reduced for the VOF model. From Fig. 32.2, it can be seen that the addition of the waves reduces the power to a mean value of approximately 440 kW ($C_p = 0.38$), a 4 % reduction.

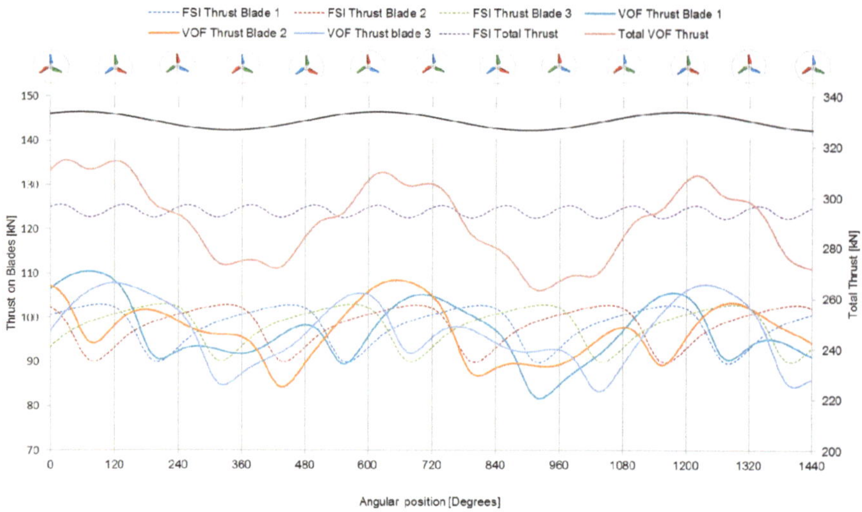

Fig. 32.3 Transient thrust curves for the VOF and FSI models. *FSI* fluid–structural interaction, *VOF* volume of fliud

32.5.2 Loading Characteristics

32.5.2.1 Thrust

Figure 32.3 shows that the transient thrust force for the FSI model is reasonably constant, save for the rotational position when the blade and stanchion have a small level of interaction. The interaction per blade is once every rotation with a frequency of 0.35 Hz. Hence, a complete rotation of the three-bladed turbine has a frequency of 1.05 Hz. The black line is a dimensionless representation of the wave position at the turbine, and the turbine images depict the blade configuration at each given time.

The amplitude of the thrust on an individual blade is 13 kN. However, the combined thrust which is transferred to the drive shaft is approximately 300 kN (C_t~0.80), with a 5 kN fluctuation. This thrust and that on the hub will ultimately be transmitted to the structural support. The hub loading is negligible compared to the blade loading.

The addition of a surface wave to the environment (VOF model) can be seen to have a large effect on the thrust loading The net thrust transferred to the drive shaft is virtually the same as that when there is no surface wave present (FSI model) as can be seen in Fig. 32.2. However, the simple repetitive fluctuation previously seen has been complicated by the addition of a second periodic event, which has a longer time period, is not a multiple of and out of phase of that of the turbine rotation. The maximum thrust has also increased to 318 kN C_t~0.85) with a 48 kN amplitude. However, there is an additional long 'enveloping' wavelength that occurs as a result of the wave–turbine rotation combination. The frequency of this wavelength is 0.14 Hz, with an amplitude of 76 kN.

32.5.2.2 Power

The power generated, with no waves present (FSI model), follows a similar trend to that of the thrust data, albeit a very slightly increased average. This is attributed to the small deflection present in the blades which have moved the blade pitch angle to the optimum value [9]. That is to say that there is a regular set of curves from each blade 120° out of phase with each other. Hence, the power generated by the whole rotor is a clean repetitive power output of approximately 460 kW (i.e. $C_p \sim 0.41$). The frequency of the oscillation in power (approximately 12 kW) matches the rotational frequency of the turbine, that is, 0.36 Hz. The introduction of waves reduces the power output to 450 kW (i.e. $C_p \sim 0.39$). The oscillation in the power, ~190 kW, has a frequency of 0.75 Hz. However, there is again a longer wavelength apparent which results from the wave–turbine interaction. This influence of the waves on the turbine also means that the peak power output does not always occur at the same rotational position of the blades or with the same magnitude. From Fig. 32.4, it can be seen that the power peaks between one and two full rotation of the turbine for the given environment. This is in contrast to the environment with no waves, where the power peaks three times every full rotation of the turbine.

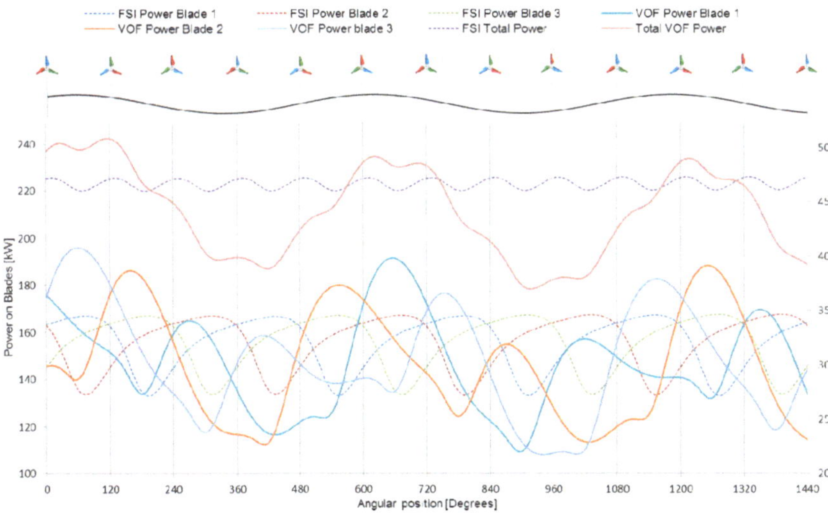

Fig. 32.4 Transient power curves for the VOF and FSI models. *FSI* fluid–structural interaction, *VOF* volume of fliud

32.5.2.3 Two-Way Coupled FSI

Over the complete rotational cycle, the blade tips deflect by approximately 23 mm. The deflection is induced by the presence of the stanchion, as each blade passes the stanchion, its thrust and power is reduced (as represented in Figs. 32.3 and 32.4), thus causing a variation in blade deflection. This deflection is particularly low due to the material used in the blade model (steel with a Modulus of Elasticity of 200 GN/m^2) being 'stiff'. However, the findings are indicative of the loading distribution through the blades and hub.

Given the constantly varying loads that the blades are exposed to due to the wave and stanchion interaction, these deflections would also be constantly changing over a rotational cycle. Although the changes in the deflections may be small, there is an increased likelihood of fatigue failure and lowering the life expectance of the component parts. This will only be exaggerated by a shearing flow through the water column.

32.5.2.4 Bending Moments

Figure 32.5 shows the resultant bending moment applied to the turbine drive shaft due to the unequal thrust loading of the tidal flow on the turbine blades. The data shown are from the VOF and the FSI model, respectively. The magnitudes of the bending moments are, on average, the same at 26 kNm. However, what is clear from the figure is that the FSI model, which is not influenced by waves, fluctuates

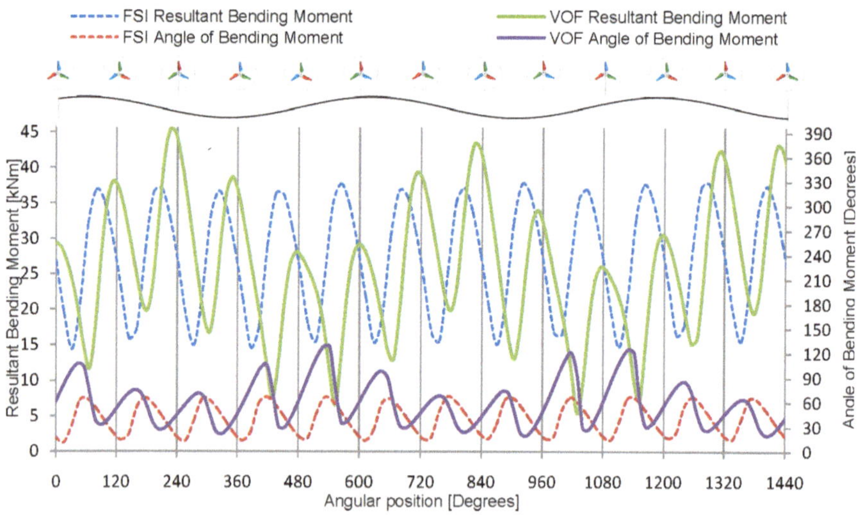

Fig. 32.5 Transient bending moment for the VOF and FSI model. *FSI* fluid–structural interaction, *VOF* volume of fliud

between 12 and 36 kNm, whereas the introduction of waves (VOF model) increases this fluctuation to between 6 and 45 kNm.

The angle of the bending moment is defined as zero when a blade is vertically upwards at 'top dead centre' and increases in the direction of the turbine's rotation. The range of the direction of the bending moment for the FSI model is between 5 and 67°, with an average of approximately 37°. Given that the velocity is uniform, the variation in angle and the fact that it remains in the upper 'quarter of the rotation' implies that the stanchion has influenced this characteristic. In contrast, the introduction of the waves (VOF model) has altered the range of the bending moment direction to between 18 and 127° and the average angle to 60°. The consequence of the narrow-angle range is that the bending moment will translate into loading being applied over a small area of the seals and bearings at the rotor end, resulting in uneven wear and possibly reduced life and increased maintenance requirements.

32.6 Discussion

The exact placement of a tidal turbine in the water column will be dependent on a number of issues. These issues include the design of the anchoring method, the seabed topography, the shipping needs, the environmental considerations, etc. What must be known is the velocity profile over a period of time including spring and neap tides, the approximate level of turbulence and a history of the wave characteristics. With this information, not only can the device characteristics be determined but also the extreme loadings under normal conditions. The fundamental effects of how the waves change the velocity of the water through the water column can be estimated from oceanographic text. However, the effects of these changes on a turbine are another matter.

The work discussed in this chapter has shown that as expected the average thrust on the turbine and power generated are only affected by a small amount. It is interesting to note that the average thrust has a small increase of approximately 5 %, whilst the average power is reduced by approximately 5 %. The major problems arise, however, from the large increase in the fluctuations in these characteristics. It should be noted that the turbine was run at a constant rotational speed and with a fixed blade pitch angle, facts that would not be true in reality, and that one or both could be varied in response to fluctuations. The structure's response time to wave-induced velocity changes would still produce large fluctuations, as the inertia effects of the rotor would play a significant role. The fluctuations will translate into increased wear and tear or fatigue on components of the turbine, so reducing the life expectancy of the device before maintenance would be required. The amplitude of the fluctuations is most likely related to the wavelength and celerity of the wave and would have to be designed for. It could be hypothesised that the waves cause low-frequency fatigue, with turbulence causing high-frequency fatigue under certain circumstances. This work has not included an acoustic Doppler current profiler (ADCP)-derived velocity profile and wave combined. However, it can be further hypothesised that given the possibility of a high shear velocity profile through the water column and the increased loading amplitude, the results discussed in this chapter are only exaggerated under such conditions.

The FSI model has provided an insight into the reaction of the turbine, specifically the blades, to the simple uniform velocity. Due to the material chosen for the blades, the deflections are relatively small at approximately 30 mm; however, there is a minor interaction with the structure as the blades pass in front of the stanchion. There are clearly fluctuations in the loading on the individual blades that cause a fluctuation in the deflection of the blades as it rotates. These fluctuations cause a 'pulsing' in the thrust transferred to the drive shaft, as well as a resultant bending moment. The consequences are likely to be a reduced life expectancy before maintenance would be required. If the results of a wave were superimposed onto the loads, then the blade roots and drive shaft will be subjected to large fluctuations in loads and bending moments.

32.7 Conclusions

This chapter has discussed the influence of waves on a tidal turbine's characteristics. What is clear is that the waves change the time-averaged values of power and thrust. However, the magnitudes and period of the changes are most likely dependent on the wave definition, that is, wavelength, amplitude and celerity.

The amplitude of the fluctuations in both the power and thrust become exaggerated with waves as the axial flow velocity is influenced by the wave's sub-surface rotational component.

There is clearly a need to include waves in the modelling of marine devices, provided the devices are within a water column depth, to the surface, of half the wavelength of a wave.

The results from the two-way coupled FSI models have shown no significant differences from the results from uncoupled CFD model. This is primarily due to the inflexibility of the blade material.

Acknowledgments The authors would like to express their gratitude for the support of Engineering and Physical Sciences Research Council (EPSRC), Welsh European Funding Office (WEFO), High Performance Computing (HPC) Wales, Fujitsu and Mabey Bridge Ltd. for their funding of the work. The support of ANSYS is also gratefully acknowledged.

References

1. European Union Committee (2008) 27th report of session 2007–08—the EU's target for renewable energy: 20% by 2020. The Stationery Office Limited, London
2. DECC (2012) UK renewable energy roadmap update 2012. Crown copyright, Department of Energy & Climate Change, London

3. Carbon Trust (2011) Accelerating marine energy: the potential for cost reduction: insights from the carbon trust marine energy accelerator
4. de Laleu V (2009) La Rance tidal power plant. 40-year operation feedback—lessons learnt BHA annual conference. www.british-hydro.org/downloads/La%20Rance-BHA-Oct%20 2009.pdf. Accessed July 2014
5. Tidal Energy Limited. www.tidalenergyltd.com. Accessed July 2014
6. Seagen Wales. http://seagenwales.co.uk/description.php. Accessed July 2014
7. Royal Haskoning (2011) SeaGen environmental monitoring programme final report. http://www.marineturbines.com/sites/default/files/SeaGen-Environmental-Monitoring-Programme-Final-Report.pdf. Accessed 2 March 2013
8. Mason-Jones A, Evans PS, O'Doherty T, O'Doherty DM (2008) Characterisation of a tidal stream turbine design using CFD and ADCP. World Renewable Energy Conference, Glasgow
9. Morris CE (2014) Influence of solidity on the performance, swirl characteristics, wake recovery and blade deflection of a horizontal axis tidal turbine. PhD Thesis. http://orca.cf.ac.uk/60952. Cardiff University
10. Pinet PR (2009) Invitation to oceanography, 5th ed. Jones & Bartlett Publishers, p. 237. ISBN 0-7637-5993–3
11. Lewis M, Neill SP, Hashemi MR (2014) Waves, wave direction and the tidal stream energy resource. TOS/ASLO/AGU Ocean sciences meeting, Honolulu, Hawaii, 23–28 Feb 2014. http://www.eposters.net/pdfs/waves-wave-direction-and-the-tidal-stream-energy-resource.pdf. Accessed July 2014
12. Nicholls-Lee RF, Turnock SR, Boyd SW (2011) A method for analysing FSI on a HATT. 9th European wave and tidal energy conference, Southampton, 5–9 Sept 2011
13. Park SW, Park S, Rhee SH (2013) Performance predictions of a horizontal axis tidal stream turbine considering the effects of blade deformation. 3rd International symposium on marine propulsors, Launceston, 5–8 May 2013
14. Mason-Jones A, O'Doherty DM, Morris CE, O'Doherty T (2012) Influence of a velocity profile & support structure on tidal stream turbine performance. Renew Energy. doi:10.1016/j.renene.2012.10.022
15. Frost C, Morris CE, Mason-Jones A, O'Doherty DM, O'Doherty T (2014) Effects of tidal directionality on tidal turbine characteristics. [Under review J of Renew Energy]
16. Mason-Jones A, O'Doherty DM, Morris CE, O'Doherty T, Byrne CB, Prickett PW, Grosvenor RI, Owen I, Tedds S, Poole RJ (2012) Non-dimensional scaling of tidal stream turbines. Energy. doi:10.1016/j.energy.2012.05.0102012
17. Tedds SC, Poole RJ, Owen I, Najafian G, Mason-Jones A, Morris CE, O'Doherty T, O'Doherty DM (2011) Experimental investigation of horizontal axis tidal stream turbines. 9th EWTEC, Southampton
18. Le Méhauté B (1976) An introduction to hydrodynamics and water waves. Springer-Verlag, New York
19 American Petroleum Institute (2000) Recommended practice for planning, designing and constructing fixed offshore platforms API-RP2A, 21st ed. API, Washington, D.C.
20. United States Army Corps of Engineers (1984) Shore protection manual. Department of the Army, Mississipi
21. Pedlosky J (2003) Waves in the ocean and atmosphere—introduction to wave dynamics. Springer-Verlag, New York

Chapter 33
Experimental Validation of Gap Leakage Flow Models in Archimedes Screw Generators

Andrew Kozyn and William D. Lubitz

Abstract Archimedes screw generators (ASGs) are a form of microhydro power generation that is increasingly being adopted as an alternative renewable energy source. ASGs operate efficiently, even as head approaches zero. A small gap must always exist between the trough and edge of the rotating screw flights to allow screw rotation. This leads to two categories of flow within the screw: the primary flow remains between the helical flights and causes screw rotation, while a secondary leakage flow occurs through the gaps at the screw edges. A high gap leakage flow reduces efficiency because this flow effectively bypasses the working parts of the screw and is lost. An accurate gap leakage model is an essential component of any ASG design model. Current gap leakage models are based on experience with Archimedes screw pumps or are derived from models assuming quasi-static flow through an opening. This experimental study measured the actual leakage flow in an operating laboratory-scale ASG. The fill heights within the buckets were measured using high-speed photography and used to determine bucket volume and therefore non-gap flow. The gap flow is determined based on the difference between the measured total flow through the system and the computed non-gap flow based on the measured fill height and screw speed. The uncertainty of the gap flow is relatively high, since it is calculated based on the difference between two variables with similar magnitude, one of which is calculated based on other measurements. The existing gap flow models did not accurately predict the experimental results. Gap flow was found to be dependent on screw rotation speed, with highest gap flows occurring at intermediate screw rotation speeds. Gap flow reduces to zero as screw rotation speed approaches the maximum possible speed. There is significant uncertainty in the existing gap flow models, and gap flow models constitute one of the largest sources of uncertainty when modeling the performance of ASGs.

Keywords Archimedes screw · Gap flow · Microhydro · Low head

W. D. Lubitz (✉) · A. Kozyn
School of Engineering, University of Guelph, Guelph, ON, Canada
e-mail: wlubitz@uoguelph.ca

A. Sayigh (ed.), *Renewable Energy in the Service of Mankind Vol I*,
DOI 10.1007/978-3-319-17777-9_33

33.1 Introduction

An Archimedes screw consists of several helical planes fastened to a central cylinder that runs the length of the screw. Archimedes screw pumps have been utilized since antiquity for pumping large volumes of water over relatively low heads [1]. Archimedes screw pumps can also be operated in reverse, using the screw to extract energy from a flow of water. Water introduced at the top of the screw causes the screw to turn as it flows through the screw. Electricity can be generated by adding a gearbox and a generator to the screw shaft. This is the basis of the Archimedes screw generator (ASG), a relatively recent microhydro power generator that converts available energy in flowing water to electrical energy. ASGs are efficient even at low heads, have minimal impact on fish, and can often be retrofit into existing dams, mill structures, and other sites. Brada [2] installed the first ASG two decades ago. Since then, hundreds of ASGs have been installed in Europe [3], although the technology is still little used elsewhere.

ASGs are external flow devices. When flowing water is introduced to the screw inlet, the water is trapped between the helical planes forming volumes of water called “buckets.” Quasi-static water pressure differences on opposite sides of the helical planes create a net torque that causes the screw to rotate. A small gap must exist between the screw and the fixed circular trough in which it rotates to allow unimpeded screw rotation. This creates two distinct flow regimes: a primary non-gap flow consisting of the bulk of the water that remains between the helical planes and a secondary leakage flow through the gaps between the screw and trough. The non-gap flow does work on the screw, while the leakage flow represents system energy losses. The leakage flow through the gap is difficult to quantify because it represents only a small portion of the total flow. The gap flow can only be directly measured in a stopped screw because then all flow through the screw is by gap flow as long as the water level within the screw is not high enough to cause overflowing over the top of the central cylinder (Fig. 33.1).

There are two types of gap flow models available in the literature. Nagel [4] presents an empirical formula for gap flow based on experience with ASG:

$$Q_{IN} = 2.5 \cdot G_w \cdot D\sqrt{D} \tag{33.1}$$

where G_w is the gap width and D is the diameter, both in meters.

Gap flow is returned in units of cubic meters per second. It is assumed that the screw is operating in a full condition at conventional rotation speed.

Gap flow has also been modeled by assuming quasi-static conditions within the screw and assuming that the gap flow is driven by the static pressure due to the difference in water depth between the two buckets adjacent to the gap. Lubitz [5] presented the following model based on this assumption, although it should be noted that this model is functionally almost identical to the model that Neurnbergk and

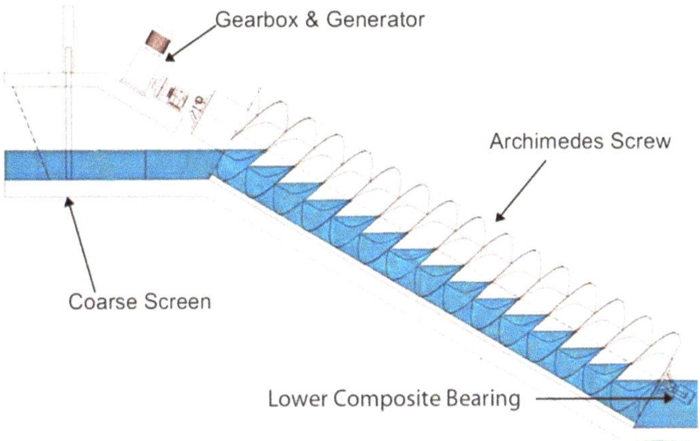

Fig. 33.1 A three flight Archimedes screw generator

Rorres [6] report was developed by Muysken in the 1930s. In this approach, leakage flow (Q_{1L}) is modeled as

$$Q_{lL} = CG_w\left(l_w + \frac{l_e}{1.5}\right)\sqrt{\frac{2gP}{N}\sin\beta} \tag{33.2}$$

where N is the number of screw flights (typically 3 or 4), P is the screw pitch, g is the gravitational constant 9.81 m/s^2 β is the screw slope angle.

The values of l_w and l_e are the wetted lengths of the gap that are submerged on both sides, and only on the upstream side, respectively (Fig. 33.2). The value of C is one if there are no minor losses in the gap, otherwise it will be less than one. A

Fig. 33.2 Lubitz model gap length definitions

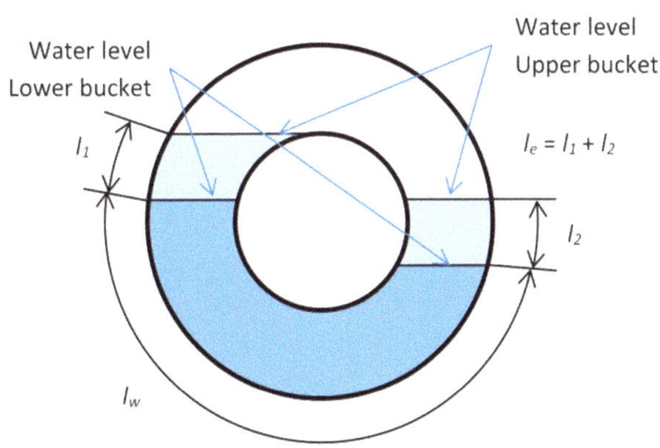

Table 33.1 Laboratory ASG dimensions

Parameter	Measurement
Inner diameter (mm)	76.26 ± 0.97
Outer diameter (mm)	144.36 ± 0.14
Screw length (mm)	584 ± 3
Angle (degree)	24.8 ± 0.3
Pitch (mm)	144.61 ± 3.67
Gap width (mm)	1.11 ± 0.15
Casing thickness (mm)	3.43 ± 0.28
Head (m)	0.24

value of $C = 0.9$ was determined from static leakage flow measurements in a small laboratory-scale Archimedes screw.

The goal of this work was to experimentally determine the leakage flow through the gap of a laboratory-scale ASG and compare the measured leakage flow to leakage predicted by the available models. The authors are not aware of prior work experimentally validating leakage models in an operating screw.

33.2 Methods

Experiments were conducted using a laboratory-scale Archimedes screw system at the University of Guelph (Table 33.1). The screw trough was a transparent acrylic tube which allowed direct observation of the internal fill heights (Fig. 33.3). The ASG was attached to two basins: an upstream basin which supplied water to the inlet and a downstream basin that collected the discharge. The downstream basin flow as routed through the flow sensor. Flow rates were regulated using an overflow weir

Fig. 33.3 Laboratory ASG

on the back of the upstream basin; raising the weir height increased the water height in the upstream basin, thereby increasing the supplied flow rate. An electric pump returned water from a lower receiving basin to the upper basin. ASG rotational speed was regulated using an adjustable Prony brake attached to the rotating shaft.

33.2.1 Experimental Procedure

Distinguishing between the primary flow and the secondary leakage flow is difficult. Leakage occurs through a gap width of approximately 1.11 mm, while the screw is rotating between 0 and 30 rad/s. The gaps themselves are constantly translating in space relative to both the entrained water and the screw casing. Since direct measurement of the flow through the gap in a turning screw is not feasible, a novel approach was taken that involved determining the gap flow based on measurements of fill height, screw rotation speed, and knowledge of the screw geometry. The fill height *(f)* is defined as the ratio of the water level in a bucket, relative to the maximum depth that can occur before the buckets overflow (Fig. 33.4). Since the geometry of the screw is well-defined, the volume of water entrapped between the planes can be calculated by determining the volume of the region in the bucket beneath the observed fill level. This was done by numerical integration [7]. Once the volume of water in a bucket is known, the volumetric flow rate, an ideal version of the screw without a gap can be determined using the measured rotation speed of the screw. Finally, the leakage flow (Q_l) is the difference between the ideal non-gap (Q_{ng}) flow and the total measured flow through the screw *(Q)*.

Three distinct runs were conducted, each corresponding to a different total flow rate through the screw. Each run consisted of approximately 60 observations distributed across the range of achievable rotation speeds controlled by adjusting the Prony brake. The rotation speed, total flow rate through the screw, and fill height were measured and averaged over 60-s time periods.

The primary total flow was measured passing the flow exiting the downstream basin through a vertical cylinder with a constricted outlet. Water depth in the cylinder was observed using a transparent sighting tube. A calibration curve relating

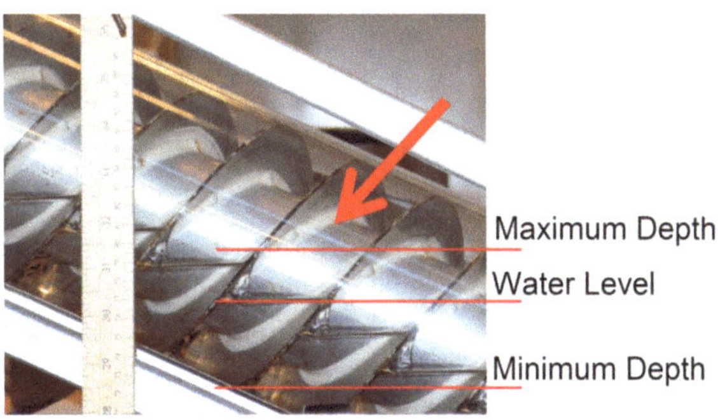

Fig. 33.4 Fill height depths, for the bucket indicated by the arrow

volume flow rate to observed water depth in the cylinder was derived and used to determine flow rate.

33.2.2 Fill Height Measurements

Fill height measurements were made using high-speed photography. A Nikon D3200 camera with a remote shutter release was used to take four pictures per run at intervals of 15 s. External lighting and shutter speeds between 1/800 and 1/1400 s were used to reduce motion blur. Image resolution was 4512 × 3000 pixels. The camera was positioned approximately 1.5 m from the screw, and maximum optical zoom was used to reduce image distortion due to lens curvature. Test images taken both with and without the acrylic casing determined that optical distortion effects due to the acrylic tube itself were negligible. Straight-edged rulers were leveled and fixed to the screw frame to include scale in the images. Observing the bottom tip of the helical planes (i.e., the minimum depth point) within the photos was difficult due to distortion effects from the water so instead measurements were made from the bottom of the acrylic casing. The thickness of the acrylic casing and the gap width were used to adjust the measurements to determine the actual bottommost point of the bucket interior. The maximum water level was impossible to see due to the curvature of the helical planes. Therefore, this distance was determined by extrapolating the screw geometry. The fill heights themselves were extracted from the photos (e.g., Fig. 33.4) using the open-source plot digitization software, Plot Digitizer.

Screw rotation speed was measured using a recessed magnetic switch fixed to a position above the ASG inner cylinder which registered each passage of a small, rare earth magnet attached to the rotating cylinder. The switch signal was digitally recorded using a National Instruments USB-6009 DAQ at a sampling frequency of 1000 Hz with custom acquisition software written in LabVIEW. Rotation speed was determined from the time between recorded pulses. Speed measurements made with the magnetic switch were validated against measurements made with a handheld optical tachometer.

Due to the inherently noisy nature of the leakage measurements, both the total flow and the modeled idealized non-gap flow were fitted with polynomials to provide a smoothed relationship between rotation speed and flow. The leakage flow rate was calculated by taking the difference between the polynomials.

33.2.3 Uncertainty Analysis

Uncertainty analyses were performed on all measurements to assess the uncertainty of the obtained leakage values. Uncertainties of all physical dimensions and sensor measurements were determined (Table 33.1). Gap width was determined by measuring tube and screw diameters using a Mitutoyo Digimatic 192-830-10 HD 12 AX digital height meter. A sensitivity and uncertainty analysis was performed on the volume calculation done with the [5] model to quantify uncertainty associated with computing bucket volume from fill height. Uncertainties were propagated through flow calculations to determine the uncertainty of the gap leakage flow measurements. The uncertainty of the flow sensor was determined directly from the flow sensor calibration equation through error propagation and found to average approximately 5.7 % of the overall measurement.

The output data from the magnetic switch used to measure the rotational speed was recorded at 1000 Hz. This measurement frequency far exceeded the actual rotational speed of the ASG screw itself, and the signal is a digital state not subject to noise. The rotational speed uncertainty was approximately ± 2 revolutions to account for the possibility of the recording interval starting and ending between signals.

The fill height measurement had the highest level of noise and most uncertainty. The uncertainties of the individual bucket fill height measurements are relatively low (less than a few percent). However, it was found that fill heights varied widely across all the buckets during a 60 s measurement period. Since the measurements were averaged over this period, it was deemed that uncertainty in the individual measurements would insufficiently underestimate the actual uncertainty of the fill height measurements. Similarly, using the maximum deviation from the measurement mean would overestimate the uncertainty because there were often single bucket measurements that deviated from the mean considerably. Therefore, the uncertainties were taken as the standard deviation of the individual fill heights for a given measurement.

There are two primary sources of uncertainty associated with the Lubitz et al. (2014) numerical model, which is a numerical computer model: error in the numerical results and error due to the sensitivity of the model itself to its input parameters. Convergence tests verified that numerical uncertainty is less than 0.1 %. Sensitivity tests were conducted to determine the uncertainty of outputs relative to inputs. This knowledge was used to propagate errors during the computation of bucket water volume based on measured fill level and rotation speed.

33.3 Results and Discussion

Figures 33.5a, b, c show the observed total flow (Q) and non-gap flow (Q_{ng}) for the three experimental runs. There is reduced flow through the screw at lower speeds during all three tests. Water entering the upstream basin has two possible paths: it can either flow through the screw or be diverted over a flow control weir situated at the rear of the basin. The amount of water that the screw can draw depends on its rotational velocity. Increasing the screw rotational velocity increases the flow through the screw. Therefore, at lower rotational speeds, more water is diverted over the back weir, causing a decrease in the total flow through the screw. This effect is seen in all three experimental runs. Fill height also decreases with speed, as the capacity of the screw increases faster than the rate at which water will enter the screw for a

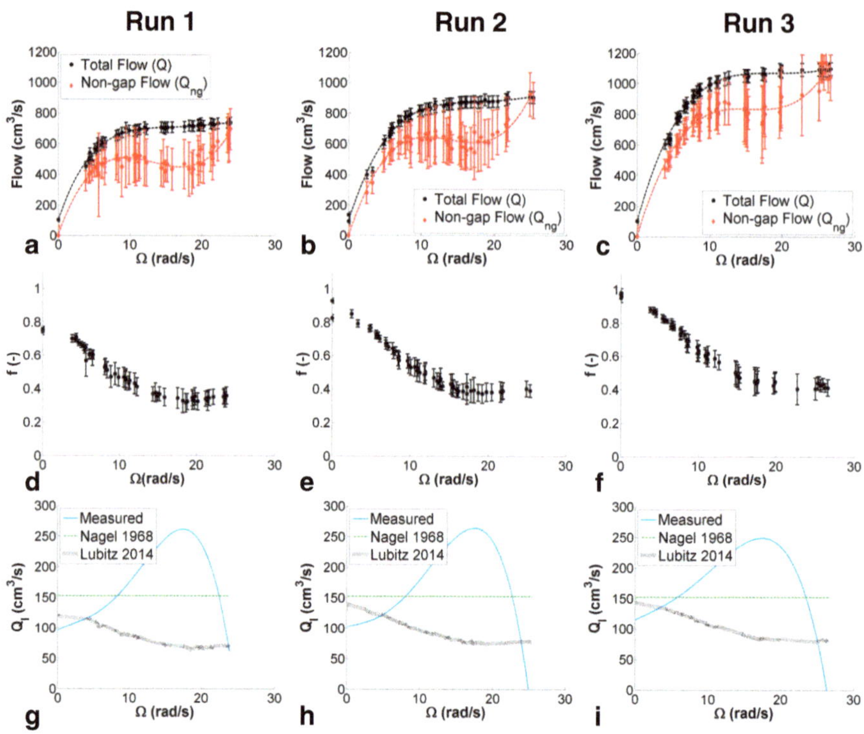

Fig. 33.5 Measured total flow and non-gap flow for **a** Run 1 **b** Run 2 **c** Run 3. Fill heights **d** Run 1 **e** Run 2 **f** Run 3. Comparison of measured leakage to Nagel and Lubitz et al. (2014) for **g** Run 1 **h** Run 2 **i** Run 3

given upstream head. At an angular velocity of 0 rad/s (i.e., screw stopped), the flow is entirely leakage flow. The maximum flow rates for experimental runs 1, 2, and 3 were 735, 906, and 1097 cm^3/s, respectively, corresponding to the maximum, or freewheeling, speed at which the screw was subjected to no braking. The non-gap (Q_{ng}) flows for each of the experimental runs are zero when the angular velocity (Ω) is zero. The non-gap flow also converges to the total flow as rotation speed approaches the freewheeling speed.

Figures 33.5d, e, f show fill height as a function of rotational speed. The fill heights can be seen decreasing with increasing angular velocity. This suggests that as the rotation speed increases and the fill heights decrease, the disparity between the non-gap flow, and the total flow is eliminated. Perhaps the most interesting result is that the non-gap flow for all runs can be seen to locally decrease at intermediate rotation speeds between 10–20 rad/s.

The measured leakage was found by fitting polynomials to the observations of flow as a function of speed for each test (Table 33.2) and then taking the difference between the fits for total flow and non-gap flow. Figures 33.5g, h, i show the ob-

Table 33.2 6th order polynomial fits for flow, Q (cm^3/s) vs. angular velocity, Ω (rad/s), for both the total and non-gap flows

Run	Polynomial
Run 1—total flow	$Q=(1.709e-5)\Omega6-0.001237\Omega5+0.02641\Omega4+0.07763\Omega3-10.21\Omega2+135.2\Omega+114.8$
Run 1—non-gap flow	$Q=(-6.973e-5)\Omega6+0.004086\Omega5-0.08554\Omega4+1.069\Omega3-14.41\Omega2+139.2\Omega+0.03932$
Run 2—total flow	$Q=(4.056e-5)\Omega6-0.003571\Omega5+0.1138\Omega4-1.436\Omega3+1.048\Omega2+124.8\Omega+90.04$
Run 2—non-gap flow	$Q=(8.020e-6)\Omega6-0.001671\Omega5+0.08217\Omega4-1.324\Omega3+1.560\Omega2+115.7\Omega+0.1914$
Run 3—total flow	$Q=(7.123e-5)\Omega6-0.006357\Omega5+0.2115\Omega4-3.070\Omega3+13.35\Omega2+107.2\Omega+114.9$
Run 3—non-gap flow	$Q=(7.385e-5)\Omega6-0.006863\Omega5+0.2399\Omega4-3.657\Omega3+18.06\Omega2+87.81\Omega+0.6349$

served leakage flow rates compared to both the leakage flow rates predicted by the Nagel [4] and Lubitz [5] leakage models for the conditions of each test run.

Figures 33.5g, h, i show the measured leakage increased with increasing angular velocity, peaked at approximately 16–18 rad/s, and then declined towards zero as speed approached the freewheeling speed. Interestingly, leakage magnitude is similar for all runs despite the varying total flow rates. Peak leakage occurs at approximately the same angular velocity (16–18 rad/s) in all runs despite the differing flow regimes. This result was not expected. The experimental process was carefully reviewed, and the authors do not believe it is an experimental error. This peak leakage is approximately 260 cm^3/s. At high speeds, leakage decreases significantly as the non-gap flow converges to the total flow.

Significant differences occurred between the measured leakage and the model predictions (Figs. 33.5g, h, i). Since Nagel's model [4] estimates leakage based solely on the diameter and gap width of the screw, it predicts a constant leakage value regardless of the flow conditions. Despite this, Nagel's leakage prediction provides a reasonable estimate of the typical leakages this laboratory screw experienced, particularly in the speed range coinciding with highest efficiency (which is 5–10 rad/s). This is also notable since the intended use of the Nagel model is for larger, full-scale Archimedes screw pumps.

The Lubitz model [5] predicts varying leakage as a function of speed, but these variations differ markedly from the measured gap flows. The Lubitz model predicts reduced leakage as fill height decreases (which coincides with higher speeds). This model does not predict the increased leakage observed at moderate speeds. A possible explanation is that at low angular rotation speeds, the static pressure at the gap may be the dominant factor driving leakage flow, but as rotational speed increases, dynamic effects due to the rotational motion of the screw itself may be playing a larger role in causing leakage. Future leakage models may have to take into account more dynamic effects to properly model ASG leakage.

33.3.1 Uncertainty

The average percentage error in total flow measurements from the flow sensor was 5.7 %. The maximum error was 22 %. The uncertainty in the non-gap flow was considerably higher, due to accumulating uncertainty from multiple inputs and calculation steps. Uncertainty in the non-gap flow was most sensitive to uncertainty associated with the fill height and angular velocity inputs. The average fill height uncertainties were approximately 7.9 % of the actual measurements, while the average angular velocity measurement uncertainty was approximately 1.1 % of the actual measurements.

The fill height uncertainty was high primarily due to the flow patterns, rather than sensor uncertainty. In practice, there is some variation in the gap width at different locations due to manufacturing tolerances, which results in varying fill height. There were noticeable differences between the fill heights in adjacent buckets even within single runs, particularly at high flows. An average fill height value was utilized for model input. It was determined that the standard deviation of the observed fill heights was a reasonable approximation for the overall uncertainty in the measurements. The average and maximum non-gap flow uncertainties, relative to the actual measurements, were approximately 17.2 and 69.3 %, respectively.

Large uncertainties in the non-gap flow measurements also resulted in high uncertainty in the resulting leakage measurements. The average uncertainty for the leakage measurements was approximately 95 % of the actual measurements. This was expected considering that the leakage flow is a small fraction of the total flow and is determined by taking the difference of two similar magnitude numbers [8]. Nonetheless, gap flow results with consistent characteristics were observed, across a range of varying flow conditions. Figure 33.6 shows the leakage measurements for all three experimental runs with associated uncertainties.

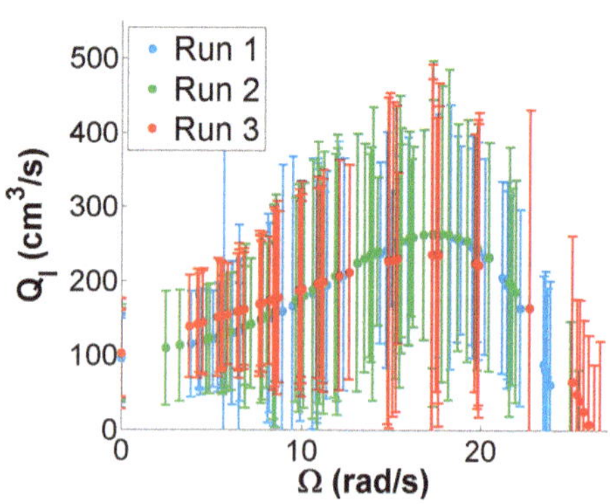

Fig. 33.6 Measured leakage and associated uncertainties for all experimental runs

33.4 Conclusion

The results presented here are the first measurements of gap flow in non-static Archimedes screws known to the authors. Predictions from two current gap flow models were compared to the observed gap flows. Gap flow in Archimedes screws was observed to approach zero as the screw approached its maximum rotation speed and reach a maximum at intermediate rotation speeds. This pattern was true across the range of volume flow rates and fill levels tested. Neither of these characteristics was predicted by either of the two gap flow models. While the current models provided reasonable predictions of gap flow at the lower operating speeds where Archimedes screws are most efficient and operated in practice, there is still significant uncertainty in the current models. Since accurate predictions of gap flow are essential to any overall performance model for an Archimedes screw, these results suggest additional study of gap flow is needed, as well as improved gap flow prediction models.

Acknowledgments This work was conducted as part of a larger research project with Greenbug Energy Inc. (Delhi, Ontario, Canada), with financial support from the Natural Science and Engineering Research Council (NSERC) Collaborative Research and Development (CRD) program. The authors also thank Sylvia Nguyen, Nicole Serrao and Alvaro Ramos de Figueiredo for their assistance with the experiments.

References

1. Rorres C (2000) The turn of the screw: optimal design of an Archimedes screw. J Hydraul Eng 126:72–80
2. Brada K (1999) Wasserkraftschnecke ermöglicht Stromerzeugung über Kleinkraftwerke [Hydraulic screw generates electricity from micro hydropower stations]. Maschinenmarkt Würzburg 14:52–56
3. Lashofer A, Hawle W, Pelikan B (2012) State of technology and design guidelines for the Archimedes screw turbine. Hydro 2012, Bilbao
4. Nagel G (1968) Archimedean screw pump handbook. RITZ Pumpenfabrik OHG, Schwabisch Gmund
5. Lubitz WD (2014 Jun) Gap flow in archimedes screws. CSME Intl. Congress, Toronto
6. Nuernbergk DM, Rorres C (2012) Analytical model for water inflow of an archimedes screw used in hydropower generation. J Hydraul Eng 139(2):213–220
7. Lubitz WD, Lyons M, Simmons S (2014) A performance model of archimedes screw hydro turbines with variable fill level. J Hydraul Eng (in press)
8. Taylor JR (1982) An introduction to error analysis: the study of uncertainties in physical measurements, 2nd edn. University Science Books, Sausalito

Part V
Low-Energy Architecture

Chapter 34
Double- or Single-Skin Façades for Low-Carbon Office Refurbishments in the UK: A Comparative Case Study

Francesco Pomponi and Poorang A.E. Piroozfar

Abstract In the UK, 75 % of non-domestic buildings will still exist in 2050, and the reduction of their emissions represents a major challenge for the UK government to meet the greenhouse gases (GHG) target. There is a growing tendency towards the use of double-skin façades (DSFs) for office refurbishments in Europe. In renovations, a DSF consists of an external glazed skin added in front of the primary building envelope and separated from it by an air cavity. DSFs have the potential to act as a thermal buffer in winter and enhance the performance of natural ventilation in summer, thus potentially reducing the energy consumption for heating and cooling in buildings.

Limited knowledge exists about DSFs for office refurbishments and, more specifically, the literature lacks comparisons between this technology and high-spec single-skin refurbishments. A case study has therefore been selected to address this gap and compare the performances of the two technologies. This chapter reports findings from the study of a cellular office building of the University of Brighton that is currently undergoing a major refurbishment. The case study is analysed with three scenarios: the building as it was, as it will be once the refurbishment is completed and as it could have been with a DSF. Detailed modelling and related energy figures are achieved by means of building energy simulation tools and compared against energy benchmarks.

Results show that DSFs are a viable option for refurbishing existing office buildings. The reduction of energy consumption with the double skin outweighs that achieved with the single-skin refurbishment. Additionally, comfort analyses for all the scenarios considered showed that the risk of overheating with the double-skin façade could be avoided by the use of shading devices to make sure that the energy reduction is not at the expense of the building users' comfort. This study contributes towards deepening and broadening the knowledge about performances of DSFs within a refurbishment context and their potential advantages over high-spec single-skin alternatives.

F. Pomponi (✉) · P. A. E. Piroozfar
Division of Built Environment and Civil Engineering, School of Environment and Technology, University of Brighton, Brighton, UK
e-mail: f.pomponi@brighton.ac.uk

A. Sayigh (ed.), *Renewable Energy in the Service of Mankind Vol I*,
DOI 10.1007/978-3-319-17777-9_34

Keywords Low-carbon refurbishments · Double-skin façade · Natural ventilation · Office buildings

34.1 Introduction

The building sector is responsible for about 40 % of the world energy consumption and its related greenhouse gases (GHG) emissions [1] and, given the vintage nature of the European building stock, the major challenge for the future has been indicated as 'to promote the sustainable refurbishment of that consolidated stock' [2, pp. 1454]. This is particularly true for the UK where 75 % of existing non-domestic buildings will still be in service in 2050, and the reduction of their emissions represents a major challenge for the UK government to meet its legally binding GHG target. Given such a context, deploying low-carbon refurbishment strategies with passive design principles can contribute to cutting back on energy consumption and GHG emissions as a resolute and concerted effort without which carbon dioxide emissions related to worldwide energy consumption will double by 2050 [3].

In this respect, naturally ventilated glazed double-skin façades (DSFs) could play an important role as a form of low-energy architectural technology since they rely only on wind and sun to exploit their potential. Moreover, DSFs are considered as one of the best options among façade technologies to reduce energy consumption and mitigate GHG emissions from the demand-side, while helping manage the efficient interaction between the outdoor and the indoor conditions [4]. A DSF consists of an external additional glazed skin installed in front of the primary building façade from which it is separated by an air cavity that acts as a ventilation channel. A review of published studies on DSF performances in moderate climates shows that this technology is capable of offering 30–60 % reduction in operational energy [5]. Furthermore, DSFs have also other benefits such as providing a thermal buffer zone, allowing for natural ventilation, solar preheating of indoor air ventilation, night cooling, protection against sound, wind and pollution (even with open windows), weather protection for shading devices, a higher degree of daylighting and an improved aesthetic of building, just to name a few [6–11]. Naturally, DSFs have some disadvantages too, the principal being the risk of overheating and a higher cost—although, the higher investment is believed to be justified by energy savings and occupants' improved comfort, hence increased productivity [12]. However, limited knowledge exists about DSFs for office refurbishments and, more specifically, the literature lacks comparisons between this technology and high-spec single-skin refurbishments.

A case study has, therefore, been developed to address this gap and to compare the performance of the two technologies. This chapter reports on findings of the study of a cellular office building of the University of Brighton that is currently undergoing a major refurbishment. The case study is analysed in three scenarios: the building as it was, pre-refurbishment; as it will be, post-refurbishment and as it could have been with a DSF. The aim of this study is to answer the following research question: Are DSFs a viable alternative option to pursue effective low-carbon refurbishments compared to advanced single-skin solutions?

34.2 Material and Methods

University of Brighton's Cockcroft building shown, with both single and double skin, in Fig. 34.1 has a cellular office layout, thus representing an excellent case study as this is the most common office type in England [13]. The building is also naturally ventilated, as the majority of existing offices in the UK are [14]. Additionally, it was constructed during the 1960s, and, in that sense, it also represents the majority of UK non-domestic refurbishments projects [14]. Its main orientation is northeast–southwest (NE–SW), and it has 10 floors, two of which (ground and mezzanine floors) have not been involved in the refurbishment and, hence, not considered in this study. However, their volume has been taken into account to correctly set other floors' heights.

Initially costing around £ 23 million, Cockcroft refurbishment is anecdotally the most expensive educational refurbishment in the UK. The main interventions in the building envelope consist of replacement of the old single glazing metal-framed windows with high-spec double glazing units (DGUs) with aluminium thermal break frames and adding of thermal insulation to the opaque parts of the façades internally. This was due to the particular geometry of the building, which incorporates projected concrete frames on the outside (Fig. 34.2). Furthermore, the roof has been reinsulated to improve the thermal performance of the building.

The major refurbishment also incorporates a new mechanical cooling system to provide comfortable summer indoor conditions. Had an appropriate DSF been selected, this objective could have theoretically been achieved through natural ventilation. Therefore, this comparative study is particularly appropriate given the recommendation to prefer, wherever possible, natural ventilation strategies over mechanical systems [14]. Additionally, mechanical cooling is energy intensive and also incorporates high-embodied energy and involves tremendous efforts; what can best justify the fact that a vast majority of existing the UK offices is still fully naturally ventilated [14]. Such an effort is likely to turn into heavy disruption to the building use and for its occupants.

Fig. 34.1 Aerial views of the Cockcroft building model in its single- *(left)* and double-skin *(right)* versions superimposed on its actual location in Google earth©

Fig. 34.2 Details of the additional internal insulation *(left)* and the external concrete frames on the south façade *(right)*

Detailed floor plans have been acquired for both layouts, the building pre- and post-refurbishment, and the models used for simulation have been generated based on the accurate drawings reproduced accordingly. In other words, there has been no simplification of the existing and layouts-to-be of the building. It is worth mentioning that the refurbished version of the building shows the presence of solar corridors on the laboratory and teaching floors. The internal layout of the DSF version of the building is identical to that of the post-refurbishment version. It should be noted that this is what was intended in the client and design briefs, but not necessarily what works to best utilise the potential benefits that a DSF has to offer for this building.

Due to the Cockcroft building characteristics, naturally ventilated with windows operable individually, and each linked to an occupied indoor space, the 'multistorey' DSF was the only practically viable choice. Specifically, it comprises a vertically and horizontally continuous cavity, which extends over the whole façade. Additionally, a narrow cavity could have not been installed due to the projected concrete frames, hence a wide cavity of 1 m has been realised. Finally, as per practice in the construction sector [15] and literature [8], the cavity extends by a height of 1.5 m over the roof level of the building to avoid, or at least, minimise the risk of overheating on the upper floors.

Energy modelling has been achieved by simulating the single and DSFs in Virtual Environment by Integrated Environmental Solutions (IESVE), a building energy software package which has previously been used in DSF studies [16, 17]. However, corrections are needed when a DSF is under study. Suggested corrections by the software vendor [18] have been applied to ensure the accuracy of the simulation. Finally, due to risk of overheating with DSF, summer thermal comfort has been assessed as per the recently published Technical Memorandum TM52 by Chartered Institution of Building Services Engineers (CIBSE) [19] which utilises an adaptive comfort approach suited to the free-running nature of the Cockcroft building. It is important to mention that TM52 is based on current European standards [20], but considers more stringent criteria than those currently in use.

34.3 Results and Discussion

As stated earlier, this research studied three different scenarios. However, preliminary assessment of the building incorporating a DSF showed serious overheating (44 %) of the total occupied rooms. Therefore, vertical shading devices were added. The most appropriate choice was light-coloured venetian blinds, placed closer to the inner layer of the outer skin as recommended in previous studies [21]. In IES, it is possible to model the shading devices to come down based on the value of the solar radiation. This aspect can be thought of as the presence of an automated system or the human action in response to heat or glare. In the specific case of the Cockcroft building, it has been found that 200 W/m^2 is the triggering value that minimises the number of uncomfortable rooms to only three. This value has also been found suitable to reduce and control glare in winter when the sun is low on the horizon.

As shown in Fig. 34.3, all three uncomfortable rooms are on the seventh floor (the last linked to the DSF), which for the benefit of clarity has been isolated from the rest.

The seventh floor is one of the two floors with a solar corridor, adopted in the single-skin refurbished version of the building to provide a buffer between the southern elevation and the teaching spaces and to reduce overheating and equalise internal temperatures [22]. When a DSF is in place, the buffer provided by the solar corridor adds on the buffer created by the DSF. While such a double buffer does not seem to create problems on the fourth floor (the other floor where a solar corridor is deployed), probably due to milder air temperatures in the cavity of the DSF, it overheats two rooms and the whole solar corridor of the seventh floor. These results are consistent with what is generally known about DSFs: shading devices are, in most cases, necessary, and overheating is likely to happen in the upper floors due to higher air temperatures in the cavity.

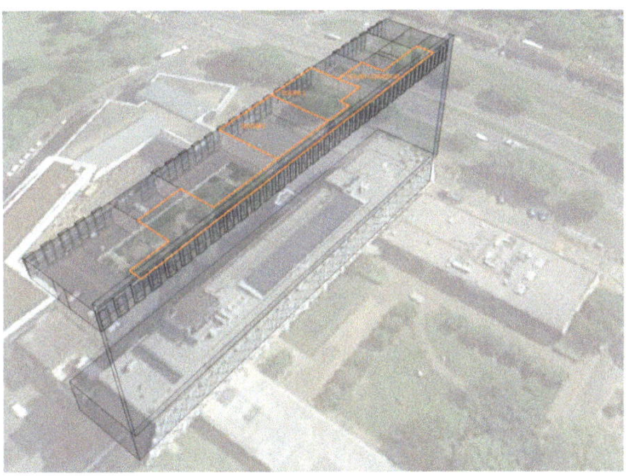

Fig. 34.3 Detail of the uncomfortable rooms

Notwithstanding the unsuitability of a DSF not equipped with shading devices in terms of internal comfort, energy results are interestingly diverse between the two DSF versions, and therefore both are reported in this study.

Figure 34.4 shows monthly data for energy consumption and carbon dioxide equivalent emissions for the four options assessed. It can be seen that DSFs have the lowest energy consumption and carbon emission values throughout the year. It is worth noting that heating of the internal spaces is not in place from May to September, as in reality. Additionally, the mechanical cooling of the post-refurbishment version of the Cockcroft building works on electricity and, therefore, there is a significant difference between energy consumption and carbon emissions, due to the higher carbon content of energy units coming from electricity from the grid. Conversion factors (kWh/kgCO_{2e} for both natural gas and electricity) used are given by [23]. These trade-offs are clearer if looked at from a cumulative point of view (Fig. 34.5).

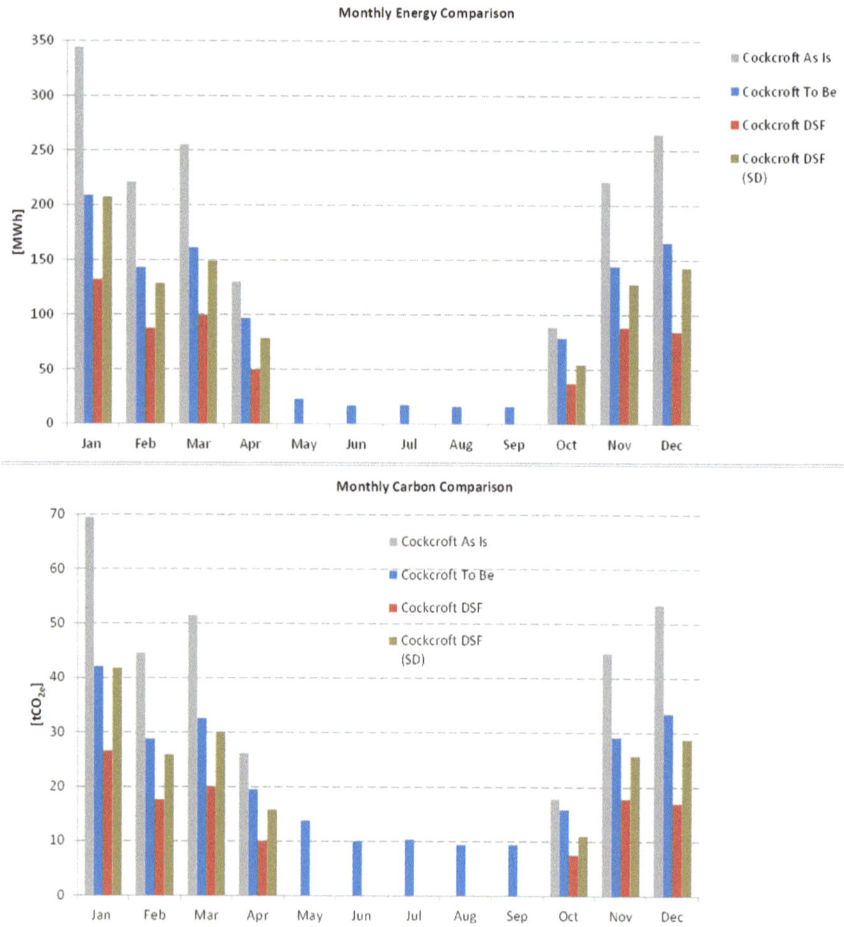

Fig. 34.4 Monthly energy and carbon comparison for the four options. *DSF* double-skin façades

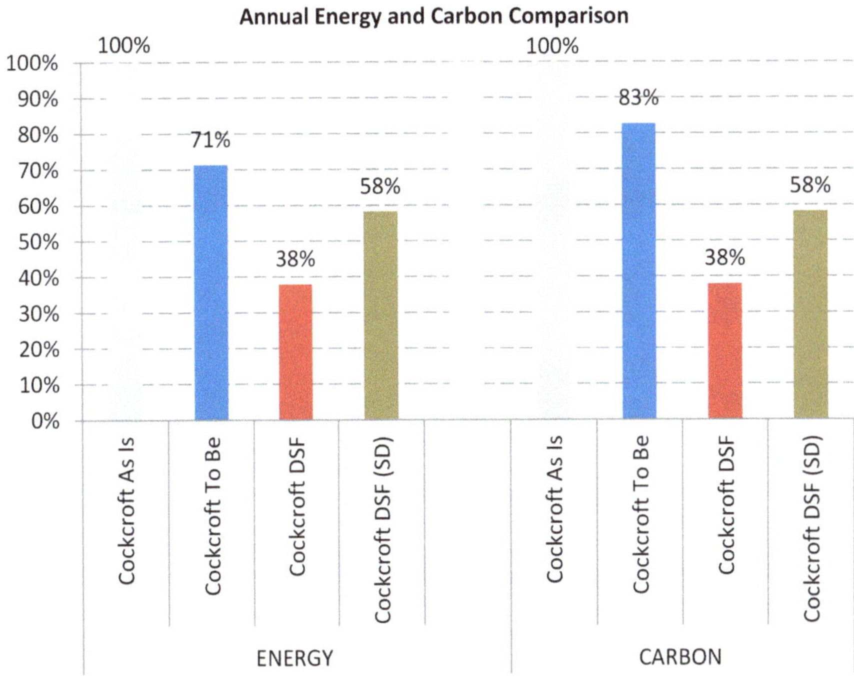

Fig. 34.5 Cumulative energy and carbon comparison (in percentage) of the four assessed options. *DSF* double-skin façades

Figure 34.5 shows the benefit of the single skin Cockcroft refurbishment towards both reducing energy consumption and cutting carbon dioxide emissions. However, while nearly a 30 % reduction is achievable towards energy, this drops to 'just' 17 % when converted into carbon due to the higher conversion factor of electricity. The DSF, instead, shows an impressive reduction potential of 42 % in both cases as its cooling strategy relies on the sole natural ventilation. Therefore, the energy saved to heat internal spaces is wholly converted into reduction in carbon emissions. It is worth noting the outstanding results that a DSF without shading devices could achieve. As said, such a choice would imply both glare and overheating issues and it is not therefore viable. However, showing such a significant difference compared to the version with the shading devices, it probably deserves further investigation to assess whether or not there are solutions with which the energy reduction potential can be obtained without affecting indoor comfort (i.e. different, high-tech glass technologies such as electro-chromic glazing).

Unfortunately, no monitoring of the building has been possible to assess—numerically—the accuracy of the model. However, not only IES VE is a validated software package but also results from the simulation are well in line with energy benchmarks available for the UK (Fig. 34.6).

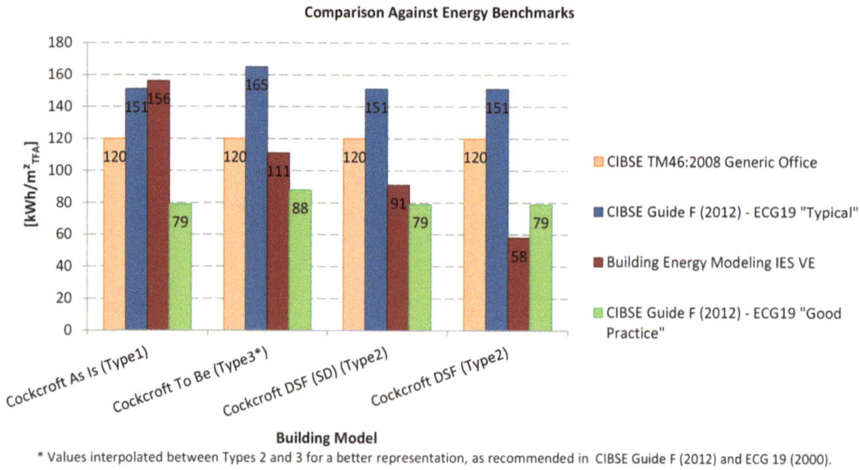

Fig. 34.6 Comparison against available energy benchmark. *DSF* double-skin façades

Figure 34.6 shows how the pre-refurbishment version of the Cockcroft building has basically the exact same value of energy benchmark for typical naturally ventilated offices. All the post-refurbishment versions show improvements, with the DSF with shading device version much closer to the 'good practice' benchmark, and the DSF without shading devices that actually goes even beyond that.

The reason why the DSF equipped with shading devices consume more energy may not be straightforward and this has been therefore investigated in more detail (Fig. 34.7).

As said, the blinds come down when solar radiation is equal to or higher than 200 W/m^2. This value prevents overheating in summer making most rooms complying with the TM52 comfort assessment and also eliminates or minimises glare in winter when the sun azimuth is low. Values of solar radiation above that threshold are frequent throughout both cold and mild season. This means the combination of a solar radiation value higher than 200 with low outside air temperature—a representative day in this respect is shown in Fig. 34.7. Solar radiation reaches 200 W/m^2 around 9.30 a.m. with the outside air temperature just above 0 °C. Since the solar radiation keeps increasing in that specific day, the shading devices stay lowered, and the air in the cavity is far colder compared to how it would be without the blinds. Such a difference peaks at a net value of 7.5 °C (16 °C without blinds, 8.5 °C with blinds). This means higher heat losses from the envelope of the building (inner skin) to the colder air in the cavity and, therefore, more energy is needed to keep the indoor temperature at its comfort level. This is confirmed in Fig. 34.7 by the boilers' energy for that specific day, which is identical as long as the solar radiation is below the triggering value for the shading devices while it starts diverging after the blinds have lowered.

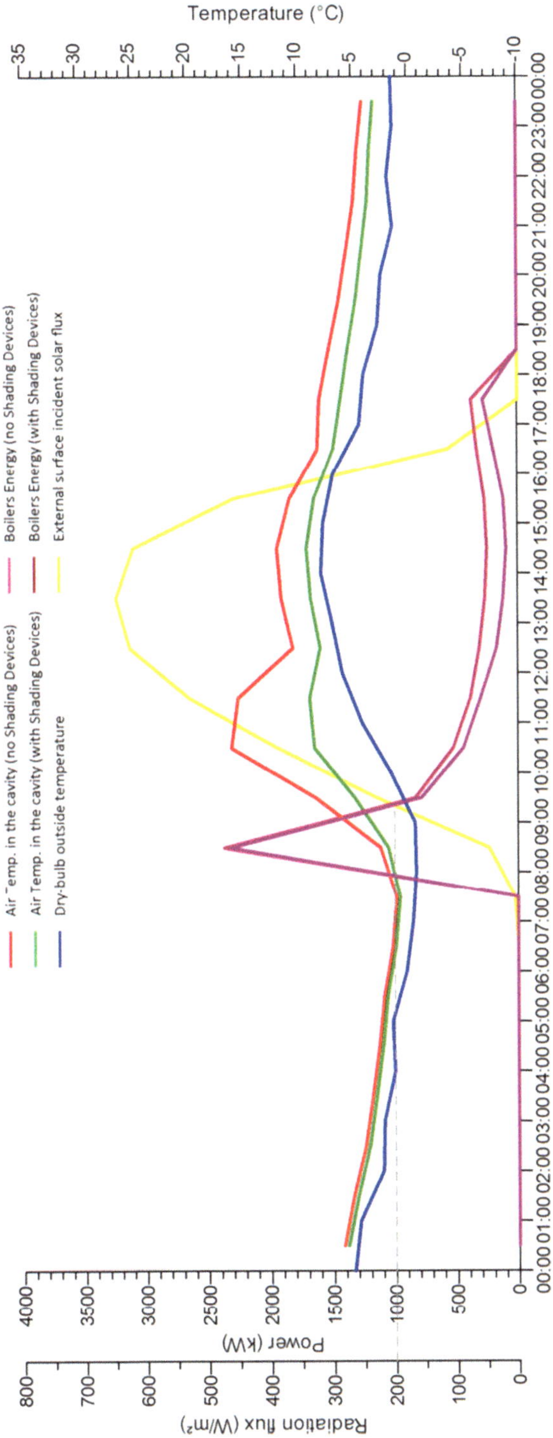

Fig. 34.7 Detailed analysis of influence of shading devices on energy consumption

34.4 Conclusions

Sustainable low-carbon refurbishments of the existing building stock represent one of the key challenges for the future if GHG targets set by all developed countries are to be met. This research focused on the use of DSFs in non-domestic buildings. A case study has been developed to assess the potential of this technology towards both reducing energy consumption and cutting carbon emissions without compromising indoor comfort. The DSF has been compared with a cutting-edge single-skin solution that has been the actual choice for the refurbishment of the University of Brighton's Cockcroft building. Results showed that DSF outperforms greatly the single-skin strategy both energy and carbon wise. Not to affect indoor comfort, shading devices had to be included within the façade cavity, and this diminished the energy reduction potential, which is, however, still very significant when compared to single-skin results. Investigations on alternative ways of preserving comfort without adopting shading devices is definitely an interesting avenue for further research. Additionally, both life-cycle energy and carbon assessment and life-cycle costing represent future research steps that could be pursued to holistically establish DSFs potential and viability for low-carbon, sustainable refurbishments.

Acknowledgements The authors gratefully acknowledge FBM Architects and Willmott Dixon for their valuable support in providing essential information to carry out this research.

References

1. Ibn-Mohammed T, Greenough R, Taylor S et al (2013) Operational vs. embodied emissions in buildings—a review of current trends. Energy Build 66:232–245
2. Ferreira J, Pinheiro MD, Brito JD (2013) Refurbishment decision support tools review—energy and life cycle as key aspects to sustainable refurbishment projects. Energy Policy 62:1453–1460
3. IEA. (2014) Technology roadmap. Energy efficient building envelopes. International Energy Agency, Paris
4. Shameri MA, Alghoul MA, Sopian K et al (2011) Perspectives of double skin facade systems in buildings and energy saving. Renew & Sustain Energy Rev 15:1468–1475
5. Pomponi F, Ip K, Piroozfar P Assessment of double skin facade technologies for office refurbishments in the United Kingdom. SB13 Conference—Graz. pp 1098–1109
6. Gratia E, De Herde A (2004) Natural ventilation in a double-skin facade. Energy Build 36: 137–146
7. Gratia E, De Herde A (2007) Greenhouse effect in double-skin facade. Energy Build 39: 199–211
8. Gratia E, De Herde A (2007) Guidelines for improving natural daytime ventilation in an office building with a double-skin facade. Solar Energy 81:435–448
9. Streicher W, Heimrath R, Hengsberger H et al (2007) On the typology, costs, energy performance, environmental quality and operational characteristics of double skin façades in European buildings. Adv Build Energy Res 1:1–28

10. Brunoro S (2008) An assessment of energetic efficiency improvement of existing building envelopes in Italy. Manag Env Quality 19:718–730
11. Kim G, Lim HS, Kim JT (2012 May) Development of a double skin facade for sustainable renovation of old residential buildings. In SHB2012—7th International symposium on sustainable healthy buildings. Seoul, Korea
12. Poirazis H. (2004) Double skin facades for office buildings—literature review. Lund University, Institute of Technology, Division of Energy and Building Design, Lund
13. Steadman P, Bruhns HR, Holtier S et al (2000) A classification of built forms. Env Plan B 27:73–91
14. CISBE (2013) Refurbishment of non-domestic buildings. Technical Memorandum—TM53. Chartered Institution of Building Services Engineers. ISBN 978-1-906846-37-4. Printed in Great Britain
15. Kragh M (2010) Facade engineering and building physics. Examples of current best practice and recent innovations. Oral presentation at the Integrated Facade Symposium, San Francisco. 21 April. http://www.cbe.berkeley.edu/research/facades/FacadeSymposium-Mikkel-Kragh.pdf
16. Poirazis H, Kragh M (2009) Establshing a simplified correlation between a proposed double skin facade and a thermally equivalent single skin facade for dynamic building energy modelling. Proceedings of the 11th International IBPSA Conference, Glasgow, Scotland. July 27–30, pp 1781–1788. International Building Performance Simulation Association. http://www.ibpsa.org/proceedings/BS2009/BS09_1781_1788.pdf
17. Kim G, Schaefer LA, Kim JT (2013) Development of a double-skin facade for sustainable renovation of old residential buildings. Indoor and Built Environ 22(1):180–190
18. IES VE (2012) MacroFlo Calculation Methods. Technical Manual. Integrated Environmental Solution Virtual Environment. http://www.iesve.com/downloads/help/Thermal/Reference/MacroFloCalculationMethods.pdf
19. CISBE (2013) The limits of thermal comfort: avoiding overheating in European buildings. Technical Memorandum—TM52. Chartered Institution of Building Services Engineers. ISBN 978-1-906846-34-3. Printed in Great Britain
20. BSI (2007) BS EN 15251:2007—Indoor environmental input parameters for design and assessment of energy performance of buildings addressing indoor air quality, thermal environment, lighting and acoustics. British Standard Institution. ISBN 978 0 580 55391 2
21. Artmann N, Sattelmayer T, Spinnler M (2004) Double-Skin Facades with Diverse Sunscreen Configurations; ISAAG 2004, Neubiberg
22. Fraser Brown MacKenna Architects (2013) Sustainable Retrofit of a 15,000sqm academic building—Cockcroft Building, University of Brighton. http://www.fbmarchitects.com/pdf/Cockcroft.pdf
23. Hill N, Walker H, Beevor J & James K (2011) Guidelines to Defra/DECC's GHG Conversion Factors for Company Reporting: Methodology Paper for Emission Factors. Published by the Department for Environment, Food and Rural Affairs (DEFRA). London. https://www.gov.uk/government/uploads/system/uploads/attachment_data/file/69314/pb13625-emission-factor-methodology-paper-110905.pdf

Chapter 35
Architectural Factors Influenced on Physical Environment in Atrium

Wei Zhao, Jian Kang and Hong Jin

Abstract In the atrium of public buildings, the physical environment is determined by architectural features in various ways; therefore, an integrated consideration of the overall physical environment is important. The aim of this chapter is to find out the common factors, which influence the daylighting, acoustics, natural ventilation, and thermal environment from the viewpoint of architectural design. Based on a systematic examination and analysis of previous studies in various aspects of physical environment, different effects of material properties, atrium geometry, atrium shape, roof structure, and adjoining spaces on the daylighting, acoustics, natural ventilation, and thermal environment have been revealed. Consequently, common factors which have influenced on two or more aspects of physical environment have been identified, along with their tendency of influence. It is expected that the results would help to optimize, at the early design stage, the physical environment.

Keywords Architectural factors · Physical environment · Atrium geometry · Roof structure · Adjoining spaces

35.1 Introduction

Atlanta Hotel by John Portman brought renaissance of classical atrium design form [1]. Currently, they are not only used for basic architectural functions but also social functions. As the core of a building, it also has its environmental influence [2]. Daylighting of an atrium decreases the requirement for artificial illumination and

J. Kang (✉) · W. Zhao · H. jin
School of Architecture, Harbin Institute of Technology, Harbin, China
e-mail: j.kang@sheffield.ac.uk

W. Zhao
zhw0904@126.com

H. jin
jinhong1963@qq.com

J. Kang
School of Architecture, University of Sheffield, Sheffield S10 2TN, UK

A. Sayigh (ed.), *Renewable Energy in the Service of Mankind Vol I*,
DOI 10.1007/978-3-319-17777-9_35

energy consumption; stack effect on natural ventilation reduces air conditioning load, and the greenhouse effect on the thermal environment reduces heating energy consumption under some design conditions; acoustic indicators affect social activities and comfort for people in the atrium [3].

For the interior design, the scale, shape, and quality are decided by geometry, measurement, form, and the materials of the ground, ceiling, and walls [4]. Not only the shape of plan and section, the roof structure system, adjoining space, and building materials but also thermal environment, acoustic environment, and light environment would be considered in atria. These researches, previously mentioned, have been researched separately. However, the comprehensive research of physical environment was limited, and it should be necessary for atrium [2].

The physical environment is influenced by climate and design in atria. In this chapter, different effects of material properties, atrium geometry, atrium shape, roof structure, and adjoining spaces on the daylighting, acoustics, natural ventilation, and thermal environment have been revealed, based on a systematic examination and analysis of previous studies of various aspects of physical environments.

35.2 Atrium Development and Architectural Factors

Atria have commonly featured in building design over the past 2000 years. The Atria in ancient Roman and Islamic buildings were not closed completely, due to the Mediterranean climate, and could be considered courtyards of brick/wood design. Following the Industrial Revolution, the development of steel and glass as building materials led to the addition of ceilings to courtyards to form atria. After 1950s, the second generation of modern atria was developed, with the atrium Atlanta Hotel, designed by John Portman, representing an archetypal atrium from the second generation. In his book, *Atrium Building—Development and Design,* Saxon Richard stated that "the motivation of the atrium for the social public space is originated from creating attractive and cheerful atmosphere rather than the aim of building function. There is no noise, smoke, or dust. In this public space, there are shade and water. People can communicate with each other casually" [1].

There are different methods for classification, such as the relative position of the atrium and main construction or the combination of atrium [1]. Atria do not exist independently, but in combinations with other spaces. From the perspective of architectural space combination [4], factors in atrium include the geometry property, surface property, and material property(Fig. 35.1).

35.3 Architectural Factors of Daylighting in Atrium

There are many factors influencing daylighting in atrium [2, 5–7]. The main factors include plan shape, cross-sectional shape and other geometry properties, roof and skylight, and surface properties of the materials.

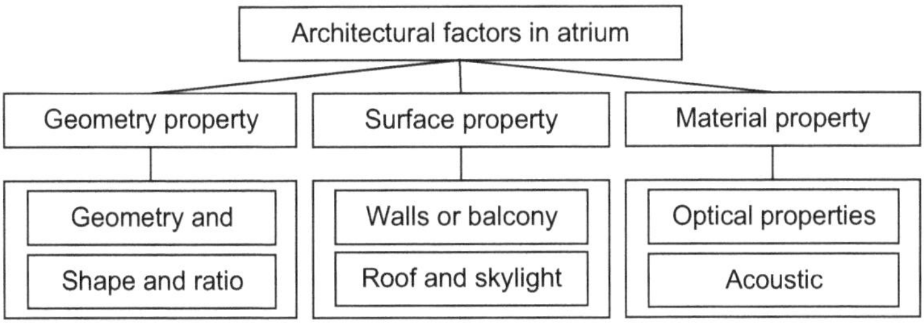

Fig. 35.1 Figure of architectural factors in atrium

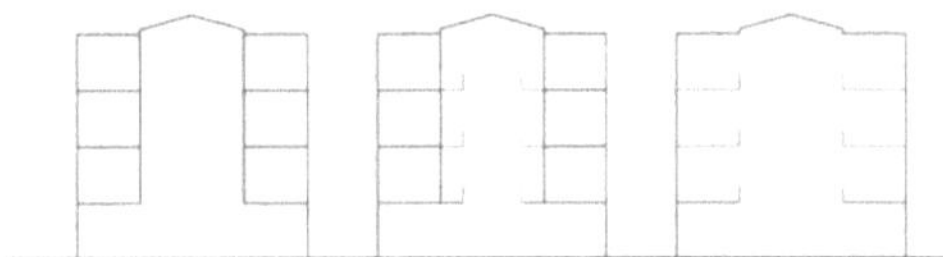

Fig. 35.2 Mutual relations between atrium and main body

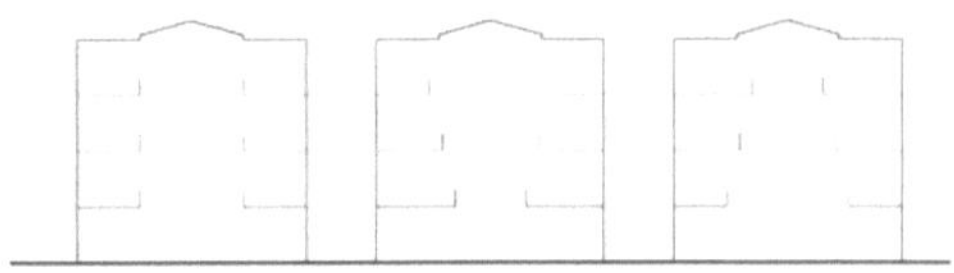

Fig. 35.3 The atrium shape of cross section

35.3.1 The Influence of Geometry Property on Daylighting

Geometry property is one of the main factors influencing daylighting distribution in atria. It includes geometry, shape and ratio of plan, and cross section.

Plan Form Plan form affects daylighting distribution. Under the four-sided atria, the equal material of glass, height of atrium and plan area, circular and rectangle atria are found to have the highest and lowest illumination levels, respectively. Rectangular or linear atria are much more suitable for vertical fenestration [7]. While keeping the height same, increasing the length of the atrium increases the light-admitting area (or reduces well index, WI) and, consequently, daylight factor (DF) [8].

Cross Section There are also different categorizations for the cross section of atria. It is divided in enclosed entity, adjoining, and open atrium according to the mutual relations between atrium and main body (Fig. 35.2). Or it is also divided in vertical form, A form, and V form according to the shape (Fig. 35.3).

The daylighting in atria is also influenced by the ratio of height to width. Square-shaped atria for a height smaller than 0.75 times the width increases the quantity of light in the adjoining spaces. Although the quantity of light decreases in atria higher than 0.75 of their widths The quality of illumination is improved in deeper atria as

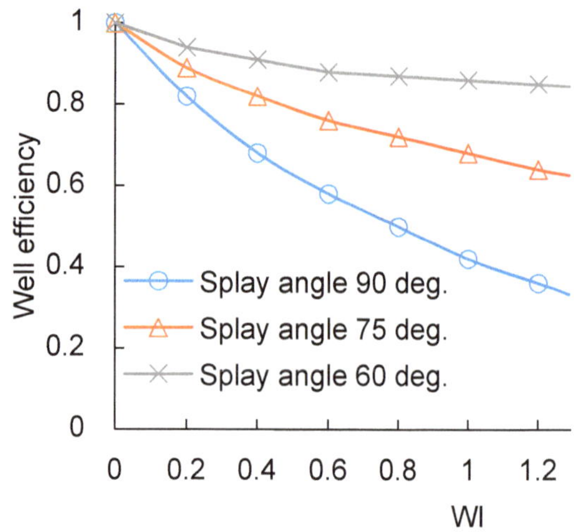

Fig. 35.4 Different well index (*WI*) makes the different well efficiency [11]

the spaces mainly receive reflected light, creating uniform illumination, improved spread across the atrium width, and reduced glare [9].

The use of stepped sections in the cross section can afford some advantages for daylight. Different forms are suited for different climates. V form could increase the daylight level and A form could improve the light penetration in adjoining spaces with better uniformity [10]. For an atrium of WI 1.0 with 60° splayed walls, the efficiency was twice that of a well with no splays at right angles (Fig. 35.4). However, the daylight performance has not been improved for the splay angles less than 60° [11]. Because of the several types of atrium construction, it is noted that the daylight performance is influenced on the form of stepped section, geometry, the size of steps, and so on.

Well Index The research shows that the DF depends on the geometry proportions, not the dimensions. The WI is a good indicator to describe the daylight availability [2]. The WI is shown as below:

$$\left[\mathrm{WI} = \frac{\text{height} \times (\text{length} + \text{width})}{2 \times \text{length} \times \text{width}} \right].$$

The formula calculating the DF differs when the reflection is changed. The general trend is that the DF becomes smaller as the WI becomes bigger. The general form is of an exponential decay [7] (Fig. 35.5).

35.3.2 The Influence of Surface Property on Daylighting

Atrium Walls or Balcony The different enclosed pattern can also influence the daylight amount. The atrium is divided in enclosed entity, adjoining and open atrium according to the mutual relations between atrium and main body. For example, the

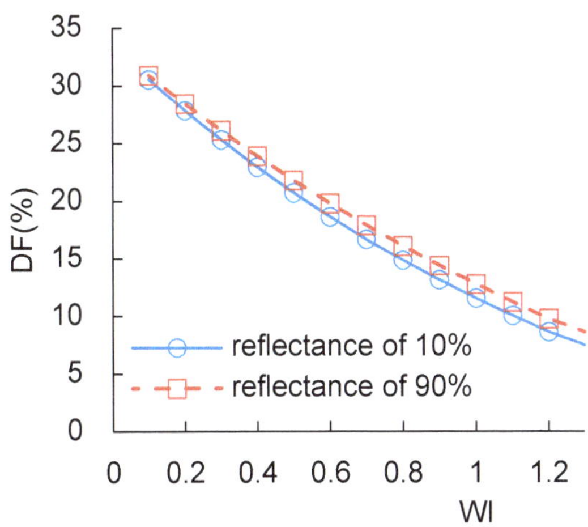

Fig. 35.5 The relationship between DF and WI is one of exponential decay. [11]. *DF* daylight factor, *WI* well index

WI and the depth dimension of adjoining space have effects on daylight. The daylight attenuation has been 70–90% on the balcony floor for the WI higher than 1.0 and the balcony depth bigger than 3 m [12]. The artificial lighting system should be considered.

Atrium Roof and Skylight The roof system has the most significant effect on the daylight intensity and direction [13]. Roof types have been chosen according to the function, plan form, span, materials, and other indicators. Figure 35.6 shows some structural forms with comparisons. The most amount of daylight was admitted by the flat top. Compared to pyramids and vaults, the sawtooth configurations have been found to have the strongest directional properties, which have the highest sensitivity to solar altitude [13]. However, the directional properties of daylight are less significant as atrium height increases. The dome is the best shape for the roof admitting low solar altitude winter sun [15]. The roof configurations would be chosen according to the usage and artistic atmosphere of atria to determine their daylight properties. For example, the sawtooth configurations would be considered

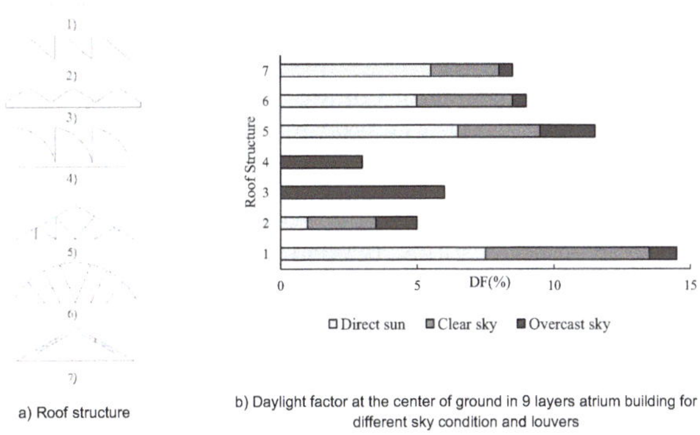

Fig. 35.6 Forms of roof structure and comparison under the different sky conditions [14]. *DF* daylight factor

Table 35.1 Summaries of papers regarding factors influencing atria daylight

–	Geometry property			Surface property		Material property
Authors and literatures	Plane form	Sectional form	Well index	Roof structure	Walls surrounded or additional space	Reflection coefficient
[2, 5–7]	√	√	√	√	√	√
[8]	√	–	–	–	–	–
[9]	–	√	–	–	–	–
[10, 11]	–	√	–	–	–	–
[12]	–	–	–	–	√	–
[13]	–	–	–	√	–	–
[14]	–	–	–	√	–	–
[16]	–	–	–	–	–	√
[8]	–	–	√	–	–	–
Peng X	√	√	–	–	–	√
Wang J	–	√	–	–	–	
Xia C	–	–	–	√	√	√
Zhou X	–	–	√	√	–	
Xu L	√	–		√	–	√
Gao H	–	–	√	√	–	√

to increase the illumination level on the vertical walls and adjoining space based on the solar altitude.

35.3.3 Influence of Material Properties on Daylighting

The reflective characteristics of atrium surfaces also influence the daylight distribution in addition to WI. Liu's research showed that the DF is much higher when the reflective coefficient is much bigger. However, this characteristic is obvious when the WI is from 1 to 2. The distribution of materials has limited influence on the daylighting environment for the deep well [2]. From Samant and Yang's research, the materials distribution has limited influence on the average DF [16].

At present, the studies of the daylight in atria are comprehensive. In some papers, other physical environmental factors affected by geometry, and other architectural elements have been considered in addition to atria daylight [2]. The elements which help to improve the daylight environment require integration, and further consideration of complex forms, beyond simple shapes, is required, such as stepped sections and irregular shapes. Table 35.1 summarizes the factors influencing atria daylighting.

35.4 Architectural Factors of Acoustic Environment in Atria

Papers focus on the daylight and thermal environment of physical environment in atria design. Limited consideration has been given to the acoustic environment. From this investigation, it was found that the lack of consideration for acoustic environment in some atria has led to very low customer satisfaction, which is a subjective assessment [17]. The objective acoustic indices including sound pressure level (SPL) and reverberation time (RT) have some effects on subject assessment [18]. The acoustic design in atria needs to be considered. Sound distribution has some special characteristics because of volume, shape, boundary conditions, and additional space. Besides the position of sound sources [19], the sound environment is influenced by architectural design elements including spatial form, volume, dimensions, shape of specific surface, and materials [3, 20–24].

35.4.1 Influence of Geometry Property on Acoustic

The acoustic characteristic is influenced by the spatial form. The small volume to area ratio and bulged specific surface have positive results to the sound field [20].

SPL and RT change with the shape of plan and number of layers. Different plan forms lead to different spatial SPL and reverberation distributions. Average SPL at all frequencies of rectangles is lower than that of squares and circles 1~1.5 dB (Fig. 35.7). The SPL distribution is stable in square and circle plan atria. At medium and high frequencies, RT is long in square form plan, short in circle, and shorter in rectangle form (Fig. 35.8). As the layers increase, RT across all frequencies becomes shorter. The variation law of RT in square or circle plan with layers is not obvious [21]. It is shown that the volume, shape, and dimensions have effects on acoustics in atria [22]. Surface geometry also has effects on sound distribution and reverberation [23].

35.4.2 Material Property Influenced on Acoustics

Sound absorption materials would improve the sound distribution. The improvement of sound distribution is a common benefit associated with absorption of ceilings. The reverberation time at medium and high frequencies would be controlled by the hanging objects [21]. Large volume and hard materials lead to long reverberation time. The different properties of hard materials, such as glass, have different effects on sound distribution and reverberation [24] (Table 35.2).

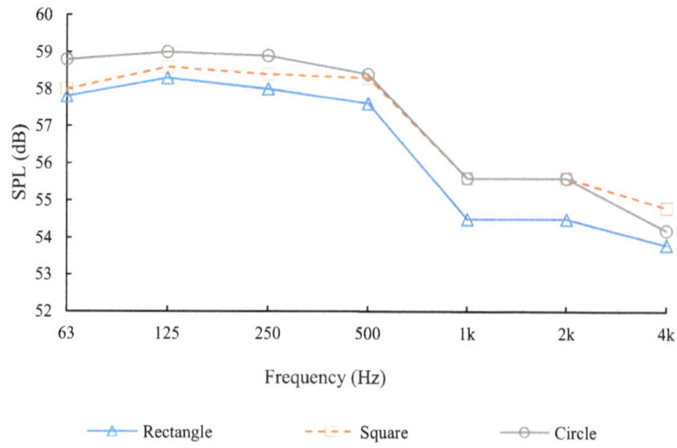

Fig. 35.7 SPL comparison of three plan forms for atria. *SPL* sound pressure level

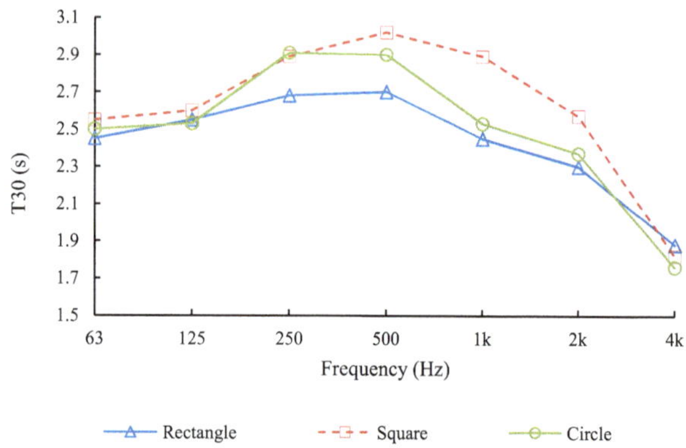

Fig. 35.8 RT comparison of three plan forms for atria. *RT* reverberation time

Table 35.2 Summaries of papers about indicators influenced on atria acoustics

–	Geometry property			Surface property	Material property
Literatures	Plane form	Sectional form	Volume	Walls surrounded or additional space	Acoustical absorption
[3]	√	–	√	√	–
[20]	–	–	–	√	–
[21]	√	√	–	√	√
[22]	√	√	√	–	–
[23]	–	–	–	√	–
[24]	–	–	√	–	√

35.5 Architectural Factors of Thermal Environment in Atrium

There is an apparent contradiction between the greenhouse effect and stack effect [1], thus, consideration should be given during design to ensure only the desired effect occurs. It is observed that the spatial form, dimension, sectional area, and top fenestration have some effect on the thermal environment.

35.5.1 Geometry Property Influenced on Thermal

In contrast to the daylight in atria, the length–width ratio is not the principal element. In high-rise residential buildings, with the increase of atrium size for the square form, the improvement in cooling effect and ventilation is much more obvious. However, the improvement becomes less significant as the size increases, which is particularly apparent for the top layer [25].

Sectional shape affects the thermal environment in atria. "A" form sectional shape can provide shade for lower layers, and at the same time, it increases the stack effect. "V" form can increase the solar radiation which could be used as atrium form in buildings in cold regions of China [26–27].

As increased height and smaller plan area are key factors for the stack effect, the buoyancy-driven ventilation is much more apparent for atria in high-rise residential buildings. However, the relationship between size and ventilation should be considered in common public buildings, as the atrium area is much bigger than it is in residential buildings. The "A" form or "V" form has the advantage of stack effect or greenhouse effect separately; however, it is not considered the sectional form influenced on the daylight or acoustics in atria.

35.5.2 Surface Property Influenced on Thermal

Stack effect depends on the pressure differences at the top and bottom of atria. The percentage of skylight opening (surface property) is the primary factor affecting ventilation, whereas, the sectional area (geometry property) has less effect [28].

Laouadi et al. [29] studied the impact of selected design alternatives on the thermal and energy performance of atria. Design alternatives focused on fenestration glazing types, fenestration surface area, skylight shape, atrium type, and interaction of the atrium with its adjacent spaces. It is observed that the skylight shape would have an impact on the energy use. For example, the pyramidal skylight increased the solar heat gain ratio by up to 25 % in the heating season for the enclosed and linear atriums. However, the shape of skylight shape did not have a significant effect in the cooling season [29] (Table 35.3).

Table 35.3 Summary of papers regarding the factors that influence thermal environment in atria

–	Geometry property		Surface property	
Literatures	Plane form and size	Sectional form and size	Form of walls surrounded	Roof structure
[25]	√	–	–	–
[26]	–	√	–	–
[27]	–	√	–	–
[28]	√	–	–	√
[30]	–	–	–	√
[31]	√	√	√	–

Table 35.4 Comprehensive analysis of factors influencing the physical environment of atria and additional spaces

–	Geometry property				Surface property		Material property	
–	Plane form	Sectional form	Size and ratio	Area and volume	Walls surrounded	Roof structure	Reflection	Sound absorption
Daylight environment	√	√	WI	–	√	√	√	–
Acoustic environment	√	–	Special shape	Volume	√	√	–	√
Thermal environment	–	√	√	√	–	√	–	–

35.6 Architectural Factors of Physical Environment in Atrium

The effects of architectural design on the physical environment of atria were also considered. Daylight in atria is affected by factors including plan shape, ratio, cross section, and skylight. The WI and skylight have significant effects on daylight. Sound distribution and reverberation are affected by factors including the shape, size, volume, and adjacent space. The size and the form of enclosed spaces have much effect on sound distribution. Ventilation has been affected by the form, size of atria, and the percentage of skylight open.

As analyzed above, effects of architectural elements on the physical environment were not consistent. For example, the greater the height of the atrium, the greater the amount of resulting ventilation. However, the increased height has detrimental effects on daylighting. The use of sound absorption materials in the atrium roof has obvious effects on the sound field of the space. However, there must be a compromise between acoustic environment and daylighting in the use of glass as a roof material, as it has low sound absorption. Boundary conditions of additional spaces could improve the sound environment but they have less effect on daylighting, which depends more on the form. The analysis of synthetic influencing factors has been displayed in Table 35.4.

35.7 Conclusions

As shown in Table 35.4, there are some discussions followed, concerning the architectural factors of atria effecting the physical environment.

35.7.1 Geometry Property

35.7.1.1 Factor 1

Form and Ratio of Plan

Daylight factor increases with the length of side, for rectangular plans. For acoustic environment, the SPL is much lower, and the RT is much shorter in rectangle than in square and circular. For thermal environment, length–width ratio has effects on ventilation for residential buildings but potentially not for public buildings.

35.7.1.2 Factor 2

Form and Ratio of Section

For daylight, atria (square shaped) with a height smaller than 0.75 times the width increases the quantity of light in the adjacent spaces. For sound environment, at all frequencies, there is a continuous decrease in SPL with increasing storeys. For thermal environment, A-shaped section is advantageous for the stack effect, and V-shaped section increases the solar radiation area.

35.7.1.3 Factor 3

Form and Ratio of Section

WI and volume for daylight, the smaller the WI, the greater the DF. For sound environment, the reverberation time increases with volume.

35.7.2 Discussion

Geometry property affecting the physical environment, including size, scale, shape, WI, and volume should be considered

- Shape and scale of plan effected on thermal and other environments should be taken into consideration

- Shape and scale of the section affecting sound distribution and RT should be taken into consideration
- WI and volume as determinant factors affecting daylight, RT, and other environments should be taken into consideration.

35.7.3 Surface Property

35.7.3.1 Factor 1

Walls Enclosed Atrium and Balcony

For daylight, the protruding balcony allows daylight attenuation. For sound environment, the protruding surfaces could improve sound diffusion. At the same time, the interaction between the atria and surrounding spaces should be considered.

35.7.3.2 Factor 2

Roof and Skylight

For daylight, the different atrium roof structures affect the daylight environment. For example, the flat top roof was found to admit the most daylight. For sound environment, the use of glass, which has low sound absorption, has little effect in improving the sound distribution. For ventilation, the opening area of skylights is an important element affecting ventilation.

35.7.4 Discussion

Surface property affecting the physical environment should be assessed including surrounding spaces and roof structure.

- Depending on the openness of surrounding spaces for enclosed, adjoining, and open type, different considerations should be given in terms of the effect on physical environment at ground level and other spaces.
- The roof structure and skylight type should be considered for the effect on sound and thermal environment.
- The geometry and surface properties should be considered for the effect on comprehensive physical environment.

Acknowledgements The authors are indebted to the research grants from China 2012BAJ06B04-02 and 2014-1-54.

References

1. Saxon R (1990) Atrium building development and design. China Architecture & Building Press, Beijing, p 7
2. Swinal Samant (2010) A critical review of articles published on atrium geometry and surface reflectances on daylighting in an atrium and its adjoining spaces. Archit Sci Rev 53(2):145–156
3. Mei H, Kang J (2012) An experimental study of the sound field in a large atrium. Build Environ 58:91–102
4. Peng Y (2004) Combination of building space, vol 13, 2nd edn. China Architecture & Building Press, Beijing, pp 43–48
5. Aizlewood ME The daylighting of atria: a critical review. ASHRAE Trans 101(2):841–857
6. Littledfair P, Aizlewood ME (1998) Daylight in atrium buildings. Building research establishment information Paper IP3. Communication Research Communicationl Ltd, London
7. Sharples S, Lash D (2007) Daylight in atrium buildings: a critical review. Archit Sci Rev 50:301–312
8. Calcagni B, Paroncini M (2004) Daylight factor prediction in atria building designs. Sol Energy 76(6):669–682
9. Willbold-Lohr G (1989) Daylighting in glazed atria. Proceedings of the second European conference on architecture. Paris, pp 16–20
10. Alraddadi TA (2004) The effect of the stepped section atrium on daylighting performance. Archit Sci Rev 47(3):303–310
11. Laouadi A (2004) Design with sky vision: a computer tool to predict daylighting performance of skylights. CIB world building congress. NRCC-IRC, Toronto, pp 1–11
12. Kim G, Kim JT (2010) Luminous impact of balcony floor at atrium spaces with different well geometries. Build Environ 45:304–310
13. Boubekri M (1995) The Effect of the cover and reflective properties of a four-sided atrium on the behavior of light. Archit Sci Rev 38(1):3–8
14. Navvab M, Selkowitz S (1984) Daylighting data for atrium design. Ninth national passive solar conference. American Solar Energy Society, Columbus, pp 495–500
15. Laouadi A, Atif M (2001) Prediction models of optical characteristics for domed skylights under standard and real sky conditions. Proceedings 7th international IBPSA conference. IBPSA, Rio de Janeiro, pp 1101–1108
16. Samant S, Yang F (2007) Daylighting in atrium: the effect of atrium geometry and reflectance distributions. Lighting Res Technol 39(2):147–158
17. Shi H, Zhou Z (2008) Investigation and assessment for atrium acoustic environment. Architecture and urban physical environment in the process of the urbanization-the 10th national conference on physical academic buildings, pp 131–134
18. Chen B, Kang J (2004) Acoustic comfort in shopping mall atrium spaces: a case study in Sheffield Meadowhall. Archit Sci Rev 47:107–114
19. Zhang Y (2012) The sound source and its influence in atrium space. Master Degree of Harbin Institute of Technology, Harbin
20. Ekrem Bahadir Caliskan. (2010) Acoustical evaluation of shopping mall typology. Master Degree of Middle East Technical University, Ankara, p 9
21. Chen L (2012) Research on acoustic environment of typical forms of atrium space and design strategy. Master Degree of Harbin Institute of Technology, Harbin
22. Meissner M (2007) Influence of wall absorption on low-frequency dependence of reverberation time in room of irregular shape. Applied Acoustics 5(3):293–312
23. Maekawa Z, Lord P (1994) Environmental and architectural acoustics. E & FN Spon, London
24. Ardeshir M, Pak J, Lechleitner J (2007) Acoustics of atria: contrasting measurement and modeling results. Proceedings: buildings simulation, pp 1219–1226

25. Li N, Zhang M, Li J, Hou S, He D (2010) Effect of sectional shape and size on ventilation for high residential buildings in the process of the urbanization-the 2010 national academic annual conference on HAVC, p 282
26. Lin C, Tian X, Fang Z (2004) Architectural design of atrium and control of its thermal amenity. Ind Constr 34(7):28–32
27. Wen C, Wang C (2006) Thermal amenity analysis of different space shape of atrium in summer. J Shandong Univ Archit Eng 21(1):36–41, 49
28. Tian P, Pi P, Chu W (2010) Influencing factors of the thermal environment of atrium based on the orthogonal analysis method. Sci Technol Rev 28(21):82–85
29. Laouadi A, Atif MR, Galasiu A (2002) Towards developing skylight design tools for thermal and energy performance of atriums in cold climates. Build Environ 37:1289–1316
30. Wang J, Zhao D (2011) Optimized roof design of lighting atrium to promote natural ventilation. Zhejiang Constr 28:43–46, 50
31. Wang C, Wen C, Zhao X (2006) Several factors influencing the thermal amenity of architectural atrium and the improvement measures. Huazhong Archit 24(3):80–83

Chapter 36
Numerical Simulation Analysis on Wind Environment of Traditional Village Courtyard in Severe Cold Regions

Xinyu Zhang, Hong Jin and Xu Dong

Abstract Outdoor wind environment has become markedly worse in recent years during winter in severe cold regions of China. The subject of the study is wind environment of the courtyard, which is an important part of rural housing and village life in China, and the quality of the wind environment, in a way, affects the using comfort of the residential indoor and outdoor thermal environment in winter. This chapter applies the computational fluid dynamics (CFD) software to simulate the wind distribution characteristics of seven cases, which have different courtyard orientation, scale, building arrangement, and the height and materials of fences. Then the design strategies are put forward which can effectively avoid the adverse effects of cold wind in winter of severe cold regions.

Keywords Severe cold regions · Traditional courtyard · Wind environment · CFD simulation

36.1 Introduction

The rural residence in cold regions faces the ordeals of piercingly cold wind, severe outdoor air temperature, and sharp indoor and outdoor temperature difference, and it is the ultimate objective of the residents to have high indoor thermal comfort degrees and low heating energy consumption in winter. In recent years, the researchers, for this purpose, frequently attach their priorities of work to the indoor thermal comfort degree, the shape factor control, the space enclosing structure insulation construction, the new type heat insulating material selection, the heating facility performance optimization of the houses, and have registered a large amount of research achievements [1, 2]. Nevertheless, different from urban residence, the traditional rural residence features the basic constitution of

X. Zhang (✉) · H. Jin · X. Dong
School of Architecture, Harbin Institute of Technology, Harbin, China
e-mail: xinyu.z@hotmail.com

A. Sayigh (ed.), *Renewable Energy in the Service of Mankind Vol I*,
DOI 10.1007/978-3-319-17777-9_36

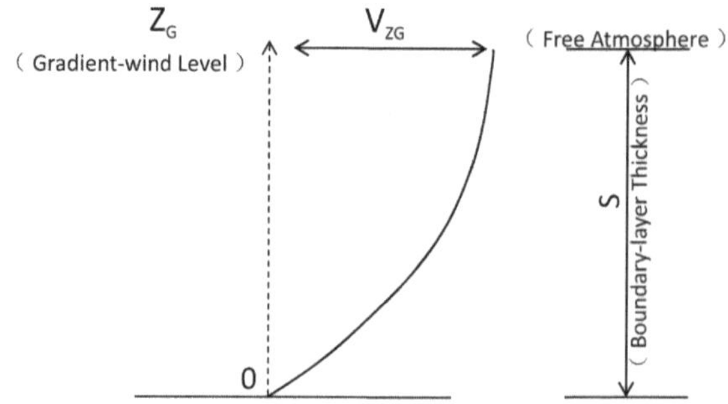

Fig. 36.1 Wind flow pattern

residence = housing + courtyard. The traditional rural residence is not a separate entity. Apart from the control of thermal insulation and energy-saving performance of itself, it is also under significant impact of the entire environment of the microclimate of the courtyard, which is often neglected by the residents and the researchers.

The velocity of wind mobility is in graded distribution due to the roughness of the underlying medium (Fig. 36.1). Meanwhile, wind is the most decisive factor among the numerous climatic factors (temperature, surface temperature, solar radiation, wind, and humidity, etc) influencing the outdoor comfort for human body [3]; for the buildings, wind is the main factor influencing the thermal exchange between the outdoor air and the external building envelope. Due to the special climatic conditions, the residential layout pattern, and the living habits, the outdoor wind environment of the cold rural regions in winter foster the important environmental aspects influencing the amount of heat dissipation, the osmotic quantity, and the indoor thermal comfort degree, etc. For this reason, it is of tremendous significance to find the flow regularity of wind around the buildings and the coupling relationship between human body comfort and building energy consumption. Computational fluid dynamics (CFD) software has been widely applied to simulate the turbulence structure, the wind velocity, and the pressure field distribution in the atmospheric boundary which have been verified[4, 5]. Nonetheless, rare available research has given due consideration to the graded distribution of wind. So, this chapter offers a comprehensive account of the roughness of the under-layer and the graded distribution of wind to truly reflect the flow of wind in the courtyard.

36.2 Factors Influencing Wind Environment in Courtyard

36.2.1 Orientation and Scale

The orientation and scale of the courtyard and buildings exert vital effects upon the wind environment in the courtyard. The orientation and the layout of the main

Fig. 36.2 Courtyard scale

entrance of rural residence in cold regions should try to avoid the prevailing wind direction in winter to reduce the area of its influence. The traditional courtyards in different cold regions boast different scales, which is controlled by the policy guidance of the local governments, the geological positions of the villages, their economic conditions and living, and production modes. For instance, the town government is close to the town center, which runs short of land and where the scale of residential courtyard is fairly small. However, the regions far away from the center are abundant in land resources, where the scales of courtyard can be appropriately expanded. Traditional courtyards in cold regions are enclosed by courtyard walls, which are mostly in the shape of a rectangle [6]. The buildings are frequently constructed parallel to the short edge of the courtyard whose scale ranges between 20×20 or 20×30 (Fig. 36.2).

36.2.2 Function Arrangement

The courtyard layout pattern influences the distribution of the wind pressure field and the wind velocity vector field in the courtyard. The function set and distribution of courtyards of rural residences hugely differ from those of courtyards of urban villas, which is decided by the living habits of the residents. The function set of the courtyard of rural residence is very complete, and it is the area to place sundries, store grains and firewood, plant vegetables for self-sufficiency, produce, and work. Generally speaking, the courtyard has the area for storage, planting, and cultivation, etc. (Fig. 36.3).

Fig. 36.3 Courtyard function arrangement in cold region

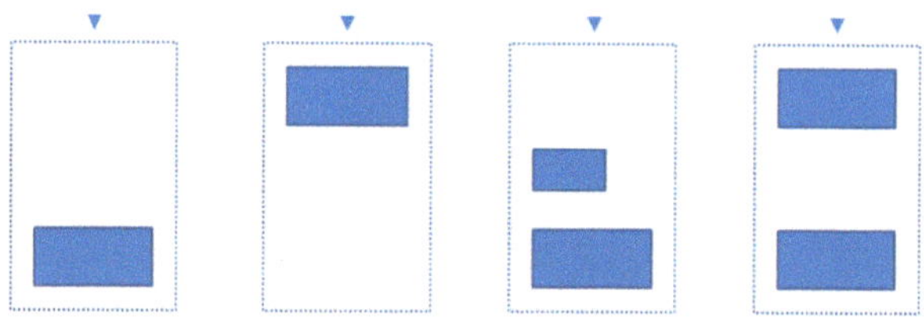

Fig. 36.4 Courtyard function arrangement pattern

Table 36.1 Basic information on courtyard walls

Village	Yinchun	Langxiang	Shengli
Height	1.3 m	1.5 ~ 1.8 m	2 ~ 2.3 m
Material	Choi steel	Wood	Wood
Space	0.1 m	0.01 m	0.02 m

Apart from these functions, the layout of courtyard should consider the climatic influence. Consequently, the layout of courtyards in cold rural regions is primarily divided into such types as forecourt, backyard, determinant, semi-enclosure, etc. (Fig. 36.4), giving full consideration to wind screening and breaking.

36.2.3 Courtyard Walls

Courtyard walls stand as the important part of rural courtyard, playing the functions of enclosing and separating the area and protecting the properties in the courtyard. Simultaneously, the height and void ratio of courtyard walls exercise impetus influence upon the airflow. The survey manifests that the villagers of different regions favor courtyard walls of different materials such as masonry, board, and wrought iron, etc. of 1.2–2.0 m, mostly 1.3, 1.8, and 2 m (Table 36.1).

36.3 Courtyard Wind Environment Simulation

36.3.1 Model Selection

The wind tunnel experiment and CFD software are usually used in the analog simulation wind environment [7]. Due to high costs and a complicated process, the wind tunnel experiment is less widely applied than the CFD software. In the simulation, the Re-Normalisation Group (RNG) k–ε two-equation model is selected after optimization [8, 9], widely tested, and successfully applied in scientific research and engineering practice.

36.3.2 Model Establishment

To simplify the simulation and achieve clearer results of comparison, the scales and layout forms of traditional courtyards in Langxiang Town are abstracted to be the ideal model of courtyard[10]. As shown in Fig. 36.5, the basic information of the ideal model is listed as follows: (1) functional sizes of the courtyards: size of courtyard, 20 × 20 m, 15 × 30 m, or 20 × 30 m; size of house, 10 × 6 × 5 m; size of warehouse, 10 × 4 × 4 m; a passageway of 3 m inside; wall height of 1.3, 1.8, or 2.3 m; distance from the house to the northern wall, 3 or 13 m; (2) the courtyards are used only for housing or permanent warehousing, rather than as simple warehouses or garages, etc.; (3) with no arbor or shrub planted in or around the traditional courtyard in Langxiang Town, and due to the low height of vegetables planted, the functions of wind shielding and stream guidance of trees and vegetables can be neglected; (4) although the courtyard walls are of different materials, the functions and effects of the courtyard wall height upon the courtyard wind are more directly displayed. The simulation is made supposing the courtyard walls are of dense and windproof materials of 0.2 m thickness. According to the different courtyard scales and courtyard wall heights, the simulation has been made principally in the seven cases as shown in Table 36.2.

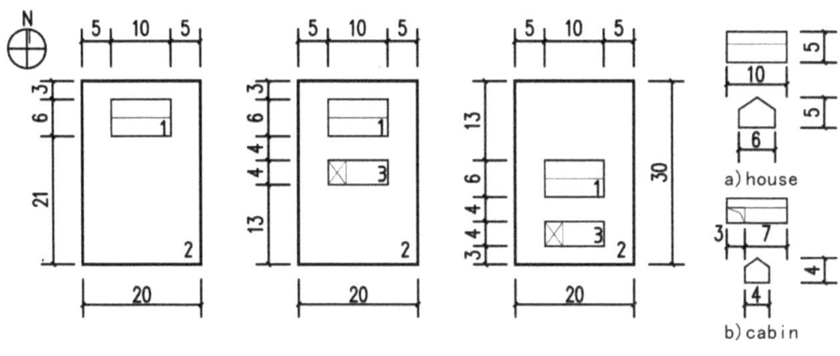

Fig. 36.5 The scale and arrangement pattern of traditional courtyard in Langxiang

Table 36.2 Numerical simulation cases

–	Size (X × Y × Z)	Building amount	Distance between builidng and North courtyard wall (m)
Case 1	20 × 20 × 1.8	1	3
Case 2	15 × 30 × 1.8	1	3
Case 3	20 × 30 × 1.3	1	3
Case 4	20 × 30 × 1.8	1	3
Case 5	20 × 30 × 2.3	1	3
Case 6	20 × 30 × 1.8	2	3
Case 7	20 × 30 × 1.8	2	13

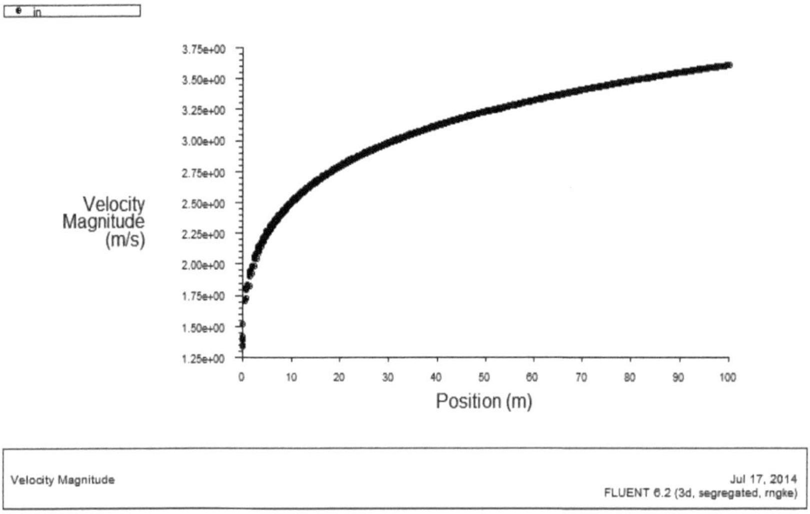

Fig. 36.6 Inlet wind velocity

36.3.3 Boundary Conditions

36.3.3.1 Wind Direction and Wind Velocity

Langxiang Town lies in the middle of Heilongjiang Province. Due to the prior study on the courtyard wind environment in the cold region, north wind (N) is primarily selected in the simulation.

According to the summary and analysis of the meteorological data of Langxiang Town, the wind velocity averages out to be 2.5 m/s in winter. Considering the roughness of the underlayer, the wind velocity is in the change of gradient with that at the inlet in the tendency shown in Fig. 36.6.

36.3.3.2 Courtyard Building Component

The layout of traditional courtyard and building is generally in the positive north–south direction. The courtyards are mostly in structured squares and are of windproof compact structure, providing areas for planting, warehousing, and cultivation, etc. Concurrently, to gain the obvious tendency of change of wind velocity, the courtyard walls also adopt the windproof compact materials of the thickness of 0.2 m.

36.3.4 Simulation Results and Analysis

Due to the disturbance of the wind from the courtyard walls and buildings, uneven distribution of wind pressure and excessive wind velocity occur in some parts of the courtyard. Taking case 4 as the benchmark comparison object, this chapter makes a comparative analysis of the wind velocity, the distribution of the wind pressure field

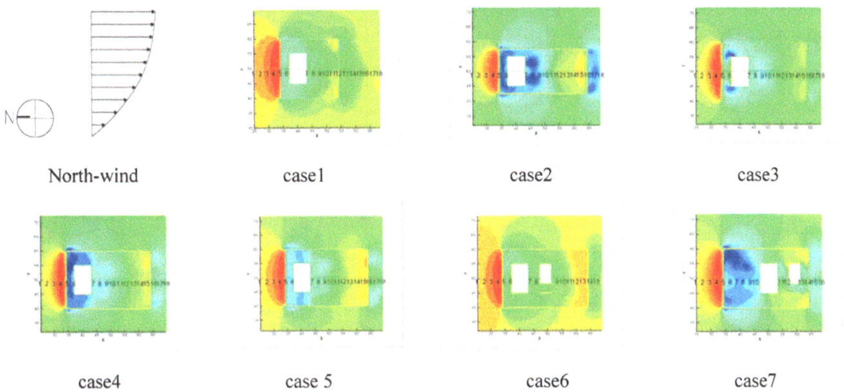

Fig. 36.7 Each dot of the axle wire in the courtyard

of different courtyards at the height of 1.5 m, and the distribution of wind velocity at each dot of the axle wire in the courtyard (as shown in Fig. 36.7, one dot is selected, each 2 m, averting from the house and the courtyard wall), analyzes the influence of each factor upon the courtyard wind environment in cold regions in winter, and proposes suitable designing strategies.

(1) Influence of the Courtyard Scale Figure 36.8 shows that, when other conditions are the same, the wind velocities in cases 1 and 4 are similar in the entire tendency of distribution. However, the courtyard in case 1 is shorter than that in case 4, and a shorter courtyard may result in untimely backflow and reflection of airflow in the courtyard and greater wind velocity at the roots of the courtyard walls on the southern and northern sides, which will severely impact the distribution of wind in the courtyard and trigger poor homogeneity.

Figure 36.9 displays that when other conditions are the same, the wind velocities near the roots of the courtyard walls on both sides in cases 2 and 4 are fundamentally identical. But the wind velocity in case 2 is remarkably higher than that in case 4, frequently fluctuating, in uneven airflow distribution, and bringing discomfort to the residents. The reasons are analyzed as follows: reducing courtyard width (namely reducing the distance from courtyard walls to the building gable) will conspicuously

Fig. 36.8 Wind velocity compared *case 1* and *case 4*

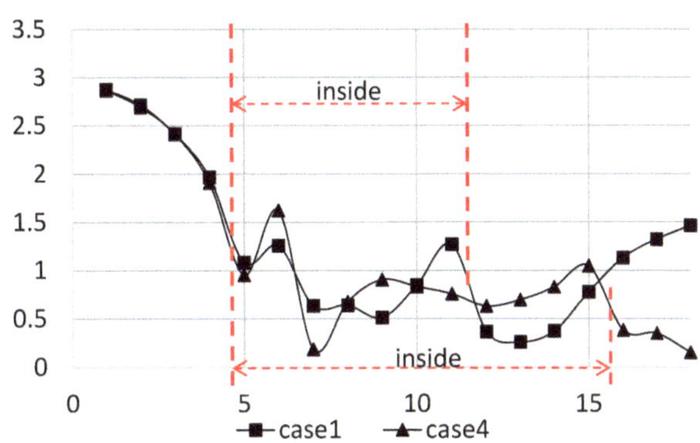

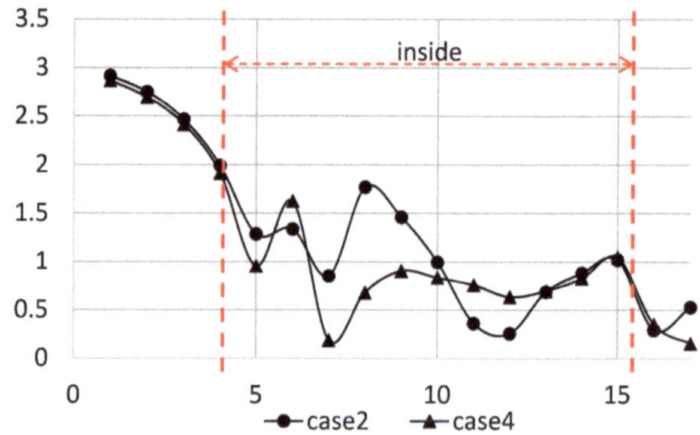

Fig. 36.9 Wind velocity compared *case 2* and *case 4*

accelerate the wind velocity at the corner of the building gable, exert influence upon the airflow in the courtyard, and generate many cyclones and whirlpools.

Consequently, the length–width ratio should be reasonably controlled to be 1:1.5 ~ 1:2, and the ratio of the edge length of building and the short edge of the courtyard wall should be suitably controlled to be 1:1.5 ~ 1:2.5, which will attain the optimal wind environment in the courtyard.

(2) Influence of the Courtyard Wall Height Figure 36.10 depicts that when other conditions are the same, the courtyard wall heights in cases 3, 4, and 5 are separately set to be 1.3, 1.8 and 2.3 m; the airflow in the courtyard in cases 4 and 5 is basically in the same tendency of distribution; since the courtyard wall height in case 3 is below the base level, the courtyard wall does not play the role of effective shielding, and the wind velocities at the southern and northern courtyard walls are evidently enhanced. The comparison of the related values in cases 4 and 5 indicates that the wind velocities in the courtyard in the two schemes are relatively equal, and the wind velocity at each point in case 5 is distinctly lower than that in case 4. Nonetheless, due to the greater height of the courtyard wall in case 5, the wind pressure outside the southern courtyard wall is great and the wind velocity is on a rise, which exercise a gigantic influence upon the wind environment on the street outside the courtyard. It can be seen that, for the inside of the courtyard, the higher courtyard

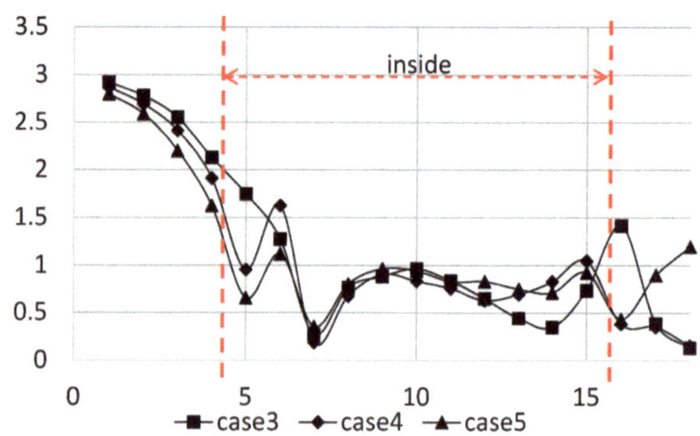

Fig. 36.10 Wind velocity compared *case 3* and *case 4* and *case 5*

wall represents the higher inside wind comfort. Nevertheless as the important structure to separate the private from the public area, the design of courtyard wall should sufficiently consider the psychological feelings of the users. For the total environment of the town, the comfort of the wind environment in the street should also be measured. Consequently, based on the further simulation, the courtyard wall height should be reasonably controlled to be 1.7–2.0 m. If the height exceeds 2.0 m, materials of high permeability should be adopted.

(3) Influence of the architectural layout The comparison of the cases 4 and 6 in Fig. 36.11 tells that when other conditions are the same, the airflow distribution in the courtyard will markedly change, and the air pressure difference on the windward and lee side of the building will clearly decrease under the influence of the warehouse constructed in front of the principal building in case 4. However, the warehouse shortens the distance from the warehouse to the southern courtyard wall. Thus, although the wind velocity in this area is heightened, it does not affect the use of main residential buildings. Therefore, it can drastically cut the thermal loss of the external building envelope in winter, and the lowered pressure difference can also effectively reduce the thermal loss caused by cold wind into the room when the residents go in and out of the building.

The comparison of the simulation results in cases 6 and 7 in Fig. 36.11 unveils that, when other conditions are the same, the wind velocity lowers and takes on the trend of decline until it gets to the windward side under the impact of weakened acceleration with the unceasingly increasing distance between the courtyard wall on the northern windward side and the building. The wind velocity rapidly dips on the lee side of the building affected by the warehouse and rises on the lee side of the warehouse affected by the courtyard wall, fluctuating frequently, and generating discomfort when the average wind velocity exceeds that in cases 4 and 6.

Hence, further simulation and data comparison reveal that, when the courtyard wall height exceeds the benchmark of 1.8 m, the distance between the courtyard wall on the windward side, and the building and the height of the building on the windward side (d/H) should be controlled to be 1:2∼2:3; under the premises of sunshine spacing, the distance between the buildings should be shortened as much as possible.

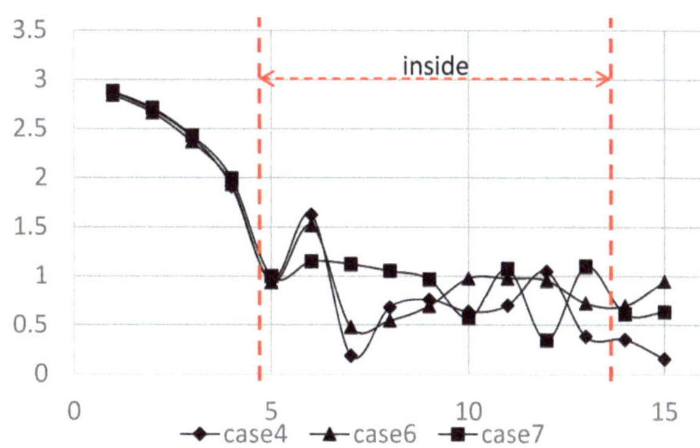

Fig. 36.11 Wind velocity compared *case 4* and *case 6* and *case 7*

36.4 Conclusions

This chapter applies fluent computer numerical simulation and practical survey, and studies the design strategies of different forms of courtyards suitable for the wind environment in cold regions in winter. Through analytical comparison, it discusses the influence of such aspects as the scale, face width, depth, wall height, and architectural layout, etc. of the courtyard upon the wind environment in the courtyard in cold regions. It predominantly gets to the following conclusions: (1) the length–width ratio and the scale of courtyard should be reasonably controlled to ensure effective improvement of the inside wind environment made by the courtyard wall; (2) appropriate courtyard wall height should be applied to insulate the influence of the external wind environment upon the inside courtyard; (3) architectural communities should be scientifically arranged, and beneficial shielding between the building and the courtyard walls and between the buildings should be made good use of to reduce the negative impact of the wind environment, so as to realize effective control of architectural energy consumption.

Acknowledgment Sponsored by the National Project of Scientific and Technical Supporting Programs Funded by Ministry of Science & Technology of China(Grant No. 2013BAJ12B02), the National Natural Science Foundation of Heilongjiang (Grant No. E201316).

References

1. Building Energy Research Centre Tsinghua University. (2009) Annual report on China building energy efficiency. China Architecture & Building Press, Beijing, pp 81–93
2. Hong J, Wei L (2010) Low-energy consumption, low-technology and low-cost. Archit J 8:14–16
3. Feng-quan YU. (2010) The summary of CFD simulation methods for architecture wind environment. J Maoming Univ 20:72–75
4. Mingfei Z, Chang X, Dong L (2010) Numerical analysis on the wind environment distribution of village buildings. Low Temp Archit Technol 6:17–19
5. Hargreaves DM, Wright NG (2007) On the use of the k-3 model in commercial CFD software to model the neutral atmospheric boundary layer. J Wind Eng Ind Aerodyn 95(5):355–369
6. Jie L (2001) Courtyard—man-made nature space. Environ Exploitation 16(3):16–17
7. Qingdang Q, Dongpeng G, Rentai Y, Yunchao W, Dan F (2011) Numerical and wind- tunnel simulations of flow and dispersion around a building. Radiat Prot 31(4):202–208
8. Yoshihide T, Ted S (2009) Numerical simulation of dispersion around an isolated cubic building:comparison of various types of k-ε models. Atmos Environ 43:3200–3210
9. Richards PJ, Hoxey RP (1993) Appropriate boundary conditions for computational wind engineering models using the k- 3 turbulence model. J Wind Eng Ind Aerodyn 47:145–153
10. Jakeman AJ, Letcher RA, Norton JP (2006) Ten iterative steps in development and evaluation of environmental models. Environ Model Softw 21(5):602–614

Chapter 37
Energy Saving and Emission Analysis via Lighting Retrofitting in a Large-Scale Hospital: Case Study in Malaysia

S Moghimi, F Azizpour, C. H Lim, E Salleh, S Mat and K Sopian

Abstract A field study on energy efficiency by lighting retrofitting has been carried out in one of the large-scale hospitals in Malaysia. The main objectives of this study were to find energy and cost savings after light-emitting diode (LED) tubes replacement in the hospital as well as the emission reduction and payback period time. The results have showed that, by applying this strategy, the energy cost saving will be almost US $ 2 million after 3 years with more than 5000 t of CO_2reduction per year as well as 2.9 years is resulted as payback period time.

Keywords Sustainable and low-energy architecture · LED lighting · Payback period · Emission reduction

37.1 Introduction

The increasing growth of energy consumption in the world has considered over limited resources, supply problem, and environmental impact including global warming, climate change, and ozone layer depletion [1, 2]. The government awareness is raised increasingly in the building sector, since buildings are huge energy consumers. Moreover, one of the most important challenges for engineers and building decision makers is to find and apply the best energy efficiency strategies in terms of energy saving and cost [1, 2]. On the other hand, population growth, enhancing comfort level in building, and inside time spending incensement determine clearly the upward trade of energy consumption and their related problems in the coming years. Due to this fact, energy efficiency in buildings becomes a primary goal in energy policy in various national and international levels. In developed countries, the contribution of both residential and commercial buildings is increased between 20 and 40 % [1, 2]. In Malaysia, there are increasing government concerns and awareness toward energy efficiency since the 2000s as well as designing and constructing

S. Moghimi (✉) · F. Azizpour · C. H. Lim · E. Salleh · S. Mat · K. Sopian
Solar Energy Research Institute, Universiti Kebangsaan Malaysia, 43600 Bangi, Selangor, Malaysia
e-mail: saeed.moghimi@gmail.com

A. Sayigh (ed.), *Renewable Energy in the Service of Mankind Vol I*,
DOI 10.1007/978-3-319-17777-9_37

Table 37.1 Energy end-use for different hospital parts

–	Energy usage percentage (%)
Air conditioning	62.0
Lighting	20.6
Medical equipments and others	17.4

a few leading energy efficient buildings as a show case of the building proficiency and to the public [3, 4]. Hospitals, in particular, are one of the most critical buildings in commercial sectors in terms of energy consumption. Hospitals are 24 h, 7-day operational buildings that may use various types of energy in air conditioning systems, lightings, medical, and office equipments. Applying energy efficiency policy in hospitals can lead to significant energy conservation. Based on a previous study, end-use energy consumption in the Malaysian case study hospital is given in the following table (Table 37.1) [5, 6]. Lighting consumes around 20 % of total energy consumption in the hospital. Present lighting systems, including tungsten filament bulbs, utilize more power in comparison to other lighting systems. Fluorescents are not eco-friendly due to their mercury, nondisposable, with a short lifetime, while light-emitting diodes (LEDs) are more power saving and eco-friendly [7].

As Table 37.1 illustrates, almost 20 % of the energy consumption in hospitals is lighting. This study focused on energy conservation in the lighting sector. The objectives of this study are as follows:

- To calculate energy saving in a large-scale hospital via replacing LED tubes
- To calculate payback period in total and in various zones separately
- To find CO_2 emission reduction via replacing LED tubes

37.2 Research Methodology

37.2.1 Site Description

This study is carried out in a large-scale hospital in Malaysia. University Kembangsaan Malaysia Medical Centre (UKMMC)was established in 1997 as a teaching hospital with 240,000 m^2built-up area. This hospital consists of various blocks including the educational block and the clinical block, which are 6 and 14 story buildings, respectively. Figure 37.1, taken from Google Earth, shows a general view of the case study [8, 9, 10].

37.2.2 Lighting Electricity Consumption

To calculate the lighting energy consumption, the field study was divided into 28 zones based on different per capita lighting (Lux/m^2), time usage, and lighting types. The data needed for hospital lighting type zoning were collected by walk-through

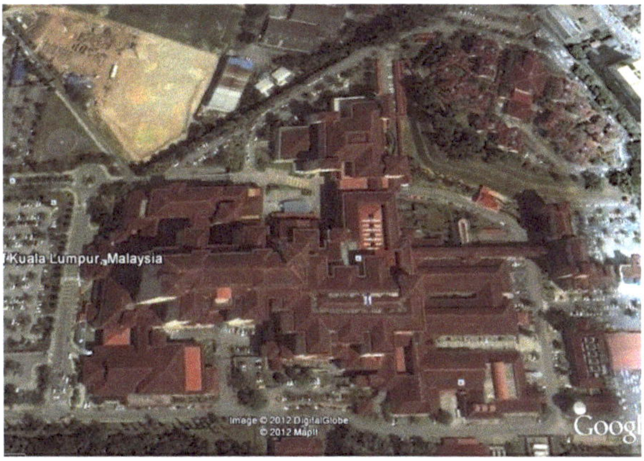

Fig. 37.1 General view of the case study (UKMMC).

Table 37.2 Quantity of different tube using in the field study

Tube types	Power consumption (W)	Quantity
2 ft tube	18	22,033
4 ft tube	36	32,235

observations, surveys, and electrical drawing maps. To light the hospital, 2 and 4 ft T8 fluorescent tubes have been used. Table 37.2 shows the quantity of utilized tubes in UKMMC.

37.2.3 Energy Saving by LED Replacement

LED tubes consume 50% less electricity in comparison to T8 fluorescent tubes used in UKMMC. Moreover, due to the lack of starter and ballast in LED tubes, the electricity usage difference between these two tubes would be more than 50%. In addition, the LED life hour is five times higher, and maintenance demand is less than T8 fluorescent. Table 37.3 illustrates the comparison between T8 fluorescent and LED tubes.

To calculate the annual energy saving (AES), formula 1 can be used as well as formula 2 for estimating the annual bill electricity saving cost.

Table 37.3 T8 fluorescent and LED tubes comparison

–	T8 fluorescent	LED
2 ft tube EC (wattage)	18+8=26	9
4 ft tube EC (wattage)	36+8=44	18
Life hours *(h)*	10,000	50,000
Yearly maintenance cost per tube (*US $)*	2	0.31
Yearly parts replacement per tube	1	0

EC electricity consumption, *LED* light-emitting diode

$$\mathrm{AES} = \frac{n \times h \times (E_{\mathrm{flu}} - E_{\mathrm{ED}})}{1000} \tag{37.1}$$

$$\mathrm{ASC} = \mathrm{AES} \times E_c, \tag{37.2}$$

where

AES annual energy saving is in kWh
n number of replaced tubes
h annual operation hours
E_{Flu} power consumption of each T8 fluorescent tube in Watt
E_{LED} power consumption of each LED tube in Watt
ASC annual saving cost in US $
E_c electricity cost per kWh in US $

37.2.4 Payback Period Calculation

One of the most important factors in applying different energy efficiency strategies is the payback period time. Equation 37.3 can be used to calculate the simple payback period time. Table 37.4 illustrates some information needed to estimate energy saving and payback period time. The LED tubes cost was quoted by P-PLUS LIGHTING SDN BHD.

$$\text{Simple payback period} = \frac{\text{Incremental cost}}{\text{Annual dollar saving}}. \tag{3}$$

37.2.5 Emission Reduction Calculation

By saving energy in buildings, energy generation from power plants can be saved which leads to reduction of CO_2 and other emissions from environment. The reduced emissions can be estimated by Eq 37.4. In addition, emission factor per unit

Table 37.4 Energy and cost saving data in UKMMC lighting

2 Fit T8 Tube power consumption (W)	18
4 Fit T8 Tube power consumption (W)	36
Ballast and starter power consumption (W)	8
2 fit LED tube cost (US $)	28.75
4 fit LED tube cost (US $)	40.62
2 fit LED tube electricity saving (%)	65
4 fit LED tube electricity saving (%)	59
Electricity price (US $/kWh)	0.10

UKMMC University Kembangsaan Malaysia Medical Centre

Table 37.5 Emission factor of fossil fuel for electricity generation

Fuels	Emission factor (kg/kWh)			
	CO_2	SO_2	NO_2	CO
Coal	1.18	0.0139	0.0052	0.0002
Petroleum	0.85	0.0164	0.0025	0.0002
Gas	0.53	0.0005	0.0009	0.0005
Hydro	0.00	0.000	0.0000	0.0000
Others	0.00	0.000	0.0000	0.0000

Table 37.6 Percentage share of fuel used to generate electricity

Coal (%)	Petroleum (%)	Gas (%)	Hydro (%)
16.76	2.44	53.2	27.6

energy and the percentage share of fuel used to generate electricity are illustrated in Tables 37.5 and 37.6 [2].

$$ER = AES \times EF, \tag{4}$$

where ER is the emission reduction in kilogram, and EF is the emission factor in kilogram.

37.3 Results and Discussions

37.3.1 Cost-Saving Calculation

Table 37.7 illustrates the energy cost, life hour, and maintenance difference between T8 fluorescent and LED tubes. The results shown are due to less energy consumption and maintenance demand of LED as well as higher lifetime, almost US $ 3 million can be saved in every 3 years.

Table 37.7 Saving cost by LED replacement in UKMMC

–	T8 Fluorescent	LED	Savings (US $)
Total energy cost per year (US $)	895,685	344,650	551,033
Tube life hours (h)	10,000	50,000	–
Yearly maintenance and parts' replacement cost (US $)	127,190	16,958	110,232
Yearly power consumption and maintenance cost (US $)	1,022,875	361,609	661,265
3 years power consumption and maintenance cost (US $)	3,068,625	1,084,829	1,983,796

LED light emitting diode, *UKMMC* University Kembangsaan Malaysia Medical Centre

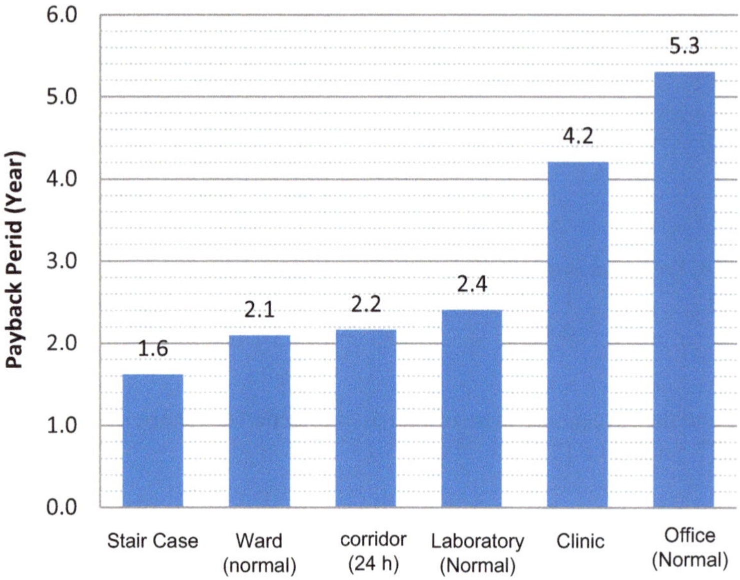

Fig. 37.2 Lighting zones payback period

37.3.2 Payback Period Times

Payback period time of lighting retrofitting for both the entire UKMMC and each lighting zone in the hospital were calculated. Results showed that for the UKMMC, the payback period is 2.9 years. Figure 37.2 demonstrates the payback period time in zones separately. According obtained results, "staircases" have the lowest rate of payback period time, which is 1.6 years and "wards" are in second place with 2.1 years. Different operation hours and usage factors were resulted in different payback period.

37.3.3 Emission Analysis

Table 37.8 illustrates the avoided emissions trough lighting retrofitting in UKMMC per year. The result shows that by LED tubes replacement in this hospital, more than 5000 tonnes of CO_2 will be reduced from the atmosphere each year.

Table 37.8 Annual avoided emission

Emission	CO_2	SO_2	NO_2	CO
Reduced (kg/year)	5,186,201	2,595,528	15,537	7319

37.4 Conclusion

A field study on energy and cost saving, payback period, and emission reduction through lighting retrofitting was carried out in a large-scale hospital. Results show almost US $ 3 million saving in 3 and 2.9 years as the initial cost payback period. This study shows that through only LED tubes replacement in this large-scale hospital, more than 5000 tonnes of CO_2 will be reduced from the environment.

Acknowledgment The authors would like to express their gratitude to the Solar Energy Research Institute (SERI), Universiti Kebangsaan Malaysia, the Ministry of Science, Technology and Innovation Malaysia for sponsoring the work under the project Science fund UKM-DLP-2011-032. Thanks as well to P-PLUS LIGHTING SDN. BHD and the UKMMC staff for their valuable assistance in this project.

References

1. Pérez-Lombard L, Ortiz J, Pout C (2008) A review on buildings energy consumption information. Energy Build 40(3):394–398
2. Saidur RÃ (2009) Energy consumption, energy savings, and emission analysis in Malaysian office buildings. Energy Policy 37(10):4104–4113
3. Lim Y-W, Kandar MZ, Ahmad MH, Ossen DR, Abdullah AM (2012 Nov) Building façade design for daylighting quality in typical government office building. Build Environ 57:194–204
4. Chan S, Reid MD, Ficek Z (2009) Entanglement evolution of two remote and non-identical Jaynes–Cummings atoms. J Phys B 42(6)
5. Moghimi S, Mat S, Lim C (2011 July 14–16) Building energy index (BEI) in large scale hospital: case study of Malaysia. In: Proceeding GEMESED'11, Proceedings of the 4th WSEAS international conference on "Recent Researches in Geography Geology, Energy, Environment and Biomedicine. Corfu Island, Greece, pp 167–170
6. Azizpour F, Moghimi S, Mat S, Lim C, Sopian K Objective and subjective assessments of thermal comfort in hot-humid region. In: Proceedings of 5th WSEAS international conferences on Recent Researches in Chemistry, Biology, Environment and Culture, Montreux, Switzerland, pp 207–210, 29–31 December
7. Thejo Kalyani N, Dhoble SJ (2012 Jun) Organic light emitting diodes: energy saving lighting technology—a review. Renew Sustain Energy Rev 16(5):2696–2723
8. Azizpour F, Moghimi S, Lim C (2011) Thermal comfort assessment in large scale hospital: case study in Malaysia. In: Proceedings of the 4th WSEAS international conference on Recent Researches in Geography, Geology, Energy, Environment and Biomedicine, pp 171–174
9. Azizpour F, Moghimi S, Lim CH, Mat S, Salleh E, Sopian K (2012) A thermal comfort investigation of a facility department of a hospital in hot-humid climate: correlation between objective and subjective measurements. Indoor and Built Environ 1420326X12460067, 2012. doi:10.1177/1420326X12460067
10. Moghimi S, Azizpour F, Mat S, Lim CH, Salleh E, Sopian K (2013) Building energy index and end-use energy analysis in large-scale hospitals—case study in Malaysia. Energy Efficiency 1–14. doi:10.1007/s12053-013-9221-y

Chapter 38
The Effect of Wind Velocity and Night Natural Ventilation on the Inside Air Temperature in Passive Cooling Ventilation in Arid Zones

H. Bencheikh

Abstract The effect of wind velocity and night natural ventilation in lowering the inside daytime air temperature in passive cooling in arid zones were investigated by numerical calculations and experimental means for different values of air change flow rate due to infiltrations and natural ventilation and different wind speed. The numerical calculations based on the inside outside air temperature, wind speed, cracks, and openings dimensions to determine the volume of air change per hour.

The experimental model was a test cell with the door facing north and the window in the opposite side facing south, the volume of the model was 9 m^3. The calculated and measured results show that 1.8 volume per hour of air change flow rate and 2 m/s wind speed show a high concordance between calculated and measured inside air temperature and can lower the inside air temperature by 3–4 °C compared to non-ventilated test cell [1].

Keywords Passive cooling · Wind velocity · Night ventilation · Arid zone

38.1 Introduction

Night natural ventilation potential for improving thermal comfort in buildings has been investigated by numerical and experimental means; Santamouris and coworkers [2, 3] introduced an interested method to calculate the energy contribution of night ventilation technique to the cooling load of a building. Many researchers investigate the improvement of space cooling by night natural ventilation such as Kolokotroni et al. [4] used temperature and humidity charts to generate a predesign tool for summer cooling evaluation of night ventilation. Geros et al. [5] carried out an experimental evolution of night ventilation effect on the inside air temperature for four buildings and simulations investigation to determine the effect of air change rate on night natural ventilation. Givoni [6] carried out experimental work

H. Bencheikh (✉)
Laboratoire de génie civil, Université Amar Telidji, Laghouat, Algeria
e-mail: h.bencheikh@mail.lagh-univ.dz

A. Sayigh (ed.), *Renewable Energy in the Service of Mankind Vol I*,
DOI 10.1007/978-3-319-17777-9_38

to investigate the effectiveness of night ventilation in lowering the indoor daytime temperature and many other researchers worked on the same subject.

This chapter focuses on numerical calculation and experimental work to validate the theoretical results on night natural ventilation effect in lowering the daytime inside air temperature.

38.2 Numerical Model

38.2.1 Airflow Rates Calculation

Airflow into buildings is either by infiltration of outside air through cracks around exterior openings or by natural ventilation through opened exterior openings. Any outdoor air that enters by way of infiltration or ventilation is assumed to be immediately mixed with inside air.

38.2.1.1 Infiltration

The determination of the amount of infiltration air is quite complicated and subject to significant uncertainty. The infiltration quantity is converted from a number of air change by hour (ACH) and unclouded in the inside air heat balance using the outside air current simulation time step.

There are three models for estimating the infiltration flow rate. The first is the design flow rate, the second is the effective leakages area based on Sherman and Grimsrud [7], and the third is the flow coefficient model based on Walker and Wilson [8].

Infiltration Design Flow Rate

The flow of air from the outside environment directly into the inside one is generally caused by cracks around exterior openings, temperature differences, and wind speed. The basic equation Coblens and Achenbach [9] used to calculate infiltration:

$$Infiltration = (Inf_{design})(F_{schedule})(A + B)(T_{ai} - T_{ae})[CV + DV^2)] \quad (38.1)$$

Infiltration Effective Leakage Area

Infiltration leakage area model is based on Sherman and Grimsrud [7]:

$$Infiltration = (F_{schedule})\frac{A_i}{1000}\sqrt{C_s \Delta T + C_w V} \quad (38.2)$$

where

$F_{schedule}$ is a value from the user-defined schedules
A_i is the effective air leakage area in cm² that corresponds to a 4Pa pressure differential
C_s is the coefficient from stack-induced infiltration in $\left(\frac{L}{S}\right)^2\left(\frac{1}{\text{cm}^4 K}\right)$
ΔT is the absolute difference between inside and outside air temperature
C_w is the coefficient for wind-induced infiltration in $\left(\frac{L}{S}\right)^2\left(\frac{1}{\text{cm}^4 K}\right)$
V is the local wind speed in m/s

Infiltration by Flow Coefficient

The flow coefficient model is based on Walker and Wilson [3] equation.

$$Infiltration = F_{schedule}\sqrt{(cC_s\Delta T)^2 + [cC_w(sV)^{2n}]^2} \tag{38.3}$$

where

c is the flow coefficient in m²/(s.Pa^n)
C_s is the coefficient from stack-induced infiltration in $\left(\frac{Pa}{k}\right)^n$
n is the pressure exponent
C_w is the coefficient for wind-induced infiltration in $\left(Pa.\frac{s^2}{m^2}\right)^2$
s is the shelter factor

38.2.1.2 Natural Ventilation

Natural ventilation is a controlled air change with the exterior environment through openings due to temperature differences, wind speed, and the opening area. The controlled natural ventilation calculation is based on three models.

Design Flow Rate

Design flow rate is based on the outside inside temperature difference, wind speed, and openings area, the basic equation to calculate the design flow rate is

$$Ventilation = (v_{design})(F_{schedule})(A + B)(T_{ai} - T_{ae})[CV + V^2)] \tag{38.4}$$

Ventilation by Wind and Stack with Open Area

The ventilation airflow rate is a function of wind speed and thermal stack effect, along with the area of the opening, the equation used to calculate the ventilation rate is (ASHRAE) [10].

$$Q_w = C_w A_{open} f_{schedule} V \tag{38.5}$$

where

Q_w volumetric airflow rate driven by wind $\left(\frac{m^3}{s}\right)$
C_w opening effectiveness dimensionless
A_{open} opening area (m^2)

$$C_w = 0.55 - \frac{(Effective\,angle - wind\,direction)}{180} * 0.25.$$

If $(Effectiveangle - winddirection) > 180^{\circ}$ the difference $= -180°$.

The equation used calculates the ventilation rate due to stack effect as given by ASHRAE 2009 handbook:

$$Q_s = C_D\, A_{open} F_{schedule} \sqrt{2g\Delta H_{NPL} (|T_{ai} - T_{ae}|) / T_{ai}}, \tag{38.6}$$

where

Q_s volumetric airflow rate due to stack effect (m^3/s)
C_D discharge coefficient for opening dimensionless
ΔH_{NPL} height from midpoint of lower opening to the neutral pressure level (m)
T_{ai}, T_{ae} the inside and outside air temperature in degree (K)

The discharge coefficient for opening dimensionless C_D is given by ASHRAE 2009 handbook:

$$C_D = 0.40 + 0.0045|T_{ai} - T_{ae}|.$$

The total ventilation rate by wind and stack airflow:

$$Q_{vt} = \sqrt{Q_s^2 - Q_w^2}.$$

38.3 Experimental Work

An experimental study of the effect of night natural ventilation on daytime inside air temperature was carried out for a typical summer day of June for Laghouat in Algeria. The first experimental work was done under clear sky and 0.25 m/s wind speed during the night-time, the second day under cloudy sky and 0.81 m/s wind speed, after 3 days of experiment the night natural ventilation had a good effect on in lowering the inside air temperature. The effect of air change flow rate and wind speed were studied and simulated for an average wind speed of 3 m/s and a variable air change flow rate (1.8, 2.7, and 3.6 V/h). Figure 38.1 shows the variations of inside air temperature for

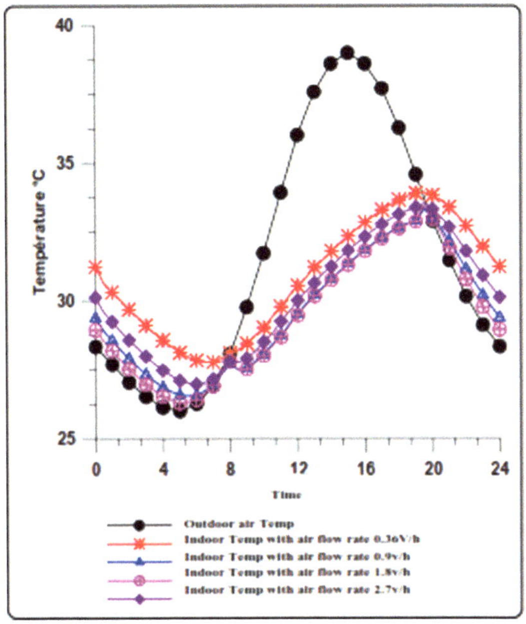

Fig. 38.1 Inside air temperature for 3 m/s wind speed and variable air change flow rate

constant wind speed 3 m/s and variable air change flow rate, the inside air temperature decreases with the increase of air change flow rate. When the air change flow rate taken to its maximum for variable wind velocities, for velocity ≥3 m/s the inside air temperature had the almost the same values as shown in Fig. 38.2.

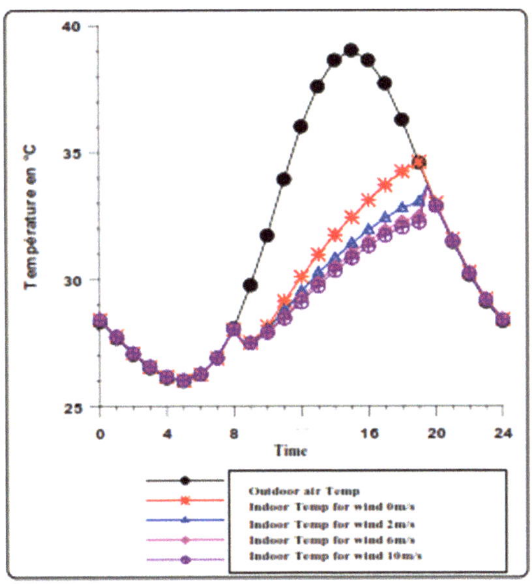

Fig. 38.2 Inside air temperature for 7.2 V/h air change flow rate and variable wind speed

38.4 Results and Discussions

Figures 38.3, 38.4, 38.5, and 38.6 shows the inside air temperatures variations for variable air change flow rate (0.9, 1.8, 3.6, and 7.2 V/h) and variable wind speed (from 0 to 10 m/s). After analyzing the results obtained by experimental and

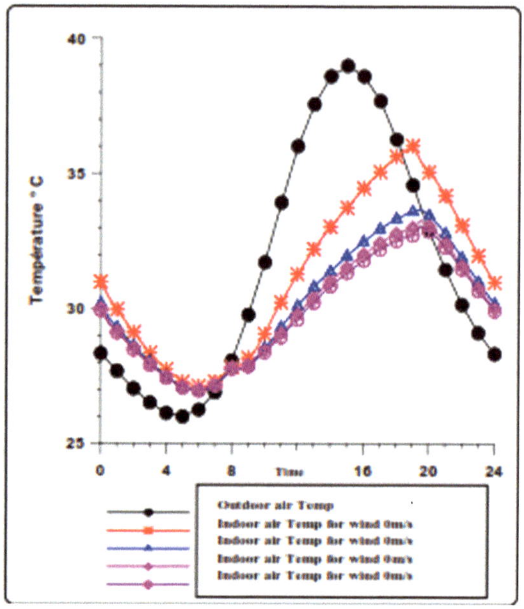

Fig. 38.3 Inside air temperature for 0.9 V/h air change flow rate and variable wind speed

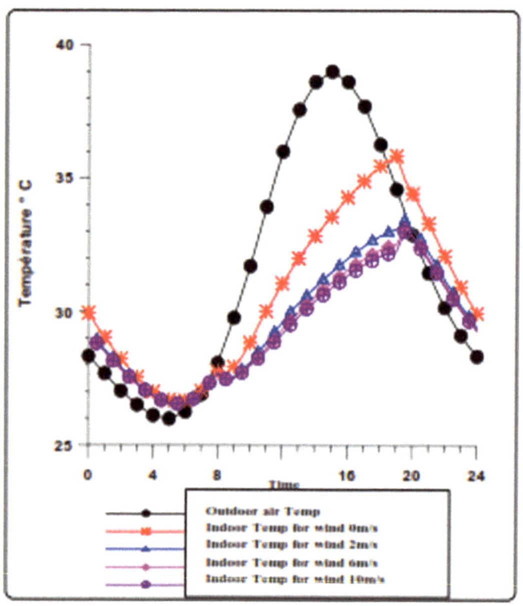

Fig. 38.4 Inside air temperature for 1.8 V/h air change flow rate and variable wind speed

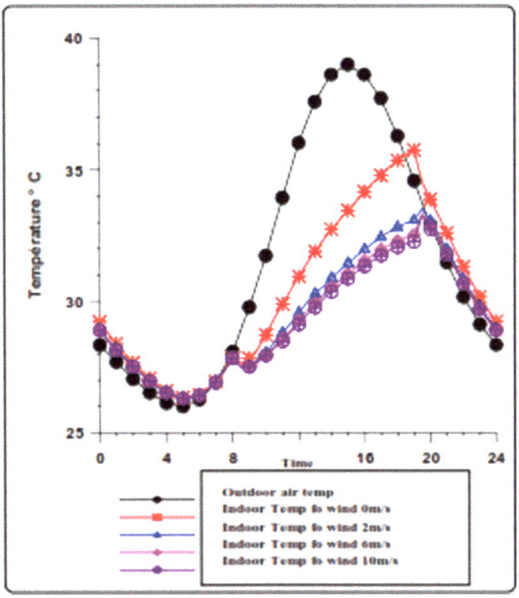

Fig. 38.5 Inside air temperature for 3.6 V/h air change flow rate and variable wind speed

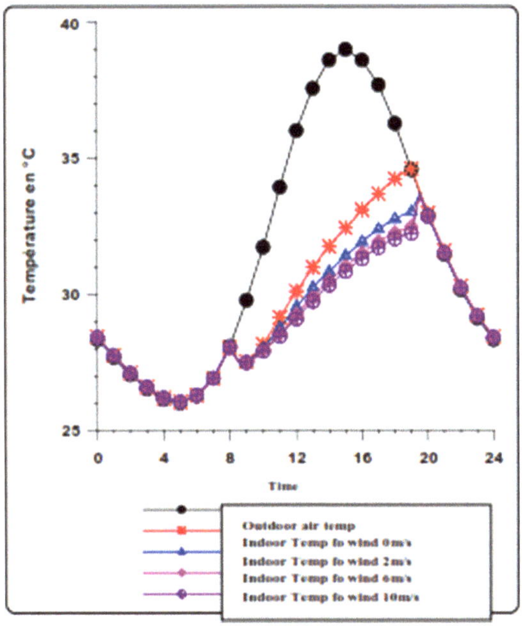

Fig. 38.6 Inside air temperature variation for 7.2 V/h air change flow rate and variable wind speed

simulation, by comparing the results, the values of 1.8 V/h of air change flow rate and 2 m/s wind speed give the smaller error between measured and simulated inside air temperature as shown in Fig. 38.7. Figure 38.8 presents the measured and simulated inside air temperature for non-ventilated space, the two curves have almost

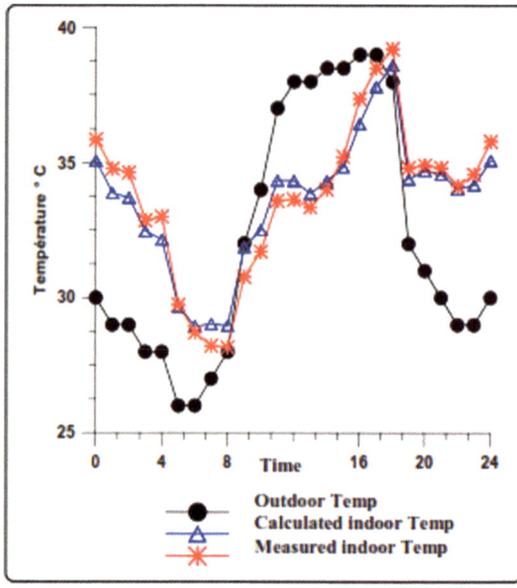

Fig. 38.7 Measured and simulated inside air temperature without night natural ventilation

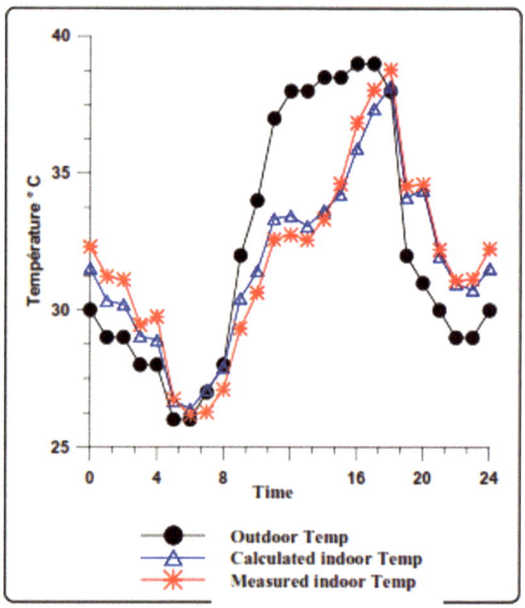

Fig. 38.8 Measured and simulated inside air temperature with night natural ventilation [11]

the same values which mean the simulations are accurate. Figure 38.9 presents the measured and simulated inside air temperature for night ventilated space for 1.8 V/h air change flow rate and 2 m/s wind speed, the two inside air temperature curves show a smaller error.

38.5 Conclusion

The results presented in Fig. 38.7 show smallest differences between the two curves of inside air temperature specially for the space without night natural ventilations; these differences were due to, wind speed which was considered a constant value during simulation period. This was different from the reality; the wind speed was very variable from time to time and sometimes presented big differences which effects directly the air change flow rate which effects the inside daytime air temperature. In the space, without night natural ventilation and when the wind speed was assumed null for space with night ventilation, the two curves of the inside air temperature measured and simulated have almost the same values as shown in Fig. 38.8. Figure 38.9 shows the inside air temperature for the space with and without night natural ventilation; so the night natural ventilation for space in arid zone participate in lowering the daytime inside air temperature from 3 to 6 °C as shown in Fig. 38.9.

Fig. 38.9 Measured inside air temperature with and without night natural ventilation for space with cooling roof [11]

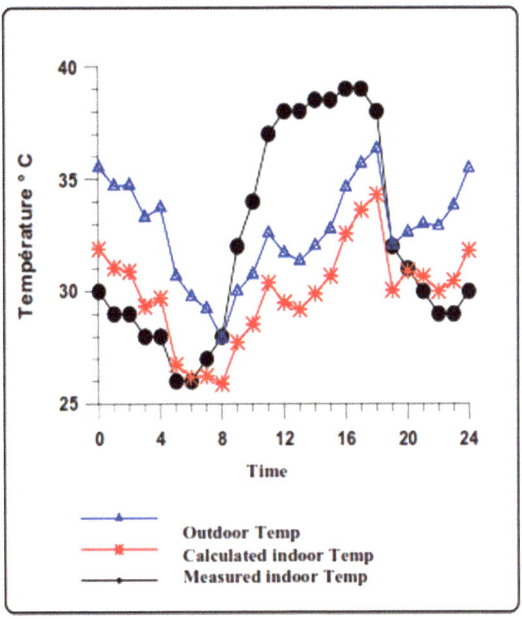

References

1. Bencheikh H, Bouchair A (2004) Passive cooling by evap-reflictive roof for hot dry climate. Renew Energy 29(11):1877–1886
2. Santamouris M, Mihalakakou G, Argiriou A, Asimakopoulos D (1996) On the efficiency of night ventilation techniques for thermostatically controlled buildings. Solar Energy 56(6):479–483
3. Santamouris M, Mihalakakou G, Asimakopoulos D (1997) On the coupling of thermostatically controlled buildings with ground and night ventilation passive dissipation techniques. Solar Energy 60(3–4)191–197
4. Kolokotroni M, Webb B, Hayes S (1998) Summer cooling with night ventilation for office buildings in moderate climates. Energy Build 27:231–237
5. Geros V, Santamouris M, Tsangrasoulis A, Guarracino G (1999) Experimental evaluation of night ventilation phenomena. Energy Build 29:141–154
6. Givoni B (1998) Effectiveness of mass and night ventilation in lowering the indoor daytime temperatures. Energy Build 28:25–32
7. Sherman MH, Grimsud DT (1980) Infiltration-pressurization correlation, simplified physical modeling. ASHRAE Trans 86(2):778
8. Walker IS, Wilson DJ (1998) Field validation of equations for stack and wind driven air infiltration calculations. Int J HVAC R Res 4(2):119–140
9. Coblenz CW, Achenbach PR (1963) Field measurement of ten electrically heated houses. ASHRAE Trans 69:358–365
10. Malcolms S Orme (2001) Air infiltration and ventilation. ASHRAE HVAC 2001 fundamentals handbook chapter 26
11. Bencheikh H (2013) Full scale experimental studies of a passive cooling roof in hot arid areas. Int J Renew Energy Technol Res 2(6):170–180

Chapter 39
The Building Energy Consumption and Outdoor Design Conditions of Severe Cold Regions Based on Climate Change

Teng Shao and Hong Jin

Abstract Climate change and its effects is a topic that attracts the attention of human beings all over the world. In recent years, the extreme climate changes frequently occurring in China's severe cold regions have led to the increase in building energy consumption and decrease in the winter indoor thermal comfort. This chapter selects rural dwellings as the research object, and based on the meteorological data of Harbin in 63 years, it applies the frequency distribution analysis to obtain the extreme temperature values under different recurrence intervals; meanwhile, taking these values as the outdoor design conditions, this study simulates and compares the building energy consumption in different conditions, so as to reveal the effects of climate change on building energy consumption of heating in winter and provide the quantitative basis for the energy-saving optimization design of rural dwellings in severe cold regions.

Keywords Climate change · Severe cold regions · Energy consumption simulation · Outdoor design conditions

39.1 Introduction

Along with the increasing deterioration of ecological environment, the extreme weather events, such as low temperature, drought, and rainstorm, have a higher frequency of occurrence, stronger degree of damage, and a wider range of influence [1]. Especially in the recent years, the extreme weather events frequently occur in winter of China's severe cold regions, leading to urban traffic paralysis, heating power supply being affected, and so on, meanwhile triggering the sustained growth of building energy consumption. The bad weather conditions presented severe challenges for China's severe cold regions; therefore, calculating the extreme low-temperature values and then taking them as outdoor design conditions have important real-world implications.

T. Shao (✉) · H. Jin
School of Architecture, Harbin Institute of Technology, NO. 66 Xidazhi Street, Nangang District, Harbin 15001, China
e-mail: 626682132@163.com

A. Sayigh (ed.), *Renewable Energy in the Service of Mankind Vol I*,
DOI 10.1007/978-3-319-17777-9_39

The probabilistic methods and distribution characteristics of extreme climate events have been carried out in some studies by domestic and overseas researchers, such as Linda O. Mearns [2] who, by taking maximum wind speed as an example, introduced the parameter estimation method and the calculation of distribution probability value in detail. They also indicated that the Gumbel distribution and Weibull distribution are two more suitable methods that can be applied for the calculation of extreme values occurrence intervals. Liu Guanghai [3], on the basis of the statistics of summer outdoor high-temperature data from 1951 to 1981, factored in the frequency analysis method and applied the Gumbel distribution to fit the summer outdoor temperature parameters, and then taking Beijing as an example, derived the summer outdoor high-temperature distribution diagram under conditions of different reliability. Jing et al. [4] used 50 year's lowest temperature data (1961–2010) of Fujian province's 67 weather stations to analyze the time and space distribution regularity of extreme low temperatures and also applied the Gumbel distribution functions to calculate the annual extreme temperatures of every station. Zengwu et al. [5] adopted the Weibull distribution and Gumbel distribution to conduct fitting experiments on the basis of Chongqing's annual minimum temperature values from 1951 to 1996 (46 years). The experiment showed that Weibull distribution is the optimal asymptotic distribution for Chongqing's annual extreme low temperatures. In conclusion, as these studies mostly aim at the special urban climate conditions, they cannot be applied to all cases.

39.2 Statistics of Meteorological Data

The commonly used sampling methods of frequency analysis comprise annual maximum value method, annual super value method, and super quantitative method; of these, the maximum value method is very simple and easy to select and often used in the statistical analysis of heavy rain and flooding in China [1]. meanwhile many studies indicate that the relationship of annual maximum values between adjacent years is very weak and can be thought of being independent, so adopting the annual maximum values method "*n*" maximum values in "*n*" years can be selected in the observational data. This chapter, according to each year's daily average temperature and daily extreme low temperature of Harbin in 63 years, presents the statistics of the annual daily average extreme low temperature and the annual extreme low temperature [6], and it applies frequency analysis to calculate the annual extreme low temperature and the annual daily average extreme low temperature of different recurrence intervals (Table 39.1).

39.3 Calculation of Extreme Temperature Under Different Recurrence Intervals

At present, Pearson-III distribution, Weibull distribution, and Gumbel distribution are recognized as excellent methods of extreme temperature analysis in China and overseas. Many researchers insisted that under extreme climatic conditions, annual

Table 39.1 Statistics of extreme temperature and daily average extreme temperature of Harbin. (Unit: °C)

Y	ET	AET	Y	ET	AET	Y	ET	AET	Y	ET	AET
1951	−34.6	−30.4	1967	−35.7	−31.2	1983	−33.2	−25.3	1999	−27.8	−22.3
1952	−32.3	−27.4	1968	−34.1	−28.4	1984	−33.7	−26.8	2000	−33.0	−27.0
1953	−35.0	−29.8	1969	−34.5	−27.5	1985	−37.7	−30.9	2001	−37.3	−30.9
1954	−29.4	−22.6	1970	−38.1	−33.0	1986	−33.3	−28.2	2002	−26.8	−21.0
1955	−32.0	−25.7	1971	−29.7	−25.1	1987	−36.1	−30.0	2003	−25.9	−22.6
1956	−32.5	−27.2	1972	−31.1	−25.6	1988	−30.1	−22.6	2004	−29.6	−24.3
1957	−30.9	−26.2	1973	−31.1	−26.4	1989	−33.0	−26.6	2005	−28.7	−22.1
1958	−37.5	−30.0	1974	−33.1	−26.0	1990	−37.5	−30.5	2006	−29.6	−24.1
1959	−35.1	−28.9	1975	−27.1	−21.1	1991	−30.9	−25.1	2007	−22.6	−17.1
1960	−33.7	−28.0	1976	−35.6	−30.9	1992	−33.0	−24.7	2008	−28.0	−23.6
1961	−36.3	−28.6	1977	−35.5	−30.8	1993	−32.2	−26.4	2009	−32.3	−27.0
1962	−31.7	−23.0	1978	−32.7	−27.0	1994	−34.6	−26.8	2010	−31.8	−27.6
1963	−34.4	−28.5	1979	−31.6	−25.9	1995	−27.3	−20.9	2011	−32.5	−28.0
1964	−32.5	−25.0	1980	−36.7	−32.0	1996	−27.7	−23.5	2012	−31.3	−26.4
1965	−33.7	−28.1	1981	−30.9	−25.6	1997	−30.3	−25.2	2013	−32.3	−28.0
1966	−33.0	−28.7	1982	−31.6	−24.8	1998	−28.6	−22.7			

Y years, *ET* annual extreme low temperature, *AET* daily average extreme low temperature

extreme temperatures comply with the Gumbel distribution and are appropriate for the frequency analysis method. In line with the statistical sample, we can obtain the accurate calculated parameters and facilitate analysis and comparison [7]. Therefore, the Gumbel distribution is applied to fit the frequency of extreme temperature distribution and four recurrence intervals—10, 20, 50, and 100 years—are selected for calculation and analysis.

39.3.1 Calculation Model

Gumbel distribution is a theoretical model for describing the extreme values distribution. The related studies show that it is reasonable to apply the Gumbel model to simulate the frequency distribution of annual extreme low temperature and annual daily average extreme low temperature. When it is used to statistically analyze the distribution of extreme low values, the distribution function is as follows [1, 7]:

$$F(X < x) = 1 - e^{-e^{a*(x+b)}} \tag{39.1}$$

where a is the scale parameter, $\frac{1}{a} = \frac{\sqrt{6}}{\pi}\sigma$, σ is the standard deviation of meteorological parameters; *b*, the number of distribution density, can be calculated by the mathematical expectation of meteorological parameters, that is $b = -\overline{x} - \frac{c}{a}$, *c* is Euler's constant, the value is 0.57722 [1]. Using *a* and *b* values given above, the Gumbel distribution function of extreme low temperature can be calculated as follows:

$$x = \frac{[\ln\ln\frac{1}{1-Fx}+0.5772]\sigma}{1.283}+\overline{x} \quad (39.2)$$

According to the statistics of extreme low temperature given in Table 39.1, the standard deviation and mathematical expectation of meteorological parameters can be calculated to sequentially get the calculation models of annual extreme low temperature and daily average extreme low temperature as follows:

Calculation model of annual extreme low temperature:

$$X = \ln\ln\frac{1}{1-Fx}\mathrm{x}2.465-30.877\,^{\circ}\mathrm{C}$$

Calculation model of annual daily mean extreme low temperature:

$$X = \ln\ln\frac{1}{1-Fx}x2.428-25.099\,^{\circ}\mathrm{C}$$

39.3.2 *Computed Result*

When considering the energy consumption differences of extreme climate conditions, the extreme low temperatures under different recurrence intervals must be calculated. For a minimal value, if the probability value F(x) is given, the minimal value X can be calculated. The corresponding probability of recurrence intervals—10 years, 20 years, 50 years, and 100 years—are respectively 10%, 5%, 2%, and 1%, and they are then substituted in the calculation model to calculate the extreme low temperatures; the results are shown in Table 39.2.

Table 39.2 Statistics of extreme low temperature under different return periods

Occurrence intervals (years)	10	20	50	100
Annual extreme low temperature	−36.8	−38.2	−40.5	−42.2
Annual daily mean extreme low temperature	−30.6	−32.3	−34.6	−36.3

39.4 Building Energy Consumption Contrast of Extreme Weather Conditions

The existing thermal design criteria only can satisfy the human comfort requirement under normal conditions. When the extreme temperatures occur, the building energy consumption rapidly increases and the heating facilities cannot meet the human comfort demand. Therefore, the chapter, taking building heating load as an evaluation standard, applies DesignBuilder software to calculate and analyze the heating load under different indoor design temperatures and different extreme temperatures of recurrence intervals.

39.4.1 Simulation Settings of Building Energy Consumption in Severe Cold Regions

39.4.1.1 Selection of Simulation Object

Rural dwellings are growing rapidly in recent years and account for a large proportion of residential buildings in China. Their low storeys, small size, and large shape coefficient lead to larger heating energy consumption than the urban buildings. Therefore, the traditional rural dwellings of China's severe cold regions were selected as the baseline building of energy consumption contrast. As shown in Fig. 39.1, the building area is 70.06 m^2 with a width of 11.3 m, depth of 6.2 m, a ceiling height of 2.8 m, and the specific building envelope construction as shown in Table 39.3.

Fig. 39.1 Typical rural dwellings of China's severe cold regions

Table 39.3 Building envelope construction of typical rural dwellings in severe cold regions

Building envelope	Construction	Heat coefficient
External wall	10 mm cement mortar + 490 mm brick + 10 mm composite mortar	1.24 W/(m² K)
Roof	Suspended ceiling + Wood joist + 100 mm plant ash + wood roof truss + tiled roof	0.93 W/(m² K)
Ground	Rammed earth + concrete cushion + 20 mm cement mortar	–
External window	Monolayer plastic steel window of single frame double glass	2.20 W/(m² K)
External door	Double wood door	2.33 W/(m² K)

39.4.1.2 Settings of Simulation Parameters

The simulation parameters mainly include two aspects: outdoor parameters and indoor parameters. Applying annual extreme low temperature and daily average extreme low temperature whose statistics are shown in Table 39.2 presents as the worst outdoor meteorological parameters. The indoor thermal environment contains design temperature, air changing, and heat source, and the specific settings are as follows: In line with the "Design standard for energy efficiency of rural residential buildings" [8], 14 °C and 16 °C were selected as the design temperature, but considering the human comfort requirement under extreme climate conditions, and combining the field test and survey, 18 °C was also selected. Air change rate is 0.5/h; equipment heat dissipation, human body clothing, solar radiation, etc. are set by the building actual use.

39.4.2 Contrast and Analysis of Simulation Results

The existing outdoor calculated temperature of heating period is the average temperature which does not guarantee 5 days on average over the years. According to the design standard, the outdoor calculated temperature of Harbin is −26 °C. This chapter, taking the heat load which is calculated to be −26 °C as base value, respectively forms contrasts and analyzes the differences between simulated heat load values and base values based on annual extreme low temperature and annual daily average extreme low temperature.

39.4.2.1 Energy Consumption Analysis Based on Extreme Temperature

The annual extreme low temperature is the minimum temperature among daily extreme low temperatures. When the interior design temperatures are 14 °C, 16 °C, and 18 °C, respectively, the building heat loads are as shown in Fig. 39.2.

It can be seen from Fig. 39.2 that with the gradual increase of indoor design temperature, there is a significant increase in building heat load; meanwhile, the energy consumption amplification under different recurrence intervals is equal, about

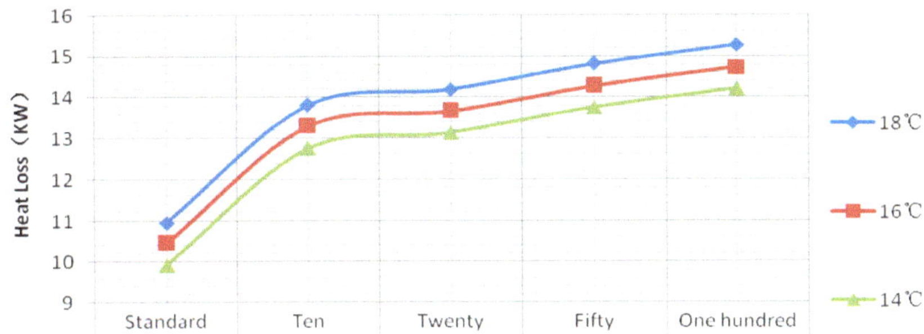

	Standard	Ten	Twenty	Fifty	One hundred
18℃	10.96	13.81	14.19	14.81	15.27
16℃	10.45	13.29	13.66	14.28	14.73
14℃	9.93	12.76	13.14	13.75	14.21

Fig. 39.2 Building heat load under different return periods and indoor temperatures

0.52 kW. With the increase of recurrence intervals, the building heat load also showed a sustained rise. Under recurrence intervals of 10, 20, 50, and 100 years, the growth rates than standard are respectively 26, 30, 35, and 40 %, making the building heat load obviously increase. This will cause the heating energy consumption to greatly augment and the indoor temperature also difficult to maintain in thermal comfort state.

39.4.2.2 Energy Consumption Analysis Based on Daily Average Extreme Temperature

The annual daily average extreme low temperature is the minimum temperature among every day average extreme low temperatures. When the interior design temperatures are 14, 16, and 18 °C, respectively, the building heat loads are as shown in Fig. 39.3.

It can be seen from Fig. 39.3 that with the gradual increase of indoor design temperature, the variation tendency of building heat load is similar to the conditions of annual extreme low temperature; with the increase of recurrence intervals, the building heat load also shows an increasing trend. Under recurrence intervals of 10, 20, 50, and 100 years, the growth rates than standard are respectively 11, 15, 21, and 和 25 %, only the amplification is reduced. In conclusion, whether in the conditions of annual extreme low temperature or annual daily average extreme low temperature, the heat loads of rural dwellings all have an obvious increase. If the building design is in line with the existing outdoor calculated temperature, the indoor thermal environment is bound to not meet human comfort requirement when the extreme weather event occurs, and leads to a significant increase in building energy consumption, causing economic and energy wastage. Therefore, the outdoor and indoor calculated temperatures should be reasonably confirmed, so as to reach the dual purposes of energy-saving and economy.

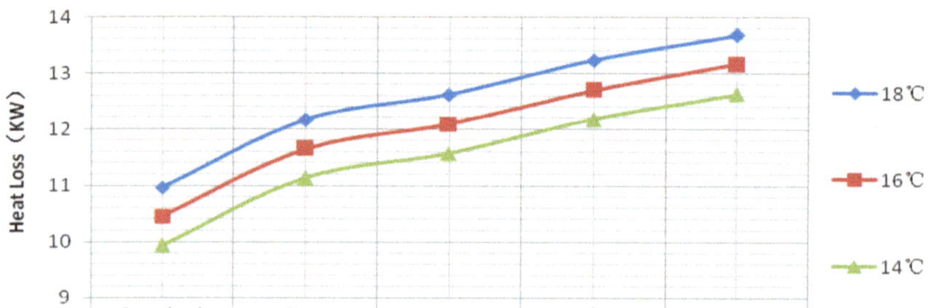

	Standard	Ten	Twenty	Fifty	One hundred
18℃	10.96	12.17	12.62	13.23	13.68
16℃	10.45	11.65	12.1	12.7	13.16
14℃	9.93	11.13	11.57	12.18	12.63

Recurrence intervals (years)

Fig. 39.3 Building heat load under different return periods and indoor temperatures

39.5 Conclusions

In this chapter, by applying the Gumbel distribution functions to statistics and analyzing the meteorological data of Harbin in 63 years, the annual extreme low temperature and annual daily average extreme low temperature under different recurrence intervals were calculated and then taking Harbin's typical rural dwellings as an example DesignBulider software was adopted to obtain the percentage difference between simulated heat load and base values of different recurrence intervals respectively under the indoor design temperature of 14, 16, and 18 °C. It indicates that building heating energy consumption will significantly increase when the extreme weather event occurs and indoor temperature also becomes difficult to maintain in the thermal comfort range. Hence, in the energy-saving design of rural dwellings,calculating the extreme temperatures and contrasting the building energy consumption are necessary.

Acknowledgements This chapter was supported by the Specialized Research Fund for the Doctoral Program of Higher Education (Grant No.: 20122302110041) and the National Natural Science Foundation of China (Grant No.: 51378136).

References

1. Kaiyu M (1993) Climate statistical principle and method. China Meteorological Press, Beijing
2. Chengdao Z Extreme high temperature phenomenon: the probability change with the average temperature variation. Meteorol Res 143:48–56
3. Guanghai L, Ruhe X (2009) Analysis and determine of the high temperature parameters model based on Gumbel distribution. J Guangzhou Univ (Nat Sci Ed) 18(4):08
4. Jing L, Hui C et al (2011) Annual extreme low temperature distribution and its parameter estimation in Fujian Province. Chin J Agro-meteorol 32(Suppl1):24–27

5. Zengwu W, Qingzhen M, Ruifeng Y, Xiaolei G (2004) The extreme value asymptotic distribution and parameter estimation of Chongqing ground minimum temperature. J Chengdu Univ Inf Technol 19(3):09
6. China Meteorological Data Sharing Service System. http://cdc.cma.gov.cn/home.do
7. Rong W (2008) Study on impact of climate change on heating/cooling energy consumption and outdoor design conditions of HVAC based on climate. Doctoral Dissertation of Xi'an University of Architecture and Technology
8. DB23/T 1537-2013 (2013) Design standard for energy efficiency of rural residential buildings in Heilongjiang Province. Urban and Rural Housing Construction in Heilongjiang Province

Chapter 40
Low-Energy Architecture: Cuban Contradictions

Dania González Couret

Abstract Cuba has a very strong tradition in low-energy architecture along the history. The aborigine architecture was a clever answer to the local environmental conditions, which have been continued by the vernacular and rural ones. Despite that the colonial cities followed the Mediterranean model, they were progressively transformed for a better adaptation to the climatic conditions. Even the modern architecture developed during the first half of the twentieth century, recovered the best former traditions, combined with the new aesthetic, generating a highly qualified built environment.

During the past half a century, research works have been systematically carried out, focused on the evaluation of architectural and urban solutions according to climatic and energy behavior. These issues were included in the curricula of architecture and urbanism since the 1960s, and several standards have been elaborated during the past 30 years to propose principles and regulations to take advantage of daylight and ventilation, as well as to reduce the thermal load in buildings.

However, most of the contemporary Cuban architecture does not apply the research results, taking into account the standards, or following the tradition of the Cuban passive low-energy architecture, but on the contrary, it use to copy foreign models coming from developed countries with cold climates. The results of study cases are presented in the chapter in order to evaluate the architectural solutions, trying to discuss the causes of the wrong practice.

Keywords Low-energy architecture · Standards · Research · Architectural tradition

---

D. G. Couret (✉)
Faculty of Architecture, Instituto Superior Poiltécnico José Antonio Acheverría, Havana, Cuba
e-mail: dania@arquitectura.cujae.edu.cu; dgcouret@yahoo.es

A. Sayigh (ed.), *Renewable Energy in the Service of Mankind Vol I*,
DOI 10.1007/978-3-319-17777-9_40

40.1 Introduction

The consumption of fossil fuels have been a determinant factor in the world ecological crisis that motivated the formulation of the sustainable development model at the end of the past century. The so-called "peak oil" means the end of the fossil fuels or, at least, of the cheap fuel oil.

Consequences of the crisis are the climate change and the loss of biodiversity. Some of the actions developed at an international scale to mitigate these effects are the movement Architecture 2030, promoted by the American architect Edward Mazria [1], the objective 20-20-20 of the European Community, as well as the neutral cities and the zero-energy buildings (or almost zero) [2]. All of them intend, among other purposes, to reduce the energy consumption in the built environment and to increase the use of renewable sources.

Despite the announced, climate change is unavoidable as a consequence of the CO_2 emissions generated by the humanity mainly during the past 50 years; it is not only needed to mitigate their effects, but to adapt to them, which requires important transformations in the present and future build environment.

However, the Cuban architecture transits in an opposite direction, characterized by the loss of the tradition and the adoption of foreign models not appropriated to the climate. The chapter presents the results of several investigations focused on the characterization and evaluation of the Cuban architecture and its evolution from the point of view of its fitness to climatic conditions.

40.2 Materials and Methods

The present investigation has been developed during several decades, applying diverse methods of the theoretical and empirical research. It departs, on the one hand, from an historical evolution of the Cuban architecture, based on the documental review, the theoretical research, and the observation of the reality. On the other hand, it is also based on the model proposed in the international literature and the precedent investigations with respect to the bioclimatic architecture recommended for the Cuban conditions as well as the evolution (in the theory and practice) of the bioclimatic design at the international scale.

The comparison between significant examples of the Cuban architecture and the reference model allowed a preliminary evaluation to detect the essential features repeated along the history, in spite of the predominant architectural style. Some quantitative results offered by precedent investigations or those achieved by measurements and empiric field work were considered in order to verify the initial qualitative evaluation.

Among the results of precedent investigations developed by other authors and taken as references in this research work, they are those carried out by Alfonso in the 1970s [3], the Group of Environmental Physics in the Ministry of Construction during the 1980s [4], as well as the studies about the microclimate of the compact city developed at the same time [5, 6].

The empiric works carried out by the author taken as basis for the present chapter, include measurements and simulations in rural and urban zones. The measurements of temperature and relative humidity in rural housing, carried out in 1987, helped to show the appropriate indoor environment in the rural housing that continued the tradition of the aborigine architecture [7], while those made in the historic center of Havana in 1990 allowed to verify the influence of the volume and orientation in the thermal indoor environment of the colonial architecture [8].

The influence of the architectural design on the thermal indoor environment and the energy consumed by artificial conditioning was also simulated in 2003 for beach hotels [9], and in 2008 for the Almendares real estate building, it was demonstrated that the cost of the shading devices could be recovered in less than 2 years, as a result of the energy saving [10]. Those effects were verified again in 2013 by measurements and simulations carried out at the University of the Informatics Sciences [11].

Studies developed in apartment buildings located in diverse urban contexts also allowed verifying the influence of the volumetric and spatial solution as well as the transitional spaces between indoors and outdoors on the thermal and visual environment in spaces without artificial conditioning. Simulations of the thermal environment were carried out in 1999 for a building projected for the historical center of Matanzas city [12], and measurements were also made during several years in interior spaces of apartment buildings with different volumetric solutions in diverse contexts of Havana city: Centro Habana, Vedadoy Miramar [13, 14].

All the referred empiric investigation support the conclusions achieved related to the bioclimatic performance of the Cuban architecture, which are discussed in the present chapter. The knowledge and evaluation of the standards were consolidated by working in the Technical Committee of Bioclimatic Design, since it was founded more than 15 years ago, and the identified study cases are also the result of the information achieved as a member of the Expert Committees of the Ministry of Construction in which these projects were evaluated.

40.3 Results and Discussion

40.3.1 Bioclimatic Design Strategies

Bioclimatic architecture optimizes the relationship between the "bios" and the "climate" [15]. Despite in theory all architecture should be bioclimatic, it does not always happen in practice, as a consequence of what additional consumes of energy are generated.

The Cuban climate, classified by Köpen [16] as tropical, is warm and humid, and also rainy. Appropriate design strategies are suggested by the bioclimatic charts as Givoni's one [17]. The oscillation of daily temperature is less than 10 °C, in such a way that it is warm during day and night, in summer or winter [18] (Fig. 40.1), and the perceived thermal sensation is warmer as a consequence of the relative humidity

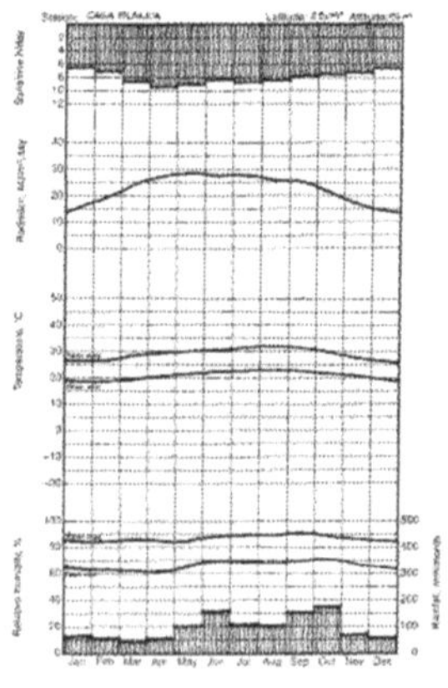

Fig. 40.1 Havana's Climogram. Casa Blanca meteorological station. Drawn by the author [22]

that make the evaporation of transpiration difficult, which is the main thermoregulation mechanism of the human body.

Bioclimatic design scheme recommendable for Cuba corresponds to the isolated building protected against the sun and permeable to the wind, what is possible in rural or suburban zones or in the open urbanizations according to the modern model (Fig. 40.2). However, the historic centres of the traditional cities developed by the Spanish colonizers followed the compact Mediterranean model more appropriate to dry climates, where instead of cross ventilation, thermal inertia of the built mass predominated allowing to generate microclimates in the courtyard (Fig. 40.3).

The traditional compact city was criticized by the hygienist movement born in the Europe of the nineteenth century and by the modern movement as a continuation of that. Then, during the first half of the twentieth century the open urbanization was developed, where narrow buildings allowed the access to the sun indoors and the distance between buildings avoided shadows [19] (Fig. 40.4).

This model generated in Europe was extended to other latitudes where the justification related to the access to the sun was substituted by the facility for ventilation. Then, buildings should be narrow to allow the wind to cross through (cross ventilation) and being distant enough one from the other for the wind to be recovered

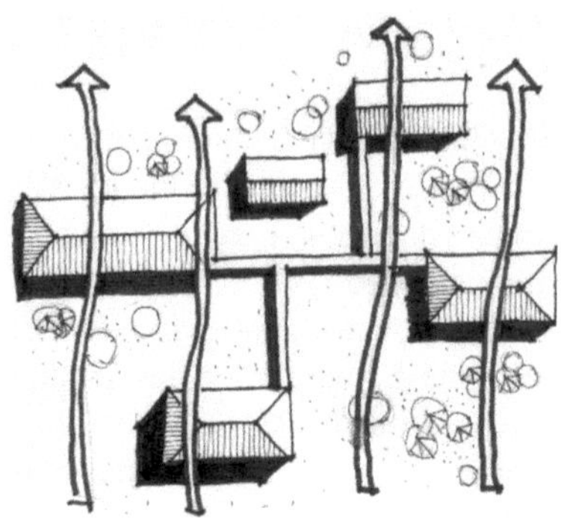

Fig. 40.2 Bioclimatic design scheme recommended for Cuba. Drawn by Pedro Rodríguez from the illustration of Konya [18]

Fig. 40.3 Compact urban morphology of the Havana's Historic Centre. From Google Earth

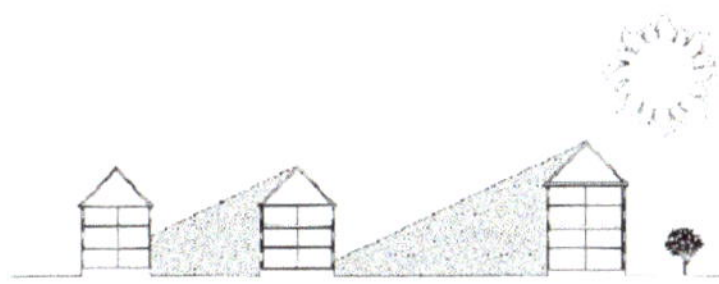

Fig. 40.4 Separación de edificios para evitar sombras arrojadas. Tomado de Butti y Perlin [19]

(avoiding wind shadow) (Fig. 40.5). These architecture and urbanism were more related to the bioclimatic design strategies recommended for warm and humid climates in every handbook than those present in the morphology of the traditional compact city.

But the handbooks of bioclimatic design produced during the 1950s, 1960s, and even 1970s copied the results of investigations carried out in models of isolated spaces or buildings exposed to laminar wind flow in fume chambers or wind tunnels, without taking into account that buildings are not isolated in cities and that the urban context modify not only the wind behavior but also temperatures, what was later on demonstrated when the "urban heat island effect" was recognized.

With the questioning of the modern urban model and the revaluation of the traditional city values produced at international scale since the end of the 1970s, the studies about urban microclimate started and many former affirmations were reconsidered. Investigations carried out in Cuba by Alfonso [6] offered new results which showed that the compact city was not as negative as it was thought and the open urbanization was not as favorable as it was believed.

The studies about urban microclimate initiated just 30 years ago, continue being an interrogation, and most of the modern buildings that follow the international style, the construction of which was possible thanks to the extensive use of artificial systems of illumination and acclimatization, are being rehabilitated to improve their energy efficiency, since in a general way, they constitute high consumers.

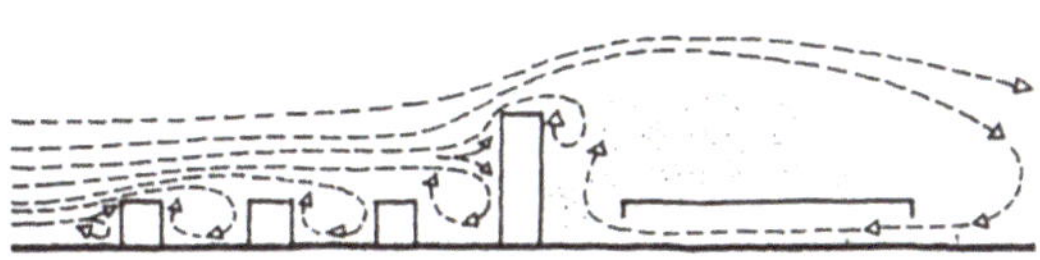

Fig. 40.5 Separación de edificios para permitir la recuperación del viento. Tomado de Konya [18]

40.3.2 Traditional Cuban Architecture

The original Cuban architecture, not affected by foreign influences, was that developed by the aborigine population before the colonial period (Fig. 40.6). This was built using natural fibers locally available, and its shape was a good answer to the climate. The tilted roofs guarantee the quick evacuation of the abundant rain; the organic materials in roofs and walls were low heat absorbent and had low thermal transference coefficient, allowing a small amount of heat to reach the interior space, and walls were permeable to the wind, favoring cross ventilation as an essential mechanism for the human thermal balance in warm and humid climates as the Cuban one.

The inhabitants of the island before the colonization lived in a development step corresponding to the primitive community. That is why this rural model is not valid as a direct reference for the urban habitat, but it has continued its development till today, by the "bohío" as the principal exponent of the Cuban rural architecture. The first cities were founded by the Spaniard colonizers according to the Mediterranean model, which shape and thermal behavior responds, as it has been said, to different thermoregulation mechanisms, traditionally recommended to dry climates. Nevertheless, colonial Cuban architecture that was very massive and less permeable at the beginning, was transformed to better fit to the climatic conditions, by increasing the height and holes, adding porches, galleries, and balconies, as well as louvers and colored glasses in order to maximize shadow and rain protection, favoring at the same time daylight and ventilation (Fig. 40.7).

Fig. 40.6 Indo-Antillean settlement. Drawing Roberto Mateizán. From Arquitectura y Urbanismo No2, 1985

Fig. 40.7 Colonial housing in Trinidad. Author's picture

Fig. 40.8 Timber housing in Varadero. Author's picture

Other manifestations of vernacular architecture, besides the rural one, also respond to the recommended model. Such is the case of the typical timber housing in the settlements associated to the sugar mills or in the coasts, using pilots, tilted roofs, large overhangs, porches, galleries, and even openings on the top for stack ventilation [20] (Fig. 40.8).

40.3.3 The Architecture of the Modern Movement in Cuba

The influence of the modern movement could be appreciated in the Cuban architecture at the end of the 1930s, when the eclectic ornamentation was lost and the horizontal elements of the "stream line" decoration appeared. The 1940s constitute a transition in which the symmetry, the massiveness, and the vertical proportion of the windows were maintained, but in the 1950s, the architecture fully assimilated the compositional principles of the modern movement and recovered the spatiality and permeability of the better former Cuban architecture (Fig. 40.9).

Despite the reinforced concrete substituted the natural building materials, its advantages were used to provide spatial continuity, wide overhangs for solar protection and large windows permeable to the wind, with movable louvers and colored glasses reinterpreted in geometric forms as light filters.

Fig. 40.9 Hermanas Farfante's housing in Nuevo Vedado. Arch. Frank Martínez, 1955. Author's picture

40.3.4 Investigation and Standards

Cuba was a pioneer in and Latin America respect to research, teaching, and standards about bioclimatic design. Examples of that constitute the first studies carried out by Joaquín Rayo in the 1960s; the investigations developed during the 1970s–1980s in the Havana's Faculty of Architecture, the Group of Environmental Physics of the Ministry of Construction, and the Institute of Hygiene and Epidemiology, as well as the inclusion of these issues in the curricula of Architecture in the 1970s.

On the other hand, the work carried out in these fields not only have been focused on researching and teaching but since the beginning of the 1980s, standards to regulate solar control, daylight, thermal behavior of the building materials, climatic variables, and others were elaborated. However, those standards were mainly focused on calculation methods, what could explain why they were generally not applied in the architectural projects.

Despite the disappearance of Group of Environmental Physics in the Ministry of Construction, some other institutions have been created, as the Technical Committee of Bioclimatic Design (CTN-40), which has been in charge of reviewing the first standards in this field. The investigation has continued in the Institute of Hygiene and Epidemiology, in the Faculty of Architecture at Havana, and in other faculties of construction along the country, in spite of the economic difficulties to acquire adequate software and measurement equipment.

40.3.5 Cuban Contemporary Architecture

As it has been exposed, Cuban architectural tradition shows a high quality in different historic periods and constitutes a good referent for the application of the essential bioclimatic design principles to a warm and humid climate. However, despite this rich tradition, the standards, as well as the research and academic work, a regression can be appreciated in the contemporary architecture respect to these issues.

A case study about projects developed in Havana between 1998 and 2009 [11], showed that despite the negative evaluation given by the expert committees, the built projects have excessive glass windows area exposed to the sun, violating the standards (NC 220.2009) [21] and generating undesirable reflections to the urban contexts as well as penetration of the direct solar radiation to the interior spaces. This provoke glare and increase the energy consumed by the artificial conditioning systems (Fig. 40.10).

The question is then, why despite the good tradition in architecture, standards, researching and teaching, the current Cuban architecture does not offer solutions fitted to the climate in order to achieve appropriate indoor environment and to reduce energy consumption?

Fig. 40.10 Colegio de San Gerónimo in the Historic Center of Havana. Author's picture

40.3.6 The Influence of Foreign Models

The described phenomena are multicausal. Based on the postgraduate teaching and the experts committees, it can be appreciated that the architects do not assimilate the knowledge offered by the curricula. The generations graduated during the 1970s–1980s use to reject the complex calculation methods that characterized the teaching of the bioclimatic issues at that time. The application of this knowledge has not always been demanded by the studio teachers, who sometimes ignore these tools and avoid the issue arguing that it is not important or that it limits the architectural aesthetic expression, about what they usually convince the students. Sometimes, those who teach these matters also contribute to the students' reaction when they affirm that those factors are determinant in the design solution and the rest are subordinated to them.

However, it is true that the architects are not the ones who really decide the design solution, but the invertors or other decision makers, who generally depart from predetermined ideas about the desired project, based on their own paradigms, usually conditioned by images coming from the architecture of the developed countries, where the climates are cool and totally different from the Cuban one. In fact, most of the projects considered in the study case were not developed in Cuba, but imported solutions imposed by the foreign partners.

And this is just the main reason behind the opinion of the teachers considering that the bioclimatic design limits the architectural expression and the student's creativity as well as the investor or decision maker dreaming with a glass box associated to the image of "development." It is a combination of disownment and lack of auto-valuation that generates a high vulnerability to the foreign paradigms, which frequently are not appropriated even in the societies where they come from.

40.4 Conclusions

Cuba exhibits a high-quality architectural tradition that maintains and reinterprets essential principles of the bioclimatic design in different historic moments, but it has been lost in the contemporary architecture, in spite of the systematic research and academic work carried out in this field, as a consequence of the importation of foreign models.

In order to change this situation, an intense work is required to promote the essential knowledge and own values, not only in professional teaching but also for the whole society, especially investors and decision makers. It is related not only to the traditions rescuing but also a basic principle for adapting to the climate change.

References

1. Mazria E (2012) Clean energy: the future is now. World Renewable Energy Forum. American Solar Energy Society, Denver
2. Montes de Oca M (2011) Edificios de energía casi nula, Una realidad inminente. In: Romera García P (ed) Curso de Arquitectura Sostenible. Escuela de Arquitectura. Las Palmas de Gran Canarias, pp 66–71
3. Alfonso A (1979) Condiciones admisibles del ambiente térmico para los edificios de vivienda. Ciencias Técnicas. Arquitectura y Urbanismo, La Habana, Facultad de Arquitectura, ISPJAE, núm. 5, pp 23–41
4. de Autores C (1989) Selección de Artículos. Centro de Información de la Construcción, Ciudad de La Habana, octubre 1989
5. Alfonso A (1985) Transpira la ciudad compacta. Arquitectura y Urbanismo, La Habana, Facultad de Arquitectura, ISPJAE, núm. 1, pp 60–69
6. Alfonso A, Díaz G, De la Peña AM (1989) Por el rescate de la tradición. La Habana, Facultad de Arquitectura, ISPJAE, núm. 2, pp 20–25
7. González Couret D (1991) Montaña, clima y arquitectura. Arquitectura y Urbanismo, La Habana, Facultad de Arquitectura, ISPJAE, núm. 3, pp 45–54
8. Castillo RM, González T, González L (1990) Tipología habitacional en Ciudad de La Habana antes de 1959. Comportamiento Climático. Directoras: Dania González y Eliana Cárdenas. Facultad de Arquitectura, ISPJAE, La Habana
9. Guzmán R, Alberto L (2003) Influencia del diseño en el consumo de energía por climatización en las habitaciones de hoteles de sol y playa en Cuba. Directora: Dania González Couret. Facultad de Arquitectura del ISPJAE, La Habana
10. Martínez R (2008) Diseño de elementos de protección solar para el ahorro energético en edificios. Directora: Dania González Couret. Facultad de Arquitectura, ISPJAE, La Habana
11. Llovet M, González N, Rodríguez E (2013) Arquitectura sustentable en clima cálido—húmedo. Caso de estudio: Universidad de las Ciencias Informáticas. Directora: Dania González Couret. Facultad de Arquitectura, ISPJAE, La Habana
12. González Couret D (2000) Design criteria for warm humid climates: a case study. Renewable energy. Renewables: the energy for the 12th century. World Renewable Energy Congress VI, Part I, Pergamón, Brighton, pp 534–540
13. González Couret D et al (2005) Vivienda apropiada para Ciudad de La Habana. Editorial CUJAE, La Habana
14. González Couret D et al (2010) El edificio de apartamentos en Cuba. Editorial CUJAE, La Habana

15. Serra R, Coch H (1994) Las energías naturales en la arquitectura. UPC, Barcelona
16. Guardiola J, María J (1974) Curso de Climatología. Instituto Cubano del Libro, La Habana
17. Givoni B (1998) Climate consideration in building and urban design. Van Nostrand Reinhold, New York
18. Konya A (1981) Diseño en climas cálidos. Editorial Blume, Madrid
19. Butti K, Perlin J (1985) Un hilo dorado. Hermann Blume, Madrid
20. González Couret D (2004) From the past to the present of bioclimatic architecture in Cuba. World Renewable Energy Congress VIII, Elsevier, Denver
21. NC 220_1, 2009 (2009) Edificaciones. Requisitos de Diseño para la eficiencia energética. Parte 1: Envolvente del edificio. CNN, La Habana
22. González Couret D (1999) Progressive social housing in the historic centre of Matanzas City. A bioclimatic analysis. Architecture, Energy and Environment, Lund University

Chapter 41
Towards Nearly Zero-Energy Buildings in 2020 in the Netherlands

Kristian Gvozdenovic, Wim Maassen and Wim Zeiler

Abstract The aim of this article is to give information and insight in nearly zero-energy building (nZEB) developments in the Netherlands that will occur in the near future and what consequences these developments have for buildings, in particular for building services. In 2020, all newly built buildings have to comply with the nZEB regulation with an energy performance coefficient ≈0. To support the future policy on energy performance improvements, this study provides a roadmap towards nZEBs with a technical and financial feasibility framework. Life-cycle cost calculations, also beyond cost optimality, are essential for determining the Dutch nZEB definition, because these determine if the energy efficient measures are profitable, cost-effective, and can be implemented in the building law.

Keywords Nearly zero-energy buildings · Life-cycle costs · Profit

41.1 Introduction

In 2010, the European Commission has launched the energy performance of building directive (EPBD) with the main targets to reduce CO_2 emissions by 90% compared to 1990 [1]. The Dutch government has set out a plan to implement nearly zero energy building (nZEB) regulation for the coming years (2015–2017) and published the "National Plan to promote nZEBs" 0 in September 2012 to implement [10] the EPBD recast requirements. The EPBD requires all newly built buildings to be nZEB in 2020 for different building functions. Existing buildings will also have to comply with this regulation towards 2050. Each European member state (MS) has to work out a plan that includes an nZEB definition for different building functions, determining specific building requirements.

W. Maassen (✉) · K. Gvozdenovic
Royal HaskoningDHV, Rotterdam, The Netherlands
e-mail: wim.maassen@rhdhv.com

K. Gvozdenovic · W. Maassen · W. Zeiler
Faculty of the Built Environment, University of Technology Eindhoven, Eindhoven, The Netherlands
e-mail: wim.maassen@rhdhv.com

A. Sayigh (ed.), *Renewable Energy in the Service of Mankind Vol I*,
DOI 10.1007/978-3-319-17777-9_41

The definition of an "nZEB" is described within the EPBD recast of the EU 0 and it is specified that by 31st December 2020 all new buildings shall be "nZEB." Governmental buildings occupied and owned by public authorities, will have to be "nZEB" by December 31, 2018 according to the EPBD recast. The actual definition of nZEB is given in Article 9 of the EPBD [9]:

> Nearly Zero Energy Building (nZEB): Technical and reasonably achievable national energy use of >0 kWh/(m^{2}a) but no more than a national limit value of non-renewable primary energy, achieved with a combination of best practice energy efficiency measures and renewable energy technologies which may or may not be cost optimal.

In the Netherlands, performance is indicated by the energy performance coefficient (EPC) which is described in the NEN 7120 norm. Currently, the EPC is 0.6 for residential buildings and will be lowered to 0.4 in 2015 according to a covenant of the new buildings sector, aimed at reducing the energy consumption of new buildings over time. In this signed agreement between the public and private sectors, a number of efforts have been agreed to reduce the energy use of new buildings by the year 2015 by at least 50 % compared to 2007 levels.

41.2 Methodology

The tightening EPC demands require a new improved cost-effectiveness methodology, therefore a practical and theoretical test have been developed for both residential and utility buildings. Focus point in this new methodology were to create a clear method for all building types, to adapt the existing method to new EU demands, and to include additional gains into the life-cycle cost (LCC) calculation. The Royal Haskoning DHV (RHDHV) study on the so-called sustainable building accelerator lies at the base of the enhance cost optimality calculation in which benefits are included next to costs [2]. Cost optimality calculations are essential for determining the Dutch nZEB definition, because they determine if the energy efficient measures are cost-effective and can be implemented in the building law.

In the near future, EPC requirements will be reduced to values that lay within the so-called cost optimal range as shown in Fig. 41.1 (green area). This range is determined by calculating the LCCs over a period of 30 years. In 2020, buildings will have to be nZEBs (blue area in Fig. 41.1). The nZEB level is determined by each EU MS based on the economic feasibility. Current calculations show that nZEBs will result in much higher LCC values than the economic optimum. Therefore, a LCC method which also takes additional gains (e.g., productivity, resale value) into account is proposed. Effectively including these gains leads to lower total LCCs and the economic optimum shifts towards nZEB requirements (blue arrow in Fig. 41.1). Focusing on gains and including these in the LCC calculation method an important foundation for the roadmap towards nZEBs.

The cost optimality is a crucial aspect for the introduction of nZEBs in the Netherlands. In 2009, the effects on lowering the EPC to 0.6 for residential buildings

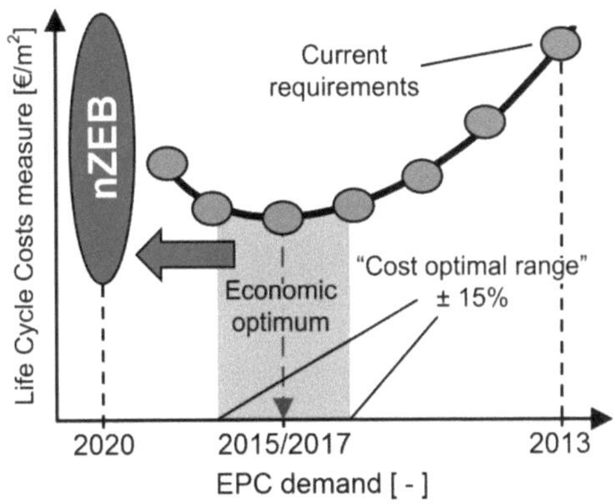

Fig. 41.1 Life-cycle costs versus the EPC-demand. *nZEBs* nearly zero energy buildings, *EPC* energy performance coefficient

in 2011 were studied [3]. The goal was to gain insight about the effect of EPC reduction on the indoor environment, energy demand, CO_2 emissions, the relation between investment costs and energy-saving measures. In 2013, a following study was done on cost optimality of energy-saving measures for residential and utility building according to the EU calculation method [4]. The results for the financial and macroeconomic calculation were quite similar, so only the results for financial cost optimality analysis will be shortly discussed. The following graphs show the additional net present values (NCC) for different packages (energy-saving measures) compared to the EPC (Q/Q) demand of different building types/functions. To satisfy the EPC demand (from the 2013), it is important that proposed measures result in a Q/Q below 1.00, see Fig. 41.2. It shows the additional net present values (NCC) for energy-saving measures for office buildings. Almost all measure for office building satisfy the EPC demand and cost of energy-saving measures prove to be cost neutral or even cost reducing! This means that all applied measures are already cost-effective for an EPC of 0.65. The goal of this study is to provide nZEB scenarios with low EPC scores and primary energy consumption in combination with low LCC. Currently, cost optimality calculation can be made for existing technologies (Fig. 41.3a).

These buildings conform to current EPC demand, and they are within the cost optimal range. Buildings that have to comply with future EPC regulation will have to be equipped with future technology. This will result in low primary energy demand (low EPC); however, these technologies are not cost-effective yet (Fig. 41.3b). To reduce LCCs, additional gains such as resale value, productivity, etc. will be added to providing a new type of graph (Fig. 41.3c) in which lower LCC and primary energy are accomplished. This new calculation method is called the LCC since it is not exactly the same method as prescribed by the EU. The LCC cost optimality calculations have been executed using the sustainable building accelerator [5] (NL: "DUBO-versneller"); an LCC calculation tool developed at RHDHV [2]. The Sustainable building accelerator can be utilized to compare LCC's of four building concept to each other. The input of the Sustainable building accelerator takes four main expenses into account:

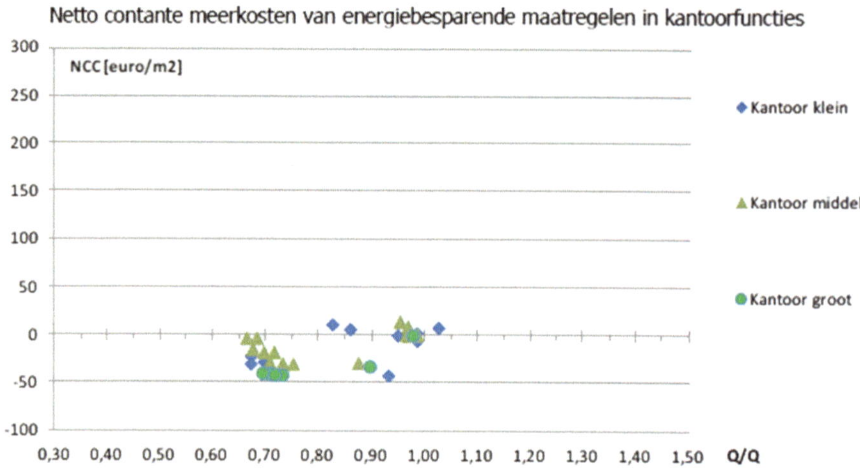

Fig. 41.2 The additional net present values (NCC) of energy-saving measures for office buildings [4]. *NCC* net constant costs

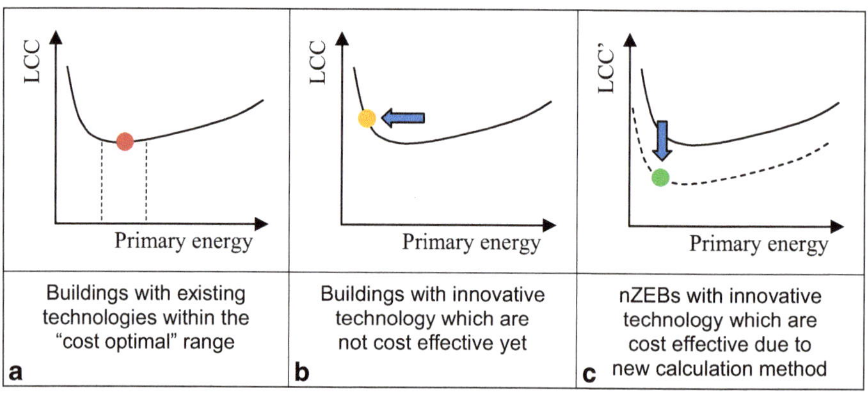

Fig. 41.3 Cost optimality trajectory **a** “standard” scenarios **b** scenario with innovative technologies **c** scenario with innovative technology and new LCC calculation method. *nZEBs* nearly zero energy buildings, *LCC* life-cycle cost

- *CAPEX* (capital expenses) in (€/m²), for example:
  Building costs
  Land costs
  Installations: mechanical and electrical installations.
  Building creators: architect, installation advisors, building manager.
- *Energy* in (€/(m²a)), for example:
  Electricity
  Gas
- *OPEX* (operational expenses) in (€/(m²a)), for example:
  Maintenance: building, mechanical, and electrical installations

Other building service: cleaning
Taxes, insurance

- *End value* in (€/m^2), for example:
  Rest value of building
  Residual value: building, land, and installations
  Dismounting and disposal

Dynamical input costs, including replacement of installation after x number of years, have also been integrated in LCC calculation. This allows taking refurbishments and overhauling costs into account.

For the cost optimality calculation, the discounted cash flow was used. The LCC calculation method also provides a possibility to perform a sensitivity analysis for all variable parameters such as interest rate, discount rates, and energy prices. A web-based lite version (NL) of the sustainable building accelerator can be found at www.duboversneller.nl.

Energy prices used in the LCC calculation have been determined using energy prices from three large energy suppliers for a middle-sized office building. These costs are specified in (€/(m^{2}a))

Gas price Energy-saving measures applied to the nZEB scenarios results in all-electric buildings, which have no gas connection. Because all buildings are all-electric, no differences appear (gas connection). When the nZEB scenarios would be compared to a gas grid connected building, cost can be further reduced. The energy demand in the cost optimality is expressed in primary energy units, meaning primary energy has to be converted to cubic meter gas and kilowatt-hour electricity. The following conversion values have been used: Natural gas: 35.17 MJ primary energy per cubic meter natural gas and electricity: 9.23 MJ primary energy per kilowatt-hour. The electricity prices for utility buildings are distinguished for an annual consumption of $< 10{,}000$, $< 50{,}000$, and $> 50{,}000$ kWh. These tariffs can affect the energy consumption when it is close to the set limit, and it will most certainly influence the PBP of PV panels.

Besides the operational expenses there are also benefits to be considered like:

- *Productivity increase*: The productivity increase was implemented using the cost reduction value of 26.50 €/(m^{2}a).
- *Sick leave reduction*: An average value was determined for sick leave reduction using the studies [6, 7]. According to the first study, sick leave by unhealthy climate works out to an average of 3.6 days per employee per year. The second study is more specific (only looking at effects of an economizer on energy and cost) and results in averaged 35 additional sick days (spread over 72 employees) when no economizer is installed. This number corresponds to 0.49 sick days per employee per year. The average value, used in the LCC calculation, was 2.05 days per employee annually. The office building in the current study is assumed to have 200 employees (15 m^2/employee) with a monthly salary of € 2000.
- *PR value*: The quantification of costs for PR value may be calculated with the budgets companies use for publicity on sustainability. Normally, money would be spent on improving a production process (making products or services more

energy efficient) which would be used for a greener image. The annual costs spending on those processes may now be spend a more sustainable building; the PR value of the building may be used for several years until regulation and other buildings have caught up to the nZEB standards.

- *Higher renting value*: This value is mostly represented by a combination of productivity increase, sick leave reduction, and PR value. The reason this is mentioned is that these cost may or may not be incorporated in the LCC calculation depending on whether the building owner is also the building user. When the building is rented, the higher renting value is most certainly lower than the combined gains (productivity, sick leave, and PR value) because it is quite difficult to charge higher rent when values on number of employees, salaries, PR budget, etc. cannot be determined. This leads to the conclusion that it is more advantageous to own an nZEB.
- *Higher rest value*: The value of building installation such as the ground source well for ground source heat pump (GSHP) or aquifer thermal energy storage (ATES) gives the building a higher end value. In the current study, all buildings concepts (U_{ref}, U_1, U_2, U_3) have wells therefore costs/gains do not alter the outcomes of the LCC calculation. However, the added value will be more significant when comparing nZEB scenarios with a conventional building with high-efficiency gas boiler (no wells).

Critical parameters of the LCC calculation, often calculated or assumed, will be tested using the sensitivity analysis. The analysis will have to satisfy minimum requirements according to the EPBD recast for different price scenarios for energy carriers (gas and electricity) and minimal two discount rates for the micro- and macroeconomic analyses. The parameters may only be changed one at the time to see the changing effects. Values from [8] were adapted according to the recalculated values of [2].

41.3 Results

An LCC calculation with a period of 30 years is performed in which four building designs for the urban area are discussed: one reference building and three nZEB scenarios with energy-saving measures; see Table 41.1. The focus will be on building installations, mainly the differences between the technologies. The calculation method is based on EU approach using Dutch principles (from previous studies) on cost optimality of energy-saving measures.

Important parameters used are inflation rate, 2.3 %; discount rate, 6.4 %; excluding CO_2 emission costs, no subsidies, energy tariffs including energy taxes, and value-added tax (VAT). The reference building has an energy performance that applies to coming EPC demands (EPC = 0.7). The nZEB scenarios are well insulated and built airtight reducing energy demand. Installations used are GSHP, ATESs and large-scale PV application. An average EPC score of 0.16 (average primary energy demand 18.5 kWh/m^{2}a) was obtained for the three scenarios (Fig. 41.4).

Table 41.1 Building installations for urban area scenarios

	Urban Area scenarios Office building			
	U_{ref}	U_1	U_2	U_3
EPC	0.70	0.20	0.14	0.15
Primary energy demand (kWh/(m^2a))	79.0	23.1	15.4	17.1
Installations				
Heating Cooling	Vertical GSHP with low temperature (30–35 °C) floor heating	Vertical GSHP with low temperature (30–35 °C) floor heating	ATES with low temperature (30–35 °C) floor heating	ATES with road collector with low temperature (30–35 °C) floor heating
Hot tap water	Small electric boiler	Small electric boiler	Small electric boiler	Small electric boiler
Ventilation	Mechanical (balanced) with heat recovery (70 %)	Mechanical (balanced) with heat recovery (95 %)	Mechanical (balanced) with heat recovery (95 %)	Mechanical (balanced) with heat recovery (95 %)
Lighting system	Efficient lighting system (8 W/m^2)	Efficient lighting system (8 W/m^2)	Efficient lighting system (8 W/m^2)	Efficient lighting system (8 W/m^2)
Electricity generation	PV cells 170 m^2 (roof)	PV cells 770 m^2 (roof) 375 m^2 (façade)	PV cells 770 m^2 (roof) 375 m^2 (façade)	PV cells 770 m^2 (roof) 375 m^2 (façade)

EPC energy performance coefficient, *GSHP* ground source heat pump, *ATES* aquifer thermal energy storage, *PV cells* photovoltaic cells

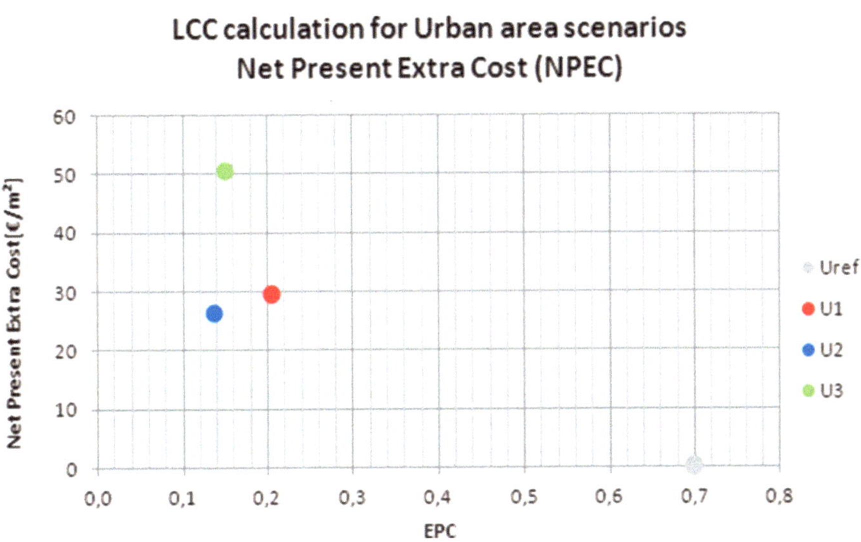

Fig. 41.4 Financial analysis without additional gains for urban area scenarios: office building

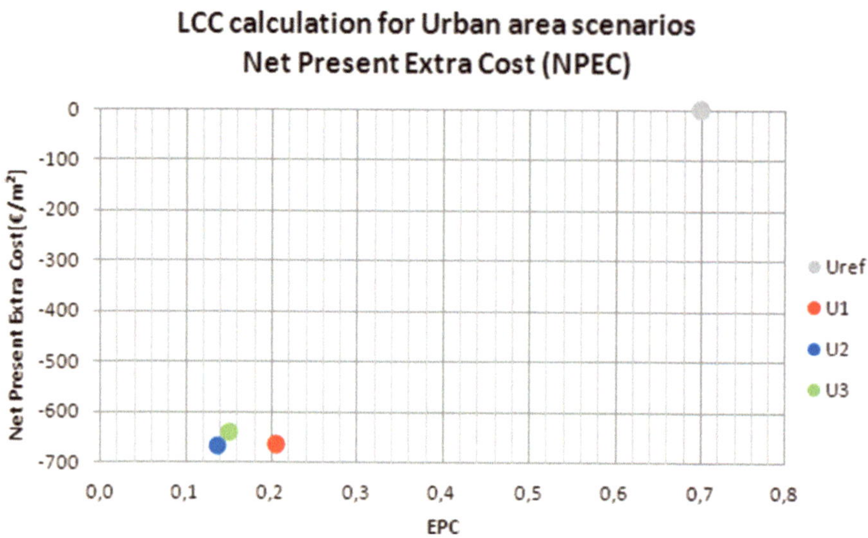

Fig. 41.5 Financial analysis with additional gains for urban area scenarios: office building

Additional gains that have been included in the LCC calculation are increased productivity and reduced sick leave. Scientific research has proven that buildings with a healthy indoor climate (high ventilation rate) increase productivity and reduce sick leave.

Results of the LCC calculation show that the nZEB scenarios are not cost-effective yet without additional gains. Average additional costs for the energy-saving measures are 15–50 €/m^2 higher compared to the reference case, and 100–140 €/m^2 higher, for the financial analysis and macroeconomic analysis, respectively. When additional gains are added, the total LCC drops significantly for both analyses (700–1100 €/m^2). The graphical representations (figures) are shown in LCC (€/m^2) versus EPC demand for the case with and without additional gains (Fig. 41.5).

41.4 Conclusions

This study on nZEBs in the Netherlands provides insight in the current situation of nZEBs and promising scenarios which are technically and financially feasible. The aim of this report was to give information on nZEB developments that will occur in the near future and what the consequences of these developments have for buildings, in particular for building services.

Examples of nZEBs (offices) show the technical capabilities of energy-saving measures: low EPC scores can already be achieved. Existing energy-saving measures have been compared to measures applied to nZEBs: It clearly shows focus should be on energy demand reduction (insulation, glazing, and air tightness) and installations (heating/cooling system, mechanical ventilation with heat recovery and application of large scale PV).

The energy infrastructure in the Netherlands needs to undergo a transformation to support the implementation of nZEB buildings to the existing grid. Currently, many different smart grid projects are being executed ranging from an energy neutral business park to a complete or partial self-sufficiency in heat and electricity network, powered by biomass combined heat and power plants (CHP) production. It is expected that smart grid (pilot) projects provide good outcomes which enable large-scale implementation of smart grids.

On the basis of this study (in line with [3] and [4]), a recommendation is done on EPC demands for offices. For a middle-sized office building, it was possible to create nZEB designs with an average energy consumption of 20 kWh/(m^2a) resulting in an EPC score of ~0.2. On the basis of this study, an EPC of 0.2 for offices is recommended since it is technically feasible, and financially (taking into account additional gains) more attractive than an office building with EPC 0.7.

Additional gains—productivity and sick leave The additional gains proved to reduce LCC of nZEB considerably. It should be noted that relatively simple methods have been applied to incorporate the gains into the LCC calculation. This method was based on available research results. For both additional gains (productivity and sick leave), values were used closest to the building performance of this study; however, a certain deviation exist. To gain more reliable results, more specialized calculations on increased productivity and reduced sick leave should be conducted for ventilation rate used in this study: $ACH_{ref}=2\ h^{-1}$ and $ACH_{nZEB}=3\ h^{-1}$. This may lead to lower additional gains; however, it is expected that the gains still lead to positive results (cost-effectiveness) for the nZEB scenarios.

Some of the additional gains were not incorporated: in this research, although mentioned are not taken into account in the LCC calculation. Gains that should be further investigated are PR value, higher renting value, and higher rest value. Further investigation on these subjects is necessary to define actual costs. It was concluded that owning and using an nZEB is more beneficial than of renting an nZEB to another party, because the additional gains of a higher renting price is probably lower than the benefits of productivity and sick leave.

Acknowledgments This project is realized by a cooperation between RHDHV and TVVL, in collaboration with the Eindhoven University of Technology.

References

1. EPBD (2010) EPBD recast, Directive 2010/31/EU of the European Parliament and of the Council of 19 May 2010 on the energy performance of buildings (recast), Official Journal of the European Union, (2010) 18/06/2010. http://eur-lex.europa.eu/LexUriServ/LexUriServ.do?uri=OJ:L:2010:153:0013:0035:EN:PDF. Accessed 12 Nov 2013
2. Maassen WH, Maaijen HN (2011) Sustainable building accelerator, international conference for enhanced building operations. Proceedings of ICEBO, New York
3. dGmR (2009) Rapport E.2009.0697.00.R002 Aanscherpingsstudie EPC-woningbouw 2011. 17 Nov 2009. http://www.rijksoverheid.nl/documenten-en-publicaties/rapporten/2013/03/20/kostenoptimaliteit-voor-eu.html. Accessed 20 Nov 2013

4. dGmR (2013) Rapport E.2012.1094.00.R001 Kostenoptimaliteit energieprestatie eisen Nederland. Verslagleg-ging kostenoptimaliteit voor EU. 11 maart 2013. http://www.rijksoverheid.nl/documenten-en-publicaties/rapporten/2013/03/20/verslagleggingkostenoptimaliteit-voor-eu-definitief.html. Accessed 25 Nov 2013
5. RHDHV (2014) www.duboversneller.nl. Accessed 1 June 2013
6. Bergs J (2010) The effect of healthy workplaces in the well-being and productivity of office workers. http://www.plants-for-people.org/eng/science/jbergs_symposium_paper.html. Accessed 1 June 2013
7. Fisk B, Seppanen O, Huang J (2004) Economics benefits of an economizer system, energy savings and reduced sick leave. ASHRAE Trans LBNL 54472
8. RHDHV (2013) EPC aanscherpingmethodiek Woningbouw en Utiliteitsbouw. 16 mei 2013. http://www.rvo.nl/sites/default/files/2014/01/EPC%20aanscherpingsmethodiek%20woningbouw%20en%20utiliteitsbouw.pdf. Accessed 21 Dec 2013
9. REHVA (2010) Journal of the European Union 18/06/2010. REHVA report—nZEB technical definition and system boundaries for nearly zero energy buildings. (Federation of European Heating, Ventilation and Air Conditioning Associations.) Report no. 4 (2013)
10. BENG (2012) Nationaal Plan voor het bevorderen van Bijna-Energie Neutrale Gebouwen (BENG) in Nederland, Sept 2012. http://www.government.nl/documents-and-publications/reports/2011/02/25/plan-of-actionenergy-saving-in-built-environment.html. Accessed 11 Nov 2013

Chapter 42
How to Reach for the Necessary Synergy Between Architecture and Engineering

Wim Zeiler

Abstract New strong demands for a more sustainable built environment led to a more complex design process. The different expertise of engineers must be used more effectively by architects especially in the conceptual design phase to reach for new solutions. To support diverse multidisciplinary building design teams, a supportive design method was developed in cooperation with the Dutch professional organizations of architects and consulting engineers. It helps architects and engineers with their new role in the conceptual design phase as it enables to structure each perspective on the design task as well as to structure the available domain knowledge by using morphological charts (MC) and morphological overviews. After testing the method in workshops as part of a training program in industry, the design method was transferred and applied at the department of architecture for master students for their multidisciplinary master project integral design (ID). In the past 9 years, master projects ID were held and there was a continuous development to optimize the cooperation between architects and engineers. The research showed that it is possible to engage engineers and let other disciplines learn from them within the conceptual building design phase, which makes it easier to achieve nearly zero-energy buildings.

Keywords Building design · Conceptual design

42.1 Introduction

Sustainability is a crucial issue for our future and architecture has an important role to direct sustainable development [1]. Sustainable building designs need to provide solutions for sustainability issues ranging from flexible use to renewable energy, energy reduction measures while maintaining, and even increasing comfort level of the users. The necessity of more sustainable buildings led to the involvement of engineers in the conceptual design phase of building design. The increased com-

W. Zeiler (✉)
University of Technology Eindhoven, Eindhoven, Netherlands
e-mail: w.zeiler@bwk.tue.nl

A. Sayigh (ed.), *Renewable Energy in the Service of Mankind Vol I*,
DOI 10.1007/978-3-319-17777-9_42

plexity of building design inevitably called for more design collaboration [2]. In the projects designed (and built) in the early 2000s, architects started to adapt their usual design process (traditional design process) by consulting engineers in an earlier stage than normally done. In sustainable building projects designed later, many architects qualified their design process as an integrated design process (IDP): The architects mentioned mostly the early engagement of engineers in the process as a clear sign of this [3, 4].

Due to the environmental problems, new building requirements require buildings to become nearly zero-energy buildings (nZEB) in Europe by 2010 (European Parliament 2010). As current energy consumption by buildings accounts for around almost 40% of the total energy in the EU and the US [5], there is a need for dramatic change. The design process of a net zero-energy solar building, compared to the design process of buildings with less ambitious energy performance targets, needs a substantial change of tasks, required information and interaction between disciplines [6]. The "Net Zero" target strongly influences the entire design process [7]. To meet the energy target, already from the very early stages of the conceptual design, not only a tight interaction between various disciplines is needed like in integrated design but also an integral design (ID) approach leading to a design team approach. At this point it is useful to define the differences between integrated design and ID more explicitly. Within integrated design two or more disciplines, all necessary and important, are treated as part of, or contained within, the whole building design approach from the early stages of a project. To put it another way, within integrated design the architectural discipline and other disciplines start separately and often in different design phases and are later made to fit, whereas within ID all necessary design disciplines start together right from the conceptual design. ID concepts are only possible by starting together and uniting various viewpoints of the different design disciplines participating in the project. Even if a designer has the ability to deploy most of these viewpoints, he or she usually does not have enough specialist knowledge to assess all of them in depth. For this reason, it is assumed that a multidiscipline design team view on design is a better way of pursuing building design synergy than a monodisciplinary individual designer view. Figure 42.1 derived from the MacLeamy Curve [8] illustrates the differences between ID, integrated design, and more traditional design.

Given the existing disparities in the construction industry, King [9] stated that in order to do anything meaningful in terms of moving to low carbon society, we need a consistent framework, design method, within which we can apply knowledge embodied in a design team. However, proper design tools and working methods are needed which could help architects in the design process [4]. Still as said by Mike Davies, one of the founding partners of the Richard Rogers Parnership, just recently said, "architectural intent, function and structural, services and environmental intent—all need to be brought together at an early stage to achieve the aesthetic, functional and civic integrity of a good building" [10].

The main body of the chapter starts (Sect. 42.2) with the development of the ID method: A design method that helps to merge the different perspective of all designers and engineers involved in the design process. The derived design method work-

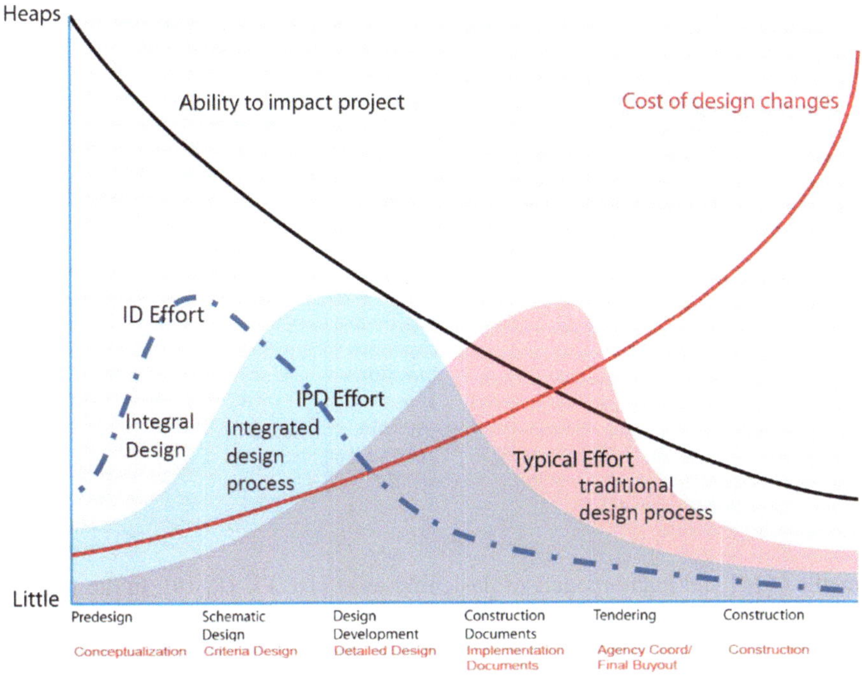

Fig. 42.1 Characteristic efforts distribution within different design process approaches. (Derived from Ref. [8])

shops for professionals, architects, and engineers in building design were tested in practice. This is described in Sect. 42.3. In Sect. 42.4, the results are presented by the application of the ID method within the workshops. A discussion about using the developed ID method to support building design teams to generate solutions for the highly complex problem of designing extreme sustainable buildings (nZEB) is given in Sect. 42.5. Finally, in Sect. 42.6, some conclusions are given about the added value of the presented approach for the generation of solutions for a low-carbon built environment.

42.2 Methodology: Integral Design (ID)

In building design, one has to work with ill-defined problems where the wanted solution and the problem itself develops almost in parallel at the early stages of the design process. Also the amount of relationships and dynamic social interactions makes it increasingly complex. Therefore, a method is needed to structure this wicked problem [11]. In the early 1960s, research started to investigate new design methods as a way to improve the outcome of design processes [12]. Despite development since than [13, 14]; there is still no clear picture about the ideal design

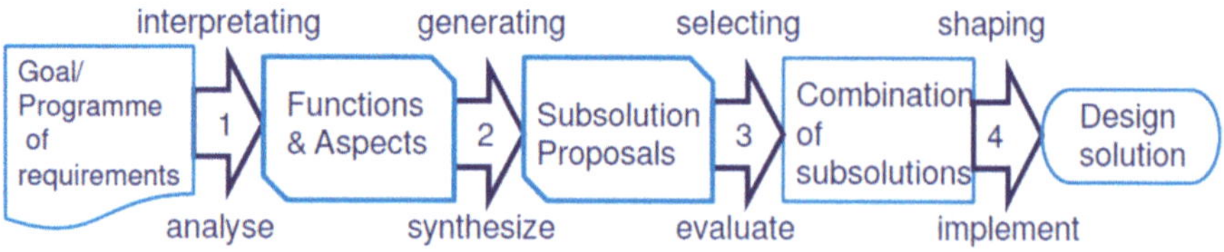

Fig. 42.2 The four-step pattern of integral design (ID)

Fig. 42.3 Concept of a morphological chart

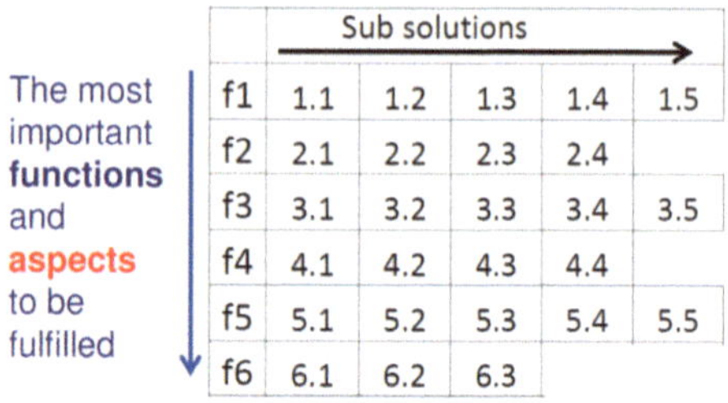

method [15] and many models of designing exist [16, 17, 18, 19]. In the Netherlands, methodical design was a quite familiar design method [20]. The methodical design process is a framework application-independent principle with its connection to the general system theory and has some exceptional characteristics [21]. It is problem-oriented and distinguishes, based on functional hierarchy, various abstractions or complexity levels during different design phase activities. This methodical design method was extended into an ID model by us by adding a fourth process step with a special focus on selection to the original three-step process. So a distinctive feature of the ID model is the four-step pattern of activities (generating, synthesizing, selecting, and shaping, see Fig. 42.2a) that occurs on each level of hierarchical functional abstraction with the design process.

Another change to the original methodical design method is the intensified use of morphological charts (MCs) developed by Zwicky [22] to support design activities in the design process [23]. A MC is a kind of matrix with columns and rows which contains the things to solve and the possible solutions connected to them. Different overall solutions are created by combining various solution principles to form a complete system combination [24]. Within the ID process a MC is formed by decomposing the main goals of the design task into functions and aspects, which are then listed on the first vertical column of the MC. The functions and aspects are derived from the program of demands. Possible solution principles for each function or aspect are then listed on the horizontal rows of the corresponding function or aspect, see Fig. 42.3 for the concept of a MC

In the first step of the ID method the individual designer has to make a list of what he thinks are the most important functions that have to be fulfilled based on the design brief and put it into the first column of the MC. In the second step of the process, the designers add possible part solutions to each aspect or function in the rows after the specific function or aspect, see Fig. 42.4.

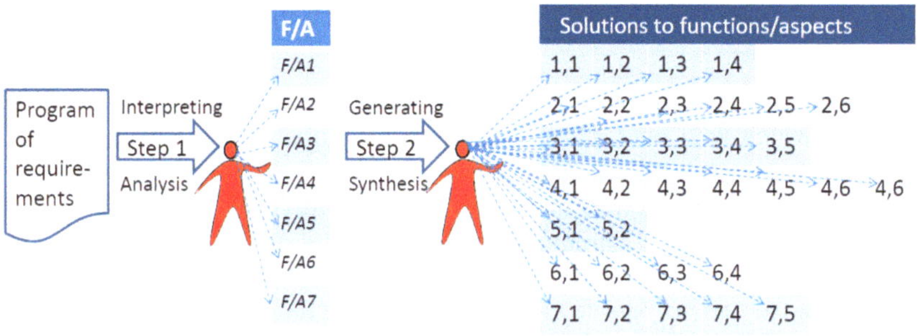

Fig. 42.4 Interpretation design brief and generating solutions

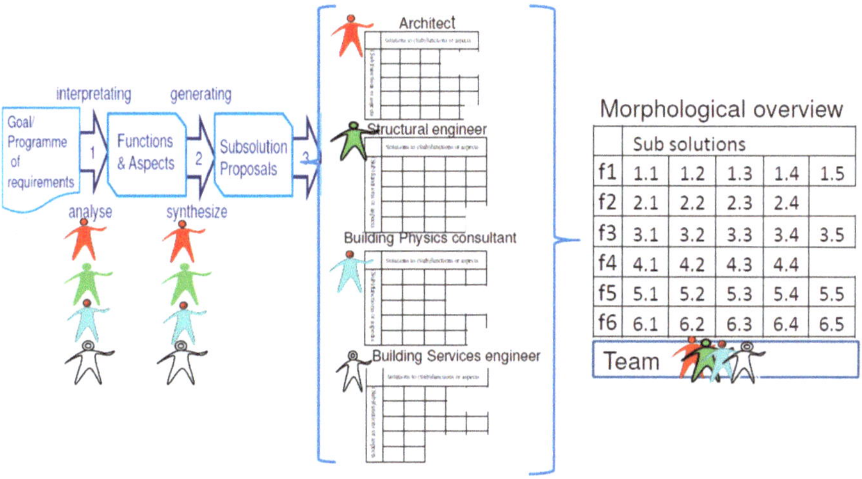

Fig. 42.5 Building the morphological overview out of the individual MCs

Using a MC is an excellent transparent way to record information about the solutions for relevant functions and as such an aid in the cognitive process of generating the system-level design solution [16, 25]. The MC to visualize subsolution alternatives plays a central role in the ID approach for design teams. Each participant of a design team develops a full MC from their own specialist point of view, see Fig. 42.5. These individual discipline-based MCs can be combined to one overall MC called morphological overview. The morphological overview of an ID team process is thus generated by combining in two steps the different MCs made by each discipline. First functions and aspects are discussed and then the team decides which functions and aspects will be placed in the morphological overview. Then after this first step, all participants of the design team can come up with their solutions for these functions and aspects to fill in the rows within the morphological overview.

42.3 Experiments

After extensive experiments with different setups for implementing the ID approach, in which well over 100 professionals participated [23], it was concluded that a good way to test our design approach was a workshop setting for professionals. Therefore, workshops were arranged as part of a training program for architects and consulting engineers (structural engineers, building services engineers, and building physics engineers) [23]. Since 2000, together with the Dutch Royal society of architects (BNA), the Dutch Association of Consulting Engineers (ONRI), the Dutch Society of Building Services Engineers (TVVL), and different roofer associations, in total 14 series of workshops were organized in which more than 200 experienced professionals from these organizations, with at least 10 years of experience, voluntarily participated. The design exercises were derived from real practice projects and as such were as close to professional practice as possible. The design tasks during the 2 days were on the same level of complexity and have been used in all workshops. In the workshops, stepwise changes were introduced in the setup of the design sessions to the traditional building design process type, in which the architect starts the process and the other designers join in later in the process. After each session the teams were rearranged to avoid learning effects between team members. The first two design sessions on day 1, provide reference values for the effectiveness of the involved of all designers from different disciplines right from the start. On the second day, the morphological overviews were introduced. The application of morphological overviews during the setup of the third design session enabled structuring of design functions/aspects and the generated (sub)solution proposals. Additionally, the third setting provided the possibility of one full learning cycle regarding the use of morphological overviews. After the feedback to the participants about their use of MCs and the morphological overview, all teams had the basic knowledge to apply them correctly in the fourth setting. During the design sessions in the 2011–2014 versions of the multidisciplinary master design project, students performed different design assignments which were based on the assignments tested in the workshops for professionals. Characteristic of the ID process is the use of MCs by individual designers which were combined into one morphological overview by the design team. All the assignments were related to aspects of nZEBs and had a similar level of complexity to make the results of the different sessions comparable. This enabled us to compare the results of the students with those of the professionals.

42.4 Results

Here, only a brief selection of the results of the professional workshops is given in relation to the later workshops for students which were derived from them, see Fig. 42.6 far right. More results and information were presented by Savanović [23]. For the comparison of the results of the different workshops of the student, work-

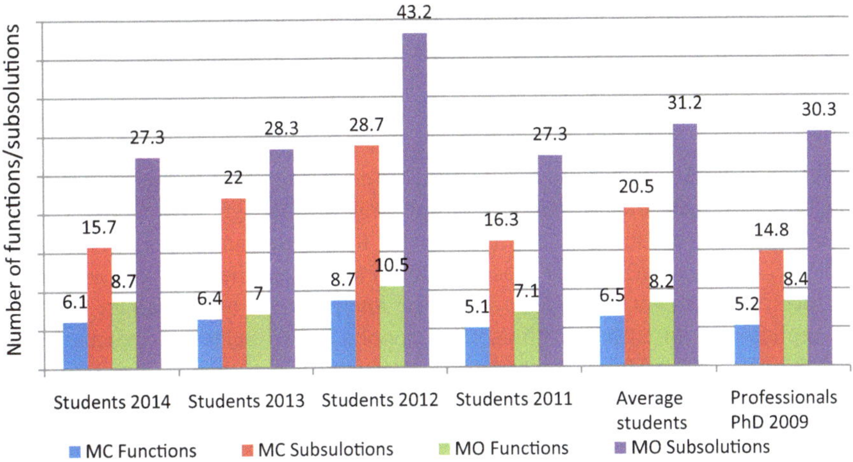

Fig. 42.6 Functions and subsolutions proposed in the morphological charts (MCs) and morphological overview (MOs) by students compared with the results from the professionals [23]

shops within the master project ID were compared with those of the professionals from the research by Savanović [23]. On average the student teams (2011–2014) produced slightly less functions (2.4 % less: students 8.2 professionals 8.4), but more subsolutions (2.6 % more: students 31.2 professionals 30.3) than the professional teams of the research by Savanović [23], see Fig. 42.6.

Applying the ID-method led to an increase of mentioned functions of 26 % (from 6.5 to 8.2) by students and 62 % (from 5.2 to 8.4) by professionals. For the number of mentioned solutions, there was an increase of 52 % (from 20.5 to 31.2) for the students and an increase of 105 % for the professionals (from 14.8 to 30.3).

42.5 Discussion and Conclusion

New design methods are needed to improve the present situation in building design. A process approach is needed in which the intentions of the client are transparently linked to the building design process by the design team of architect and engineers. Support for both stakeholders and design team members by supplying information more transparent will greatly enhance the understanding of the design process and the outcome of the combined efforts of the design team members. At the technological university (TU) Eindhoven an ID method has been developed. In workshops with experienced professionals, a first prototype of the ID method was developed and tested in practice in cooperation with the BNA and the Dutch society of consulting engineers (NLIngenieurs).

Although the quantitative results of applying the ID-method look very promising, there might not be such a positive and direct effect on the quality of the end result of the design process. However, this is almost impossible to study, but given the fact that a more transparent exchange of ideas and interpretations took place within the design teams which should lead to improvements of the further process phases during the building design process.

The primary goal of this research was to find a way to integrate architecture with different engineering disciplines into an ID process for nZEB design. Right at the beginning of such integral building design project, a design method was offered to integrate in a meaningful way engineering knowledge and experience. Putting the MCs together with the ID process enables to "put on the table" the individual perspectives from each discipline about the interpretation of the design brief and its implications for each discipline. This supports and stimulates the discussion on and the selection of functions and aspects of importance for the specific design task. Structuring the process communication between architects and engineering consultants based on abstraction of functions and subsolutions, the building model is gradually described in an increasingly more detailed manner.

For the client, such ordering of the design process is made more clear by presenting it in part results directly related to the interpretation of the client's intentions. The results show that the ID method is relevant for increased insight between design team members, which makes it easier to create sustainable concepts like net or nZEBs. As such, we conclude that ID method supports architects, structural engineers, building physics engineers, and heating, ventialtion, and air-conditioning (HVAC) engineers to develop new necessary professional skills to cope with their new roles for sustainable development.

References

1. Taleghani M, Ansari HR, Jennings P (2010) Renewable energy education for architects: lessons from developed and developing countries. Int J Sustain Energy 29(2):105–115
2. Lee J, Jeong Y (2012) User-centric knowledge representations based on ontology for AEC design collaboration. Comput Aided Des 44:735–748
3. Kanters J, Horvan M (2012) The design process known as IDP: a discussion. Energy Procedia 30:1153–1162
4. Kanters J, Horvat M, Dubois M (2012) Tools and methods used by architects for solar design. Energy Build 68:721–731
5. Juan YK, Gao P, Wang J (2010) A hybrid decision support sytem for sustainable office building renovation and energy performance improvement. Energy Build 42:290–297
6. Larsson N, Poel B (2003) Solar low energy buildings and the integrated design process—an introduction. http://www.iisbe.org/system/files/Task23_introduction.pdf. Accessed 7 July 2015.
7. Biesbroeck K, Klein R, Versele A, Breesch H (2010) Design process for Net-Zero Energy Buildings. Proceedings conference: Towards 0-impact Buildings and Built Environment SB10, Maastricht
8. CURT (2004) Construction users roundtable (CURT) architectural/engineering productivity committee. Collaboration, integrated information and the project life cycle in building design, Construction and Operation (WP-1202)
9. King D (2012) Holistic approach. CIBSE J January 47–49

10. Young L (2014) I cannot think of a better career than services engineering. Inteverview with Mike Davis. CIBSE J April:29–31
11. Simon H (1969) Sciences of the artificial. MIT Press, Cambridge
12. Frankel L, Racine M (2010) The complex field of research for design, through design, and about design. Proceedings Design Research Society conference Design & Complexity, 7–9 June 2010, Montreal
13. Chai K-H, Xiao X (2012) Understanding design research: a bibliometric analysis of design studies (1996–2010). Des Stud 33(1):24–43
14. Cross N (2007) Editorial forty years of design research. Des Stud 28(1):1–4
15. Bayazit N (2004) Investigating design: a review of forty years of design research. Des Issues 20(1):16–29
16. Wynn D, Clarkson J (2005) Models of designing. In: Clarkson J, Eckert C (eds) Designing process improvement—a review of current practice. Springer, London
17. Pahl G, Beitz W, Feldhusen J, Grote KH (2006) Engineering design. A systematic approach, 3rd edn. Springer, London
18. Ranjan BSRC, Srinivasan V, Chakrabarti A (2012) An extended, integrated model of designing. Proceedings of TMCE, Karlsruhe
19. Gericke K, Blessing L (2012) An analysis of design process models across disciplines. Proceedings International Design conference Design 2012, Dubrovnik
20. Zeiler W, Savanović P (2009) General systems theory based integral design method. ICED'09, Stanford
21. Blessing LTM (1994) A process-based approach to computer supported engineering design. PhD-thesis, Universiteit Twente
22. Zwicky F (1948) Morphological astronomy. The Observatory 68(845):121–143
23. Savanović P (2009) Integral design method in the context of sustainable building design. PhD thesis, Technische Universiteit Eindhoven
24. Ölvander J, Lundén B, Gavel H (2008) A computerized optimization framework for the morphological matrix applied to aircraft conceptual design. Comput Aided Des 41(2):187–196
25. Ritchey T (2010) Wicked problems social messes, decision support modelling with moprhological analysis, Swedish morphological society. Stockholm

Chapter 43
Study of the Aeraulic Flows in the Building of the Valve Halls Mandarins, France

N. Laaroussi, L-V. Bénet, F. Lacroux and M. Garoum

Abstract The site object of this work connects Sellindge, England, and Les Mandarins by a National Grid and Réseau de Electricity Transmission Network, usually known as RTE, to allow the transfer of high-voltage electricity from England to France and vice versa. The aim of the project studied is the renovation of two converter stations having a capacity of 2 × 1000 MW high-voltage direct current (HVDC) with an electric potential of 270 KV, and the link connecting the Great Britain (GB) and French transmission systems was attributed to AREVA T&D (PC 8892 A AREVA, IFA 2000 Valves & Controls Replacement Project Sellindge and Les, 2009). After this, the electricity is produced at a power station, where it will be distributed to the customers.

This study concerns the building of the valve halls mandarins, France. The valves have to be replaced because the existing valves discharge all of their heat into the halls, whereas the new valves will have water cooling systems to remove most of the heat at source. However, some heat will still be discharged into the halls, so the ventilation systems must be changed to assure the requirements of the new valves and to maintain the halls at the suitable conditions.

The produced electricity is transferred towards the outside via the overhead conductors which operate at high voltage and connect the valves with the bushings. The maximum air temperature around bushings must be lower than 40 °C. The high degree of temperature is not supported by the bushings and will be damaged.

N. Laaroussi (✉) · M. Garoum
Laboratory of Energy, Materials and Environment (LEME),
University of Mohammed V Rabat, EST of Sale,
Avenue Prince Héritier, BP. 227, Salé Medina, Marocco
e-mail: n.laaroussi@est.um5a.ac.ma

L.-V. Bénet
Socotec Industries, 1, avenue du Parc, 78180 Montigny-le-Bretonneux, France

F. Lacroux
AREVA T&D, Tour AREVA—1, place Jean Millier, 92084 Paris La défense cedex, France

A. Sayigh (ed.), *Renewable Energy in the Service of Mankind Vol I*,
DOI 10.1007/978-3-319-17777-9_43

Keywords Aeraulic · Thermal insulating · Building materials · Thermal conductivity

Nomenclature

A	thermal diffusivity $a=\lambda/\rho c_p$ ($m^2\ s^{-1}$)
H	high of Hall (m)
m	mass (kg)
M_a	molecular weight of air ($kg\ kmol^{-1}$)
P	thermodynamic pressure (Pa)
C_p	thermal capacity ($J\ kg^{-1}\ K^{-1}$)
q_v	flow rate ($m^3\ s^{-1}$)
S	surface (m^2)
T	temperature (K)
$\bar{T}$	mean temperature (K)
Ra	thermal Rayleigh number, $Ra=\rho g\beta\Delta TH^3/\mu.a$
Gr	Grashof number, $Gr=g\beta\Delta T\,H^3/v^2$
C_1, C_2, C_3	coefficients in (k–ε) models

Greek Symbols

ρ	density ($kg\ m^{-3}$)
λ	thermal conductivity ($W\ m^{-1}\ K^{-1}$)
μ	dynamic viscosity ($kg\ m^{-1}\ s^{-1}$)
μ_t	turbulent viscosity ($kg\ m^{-1}\ s^{-1}$)
ε	dissipation rate ($m^2\ s^{-3}$)
K	turbulent kinetic energy ($m^2\ s^{-2}$)
β	coefficient of thermal expansion (k^{-1})

43.1 Introduction

The buoyancy forces of heat transfer resulting from temperature gradient is shown to have large influences on heat transfer of fluid in many engineering systems, for example, drying processes and building technologies.

Two-equation turbulence models allow the determination of both a turbulent length and time scale by solving two separate transport equations. The standard *k*–ε model in ANSYS [2] falls within this class of models and has become the work-horse of practical engineering flow calculations in the time since it was proposed by Launder and Spalding [3]. The standard *k*–ε model is a model based on model transport equations for the turbulence kinetic energy *(k)* and its dissipation rate *(ε)*. The model transport equation for *k* is derived from the exact equation, while

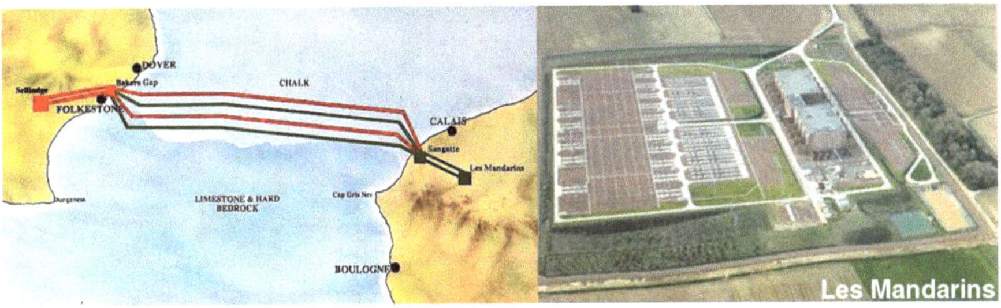

Fig. 43.1 The located site at "Les Mandarins," France

the model transport equation for ε is obtained using physical reasoning and bears little resemblance to its mathematically exact counterpart. In the derivation of the k–ε model, the assumption is that the flow is fully turbulent, and the effects of molecular viscosity are negligible. For a much more practical approach, the standard k–ε turbulence model is used which is based on our best understanding of the relevant processes, thus minimizing unknowns and presenting a set of equations which can be applied to a large number of turbulent applications. Therefore, in our study, we employ a standard k–ε turbulence model [3]. In fact, the standard k–ε model [4, 5] is, till now, by far the most widely used and validated of the classical models.

The computational fluid dynamics (CFD) method is used to investigate the temperature fields in a mechanically ventilated hall. Heat transfer in the hall is numerically simulated using the standard k–ε model of turbulence. For a very hot weather, this study aims to evaluate the local temperatures in the ambient air of the hall, with assuming running valves and air conditioning device in open loop with a 35 °C inlet temperature. The study shows that the air conditioning is efficient enough to maintain low levels of temperature disparity Fig. 43.1.

43.2 Mathematical Formulation

Computations are made on the three-dimensional Cartesian coordinate system using the standard k–ε model of turbulence in incompressible form. There are no chemical reactions, and the radiative heat transfer in the hall is neglected. The air is invoked as an ideal gas $\left(\rho = \dfrac{PM_{\mathrm{a}}}{RT}\right)$, and the thermophysical properties of air are assumed to be constant.

In Reynolds averaging, the solution variables in the Navier–Stokes equations are decomposed into the mean ϕ and fluctuating ϕ' components.

The fluid is considered as incompressible with an ideal gas law depending only on temperature. The filtering equations for continuity, momentum, and energy are the following:

The continuity equation is:

$$\frac{\partial u_i}{\partial x_i} = 0. \tag{43.1}$$

The momentum equation is written as:

$$\rho \frac{\partial}{\partial x_j}\left(u_i u_j\right) = -\frac{\partial p}{\partial x_i} + \frac{\partial}{\partial x_j}\left[\mu\left(\frac{\partial u_i}{\partial x_j} + \frac{\partial u_j}{\partial x_i}\right)\right] + \rho \frac{\partial}{\partial x_j}\left(-\overline{u_i' u_j'}\right) + \rho g_i. \tag{43.2}$$

The energy equation is written as:

$$\rho c_{\mathrm{p}} \frac{\partial}{\partial x_i}(u_i T) = -\frac{\partial}{\partial x_i}\left(\lambda \frac{\partial T}{\partial x_i}\right) + \rho c_{\mathrm{p}} \frac{\partial}{\partial x_i}\left(-\overline{u_i' T'}\right) + S_h. \tag{43.3}$$

The source term S_h, introduced in the energy equation, is for the volumetric heat flux produced within the valves.

The turbulence kinetic energy, k, and its rate of dissipation, ε, are obtained from the following transport equations:

$$\rho \frac{\partial}{\partial x_i}\left(k u_i\right) = \frac{\partial}{\partial x_j}\left[\left(\mu + \frac{\mu_t}{\sigma_k}\right)\frac{\partial k}{\partial x_j}\right] + G_k + G_{\mathrm{b}} - \rho\varepsilon, \tag{43.4}$$

and

$$\rho \frac{\partial}{\partial x_i}\left(\varepsilon u_i\right) = \frac{\partial}{\partial x_j}\left[\left(\mu + \frac{\mu_t}{\sigma_\varepsilon}\right)\frac{\partial \varepsilon}{\partial x_j}\right] + \frac{\varepsilon}{k}(C_1 G_{\mathrm{k}} + C_3 G_{\mathrm{b}}) - C_2 \rho \frac{\varepsilon^2}{k}. \tag{43.5}$$

In these equations, G_k represents the generation of turbulence kinetic energy due to the mean velocity gradients. G_{b} is the generation of turbulence kinetic energy due to buoyancy. C_1, C_2, and C_3 are constants. σ_k and σ_ε are the turbulent Prandtl numbers for k and ε, respectively.

$$G_k = -\overline{u_i' u_j'} \frac{\partial u_j}{\partial x_i} \quad \text{with} \quad \overline{u_i' u_j'} = -\nu_t\left(\frac{\partial u_i}{\partial x_j} + \frac{\partial u_j}{\partial x_i}\right), \tag{43.6}$$

$$G_{\mathrm{b}} = \beta g_i \frac{\mu_t}{\mathrm{Pr}_t} \frac{\partial T}{\partial x_i} \tag{43.7}$$

The coefficient of thermal expansion, β, is defined as: $\beta = -\frac{1}{\rho}\left(\frac{\partial \rho}{\partial T}\right)_p$. For ideal gas, Eq. (43.7) reduces to:

$$G_{\mathrm{b}} = -g_i \frac{\mu_t}{\rho \mathrm{Pr}_t} \frac{\partial \rho}{\partial x_i}, \tag{43.8}$$

where Pr_t is the turbulent Prandtl number for energy, and g_i is the component of the gravitational vector in the i direction. For the standard k–ε models, the default value of Pr_t is 0.85.

The turbulent viscosity, μ_t, is computed by combining k and ε as follows:

$$\mu_t = \rho C_\mu \frac{k^2}{\varepsilon}, \tag{43.9}$$

where ρ represents the density. k and ε correspond to the turbulence fields: turbulent kinetic energy and its rate of dissipation, respectively. $C_\mu = 0.09$ is the empirical value.

The model constants C_1, C_2, σ_k, and σ_ε have the following default values determined from experiments for fundamental turbulent flows: $C_1 = 1.44$, $C_2 = 1.92$, $\sigma_k = 1$, and $\sigma_\varepsilon = 1.3$.

Inclusion of the production term due to buoyancy in the k–ε model was first proposed by Rodi [6]. Cox et al. [7] conducted a detailed study to determine the influence of C_3 to the k–ε equations. Viollet has demonstrated the effectiveness of G_k, C_1, and C_3 in flows through reactor coolant analysis as mentioned above [8]. Viollet [8] has adopted the following procedure for determining the value of $C3$ based on the sign of Gk—the term denoting the buoyancy production of turbulence in k. If $Gk > 0$, $C3 = C1 = 1.44$ and $C3 = 0$ if $Gk < 0$. Note that the sign of Gk is related to the stability of flow.

Finally, the two equations for the kinetic energy and dissipation rate of the standard k–ε turbulence model are added to the continuity equation, the Navier–Stokes momentum equations, and the thermal energy equation in order to capture the air flow in the hall.

43.3 Solution Procedure

43.3.1 Computational Conditions

Thermal heat transfer within the valve halls is numerically simulated using the standard k–ε model of turbulence. Considering the geometrical symmetry of the valve halls, only one half of the entire field is used as the solution domain, as shown in Fig. 43.2.

The dimensions of each valve hall are 18.3 m wide, 60.5 m length, and 18.1 m high. The construction of the buildings is assumed to be as follows:

- Valve hall floor, 200 mm reinforced concrete with a density of 2500 kg m^{-3}
- Valve hall walls, 350 mm reinforced concrete with a density of 2500 kg m^{-3}
- Valve hall roof, 150 mm reinforced concrete with a density of 2500 kg m^{-3} with bitumen felt on top. The insulation under roof felt is 60-mm-thick foam glass with a thermal conductivity of 0.042 W m^{-1} K^{-1}.

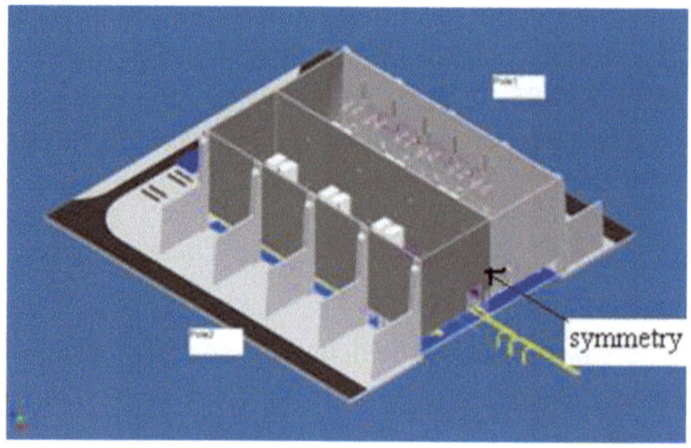

Fig. 43.2 The building of the valve halls mandarins, France

The thermophysical properties of air were evaluated at reference temperature $T_0 = 313$ K (Table 43.1).

Calculations were carried out by utilizing the commercial, control-volume-based code Ansys13. Results of the simulations were collected and processed by employing in-house softwares. A second-order upwind scheme was used for the advective and transport terms. The velocity–pressure coupling was solved with simple algorithm, and the pressure was calculated with a body-force-weighted scheme. The convergence criterion required that the scaled residuals be smaller than 10^{-5} for mass and momentum equations and smaller than 10^{-6} for energy and transport equations.

43.3.2 Boundary Conditions

The studied hall contains three valves. When the valves are operating, the internal heat gain is 40 kW emitting by the three valves and the temperature is of 50 °C local to the valves. However, some heat will still be discharged into the hall, and the air conditioning systems are designed to remove this heat and maintain the hall at the correct conditions. The system can run continuously for 1 year without maintenance necessitating a shutdown, apart from filter changing. The ventilation shall operate 24 h per day to maintain conditions.

Ventilation includes both the exchange of air to the outside and circulation of air within the hall. It is one of the most important factors for maintaining acceptable condition in the hall. To maintain the area around the valve hall at a reasonable temperature, we have assumed that several fans would be installed in the external walls

Table 43.1 Thermophysical properties of air at $T_0 = 313$ K and atmospheric pressure

ρ_a (kg/m^3)	C_{pa} (J/kg K)	μ_a (kg/m s)	k_a (W/m K)	Pr	Ra	Gr
1.132	1006	2.006×10^{-5}	0.0272	0.71	4.39×10^{12}	5.92×10^{12}

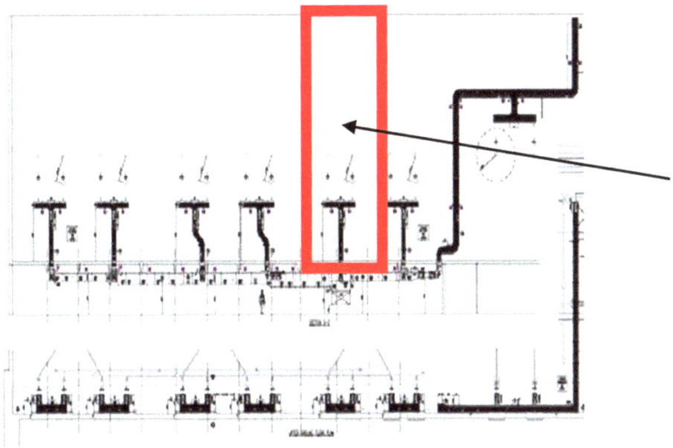

Fig. 43.3 The ventilation plan of the valve hall

and roof around each hall. The ventilation rate is expressed by the volumetric flow rate of outside air being introduced to the hall. The total flow rate delivered by the ventilator during inspiration is $q_v = 1.53\ m^3/s$.

At Les Mandarins, to achieve the air temperature around the bushings at 40 °C, air will be directed to the bushings at approximately 35 °C or lower. This will generally be achieved using ambient air; however, when the ambient temperature rises above 35 °C, the cooling system will be activated. Also, the natural ventilation of the hall is achieved through openings in the lower areas. The jet is maintained at a fixed temperature T_e by an air-treatment system. Also, the ventilation system allows us to impose inlet flow rate q_m (Fig. 43.3).

The simplified model was composed by two bushings, one valve, and ventilation systems with three rectangular openings.

These elements are included in a volume of air including a wall and three vertical plans of symmetries. The computational conditions are summarized in Table 43.2. The cases computed are shown in Fig. 43.4.

Table 43.2 Boundary conditions of the simulation

Heat source	$V_{valve} = 178\ m^3$ and $S_h = 74.7\ W\ m^{-3}$
Mass flow rate and temperature of ventilation	$q_m = 3 \times 0.577\ kg\ s^{-1}$ and $T_e = 35\ °C$
Opening of aeration	Outflow
Floor	Adiabatic for temperature and $\vec{u} = 0$
Roof	$h_T = 0.67\ W\ m^{-2}K^{-1}$, $T_{ex} = 45\ °C$, and $\vec{u} = 0$
External vertical walls	$h_{ex} = 25\ W\ m^{-2}K^{-1}$, $T_{ex} = 45\ °C$, and $\vec{u} = 0$
Symmetry plans	Normal velocity to the plan is set at 0 and free slip, otherwise (normal gradient of zero)

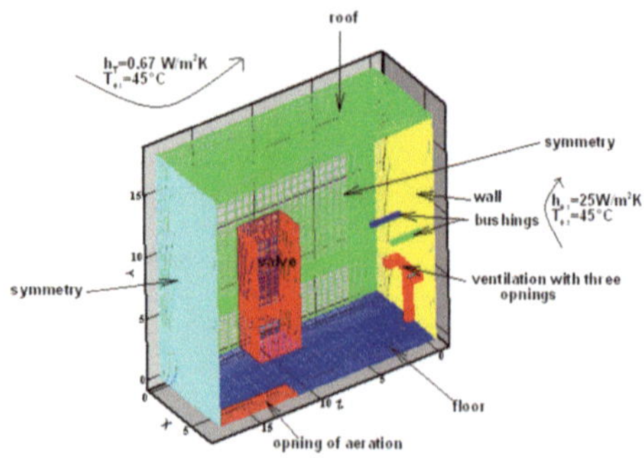

Fig. 43.4 The simplified model of the valve hall

43.4 Computational Results and Discussions

The section presents the numerical results for the simplified and the complete models as shown in Figs. 43.4 and 43.7. The comparison concerns the mean temperature fields. Several plans have been chosen for the analysis: the longitudinal plans according to the bushings having a diameter of 0.356 m and length of 2.46 m.

43.4.1 Simplified Model

To conduct our study, we start with the simplified model; this approach seems attractive because the increase in size of the computational domain proves to be expensive, in both memory and computational time. For these reasons, the domain extensions are often either relatively reduced or large but coarsely discretized. In order to evaluate the accuracy of our models, we then compare the mean temperature and magnitude of the velocity in the jet zone. In all cases, the simulation models prove that the expansion of the jet, and then the dynamic of the fluid, can be correctly predicted.

Figure 43.5 indicates that the maximum valve hall temperature is about 50 °C, and the maximum air temperature around bushings is about 40 °C. In the ventilated hall, the dynamic of the flow is quite simple because there are two principal zones: the jet zone and the nonmoving fluid zone, where the fluid velocity is less than 0.05 m/s, as shown in Fig. 43.6.

The thermal convection leads to a vertical temperature stratification within the hall with the lowest temperatures along the bushings estimated between 38 and 40 °C, fed by the air of ventilation. The temperature-high region has approximately 42 °C fed by discharged heat of the valves into the hall. The changes in the shapes of the isotherms indicate that thermal buoyancy, acting downward opposite to forced convection, is due to air conditioning.

Figure 43.6 provides a representation of the evolution of the two jets simulated. Colored velocity vectors represented along vertical planes illustrate the growth of each jet along its centerline. Close to the opening, the centerline velocity of jets remains equal to the initial velocity, denoting the core region of the jets. An axisym-

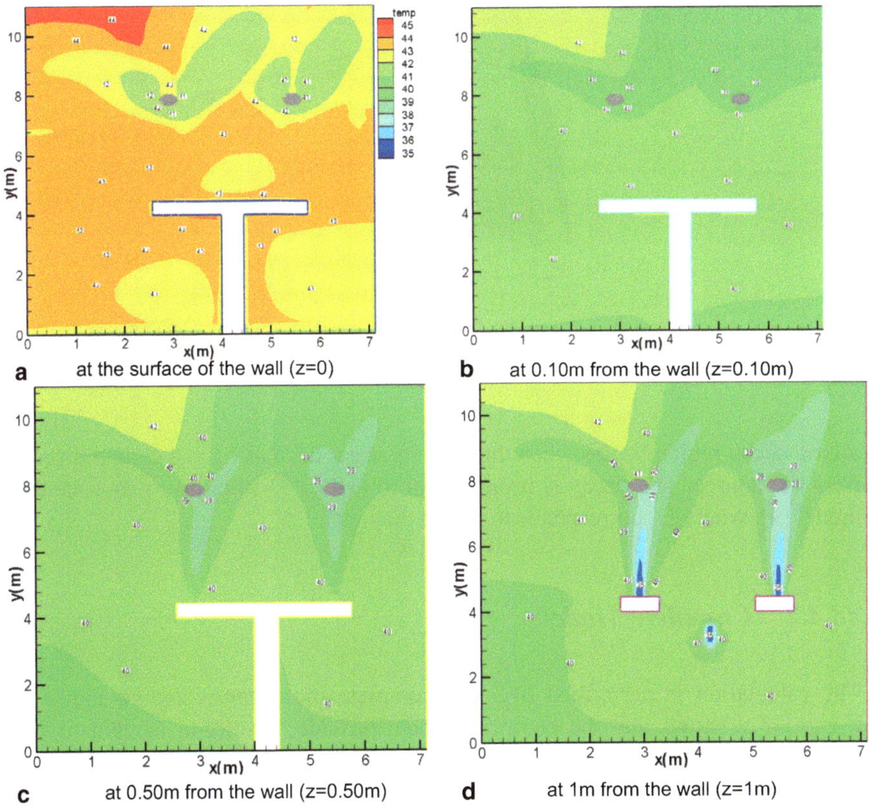

Fig. 43.5 The mean temperature fields **a** at the surface of the wall ($z=0$), **b** at 0.10 m from the wall ($z=0.10$ m), **c** at 0.50 m from the wall ($z=0.50$ m), and **d** at 1 m from the wall ($z=1$ m)

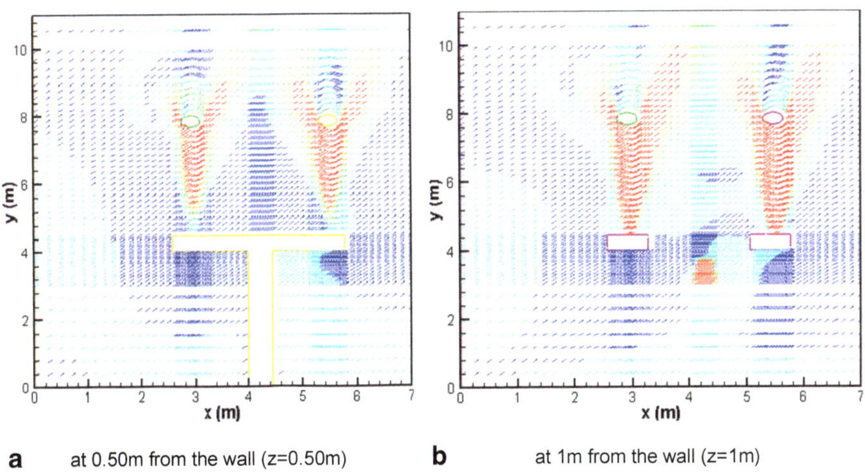

Fig. 43.6 The mean velocity vectors **a** at 0.50 m from the wall ($z=0.50$ m) and **b** at 1 m from the wall ($z=1$ m)

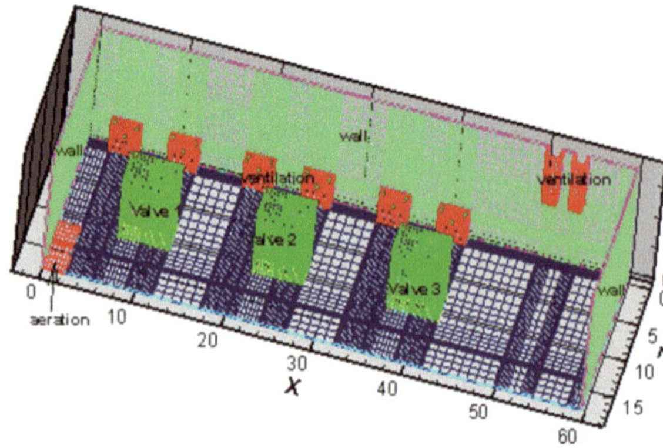

Fig. 43.7 The complete model of the valve hall

metric decay region begins when the velocity starts decreasing at a certain distance from the source. The decay is proportional to $(x/D_0)^{-0.5}$ when using k–ε standard model [9], where D_0 corresponds to the jet diameter.

43.4.2 Complete Model

The calculation is performed over the complete model presented in Fig. 43.7. Figure 43.9 shows that the complete model provides results in agreement with the simplified model. An analysis of the results obtained for the simplified model proves to exhibit very similar properties and coincides very well with those of the complete model based on the simulation of the entire hall (Fig. 43.8), even if for slightly different parameters. Hence, the simplified model presents the advantage of having fewer grids, which required a half-day computing time on a workstation.

The temperature distributions are displayed in Fig. 43.9 for plans at $z=0$ and $z=1$ m from the wall, respectively. The possible symmetry of the flow fields about the vertical middle plans is evidenced at steady state.

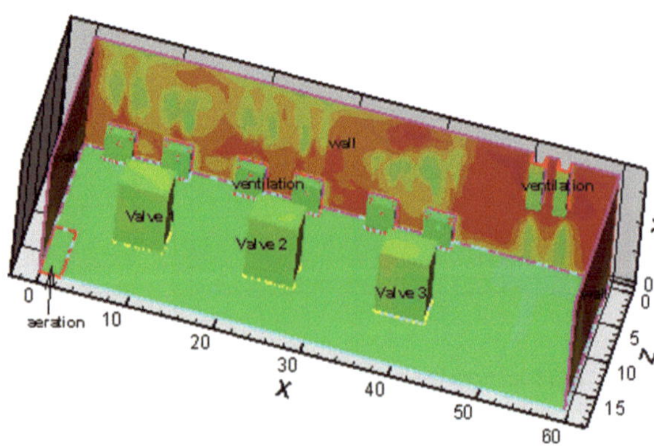

Fig. 43.8 The surface temperatures of the valve hall

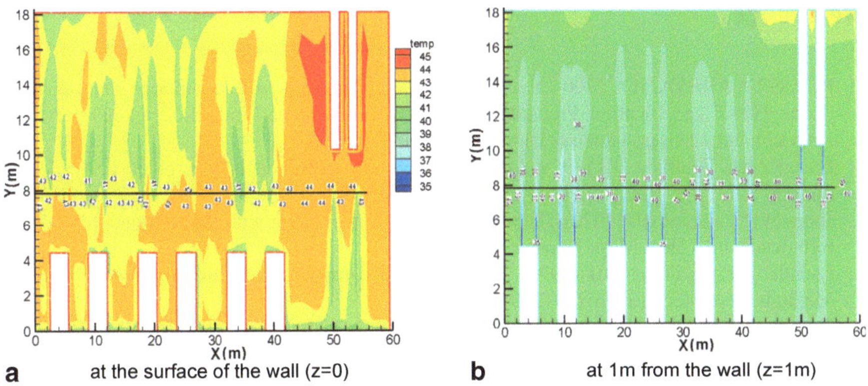

Fig. 43.9 The mean temperature fields **a** at the surface of the wall ($z=0$) and **b** at 1 m from the wall ($z=1$ m)

43.5 Conclusion

The simulations presented were carried out for both simplified and complete models. The standard k–ε model seems to predict well the velocity and temperature fields in the valve hall. So, the global comportment of the fluid is correctly solved. The prediction of the airflow and the slope of the jet are well identified. However, this parameter is very important for the thermal comfort conditions of a building hall.

Knowing that the design of all thermal loads and selections of equipment will be based on the external design conditions, which indicate that during summer, the average maximum of temperature reaches 45 °C. The calculated mean air temperatures reach 39.5 °C inside the valve hall. The local air temperature variation around the double bushing is about ± 1 °C from mean value, but the simple bushing located at 16.85 m high is more exposed to warm air at 42 °C than the doubled bushing. The study indicates that the staff will wait 2 h before entering the valve hall with the air temperatures becoming lower than 25 °C when the valves are stopped and air conditioning loop is closed.

It should be noted that the cost of air conditioning of the valve hall is rapidly increasing. The best way to reduce the cooling cost is to add insulation to walls of the hall that is below recommended levels.

Finally, the aim of this study was to ensure a satisfactory installation and the feasibility of the project.

References

1. PC 8892 A AREVA, IFA 2000 valves & controls replacement project sellindge and Les Mandarins, HVAC specification. Prepared by RB Plant Construction Limited, Sept 2009, Rev2
2. Ansys Fluent 12 Users Guide (n.d), www.fluentusers.com
3. Launder BE, Spalding DB (1972) Lectures in mathematical models of turbulence. Academic, London
4. Jones WP, Launder BE (1972) The prediction of laminarization with a two-equation model of turbulence. Int J Heat Mass Transfer 15:301–314
5. Launder BE, Spalding DB (1974) The numerical computation of turbulent 6ows. Comput Methods Appl Mech Eng 3:269–289
6. Rodi W (1978) Turbulence models, their application in hydraulics—a state of art review. University of Karlsruhe, SFB/80/T/127, 128
7. Markatos NC, Malin MR, Cox G (1982) Mathematical modeling of buoyancy-induced smoke flow in enclosure. Int J Heat Mass Transfer 25:63–75
8. Viollet PL (1987) The modeling of turbulent recirculating flows for the purpose of reactor thermal-hydraulic analysis. Nucl Eng Des 99:365–377
9. Menchaca Brandan M (June 2005) Study of airflow and thermal stratification in naturally ventilated rooms. Thesis, Massachusetts Institute of Technology

Chapter 44
Experimental and Theoretical Study for the Performance of New Local Thermal Insulation in Iraqi Building

Ghanim Kadhim Abdulsada and Tawfeeq Wasmi M. Salih

Abstract Iraq is located in a hot-arid region, where the hot summer season is very long and the ambient temperature is about 50 °C, while in the winter the temperature is about 10 °C. Therefore, cooling of buildings needs A/C system, which consumes 60 % of the total electricity supplied to residential building. So, there is really a need to minimize the energy consumption by enhancing the performance of Iraqi buildings. To minimize the effect of the environment upon the building, one must look for methods to increase the thermal resistance of walls and roof. This chapter deals with the experimental and theoretical study using new local thermal insulation on walls and roof with an air gap. This study takes into consideration several parameters like ambient temperature, solar radiation, orientation, and duration. The results of the work give good indication that using this insulating system in residential building has great effect on the energy consumed. Also, the results reveal that a system with sufficient insulation can save the annual cooling load for the room by 6 MW and that the indoor temperature can reach 35 °C, while the ambient temperature is about 45 °C.

Keywords Energy saving · Passive system · Thermal insulation · Building performance

G. K. Abdulsada (✉) · T. W. M. Salih
Department of Mechanical Engineering, Al-Mustansiriyah University, Baghdad, Iraq
e-mail: ghanimkadhim@yahoo.com

T. W. M. Salih
e-mail: tawfeeq1978@yahoo.com

A. Sayigh (ed.), *Renewable Energy in the Service of Mankind Vol I*,
DOI 10.1007/978-3-319-17777-9_44

44.1 Introduction

The local severe summer weather conditions in Iraq are a good example of arid climatic conditions with high levels of solar radiation and large fluctuations in air temperature. In order to reduce the climatic effects upon buildings, it is important to use insulating materials for the exposed walls and roof. The various thermal insulation materials and component systems for the building envelope are reviewed and assessed including mineral and glass wool, cellulose, and polymers. The choice of the thermal insulation material depends on the thermal performance of it. To evaluate the thermal performance of insulation, heat transmission through walls and roof should be analyzed. Three modes of heat transfer are presented: conduction, convection, and radiation, and to minimize the effect of these quantities, it should seek for [1]:

- Low-thermal conductivity to minimize the conduction
- Evacuated portion to minimize the convection
- Low emittance high-reflective surface to attenuate radiation

A theoretical study carried out by Almusaed et al. [2] shows that improving energy efficiency in the performance of Iraqi building designs can be satisfied by means of natural elements like use of insulation, shadow, air flow, and rooms planning. The U value of the optimized wall must not be exceeding 0.12 W/m^2 K. For an efficient roof, it needs to be developed in ways to get a U value of 0.082 W/m^2 K. A complete thermal bridge-free construction is needed with linear thermal transmittance below 0.01 W/m K. On a summer day, the temperature of a gravel roof can increase to 40 °C, so using 15 cm of mineral wool is required. Also, shading of the roof can reduce air temperature as much as to 5 °C. An experimental study carried out by Abdulsada [3] deals with the reducing of energy consumption in buildings by eliminating the effect of direct solar radiation in summer season. Roof surface is the most exposed area to solar radiation, and most of the solar gain absorbed by roofs is transmitted down to the inside living space because the roofs in Iraqi buildings work as solar collectors. In that study, a system is stated to spray water upon the external roof surface. It consists of network piping, control valves, spray nozzles, and thermal sensors. The results of this study showed that the energy transmission through the roof surface is reduced to only 4 W/m^2, while, for the case of no-spray water system, the energy transmission through the roof surface rises up to 40 W/m^2. Another study carried out by Kalif et al. [4] tries to calculate the effect of packaging materials on the cooling load in Baghdad. The study conducted a survey of those materials, and it also includes testing and thermal analysis. Computer programming is done to estimate the amount of cooling load through the walls by using CLTD (cooling load temperature differences) style method described by the American Society of Heating (ASHRAE). The U value of materials is determined depending on the resistance method, where measured values of thermal conductivity of materials are taken. The results showed that the percentage reduction in cooling load when using internal packaging materials ranges between 5 and 12%.

44.2 Experimental Study

The aim of this study is to investigate the actual effect of insulated material on the building performance. The experimental study investigated the thermal behavior across two rooms in Baghdad (33.3°, 44.4°) at Al-Mustansiriyah University, Mechanical Engineering Department. These rooms are located at the terrace floor. The first room is located at the southwest corner and the second at the southeast corner. Both rooms have the dimensions 4 m × 4 m × 3 m. The walls of both rooms are constructed from 24 cm common brick, 2 cm external cement plaster, and 2 cm internal gypsum plaster, while the roof is constructed from 15 cm steel-reinforced concrete, 5 cm external cement tiles with soil, and 2 cm internal gypsum plaster. Each exposed surface is covered by insulation materials with an air gap of 5 cm as shown in Fig. 44.1. The insulation is formed locally from sheets of soot (k=0.04 W/m K) and of 1.5 cm thickness and packed into reflective Al foils as shown in Fig. 44.2. The soot is collected locally and cracked into small pieces and then grinded. This mechanism (insulation material and air gap) is selected to minimize all three modes of heat transfer. The experimental study required several instruments. Surface temperature sensors are connected to the data logger system (Lutron-Model BTM-4208SD), which records the data at every 15 min during the daytime for the period 28 April–25 June. Indoor and outdoor temperatures are also recorded. Also, the total heat solar radiation is measured using solar power meter (Lutron-Model SPM-1116SD). An infrared

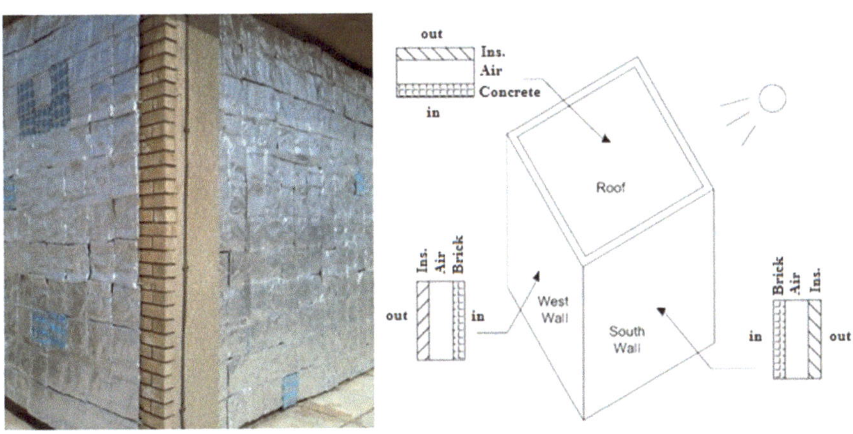

a Insulation layers b Compositions of walls and roof

Fig. 44.1 Insulating system in room (1)

Fig. 44.2 Local-made insulation from soot and covered by Al foil

thermographic camera (Fluke-Model Ti32) is used to capture the thermal images on the building surface.

44.3 Theoretical Study

The theoretical approach has been presented in a computer program which calculates the rate of heat transmission through walls and roof in all orientations. Data of air temperature and solar radiation for Baghdad have been taken from references [5–7]. The analysis has been carried out for most common types of walls and roof in Iraq by taking into account the values of physical properties of building materials including thermal conductivity, specific heat, density, and surface absorbance. In summer time, the absorbed radiation increases the temperature of the outside surface to a value that is greater than outside air temperature. This outside surface temperature is called sol-air temperature. The solar radiation amount depends on the orientation of the surface, solar altitude angle, and solar azimuth angle [8]. The heat transmission through a wall or roof involves the solution of one-dimensional energy equation and finding values of outside surface temperature (T_o) and inside surface temperature (T_i). The energy balance via walls could be presented as:

$$Q_{\text{sol}} = Q_{\text{cond}} + Q_{\text{conv}} + Q_{\text{rad}} + Q_{\text{acc}}, \tag{44.1}$$

where Q_{sol} is the solar power incident upon the surface.

And the conduction heat transfer is:

$$Q_{cond} = UCLTD = U(T_o - T_i) \tag{44.2}$$

The convection loss to the ambient is:

$$Q_{\mathrm{conv}} = h\,(T_{\mathrm{o}} - T_{\infty}). \tag{44.3}$$

The radiation loss to the ambient is:

$$Q_{\mathrm{rad}} = \sigma \epsilon \left(T_{o}^{4} - T_{\infty}^{4}\right). \tag{44.4}$$

Finally, the accumulated heat inside the wall is:

$$Q_{\mathrm{acc}} = m\,cp \frac{\partial T}{\partial \tau}. \tag{44.5}$$

The use of these equations and appropriate boundary and initially conditions, including the calculation of overall heat transfer coefficients, gives confident results. For example, it is noted that the temperature rate inside a certain wall $\left(\frac{\partial T}{\partial \tau}\right)$ is about 0.5°C/h for the summer time.

44.4 Results and Discussion

There were many runs upon the building during the period 28 April–25 June which included several case studies such as:

- Case (1) in room (1) where insulation is at south wall and west wall (28 April–02 May)
- Case (2) in room (1) where insulation is at south wall, west wall, and roof (19 May–28 May)
- Case (3) in room (1) where insulation is at south wall and roof (29 May–01 June)
- Case (4) in room (1) where insulation is at roof only (02 June–04 June)
- Case (5) in room (2) where insulation is at south wall and roof (10 June–13 June)
- Case (6) in room (2) where insulation is at south wall, east wall, and roof (17 June–20 June)
- Case (7) in room (2) where insulation is at south wall and east wall (22 June–25 June)

The experimental results are presented for both with and without insulation as shown in Figs. 44.3, 44.4, 44.5, 44.6, 44.7, 44.8 and Fig. 44.9. Generally, these results show the decrease in the indoor temperature for the case of insulation compared with that of non-insulation. It is clear that there is a relaxation change (about 3 °C) in indoor temperature for insulated case compared to the non-insulated case which

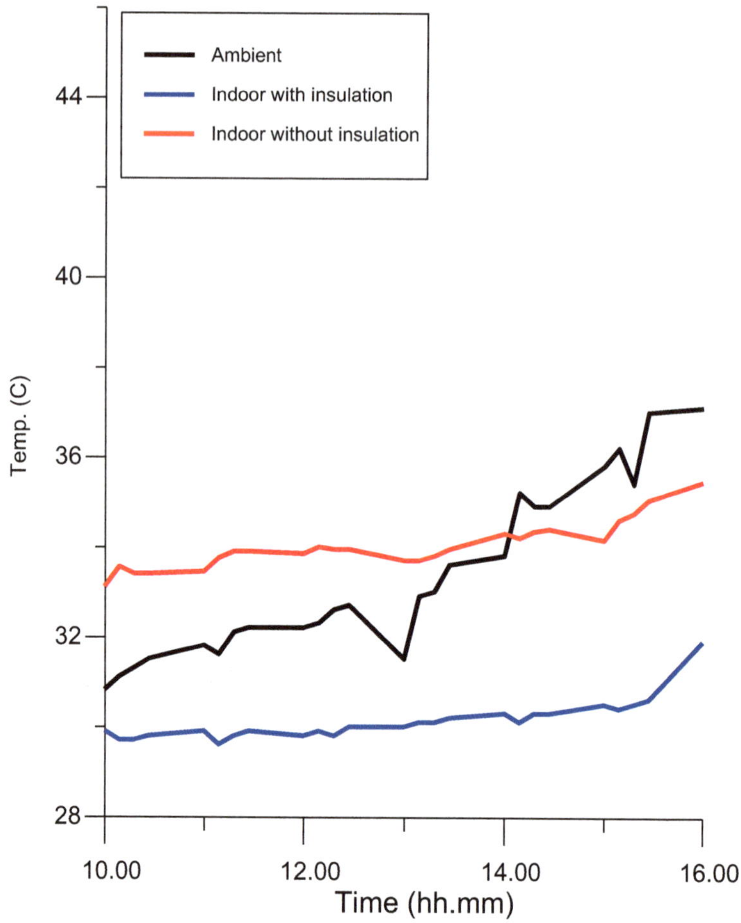

Fig. 44.3 Data in room (1) where insulation at south wall and west wall on April 28, 2013 (case 1)

has fluctuated variations (about 6 °C). The results also show that the maximum indoor temperature occurred after 2:00 p.m. due to high-accumulated heat. In case (3), there is a rapid increase in temperature after noontime because the west wall is exposed to direct solar radiation as shown in Fig. 44.5. This behavior becomes reverse for case (5), where temperature increment became so fast in the morning time because the east wall is exposed to direct solar radiation; on the other hand, temperature increment became less after noontime due to internal west wall as shown in Fig. 44.7. In Figs. 44.4 and 44.8, the indoor temperature is quite steady due to sufficient insulation for these cases. It should be noted that these cases are tested at different durations in summer, where the solar impact is variable. Figure 44.10

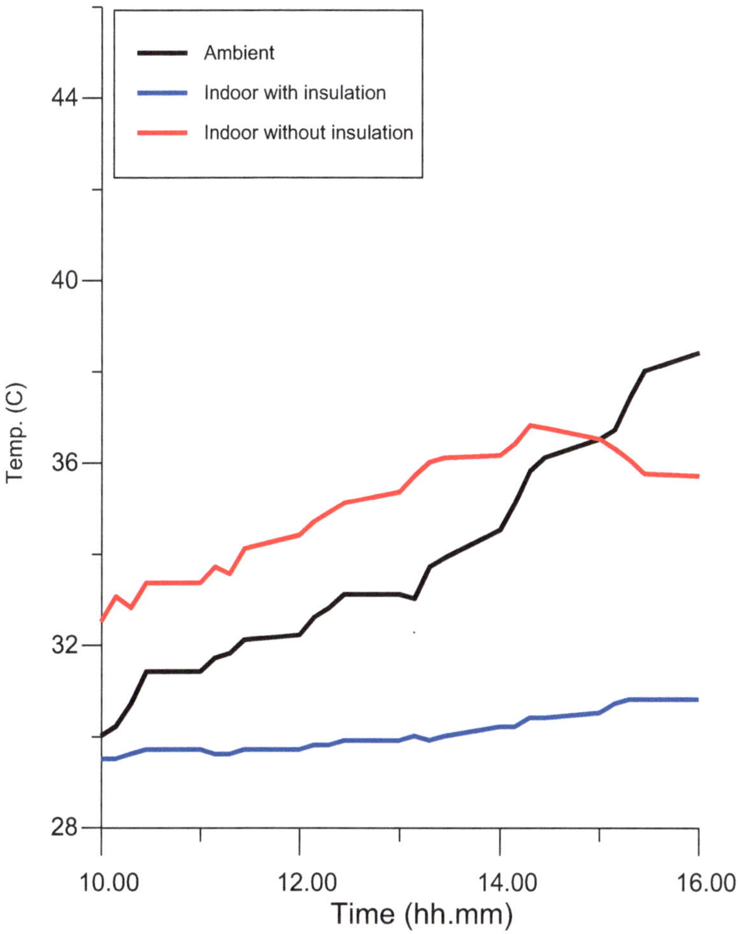

Fig. 44.4 Data in room (1) where insulation at south wall, west wall, and roof on May 23, 2013 (case 2)

shows the recorded solar radiation. It is a real need to avoid radiation in the east and west orientation because the heat transmitted through the east and west walls is really much than that of the south wall. Further, in the morning and evening, the sun has a low angle, so the solar radiation focuses directly on the walls as compared with the mid-daytime period where the sun is highest and the rays are incident directly upon the roof. Furthermore, it is noticed (experimentally) that the intensity of the solar radiation in the morning is more than that in the evening because of clear weather in the morning against the dusty weather in the evening. Hence, the heat transmission via east wall is the highest.

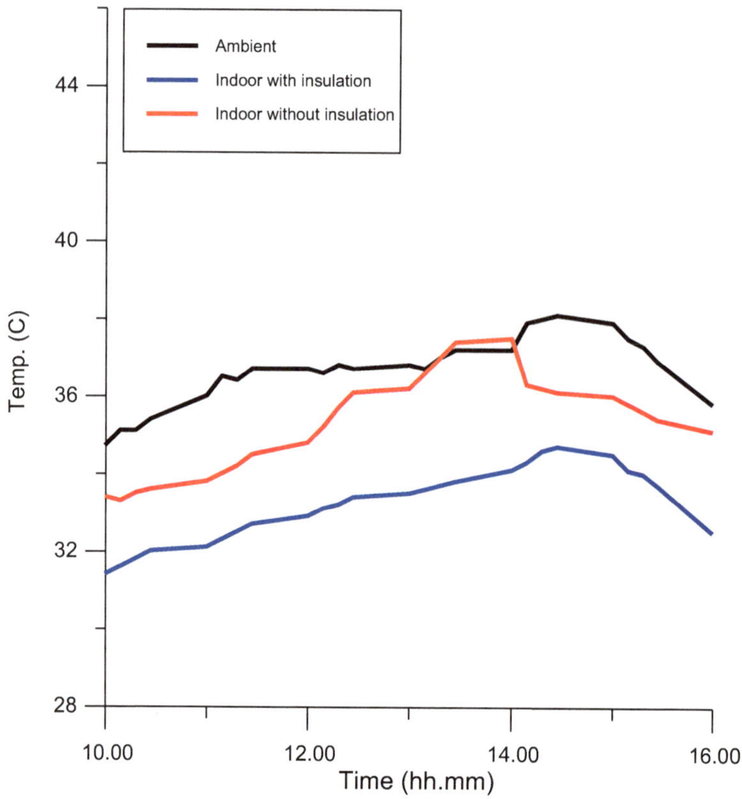

Fig. 44.5 Data in room (1) where insulation at south wall and roof on May 30, 2013 (case 3)

As shown in Fig. 44.11, the percentage of reduction in average indoor temperature for the case (2) is 14 % and for case (6) is 12 % which are the highest among all cases. This ratio is satisfied when all exposed walls and the roof are insulated, and the air change per hour is less than 1. The calculation of energy consumption depends on the average indoor temperature of the daytime for June and July.

Table 44.1 shows the energy consumption and average indoor temperature where these values are calculated using the simulated program, which is also based on the results extracted from the experimental work. The results show that average indoor temperature reaches about 41 °C for conventional case (no insulation) while it can reach up to 35 °C with sufficient insulation (insulation for all exposed walls and roof). The results also showed that the annual saving in cooling energy could reach up to 362 kWh/m^2, and the annual saving in cooling energy for the testing room (16 m^2) could reach up to 6000 kWh.

Thermal images are captured for room (1) during noontime (1230–1330 h) on May 27, 2013, at the period when the sun bath starts to leave the south façade (with

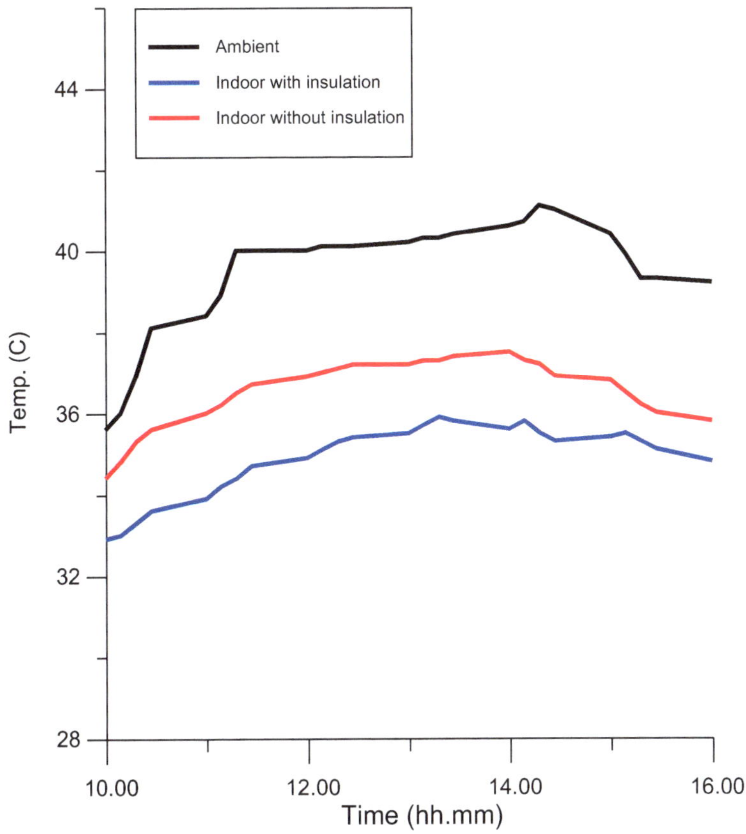

Fig. 44.6 Data in room (1) where insulation at roof only on June 3, 2013 (case 4)

its high angle) and faces the west (with low angle). As shown in Fig. 44.12, the temperature of the exposed surface (south) is about 50 °C except for the upper part which is about 40 °C due to overhang shading effect. It is also noticed that temperature values are not unique over the entire wall due to some regions where the insulation is distributed nonhomogenously. Therefore, more heat can be transmitted inside, and that is cleared wherever temperatures are lower. On the other hand, the temperature of the western exposed surface is about 46 °C. Also, the effect of nonhomogenous distribution of the insulation is cleared for both facades. On the south wall, there is a central region of lower temperature (referred by blue color), this represents the region of covered window by insulation. Hence, heat behind the insulation penetrates easily through the glass because of its low resistance comparing with wall resistance. Temperature near the bottom is higher by 4 °C than the average wall temperature due to thermal bridge effect with the hot nearby basement (concrete tiles).

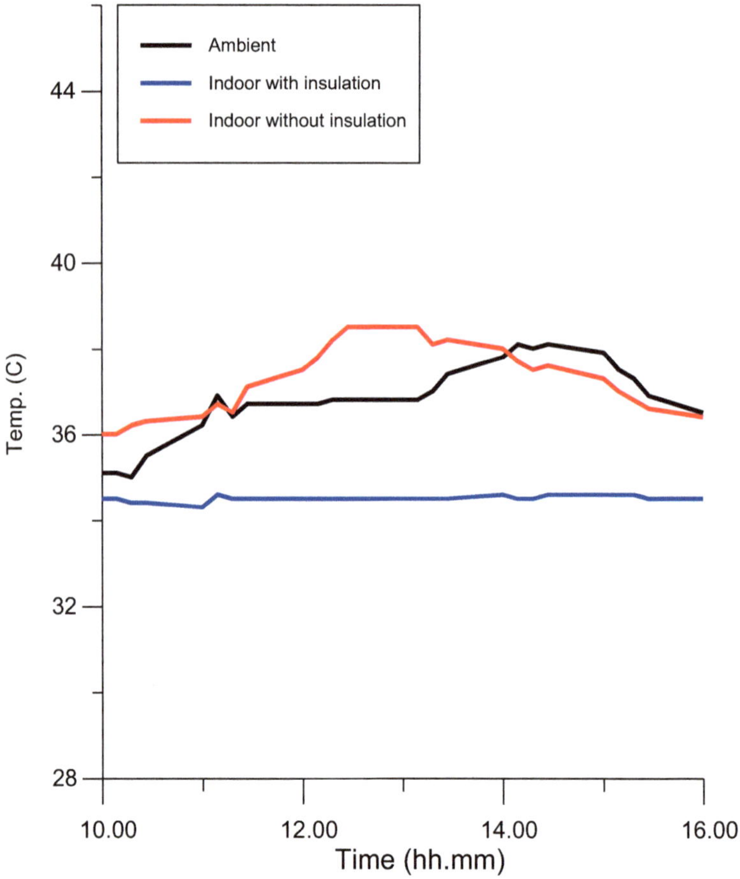

Fig. 44.7 Data in room (2) where insulation at south wall and roof on June 12, 2013 (case 5)

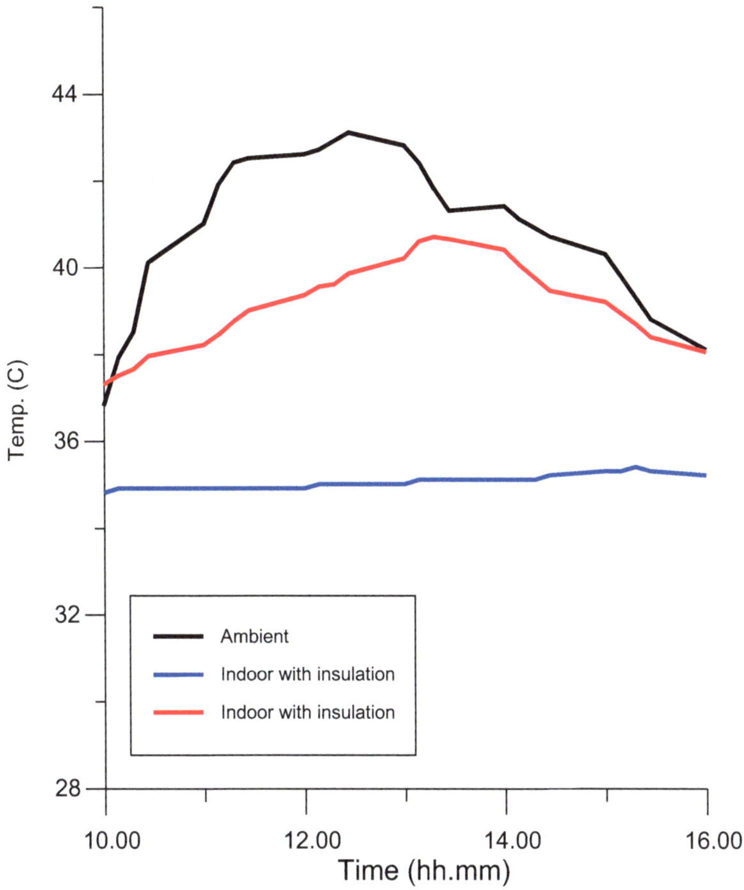

Fig. 44.8 Data in room (2) where insulation at south wall, east wall, and roof on June 20, 2013 (case 6)

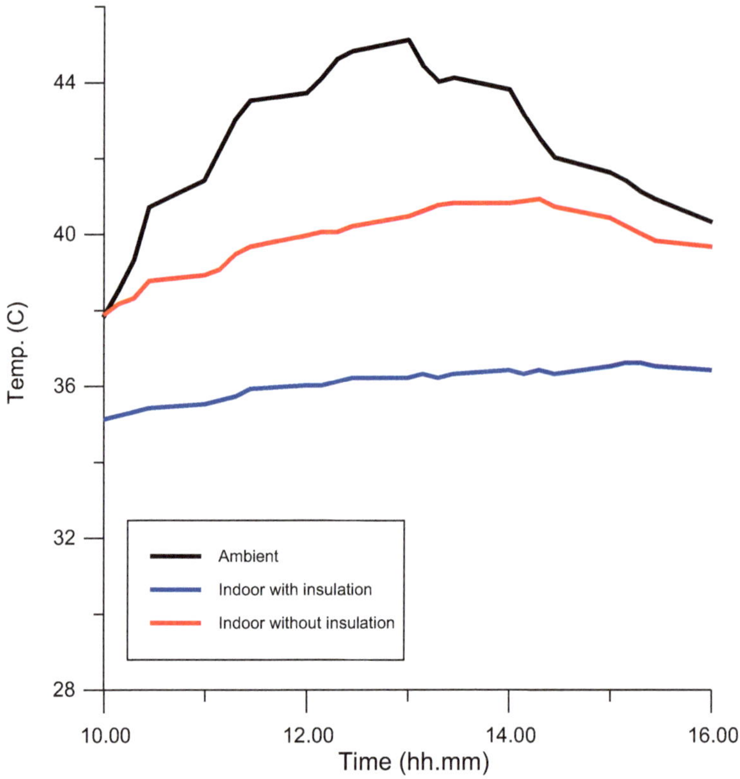

Fig. 44.9 Data in room (2) where insulation at south wall and east wall on June 25, 2013 (case 7)

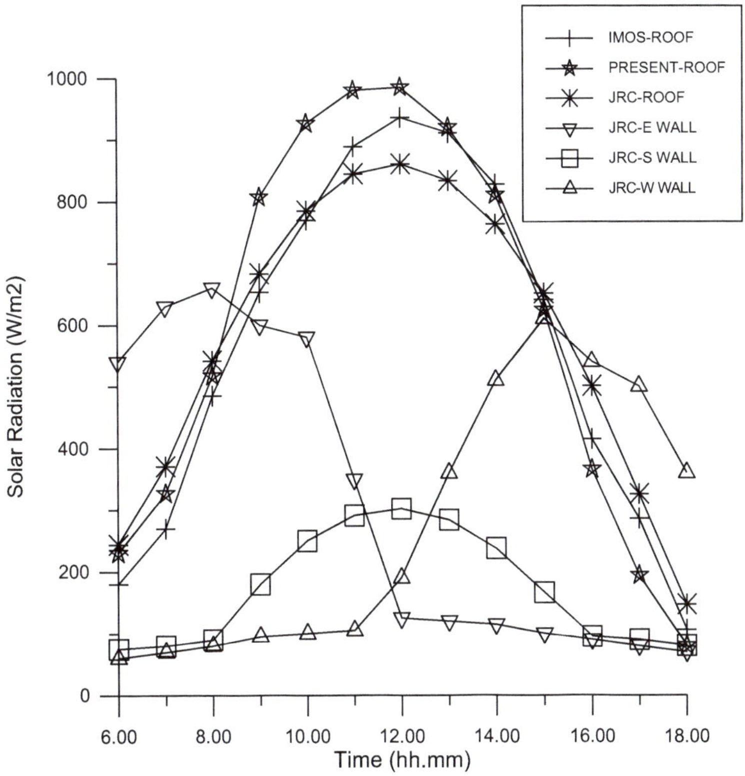

Fig. 44.10 Data for solar radiation on June 15, 2013

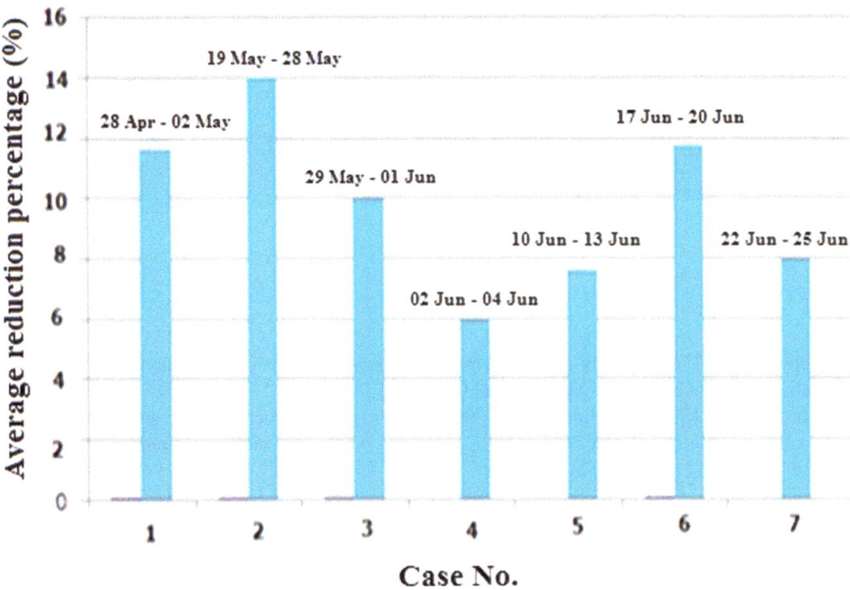

Fig. 44.11 Average reduction percentage in indoor temperature for all cases

Table 44.1 Energy consumption for various cases of insulation

Description	Room cooling load (kW)	Annual cooling consumption (kWh/m^2 year)	Cooling energy saving (kWh/m^2 year)	Average indoor temp. for June and July
Room (1) or (2) without insulation	4.8	725	–	40.8
Room (1) or (2) with insulation for roof only	3.8	552	173	38
Room (1) or (2) with insulation for all exposed walls without roof	3.6	521	204	37.6
Room (1) or (2) with insulation for south wall only and roof	3.3	475	250	37.4
Room (1) or (2) with insulation for west wall or east wall and roof	3.1	448	277	37.2
Room (1) or (2) with insulation for all exposed walls and roof	2.7	363	362	35.7

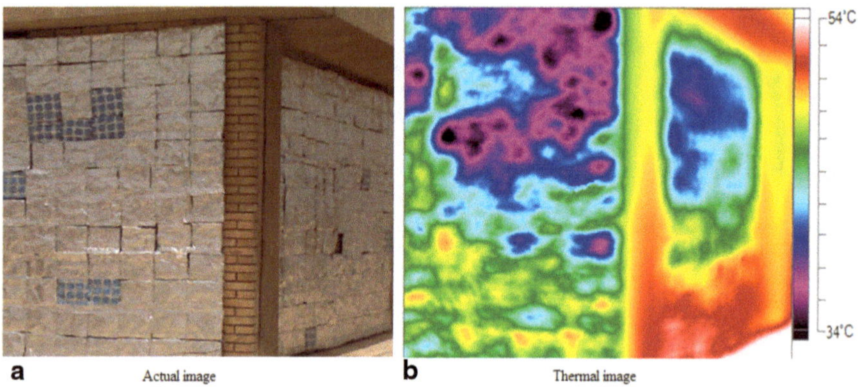

Fig. 44.12 Thermal images captured for room (1) at noontime

References

1. Ecovolve (2007) Principles of thermal insulation: heat transfer via conduction, convection and radiation. http://beodom.com. Accessed 06 Jul 2013.
2. Almusaed A, Khalil A, Almusaed A (2008) Towards a zero energy house strategy fitting for South Iraq climate. PLEA 2008–25th Conference on Passive and Low Energy Architecture, Dublin, 22–24 October 2008
3. Abdulsada GK (2013) Techniques to reduce the energy consumption in building in hot arid rejoin (Iraq). J Energy Power Eng 7:1052–1057.

4. Kalif A, Ismael A, Hussein N (2013) Comparison of the thermal behavior for natural and industrial packaging materials on the cooling load in Baghdad. J Babylon Univ Eng Sci 21(4):1450–1463
5. IMOS (2013) Iraqi Meteorological Organization and Seismology, Ministry of Transportation, Iraq. http://www.meteoseism.gov.iq. Accessed 10 Oct 2013
6. Weather Spark Website (2013) http://weatherspark.com. Accessed 14 Dec 2013
7. JRC (2012) Joint Research Center, The European Commission's In-House Science Service. http://ec.europa.eu. Accessed 08 Mar 2013
8. Thomas V (2003) Energy efficient building design—heat gains and losses: roofs and walls. (Syllabus Notes). College of Architecture, Illinois Institute of Technology, Chicago

Chapter 45
Simulation-Based Optimization for Energy and Buildings

Ala Hasan, Matti Palonen and Mohamed Hamdy

Abstract This chapter presents implementations of simulation-based optimization by giving two examples for finding optimum values of design/decision parameters in the building envelope, building heating/cooling systems, and energy systems. In addition, the features of the multi-objective building performance optimization program (**MOBO**®) are presented. MOBO is a new generic freeware able to handle single- and multi-objective optimization problems including continuous and/or discrete variables. It can be coupled to many simulation programs and has a library of different types of algorithms (evolutionary, deterministic, hybrid, exhaustive, and random).

It is concluded that simulation-based optimization is an efficient method for finding optimum values of design/decision variables in the building envelope, heating/cooling systems, and energy generation systems. While an exhaustive search method will need a huge number of simulations to find optimal solutions, the optimization algorithm will need reasonable time and number of simulations to find comparable solutions. It can be used in finding optimal solutions for the fulfillment of zero, net- or nearly-zero energy/emission buildings, where minimization of energy emissions and cost and maximization of indoor air quality are the targets.

Keywords Building performance simulation · Optimization · MOBO software · Zero-energy buildings

A. Hasan (✉)
VTT Technical Research Centre of Finland, Tekniikantie 4A, Espoo, Finland
e-mail: ala.hasan@vtt.fi

M. Palonen · M. Hamdy
Department of Energy Technology, Aalto University, Espoo, Finland

M. Hamdy
Building Physics and System Unit, Eindhoven University of Technology, Eindhoven, The Netherlands

A. Sayigh (ed.), *Renewable Energy in the Service of Mankind Vol I*,
DOI 10.1007/978-3-319-17777-9_45

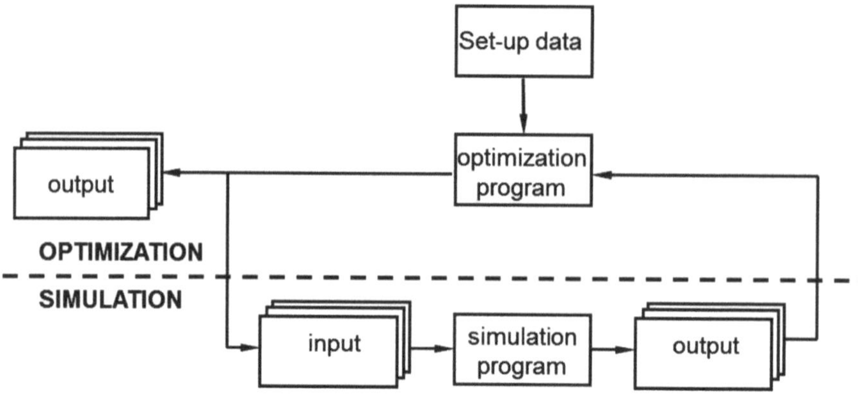

Fig. 45.1 Simulation-based optimization: combined simulation and optimization

45.1 Introduction

Simulation tools are increasingly used in the design and performance analysis of buildings and their energy systems. In order to find better designs or energy system operation in buildings, longtime and excessive efforts are needed to explore huge numbers of new designs through trial and error simulations, which may not necessarily lead to finding better results. The task will be more demanding if optimal values are required to be found in the building envelope, heating, ventilating, and air conditioning (HVAC) system, and the energy generation systems. Combined simulation–optimization using search techniques can be used for this purpose.

45.1.1 Simulation-Based Optimization

Optimization tools require much less time to reach optimal/near-optimal solutions compared to exhaustive search. By simulation-based optimization, we are able to combine simulation tools with optimization tools. Figure 45.1 depicts such a combination. Using simulation-based optimization, it is possible to find optimum values of design/decision variables that achieve defined objective functions. Various decision variables can be searched in the building envelope, heating/cooling system, energy generation/storage systems, etc. The objective(s) can be minimization of energy, emissions, cost, equipment size, etc. and/or maximization of indoor air quality, energy efficiency, etc. The optimization problem can be single- or multi-objective problem. There may also be implementation of constraint functions, for example, satisfaction of thermal comfort levels, limitations on investment cost, etc.

In this chapter, two examples for the implementation of simulation-based optimization are presented including one using the multi-objective building performance optimization (**MOBO®**) software.

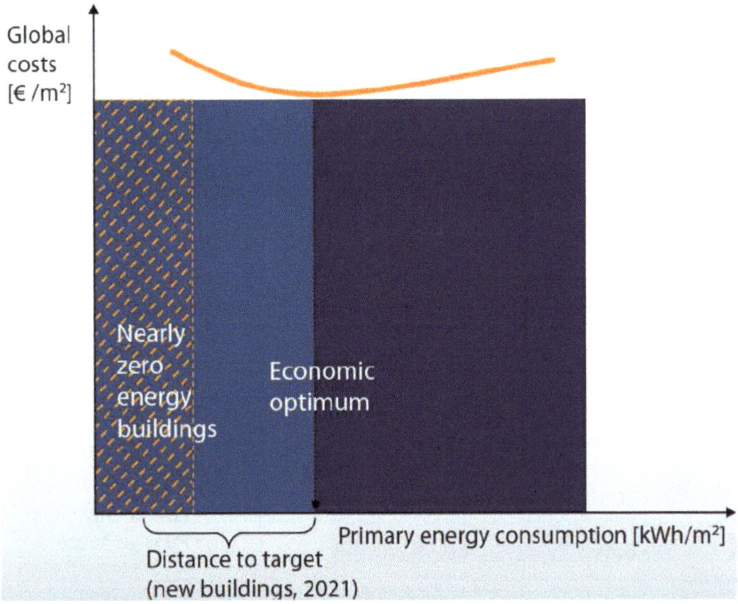

Fig. 45.2 Cost optimum curve [1]

45.2 Implementations of Simulation-Based Optimization

45.2.1 Finding Cost-Optimal and Nearly-Zero Energy Building Solutions

According to the recast of Energy Performance of Buildings Directive (EPBD) issued by the European Commission [1], all new buildings shall be "nearly zero-energy buildings" (nZEB) as of the end of 2020, and 2 years prior to that for the public sector. For each building, a cost curve (Fig. 45.2; global costs vs. primary energy consumption) has to be derived from results of different packages of measures (combinations of compatible energy efficiency and energy supply measures). The lowest part of the curve is the economic optimum, while the left part of the curve that has the lowest energy consumption is the nZEB solution.

An example is presented here that shows an implementation of simulation-based optimization for finding the cost-optimal and nearly zero-energy solutions for a house having different design options in the building envelope, heating systems, and energy generation. Figure 45.3 gives an overview of the studied options. The task is to find optimal combinations of different design variables for the electricity and heating energy generations (on-site micro-generation or district connection), different sizes of primary heating unit (HU1) and auxiliary unit (HU2), and different constructions. These are described in Table 45.1, where all the values of the design variables are indicated. The house is located in Helsinki, Finland, and it

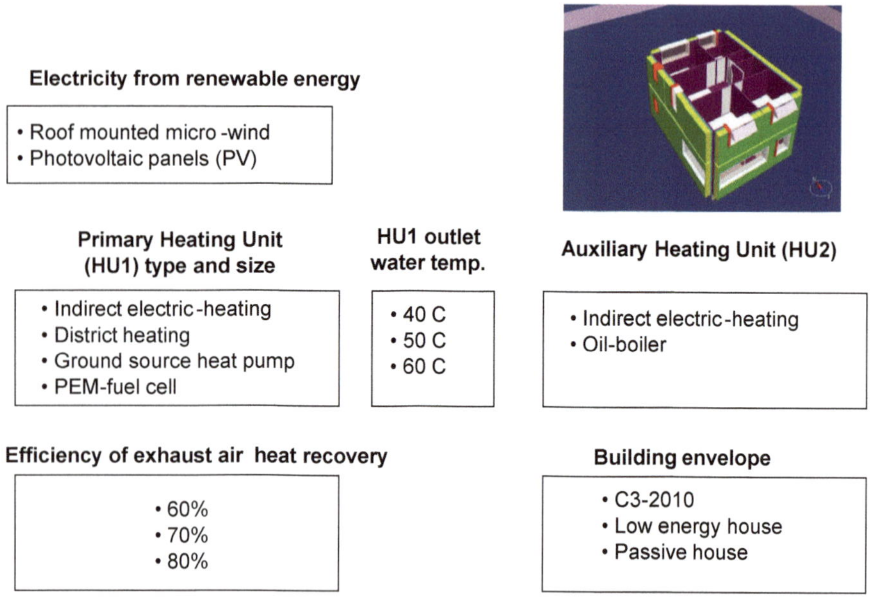

Fig. 45.3 Design alternatives. *PEM* proton exchange membrane

Table 45.1 Design variables and given options

Design variable	Description	Options
Building envelope	C3-2010, low-energy, or passive house	3
Heating unit (HU1; primary unit)	Indirect electric heating, district heating, GSHP, or PEM fuel-cell	4
Heating unit (HU2; auxiliary unit)	Oil boiler or electric heater	3
Micro-wind turbine (maximum power, W)	120, 400, or 950	4
Photovoltaic (area, m^2)	0–60 (1 m^2 step)	61
Size of HU1 (primary unit; %)	Percentage of peak: 50,75, or 100	3
Outlet water temperature of HU1 (C)	40, 50, or 60	3
Size of heating coil inside the DHW storage tank (%)	Percentage of peak: 0, 50, 75, or 100	4
Efficiency of heat recovery unit (%)	60, 70, or 80	3

GSHP ground source heat pump, *PEM* proton exchange membrane, *DHW* domestic hot water

has two floors with a total floor area of 156 m^2. There are three options for the house construction (according to the Finnish C3-2010, low energy or passive house regulations). Heat can be produced in the HU1 by indirect electric heating, district heating, ground source heat pump (GSHP), or proton exchange membrane (PEM) fuel cell. The size of any of these units may cover 50, 75, or 100 % of the total peak heating load for both space heating (under-floor heating and ventilation heating) and domestic hot water. HU1 can supply hot water at a temperature of 40, 50, or 60 °C to a storage tank. However, in case HU1 is not able to meet the required peak

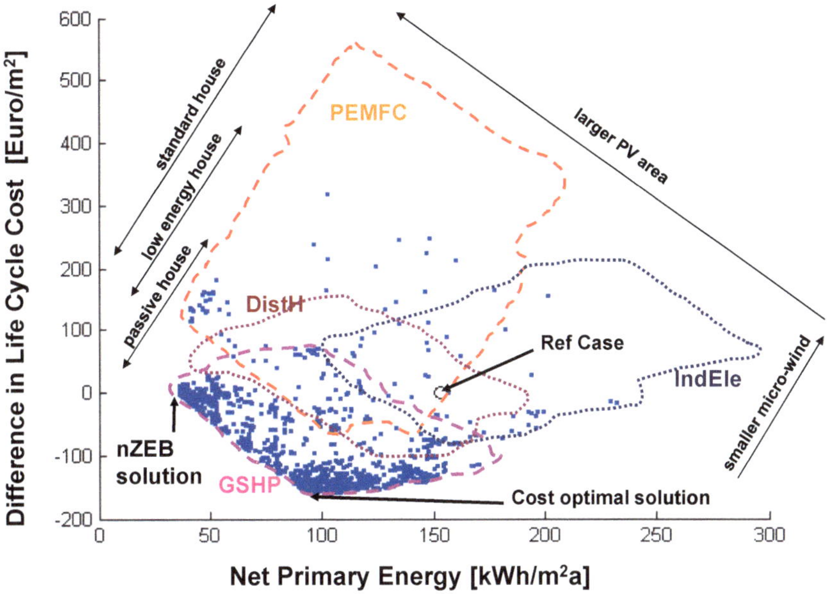

Fig. 45.4 Optimization results. *LCC* life-cycle cost

heat load or supply temperature, the auxiliary heating unit (HU2), which has two options: oil boiler or electric heater, can come into operation. The on-site electricity generation can be by a roof-mounted micro-wind, photovoltaic (PV) panel, or the PEM fuel cell. There are different sizes of the micro-wind and areas of the PV panels, as indicated in Table 45.1. The air ventilation system is a supply and exhaust system. There is an exhaust air heat recovery unit with three options (60, 70, or 80 % efficiency).

The optimization procedure for this problem was made into two steps min–min and min–max for the two objective functions: difference in the life-cycle cost (LCC) and primary energy consumption. The LCC difference refers to the difference between the LCC of any combination of the design variables and that of a reference case. For the building performance simulation calculations, the IDA-ICE software was used. The optimization algorithm made 3000 iterations. The results of the optimization are presented in Fig. 45.4. The whole history of the optimization is indicated by the points on this figure. From these results, we can note that the optimization algorithm was able to construct the cost curve and find the cost optimal and nZEB solutions. Most of the points on the cost curve have lower LCC when compared to the reference case. The lowest is that for the cost optimum solution, which makes 37 % reduction in the primary energy consumption. The cost-optimal points are composed of solutions that are mainly having combinations of GSHP, the larger size of the micro-wind, the higher efficiency of the heat recovery unit, the oil boiler in HU2, and the passive house. This is also valid for the points on the cost curve to the left of the cost-optimal, but with increased contribution of the PV when going for better alternatives toward nZEB solutions. The solutions on the right-hand side of Fig. 45.4 have the highest environmental impacts and are mainly composed of combinations of indirect electric heating and standard house construction (C3-2010). The district heating and the fuel cell do not appear on the front of the cost curve due to their higher LCC values.

45.2.2 Multi-Objective Building Performance Optimization Software

There are many building optimization tools that are currently available with different features. However, we think that there is still a need for a new tool that should be a generic freeware and can fill the shortages recognized in the available tools. For these reasons, we have developed **MOBO®** (http://www.ibpsa-nordic.org/tools.php).

45.2.2.1 Main Features of MOBO

MOBO is able to handle single- and multiple-objective optimization problems. MOBO can be used to run external (simulation) programs. Most of its algorithms include automatic constraint handling. However, this does not limit defining penalty functions if the user wishes. It has a graphical user interface (GUI) for defining the optimization problem. The interface can be used with several tools such as IDA-ICE, TRNSYS, EnergyPlus, etc. It handles the input and output through text files. The user needs to define the command that starts the external program, template file, and the names of the input and the output files of the external program. Since the interface is generic, that is, it can be used with different programs, the user needs to specify where to write the values of the variables (in the input file) and where to read the function values in the output file. MOBO can handle both continuous and discrete variables. For both variables, a preprocessing function can be easily added by using standard algebraic symbols. The software supports approximately 30 functions that can be used in the formulas (e.g., *sin*, *cos*, *sqrt*, *exponent*, etc.). The input (optimization problem and simulation software parameters) is given through the GUI, which interactively checks if there is any error in the input. The progress of the optimization can be viewed with two different graphs: the first graph shows the progress of the decision variables values as a function of the simulation number, and the second graph shows a bi-objective function space.

MOBO has the feature of running simulations in parallel. Before starting the optimization, MOBO will automatically detect the number of threads (or processors) in the machine. Then, the optimization procedure will run multiple simulations (threads) in parallel. This parallel computing feature will lower the optimization time with a factor equal to the number of threads available. All the algorithms in MOBO, excluding the algorithm of Hooke and Jeeves, can use this parallel computing feature.

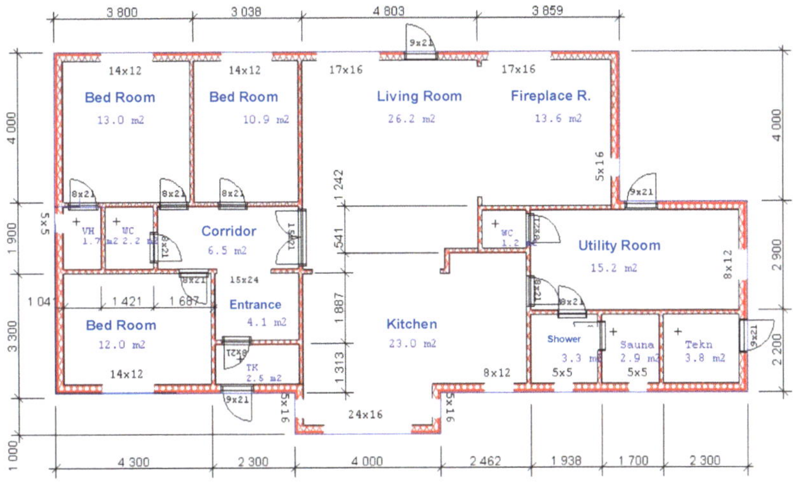

Fig. 45.5 The detached house

Table 45.2 Design variables

Design variables	Type	Minimum value	Maximum value
Additional insulation thickness in external walls (m)	Continuous	0	1
Additional insulation thickness in roof (m)	Continuous	0	1
Additional insulation thickness in floor (m)	Continuous	0	1
Window's *U*-values (W/m²K)	Discrete (two options)	1	1.4
Heat recovery efficiency (%)	Discrete (two options)	70	80

MOBO has a library of different types of algorithms: (1) genetic algorithms GA (NSGA-II Deb et al. [2], Pareto Archive NSGA-II Hamdy et al. [3], and OMNI-optimizer Deb and Tiwari [4]), (2) deterministic algorithm (the algorithm of Hooke and Jeeves), (3) hybrid algorithm, which first uses GA to find a good initial point for the Hooke and Jeeves algorithm, (4) brute-force, which is an exhaustive search that can sample the whole solution space, and (5) random search method.

45.2.2.2 Example of MOBO Implementation

As a demonstration of the implementation, MOBO is used to handle a multi-objective optimization problem of a single-family detached house located in Helsinki, Finland (Fig. 45.5, Shemeikka and Laitinen [5]). The problem was solved in our previous studies using different optimization algorithms (Hasan et al. [6], Hamdy et al. [7], and Palonen et al. [8]). The problem includes five design variables (Table 45.2): three continuous variables (additional insulation thickness in the

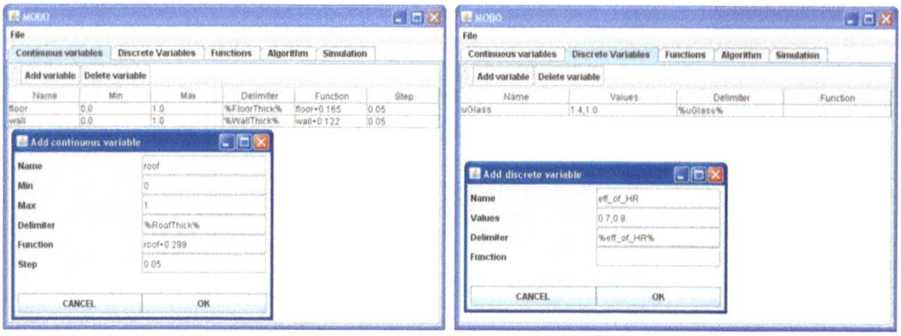

Fig. 45.6 Set-up of the continuous and discrete variables

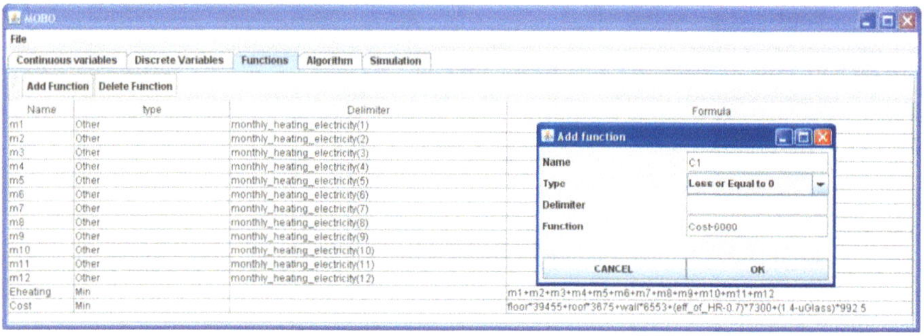

Fig. 45.7 Set-up of the objective and constraints function

external wall, roof, and floor), and two discrete variables (types of windows and heat recovery). The space heating energy and the additional investment cost are the two objective functions to be minimized. The house is considered as a single zone in the energy calculations. The initial U-values are in accordance with the Finnish National Building Code C3-2003 requirements. The house is heated by direct electricity (electric radiators inside the house and an electric heater in the air-handling unit).

The problem is solved here using MOBO. First, the set-up of the design variables and their bounds are made (Fig. 45.6), then the two objective functions and the constraint function are added (Fig. 45.7). The two objective functions to be minimized are *Eheating*, which is the heating energy consumption for the whole year and *Cost,* which is the additional investment cost. As shown in Fig. 45.7, *Eheating* is represented by the summation of the 12 functions of type *other* used to read the monthly energy consumptions, and *Cost* is represented by the summation of the increase in the component's costs. The constraint functions *C1* is defined such that *Cost- 6000*≤ 0. After that, the location of the model file, names of the input- and output files of the simulation software, and the command that starts the simulation are given.

The problem is optimized using three algorithms: the brute-force algorithm, random search algorithm, and Pareto-archive aNSGA-II algorithm. The brute-force

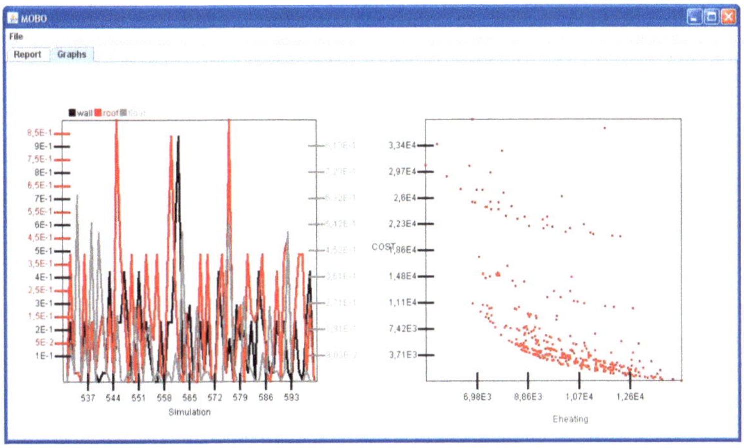

Fig. 45.8 Online output of MOBO

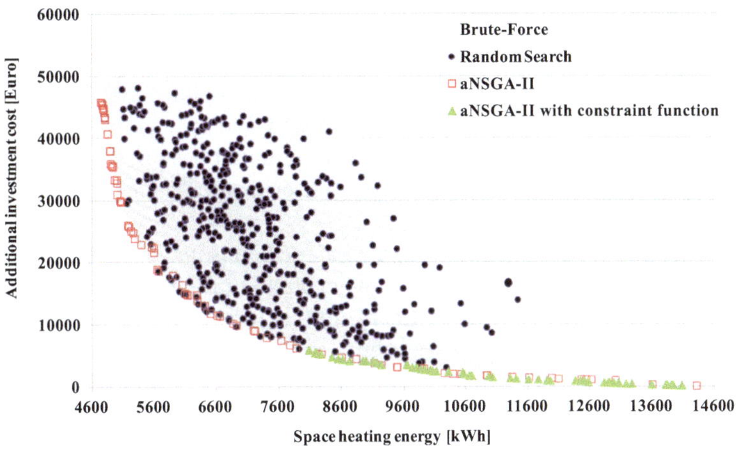

Fig. 45.9 MOBO results of the four search/optimization runs

algorithm used a step of 0.05 m for the continuous variables (insulation thickness). This resulted in 32,000 simulations. The random search algorithm applied 600 simulations. The binary aNSGA-II made two runs. In the first run, no constraint function was included, while in the second run a maximum additional investment cost of € 6000 was considered as a constraint function. The population size was set to six and the number of generations to 100, making a total of 600 simulations. For the building performance simulation calculations, the IDA-ICE software was used.

The online progress of the output results during the optimization run can be seen in Fig. 45.8. The results of the four runs are indicated in Fig. 45.9. We can note that the results of the brute-force algorithms cover the solution space and determine the global pareto front. The whole history of the random search solutions is also shown.

It can be noted that some of these solutions converge close to the global pareto front in its middle section. However, it could not generate many points on the right-hand side of the search space. In its both runs, the aNSGA-II algorithm generated a diverse set of solutions close or on the global pareto front. Only the non-dominated solutions of these two runs are shown in the figure. It can be noted that solutions were generated that are very close to the extreme points on both the right-hand and left-hand sides. However, in the first run, there was room for little improvement to the maximum value of the investment cost. It is interesting to see that the random search found few points that actually dominate the aNSGA-II points in the middle section. It is worthwhile mentioning that these are results of running these stochastic algorithms only once. However, the results are of high quality.

45.3 Conclusions

Simulation-based optimization is an efficient method for finding optimum values of design/decision variables in the building envelope, heating/cooling systems, and energy generation systems. While an exhaustive search method will need a huge number of simulations to find optimal solutions, the optimization algorithm will need reasonable time and number of simulations to find comparable solutions. Better designs can be found that are better than the initial designs in terms of lower energy requirements, lesser CO_2 emissions, lower investment and operating costs, and more comfortable indoor air conditions.

The combined simulation and optimization method is implemented in this paper in two examples. The first one is an implementation of the new recast of the European EPBD-2010 that calls for nZEB by 2021. From the optimization results for the two objective functions (primary energy consumption and life and cycle cost), the whole cost curve is constructed, and the cost-optimal and nZEB solutions are found.

Besides, the features of the new building optimization freeware MOBO are presented. MOBO can handle single- and multi-objective optimization problems with automatic constraint handling. It is a generic optimization tool that can be coupled to many building simulation programs. It tries to minimize associated efforts when making the coupling. It has a GUI by which the coupling can be established, the input data is given, and the output results are viewed. Through the GUI, the user is able to use simple algebraic symbols and calculator functions to describe the continuous and discrete variables, as well as the objective and constraint functions. There is a library including different optimization algorithm types: evolutionary, deterministic, hybrid, exhaustive, and random. The implementation of MOBO is demonstrated by solving a building performance optimization problem using three algorithms. The MOBO software is available for free download for Academic non-commercial use from the following link http://www.ibpsa-nordic.org/tools.php.

Acknowledgments Acknowledgements are due to the Academy of Finland for funding this work.

References

1. EU Directive 2010/31/EU (2010) the European Parliament and of the Council of 19 May 2010 on the energy performance of buildings. Official J Eur Union 53
2. Deb K, Pratap A, Agarwal S, Meyarivan T (2002) A fast and elitist multiobjective genetic algorithm: NSGA-II. IEEE Trans Evol Comput 6(2):182–197
3. Hamdy M, Palonen M, Hasan A (2012) Implementation of Pareto-Archive NSGA-II Algorithms to a nearly-Zero-Energy Building Optimization Problem. 1st Building Simulation and Optimization conference (BSO12), IBPSA-England, Loughborough University, pp. 181–188
4. Deb K, Tiwari S (2005) Omni-optimizer: a procedure for single and multi-objective optimization. In Proc EMO, Mexico, pp 47–61
5. Shemeikka J, Laitinen A (2005) Specification of RET-single family house. Technical Research Centre of Finland VTT, Building and Transport. Version 2.0, 2 March 2005 [in Finnish]
6. Hasan A, Vuolle M, Siren K (2008) Minimisation of life cycle cost of a detached house using combined simulation and optimization. Build Environ 43(12):2022–2034
7. Hamdy M, Hasan A, Sirén K (2009) Combination of optimization algorithms for a multi-objective building design problem. 11th IBPSA conference, Glasgow-UK
8. Palonen M, Hasan A, Siren K (2009) A genetic algorithm for optimization of building envelope and HVAC system parameters. 11th IBPSA conference, Glasgow-UK

Chapter 46
Low-Energy Earth–Air Heat Exchanger Cooling System for Buildings in Hot and Humid Malaysia

Aliyah N. Z Sanusi, Li Shao and Nila I Keumala

Abstract This chapter presents an extended exploration of the performance of low-energy earth–air heat exchanger (EAHE) cooling in hot and humid Malaysia from the published work of Sanusi (2013). The increasing demand of air-conditioning for cooling purposes motivates this investigation in search for better cooling alternatives. The passive technology, where soil underground was used as a heat sink to produce cooler air, has yet to be investigated further in hot and humid countries. This work, the passive technology, is tested by carrying out two interconnected field investigations: soil temperature measurement and the earth pipe cooling experiment for 1 year. In 1 year, soil temperature distribution at 1 m depth fluctuates, influenced by the different seasons in a year. However, the soil at 1 m depth is cooler than deeper soil during the wet season. Therefore, for investigating the EAHE cooling performance for 1 year, the pipe was buried at 1 m depth at the same test site. A significant temperature drop was found in the 1-m-depth-buried pipe: up to 6.3 and 8.6 °C, respective to the seasons of the year. Thermal comfort of air temperature at the buried pipe outlet was analyzed with Khedari thermal comfort chart, and the result has shown that the buried pipe outlet air temperatures are within the range of thermal comfort conditions for hot/humid countries. This study has showed a potential of earth pipe cooling technology in providing low-energy cooling in buildings in Malaysia.

Keywords Soil temperature · Hot and humid Malaysia · Earth pipe cooling · Passive · Low energy · Thermal comfort · Ground cooling

---

A. N. Z. Sanusi (✉)
Department of Architecture, Kulliyyah of Architecture and Environmental Design, International Islamic University Malaysia, Kuala Lumpur, Malaysia
e-mail: aliyahsanusi@gmail.com

L. Shao
Division Institute of Energy and Sustainable Development, University of Reading, Reading, UK
e-mail: l.shao@reading.ac.uk

N. I. Keumala
Department of Architecture, Faculty of Built Environment, University of Malaya, Kuala Lumpur, Malaysia
e-mail: nimk@um.edu.my

A. Sayigh (ed.), *Renewable Energy in the Service of Mankind Vol I*,
DOI 10.1007/978-3-319-17777-9_46

46.1 Introduction

This research was intended to seek for a better cooling system in terms of sustainability than air-conditioning, which has significant contribution to the vast increment in energy consumption in Malaysia [1]. The current energy issue was tackled in this research by conducting an investigation on passive earth pipe cooling technology at an experimental site in Malaysia.

Earth pipe cooling technology works as a passive cooling system that utilizes the soil underground as a heat sink. In the system, ambient air is channeled through pipes, which are buried underground. Within the soil underground, heat dissipation occurred between the air inside the pipe and soil surrounding the pipe via conduction. Since the earth pipe technology uses soil as a heat sink, its efficiency is largely influenced by temperature difference between ambient air and soil temperature and the soil thermal conductivity. Other factors that usually influence the efficiency of this technology are air flow inside the pipe, pipe length, and pipe diameter. The conditioned air is then channeled into the desired space.

Many researchers have explored this earth pipe cooling technology and building designers have used it as a cooling means for various building types in temperate countries as well as hot and arid countries. The various performances of the technology have been significant and positive [2–5]. In many cases, previous investigations found that the optimum pipe burial depth is at 4 m underground. However, more investigation needed to be carried out to find the optimum pipe burial depth for Malaysian climate.

46.1.1 Statement of Problem

The research was driven by the current energy demand issue in Malaysia. First, there is an increasing demand of air-conditioning in buildings in Malaysia. Occupants solve thermal comfort problem with building add-ons such as air-conditioning that consumes 64 % of energy consumption in one typical office [1]. Second, there is an increase in air temperature in the current climate of Malaysia. Putrajaya, the youngest state (36 km from Kuala Lumpur) is 5 °C hotter than other cities in Malaysia. It was revealed that Putrajaya has become an urban heat island (UHI) with temperatures up to 40 °C [6]. The Malaysian government aims to reduce CO_2 emissions to 40 % by the year 2020 [7].

46.1.2 Research Objective

The research objectives were first to ascertain air temperature reduction performances between the inlet or outlet of the earth–air heat exchanger (EAHE) system buried pipes. The next objective was to achieve temperature reduction from the inlet to the outlet EAHE system pipe by at least 3 °C. The third objective was to evalu-

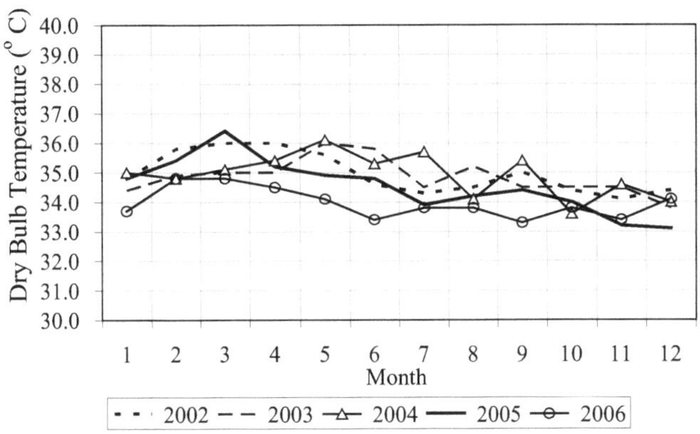

Fig. 46.1 Trend of monthly absolute maximum dry bulb temperature from the year 2002 to 2006 in Kuala Lumpur, Malaysia. (Source: Meteorology Department 2007)

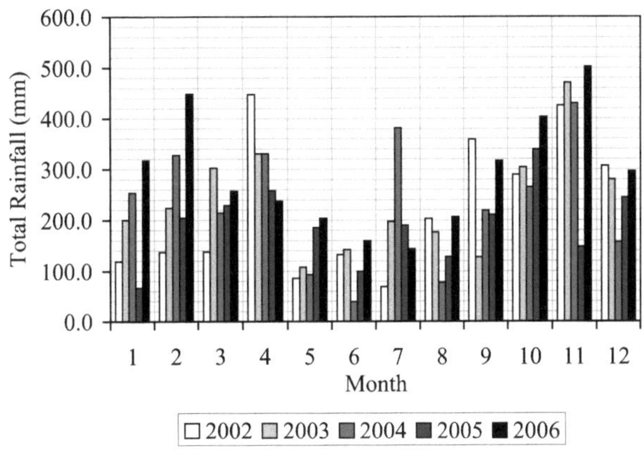

Fig. 46.2 Monthly total rainfall in 5 years from 2002 to 2006. (Source: Meteorology Department 2007)

ate thermal comfort performance of the EAHE in buildings in the hot and humid climate of Malaysia.

46.1.3 Malaysian Weather

Figure 46.1 shows the trend of monthly absolute maximum dry bulb temperature (DBT) within 5 years. Malaysia experiences warm and humid weather throughout the year, comprising a wet season from October to December and hot/dry season from May to June. The air temperature normally ranges from 21.5 to 36.1 °C.

Figure 46.2 shows the rainfall trend within 5 years. It is shown that within those 5 years, the weather is dry mostly in May and June but turns wet at the end of the year in most years

Literatures regarding the Malaysian thermal comfort zone have been reviewed and found that, for naturally ventilated building, occupants felt comfortable in temperature ranges from 22.7–31.6 °C. Among the findings from various researches on the Malaysian thermal comfort zone for naturally ventilated buildings, the average of the upper limit of the comfort zone is found to be 30.1 °C.

46.2 Methodology

With the basic knowledge of earth pipe cooling, the Malaysian weather and the Malaysian comfort zone, a research methodology was formed. Field experiments became the main investigation of this research. It is divided into two interrelated field investigations: soil temperature measurement and earth pipe cooling experiments.

46.2.1 Research Limitation

The fact that challenges the earth pipe cooling experiment in Malaysia is that there is no systematic record of the Malaysian soil temperature hourly data at up to 4 m depth underground. Furthermore, there were very few records of earth pipe cooling technology being used in Malaysia or other hot and humid countries.

46.2.2 Experiment Site

The site is located within the International Islamic University Malaysia (IIUM) campus in Kuala Lumpur, Malaysia. It is in an exposed space, and the ground is covered with short grass. Very few trees are found scattered around the site (Fig. 46.3).

46.2.3 Soil Temperature Measurement

An initial investigation on soil temperature measurement was carried out during the wet season, in October and November, when the weather is cooler. It comprises outdoor DBT, ground surface temperature, and underground soil temperature at 1, 2, 3, 4, and 5 m depth.

Soil temperature measurement was conducted again focusing on the soil around 1 m depth, in which the data taken was outdoor DBT and soil temperature at 0.5, 1.0, and 1.5 m underground. The second soil temperature measurement was conducted for approximately 1-year duration.

46.2.4 Earth Pipe Cooling Experiment

The 1-year earth pipe cooling experiment is work resumed from the field experiment presented in Sanusi [9], where the experiment was carried out for 1 month,

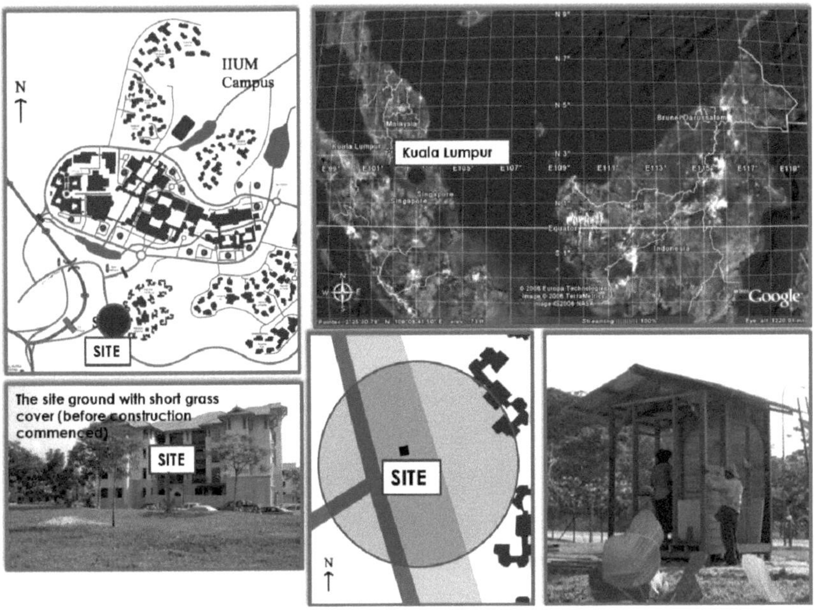

Fig. 46.3 Location of experiment site and site condition. *IIUM* International Islamic University Malaysia. (Source: Google map 2007; IIUM Development Division 2007 and [8])

each during the wet season and the hot and dry season. In both seasons, based on the results obtained in soil temperature measurement, earth pipe cooling experiment was conducted with three 3″ polyethylene pipes buried separately, at 0.5, 1.0, and 1.5 m deep underground. All the pipes were 25 m long. There was a fan blower connected to all pipe inlets, which provided air flow into the buried pipes at 5.6 m/s. The fan was switched on daily from 10 a.m. until 6 p.m. to allow night ventilation. The pipe inlets and outlets were sheltered by an experimental shed (Fig. 46.4), which has 2 × 2 × 2 m dimensions. Data of each buried pipes were recorded for 48 h at 10 min interval. The tests were carried out in sequence rather than in parallel to avoid thermal interference between the pipes.

Due to significant differences in buried pipe outlet temperature between wet season and hot and dry season, a 1-year investigation of earth pipe cooling experiment was conducted, but with only one buried pipe, at 1.0 m depth to avoid thermal interference. In the annual investigation, the fan blower operated for 24 h, 7 days a week. Data were recorded at 10-min interval. Soil temperature was also measured for 1 year alongside the earth pipe cooling experiment.

Table 46.1 and Fig. 46.5 shows the equipment used during the field investigation of soil measurement and earth pipe cooling experiment.

46.3 Results

All field data results were compiled into two main interrelated investigations: the soil temperature measurement and earth pipe cooling experiments.

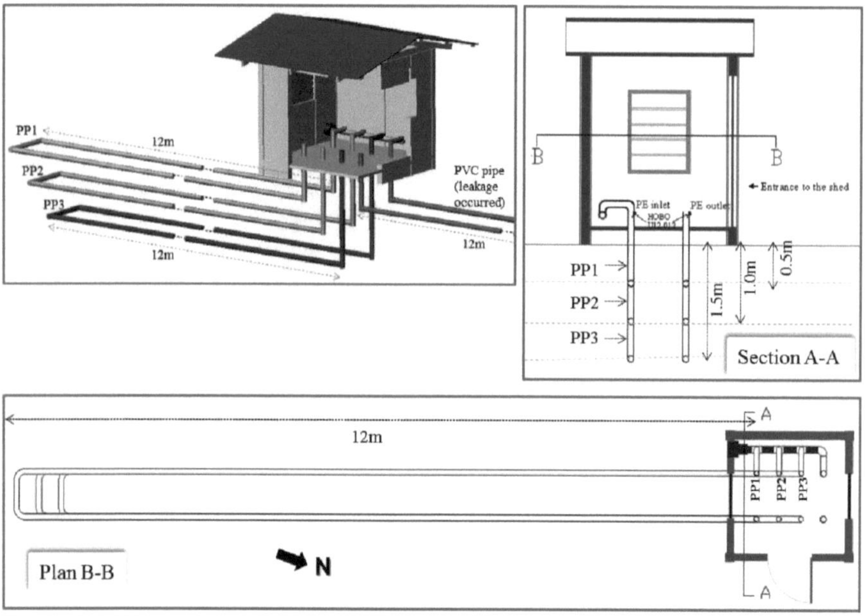

Fig. 46.4 Plan, section, and axonometric drawings of the experiment setup. *PE* polyethylene, *PVC* polyvinylchloride

Table 46.1 List of equipment used during field investigation

Figure	Equipments	Functions/measurements	Units
46.5a	Fan blower	Supply air flow	
46.5b	USB TC-08 thermocouple data logger with 7 K-type thermocouple wires	Soil and air dry bulb temperature	°C
46.5c	HOBO pendant temperature logger (UA-001-64)	Dry bulb temperature	°C
46.5d	HOBO U12-012 Temp/RH/Light/Ext	Dry bulb temperature/relative humidity	°C/%
46.5e	HOBO external sensor TMC6-HD	Dry bulb temperature (initially measure buried pipe inlet/outlet)	°C
46.5f	Air flow meter from Testo	Air velocity	m/s

Fig. 46.5 Pictures of equipments listed in Table 46.1

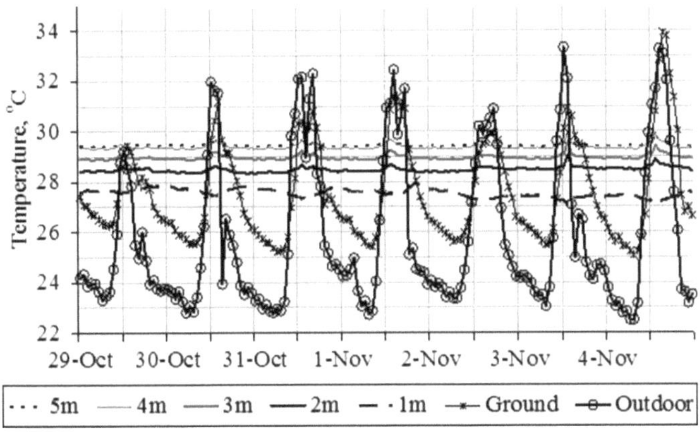

Fig. 46.6 Ambient air and ground soil temperature and underground soil temperature at 1, 2, 3, 4, and 5 m depths. (Source: Sanusi and Shao [10])

46.3.1 Soil Temperature Measurement

The result of soil temperature measurement has shown the possibility of making the soil underground as heat sink to reduce extreme solar heat gain especially during 12:00–14:00 h of the day. Figure 46.6 shows the results obtained from the initial ambient air and soil temperature measurement during the wet season in October 2007.

With reference to Fig. 46.6, it is shown that the soil temperature is fairly stable at 1 m depth underground with reduced daily amplitude compared to the ground-level soil temperature. It also shows that the soil temperature increases with increasing depths underground, and the soil temperature changes a little beyond 4 m deep underground.

Figures 46.7 and 46.8 present the result of soil temperature measured at 0.5, 1.0, and 1.5 m depth underground, which were measured for approximately 1 year.

Figures 46.7 and 46.8 show that in 1-year measurement, the soil temperature at 1 m depth fluctuates, but it takes advantage of the cool temperature in the wet season.

46.3.2 Earth Pipe Cooling Experiments

Throughout the earth pipe cooling experiment, it is evident that temperature reduction does occur within the inlet and outlet of the buried pipes.

For 1-year data, a 1-m deep buried pipe was selected for an extensive investigation. In 1 year, at 1 m depth, the maximum temperature reduction through the buried pipe was 8.6 °C, which occurred in January (Fig. 46.10). Meanwhile, the average daily maximum temperature reduction through the buried pipe was 5.7 °C (Table 46.2).

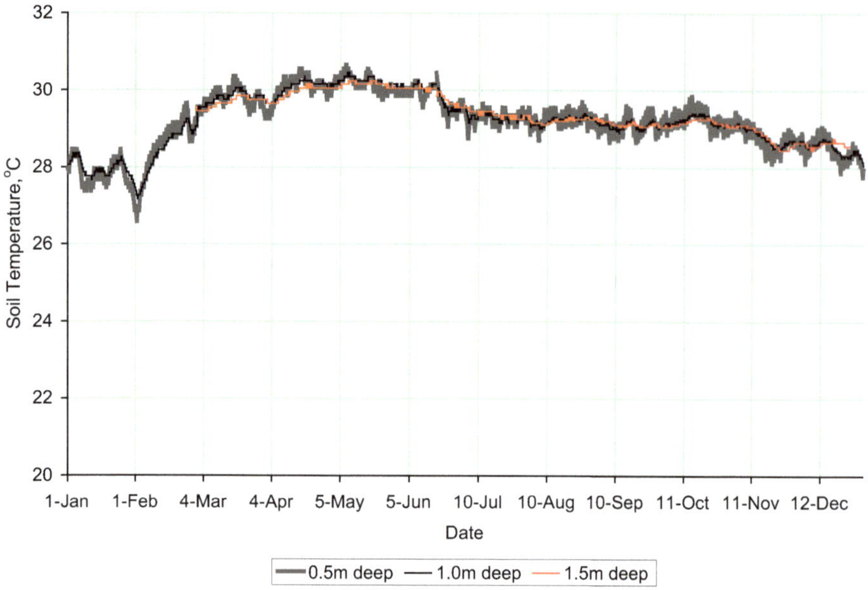

Fig. 46.7 Hourly distribution of soil temperature (°C) at 0.5, 1.0, and 1.5 m depths within a year at the field work site in Gombak, Kuala Lumpur, Malaysia. (Source: Sanusi and Shao [10])

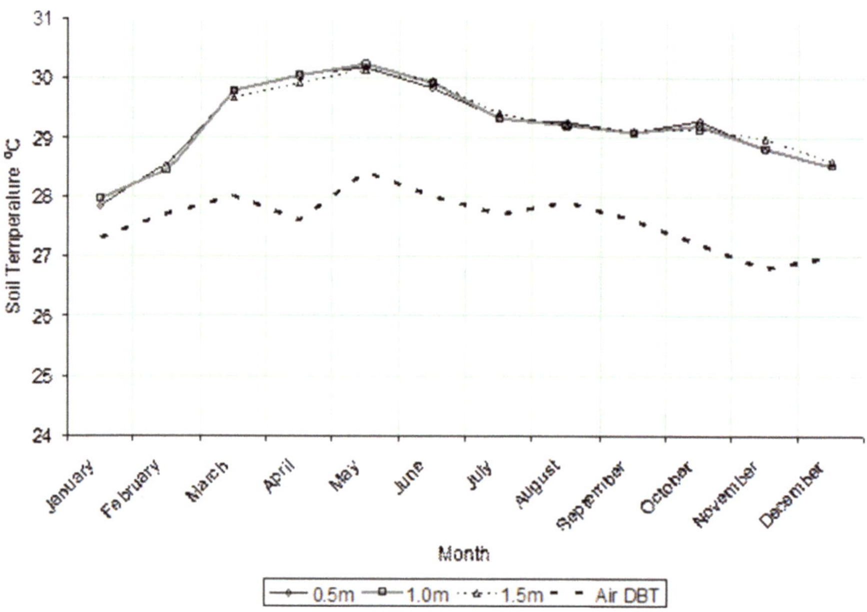

Fig. 46.8 Monthly average of ambient air and underground soil temperature at 0.5,1.0, and 1.5 m depths at Gombak, Kuala Lumpur. *DBT* dry bulb temperature. (Source: Sanusi and Shao [10])

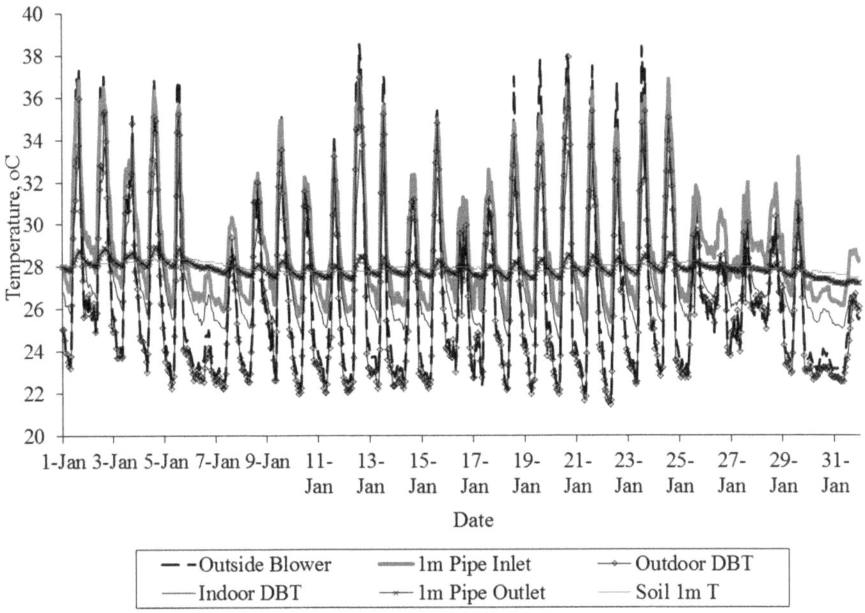

Fig. 46.9 Temperature data of earth pipe cooling experiment in January (pipe buried at 1 m depth underground). *DBT* dry bulb temperature

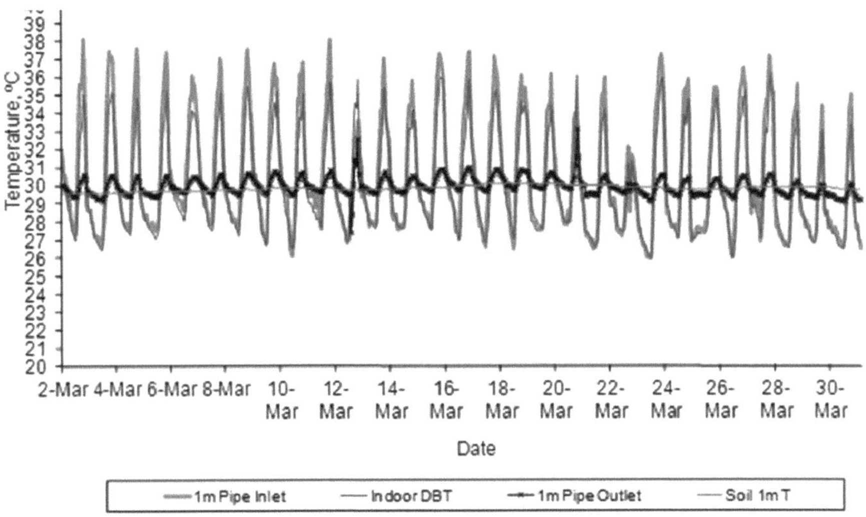

Fig. 46.10 Temperature data of earth pipe cooling experiment in March (pipe buried at 1 m depth underground). *DBT* dry bulb temperature

Table 46.2 The summary of earth pipe cooling system performance with respect to temperature

Month	Jan	Feb	Mar	April	May	June	July	Aug	Sep	Oct	Nov	Dec
Unit	°C											
Max TR	8.6	8.1	7.5	6.3	6.4	6.3	Data not available	Data not available	6.6	7.0	8.0	8.0
AveMx TR	5.3	7.6	5.7	5.4	5.2	5.0			5.2	5.9	5.9	5.8
Outlet	27.1–28.9	26.9–29.0	27.4–30.9 (33.1)[a]	28.9–30.7	29.0–30.9	28.7–30.5			28.1–29.5	28.5–29.5	28.1–29.7	27.7–29.1
Inlet	25.4–37.0	25.1–36.7	25.9–38.1	26.3–36.9	25.7–37.1	25.8–36.8			26.5–36.1	25.7–36.5	25.8–37.2	24.4–37.1

MaxTR maximum temperature reduction between pipe inlet and outlet, *AveMx* average daily maximum, *Ave* average

[a] The value in a bracket is an error found which only occurred twice in the month

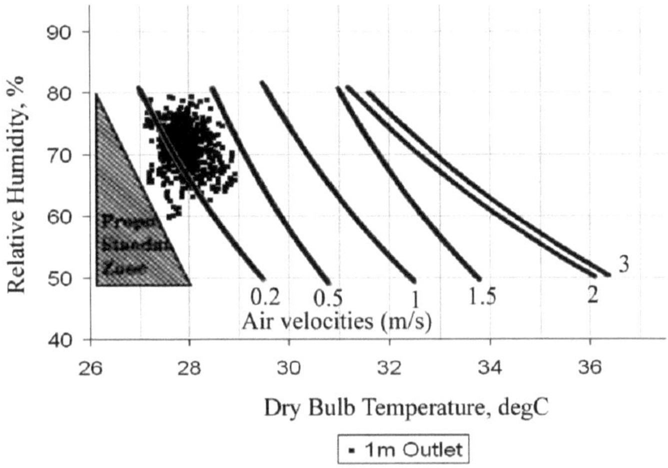

Fig. 46.11 Field work data in January are plotted in Khedari's thermal comfort chart

Table 46.2 shows the range of buried pipe inlet and outlet temperature for each month. As mentioned above, the maximum temperature reduction between the buried pipe inlet and outlet in January is 8.6 °C, and its average daily maximum temperature reduction is 5.3 °C. Meanwhile, the outdoor DBT ranges from 21.5 to 37.9 °C in January (Fig. 46.9). With regard to the soil property, the soil temperature at the same depth of 1 m underground ranges from 27.5 to 28.4 °C.

Meanwhile, in March, which is a hot and dry season, the pipe outlet air temperature was reduced to 30.1 °C, and the EAHE system achieved 7.5 °C of temperature reduction (Table 46.2 and Fig. 46.10)

46.3.3 Thermal Comfort Discussion

An analysis of comfort cooling with the cooled air from the buried pipe outlet was carried with the aid of a thermal comfort chart, formed by Khedari using Thailand data and samples. The field work results of buried pipe outlet temperature were plotted against the Khedari Thailand thermal comfort chart for each month with available field work data. It was found that all buried pipe outlet data fall within Khedari thermal comfort zone, provided there is an adequate air velocity, which is either 0.5, 1/0, or 1.5 m/s, depending on the weather conditions of specific months considered. Figures 46.11, 46.12, and 46.13 show an example of plotted Khedari thermal comfort chart referring to the months January, March, and December data.

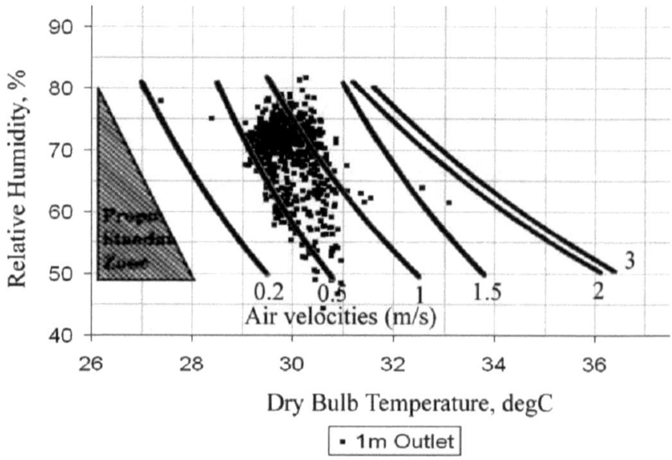

Fig. 46.12 Field work data in March are plotted in Khedari's thermal comfort chart

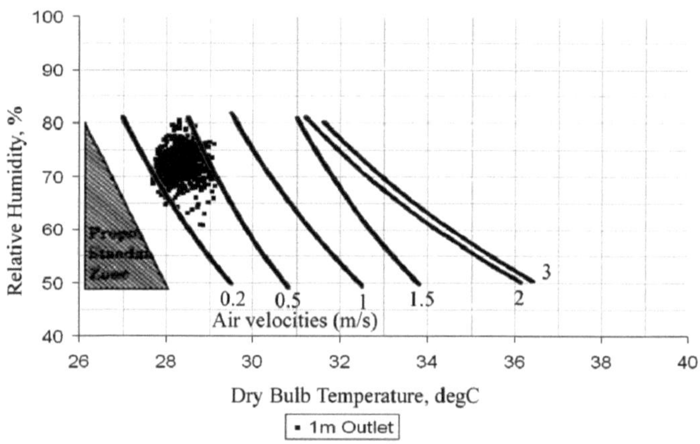

Fig. 46.13 Field work data in December are plotted in Khedari's thermal comfort chart

46.4 Conclusion

In conclusion, the research objectives were achieved through research findings. Evidently, in field work data, there are significant temperature reductions from the ambient air temperature and buried pipe inlet to the buried pipe outlet of the EAHE cooling system. The monthly average temperature reduction between the buried pipe inlet and outlet ranges from 5 to 7.6 °C. Meanwhile, the monthly maximum temperature reduction ranges from 6.3 to 8.6 °C. Therefore, during cooling period, the EAHE system achieved temperature reduction more than 3 °C. The last findings concluded that all pipe outlet data fall within Khedari thermal comfort zone [11], where the adequate air velocity required is 0.5, 1.0, or 1.5 m/s depending on the weather conditions of specific months considered.

The applicability of earth pipe cooling for hot and humid Malaysia as an alternative to air-conditioning was shown to be possible, based on thermal comfort analysis using Khedari comfort chart. This alternative cooling system can contribute to achieve the aim of Najib Tun Razak (the Prime Minister of Malaysia) in reducing CO_2 emission to 40 % by the year 2020.

46.5 Recommendation for Future Research

The field investigation findings show similarity in average between the buried pipe outlet temperature and the relative soil temperature at most times. This demonstrates the high influence of soil temperature on the performance of earth pipe cooling technology. This suggests the treatment of the soil, having it shaded from the daytime solar radiation with either plantation or lightweight structure. According to a study by Givoni, as mentioned in Sect. 3, shaded soil has lower average temperature. Since the soil temperature affects the pipe outlet temperature, a lower soil temperature could reduce the pipe outlet temperature even more than recorded in this research finding. Furthermore, with more time and funding, the EAHE system investigation could be extended with different building forms and construction, soil, and climate.

Acknowledgments I would like to express sincere gratitude to the IIUM and De Montfort University for funding the research work and providing useful equipments. Thanks to the Development Division of IIUM for giving the permission to use the IIUM land for my experiment setup. Thanks are also extended to my research assistant, Nik Azmin Riza Nik Ahmad, for his full dedication.

References

1. Chan SA (Ar.) (2004) Energy efficiency—designing low energy buildings using energy 10. CPD seminar, Pertubuhan Arkitek Malaysia, Kuala Lumpur. Pertubuhan Arkitek Malaysia. 7 Aug, Malaysia
2. Ghosal MK, Thiwari GN (2006) Modelling and parametric studies for thermal performance of an earth to air heat exchanger integrated with a greenhouse. Energy Convers Manage 47:1779–1798
3. Santamouris M, Mihalakalou G, Balaras CA, Argiriou A, Asimakopoulos D, Vallindra M (1995) Use of buried pipes for energy conservation in cooling of agricultural greenhouses. Solar Energy 55:111–124
4. Solaini G, Dall'O' G, Scansani S (1998) Simultaneous application of different natural cooling technologies to an experimental building. Renewable Energy 15:277–282
5. Thanu NM, Sawhney RL, Khare RN, Buddhi D (2001) An experimental study of the thermal performance of an earth-air-pipe system in single pass mode. Solar Energy 71:353–364
6. Krishnan G (2007) Putrajaya i5 5 °C hotter than other local cities. The Star Archives. http://www.thestar.com.my/story.aspx/?file=%2f2007%2f6%2f19%2fcentral%2f17992083&sec=central. Accessed 20 Sept 2013
7. Bernama (2009) Malaysia announces conditional 40 % cut in emissions. News. Copenhagen http://www.pmo.gov.my/?menu=newslist&news_id=900&page=1731&news_cat=13. Accessed 20 Sept 2013

8. Sanusi ANZ, Shao L, Ibrahim N (2013) Passive ground cooling system for low energy buildings in Malaysia (hot and humid climates). Renewable Energy 49:193–196
9. Sanusi ANZ (2012) Low energy ground cooling system for buildings in hot and humid Malaysia. PhD thesis. De Montfort University
10. Sanusi ANZ, Shao L (2014) Seeking underground for potential heat sink in Malaysia for Earth Air Heat Exchanger (EAHE) application. Aust J Basic Appl Sci 8(8):54–57(Special 2014)
11. Khedari J, Yamtraipat N, Pratintong N, Hirunlabh J (2000) Thailand ventilation comfort chart. Energy Buildings 32:245–249

Chapter 47
Analysis of Passive Solar House to Improve the Indoor Thermal Environment in Winter in Lhasa, China

Ming Zhang, Wei Yu and Baizhan Li

Abstract The application of solar heating in winter was studied based on the solar resources in Lhasa area of China. The difference of indoor thermal environment and residents' thermal sensation vote between the passive solar house and typical Tibetan residential buildings was compared in winter in Lhasa, through the methods of subjective questionnaire and actual test. The results show that the winter indoor thermal environment in passive solar houses is better than that in typical Tibetan residential house. First, the residents' comfort satisfaction in solar houses was higher than that in typical Tibetan residential houses. Second, the indoor temperature range, which was accepted by more than 80 % of the residents ranged from 11.3 to 22.5 °C in passive solar houses, while the range was from 13.6 to 23 °C in typical Tibetan residential buildings. Third, the measured winter thermal neutral temperature in passive solar houses was 16.44 °C, 1 °C lower than typical Tibetan residential buildings. At the same time, the temperature in passive solar houses is higher than in the traditional Tibetan residences, but it changes greatly during the day and night. This chapter shows the effectiveness of passive solar houses in improving the indoor thermal comfort by using solar energy in winter in Lhasa. The research findings have an important role in promoting the use of passive solar house or other energy-saving buildings in Lhasa and the Qinghai–Tibetan Plateau area.

Keywords Passive solar house · Indoor thermal environment · Thermal comfort · Traditional Tibetan house · Energy efficiency · Lhasa

B. Li (✉) · M. Zhang · W. Yu
Key Laboratory of Eco-environments in Three Gorges Reservoir Region, Ministry of Education, Chongqing University, 400045 Chongqing, China
e-mail: baizhanli09@gmail.com

National Centre for International Research of Low-Carbon and Green Buildings, 400045 Chongqing, China

A. Sayigh (ed.), *Renewable Energy in the Service of Mankind Vol I*,
DOI 10.1007/978-3-319-17777-9_47

47.1 Introduction

With the improvement of peoples' living standard, more and more people pay attention to improving the winter indoor thermal environment in the Tibetan plateau area of western China. The central government of China also attaches great importance to the social development in this area [1]. It is particularly important to improve the indoor thermal environment of residential buildings in Lhasa, a traditional, representative city of high altitude areas in China, with a special climate in winter.

However, which kind of control strategies should be taken to improve the indoor thermal environment in plateau areas such as Lhasa? There is a big difference in the natural energy resources of the Tibetan plateau area and mainland China. Lhasa, for example, lacks coal, oil, gas, and other conventional fossil fuels [2, 3]. In addition, the ecological environment is fragile in this region [4]. So it is not suitable to use coal, oil, natural gas, and other conventional energy sources. However, the region is very rich in solar energy resources and Lhasa is called the "sun city" of the world. This creates excellent conditions for using passive control strategies to improve the indoor thermal environment in this area [5]. The passive control strategy may not only reduce energy consumption but also create a more comfortable, natural and harmonious indoor thermal environment [6–8].

This chapter analyzes the differences in indoor thermal environment between the passive solar house and traditional Tibetan buildings in Lhasa in winter, and proposes energy-saving control strategies for the indoor thermal environment. The findings have an important significance in promoting the use of the passive solar house in Lhasa and the Tibetan plateau area.

47.2 Analysis of the Climate Environment in Winter in Lhasa

Lhasa, a plateau area with low latitude and high altitude, belongs to the cold and severe cold regions of China [9]. The climate and environment is unique in this region, especially in winter, with strong radiation, long hours of sunshine, low average temperatures, and a large diurnal temperature range. Data for a traditional meteorological year in Lhasa [10] and the outdoor temperature parameters in winter are shown in Fig. 47.1.

As shown in Fig. 47.1, the average temperature from November to the following February is below 5 °C, and it reaches the lowest temperature in January. The temperature in winter fails to reach the thermal comfort temperature range recommended in the American Society of Heating, Refrigerating, and Air-Conditioning Engineers (ASHRAE) standard. On the coldest day, the temperature is almost continually below 0 °C, and the lowest temperature reaches minus 13.2 °C.

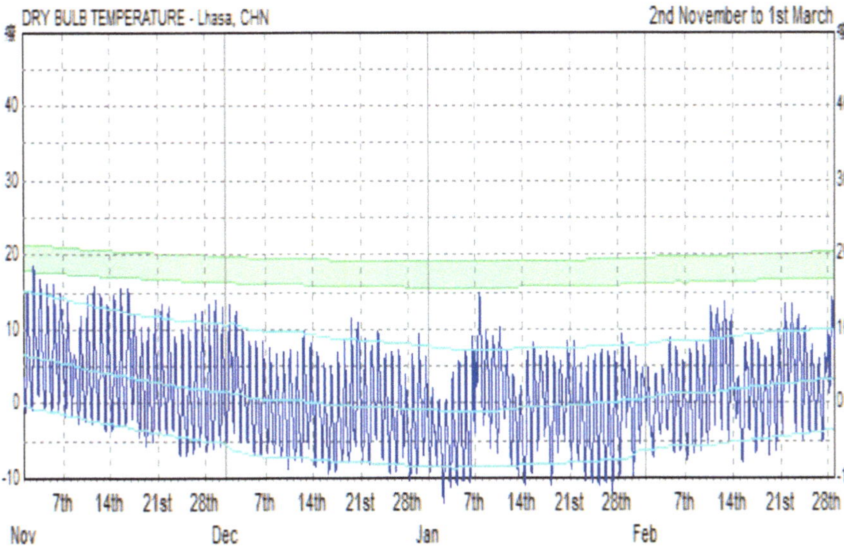

Fig. 47.1 The weather data in winter of traditional year in Lhasa. The *light blue bar* represents the dry bulb temperature change in a typical year between November and February. The *blue line* represents the average daily winter temperature. The *green line* area shows the thermal comfort temperature range recommended by the ASHRAE standard. *ASHRAE* American Society of Heating, Refrigerating, and Air-Conditioning Engineers [11]

However, solar resource is very rich in Lhasa and the sunshine rate is generally higher than 70% in winter. In January, for example, the average quantity of solar irradiation is 16.556 MJ/(m^2.d), which ranks the region first in China. Comparing outdoor design parameters in the coldest month with Beijing and Xi'an, shows that Lhasa is very rich in solar energy resources [12]. This creates good natural conditions for the application of solar heating [13].

47.3 Methods

47.3.1 The Selection of Two Kinds of Residential Buildings for the Study

The indoor thermal environment among different kinds of buildings in the same area presents big differences and differences in the residents' indoor thermal sensation. These differences are obvious in winter. At present, the traditional Tibetan dwellings and passive solar house are the most common residential buildings in Lhasa. The two kinds of buildings are shown in Figs. 47.2 and 47.3.

Fig. 47.2 The sample of tested traditional Tibetan house

47.3.2 Survey Method and Measurements

The indoor thermal environment of these two types of buildings was studied for more than 20 days from January to February, 2014. The methods of subjective questionnaire and objective testing were adopted to survey residents and the testing process was strictly in accordance with related standards. The residents' subjective evaluation of their thermal environment were obtained by the questionnaire, including details of their sense of cold or hot and moistness across the range of seven indicators adopted from the ASHRAE standard [14]. In total, 495 responses to the survey were obtained. The indoor environment parameters were measured through objective testing, including the indoor temperature and humidity, black temperature, and wind speed.

Fig. 47.3 The sample of tested passive solar house

Moreover, the research survey data also included the basic situation of residential buildings, family backgrounds, dress, active status, living habits, and the ways to improve indoor thermal environment. It was found that the residents' homes had no additional heating systems apart from the heating required for cooking.

47.3.3 *Data Statistical Analyses*

In the survey, there were 289 valid responses to the survey document about traditional Tibetan house and 206 valid responses relating to the passive solar houses. Both the heat resistance of clothing and the levels of active levels are important factors to affect residents' thermal sensation [15]. From the survey investigation, the active states of the residents could be assessed. These were sitting and resting, sitting, standing up, and so on. Their winter metabolic rate was from 0.8 to 1.4 m [16]. Different residents had different habits for the selection of clothing, but most of people wore clothes which included thermal underwear, long woolen underwear, sweaters, and a thick coat. The average winter thermal resistance was 2.6 clo. Combined with the test data, the differences of the residents' vote of thermal sensation together with the test data were compared between the two different types of residential buildings.

47.4 Results of Residents' Thermal Sensation Vote in Buildings

The method of temperature frequency [17] was used to process the thermal comfort data and the operating temperature was divided into intervals of 0.5 °C. This chapter discusses the indoor thermal sensation unsatisfactory rate (predicted percentage dissatisfied, PPD) and the mean thermal sensation vote (MSV) in the two types of residential buildings. In the data processing, the operating temperature is replaced by the indoor air temperature which is similar to the operating temperature [18].

47.4.1 *The Comparison of Residents' Evaluation of Indoor Thermal Environment*

The overall evaluation on indoor thermal environment was obtained by the analysis of residents' thermal sensation vote in the two types of residential buildings.

As shown in Fig. 47.4, the residents' satisfaction with their level of comfort in the passive solar house is higher than that in the traditional Tibetan houses. In the passive solar houses nearly 50 % of the residents felt comfortable with the winter indoor environment, and about 38 % of the residents felt a little uncomfortable. Fewer residents felt comfortable in the traditional Tibetan houses.

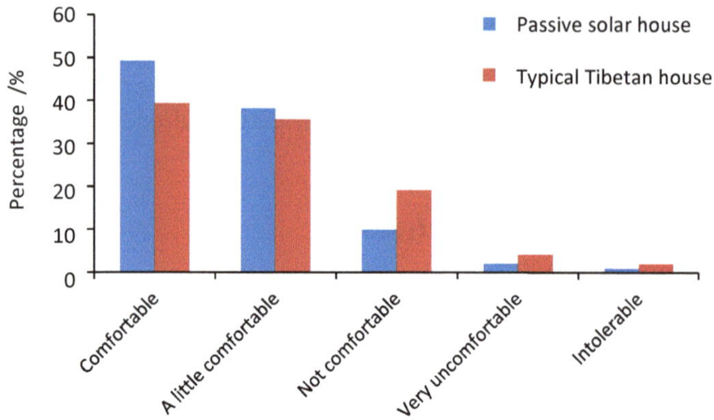

Fig. 47.4 The comparison of residents' thermal comfort satisfaction

47.4.2 The Comparison and Analysis of Residents' Acceptable Indoor Thermal Environment

According to the provisions of the ASHRAE standard and ISO7730, the temperature range which meets the need of more than 80 % residents is obtained in the two different residential buildings. The relationship between PPD and indoor operating temperature is shown in Fig. 47.5.

The regression equations of the curves are as follows:

For the passive solar house,

$$y = 0.4805x^2 - 16.243x + 141.56(R^2 0.9074) \tag{47.1}$$

For the traditional Tibetan houses,

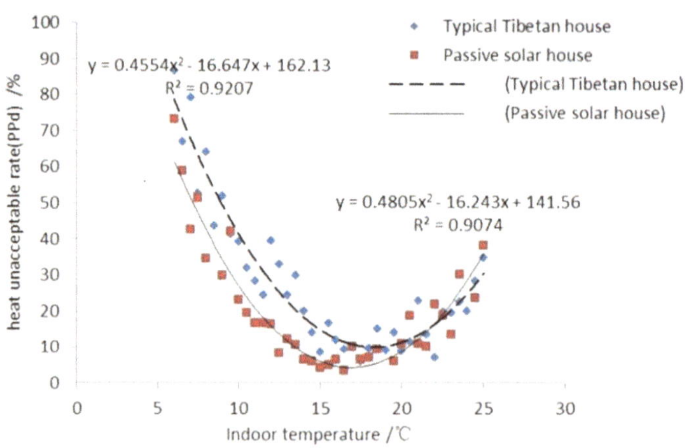

Fig. 47.5 The match between heat unacceptable rate and indoor temperature

$$y = 0.4554x^2 - 16.647x + 162.13 (R^2 = 0.9207) \quad (47.2)$$

The range of acceptable temperature in passive solar house is from 11.3 to 22.5 °C, which is more broad and lower than that in traditional Tibetan houses which is from 13.6 to 23 °C. Meanwhile, less residents feel unacceptable about the indoor thermal environment in passive solar houses than the traditional Tibetan houses when the indoor temperature is the same and lower than 22 °C.

47.4.3 The Comparison and Analysis of the Predicted Thermal Neutral Temperature

By using the method of regression analysis, the thermal neutral temperatures of the two kinds of residential buildings were obtained by curve fitting the measured data.

The regression equations of two kinds of residential buildings can be seen from Fig. 47.6.

For the passive solar houses,

$$y = 0.2019x - 3.3191 (R^2 = 0.8179). \quad (47.3)$$

For the traditional Tibetan houses,

$$y = 0.2458x - 4.2867 (R^2 = 0.8852). \quad (47.4)$$

From the equation, the calculated indoor thermal neutral temperature in the passive solar house is 16.44 °C in winter which is 1 °C lower than that in the traditional Tibetan houses. It can be seen that the residents may feel warmer in the solar houses.

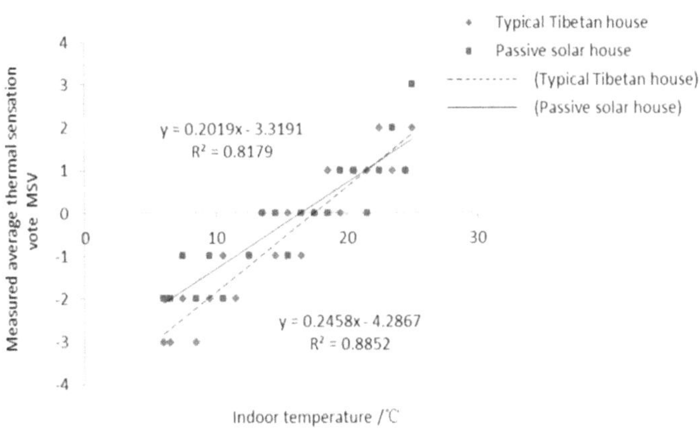

Fig. 47.6 The linear relationship between MSV and the operating temperature

Most of residents feel more comfortable in a passive solar house than a traditional Tibetan house.

In fact, because of the sun radiation effects in winter, it is not difficult to understand the difference of thermal environment between two kinds of residential buildings. A large area of wall in the passive solar house is made of transparent glass. There is more solar radiation heat transferred into the room through the glass, so the temperature of the surface of wall and indoor furniture will rise [20]. Residents will feel more comfortable when experiencing the radiative heat from the surface of the wall and furniture. So, a more comfortable indoor thermal environment can be created in the passive solar house. Even when the indoor temperature is low, residents will not feel the cold because of the solar radiation.

47.5 The Verifying Case

In order to confirm that the passive solar house can improve the indoor thermal environment, the members of the project group measured the indoor temperature for five continuous days in the two kinds of residential buildings. The data were summarized in Fig. 47.7.

As is shown in Fig. 47.7, the difference of the temperature in traditional Tibetan houses, passive solar houses, and outside can be seen vividly. Figure 47.7 shows three key points.

First, the temperature in the passive solar house is clearly higher than the traditional Tibetan house, and there is a big difference in temperature between day and night. The lowest temperature in the passive solar house is 6.4 °C in the morning.

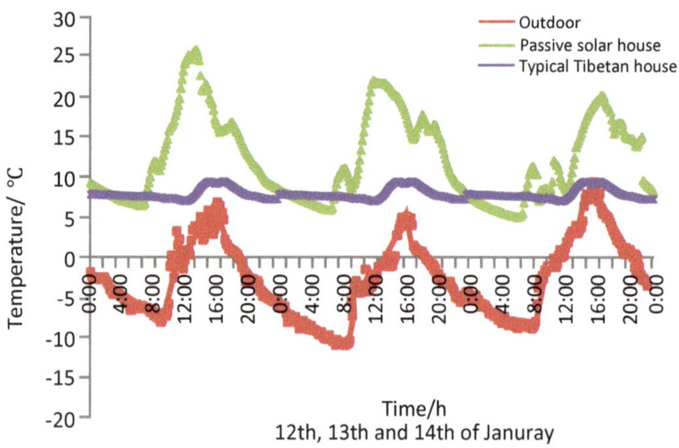

Fig. 47.7 The comparison of temperature in passive solar house, traditional Tibetan house, and outside

The highest temperature can reach up to 25.9 °C. The range of temperature between day and night is up to 19.5 °C in solar house.

Second, the temperature in traditional Tibetan house is low and stable. The minimum test temperature in the Tibetan house is 7 °C, and the highest test temperature is only 9.5 °C which is significantly lower than that in the passive solar house. This shows that the temperature cannot meet most residents' needs for indoor thermal comfort environment.

Third, the indoor temperature in both of the buildings is significantly higher than the outdoor temperature. The lowest outside temperature is −11.5 °C and the maximum temperature is 8.7 °C in winter. The outdoor temperature change between day and night can be up to 20.2 °C. Therefore people may feel very cold at night or in the morning, while feeling hot at noon.

47.6 Discussion of Passive Solar Heating to Improve Indoor Thermal Environment

The method of improving the indoor thermal environment by reasonable utilization of local climatic resources can be called "passive control strategy," which not only improves the indoor thermal environment but can also save energy [8, 21]. Solar radiation is one kind of natural energy for increasing the indoor temperature in winter [22]. The solar energy resource is very rich in Lhasa, so it can be utilized to improve the indoor thermal environment. It can be seen from this study that the indoor thermal environment in the passive solar house is more comfortable.This is verified by sample data. Passive solar heating can improve the indoor thermal environment in Lhasa.

According to established literature [6], combined with the typical weather data in Lhasa, the energy metabolism rate (M) of indoor personnel is 1.1 m, namely people in the level of sitting and relaxed. Analysis shows that using passive solar heating in winter in Lhasa area the comfort zone is extended to a wider temperature range. This is shown in Fig. 47.8.

Figure 47.8 shows that the meteorological parameters in winter in Lhasa are lower than the comfortable temperature range for most of time, and if heating facilities are not used residents live in an uncomfortable (cold) indoor thermal environment. However, through the analysis of passive control strategies in Lhasa, the thermal comfort zone can be expanded by using passive solar heating, and the comfortable temperature range can be achieved for most of time in winter. The passive solar house is one of the best methods to improve indoor thermal environment. This has important practical significance in Lhasa.

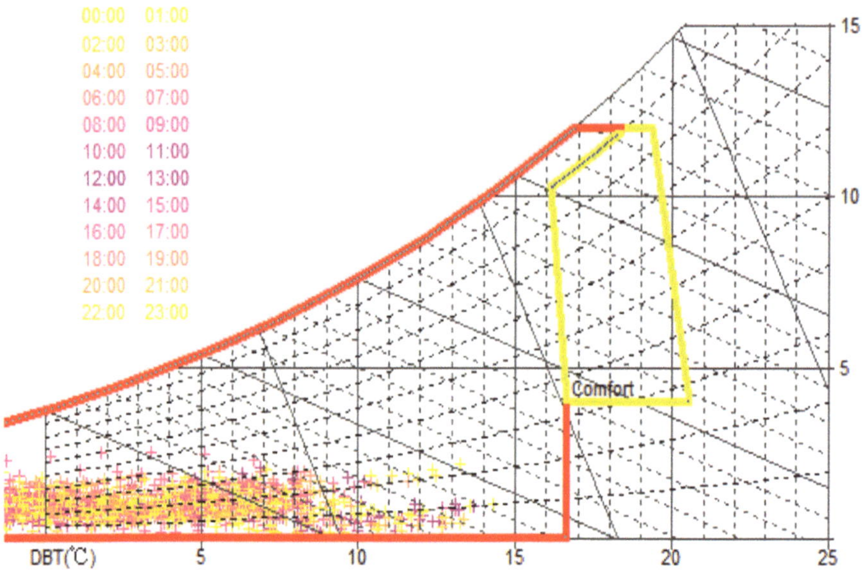

Fig. 47.8 The thermal comfort zone by using passive solar heating in winter in Lhasa. The *yellow area* represents the winter thermal comfort region in Lhasa, which includes the parameters of the air temperature, relative humidity, and the average radiation temperature. The *red area* represents the region where residents can improve the indoor thermal comfort environment by using passive solar heating. The points with different colors represent the meteorological parameters at different times in winter

47.7 Conclusions and Suggestions

The following conclusions can be drawn through the above analysis:

First, residents' thermal satisfaction in the passive solar houses is higher than that in the traditional Tibetan houses in winter. This shows that the majority of residents accept the indoor thermal environment in passive solar houses. Second, the indoor thermal environment in the passive solar house is better than traditional Tibetan residential buildings in winter. The temperature in passive solar house is generally higher, which is in the comfortable temperature range in most of the time.

Throughout the analysis, suggestions are given to improve indoor thermal environment in winter. Solar heating has a wide applicability in Lhasa, and the passive solar house, one of the best examples of using solar heating, can be widely applied in this area. While there are still some problems in the passive solar house, involving the large temperature range between day and night, and the small effect on improving humidity, these are not significant. Therefore, the design of passive solar houses should be improved, making the buildings collection, storage and distribution of solar energy reasonable, which can further improve the indoor thermal environment in winter in Lhasa.

Acknowledgments A sincere thanks to tutors who gave much help to revise the article. At the same time, thanks to the classmates who helped the author to do the investigation. This chapter is based on research project (2013BAJ03B05) by the National Key Technologies R&D Program of China and Project (2013CDJZR210008) supported by the Fundamental Research Funds for the Central Universities.

References

1. Yi J, Ya F, Xiangyang R, Yungang P, Xiangzhao F (2013) Discussion on urban heating modes for Lhasa. HV & AC (Tsinghua University) 43(6):1–7
2. Science and Technology Commission Nakchu Tibet (1996) Overview of new energy development and utilization in Tibet. Energy China
3. Zhen C (1996) Overview of new energy development and utilization in Tibet China. Energy Eng
4. Lei W (2008) Study on the passive solar heating building in Tibet. Southwest Jiaotong University, Chengdu
5. Ya F, Xudong Y, Huizhi Z (2013) Heating potential of passive solar building in Lhasa. HV & AC (China Southwest Architectural and Research Institute, Tsinghua University) 43(6):31–34
6. Li B-Z, Yu W, Liu M, Li N (2011) Climatic strategies of indoor thermal environment for residential buildings in Yangtze River Region, China. Indoor Built Environ 101–111
7. Yao R, Li B, Steemers K, Short A (2009) Assessing the natural ventilation cooling potential of office buildings in different climate zones in China. Renew Energy 34:2697–2705
8. Runming Y, Li B, Liu J (2009) A theoretical adaptive model of thermal comfort—Adaptive Predicted Mean Vote (aPMV). Build Environ 44:2089–2096
9. China Academy of Building Research GB50176-93 (1993) Thermal design code for civil building. China Planning Press, Beijing
10. Chinese Meteorological Information Center meteorological data, Building Technology Research Center of Tsinghua University (2005) Meteorological data on analysis Chinese building thermal environment. China Building Industry Press, Beijing
11. American Society of Heating, Refrigerating, and Air-Conditioning Engineers, ANSI/ASHRAE Standard55-2004. Thermal environmental conditions for human occupancy. ASHRAE, Atlanta. ISSN 1041-2336
12. Chinese architectural design and Research Institute (2012) JGJ/T267-2012. Technical code for passive solar energy building. China Building Industry Press, Beijing
13. Xiangyang R, Xiaodan M, Pengfei S, Qinglong G (2013) Suitability analysis for Lhasa heating techniques. HV & AC 43(6):23–30
14. ASHRAE 55-1992 (1992) Thermal environmental conditions for human occupancy. American Society of Heating, Refrigerating and Air Conditioning Engineers, Inc., Atlanta
15. Zhaofen J, Yingxin Z (2003) Building environment theory. China Building Industry Press, Beijing
16. Chongqing University, China Academy of Building Research GB/T 50785-2012 (2012) Evaluation standard for indoor thermal environment in civil buildings. China Building Industry Press, Beijing
17. Zhaojun W (2004) Selection of thermal comfort indexes in the field study. HV & AC 34(12):39–42
18. Yi W, Jiaping L (2000) Thermal comfort properties for air conditioned space in winter. HV & AC 30(3):12–14
19. ISO 7730 (1984) Moderate thermal environment-determination of the PMV and PPD indices and specification of the conditions for thermal comfort. International Standard Organization, Geneva

20. Tangshi Y, Fusheng G (2000) Effects of solar radiation on human thermal comfort in winter. National HV & AC Annual Conference in 2000
21. Runming Y, Li B (2002) Simulation for passive building design and thermal comfort in South China. PLEA 2000: Design with the Environment International Conference on Passive and Low Energy Architecture, France, 22–24 July 2002
22. Nianxiong L, Youguo Q (2005) Building thermal environment. Tsinghua University Press, Beijing

Chapter 48
Building Energy Index and Students' Perceived Performance in Public University Buildings

S. N. N. Syed Yahya, A. R. M. Ariffin and Muhammad Azzam Ismail

Abstract Numerous studies have shown that students' learning experience is closely associated with the physical comfort level of their teaching and learning environment. The different strategies or allocation of air-conditioning, mechanical ventilation (ACMV) and lighting systems contribute greatly to the energy performances in the buildings. This study explores the relationship between electricity consumption of the academic buildings of a public university in an urban context and its students' perceived performance. It seeks to find the answer patterns from unsuspected subjects; whether there is a difference between a lower energy-use building and a higher energy-use building. To achieve the objective, the study adopts the quantitative method of assessing students' perceived performance through questionnaire survey. The questionnaires, adopted from Building Use Studies, UK, were distributed randomly through convenience sampling to students from two academic buildings in the campus. Both buildings were selected through purposive sampling method with specific inclusion and exclusion criteria. Subsequently, power and energy logger was installed into the same buildings to monitor electricity consumption at specific intervals. For comparison, building energy index (BEI) for each building was calculated. The study found that the building that scored higher in students' performance, also has higher calculated BEI. As control for indoor comfort account for more than half of the total electricity consumption, the result suggested that electricity consumption influences students' performance positively. The study also revealed that both buildings' calculated BEI were lower than recommended by many standards. This suggests that these buildings have extremely high potential of achieving green building status. If executed properly, the university campus, which equates the size of a small city, may achieve green campus status sooner than expected and may lead others in the flagship project towards a low-carbon university campus.

Keywords Building energy index (BEI) · Students' performance · Academic buildings · Public university

S. N. N. Syed Yahya (✉) · A. R. M. Ariffin · M. A. Ismail
Department of Architecture, University of Malaya, Kuala Lumpur, Malaysia
e-mail: shipah79@gmail.com

A. Sayigh (ed.), *Renewable Energy in the Service of Mankind Vol I*,
DOI 10.1007/978-3-319-17777-9_48

48.1 Introduction

Where numerous studies have shown that students' learning experience is closely associated with the physical comfort level of their teaching and learning environment [1–5], little have shown the indirect relationship between students' performance and the energy consumption. The maintenance of this teaching and learning environment, that is, building envelope, facilities and the indoor air quality depends highly on energy supply [6, 7]. For example, the choice of strategies to cool and ventilate the air and to illuminate the functional spaces determine the energy usage in the buildings [8, 9]. Therefore, it can be assumed that the students' performance is also associated with their environment's (i.e. the building) energy performance.

Since the initiation of the sustainability movement in the late 1970s, new buildings, particularly public buildings have been encouraged to increase energy performance and energy efficiency. For a university building, regardless of the enthusiasm to pursue energy-efficient building status, it must not compromise the quality of its core business which is teaching and learning. Reducing energy consumption should not risk reducing students' performance.

The indirect relationship between energy consumption and students' performance has motivated this study to explore the relationship between the two variables. More specifically, this study compared the electricity consumption and students' perceived performance between two buildings in a renowned public university in Malaysia. The outcome of this study answered whether the students' perceived performance in low energy-use building was better or worse compared with high energy-use buildings.

48.2 Research Methodology

The chapter achieved the objective of the study by collecting data from two subjects. This method is necessary to explore the relationship between two variables, which are the energy consumption and students' performance. The first subject is the study buildings, while the second subject is the occupants of the buildings, more specifically, the students. The study buildings were selected through purposive sampling, while the students through convenience sampling.

A set of criteria was established to isolate buildings to study. The criteria restricted only multifunctioning academic buildings to be included in the study. This exclusion criterion managed to discard 310 buildings, leaving only 8 to remain in the sampling frame. The research team did a 1-day observation on each building shortlisted. The observation discovered that two buildings from the list were barely occupied. Aside from less than a dozen permanent occupant, the buildings intermittently received visitors who remained only for short periods. The two buildings were removed from the list of buildings to study, foreseeing that the potential respondents may not be able to express their perception of the building for their lack of experience occupying the buildings. One hundred and fifty questionnaires were distributed to each of the remaining six buildings.

The Building Use Studies (BUS) occupant survey was found most suitable and was utilised under license from BUS methodology 2012 for this study. The BUS occupant survey is one of two large-scale studies that have been initiated in 2002 and widely used globally to study building performance [10]. BUS methodology has also developed a database and "benchmarks" based on most recent 50 buildings surveyed from 17 countries [11]. Although the BUS questionnaire form is available in print and electronic forms, only the former is used for this study because as reported in previous research, print form provides higher response rate than the electronic [12].

The questionnaire asks the respondents to state their perception of how the building environment influences their performance, behaviour and health. Respondents were also encouraged to elaborate their answers in the "comments" column at the end of each section. There have been arguments that performance was a subjective variable that can be biased if not measured properly and supported by evidence [13]. Only "perceived" performance is used as indicator for this study because it can be recorded quickly from the survey. Actual students' performance requires access to classified records and may not be achievable in the time of research.

The response rate for each building varied where the lowest response rate was 4.67%, while the highest response rate was 72.67%. Only the two buildings (Table 48.1, Fig. 48.1 and Table 48.2) with the highest response rates were selected

Table 48.1 Study building description and characteristics

Building	A1 (Arts Faculty annexe building)	A2 (Faculty of Languages and Linguistics)
Building facade		
Year built	1999	1979
Built-up area	8314.01 m^2	10,532.49 m^2
Building form and orientation	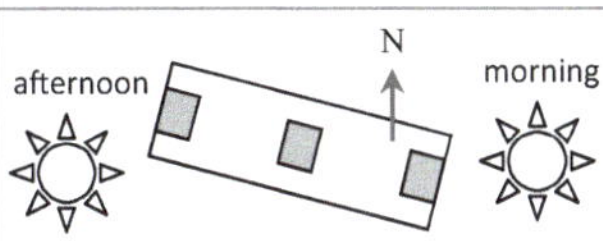	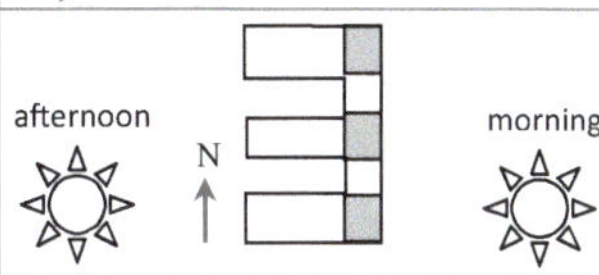
Building height	Three storeys (split levels)	Two storeys (split levels)
Roof properties	Pitched roof with concrete ceiling and parapet	Pitched roof with plasterboard ceiling (concrete roof tiles)
Glazing properties	Float glass, tinted film casement windows	Frosted jalousie windows
External wall properties	Brickwall	Brickwall
Internal wall properties	Plaster and paint, rooms too small to have internal partitions	Plaster and paint, gypsum
Ventilation type	Mechanical cooling—split unit	Mechanical cooling—split unit and air-cooled package and water-cooled package

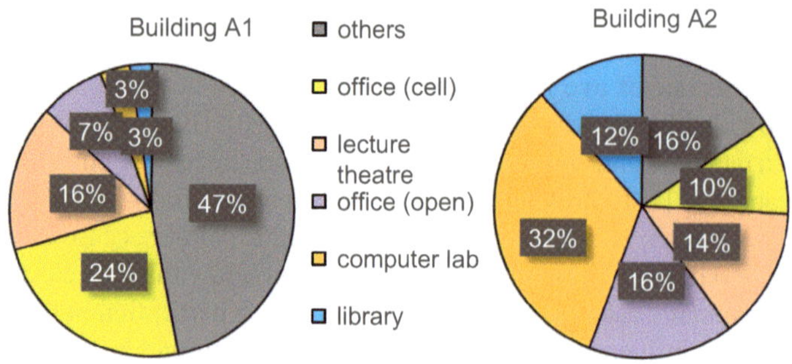

Fig. 48.1 Space function for study buildings by area

Table 48.2 Position and type of ventilation for internal spaces

Building	A1	A2
Teaching and learning spaces	Located along external wall with mechanical ventilation	Located along external wall with mechanical ventilation
Administration	Located along external wall with mechanical ventilation	Located along external wall with mechanical ventilation
Lobby	Open and naturally ventilated	Open and naturally ventilated
Walkway/corridor	Located along interior wall naturally ventilated	Open and naturally ventilated
Staircase	Open and naturally ventilated	Open and naturally ventilated
Toilet	Located at external wall with natural ventilation	Located at external wall with natural ventilation
Observations and comments	Vernacular design allows natural lighting. Mechanical ventilation is still needed because air is still	Renovation on-going at time of observation Doors and windows grilled for security Room layout has the advantage of natural lighting but external windows were fitted with black (almost opaque) blinds

to be further studied due to availability of only one power and energy logger (PEL). Based on research by others, the PEL was among the best instrument to be used to measure energy consumption [14]. The availability of only one PEL forced the data to be collected consecutively instead of concurrently. Due to the short time of the research, only two buildings managed to be studied. The PEL was set to record energy consumption at the interval of 10 minutes for 14 days in each building. Upon completion of 14 days, the PEL was dismantled from the first study building (A1) and relocated to the second study building (A2). The installation, dismantling and reinstallation were conducted by a certified and competent chargeman because the PEL needs to be installed on the main switch board (MSB). Recorded data were downloaded to the dedicated software supplied by the PEL manufacturer and then analysed.

The PEL succeeded in recording the energy consumption value for a period of 14 days; it also recorded the consumption trend, among others. For this study, only the energy consumption value is used to determine energy performance. Building energy performances were calculated for both buildings and later were analysed together with the result from the questionnaire survey.

48.3 Building Energy Performance

A building's energy performance is commonly used to determine its energy consumption against its building size for the purpose of comparison [15, 16]. A look at the energy consumption alone is unable to determine whether the building overconsumes or under-consumes energy. Energy consumption has to be analysed relative to its building size. Calculating the energy consumption over building size is known as building energy index (BEI) or sometimes as energy performance index (EUI) [17–19]. BEI is calculated simply by dividing the total annual energy consumption of the building (kilowatt-hour per year) with its total occupied floor area (m^2) as follows:

$$\mathrm{BEI}\left(\mathrm{kWh/m^2/year}\right) = \frac{\sum \text{Annual energy consumption}\left(\mathrm{kWh/year}\right)}{\sum \text{Net floor area}\left(\mathrm{m^2}\right)}.$$

The above BEI calculation requires the actual annual energy consumption for the building. However, given that only data for 14 days were available, the data were aggregated to obtain data for 365 days. The aggregated data were then used in the equation. Compared to BEI, energy benchmarking is a relatively more accurate method that is widely used to compare energy performance between buildings [15]. This method adds another variable to the quantitative energy consumption and building size, which is the building function or type. This chapter adopts the UK's Chartered Institution of Building Services Engineers (CIBSE) energy benchmarking system for the purpose of comparison. The system has established "good standard" and "typical standard" benchmark as per Table 48.3.

For this chapter, the "good standard" and "typical standard" for each study building is calculated by multiplying the area of each functional space with the respective benchmark and dividing it by the total floor area of the building. For example, the "good standard" index for building A1 is shown below (Table 48.4):

Table 48.3 CIBSE benchmarking system for energy performance (Adopted from [15])

Function	CIBSE good standard ($kWh/m^2/year$)	CIBSE typical standard ($kWh/m^2/year$)
Others	54	85
Office (cell)	33	54
Lecture theatre	67	76
Office (open)	54	85
Computer lab	155	175
Library	46	64

CIBSE Chartered Institution of Building Services Engineers

Table 48.4 CIBSE good standard index calculation

	CIBSE good standard (kWh/m²/year)	Area (m²)	A1 good standard (kWh/year)
Others	54	2493.79	134,664.66
Office (cell)	33	1257.30	41,490.90
Lecture theatre	67	859.72	57,601.24
Office (open)	54	391.21	21,125.34
Computer lab	155	180.19	27,929.45
Library	46	131.23	6036.58
Total		5313.44	288,848.17

CIBSE Chartered Institution of Building Services Engineers

$$\text{Good standard index}_{\text{A1}} = \frac{\sum\left(\text{area for each function}\left(\text{m}^2\right) \times \text{good standard index}_{\text{CIBSE}}\left(\text{kWh/m}^2\text{/year}\right)\right)}{\sum \text{nett area of A1}\left(m^2\right)}.$$

$$\text{Therefore, good standard index}_{\text{A1}} = \frac{288{,}848.17\,\text{kWh/year}}{5313.44\,\text{m}^2} = 54.36\,\text{kWh/m}^2\,\text{/year}.$$

For comparison, the chapter uses energy index from the University of Malaya main campus (92.25 kWh/m²/year), Universiti Kebangsaan Malaysia Hospital (244.81 kWh/m²/year) [17], the recommended Malaysian Standard for Energy Efficiency, MS1525:2001 (130 kWh/m²/year) [20] and both the CIBSE benchmarking for good (54.36 kWh/m²/year for A1 and 63.34 kWh/m²/year for A2) and typical standard (78.74 kWh/m²/year for A1 and 80.72 kWh/m²/year for A2). Apart from being compared with each other, these study buildings were also compared with other buildings.

48.4 Study Building A1

The result section is reported according to the study building so that comparisons can be discussed in the discussion section. The distribution of questionnaire survey was done randomly through convenience sampling and was not restricted to just students. Therefore, there was a mix of staff, students and visitors who responded to the survey. Thirty-three out of 102 respondents for building A1 were students. Thirty of them were aged less than 30 years old and 23 occupied the building for at least 1 year. Only responses from students were used for this chapter. The respondents mean score for perceived performance was 5.67 compared to 4.13 of the BUS benchmark. Fourteen respondents (42.4 %) perceived that the building environment increases their performance by at least 10 % while the majority of 16 respondents believed that their performance is not affected by the building environment.

The trend for energy consumption of building A1 showed a maximum consumption of 18.72 kWh and was recorded on a Wednesday, 12.40 p.m. At the end of the 14 days,

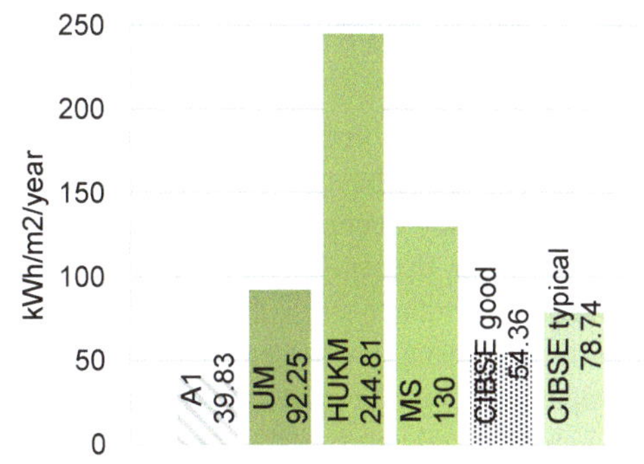

Fig. 48.2 BEI for building A1 and other standards *HUKM* Hospital Universiti Kebangsaan Malaysia, *UM* University of Malaya, *MS* Malaysian Standard for Energy Efficiency, *CIBSE* Chartered Institution of Building Services Engineers

the PEL recorded that building A1 has consumed 12.71 MWh or 12,710.00 kWh of energy. Using the BEI equation, the calculated BEI for building A1 is 39.83 kWh/m^2/year. Figure 48.2 shows the BEI for building A1 against other standards.

48.5 Study Building A2

As with building A1, the respondents for building A2 came from different backgrounds where only 49 out of 109 respondents were students. Thirty-six students have been occupants of the building for at least a year and 41 aged less than 30 years. Responses from other respondents were discarded and only responses from students were analysed from here on. The mean score obtained from the responses for perceived performance were 5.73 compared with 4.13 for BUS benchmark. Twenty-eight respondents, equivalent to 58.3 % reported that their productivity were increased by at least 10 % due to their building environment, while only 12 respondents perceived that their productivity is not affected by their building environment.

The PEL recorded maximum energy consumption of 52.88 kWh on a Monday at 10.50 a.m. On the 14th day of the equipment installation on building A2, the PEL recorded that 31.75 MWh (31,750.00 kWh) of energy has been used. The calculated BEI for building A2 is 78.59. BEI for building A2 and other standards are plotted in Fig. 48.3.

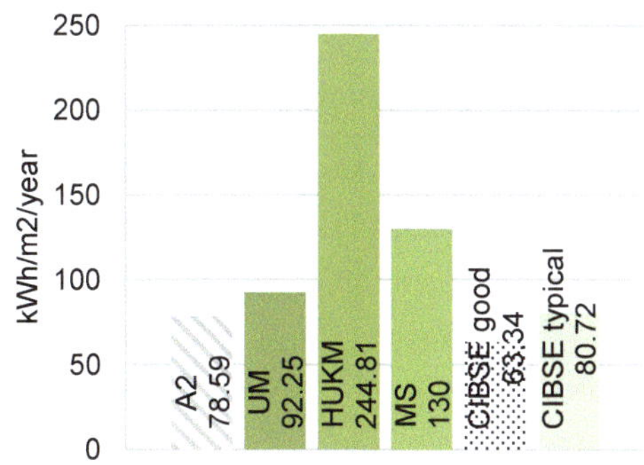

Fig. 48.3 BEI for building A2 and other standards *HUKM* Hospital Universiti Kebangsaan Malaysia, *UM* University of Malaya, *MS* Malaysian Standard for Energy Efficiency, *CIBSE* Chartered Institution of Building Services Engineers

48.6 Discussion

The demographic profile of the respondents shows that more than 60% of student respondents have been occupying the building for more than 1 year. It can be assumed that they occupy the building permanently and are not transient occupants. This signifies that their responses are genuine and reliable. Mean scores for perceived performance show that the occupants for both buildings perceived that their building environment increases their performance by at least 10%. These scores are comparatively better than the BUS benchmark score.

Similarly, when plotted against BEI from other standards, both study buildings' energy performances are considerably better. Building A1's energy performance is substantially less than CIBSE good standard index, while building A2 has a very high potential of meeting the CIBSE good standard benchmark. By combining the data from the two subjects, they answered the research question that the building with higher electricity consumption scores higher mean score for perceived performance. As reviewed by Yang, Yan and Lam [7], control for indoor comfort account for more than half of the total electricity consumption, the result suggested that electricity consumption influences students' performance positively. In other words, more energy is needed to adjust the building environment to suit the occupants' comfort so they can perform better. Although building A2 scores higher in terms of students' perceived performance and energy index, a further analysis revealed that building A1 uses energy more efficiently in terms of students' performance. This is shown in Table 48.5 below. The table shows that although BEI for building A2 is doubled from building A1, the reported perceived performance is small comparatively. The table also calculates that on average, every 1 kWh/m^2/year only contributes to 0.74% of performance for building A2, while 1 kWh/m^2/year only contribute to 1.07% of students' performance.

Looking at the possible causes of the large difference in electricity consumption for buildings A1 and A2, there is a considerable difference in space usage; where building A2 comprised more computer labs and lecture theatres, building A1 is made up of offices and other spaces that consume less energy. As established by CIBSE, computer labs and lecture theatres contribute to higher energy indices. Another factor that is known to cause the difference in energy usage and performance is the passive designs observed on the two buildings. Building A2 permits natural light and ventilation while the other does not. Although it is usual that passive features reduce energy usage, it is not the case for building A2. Natural light

Table 48.5 Percentage of performance increase every 1 kWh/m^2/year

Code	Mean score for perceived performance (a)	Reports increase 10% or more (b) (%)	BEI (kWh/m^2/year) (c)	Perceived performance every 1 kWh/m^2/year (b)/(c) (%)
A1	5.67	42.4	39.83	+1.07
A2	5.73	58.3	78.59	+0.74

that penetrates the building permits high heat indoors while the natural vents cause air-conditioner leakages and make them less effective.

48.7 Research Limitations

The original design for this study was to concurrently measure the electricity consumption for both buildings and conduct the questionnaire survey. Unfortunately, funding received was only able to accommodate one power logger, therefore measurement cannot be done concurrently. In addition, although originally, two similar buildings in size were finalized to be studied, it was discovered that one building was operated by two separate main switch boards (MSB), and there was only one energy logger. Therefore, building A2 was chosen to replace the originally shortlisted building. Another limitation faced by the researchers is the short research period. Therefore, actual electricity usage for 365 days is impossible to record. In fact, only 14 days were allocated for each building. Recorded data had to be aggregated to estimate consumption for 365 days. It would also be less difficult for the researchers if occupancy profile for both buildings were available. Although student enrolment records are available, the records do not reflect the actual occupancy of the buildings. The only occupancy profile available for these buildings is for administration offices.

48.8 Recommendations and Conclusion

The results of the study have answered the research question that the building with higher electricity consumption scores higher mean score for perceived performance. Although it has been discussed that students perform better in buildings with higher electricity usage, the significance of the difference should be further looked into. Is a 10% increase in students' performance worth double the energy usage and electricity bill? The next phase for the study is the potential to pursue green building status for the buildings studied. From the study, it is evident that both buildings have very high potential in being energy efficient. The low BEI of both buildings compared to other standards should be an encouragement towards a greener campus. If executed properly, the university campus, which equates the size of a small city [21, 22], may achieve green campus status sooner than expected and may lead others in the flagship project towards a low-carbon university campus.

Acknowledgements This research is funded by the "Low Carbon Assessment and Strategy: University of Malaya towards a Low Carbon University Campus—A Showcase" (Project No.: FL027-2012) Flagship Program under the University of Malaya Sustainability Science Cluster.

References

1. Yang Z, Becerik-Gerber B, Mino L (2013) A study on student perceptions of higher education classrooms: impact of classroom attributes on student satisfaction and performance. Build Environ 70(0):171–188
2. Plympton P, Conway S, Epstein K (2000) Daylighting in schools: improving student performance and health at a price schools can afford. In: proceedings of the American Solar Energy Society Conference; Wisconsin, USA
3. Lee MC et al (2012) Student learning performance and indoor environmental quality (IEQ) in air-conditioned university teaching rooms. Build Environ 49(0):238–244
4. Hill MC, Epps KK (2010) The impact of physical classroom environment on student satisfaction and student evaluation of teaching in the university environment. Acad Educ Leadersh J 14(4):65–79
5. Choi S et al (2014) Indoor environmental quality of classrooms and student outcomes: a path analysis approach. J Learn Spaces 2(2)
6. Olanrewaju AA (2011) Green maintenance management initiative for university buildings. Built Environ J 8(1):17–24
7. Yang L, Yan H, Lam JC (2014) Thermal comfort and building energy consumption implications—a review. Appl Energy 115(0):164–173
8. Bülow-Hübe H, Blomsterberg Ã (2011) Important design aspects of a very low—energy building. Intelligent Energy Europe—NorthPass: promotion of the very low-energy house concept to the North European Building Market, Sweden
9. Shen H, Tzempelikos A (2012) Daylighting and energy analysis of private offices with automated interior roller shades. Solar Energy 86(2):681–704
10. Gou Z, Prasad D, Siu-Yu Lau S (2013) Are green buildings more satisfactory and comfortable? Habitat Int 39(0):156–161
11. Building Use Studies (2011) The Building Use Studies (BUS) occupant survey: origins and approach Q & A. Building Use Studies, Editor, United Kingdom
12. Nulty DD (2008) The adequacy of response rates to online and paper surveys: what can be done. Assess Eval Higher Educ 33(3):301–314
13. Palmer J (2013) Post-occupancy evaluation of buildings. In: Mumovic D, Santamouris M (eds) A handbook of sustainable building design and engineering: an integrated approach to energy, health and operational performance. Routledge, London, pp 347–354
14. Alajmi A (2012) energy audit of an educational building in a hot summer climate. Energy Build 47:122–130
15. Altan H, Douglas J, Kim Y (2014) Energy performance analysis of university buildings: case studies at Sheffield University, UK. J Architect Eng Technol 3(129):2
16. Abdul-Rahman H, Wang C, Kho MY (2011) Potentials for sustainable improvement in building energy efficiency: case studies in tropical zone. Int J Phys Sci 6(2):325–339
17. Moghimi S et al (2011) Building Energy Index (BEI) in large scale hospital: case study of Malaysia. In: 4th WSEAS international conference on recent researches in geography geology, energy, environment and biomedicine, Corfu Island
18. GreenTech Malaysia and SEDA (2013) Method to Identify BEI, Net BEI, GFA, NFA, ACA. Seminar by building consumption input system, Malaysia
19. Bishop R (2012) Why some buildings have very high or low EUIs. BEES seminar by BRANZ Ltd, New Zealand
20. Aun CS (2004) Energy efficiency: designing low energy buildings using energy 10. CPD Seminar 2004 by Pertubuhan Arkitek Malaysia, Malaysia
21. Hoe YK (2011) Achieving sustainable campus in Malaysia University, in Department of Civil Engineering. University Teknologi Malaysia, Malaysia
22. Mat S et al (2009) Managing sustainable campus in Malaysia—organisational approach and measures. Eur J Soc Sci 8(2):201–214

Chapter 49
Energy-Efficient Refurbishment of Existing Buildings: A Multiple Case Study of Terraced Family Housing

D.K. Serghides, N. Saboohi, T. Koutra, M.C. Katafygiotou and M. Markides

Abstract The building sector in Europe consumes an estimated 40% of the total energy. It is well established that the reduction of building energy consumption necessitates energy-efficient renovation of the existing buildings. To achieve this objective, primarily, it is necessary to document the existing building stock.

This chapter focuses on multiple case studies of terraced family houses. Representative terraced family houses from different chronological periods, which are typical and representative of the national residential building stock typologies, act as model buildings and are used for demonstrating the energy performance and the potential energy savings resulting from energy conservation measures (ECMs) and interventions made on both the building envelope and the heat supply systems.

Successful strategies and techniques toward reducing the energy consumption and CO_2 emissions are simulated using the Tabula.xls tool, which is an excel spreadsheet algorithm-based tool. A database with the appropriate information concerning the energy-related characteristics of the terraced family housing has been developed. The research aims to fill in the gap of the current knowledge and the building typologies and it could be a useful instrument to facilitate the energy performance assessment of the building stock and to highlight the potential of renewable energy use in terraced family housing.

D. K. Serghides (✉) · N. Saboohi · T. Koutra · M. C. Katafygiotou · M. Markides
Department of Environmental Science and Technology, Cyprus University of Technology, Anexartisias 33, 3036 Limassol, Cyprus
e-mail: despina.serghides@cut.ac.cy

N. Saboohi
e-mail: sabuhinegar@gmail.com

T. Koutra
e-mail: koutra.ni@gmail.com

M. C. Katafygiotou
e-mail: martha.katafygiotou@cut.ac.cy

M. Markides
e-mail: marina.markides@cut.ac.cy

A. Sayigh (ed.), *Renewable Energy in the Service of Mankind Vol I*,
DOI 10.1007/978-3-319-17777-9_49

Keywords Energy performance · Building typology · Terraced housing · Existing building stock

49.1 Introduction

The increasing energy demand already threatens the future of the planet, as researches show that 10 % of the world population exploits 90 % of energy resources [1]. Under this threat, the European Union aims to achieve a more sustainable future and therefore focuses on the existing building stock by identifying the potential energy conservation of the building sector. However, this venture presents many obstacles and it seems very difficult to find data on European or on national levels [2].

The building sector consumes 40 % of the world energy. The existing residential building stock exceeds the number of newly built dwellings in most developed countries. Moreover residential buildings are responsible for approximate of two third of the energy consumption of the building sector [3]. While new constructions add at most 1 % a year to the existing stock, the other 99% of buildings are already built and produce about 26 % of the energy-use-induced carbon emissions [4]. Thus, greater potential of energy savings can be achieved in the existing residential buildings than the newly built structures.

The improvement of the energy performance of existing residential stock in every country is essential since the operational cost, the energy consumption and the carbon dioxide emissions are major issues worldwide. The chapter is based on information laid out by the European, IEE Episcope Project [5]. The study focuses on the terraced family house (TFH) typology, which is one of the main residential typologies in Cyprus; three representative buildings from the national stock were selected to conduct this research. This chapter is a multiple case study and aims at reducing the energy consumption of the dwelling through energy refurbishment measures.

49.2 Methodology

According to previous studies by the EU, IEE Project EPISCOPE [5], nine residential building stock typologies were established as typical and representative of the national residential building stock. These are classified according to their chronological period of construction and their architectural characteristics. The three building typologies consist of: the multi-storey family houses (MFH), the TFH, and the single family houses (SFH). These are divided into three different chronological periods, supported by the data collected by the Cyprus Statistical Service [6]. Each chronological division was defined based on the different constructional regulations and techniques that were applicable throughout the years, formulating the three distinctive chronological categories, before 1980, between 1981 and 2006, and after 2007. The three selected buildings vary in constructional elements, energy consumption demand, and electromechanical systems, and they are typical and representative samples for their construction period.

The modeling tool used throughout this chapter is the calculation tool Tabula.xls [7]. This is an advance excel workbook, based on algorithmic equations and encompasses the climatic data for each country along with the constructional data of the typical buildings, the performance of the supply systems, the frequency of building and system types, providing results regarding the primary energy consumption, CO_2 emissions, and operational costs for each building [8]. Furthermore, it performs comparative simulations for existing and refurbishment scenarios once all the data have been imported. Construction information and heat supply systems as well as the energy carrier (EAC—Electricity Authority Cyprus) for the houses were gathered, coded, and imported in the Tabula.xls tool. The total area of each building, the heated living volume, and the constructional elements that form the building envelope were collected. The characteristics of the roof, the wall, the floor slab, the structure, and the openings were recorded and their U value as well as their thermal capacity was calculated. The installed electromechanical systems were documented and their efficiency was measured. Due to the lack of available data mainly concerning the existing electromechanical systems installed in the representative buildings, certain assumptions were (length and type of pipes, condition of systems) made when this study was contacted.

49.3 Case Studies and Results

Representative buildings concerning the TFH typology in the three different chronological periods were identified. Each building was extensively studied and information regarding the size, the heated living volume, the thermal characteristics of all its constructional elements, and the heating system were recorded. The sample buildings that are used as representative dwellings for this research are presented in Fig. 49.1.

Terraced Family House (TFH01) **Chronological period: pre 1980** | **Terraced Family House (TFH02)** **Chronological period: 1981-2006** | **Terraced Family House (TFH03)** **Chronological period: after 2007**

Fig. 49.1 Images of the typical dwellings for the TFH typology for all chronological periods

49.3.1 Existing Condition of Buildings

TFH01 is located in a village in Troodos, mountainous area. It belongs to the first chronological period prior to 1980 and it was built in the late 1970s. It is an average size house of 120 $m^2$with heated living volume of 720 m^3, in the middle of the row. This terraced house has a non-insulated, tiled, pitched roof with horizontal ceiling, rendered brick walls, floor slab in contact with the ground, and wooden frame single-glazed windows. The house has constant temperature non-condensing oil boiler, without thermal insulation. There are no solar thermal panels on the roof for domestic hot water (DMH); however, there are main distribution water pipes and a central system with natural circulation with an old internal backup electric resistance. The thermal characteristics (*U* value) of the building envelope for the TFH01 are calculated as shown on Table 49.1.

The resulting heat losses for the existing TFH01 during the heating period are mainly through the exterior walls and rise up to 70.8 kWh/m^2a. There are further heat losses through the roof at 43.9 kWh/m^2a and floor at 14.7 kWh/m^2a. Results indicate that the required energy for space heating is 138.9 kWh/m^2a, as shown on Fig. 49.2.

Figure 49.3 shows the energy consumption results calculated for the existing condition of TFH01. The primary energy consumption was calculated at 349.2 KWh/m^2a. The total thermal energy consumption was calculated at 138.9 kWh/m^2a (305.2 KWh/m^2a for fuel (oil) consumption), the electricity consumption was 5 kWh/m^2a. The annual operating cost was estimated at 49.5 €/m^2a and the total annual CO_2 emissions reach the 0.09 t.

TFH02 is located in Nicosia, inland. It belongs to the second chronological period 1981–2006 and was built in the late 1990s. It is 126 m^2 with heated volume of 756 m^3, in the middle of the row of houses. It has a non-insulated, tiled, pitched roof

Table 49.1 Thermal characteristics of the building envelope for TFH01

Construction element	*U* value W/(m^2k)
Pitched roof with horizontal ceiling	3.19
External walls	1.39
Floor in contact with the ground	0.91
Windows	6.1

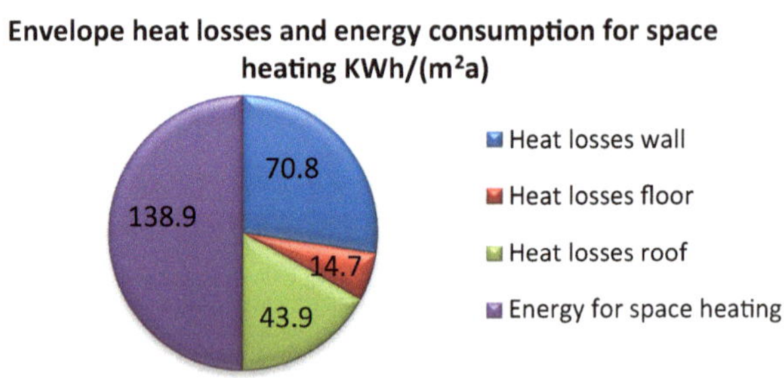

Fig. 49.2 TFH01, Building envelope heat losses and space heating consumption energy, kWh/m^2a

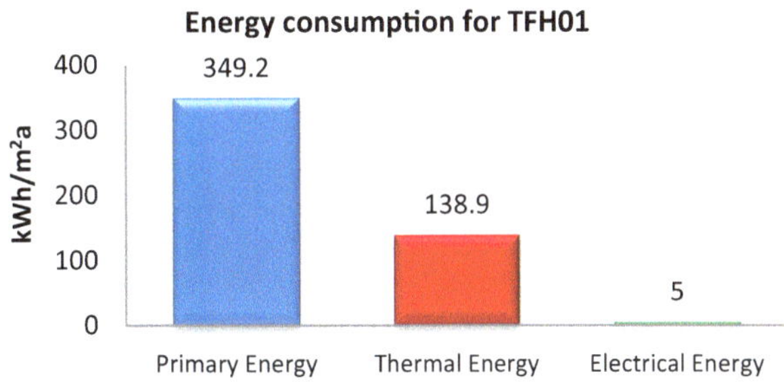

Fig. 49.3 Energy consumption for TFH01

with horizontal ceiling, rendered brick walls, floor slab in contact with the ground, and aluminum frame single-glazed windows. The house has a constant temperature oil boiler, without thermal insulation with a water storage tank that connects to solar thermal panels at the roof for domestic hot water, backed up by an electric element. The thermal characteristics of the building envelope of the TFH02 are calculated as shown on Table 49.2.

The existing heat losses for the second TFH during the heating period, as shown on Fig. 49.4, are mainly through the exterior walls at 28.5 kWh/m^2a. Further heat losses escape through the roof at 26.6 kWh/m^2a, and the floor at 14.5 kWh/m^2a. Results also indicate that the required energy for space heating is 79 kWh/m^2a.

Figure 49.5 presents the results calculated for the existing condition of the second TFH concerning the primary energy consumption which is 168.4 kWh/m^2a, the total thermal energy which is 79 kWh/m^2a (141.9 kWh/m^2a for fuel (oil) consumption) and the electricity used is 4.6 kWh/m^2a. The annual operating cost is estimated at 23.3 €/m^2a with the total annual CO_2 emissions reaching the 0.04 t.

Table 49.2 Thermal characteristics of the building envelope for TFH02

Construction element	U value $W/(m^2k)$
Pitched roof with horizontal ceiling	3.42
External walls	1.4
Floor in contact with the ground	0.91
Single-glazed windows	6.1

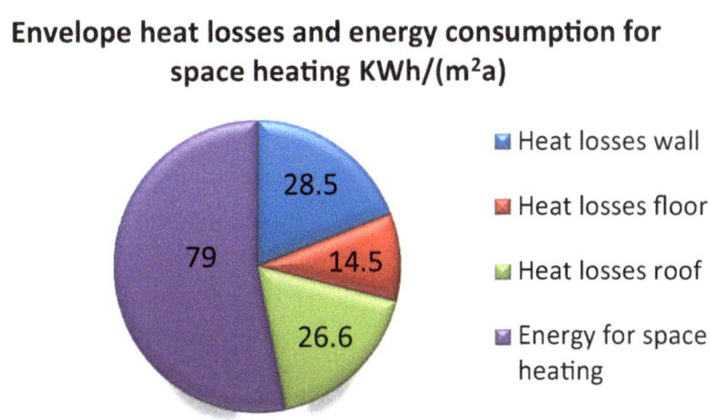

Fig. 49.4 Heat losses and space heating requirements, kWh/m^2a for TFH02

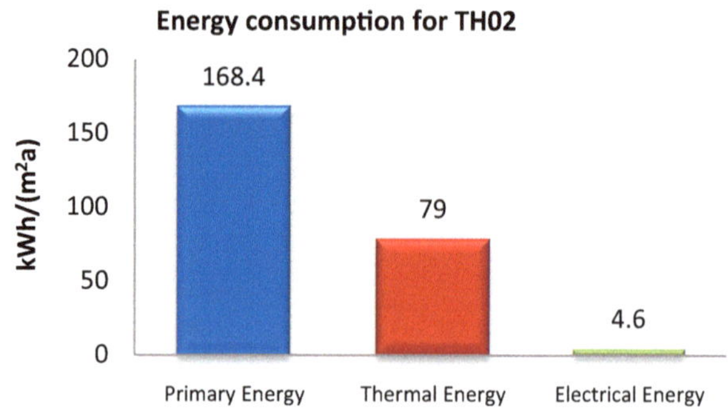

Fig. 49.5 Energy consumption for TH02

TFH03 is located in Limassol, coastal. It belongs to the third chronological period after 2007 and it was built in 2012 following the Cyprus minimum requirements of thermal performance developed for EU energy performance building directive (EPBD) [9]. It is 226.1 m^2 with a heated volume of 1356.6 m^3, in the middle of this row of houses. It has a flat thermally insulated roof (30 mm expanded polystyrene), rendered thermally insulated (20 mm expanded polystyrene) brick walls, and double-glazed windows. It has a well-maintained constant temperature non-condensing natural gas central boiler with distribution pipes within the envelope, thermally insulated with a water storage tank combined with solar collectors that covers 60 % of domestic hot water. The thermal characteristics of the building envelope of the TFH03 are calculated as shown on Table 49.3.

The resulting heat losses for the third TFH during the heating period are mainly through the wall at (9.1 kWh/m^2a). Further heat losses are through the floor slab at (7.1 kWh/m^2a) and through ventilation (5.6 kWh/m^2a). Results indicate that the required energy for space heating is (25 kWh/m^2a), as shown on graph 5 (Fig. 49.6).

The results, as calculated for the existing conditions of the third TFH are shown on Fig. 49.7. Primary energy consumption is estimated on 59.3 kWh/m^2a, the total thermal energy used is 25 kWh/m^2a, (44.6 l for fuel (natural gas) consumption), and electricity use is 4.8 kWh/m^2a. The annual operating cost is estimated at 4.5 €/m^2a. The total annual CO_2 emissions reach 0.01 t.

Through comparative studies of the above results, it is clearly shown that the primary energy consumption of the first TFH is the highest followed by the second TFH. Although the first and second houses have very similar heat transfer characteristics (the *U* value of the building elements are almost the same), the house which

Table 49.3 Thermal characteristics of the building envelope for TFH03

Construction element	*U* value W/(m^2k)
Flat concrete roof slab with 30 mm (expanded polystyrene) thermal insulation	0.77
External walls 20 mm (expanded polystyrene) thermal insulation	0.828
Floor in contact with the ground	0.91
Double-glazed windows	3.2

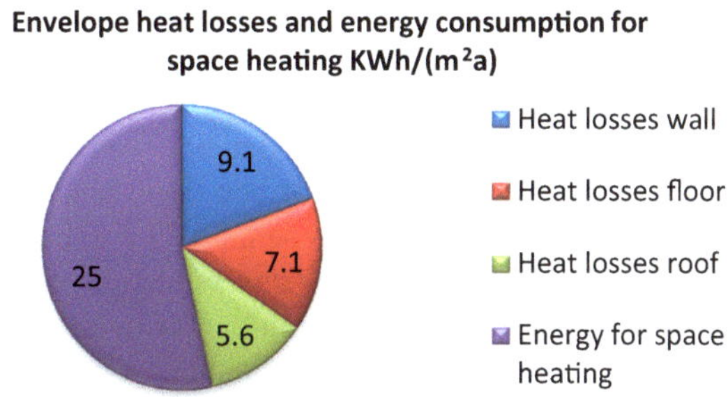

Fig. 49.6 Heat losses and space heating requirements, kWh/m^{2}a for TFH03

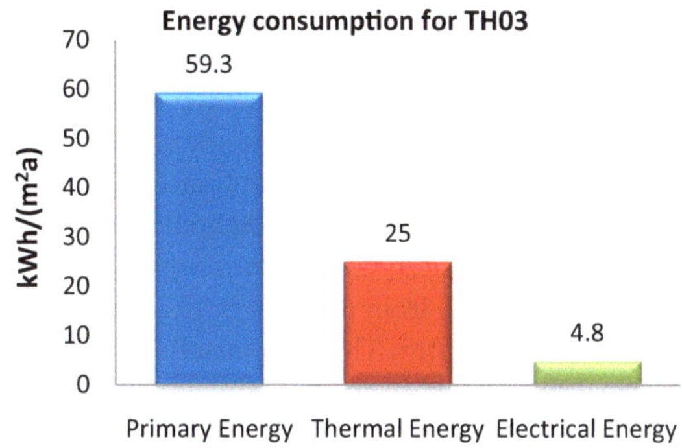

Fig. 49.7 Energy consumption for TFH03, kWh/m^{2}a

is located on the mountains has significantly higher energy consumption, because of its location and therefore the much colder winters, which demand higher energy consumption for heating; also energy demanding is its old-dated heating system without any solar collectors. In addition, the second TFH has a newer more efficient heating system, than the first one and has solar collectors installed for domestic hot water. The energy consumption for both houses is significantly higher than the one built according to the European directive, in the latter chronological period (TFH03). This is mainly due to the lack of thermal insulation on the two houses and therefore due to the increased heat losses through their building envelope and also their old-dated heating systems.

49.3.2 Refurbishment Scenarios

Two retrofitting scenarios were adopted for the simulations; the energy conservation measures (ECMs) are based on the requirements of the EPBD Directive [9], applied on the two houses built before the enforcement of the European directive and a more advance one applied on the house built afterward.

In the first, TFH, 30 mm of thermal insulation (expanded polystyrene) is added externally on the walls and 50 mm internally on the horizontal ceiling. Further retrofitting simulations were carried out; concerning the replacement of the single-glazed timber frame windows with doubled-glazed metal frame ones. To minimize the thermal energy used for heating and the hot water system, the condensing oil boiler is thermally insulated and maintained and the existing hot water system was replaced with a more efficient one (with natural circulation and electric element, thermally insulated single pipeline). The Tabula.xls tool calculated the potential energy savings and concluded on a reduction of 69.5 % in primary energy consumption from 349.2 to 106.4 kWh/m^{2}a, 69.4 % on CO_2 emissions, from 0.085 to 0.026 t and 71.1 % saving on the operational cost from 49.5 to 14.3 €/m^2. The total installation and construction cost for this refurbishment scenario reaches an approximate of 6380.00 €. This cost can be offset from the annual savings of the operating cost in 1.5 years.

The same concept is applied for the second TFH, 30 mm of thermal insulation (expanded polystyrene) is added externally on the walls and 50 mm internally on the horizontal ceiling of the tiled pitched roof. Furthermore, all the single-glazed windows were replaced with double-glazed and the existing heating system remained the same with the only intervention of thermal insulation and maintenance. The Tabula.xls tool calculated the potential energy savings and concluded on a reduction of 40.32 % in primary energy consumption from 168.4 to 100.5 kWh/m^{2}a, 46.37 % on CO_2 emissions from 0.041 to 0.022 t and 60.08 % saving on the operational cost from 23.3 to 14 €/m^2. The total installation and construction cost for this refurbishment scenario reaches an approximate of 7790.00 €. This cost can be offset from the annual savings of the operating cost in 7 years.

The TFH01 has greater savings and lower payback period than the TFH02. This is attributed to the mountainous location of the TFH01 and therefore to the greater effectiveness of the refurbishment employed measures, which focus on the heating aspects of the houses. Furthermore, the TFH01 has the smallest wall, roof, and window surface area than the other two sample houses, thus it has the lowest refurbishment construction cost.

The third TFH was refurbished with stricter ECMs, since the European Union is continuously striving toward better energy performance of buildings. Therefore, 20 mm thermal insulation was added on the existing thermally insulated external walls (amounting to a total of 40 mm expanded polystyrene) and an extra 20 mm of thermal insulation (expanded polystyrene) was applied on the roof slab, amounting to a total of 50 mm insulation. In addition, all the double-glazed aluminum frame windows were replaced with low-e doubled-glazed aluminum frame windows. The heating system was new, efficient, and in a very good condition, hence no further measures were taken. The Tabula.xls tool calculated the potential energy savings and concluded on a reduction of 0.84 % in primary energy consumption from 59.8 to 59.3 kWh/m^{2}a, 1.83 % on CO_2 emissions, from 0.0109 to 0.00107 t and 2.22 % saving on the operational cost from 4.5 to 4.4 €/m^2. This simple refurbishment scenario has very little impact on primary energy demand, CO_2 emissions, and operational cost, due to the fact that the building envelope is already thermally insulated with an efficient heat supply system, with very small margins for savings. Furthermore, this third TFH has almost double wall, floor, and window area of the other two dwellings. Additionally, the windows were replaced with low-e double-glazed, which are far more expensive than the conventional double-glazed ones; therefore, it has the highest total refurbishment cost, approximately 10355.75 €. The high retrofitting cost and the low savings, make the conservation measures non-cost effective. The low amount of energy savings is attributed to the fact that there was less to save on an already insulated house and the extra insulation cost of the added thickness of insulation balances out the resulting savings and so concludes to very high payback period [10].

Even though the interventions for these refurbishment scenarios were limited, the results indicate clearly the effectiveness of the insulation, the type of heating system and solar collectors contribution toward domestic hot water have great impact toward primary energy demand for space heating and a significant reduction to the annual operating cost for the TFH. As it is shown in the Fig. 49.8, the primary energy demand for all three representative buildings has a significant decrease, 69.5 % on the first, and 40.32 % on the second, and 0.84 % on the third TFH.

49.4 Conclusions

The study was performed to model the energy performance on the TFH national typologies. Various energy-efficient retrofitting scenarios simulations were performed to upgrade their energy performance.

The results of this chapter indicate that the application of thermal insulation and the proper thermal protection of the building envelope could have a significant impact on the buildings. By introducing simple retrofitting interventions, such as external thermal insulation and double glazing, and a basic upgrade of the existing heat supply systems, the first TFH achieved a reduction of 69.5 % in primary energy demand, 69.4 % on CO_2 emissions, and had 71.11 % saving on the operational cost. With the same interventions, the second TFH had a calculated decrease of 40.32 % in primary energy demand, 46.37 % on CO_2 emissions, with a saving of 60.08 %

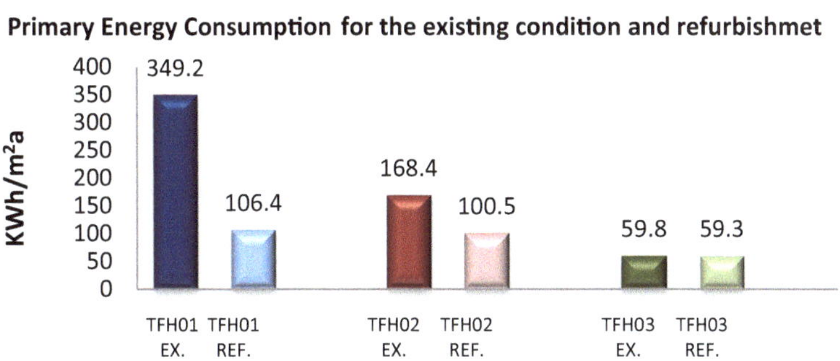

Fig. 49.8 Calculated primary energy consumption for the existing (ex.) condition and the refurbishment (ref.) for TFH01, TFH02, and TFH03

on the operational cost. The application of the ECMs on the third TFH had a slight decrease on the primary energy demand by 0.084, 1.83 % on the CO_2 emissions, and 2.22 % on the operational costs. It is obvious from the results of the examples, TFH constructed before 1981 and TFH constructed between 1981 and 2006, that the upgrading of the heating supply system is also important to further reduce the operating cost and the CO_2 emissions. It is evident that these targeted retrofitting scenarios have greater benefit on the older generation houses, which were built before 2007. This is reinforced by the payback period for each TFH; the first one equalizes the refurbishment expenses in 1.5 years and the second in 7 years. It was also clear from the results that the added thickness of insulation on the envelope of the already insulated TFH does not incur significant energy savings.

The results emphasize the poor energy performance of the existing national residential building stock, especially concerning the dwellings that were built before the enforcement of the European directives, in contrast with the third TFH, which conforms to the directive. The urgency for immediate actions and simple retrofitting guidelines to improve the performance of the buildings' envelope and their energy consumption is confirmed.

References

1. NATO Science Series (2005) Thermal energy storage for sustainable energy fundamentals, case studies and design. Volume 234 pp. Preface ix
2. Meijer F, Itard L, Sunikka-Blank M (2009) Comparing European residential building stocks: performance renovation. Build Res Inf 533–551
3. Konstantinou T, Knaack U (2013) An approach to integrate energy efficiency upgrade into refurbishment design process, applied in two case study buildings in Northern European climate. Energy Build 59:301–309
4. Konstantinou T, Knaack U (2011) Refurbishment of residential buildings: a design approach to energy-efficiency upgrades. Procedia Eng 21:666–675
5. IEE EPISCOPE Project: http://episcope.eu/index.php?id=97
6. Typology of the building stock in Cyprus Statistical Service of Cyprus published May 2012
7. Loga T (2013) Introduction to the excel workbook tabula.xlsm. Tabula Project Team, Darmstadt. http://www.building-typology.eu/downloads/intern/development/TABULA-XLS_Introduction.pdf. Accessed 20 April 2014
8. Dascalaki EG, Droutsa KG, Balaras CA, Kontoyiannidis S (2011) Building typologies as a tool for assessing the energy performance of residential buildings—a case for Hellenic building stock. Energy Build 43:3400–3409
9. Directive 2002/91/EC http://eur-lex.europa.eu/legal-content/EN/TXT/?uri=CELEX:32002L0091
10. Serghides DK, Georgakis CG (2012) The building envelope of Mediterranean houses: optimization of mass and insulation. Build Phys 36(I):83–98

Chapter 50
Low-Energy Architecture: From Theory to Design

Despina K. Serghides

Abstract This introduction projects the potential application of bioclimatic design interwoven with the objective of energy efficiency and the utilization of renewable energy resources for indoor comfort. This is, nowadays, of particular interest for all countries, since architecture is currently influenced beyond the usual functional, cultural and technological constraints by the incidence of rapid urban development and huge tourist inflows. This often results to standardized "international architecture" with high-technology services and catastrophic ecological and cultural consequences. Sadly, these international designs rely mostly on mechanical means, with high consumption of energy, ignoring the local tradition of climatic design, which respects the environment and reflects the thread of continuity, coherence and local flavour. Bioclimatic architecture considers the building totally from the stage of its inception as a place of energy exchange between the indoor and the external environment, natural and climatic. It considers the building as a living organism; a dynamic structure which utilizes the beneficial climatic parameters (solar radiation for winter, sea breezes for summer) whilst avoiding the adverse climatic effects (cold winds for winter, solar radiation for summer).

It is precisely the approach to this challenge of bioclimatic, energy-efficient architecture, posed to every designer as the coordinator of multiple considerations that is to be presented and discussed during this introduction. The main heating and cooling strategies are outlined and their adoption in the design process is approached in four stages from site planning, orientation and shape, layout and envelope of the building. These aspects are illustrated with the "Bioclimatic Designs for the Student Housing of the New University Campus of Cyprus", for which the author was the bioclimatic consultant, and the first phase of buildings for the University of Cyprus (Architect A. Kyprianou & Associates), to indicate how bioclimatic techniques address the problems of thermal and optical control.

Keywords Low-energy architecture · Bioclimatic strategies · Climate moderation · Optimized design · Integrated approach

D. K. Serghides (✉)
Department of Environmental Science and Technology,
Cyprus University of Technology, Anexartisias 33, 3036 Limassol, Cyprus
e-mail: despina.serghides@cut.ac.cy

A. Sayigh (ed.), *Renewable Energy in the Service of Mankind Vol I*,
DOI 10.1007/978-3-319-17777-9_50

50.1 Introduction

Through the highly complex set of design criteria of conventional practice, the basic objective of low-energy architecture is to derive recommendations for the efficient utilization of energy in buildings. For the attainment of this aim and to assess the energy demands for heating and cooling in buildings, primarily it is imperative to evaluate the free energy systems available to contribute to these requirements. For this, in the first stage, the comfort criteria and standards, through investigation of current thermostat settings and reviews of thermal comfort studies, must be established [1]. This is followed by a careful consideration and analysis of the prevailing local climatic conditions. Through bioclimatic chart analysis, it is then possible to outline the appropriate architectural strategies that could result to indoors thermal comfort for that location during winter and summer.

Such bioclimatic analysis is important for preliminary architectural designs. The selection of the appropriate design strategies, derived from a bioclimatic analysis, compatible with each other and other architectural aspects, could considerably reduce the cost of a building by minimizing the mechanical means for cooling and heating [2].

50.2 Siting of the Building for Bioclimatic Design

It is natural that different sites present totally different constraints and opportunities. Therefore, prior to the siting of the building a study of the site must be carried out, both under existing built conditions and taking into account extensions and development which may be governed by regulations to determine the optimum solar gain in winter. The microclimate, the predominant wind patterns, particularly the wind direction and intensity and the solar radiation are affected by the topography of a location, the presence of water, the vegetation and the man-made features. Appropriate landscape may offer buffering to the cold winter winds shading and generally cooling effect on the building by channelling the cool summer breezes and causing reduction of air and ground temperature (Fig. 50.1) [3].

50.3 Shape, Volume and Orientation

For the optimal seasonal performance of the building, its shape, volume and orientation are determined by the strategies which must be adopted for the local climate. The main aim of the strategies for winter is to maximize solar gains and minimize heat losses. Reversely, in the summer the aim is to minimize solar gains and maximize heat losses (Fig. 50.2) [4].

The strategies to achieve this in winter regarding shape, volume and orientation could be outlined as follows:

Fig. 50.1 Courtyard defined by circular wall with wetted rubble stone columns for summer evaporative cooling and enhancement of Western breezes. In winter, the courtyard becomes a pleasant sunny place, warmed by sun and mass

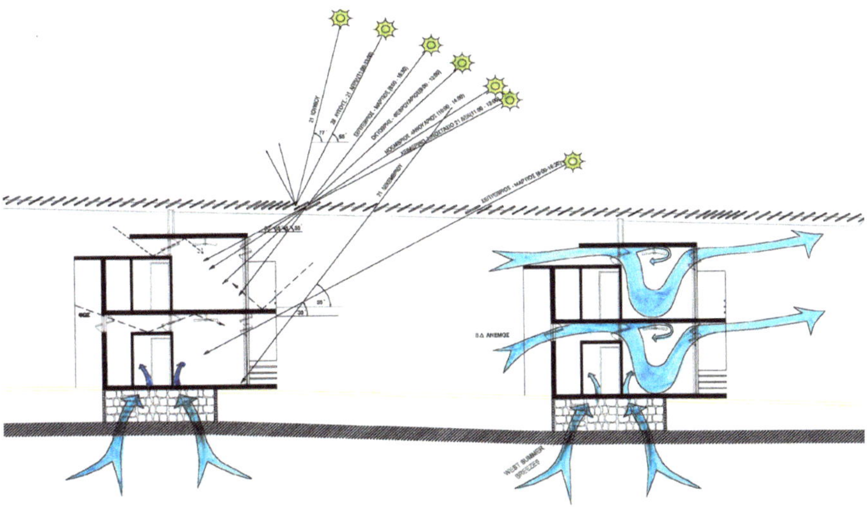

Fig. 50.2 Section views showing, in winter, the sun penetration in the centre of the day through the properly angled canopy slats. In summer, direct sun penetration is prevented by the slat angle and spacing. The narrow shape and placement of openings promote natural ventilation. Evaporative cooling from the underfloor wetted rubble stones provides additional summer comfort

- To minimize the outside wall and roof areas (ratio of exterior surface to enclosed volume). For a given volume, the more compact the shape the less wasteful it is in losing heat.
- Elongating the building with long east–west axis to maximize exposure to winter sun. This maximizes exposure to winter sun and minimizes heat in summer.

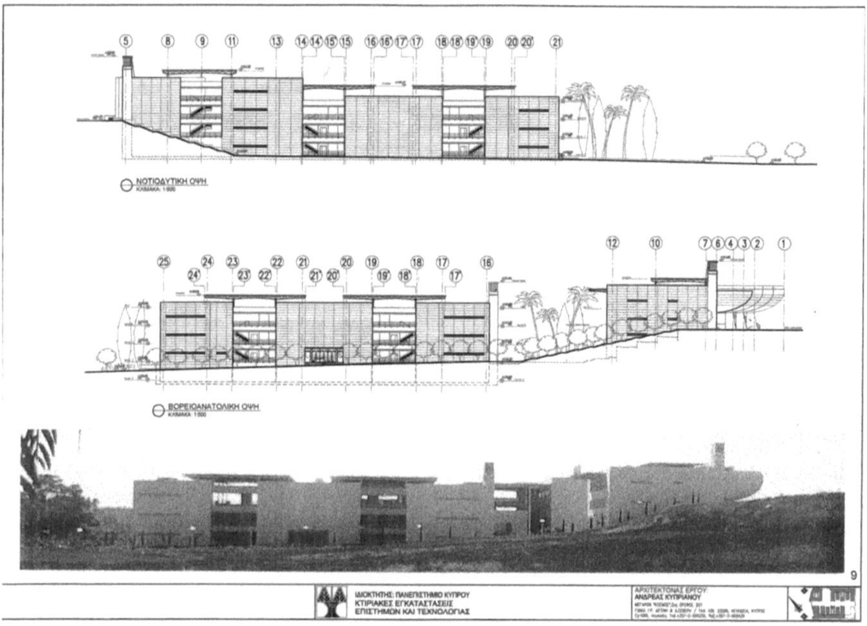

Fig. 50.3 Partitioning the interior of the building into different heating and cooling zones is significant in adjusting the indoor spaces to meet their daily living and thermal needs

- Streamlining the geometric form of the building to minimize winter wind turbulence. This could deflect air flow over the house avoiding air dam effect that its height would otherwise have.
- To recess structure below grade or raise existing grade for earth sheltering effect. This eliminates infiltration and reduces conduction–convection losses.
- Providing outdoor semi-protected areas.
- Recessing structure below grade to avoid direct exposure to the sun and provide a source of cooling.
- Building simulation studies for Mediterranean houses regarding the building shape explicitly indicate that a change of one parameter can frequently be compensated for by changes in the others, thus emphasizing the vital importance of an integrated design approach [5].

For the summer, exposure to the sun must be minimized and utilization of the summer breezes must be maximized with the following strategies:

- The proportions of the plan should be selected to equalize the sum of gains at north and south elevations with those received at the east and west facades. This minimizes exposure to summer sun.
- Using split levels, high building facades and orientation of the longer ones perpendicular to prevailing winds enhances ventilation.

Outdoor semi-protected areas must be provided for a year-round climate moderation (Figs. 50.3 and 50.4).

Fig. 50.4 The elongated shape of the structures, with an east–west long axis, allows favourable solar orientation to all student living units and common spaces, in all buildings. Shown is the unobstructed solar access, sunrise to sunset, from the beginning of October to mid-March

50.4 Building Indoor Planning

The orientation, organization and location of the indoor spaces are complex [6]. Besides their thermal requirements, a large number of various interrelated factors (safety, privacy, function, view, etc.) must be considered for the appropriate choice of the location of the spaces and their openings.

In approaching the indoor planning of the building five major strategies are identified:

- Zoning the indoor building volume
- Orienting indoor spaces for energy efficiency
- Locating and organizing climatic buffers
- Planning for heat recovery, distribution and storage
- Planning for airflow and cooling

A linear layout with the long axis stretching east–west in one room depth will offer an optimum south orientation for most of the rooms which presents interesting solar and architectural possibilities. In addition, the building will enjoy daylight from two sides and cross ventilation [7].

Nevertheless, when orienting the rooms for favourable heating in winter and cooling in summer, in addition to the sun path the following must be considered:

- The heating and cooling requirements
- The levels of the internal gains
- The size exposure and function of the room

The height and the number of floors are determining factors for the optimal, vertical thermal distribution. Two-storey spaces and tall interior spaces create pressure differentials and cause vertical air movement.

50.5 The Building Envelope

The main considerations for the design of the building envelope could be outlined as:

- The heat transfer through the building envelope aiming to restrict heat losses in winter and to enhance solar gains, whilst limiting them in the summer and promoting cooling with various techniques.
- The thermal capacity, which is an important property of the building envelope for energy conservation, since excess heat is stored in it and dissipated at a later stage when needed. In this way, the indoor temperature fluctuations are regulated and overheating is avoided [8].
- Control of moisture migration and vapour formation

External insulation is the most effective for hot and moderate climate. Yet, the insulation positioning depends also on the type of building and the air-conditioning used. For these climates, application of insulation on the roof is the most cost-effective energy saving design measure [9].

In the Mediterranean countries, which have mostly climates with large diurnal fluctuations, the aspects relating to the thermal mass are of particular significance for the energy conservation in buildings. The enhanced thermal capacity and hence its greater capacity of retaining within it the solar heat of the sunny winter days reduces temperature fluctuations and results to more heating energy savings. Also, the potential of the mass to retain the coolness of the summer nights contributes to the reduction of cooling energy load [9]. (Fig. 50.5).

50.6 The Occupants of the Buildings

It is possible to achieve indoors comfort conditions by many different combinations of optimized and effective variables in the building design, such as compact shape and the optimization of insulation, mass and fenestration designs. However, peoples' responses are a very important consideration in the creation of indoor comfort conditions, but the occupants' designed comfort level is not an input factor which may be manipulated. Moreover, a house is not merely a container in which people act like robots and are placed to receive its thermal effects. There is a dynamic dialogue between building controls and building use. Furthermore, it is necessary that some of the employed passive systems must be activated by the users to be effective [10].

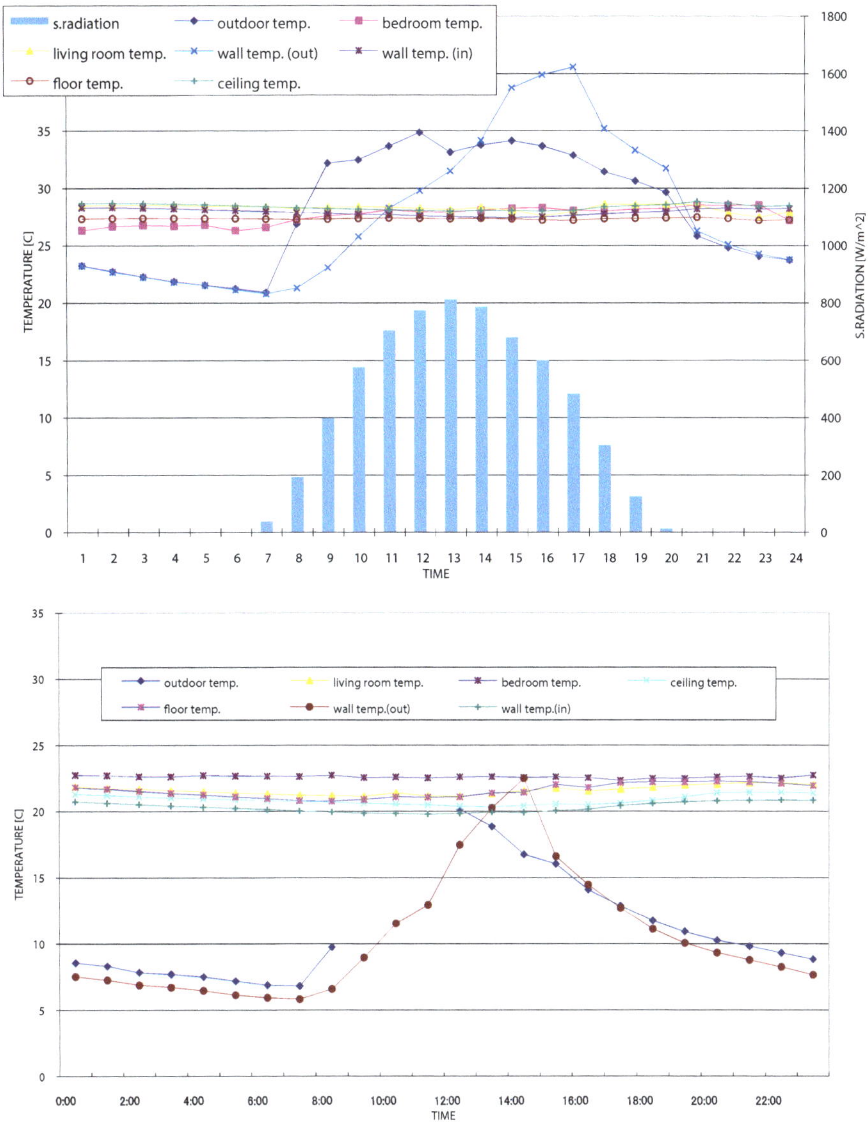

Fig. 50.5 Well-insulated house: average temperature measurement results for August 1997

50.7 Conclusions

It is concluded that during the optimization of the regulatory building systems, to achieve its fine tuning and become a successful climatic moderator, evaluation of the building performance and ultimately analysis of cost-effectiveness, necessitates detailed and at the same time robust, dynamic and interactive design approach.

This is nowadays possible with the use of computer analogues, a well-established practice, but at the same time on continuous development. No doubt the potential of bioclimatic design is dependent on a multidisciplinary design approach. However, the thought that the buildings could be permanent energy savers, demands building designers to consider carefully the practical options available to them in an integrated mode.

References

1. Serghides DK (1996) Prototype solar house for cyprus. EuroSun '96 Internationales Sonnenforum 10:1128–1130
2. Serghides DK (1993) Zero energy of the Mediterranean houses. ISES '93 solar world congress. Solar Archit 7(2):48–256
3. Serghides DK (2010) Mediterranean traditional architecture versus contemporary architecture—the energy challenge—the open construction & building technology. Journal 4:29–38. ISSN: 1874-8368
4. Serghides DK (2011) Low energy design solutions for the Mediterranean area. Contr. In: TEENERGY Province of Lucca (ed) High energy efficient schools in the Mediterranean area TEENERGY schools (pp. 80–83). Tipografia San Marco Litotipo
5. Serghides DK (1997) Comparative assessment of shape variation for Mediterranean houses. ISES '97 solar world congress proceedings. Taejon, Korea, KSES, D4, p 300
6. Serghides DK (2008) Optimisation of fenestration shading for low energy building design-Contr. In: "Renewable energy innovative technologies and new ideas" Nr Zamowienia 428/08 pp. 331–339
7. Serghides DK (2010) The dynamics of the building envelope—Fenestration, the human factor, indoor comfort and energy efficiency Session 15: progress on ventilation, indoor air quality and lighting, PALENC2010 International Conference proceedings "Cooling the Cities". September 28–October 1. Rhodes, Greece
8. Serghides DK (2007) Low energy building design—the effectiveness of mass. Proceedings of 2nd PALENC Conference and 28th AIVC Conference on Building Low Energy Cooling and Advanced Ventilation Technologies in the 21st Century September 2007, Crete Island, Greece 1:936–940
9. Serghides DK (2009) Optimisation of insulation on Mediterranean houses. ICPSR J 5(8):79–83
10. Serghides DK (2008) Integrated design for the zero energy house and the human factor–PLEA towards zero energy building PLEA 2008. ISBN: 78-1-905254-34-7, Paper ID. 745

Chapter 51
Development of a Luminous Efficacy Model Using Ground and Satellite-Based Data from the Tropics

Rungrat Wattan and Serm Janjai

Abstract In this work, a global luminous efficacy model was developed. The formulation of the model was based on global illuminance and global irradiance measured at four tropical sites in Thailand. The model expresses luminous efficacy as an empirical function of aerosol optical depth, precipitable water, satellite-derived cloud index and cosine of solar zenith angle. The aerosol optical depth and precipitable water were obtained from Aerosol Robotic Network (AERONET) of NASA, and cloud index was derived from multifunctional transport satellite (MTSAT)-1R satellite. The model coefficients were calculated using a 4-year period of hourly data (2007–2010), and the model was validated against an independent data set for 2011. The model predicted well the global illuminance with a root mean square difference (RMSD) of 3.6 % and a mean bias difference (MBD) of 0.3 %. In addition, the model was compared favourably with most existing models when tested against this independent data set.

Keywords Daylight · Luminous efficacy · Model · Tropics

51.1 Introduction

Daylight-integrated buildings can bring substantial savings in lighting cost [1, 2]. To design such buildings at a location, information on daylight environment of that location should be known, and daylight illuminance is part of such information. Although illuminance can be obtained from daylight monitoring stations, the number of the stations where illuminance is routinely measured is sparse, especially in developing countries. As illuminance is part of solar spectrum, there is a relationship between illuminance and broadband solar irradiance. This relationship is generally called "luminous efficacy", and it can be used to estimate the illuminance from more widely acquired irradiance.

S. Janjai (✉) · R. Wattan
Department of Physics, Solar Energy Research Laboratory, Faculty of Science, Silpakorn University, Nakhon Pathom 73000, Thailand
e-mail: serm.janjai@gmail.com

A. Sayigh (ed.), *Renewable Energy in the Service of Mankind Vol I*,
DOI 10.1007/978-3-319-17777-9_51

A number of authors have proposed luminous efficacy models with a wide range of degree of complicity using illuminance and irradiance data collected from many parts of the world [3–10]. The simplest models express luminous efficacy as constant [11, 12] whereas some models are written as functions of atmospheric parameters [5–10], and these models predicted illuminance with different accuracies.

As the absorption by various atmospheric constituents in photopic band (380–770 nm) due to water vapour, aerosol and cloud is different from that in broadband (300–3000 nm), luminous efficacy depends on the air mass and radiative properties of these atmospheric constituents.

The objective of this work is to develop a global luminous efficacy model which includes atmospheric parameters derived from both ground- and satellite-based parameters as independent variables. The modelling process is based on illuminance, irradiance and atmospheric parameters taken at four tropical sites in Thailand.

51.2 Measurements and Data

51.2.1 Ground-Based Measurements

Radiation monitoring stations have been deployed in each of the main geographical regions of Thailand. Chiang Mai (18.78 °N, 98.98 °E) is the northernmost station which is situated in the mountain north and characterized by cool winter. The station located in Ubon Ratchathani (15.25 °N, 104.87 °E) represents the dry northeastern highland, followed by Nakhon Pathom (13.82 °N, 100.04 °E) in the lowland central region near Bangkok and Songkhla (7.20 °N, 100.60 °E) in the peninsular southern region with low pollution and longer period (8 months) of the rainy season.

At each station, the solar radiation in different wavelength bands is routinely monitored, including global irradiance and illuminance. For each station, a luxmeter (EKO, model ML-020S-O) is used to measure global illuminance on a horizontal surface. Global irradiance is measured using a Kipp and Zonen pyranometer (model CM21) at Chiang Mai, Ubon Ratchathani and Songkhla, and a Kipp and Zonen pyranometer (model CM11) at Nakhon Pathom.

All recorded data are sent to our laboratory at Silpakorn University in Nakhon Pathom by email once a month. At the laboratory, voltage data from the pyranometers and luxmeters are converted into solar irradiance and illuminance using the corresponding responsivities of these instruments. All irradiance and illuminance data are subjected to quality control process. The abnormal data and data which violate physical laws are discarded from the data sets. The pyranometers and luxmeters at the four stations are calibrated yearly using a travelling pyranometer and luxmeter recently calibrated at the manufacturers.

The four stations are equipped with Cimel sunphotometer (model CE318) which are used to measure the solar spectral radiance at wavelengths of 340, 380, 440, 500, 675, 870 and 1020 nm. These sunphotometers belong to our laboratory and form part of the Aerosol Robotic Network (AERONET) of NASA [13]. The spectral data from these sunphotometers are routinely processed by AERONET to provide aerosol optical properties and precipitable water. These parameters are used in the modelling of luminous efficacy. The sunphotometers are regularly calibrated by AERONET. All instruments are maintained by well-trained officers at the stations. The location of the four stations and the pictorial views of the station instruments are shown in Fig. 51.1.

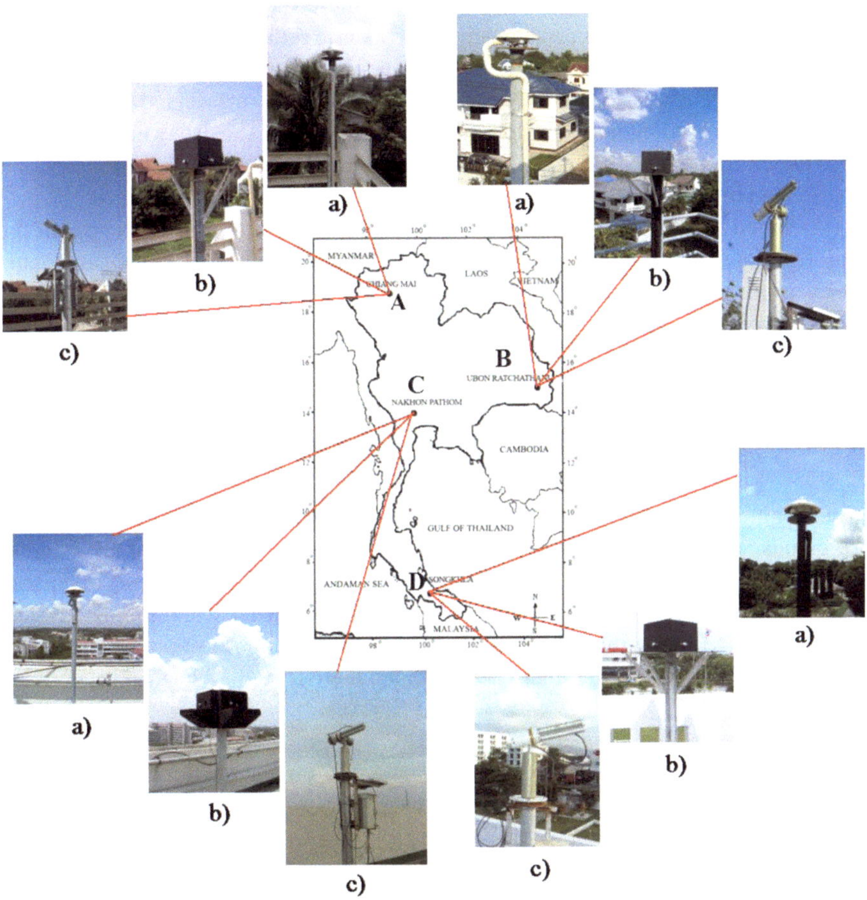

Fig. 51.1 Position of the four measuring stations and pictorial views of the equipment used in this work. **a** Pyranometer for measuring global irradiance, **b** luxmeter for measuring global illuminance and **c** sunphotometer for measuring aerosol optical depth and precipitable water. A, B, C and D indicate, respectively, the northern region, northeastern region, central region and the southern region of Thailand

51.2.2 Satellite-Based Data

Cloud information is obtained from the digital data retrieved from the visible channel of the multifunctional transport satellite (MTSAT)-1R satellite for the period 2007–2011. The data are displayed as images covering the entire area of Thailand, with a spatial resolution of 3×3 km^2 at the subsatellite point. These images are transformed to cylindrical projection and then navigated using the coastline as reference. Nine pixels centred at the position of each station are selected from the image and averaged to represent the grey level of these stations (Fig. 51.2). The grey level is further converted into the earth-atmospheric reflectivity (ρ_{EA}) and is used to derive the cloud index as explained in the next section.

The data sets consisted of 5 years (2007–2011) of both surface and satellite data, and they were used in model development and validation. The data were partitioned into two sections. The first one (2007–2010) was used for model development, and the second one (2011) for validation.

As stated in the Sect. 51.1, monthly averaging was done according to hour of the day from the hours of 8–16. This figure translates to a total of 108 measurements per year for each variable, or 540 for the entire 5-year study at each of the four stations.

51.3 Formulation of the Efficacy Models

The luminous efficacy models proposed in this work are based on the properties of clouds, aerosol and water vapour as important depletion agents of solar radiation, exhibiting different extinction in the photopic band (380–770 nm) compared to the broadband (300–3000 nm) region of the solar spectrum. Clouds tend to reflect evenly solar radiation throughout the solar spectrum, but absorption is confined to the near infrared part of the solar spectrum, and there is negligible absorption in the

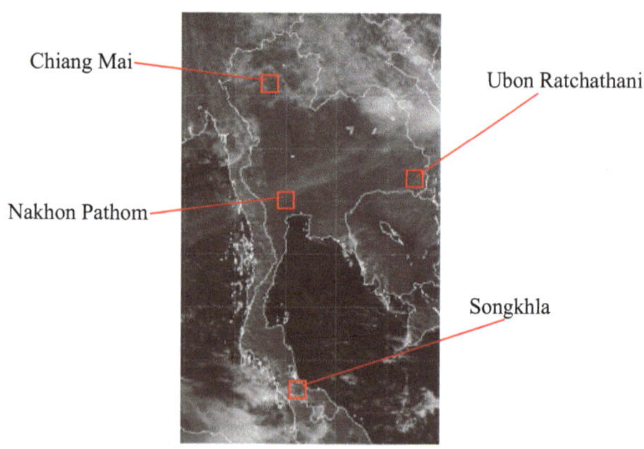

Fig. 51.2 Example of a satellite image and the position of the pixels selected for calculating cloud index

visible part [14, 15]. In this work, a satellite-derived cloud index is used to quantify the cloudiness, as it can be derived from satellite data, available for most parts of the world with good spatial and temporal resolutions. This cloud index is defined as [16]:

$$n = \frac{\rho_{EA} - \rho_G}{\rho_C - \rho_G} \tag{51.1}$$

where, ρ_{EA}, ρ_G and ρ_C are the reflectivity of the earth-atmospheric system, the reflectivity of the ground and the maximum reflectivity of cloud, respectively. The cloud index n is an indirect measure of cloudiness through the reflection properties of the ground, cloud and atmosphere. In the absence of cloud cover, ρ_{EA} approximates ρ_G and n is zero. By contrast, in a pixel which is totally covered by cloud, ρ_{EA} is equal to ρ_C and n is equal to one. For the case of partly cloudy pixel, the values of n will be in the interval: $0<n<1$.

As aerosol scatters more shortwave radiation of the visible photopic band than longer wave of near infrared radiation [17], aerosol load is considered to be a parameter in the proposed efficacy models. In this work, aerosol optical depth (AOD) at 500 nm is employed to represent the effect of the aerosol, as AOD is now available from both ground- and satellite-based measurements (e.g. AERONET and MODIS).

The amount of water vapour is generally high in the tropics due to high availability of surface water and high solar radiation. As water vapour has a low absorption in the photopic band as compared to the infrared parts of the solar spectrum, the amount of water vapour is included in the efficacy models. Precipitable water is used to quantify water vapour.

As the depletion of solar radiation by the above-mentioned atmospheric parameters depends on the path length of the atmospheric through which solar radiation travels, the cosine of the solar zenith angle is also included in the model.

Preliminary examination of the data revealed that each of the variables being modelled varied as a linear function of the independent variables. These observations prompted us to construct the regression model for global luminous efficacy in the following form:

$$K_g = a_0 + a_1 \cos\theta_z + a_2 w + a_3 \mathrm{AOD} + a_4 n \tag{51.2}$$

where K_g is the global luminous efficacy (lm/W), θ_z is the solar zenith angle (degrees), w is the precipitable water (cm), AOD is the aerosol optical depth, n is the cloud index, and a_0, a_1, a_2, a_3 and a_4 are the empirical coefficients.

To determine the empirical constants, Eq. (51.2) was fitted with the luminous efficacy and the data of AOD, w and n from Chiang Mai, Ubon Ratchathani, Nakhon Pathom and Songkhla for the period from 2007 to 2010 by using a non-linear multivariable regression technique. The results are: $a_0 = 127.097$, $a_1 = -5.093$, $a_2 = 1.100$, $a_3 = -2.226$ and $a_4 = 14.397$.

51.4 Validation of the Models

To investigate the model performance, the model was used to calculate global illuminance at four stations: Chiang Mai, Ubon Ratchathani, Nakhon Pathom and Songkhla by using values of AOD; *w* from AERONET; and *n* from MTSAT-1R satellite at these stations in 2011. As these data were not involved in the formulation of the model, they are an independent data set. The result is graphically shown in Fig. 51.2.

It was found that the proposed model predicted well the illuminance for all stations with root mean square difference (RMSD) of 3.6 % and mean bias difference (MBD) of 0.3 %.

The performance of the proposed model was also compared with that of models available in literatures [4–6], and the results are shown in Table 51.1. It is observed that the proposed model performance is better than most of the other models. This is likely due to the fact that the proposed model takes cloud, aerosols and water vapour into account, and these atmospheric constituents have important effects of solar radiation in this region, while most of other models do not include these effects (Fig. 51.3).

51.5 Conclusions

In this work, a global luminous efficacy has been modelled by using solar radiation and atmospheric data in the tropics. The model expressed the efficacy as a function of satellite-derived cloud index, aerosol optical depth, precipitable water and solar zenith angle. The performance of the model has been tested against independent

Table 51.1 Comparison of the performance of the proposed model for global illuminance and that of other models

Model	Chiang Mai		Ubon Ratchathani		Nakhon Pathom		Songkhla		Combined data from four stations	
	MBD (%)	RMSD (%)	MBD (%)	RMSD (%)	MBD (%)	RMSD (%)	MBD (%)	RMSD (%)	MBD (%)	RMSD (%)
Proposed model	−0.2	4.1	−0.7	2.7	2.0	3.9	0.7	3.4	0.3	3.6
Muneer and Kinghorn [4]	−10.4	12.2	−11.4	12.3	−9.2	9.9	−9.6	10.8	−10.1	11.3
Ruiz et al. [5]	−14.2	14.2	−13.0	14.2	−11.1	11.8	−10.8	12.3	−12.2	13.9
Perez et al. [6]	−13.2	15.2	−16.7	18.9	−15.5	17.0	−16.4	18.6	−13.7	16.7

MBD mean bias difference, *RMSD* root mean square difference

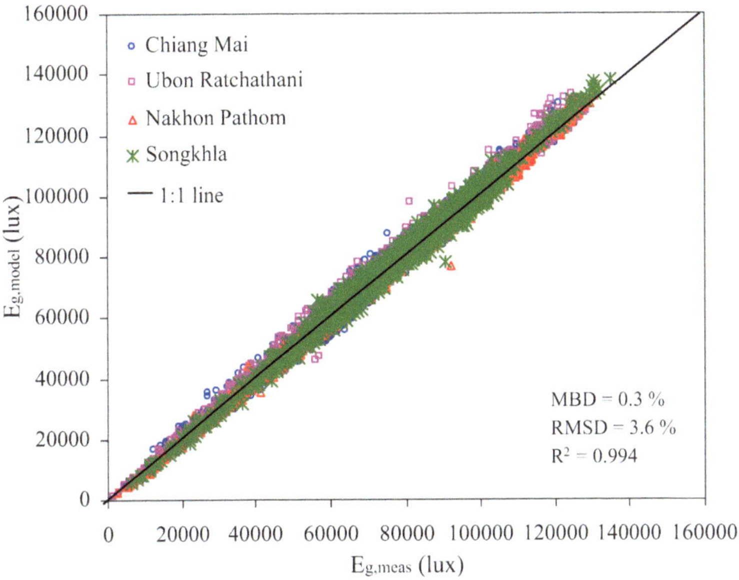

Fig. 51.3 Comparison between global illuminance calculated from the proposed model ($E_{g,model}$) and that obtained from the measurements ($E_{g,meas}$) for **a** Chiang Mai, **b** Ubon Ratchathani, **c** Nakhon Pathom and **d** Songkhla. *R* is the correlation coefficient

data sets. Global illuminance estimated from the model and those obtained from measurement are in good agreement. In addition, the model performs favourably, when compared with other existing models.

References

1. Chirarattananon S (2005) Building for energy efficiency. Asian Institute of Technology, Bankgok
2. Li DHW, Lam JC (2001) Evaluation of lighting performance in office buildings with daylighting controls. Energy Build 33:793–803
3. Tsikaloudaki K (2005) A study of luminous efficacy of global radiation under clear sky conditions in Athens, Greece. Renew Energy 30:551–563
4. Muneer T, Kinghorn D (1998) Luminous efficacy model: evaluation against UK data. J IES 27(1):163–170
5. Ruiz E, Soler A, Robledo L (2001) Assessment of muneer's luminous efficacy models in madrid and a proposal for new models based on his approach. J Solar Energy Eng 123:220–224
6. Perez R, Ineichen P, Seals R, Michalsky J, Stewart R (1990) Modeling daylight availability and irradiance components from direct and global irradiance. Solar Energy 44(5):271–289
7. Littlefair PJ (1988) Measurement of the luminous efficacy of daylight. Light Res Technol 20:177–188
8. Chung TM (1992) A study of luminous efficacy of daylight in Hong Kong. Energy Build 19:45–50

9. Kong HJ, Kim JT (2013) Modeling luminous efficacy of daylight for Yongin, South Korea. Energy Build 62:550–558
10. Robledo L, Soler A (2000) Luminous efficacy of global solar radiation for clear skies. Energy Convers Manage 41:1769–1779
11. De Rosa A, Ferraro V, Kaliakatsos D, Marinelli V (2008) Simplified correlations of global, direct and diffuse luminous efficacy on horizontal and vertical surfaces. Energy Build 40:1991–2001
12. Cucumo M, De Rosa A, Ferraro V, Kaliakatsos D, Marinelli V (2008) Correlations of global and diffuse solar luminous efficacy for all sky conditions and comparisons with experimental data of five localities. Renew Energy 33:2036–2047
13. Holben BN, Eck TF, Slutsker I, Tanre D, Buis JP, Setzer A, Vermote E, Reagan JA, Kaufman YJ, Nakajima T, Lavenu F, Jankowiak I, Smirnov A (1998) AERONET-A federated instrument network and data archive for aerosol characterization. Remote Sens Environ 66:1–16
14. Paltridge GW, Platt CMR (1976) Radiative process in meteorology and climatology. Elsevier, New York
15. Weymouth G, Le Marshall JF (2001) Estimation of daily surface solar exposure using GMS-5 stretched-VISSR observations: the system and basic results. Aust Meteorol Mag 50: 263–278
16. Cano D, Monget JM, Albuisson M, Gurnard H, Regas N, Wald LA (1986) A method for the determination of the global solar radiation from meteorological satellite data. Solar Energy 37:31–39
17. Iqbal M (1983) An introduction to solar radiation. Academic Press, New York

Chapter 52
Investigation on the Existing Circumstances and Contributing Factors of Thermal Environment of Rural Housings in Severe Cold Zones of China in Winter

Hong Jin and Kai Chen

Abstract In order to improve the dwelling conditions and living standards of peasants in severe cold zones of China, a comprehensive and in-depth survey is taken on the residential buildings in severe cold zones. Through a combination of field survey and questionnaire, a number of investigations are conducted including the existing status of residential buildings in terms of house types, spatial arrangements, heating facility, construction levels, material and structure system, fuel consumption, and the utilization of local renewable resources as well as the economic conditions, demographic structures, and indoor thermal perceptions of peasants. The indoor thermal environment of residential buildings in winter is also measured. By investigating and summarizing the existing circumstances of thermal environment in rural housings in winter, as well as analyzing the main factors contributing to the poor indoor thermal environment, adaptive approaches are explored to lay foundations for further in-depth investigation and construction of the energy-efficient green residential buildings.

Keywords Severe cold zone · Rural housing · Thermal environment in winter · Investigation and analysis

52.1 Introduction

China, a major agricultural country, has a 900 million rural population among the total 1.3 billion and a rural construction area as high as 24 billion m^2. However, energy shortage, low fuel utilization rate and poor living environment quality are very common issues in the existing rural living conditions of China, especially in severe cold zones, where the outdoor air temperature in winter can be as low as −20

H. Jin (✉) · K. Chen
School of Architecture, Harbin Institute of Technology, NO.66 Xidazhi Street, Nangang District, 150001 Harbin, China
e-mail: jinhong1963@qq.com

A. Sayigh (ed.), *Renewable Energy in the Service of Mankind Vol I*,
DOI 10.1007/978-3-319-17777-9_52

to −30 °C, heating period in the range of 131–210 days, and degree days of heating period as high as 2568–6825(°C·days). Residences suffer a lot from the extreme cold conditions. Therefore, investigating existing circumstances of thermal environment in rural housings in severe cold zones, analyzing the contributing factors and then exploring the possible adaptive approaches have great significances in terms of improving the living quality of peasants in severe zones, enhancing their living standards, and creating a harmonious society.

52.2 Investigations on Thermal Environment of Rural Housings in Severe Cold Zones in Winter

Through a combination of field survey and questionnaire, a number of investigations were conducted, including the existing status of residential buildings in terms of house types, spatial arrangements, heating facility, construction levels, material and structure system, fuel consumption, the utilization of local renewable resources, the economic conditions, demographic structures, and indoor thermal perceptions of peasants. The indoor thermal environment of residential buildings in winter was also measured.

52.2.1 Measurements of Indoor Temperature and Humidity

Team members of the project went into the countryside in the coldest month of the year (January) in order to conduct first-hand field surveys and measurements on the typical residential buildings in severe cold zones. They experienced the living condition and indoor thermal environment of peasants personally, while taking tests of the indoor thermal environment. Figure 52.1 shows the test results of temperature and humidity of three typical rural housing systems in the Heilongjiang province, the most northern province of China.

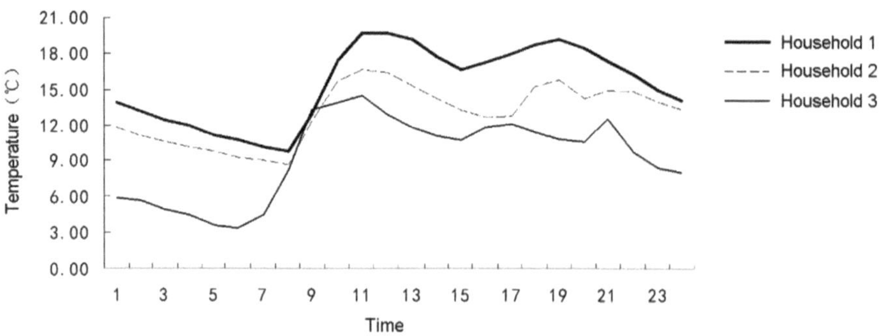

Fig. 52.1 The hourly variation curve of temperature in the bedroom of a typical rural housing system

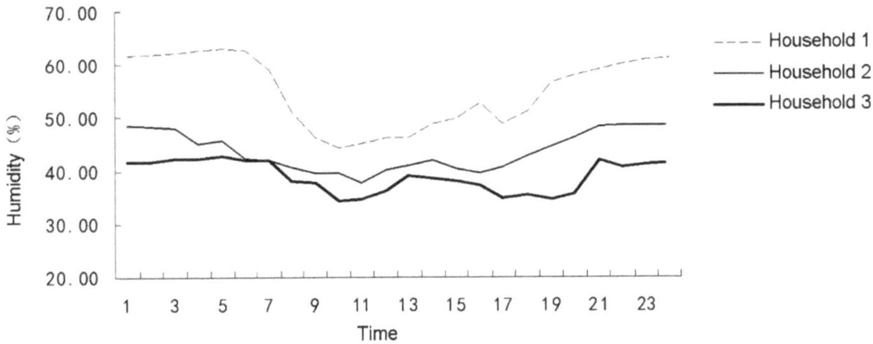

Fig. 52.2 The hourly variation curve of humidity in the bedroom of a typical rural housing system

It can be seen from Figs. 52.1 and 52.2 that the fluctuations of indoor temperature of rural housings in severe cold zones in winter are relatively great. The differences between the maximum daytime temperature and minimum night temperature can be as large as 10 °C; there are also great variations among households with different economic conditions, which means the richer the house owner is, the higher indoor temperature could achieve. The mean indoor air temperatures of bedrooms among the three households are 13.01, 9.37 and 15.43 °C, respectively, and the maximum difference among them can be as high as 6.06 °C. With relatively low indoor temperatures, the users of household 1 and 2 feel quite cold, while the user of household 3, who has a comparatively fine economic condition, feels more comfortable with a higher indoor temperature, but may still feel a bit cold at night.

In terms of humidity, it has been shown by researches that within the comfort temperature range, the effects of humidity on thermal perception of humans are relatively small, and the human body almost could not feel the humidity change when it fluctuates between 20 and 70 %. The humidity range of bedroom of rural housings in severe cold zones in winter, according to a number of investigations, is normally 30–65 % (Fig. 52.2), while it varies even more significantly in the kitchen, with the humidity as high as 80 % sometimes. High level of humidity is one of the main causes of the condensation of the inner surface of the exterior wall.

52.2.2 Subjective Survey of Thermal Environment

The investigation on the perspective comfortable temperature of occupiers is conducted in the representative dwellings in severe cold zones, according to which, the majority of subjects are not satisfied with the indoor thermal environment, among which 50.38 % of the peasants feel relatively cold (Fig. 52.3). It is also shown that people of the same age group tend to have the same perception on thermal comfort, within the same thermal environment. The perception on thermal comfort is quantified as temperature interval, which is shown in Table 52.1, from which it can be seen that the range of thermal comfort among peasants in winter is 15–17 °C, which

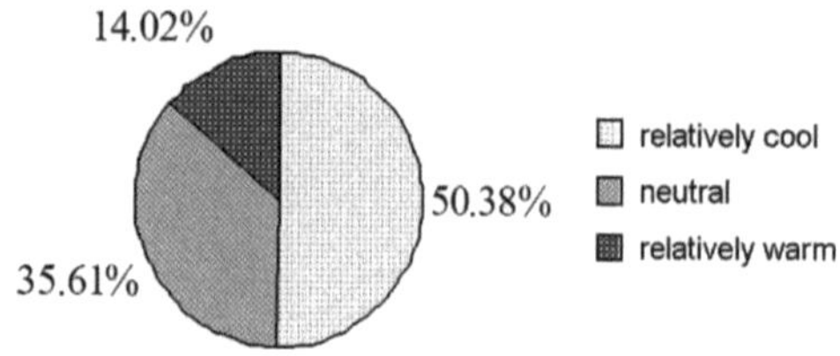

Fig. 52.3 Perceptions of peasants towards thermal environment of bedrooms in severe cold zones in winter

is lower than that of city dwellings. The main factors contributing to this can be concluded as follows: (1) unlike the citizens, northern peasants have their specific living modes, they would like to drop around and visit neighborhoods in slack season, as well as go out to the village-type dry toilet, or fetch something from outside to treat visitors. Considering as an easy access, the occupiers normally wear more clothes at home. Such living modes together with heavy clothes lead to the reducing requirement of human body in terms of thermally comfortable temperature. (2) Peasants in severe cold zones usually like to gather on the "kang" in winter to do housework or chat. Kang, as a major place of indoor activities for peasants in winter, provides a space for dining, studying, laboring, meeting, and sleeping of the whole family and visitors. Normally, people tend to go to kang as soon as they enter the room, and accordingly there are some specific furniture to match the kang, such as kang table, kang bench, and kang cabinet, etc. The temperature of the surface of kang, according to tests, can be maintained between 33 and 45°C all the time in spite of the –30°C outdoor temperature, and thus forming a comfortable microclimate zone around it [1]. Therefore, although the indoor temperature in winter is rather low and fluctuates constantly, the kang zone is always kept in a relatively comfortable warm status.

52.2.3 Existing Circumstances of Heating Facilities in Rural Constructions

The main heating facilities of rural areas in severe cold zones, according to a number of investigations, can be summarized as heated kang, hot wall, the ground kang, and homemade heating (Fig. 52.4), and they are usually used associated with each other, such as heated kang + heated wall, heated kang + homemade heating, etc. Among the two, heated kang is commonly used in the rural housing system in severe cold

Table 52.1 Inquisition on the perceptive comfortable temperature of peasants in severe cold zones

Age group	Proportion(%)	Comfortable temperature(°C)
Children	23	14–17
Young and middle-aged people	58	15–16
Aged people	19	15–17

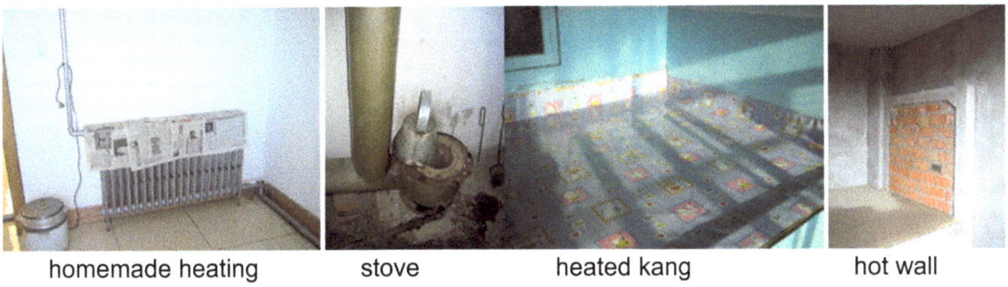

Fig. 52.4 Heating facilities of rural housings in severe cold zones

zones, accounting for 90 % out of all the main heating facilities in the households investigated. The surface of kang as well as indoor air is normally heated by the waste heat of cooking, which provides energy for both cooking and warming, but it needs reheating at night during the coldest period in winter. Such heating facility has fairly high thermal efficiency and energy utilization; however, the fluctuations of indoor temperature are quite wide.

52.3 Analysis of the Contributing Factors

Based on a comprehensive analysis of the survey data, the main factors contributing to the poor indoor thermal environment of rural housings in severe cold zones in winter can be listed as follows:

1. The overall design and layout of residential buildings are unreasonable, and the shape coefficient is too large

Shape coefficient is the most significant factor accounting for building energy consumption, which is defined as the ratio of exterior surface area between building and outdoor air to the volume of the building, it is therefore obvious that a higher shape coefficient means larger heat loss area per spatial unit and greater energy consumption. It is shown by investigations that under the climate circumstances of severe cold zones, based on the shape coefficient of 0.3, the energy consumption increases about 2.4–2.8 % with the shape coefficient rising 0.01, and decreases around 2.3–3 % with shape coefficient dropping to 0.01 [2]. Therefore, the Design Standard for Energy Efficiency of Residential buildings in Severe Cold and Cold Zones (JGJ26-2010) states that the shape coefficient of residential buildings less than three stories in severe cold zones of China should be confined within 0.50. Most of the rural housings in severe cold zones of China, according to numerous surveys conducted by the author, are single-story detached dwellings (Fig. 52.5), with the built-up area fluctuating between 60 and 180 m^2 and shape coefficient between 0.70 and 0.88 by calculation (Fig. 52.6). It is therefore obvious that the energy consumption of rural housings is much greater than that of the urban constructions.

Fig. 52.5 Full view of a certain village in severe cold zones

2. Most of the residential buildings still continue to adopt red bricks as good materials, without any thermal insulation measurements

It is shown by investigations that residential buildings whose walls are made of clay bricks without thermal insulation measurements occupy 83.1 % of the total rural households investigated in severe cold zones. The heat transfer coefficient of such a wall, according to a number of tests, is merely 0.833 w/m^2·K, which is far higher than the regulations of Design Standards of Energy-Efficient Rural housing GB/T 50824-2013[3]. Even after 2000, residential buildings, whose walls are equipped with thermal insulation construction, occupy no more than 24 % of the total annual construction (Fig. 52.7).

3. Huge amount of cold air infiltration through doors and windows causes great heat loss

The exterior door and windows are weak points of buildings in terms of thermal insulation. First of all, as one of the main openings of buildings, entrance is the

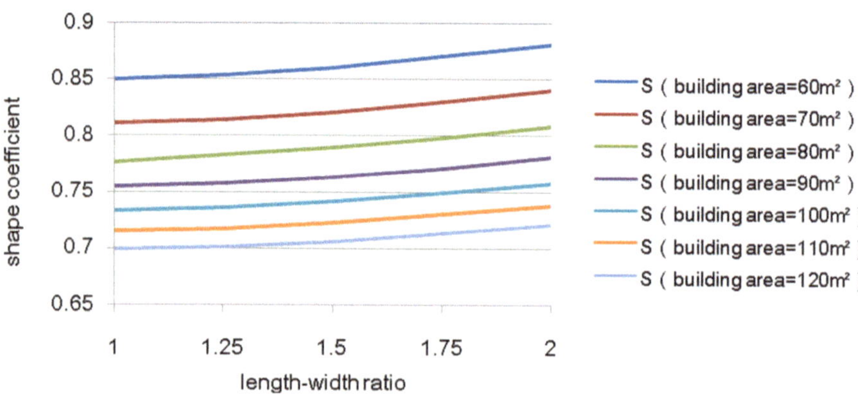

Fig. 52.6 The distribution curve of shape coefficient of rural housings in severe cold zones

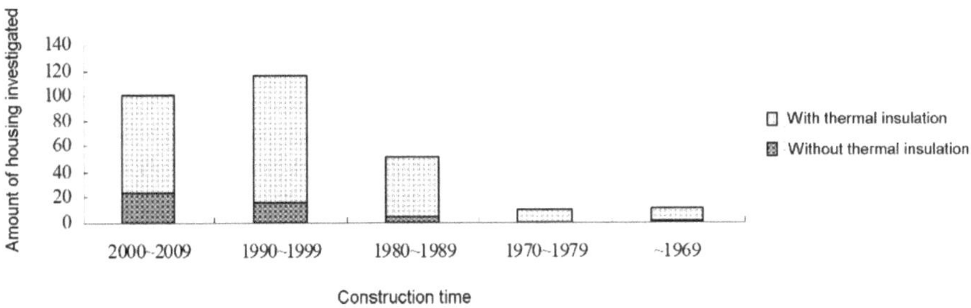

Fig. 52.7 The thermal insulation circumstances of walls according to construction time of rural housing system in severe cold zones

most frequently used part. Especially in winter, entrance serves as the only opening of rural housing in severe cold zones. As a result of the low temperature and strong wind in rural areas in severe cold zones in winter, huge amounts of cold air infiltrate with the opening of main entrance. Further, most of the main entrances in residential buildings are single-layer wooden doors, whose thermal insulation performance is far less than that of the wall; second, the heat transfer coefficient of windows is much greater than that of the wall, and the bigger the area of the window, the higher is the heat transfer of construction. It is regulated by the Design Standards of Energy-efficient Rural housing GB/T 50824-2013[3]that the window to wall ratio of south window, west and east window, and north window should be confined within 0.40, 0.30, and 0.25, respectively. However, the window to wall ratio of south window in many of the rural housing systems in severe cold zones, according to a number of investigations, achieves 0.45 or even exceeds 0.50, which, combining with the general north windows of the same window to wall ratio, causes a great heat loss in rural residential buildings.

4. The poor construction quality of residential buildings leads to a thermal performance lower than the theoretical value

The rural housing systems are mostly designed and prepared by peasants themselves and constructed with the help of relatives, friends, and villagers. The quality of buildings is difficult to guarantee because of the lack of supervision and guidance from professionals. The absence of standard in terms of quality of materials and construction processes leads to the relatively high heat transfer coefficient as well as the poor seal property in the exterior envelopes of buildings.

5. The energy shortage forces some peasants to save energy by undermining the thermal comfort

The demands of heating energy consumption of rural areas in severe cold zones are fairly large, with the energy consumption per unit built-up area exceeding 30 kgce/a. As a result of the less developed economy of rural areas in severe zones, the heating and cooking energy of rural housing still mainly depends on fuel wood such as straw and hay, thus inducing huge demands on plant fuel. However, the low effective utilization rate of burning fuel wood directly leads to a great wastage of resources. In areas with barren lands and shortage of fuel wood, peasants have no choice but to get fuel by deforestation and weeding, which causes forest reduction, water loss and soil erosion, and climatic deterioration, thus leading to the imbalance of agricultural ecology and vicious circle of geographical environment. It is calculated that based on the minimum subsistence, there are still about 22 % of shortage in terms of domestic energy of peasants. Therefore, under such circumstances of energy shortage, the peasants are forced to save energy by lowering the demands on thermal environment.

52.4 Conclusions

To sum up, the circumstance of thermal environment of rural housing system in severe cold zones in winter is far from reaching the demands of thermal comfort and urgently needs to be improved. Through a systematic analysis of the contributing factors, it is recommended that adaptive approaches should be explored in terms of integrated design, plan layout, envelope structure, construction process, utilization of renewable energy, etc. In order to improve the indoor thermal environment of rural housing in winter, appropriate technologies should be developed in depth to suite the local circumstances, thus achieving the aim of lifting the living standards of peasants as well as saving energy.

Acknowledgments This research is supported by the National Project of Scientific and Technical Supporting Programs Funded by Ministry of Science & Technology of China (2011BAJ08B06-2)

References

1. Jin Hong, Zhao Hua (2001) The thinking about rural energy-saving house design of cold regions. J Harbin Univ Civil Eng Archit 3: 96–100
2. JGJ26-2010 (2010) Design standard for energy efficiency of residential buildings in severe cold and cold zones. Ministry of Housing and Urban-Rural Development of the People's Republic of China (MOHURD)
3. GB/T 50824-2013 (2013) Building energy-saving design standard for rural residential buildings. Ministry of Housing and Urban-Rural Development of the People's Republic of China (MOHURD)

Chapter 53
Simulation Analysis and Planning Strategies for the Wind Environment of Residential Quarter in Harbin

Ming Li, Hong Jin and Teng Shao

Abstract Currently, a residential quarter is the main form of residential buildings in China, whose surrounding microclimate, especially the wind environment, has a great impact on the living quality of residents. Based on the climate conditions of Harbin, which is located in the severe cold zone in China, this chapter selects five typical layouts of residential communities, and analyses the characters of wind environment in winter within each layout mode using Fluent software. The optimal distribution mode and planning strategies are hopefully proposed, as well as the corresponding improvement methods according to the existing problems. It is suggested that the prediction and evaluation of wind environment of residential communities should be carried on from the stage of planning and design, in order to eliminate the negative factors in time.

Keywords Residential quarter · Wind environment · Simulation analysis · Layout planning strategies

53.1 Introduction

With the increasing urban population density, the construction scale of residential quarter is becoming larger, and the distribution of buildings is more and more complicated which, to some extent, brings certain difficulties to the wind environment design of residential communities. Moreover, designers of current residential communities mostly concentrate on the form and layout of buildings and landscape planning, rather than the impact of physical environment within residential quarters on human activities and buildings. The situation is more serious in the severe cold zone of China, where the winter is extremely cold and long, and unreasonable distribution of buildings will cause huge wind locally, which will greatly affect the daily travel and outdoor activities, as well as add energy consumption and reduce

H. Jin (✉) · M. Li · T. Shao
School of Architecture, Harbin Institute of Technology, No. 66 Xidazhi Street, Nangang District, Harbin 150001, China
e-mail: jinhong1963@qq.com

A. Sayigh (ed.), *Renewable Energy in the Service of Mankind Vol I*,
DOI 10.1007/978-3-319-17777-9_53

indoor thermal comfort by increasing cold wind penetration. Therefore, simulation as well as prediction and evaluation of wind environment should be conducted at the planning stage of residential quarter, in order to avoid adverse cases such as corner wind, wind tunnel and static wind, and guide the layout planning of residential quarter in the severe cold zone.

53.2 Simulation Settings of Wind Environment Within Residential Quarter

Approaches to explore wind environment in outdoor spaces include wind tunnel test, numerical simulation and field survey. In view of the long cycle and high cost of wind tunnel test and field survey, it is not suitable to apply them in scenario prediction and evaluation at the design stage; furthermore, with the development of computational fluid method (CFD), as well as the operation ability of computer, CFD simulation and analysis method have been widely used in the evaluation of outdoor wind environment, which can be shown in a more vivid and visualized way, and its feasibility has been proved by previous research [1].

53.2.1 Numerical Model

The airflow within a residential quarter is not compressed with low turbulence. The contact between the airflow and construction forms limited flow and the standard k– ε model has good effect on limited flow (wall constraint) [2]; moreover, the standard k–εmodel has low calculation cost and relatively accurate prediction, and therefore is chosen to simulate the wind environment of the residential quarter [3].

53.2.2 Calculation Domain and Grid Distribution

In order to facilitate numerical analysis and improve calculating speed, constructions are shrunk to a certain scale according to the principle of similarity and are idealized as cuboids; a larger calculating region is selected when taking into account the big actuating range of wind field. According to relevant data, when the overall external dimension of residential quarter is 1, the upwind side length of the simulating region is five times the community length, the downwind side length is ten times the community length, the width of both sides is five times that of the community and height is five times the highest construction in community[3].

The geometric model is divided in grids by fore treatment software GAMBIT, the edge of the region and the outline of the building is partitioned in line, with the solid region of constructions that do not participate in the calculation being deleted

[4], and then the fluid region is partitioned in line and surface. Considering the region around the building is the main research target of wind environment, so the grids near the facade of buildings are more intensive, while the grids along the region boundary are sparser. The surface is divided into a triangle mesh, which forms an intensive interior and sparse exterior mesh frame.

53.2.3 Settings of Boundary Condition

53.2.3.1 Determination of Wind Direction

Harbin belongs to the cold temperate continental monsoon climate, the winter is cold and long, and the summer is cool and short, with an average annual temperature of 3.6 °C, and the hottest month average temperature and coldest average temperature are 22.8 and −19.4 °C, respectively. The residential quarter in Harbin is mostly affected by the physical environment of winter, so the main concern of wind environment is the resistance of wind in winter. As shown in Fig. 53.1, according to the wind-rose diagram of Harbin, a typical city in the severe cold zone, the most frequent wind directions of Harbin in winter are west (W), southwest (WS) and south (S). The southwest wind direction is chosen in this chapter as the boundary condition of simulation and analysis.

53.2.3.2 Determination of Inlet Velocity

The boundary condition of inlet surface within the horizontal-level fluid inlet conditions, the inlet velocity is 3.58 m/s (average wind speed of the third-level wind).

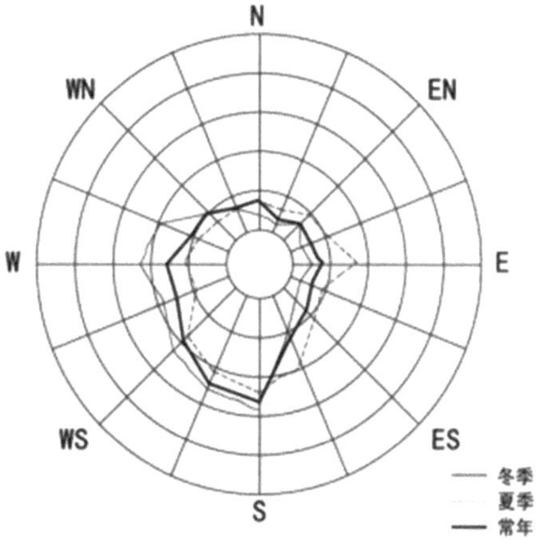

Fig. 53.1 Wind-rose diagram of Harbin

53.2.3.3 Boundary Conditions of the Pressure Outlet

The boundary conditions can be used to solve the reflux problem; a reasonable outlet reflux condition is conductive to solving the convergence difficulties with reflux outlet problem, given the airflow intensity is 2 % and viscosity ratio is 5 [5].

53.2.3.4 Boundary Conditions of the Upper and Side Surface

Because the selected computational domain is large, the airflow in the upper and side surface is hardly affected by the building, so it can be set to free-slip surface.

53.2.3.5 Boundary Conditions of Building facade and Underlying Surface

As the standard $k\text{–}\varepsilon$ model is only for the fully turbulent region apart from a certain distance from the facade of construction, the laminar viscous effects are strengthening near the solid facade of construction, so the standard $k\text{–}\varepsilon$ model must be revised. This chapter uses the facade function method to modify the building border region[5].

53.3 Wind Environment Analysis of Typical Layout Mode in Harbin

Among the current layout forms of residential communities in Harbin, the types of determinant, peripheral and cluster are the three basic prototypes in terms of the combination of residential communities; in addition, there is a type called hybrid, which is a combination of the three archetypes above, and another type called free as a result of the limited land or topographical conditions. The latter two types can be seen as derived types from the former three basic ones, with various forms. The wind environment based on the five typical combination types is analysed in order to explore the best combination mode in terms of the resistance of wind in winter.

53.3.1 Simulation and Analysis of Wind Environment of the Determinant Layout

As shown in Fig. 53.2, the wind velocity increases significantly at the entrance of the cluster, and there is high-speed wind funnels in front of the windward side of the building; as a result of the shelter of the building ahead and the relatively spacious distance between two constructions, the wind velocity near the rear part of the central greenery is quite small.

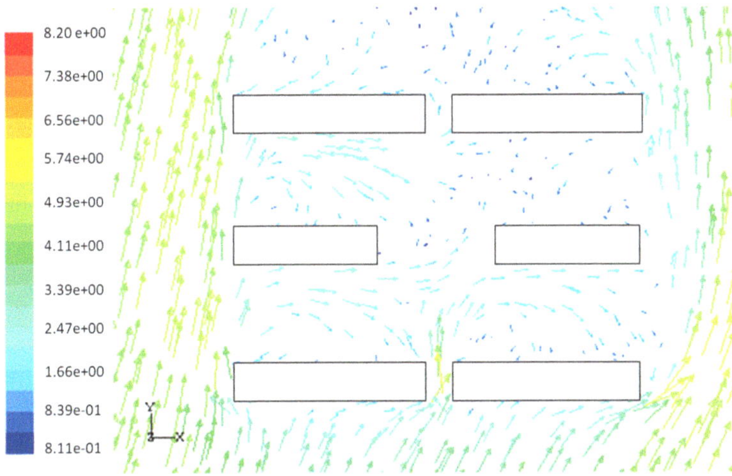

Fig. 53.2 Wind velocity vector diagram of determinant layout

53.3.2 Analysis of the Simulated Wind Environment of Staggered Layout

As a result of the staggered gable, the defence capability against wind decreases, and the wind rises suddenly at the corner because of the effect of corner wind, which is fairly unfavourable in winter. As can be seen in Fig. 53.3, there are wind funnels between two rows of buildings, especially in the centre part of the site. Therefore, it can be concluded that this type of layout has no improvement to the outdoor wind environment in winter, at least not as good as the determinant layout.

53.3.3 Analysis of the Simulated Wind Environment of Staggered Layout

As can be seen in Fig. 53.4, blocked by the second row of buildings, there are significant wind funnels behind the first row of buildings; as a result of the narrow road wind generated by front openings, there is notably increasing wind in front of the second row of buildings, which is not a preferable location for entrance considering the wind chill in winter. There is considerably huge wind in the vacant lot inside the cluster, which is not suitable for outdoor activities in winter.

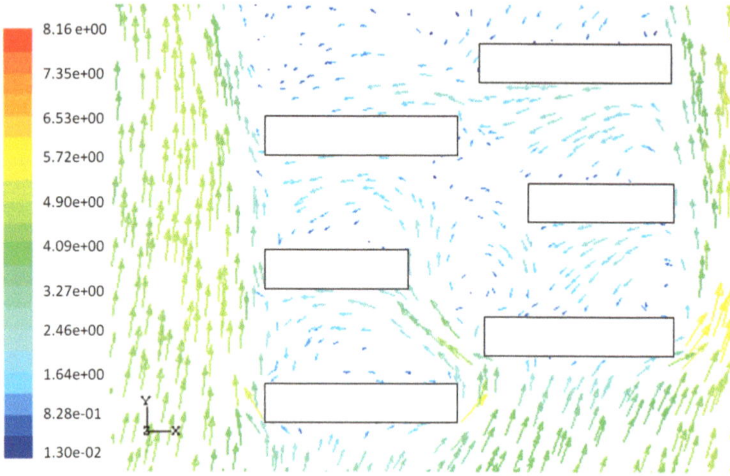

Fig. 53.3 Wind velocity vector diagram of staggered layout

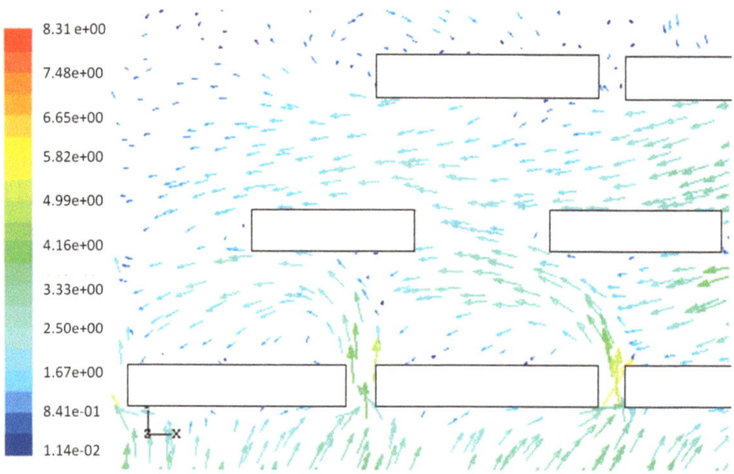

Fig. 53.4 Wind velocity vector diagram of staggered layout

53.3.4 Analysis of Simulated Wind Environment of the Determinant Layout with Increased Distance

As can be seen in Fig. 53.5, the distribution of wind environment of such a layout is relatively similar to that of the general determinant layout. There are high-speed wind and significant wind funnels in the windward side of the back row of buildings, and the wind velocity at the central greenery decreases a bit because of the

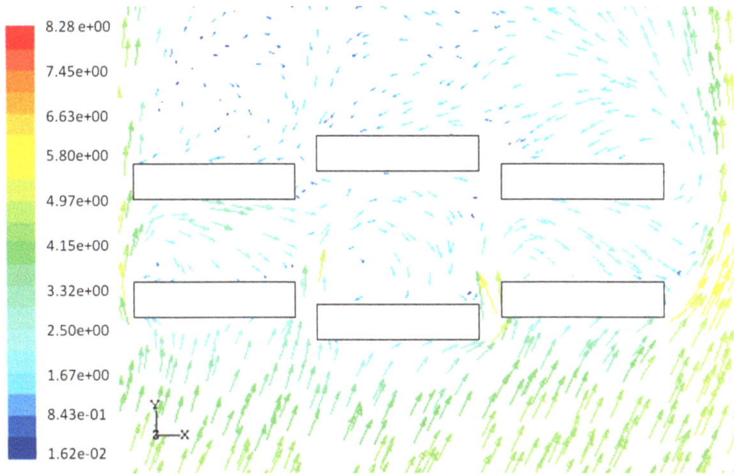

Fig. 53.5 Wind velocity vector diagram of determinant layout with increased distance

increased distance between two rows of buildings, but there are still notable wind funnels.

53.3.5 Analysis of Simulated Wind Environment of Single Peripheral Layout

As can be seen in Fig. 53.6, as a result of narrow-way effect, the wind velocity increases when entering from the front entrance, with a velocity ratio of around 1.12, and keeps high until going out from the back opening; the wind funnels are generated during the process of flowing, which, however, does not have significant effects on the wind environment, as a result of the relatively less incoming airflow as well as the comparatively low-speed wind funnel inside the cluster.

After comparing the five basic types of layouts, it can be concluded that the peripheral layout is the most significant type to resist wind, which can sustain a relatively comfortable physical environment within the residential quarter in winter and avoid the effects brought by the adverse-wind case, and thus can be applied to the residential quarter plan and design in severe cold zones.

53.4 Analysis and Strategies of the Wind Environment of Typical Residential Communities in Harbin

In order to analyse and solve the existing issues on wind environment of residential quarters in Harbin, the Min Jiang residential quarter is selected as the object of simulation and analysis, which mainly belongs to the peripheral layout, and is

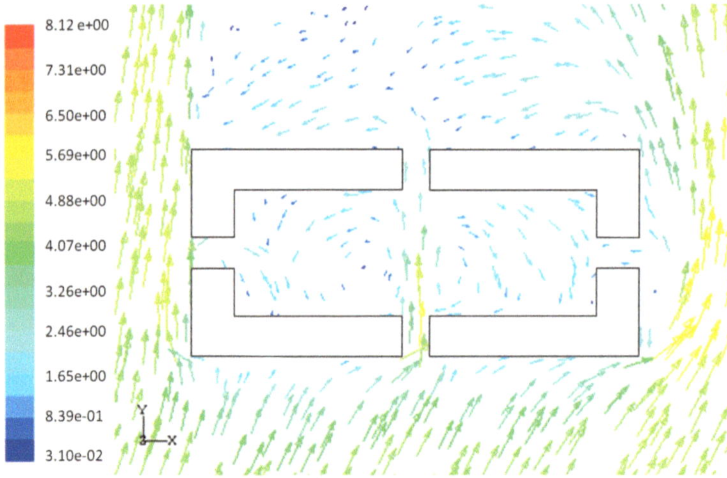

Fig. 53.6 Wind velocity vector diagram of single peripheral layout

Fig. 53.7 General plan of Min Jiang residential quarter

composed of four enclosed clusters, with the buildings of multistorey residential (six floors). Besides, there is a square in the centre of the quarter, which is used for entertainment of the residents, and is the typical mode of residential quarter in Harbin at that time (see Fig. 53.7). The simulation results are shown in Figs. 53.8 and 53.9, and the wind environment can be described as follows:

a. From the picture of wind velocity, as a result of the corner wind at the front entrance, the wind velocity rises significantly, with a wind ratio of about 1.13, and there are wind funnels by the entrance, which seriously affects the wind

Fig. 53.8 Wind velocity distribution diagram of Min Jiang residential quarter

Fig. 53.9 Wind pressure distribution diagram of Min Jing residential quarter

comfort nearby; as a result of the double block of buildings, the incoming airflow through the left opening decreases significantly, which forms a relatively fine wind environment in the zone on the right.

b. In view of the residential quarter, every cluster can enclose a considerably good wind environment, there is even calm wind shown in some of the zones in the northeastern and southeastern clusters, but the continuous airflow in the path between clusters can form a certain kind of draft sensation to humans; however, greenery, water and landscape sketch located at the centre of the quarter, to some extent, can relieve and guide the airflow passing by. So in reality, there will not be airflow running through the central zone as shown in the wind velocity diagram.

Fig. 53.10 Wind velocity distribution diagram of Min Jiang residential quarter after optimization

c. From the point of wind pressure, the wind pressure is relatively high in the left and front sides (the windward side) of the building; therefore, the thermal insulation and air tightness of the exterior facade should be strengthened; inside the community, the wind pressure is highest in the southwestern clusters and relatively low in the central zone and spaces around the front entrance, the wind pressure in the rest of the parts are comparatively well distributed, which cannot form adverse effects to buildings.

In general, the wind environment of the whole residential quarter is relatively good. The cluster mode, greenery distribution and wind barrier are all sufficient to resist wind in winter; however, considering the prevalent south and southwest wind directions in Harbin, the air ventilation in summer as well as the wind resistance in winter should also be taken into account, so calm wind should be avoided in some zones, which not only has adverse effects on ventilation in summer but also is not conductive to pollution diffusion. Therefore, the optimization measurements towards residential quarters are listed as follows:

1. Wind barrier and greenery should be located in the front entrance of the residential quarters in order to slow down the airflow.
2. To satisfy the requirement of ventilation in summer, openings should be set to the left U-shape buildings of the northwest and southwest clusters, or the buildings can be split into two L-shape ones.
3. To avoid calm zones, the scale and location of openings in some zones of the northeast and northwest clusters should be adjusted. The distribution of wind environment after optimization is shown in Figs. 53.10 and 53.11:

    a. As shown in the diagram of wind velocity, the incoming airflow deceases significantly after the wind barrier being located at the front entrance of quarter, with the wind velocity of around 0.7, only some low-speed airflow runs through the quarter, and the intensity of wind funnels also decreases. In terms of wind pressure, the wind pressure increases slightly compared to that before optimization.

Fig. 53.11 Wind pressure distribution diagram of Min Jiang residential quarter after optimization

b. After adding openings to the left buildings, the ventilation in summer is improved because of some airflow through openings into the cluster; as the direction of airflow is vertical to that of openings, only some low-speed airflow is allowed to access the cluster, which will not have a great impact on the wind environment in winter.
c. After adjusting location and scale of some openings in the northeast and southeast clusters, the calm zones have been improved, thus forming a relatively fine wind environment. However, in reality, there is greenery in both the central zone and along the road, which can effectively control the high-speed airflow within and between clusters.
d. From the wind pressure diagram, the wind pressure within the whole quarter has been raised and distributed evenly, except for the drop around the front entrance of the quarter. Improvement can be shown in the residential plan and design in terms of wind velocity and pressure.

53.5 Conclusion

This chapter compares and analyses five typical layout modes in Harbin using Fluent software, and the optimum layout mode has been discussed, based on which the existing residential communities in Harbin have been simulated and analysed, and optimization measures in terms of wind barrier, opening scale adjustment and building on silt have been opposed, the availability of improvement strategies has been verified through comparison and analysis. The results show that when peripheral layout is applied in residential quarter of severe cold zones, the continuous length of buildings should be controlled properly and openings should be located appropriately; the amount, scale and location of openings inside the quarter should be adjusted effectively to control the direction and velocity of airflow; wind barriers within quarter or at the entrance can be helpful to guide and adjust the airflow. Therefore, survey on the wind environment should be conducted at the initial stage of residential quarter plan and design, in order to solve the potential wind environment problems, as well as create fantastic outdoor space for human activities in winter.

Acknowledgments This chapter was supported by the National Natural Science Foundation of China(Grant No.: 51378136)and the Specialized Research Fund for the Doctoral Program of Higher Education(Grant No.: 20122302110041).

References

1. Asfour OS, Gadi MB (2007) A comparison between CFD and network models for predicting wind-driven ventilation in buildings. Build Environ 42:4079–4085
2. Li Wanping (2004) Computational fluid mechanics. Huazhong University of Science and Technology Press, Wuhan, pp 58–59
3. Yang Jie, Tu Guangbei, Yi Chuanxiong (2004) Natural ventilation of high-rise residential buildings with sky garden. Heat Vent Air Cond 34(3):1–5
4. Shao Teng (2013) Optimal design for wind environment of residential areas in severe cold regions. Master Dissertation, Harbin Institute of Technology, Jan 2013
5. Li Yunping (2007) Study on simulating analysis of high-rise settlements' wind environment in chill area and design strategy. Master Dissertation, Harbin Institute of Technology, July 2007

Chapter 54
Energy Efficiency Building Codes and Green Pyramid Rating System

George Bassili Hanna

Abstract Three building energy codes were introduced in Egypt between 2005 and 2010. They impose mandatory energy performance requirements for residential, commercial, and governmental buildings. Another code was presented (2013) to improve the indoor air quality and ventilation requirements and system. This chapter addresses the current states and the potential of energy efficiency, renewable energy, and green buildings to reduce the dependence on the fossil fuel. These codes were the base line for green building code. The first edition of the Green Pyramid Rating System was published in December 2010. The Green Pyramid Rating System is a national environmental rating system for buildings. It provides definitive criteria by which the environmental credentials of buildings can be evaluated, and the buildings themselves can be rated. The Egyptian government has an interest in promoting green buildings as part of Egyptian policies of overall sustainable development. The Housing and Building Research Centre (HBRC) established the Egyptian Green Building Council (EGBC) at the beginning of 2009. Another code is under preparation for Green Hotel Rating System in Egypt to improve the Egyptian Touristic sector. The essence of this chapter is to introduce the efforts being undertaken by Egypt to improve the quality of life and highlight the barriers and overcome the challenges that Egypt faces, and these are summarized in three main parameters, namely, energy, water, and housing.

Keywords Energy in Egypt · Passive technique · Energy Codes · Renewable Energy · Green Energy · Green Buildings Code · Green Pyramid Rating System

54.1 Egypt: Background and Energy Consumption

About two thirds of Egypt's total population of 90 million (2014) are living in urban areas. Egypt's population is concentrated in only 6% of the territory. About 30% of the Egyptians are living in Cairo and Alexandria, and about 65% of the total

G. B. Hanna (✉)
Housing & Building National Research Center, P.O. Box 1770, 12311 Cairo, Egypt
e-mail: gbhanna44@yahoo.com; gbhanna44@hotmail.com

A. Sayigh (ed.), *Renewable Energy in the Service of Mankind Vol I*,
DOI 10.1007/978-3-319-17777-9_54

population are living in Greater Cairo and the Nile Delta. The Egyptian building stock comprises about 12 million buildings, and 60 % of building units are in the residential sector. The number of residential units are about 18 million [1], and it is expected to reach 23 million in 2030, while the number of nonresidential units is about 12 million and is expected to reach 17 million in 2030 [2]. In Egypt, more than 60 % of the total electricity consumption is attributed to residential, commercial, and institutional buildings. Artificial lighting is estimated to account for 36 % of the electricity used in the nonresidential sector and 35 % of the electricity used for heating, ventilating, and air conditioning (HVAC) system. A significant increase in electricity demand is expected over the next 5 years with a growth rate of 8 %. To improve the energy efficiency of buildings, energy efficiency codes have been developed for new residential [3], nonresidential [4] and governmental buildings [5] in Egypt. Energy subsidies are about 147 million which is only 6 % of the total subsidies that go to the residential sector. Energy consumption in Egypt is increasing very fast, at rates more than 5 %. Estimates indicate that energy consumption could be more than double in 2030 and triple in the building sector. Energy consumption in the building sector increases due to growing populations, comfort demand, and electricity use of low-income inhabitations. The household sector is the biggest consumer of electric energy and accounts for 41.1 % [1]. The Egyptian energy tariffs are still very low with respect to other developing countries, but there is a plan to increase the electricity tariff for housing for consumers using more than 650 kWh/m. The current national energy supply mix in Egypt is 96 % from fossil fuel (petroleum products and natural gas) and 4 % from renewable recourses (mainly hydro and limited wind).

54.1.1 Key Issues and Challenges

The energy sector plays an important role in achieving the economical and social development in Egypt, through supplying the energy resources mix required for different sectors. In 2009/2010, oil accounted for 40 % of the total primary energy consumption, while natural gas “NG” accounted for 56 % (about 70 % goes for electricity), and only 4 % was from renewable energy “RE” including hydro recourses. The industrial and transport sectors are the largest end users of primary energy sources, accounting for 34 %, respectively, in 2009/2010. The electricity consumption for the domestic sector was the highest (39 %), followed by the industrial sector (32 %), as shown in Fig. 54.1. The passive design is the most cost-effective to reduce the energy consumption through the building envelope. Figure 54.2 shows [6] the design level from urban planning to architectural design [7], and it is cost-effective energy-efficient buildings. A good design can reduce the investment cost of a building, when considering compacts, efficient layouts, and orientation (Figs. 54.3 and 54.4).

Analysis of the Egyptian population quintiles [8] (20 % population segments based on income, low–high) found that the richest urban quintile received 46 % of the total benefits of petroleum subsidies, while the poorest urban quintile received

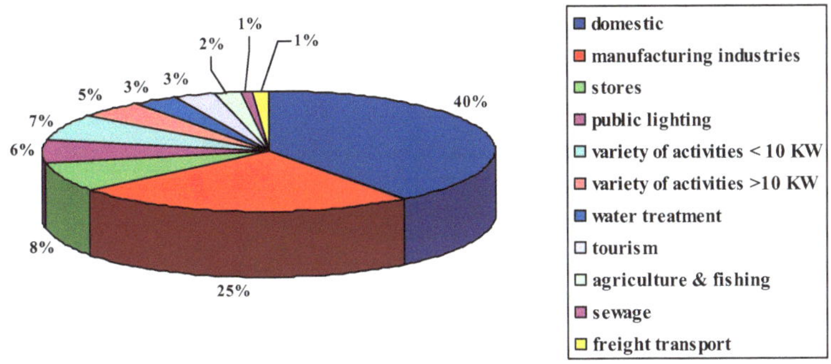

Fig. 54.1 Distribution of electricity consumption by sector

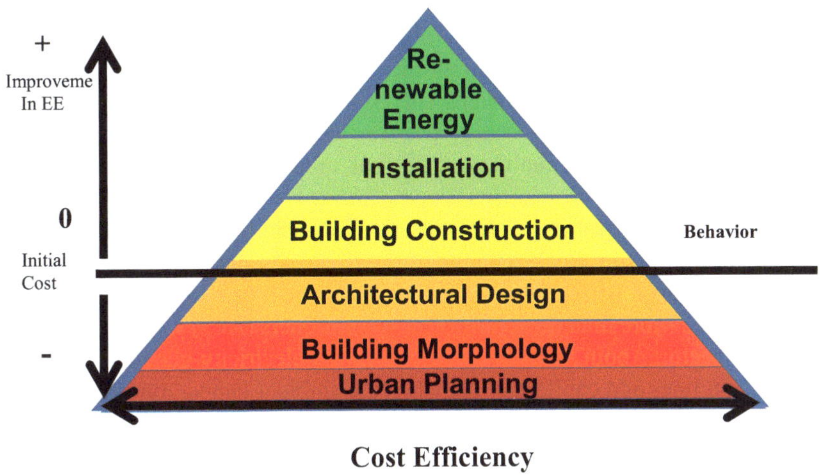

Fig. 54.2 Strategy for cost-effective energy-efficient buildings

Fig. 54.3 Primary energy consumption 2011/2012 = 81.7 Mtoe. *NG* natural gas, *Mtoe* million tons of oil equivalent

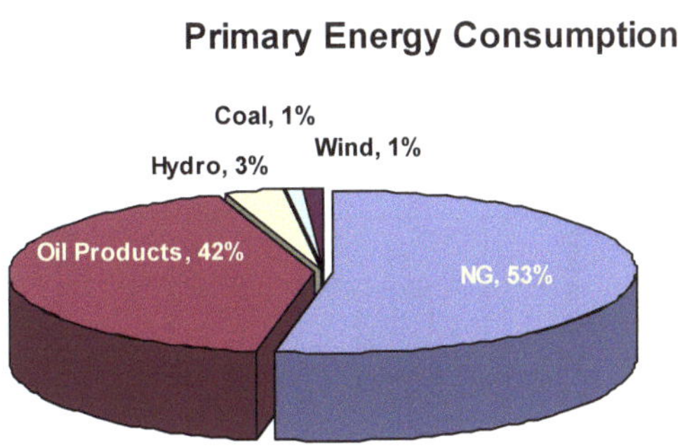

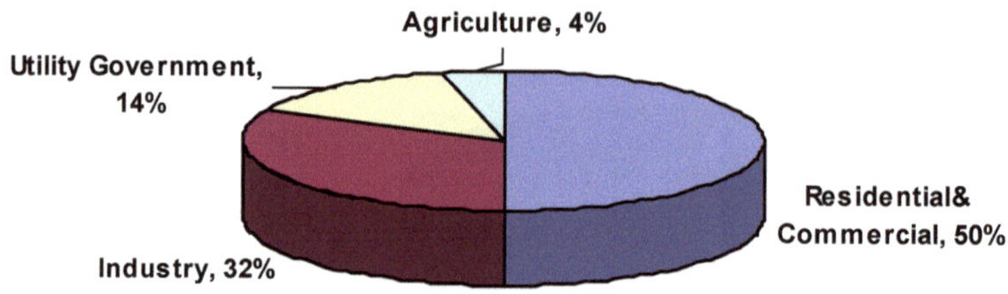

Fig. 54.4 Total electricity consumption, 2011/2012 = 134×10^9 kWh

only 9 %. The stark divide illustrates the division in benefits of subsidies between the rich and the poor. A similar divide also exists within both urban and rural areas as well as between them, with benefits primarily going to the urban rich, rather than to the rural poor. These figures give strong evidence of the common reality of inequitable distribution of universal energy subsidies (Fig. 54.5).

The electricity generation activity utilizes around 30 % of the fossil fuel and natural gas resources in addition to all the hydro and wind energy resources. The industrial activities in Egypt consume around 40 % of the overall energy available. The average annual gross for energy use was less than 6 % over the past decade, and it is expected to continue if the GOE not solve the energy problem. The actual data of electricity consumption are very limited. Surveys indicate that lighting and cooling are the most important end use of electricity in the residential sector. All account about two thirds of the consumption. The increased use of electricity for cooling adds sustainability to the peak load during summer, requiring new power stations. More than three million air-conditioners (ACs) were installed in the residential sector. About 13 % of electricity of the electricity-generated lost in the transmission lines and cables are lost when the ambient temperature reaches 40 °C or above in summer. The overheating period reaches about 5–6 months. The Ministry of Electricity and Energy must add about 10,000 MWh in the next few years to overcome the cut in the supporting the houses with electricity. The second problem is housing for low-income inhabitants. More than half of the new housing stock is

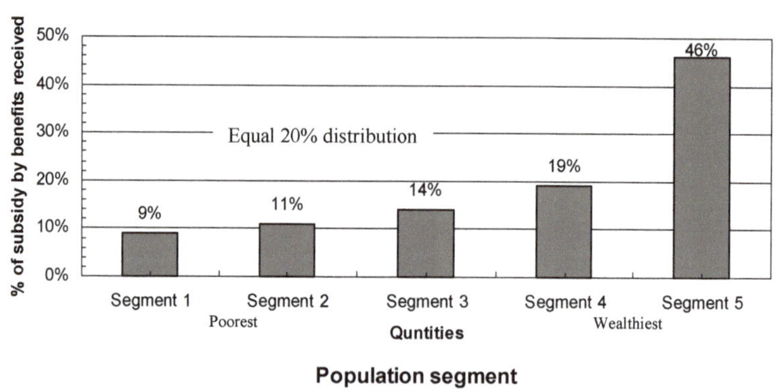

Fig. 54.5 Distribution of subsidy in Egypt

Table 54.1 Indicator of primary energy (Mtoe) and electricity consumption (GWh)

Description (sectors)	Energy (Mtoe)	Electricity (GWh)
Industries	44%	32.5%
Agriculture	1%	4%
Transport	27%	0
Governmental utilities	3%	10.5%
Residential, governmental, and commercial	25%	52.6%
Total (Mtoe)	100%	99.7%

built informally; the rest is shared by the government and the private sector. Energy subsidies more than 140 B. Egyptian pounds which affects the other matter of life. The third problem is water, Fig. 54.2. The Egyptian chair from the river Nile is limited by 55.5 B m^3/year. Table 54.1 shows the primary energy indictor (million tons of oil equivalent) and electricity consumption (giga watt hour; Fig. 54.6).

The new government of Egypt (GOE), after June 30, 2013, signed an agreement with the United Arab Emirates represented by the Arabtec company and the engineering division in the Ministry of Defence to build one million residential units in the next 5 years, that is, 200,000 units per year for low-income group. Another million residential units will be established from the Egyptian private investor's sector and GOE. These buildings must consider and apply the Egyptian building codes for energy efficiency and the green building concept (envelope requirements; ventilation; window to wall ratio, WWR; solar heat gain coefficient, SHGC; and solar water heating, SWH, in addition to energy-labelled appliances) could save more than 50 % from electricity consumed in the residential sector (Fig. 54.7).

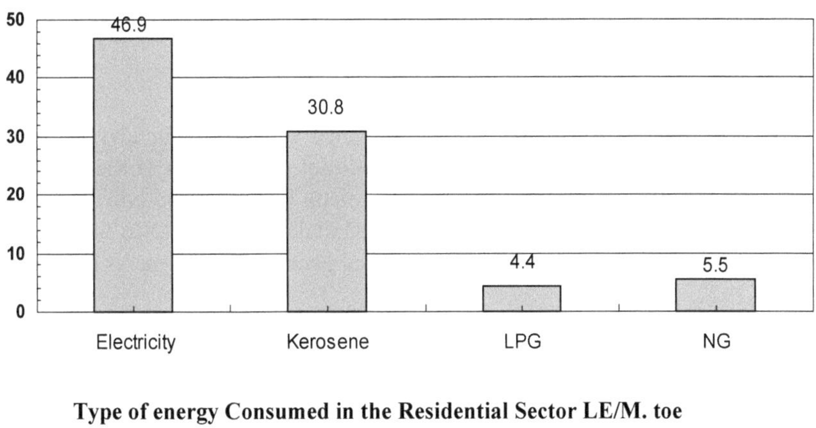

Fig. 54.6 Energy subsidy in Egypt. *LPG* liquefied petroleum gas, *NG* natural gas

Fig. 54.7 Egyptian parries

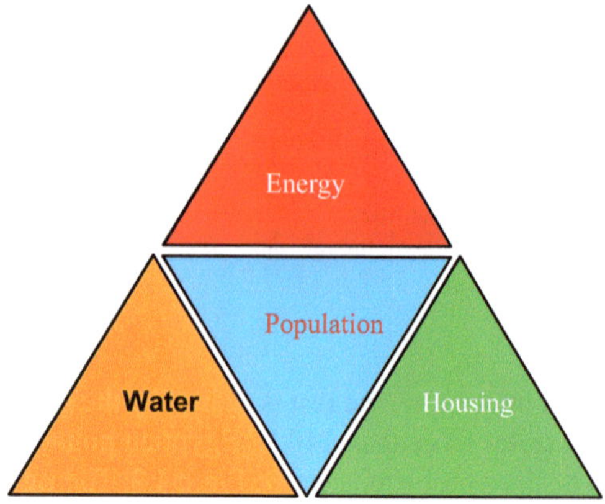

54.1.2 Green Energy and Green Buildings

Egypt was among the first nations to use renewable energy resources, especially in solar architecture, crop dehydration, pumping, and in using animals and agricultural wastes. The Egyptian renewable energy strategy [10] aims to generate 3% of the total electricity demand using renewable energy by 2014. The estimated energy saving in some households and industrial applications is 90,000 toe and 60000 toe from electricity generation using wind farms, in addition to 3.6 Mtoe biomass. A new and renewable energy resource includes solar, wind, and biomass. The energy utilization from such resources was estimated to be 150,000 toe in 2003 plus 3.6 Mtoe biomass.

54.1.2.1 Hydropower

Hydropower potential resources were established at Nagah Hamady and Assuit, as well as some other small available capacities at main canals of the river Nile, in delta, and at the Zefta barrages on the Damietta branch of the Nile. A feasibility station (pump and store) at Attaka and El-Gallala on the Red Sea is now being made. The share of hydro generation to the total generations represents about 8.2% in 2010–2012.

54.1.2.2 Solar Energy

Egypt lies between latitudes 22 and 32 with a daily sunshine of 9–11 h. Solar energy is available in all regions with an average total solar radiation of 1900–2600 kWh/m^2/year, while the direct solar radiation is 1970–3200 kWh/m^2/year. The first thermal power plant was constructed at El-Kuraymat, 50 km southwest of Cairo. The installed capacity was 140 MW out of which 20 MW was the capacity of solar component and it used hybrid solar, combined cycle *(CC)* before the steam turbine *(ST)* was introduced to generate electricity. The power plant is financed by the Global Environmental Facility *(GEF)* and Japan Bank (Japan International Cooperation Agency, *JICA*) for International Development. The project started on 30/6/2011 with an estimated total energy generation of 852 GWh/year. This station will save about 10,000 t toe and reduce the CO_2 emission by 20,000 t/year. A Solar Energy Programme is implemented during the five years started 2012–2017 for the concentrated solar power plants, with total capacity of 100 MW at Kom Ombo south of Egypt. The New and Renewable Energy Authority (NREA) built and operated 547 MW of wind farm capacities and 140 MW solar thermal power plant using parabolic concentrators through technology integrated with combined cycle power plant. A total of 25 governmental office buildings will install (PV) systems on their rooftops as the one installed in the main office of the electricity office building in Cairo.

54.1.2.3 Wind Energy

Egypt is endowed with an abundance of wind energy resources, especially in Suez Gulf area which is considered as one of the best sites in the world due to high and stable wind speeds. The west of Suez Gulf is the most promising sites in the Middle East and Africa. The wind speed ranges from 8 to 10.5 m/s on average, and also due the arability of large inhabitant desert area. Egypt enjoys considerable wind energy resources, with an average wind speed of 10 m/s in the Gulf of Suez area and 7 m/s in the East Owainat area and the Zafrarana region with about 9.5; a wind atlas has been issued for the Gulf and is being developed for Egypt. A pilot wind farm was built at Hourghada; it consists of 42 wind units (100–300 KW each) with 5.4 MW capacities. Its output has been connected to local electricity grid since 1993.

Two successive stages of wind farm at Zafrana (140 MW) have been executed as of 2001. The estimated electrical energy consumption is more than 570 TWh/year. The energy saving will be more than 125,000 toe/year. The target for total installed capacity of wind energy is about 650 MW in 2010. A new project was started at Gabal El-Zeit in the Suez Governorate to establish a new wind farm to produce 220 MW.

54.1.2.4 Biomass

Production of biomass energy using agriculture, animal, human, and solid wastes has high potential in Egypt. Power generation from gasification of sewage slued in waste water treatment plants (El-Gabal El-Asfer 23 MW plant) has already been used. High potential projects for power generation based on gasification or direct combustion of organic solid wastes or agricultural wastes are under consideration. A potential of 1000 M could be generated from agriculture waste.

54.1.2.5 Solar Distillation

The maximum water allowed from the river Nile is only 55.5×10^9 m^3 since 30 years ago, and the population was doubled in this period (less than 45 million), and the needs for fresh water became vital. The Arab Organization for Industrialization in Egypt manufacture and apply 168 projects in the Red Sea Region to provide this area by fresh water for domestic and touristic resorts away from the water pipe lines.

54.1.3 Egyptian Building Energy Codes

For the past two decades, the GOE has published three energy-efficiency building codes for residential [3], commercial [4], and governmental buildings [5] and another code was approved for ventilation [6]. Five energy label was approved and used in life practice with testing facilities to reduce the electricity consumption in the residential sector. At the time, detailed studies had shown the close line between population growth and energy. The thermal elements for eight climatic regions are summarized and shown in Table 54.2. The thermal envelope is very important which includes wall, roof, insulation, windows (ventilation and daylighting), doors, finishes, weather strip, and air/vapour retarders.

54.1.4 Egyptian Green Building Code

The second draft of the Egyptian building code was introduced to the Egyptian Green Building Council (EGBC) for discussion [9]. The main items that were considered in the green building draft code for new buildings were energy efficiency (EE), water efficiency (WE), and indoor air quality (IA).

Table 54.2 Design of an overall thermal transfer values for residential buildings

Description	Roof		Walls			
Thermal Indices	U r (W/m^2°C)	OTTV r (W/m^2)	U w (W/m^2°C)	OTTV w(W/m^2)	WWR (%)	SHGC[a]
North coast	≤0.40	≤20	≤1.4	≤95	>10–<30%	>0.4–<0.7
Cairo and Delta	≤0.50	≤25	≤1.0	≤90	>10–<20%	>0.45–<0.7
North upper region of Egypt	≤0.35	≤20	≤0.6	≤95	>10–<20%	>0.42–<0.68
South upper region of Egypt	≤0.333	≤25	0.566–1.111	≤55	>10–<20%	>0.4–<0.65
Eastern coast	≤0.357	≤20	0.909–2.222	≤45	>10–<20%	>0.4–<0.68
Highland Hhlls	≤0.500	–	0.83–1.0	–	>10–<30%	–
Western desert	≤0.333	≤25	≤55	0.71–1.25	>10–<20%	>0.55–<0.71
Southern Egypt	≤0.333	≤25	0.667–1.000	≤60	>10–<20%	>0.4–<0.7

OTTV overall thermal transfer value, *SHGC* solar heat gain coefficient, *WWR* window to wall ratio
[a]Depends on orientation of the main façade

54.1.5 The Green Pyramid Rating System

The energy-efficiency code was the first step to go further to develop the green building code. The Egyptian Green Building Council (*EGBC*) was established in 2009 and consists of both national and international personalities, including governmental ministers from cabinet-level agencies, officers from respected NGOs, and others. One of the objectives for establishing this council is to provide a mechanism to encourage building investors to adopt the BEECs as well as other sections of existing codes that satisfy both energy efficiency and environmental conservation.

Climate change is the greatest challenge facing humanity, and research shows that the phenomena are a result of increased levels of greenhouse gas emissions resulting from human activity. It is estimated that half of the total carbon-related emission comes from buildings and their use, sustainable building development, and green building. The GOE, represented by the Ministry of Housing has an interest in promoting green buildings as part of the ministry's overall sustainable development policies. As a result, Green Pyramid Rating System (GPRS) has been developed. GPRS is a rating system that can redefine the way we think about projects we design in Egypt [11]. Contemporary design, architecture design firms make sure their buildings are green, and seek to award the GPRS certification. Green building should reduce pollution and enhance the efficiency of energy and water use.

54.1.5.1 Category Weightings

The GPRS system [12] consists of the main topics, such as EE, WE, and indoor air quality and stmosphere, and other topics called "credit categories". The building performance regarding each of these credit categories is evaluated. Each category in turn contains several elements that need to be assessed in the building. These elements are credit Green Pyramid Category Weightings are as follows:

No	Green pyramid category	Category weighting (%)
1	Sustainable site, accessibility, ecology	10
2	Energy efficiency	25
3	Water efficiency	30
4	Materials and resources	10
5	Indoor environmental quality	10
6	Management	10
7	Innovate on added value	5

It is clear from the above table that energy (50 credits) and water (70 credits) efficiency are very important categories for Egypt. The objective of the EE are (1) to reduce energy consumption and carbon emissions by cooperating passive design stage, (2) to optimize the choice of electrical and mechanical equipment and evaluate the inventory of energy and carbon for each development mechanical/electrical/plumbing (MEP) system, and to minimize their impact on the environment, (3) to reduce energy demand for loads at peak use times through efficient building and services design and site based through renewable energy generation, (4) to encourage the provision of metering facilities that allow energy performance of the building, and (5) to minimize the energy consumed. The maximum credits in this category at 12 points for RE sources and 6 points to reduce the peak load, followed by 5 points for passive external heat gain reduction, that is, more than 50%. The objective of water efficiency are summarized as follows: (1) Passive distillation systems is 12 points for storm water harvesting, 8 points for water efficiency improvements, 8 points for outdoor water efficiency improvements, 8 points for waste water management, 6 points for water leakage detection, and 6 points for water feature efficiency with a total 70 pints credit.

54.1.5.2 Certification and Levels of Rating

To earn the pyramid certification a project must satisfy all the stated mandatory minimum requirements and may obtain credit points by meeting certain criteria. Projects will be treated based on credit points accumulated, according the following rating system:

- GPRS certified: 40–49 credits
- Silver pyramid: 50–59 credits
- Gold pyramid:60–79 credits
- Green pyramid:80 credits and above

Projects with less than 40 credits will be classified as "uncertified".

54.1.5.3 Process of assessment

The assessment used a spreadsheet for evaluation process as follows:

1. For each category the number of credits awarded will be determined by a green pyramid.
2. The credit achieved for each green pyramid category is calculated.
3. The percentage of credits achieved is then multiplied by the corresponding category weighting.
4. This section score.
5. The scores for each category are then added to give the overall green pyramid rating, see Table 54.3

54.1.5.4 Green Star Hotels

Tourism is one of the major contributions to Egypt's gross domestic product (GDP). It is considered as one of the many sources of foreign currency, about 15% from the

Table 54.3 Green Pyramid Rating Calculation

Green pyramid category	A	B	C=A/B×100%	D	E=C×D
	Credits available	Credits available	% Credits achieved	Category weight	Category score
1. Sustainable site, accessibility, ecology	10	5	50%	10%	5
2. Energy efficiency	50	40	80%	25%	20
3. Water efficiency	70	35	50%	30%	15
4. Materials and resources	20	15	75%	10%	7.5
5. Indoor environmental quality	30	15	50%	10%	5
6. Management	20	10	50%	10%	5
7. Innovation and added value	10	0	0%	5%	0
Total	57.5				
Green pyramid rating	SILVR				

national income after Suez Canal. Green Hotel Rating System [13] became necessary for the Egyptian touristic sector. A case study [14] focussed on hotels in the Red Sea costal area to estimate the energy efficiency base line and compared that data with the energy performance of some typical hotels in the Red Sea area according to a protocol signed between Gesellschaft für Internationale Zusammenarbeit (GIZ) and HBRC in 2010. The study is in progress and has not been completed yet.

54.2 Conclusion

This chapter presents the potential of EE and RE sources with the potential of GPRS in Egypt. The main conclusions derived from this chapter may be summarized as follows:

1. Hydro power from the high dam and other stations contribute about 9% of the total power.
2. Wind energy is very promising and needs more chair from the private sector.
3. Solar energy has not received much attention, since the electricity is subsidised and must be given to SWH for domestic use but, with incentives and rebates, many projects have been started in different locations.
4. As long as all petroleum products and natural gas are priced far below value, there will be no level playing field for RE (RE appears too costly) in power generation and decentralized application.
5. The only way out is to change the instruments, restrict misappropriation, smuggling, and wasteful low energy prices to basic energy (and public consumption transportation, and increase the incentives.
6. Egypt decided to reduce the value added tax (VAT) for efficiency and disincentives for renewable energy equipment.

The energy crisis threatens the whole world and specially Egypt. It can be summarized in that most of the energy needs in Egypt is dependent upon fossil fuel, whose supply is constantly decreasing, while the demand is greatly increasing. Another problem is the pollution growth which greatly threatens the health of all living creatures. Third, global warming due to the emission of green house gases as by-products of energy production, a problem that causes the temperature to rise constantly which threatens the life of many species, leads to more energy consumption in cooling, and causes floods in areas and droughts in others. As a response to these problems, and as buildings are important contributors to the problems, green buildings have emerged, and green building rating systems like GPRShave been developed.

References

1. Annual Report (2011/2012) EEHC
2. Liu F (2011) Mainstreaming energy efficiency code in developing countries, global experiences and lessons from early adopters. World Bank, No. 204
3. Energy Efficiency Residential Building Code (2005) Ministry of Housing, HBRC, Egypt
4. Energy Efficiency Commercial Building Code (2007) Ministry of Housing, HBRC, Egypt
5. Energy Efficiency Governmental Building Code (2010) Ministry of Housing, HBRC, Egypt
6. Natural and Mechanical Ventilation Energy Code (2013) Ministry of Housing, HBRC, Egypt.
7. Energy Efficient Buildings Guidelines (2013) MED–ENEC
8. Energy Subsidies "A Road Map for Reforms" (2013), MED–ENEC
9. Green Building Code, Second Draft (2013) HBRC, Egypt
10. Hanna GB (2013) Green energy and green buildings in Egypt. Int J Eng Res Appl 3(4):466–470
11. Hanna GB (2013) Sustainable energy investment potential in residential sector in Egypt. J Environ Sci Energy 2(6)369–377
12. The Green Pyramid Rating System (2011) Second draft
13. Hanna GB, Farouh H (2012) Energy efficient hotels as a step towards sustainable tourism in Egypt. 6th international conference of Military Technical College, 29–31 May 2012, Cairo, Egypt
14. Medhat AM (2013) Energy-base-line for green rating system in Egypt. International conference building simulation Cairo 2013—towards sustainable & green built environment, Cairo, 23–24 June, 2013

Chapter 55
Simulation Comparison Between Natural and Hybrid Ventilation by Fans at Nighttime for Severe Hot Climate (Aswan, Egypt)

A. Rizk, A. El-Deberky and Nabil M Guirguis

Abstract The computer simulation measures the effect of ceiling and wall fans in decreasing air temperature during nighttime. The indoor temperature exceeds the thermal comfort through the overheated period in Aswan. A high time lag of envelope in the daytime increases the indoor temperature at nighttime. As a result, the indoor temperature becomes higher than the outdoor temperature at nighttime of the overheated period.

The simulation study consists of three models according to ventilation conditions (types). The first model that depends only on cross ventilation is called the natural ventilation model. The second model is the hybrid ventilation that consists of natural ventilation (cross ventilation) and mechanical ventilation (ceiling fan). The third model uses wall (side) fan instead of the ceiling fan. The boundary conditions are placed according to the weather conditions in Aswan in the overheated period (July).

Simulation results of hybrid ventilation models decreased the indoor temperature from 2 to 4 °C. The wall (side)—average indoor air velocity in the case of wall fan is 1 m/s—fan is better than the ceiling fan—average indoor air velocity in the case of ceiling fan is 0.86 m/s) according to decreasing indoor temperature (4 °C), and airflow is well distributed inside the room at the sitting level.

Keywords Comfort ventilation · Nocturnal Ventilation · Night cooling · Cross ventilation · Fans: ceiling and side · CFD models

---

N. M. Guirguis (✉)
Housing and Building Research Center, Cairo, Egypt
e-mail: nabilmilad@yahoo.com

A. Rizk
Department of Architectural Engineering, Tanta University, Tanta, Egypt

A. El-Deberky
Department of Architecture, Minya University, Minya, Egypt

A. Sayigh (ed.), *Renewable Energy in the Service of Mankind Vol I*,
DOI 10.1007/978-3-319-17777-9_55

55.1 Introduction

55.1.1 Previous Studies

In the case of ceiling fan, according to Kadiri's study [1], the main factors of increasing airflow indoors are the distance from the fan and the velocity of the fan. At the distance of 1 m from the ceiling fan, the air velocity of the ceiling fan is 2.5 m/s, the air velocity at a distance of 1 m from the ceiling fan has the same velocity of the fan at 2.5 m/s. At a distance of 5 m from the ceiling fan, the air velocity of the ceiling fan is 2.5 m/s; the air velocity at 5 m from the ceiling fan decreases up to 1 m/s (40 % from the fan's velocity).

55.1.2 Aswan Problem

55.1.2.1 Visual DOE Application at Aswan

Visual DOE is used to calculate the rate of energy consumption in different ventilation conditions (natural ventilation, natural ventilation and mechanical ventilation by using the fan, and air-conditioning). Table 55.1 shows that the air-conditioning achieves comfort temperature (outdoor temperature in the sun, 48 °C, indoor temperature, 25 C) with extreme high-energy consumption (1879 kW/h), while the fan decreases around 5 °C with low energy (165 kW/h) in the cross-ventilation condition [2].

55.1.3 Air Temperature

Giovanni's equation (55.1) ([3], p. 60) suggests the indoor temperature according to outdoor temperature.

Table 55.1 Aswan outdoor temperature with ventilation types

No.	Ventilation condition	Air temperature (°C)											
		Jan	Feb	Mar	Apr	May	June	July	Aug	Sep	Oct	Nov	Dec
1	Outdoor temperature in the sun	31.2	34.3	37.8	40.9	45	48	47.5	47.3	46.5	46	42	36.3
2	Cross ventilation	26.3	27.8	32.7	35.6	40.2	43.2	42.8	43	40.6	39.1	35.6	32.2
3	Fan + cross ventilation	26.1	27.6	32.5	35.4	40	43	42.6	42.8	40.5	39	35.3	32.1
4	Air-conditioning	22.5	22.5	22.6	23	24.7	24.9	24.6	25.1	24.9	24.9	22.1	22

$$T_{max} = GT_{avg.} + \mathrm{Del}T + K\left(T_{avg.} - GT_{avg.}\right) \tag{55.1}$$

T_{max} Maximum indoor temperature (°C)

$GT_{avg.}$ Grand average of outdoor temperature is 4 °C in summer and 5 °C in winter.

Del T Depends on mass, shading, and ventilation condition (type), equals 12 °C in non-ventilation condition, 8–10 C in ventilation condition.

K Depends on mass level, equals 0.5–0.8.

$T_{avg.}$ Maximum outdoor temperature + minimum indoor temperature/2

T_{max} at Aswan = 4 + 12 + 0.7 (41 + 28/2 − 4) = 38 °C.

Figure 55.1 shows the chart of outdoor air temperature in the shade and suggested indoor temperature according to Eq. (55.1) through a full day of hot July at Aswan; the chart also shows the hours that are effective according to cooling natural ventilation.

Table 55.2 shows outdoor temperature in the shade through a day throughout the year. A total of 50 % of hot period is suitable for ventilation, especially at nighttime because of low temperature that is not more than 32 °C; night ventilation can be achieved when night outdoor temperature is lower than the indoor temperature, and night outdoor temperature is not more than 32 °C according to the thermal comfort of people living in hot climate [3, 4].

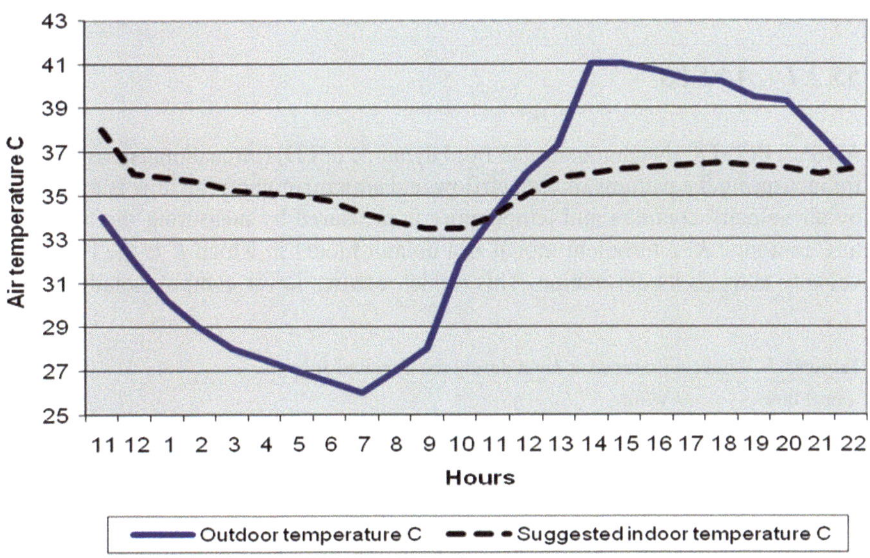

Fig. 55.1 Outdoor and suggested indoor temperature

Table 55.2 Maximum/minimum outdoor air temperature, the critical ventilation hours, and the night ventilation hours for the upper Egypt region (Aswan)

Time	Hours											
	Jan.	Feb.	Mar.	April	May	June	July	Aug.	Sep.	Oct.	Nov.	Dec.
0	13.7	13.8	19	25.4	28.4	30.1	31.9	31	29	26	20	14.4
2	12	12.5	18	23.6	27	28.1	30.2	29.2	27.3	24.4	18	13
4	11.4	11.8	16.6	22.5	25.4	26.6	28.9	28.2	26.1	23.1	17.1	12.5
6	11.6	11.6	16.5	22.8	25.8	27	29	28.3	26.4	22.5	16.8	12.1
8	11.4	12.3	18.3	23.8	28.6	30.6	32.3	31.4	28.7	24.7	17.9	13.3
10	15.3	16.2	22.8	28.3	32.2	34.4	36.1	35.2	32.1	29.3	22.1	18.2
12	19.6	19.5	26.1	31.4	35.4	38.1	38.2	38.8	36.5	32.5	25	21.2
14	22	21	28.4	33	37.2	39.8	40.7	40.5	38.7	34.8	26.7	23.4
16	21	21	28	32.7	36.2	39.5	40.3	40	38.5	34.5	26.1	22.2
18	20	19.7	26.3	31.6	35.4	38.1	39.5	38.8	36.7	32	25.2	20
20	17.4	17.7	23.5	28.9	32.1	35.1	36.5	35.5	33.5	29	22.7	17.7
22	15.5	15.6	21.4	27.2	30.3	32.7	33.9	33	31	27.2	22.1	16

55.1.3.1 Wind

Table 55.3 shows wind velocity, direction, and temperature (wind chill) at nighttime in July, Freemeto.com [5]. Wind velocity ranges between 0.8 and 2 m/s at 5 m height (low wind velocity). Wind direction ranges between north and west (constant direction). Wind temperature ranges between 21 and 29 °C.

55.2 Simulation Experiments

55.2.1 ANSYS

ANSYS FLOTRAN computational fluid dynamic (CFD) (three dimensions) is used for measuring the patterns of both airflow and air temperature; the flow is measured by air velocity contours and temperature is measured by measuring the temperature contours. *K*–*e* turbulent model is a default model in which *K* is the turbulent velocity and *e* is its dissipation. This model requires lower computational domain

Table 55.3 Wind environment at night during the month of July

Local time	Wind		
	Velocity (m/s)	Direction	Temperature (°C) (chill)
8 p.m.	0.8	NNW	27
11 p.m.	2	WNW	29
2 a.m.	1.4	N	25
5 a.m.	0.8	NNW	21

NNW north–northwest, *WNW* west–north west, *N* north

Table 55.4 Experimental conditions

No. of models	Models due to ventilation	Description of models by drawings	Description of models	Boundary conditions
Model no 1	Natural cross ventilation due to wind pressure through inlet and outlet openings	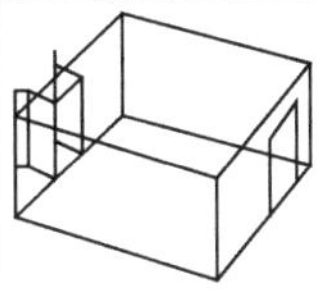	Models contain two openings (inlet and outlet). Inlet equals outlet. Inlet is characterized by wings	Intake wind velocity = 1 m/s Outdoor Intake wind temperature = 301 and 315 K Indoor temperature = 312 K
Model no 2	Natural cross ventilation + mechanical ventilation by ceiling fan	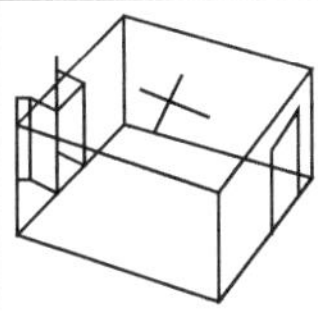	As per previous description, the model contains the ceiling fan at mid-roof and at 2.5 m from the ground level	Type of fan: ceiling or central fan Diameter of fan = 1 m Velocity of fan = 1.5 m/s
Model no 3	Natural cross ventilation + mechanical ventilation by side fan	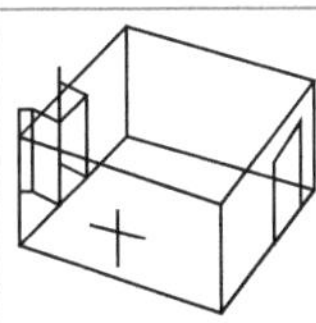	As the basic model side, adding the fan at mid-wall instead of the ceiling fan and at a height of 1 m from the ground level	Type of fan: wall or side fan Diameter of fan = 1 m Velocity of fan = 1.5 m/s

and is more applicable to the wide-ranging flow. According to processor menu, the number of solutions is 100 once at least in two main cases: thermal or adiabatic. The measurement unit is metric (Air.S1). The model is under forces: incompressible. Inputs (boundary conditions) are shown in Table 55.4. Outputs are contours from minimum value that are characterized by *blue color* and maximum value by *red color*.

55.2.2 Air Properties

Air properties at 305 °K (32 °C) are: Specific heat = 1004 GL/K, thermal conductivity = 0.025 W/m^2, dynamic viscosity = 1.87×10^{-5}, density = 1.1614 KG/m^2, static pressure = 10,133 Pa, gravitational acceleration = 9.81 m/s^2, pressure coefficient at windward side = 1.4, and loss coefficient 1.5 m/s.

55.2.3 Case Study

The test model is a single room with dimensions that are 6 m length, 4 m width, and 3.5 m height as shown in Fig. 55.2. The inlet and outlet opening dimensions are

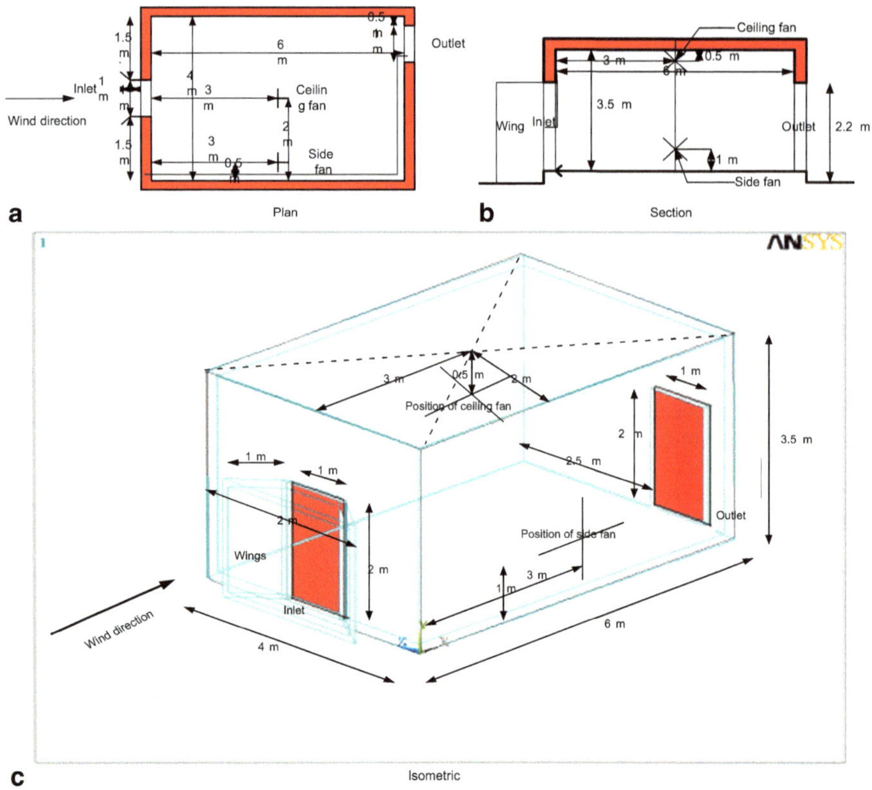

Fig. 55.2 Description of the test model

1 m width and 2 m height. The model consists of two types of fans, the first fan is central (ceiling) fan and the other is side (wall) fan. The diameter of the fan is 1 m. The position of the ceiling fan is at the center of the roof, and its height is at 3.2 m from the ground level. The position of the side fan is at the center of the sidewall, and its height is at 1 m from the ground level. Temperature and flow are measured at 1 m from the ground level (sitting level).

55.2.4 Case Studies and Ventilation Conditions

Table 55.3 shows the three main types of analyzed model due to ventilation characteristics.

55.2.4.1 First Model (Cross-Ventilation Model)

First model (no. 1) basic model depends on inlet at the center of windward side that can achieve maximum positive pressure according to wind pressure [6].

55.2.4.2 Second Model (Ceiling (Central) Fan Model)

Second model (no. 2) is to measure the effect of adding ceiling fan to enhance indoor ventilation and decrease the temperature.

55.2.4.3 Third Model (Wall (Side) Fan Model)

Third model (no. 3) is to compare the effect of ceiling (central) fan and wall (side) fan in increasing velocity and decreasing temperature.

55.2.4.4 Fan Height and Velocity

Ceiling fan is at 2.5 m, and wall fan is at 1 m from the ground level in order to enhance ventilation rate at sitting level (1 m height from the ground level). Kadiri [2] shows that the effectiveness velocity of air would be at the first meter from fan's position, the velocity decreased when the subject position fared from fan to achieve half of fans velocity at the second meter. Both ceiling and wall fans' velocity is 1.5 m/s.

55.2.4.5 Boundary Conditions

Wind velocity at the inlet opening is 1 m/s, and the night wind temperature is 300 K (28 °C). Night indoor temperature is 312 K (38 °C) due to the outdoor temperature in the day at 315 K (43 °C)—weather conditions, Giovanni's equation (55.2), and Fig. 55.1.

55.3 Results and Discussions

Figures 55.3 and 55.4 show the plan of the air temperature and velocity patterns. According to decreasing reference indoor temperature (312 K), Figs. 55.3 and 55.5 show the performance of the indoor temperature pattern in three models: Model 1—cross-ventilation model—decreases 5 K as shown in Fig. 55.3a. Model 2—ceiling (central) fan model—decreases 7 K as shown in Fig. 55.3b. Model 3—wall (side)

Fig. 55.3 Patterns of air temperature: **a** Model 1: pattern of air temperature. **b** Model 2: pattern of air temperature. **c** Model 3: pattern of air temperature

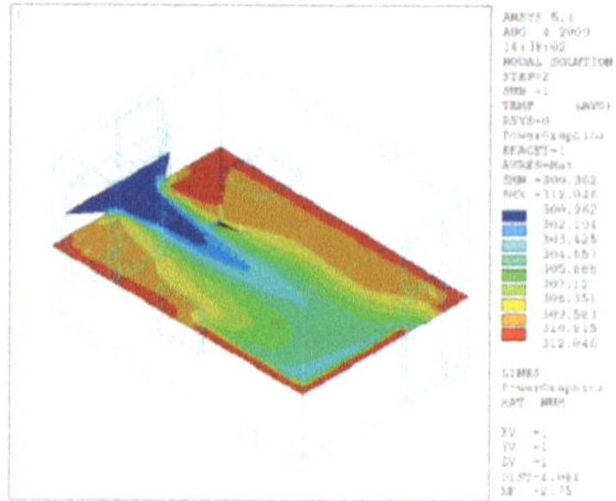

Model 1 pattern of air temperature

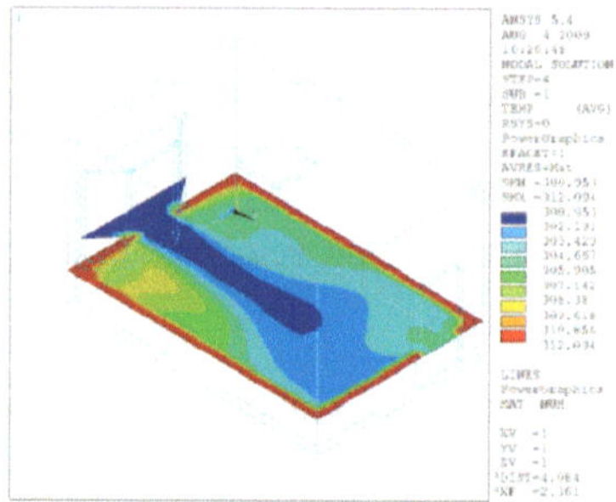

Model 2 pattern of air temperature

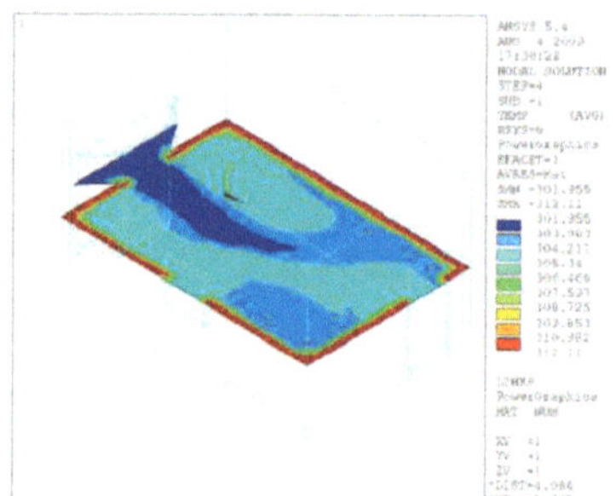

Model 3 pattern of air temperature

fan model—decreases 8 K as shown in Fig. 55.3c. According to reference outdoor air velocity (1 m/s), Figs. 55.4 and 55.5 show average indoor airflow (velocity) pattern in three models: Model 1—cross ventilation model—achieves 45 % (0.45 m/s). Model 2—ceiling (central) fan model—achieves 86 % (0.86 m/s). Model 3—wall (side) fan model—achieves 100 % (1 m/s). Figure 55.5 shows both temperature and flow patterns of nine measurement points inside the room. Both the minimum air temperature and maximum air velocity are at the center of the room's long axis. Increasing the temperature and decreasing the velocity, the long axis's measurement point is fair from the inlet opening. According to high-performance patterns of low air temperature and high air velocity (flow), the left corner that has no outlet opening

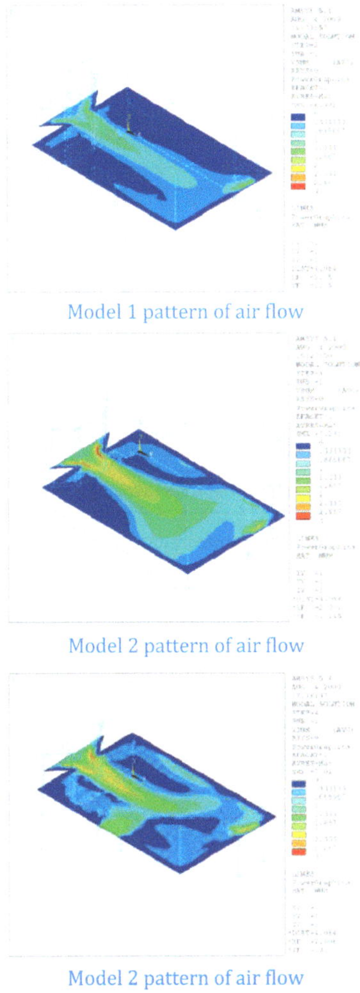

Fig. 55.4 Patterns of air flow: **a** Model 1: pattern of airflow. **b** Model 2: pattern of airflow. **c** Model 3: pattern of airflow

is better than the right corner that has the outlet opening. According to night ventilation cooling, high indoor airflow (velocity) can achieve low indoor temperature. The maximum air velocity (1 m/s) in model 3 can decrease 8 K. The minimum air velocity (0.45) in model 1 can decrease 5 K. Figure 55.6 shows the comparison between air velocity and air temperature of three models. Results can be summarized as follows: The Aswan climate problem is increasing the air temperature to 312 K (38 °C) in heavy mass of buildings at nighttime. Cross ventilation can decrease 5 K. Adding the fan can decrease cross ventilation (7–8 K). Decreasing the indoor temperature depends on the position of the fan according to the sitting level. Table 55.5 and Fig. 55.7 show the difference between night ventilation (low-intake

Number of model	Temp.(°K) & flow (m/s)	Measured points									
		1	2	3	4	5	6	7	8	9	Avg.
Model 1	Temp.	311.43	301.57	310.2	310.2	302.8	309	310.2	304	305.2	307.17
	Flow	0.45	1.15	0.15	0.15	0.88	0.2	0.2	0.8	0.45	0.45
Model 2	Temp.	306.5	301.57	307.67	305.3	301.57	307.67	304	302.8	302.8	304.43
	Flow	0.15	2	0.47	0.3	1.8	0.3	0.8	1.15	0.8	0.86
Model 3	Temp.	304.5	302	304.5	303.5	302.5	304.5	307.6	303.5	303.5	303.5
	Flow	0.45	2.15	0.45	0.83	1.85	1.8	0.45	0.45	0.45	1

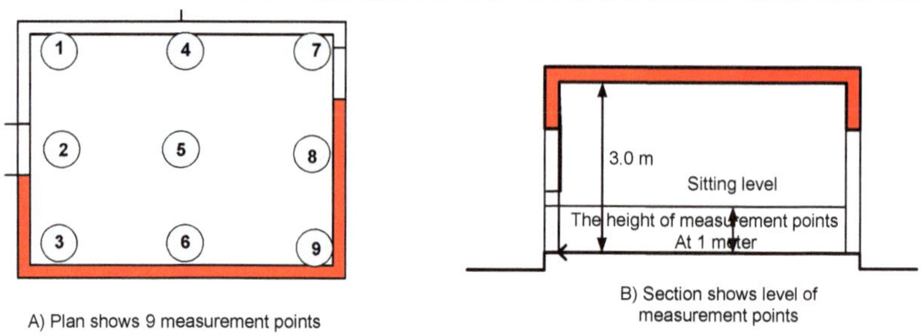

Fig. 55.5 Air temperature and airflow patterns of nine measurement points inside the room

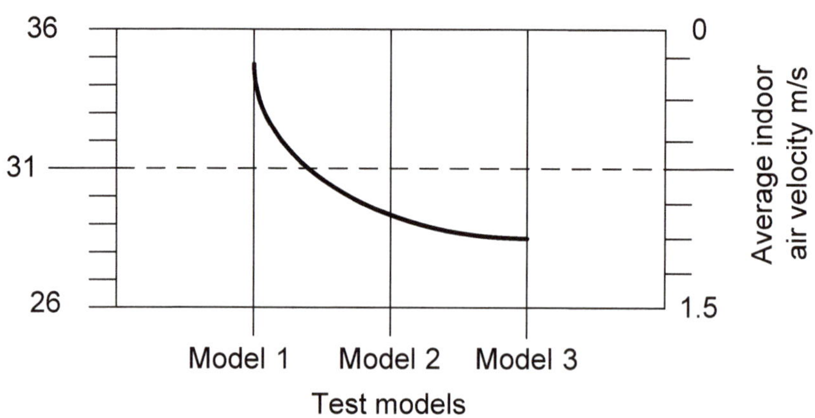

Fig. 55.6 Comparison between air velocity and air temperature of three models. **a, c** Model 1—cross-ventilation model. **b, d** Model 2—ceiling fan model

wind temperature: 300 K (28 °C)) and day ventilation (high-intake wind temperature: 315 K (43 °C)) on air temperature pattern. Night ventilation is preferable at nighttime and is not preferable during the day. At the center of the long axis, night ventilation decreases indoor temperature (10 K) and the day ventilation increases the indoor temperature (4 K) according to the reference indoor temperature (312 K; 38 °C). The fan is more effective at nighttime (13 K) than in the daytime (1 K). These results are only in the overheated period at hot dry climate such as Aswan that is characterized by high temperature (315 K, 43 °C) and low temperature (300 K, 28 °C).

Table 55.5 Comparison between low and high-intake wind temperature (night and day ventilation)

Model	Boundary conditions	Measurement points at long axis of room (K)				Reference indoor temperature (K)
		2	5	8	Avg.	
Model 1	Low-intake wind temperature (300 K)	301.57	302.8	304	302.79	312
	High intake wind temperature (300 K)	316	316	316	316	312
Model 2	Low-intake wind temperature (300 K)	302	302.5	303.5	302.66	312
	High-intake wind temperature (300 K)	316	316	314.65	315.5	312

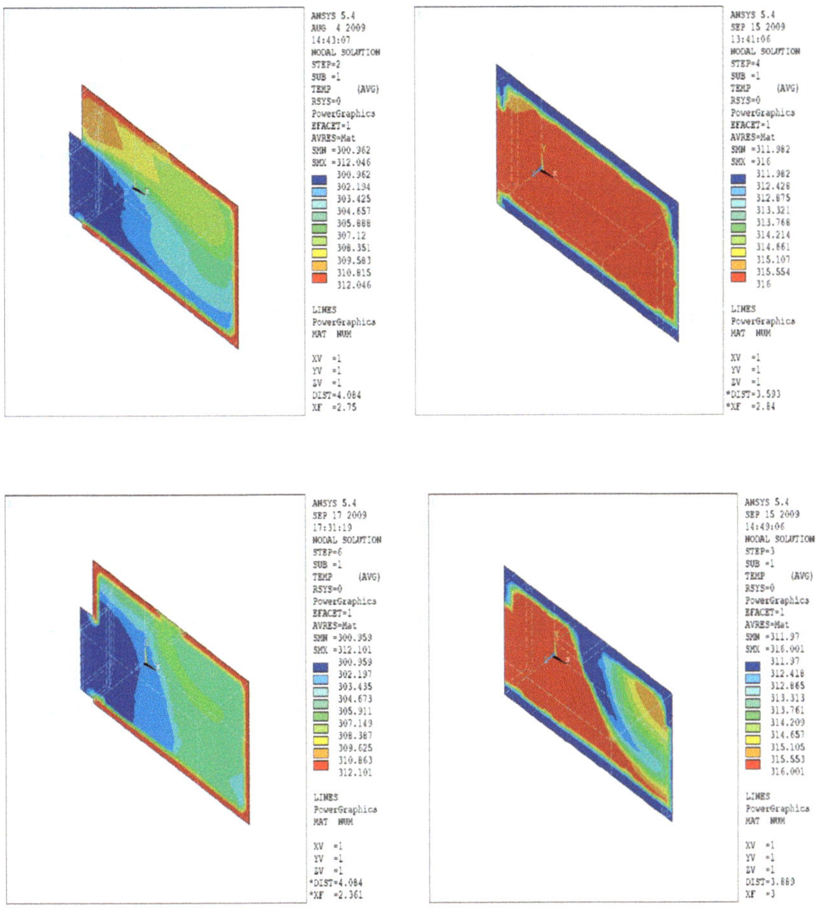

Fig. 55.7 Comparison between low-intake wind temperature and high-intake wind temperature

55.4 Conclusion

1. Internal air velocity that is equal to 0.45 m/s can decrease 5 K in the only condition that outdoor temperature is less than indoor temperature and night outdoor is not more than 32 °C.
2. Hybrid ventilation by fan is more effective according to decreased temperature regularity inside the room than only natural cross ventilation—paper's weather conditions.
3. Side fan is better than central fan.
4. The strong relationship is between increasing indoor air velocity and decreasing indoor air velocity. Well-ventilated model—model 3—decreases indoor air temperature as shown in Fig. 55.3.
5. Fan is not effective in the overheated period. According to this chapter, the fan depends on not only velocity and distance but also intake temperature.

References

1. Kadiri KO (2006) The effectiveness of fan in enhancing comfort. J Appl Sci Res 2:904–908
2. VISUAL DOE software
3. Energy-plus, weather data
4. Giovini B (1994) Passive and low energy cooling of buildings. Van Nostrand Reinhold, New York
5. Freemeto.com
6. Lstiburek (2006) Pressure in buildings. Sci Dig 109

Chapter 56
Green Building and Energy Saving

Mahmoud A Hassan and Nabil M Guirguis

Abstract The colored glass has great effects to prevent the solar heat gains and consequences to minimize the air-conditioning loads which mean low energy consumption. In this chapter, the effect of glass color on the energy saving in a flat was carried out in two climate zones Alexandria and Aswan in Egypt to know the suitable colored glass for each zone. The investigation was done using the computational fluid dynamics (CFD) called "Visual DOE" which is one of the famous energy programs. The results indicated that the light-colored glass type has a great effect on the energy saving in building.

Keywords Color · Glass · Energy · Visual

56.1 Introduction

In warm and sunny climates, the penetration of solar energy to the building interior is generally not desirable from a thermal perspective. The problem is thus one of providing effective natural lighting without imposing a penalty in terms of unwanted heat and of allowing a view outdoors. The approach to solving these seemingly conflicting requirements has been to develop spectrally selective glazings [1]. Smart windows are characterized by their ability to vary the throughput of radiant energy—visible light as well as solar radiation. Window glasses with coatings for the limitation of solar irradiation partially reflect or absorb incident solar radiation and usually have viewing transparency. They offer shielding against excessive solar heat input in summer and also superior insulating properties against heat losses in winter. However, the design of these window glasses has to be a compromise in fulfilling contradictory requirements, as the shielding of the excessive solar irradiation in summer causes a disadvantageous reduction of daylight illumination and solar heat gain in winter. Windows and facades with transparent insulation materials are highly translucent and generally do not have viewing transparency [2]. Chow et al. [3] compared thermal performance of water-flow window with conventional

M. A. Hassan (✉) · N. M. Guirguis
Housing and Building Research Center, Dokki, Giza, Egypt
e-mail: waghassan@yahoo.com

A. Sayigh (ed.), *Renewable Energy in the Service of Mankind Vol I*,
DOI 10.1007/978-3-319-17777-9_56

single-pane and double-pane design. The simulation results show that the reduction in annual room heat gains is around 32 and 52 %, respectively, in comparison to the conventional double- and single-pane design. Chunying and Chow [4] presented a novel design of water-filled double-reflective glazing system and predicted its annual performance through computational simulation with the use of Hong Kong weather data. The results show that the indoor room heat gain can be largely reduced by the novel glazing system. Wong et al. [5] investigate the effects of double-glazed facade with ventilation system on the energy consumption, thermal comfort, and condensation. Carlos and Capeluto [6] explore the influence of incorporating intelligence in building in hot climates through the perspective of energy consumption and user comfort.

56.2 Modeling Approaches

An energy computerized program called "Visual DOE" was used to conduct this investigation in order to estimate the effect of the different glazing types on the cooling electrical consumption for a flat of areas 168 m^2 as shown in Fig. 56.1; the flat has eight zones, 14 facades, the floor-to-floor height was 3.5 m, and the average light power density (LPD) and electrical power density (EPD) were fixed at 3 and 4 W/m^2, respectively. The U factor of different building materials which are used in the block construction is illustrated in Table 56.1. All windows have an area of 2 m height × 2 m width which make a window-to-wall ratio of 25 %. Table 56.2 illustrates the construction summary.

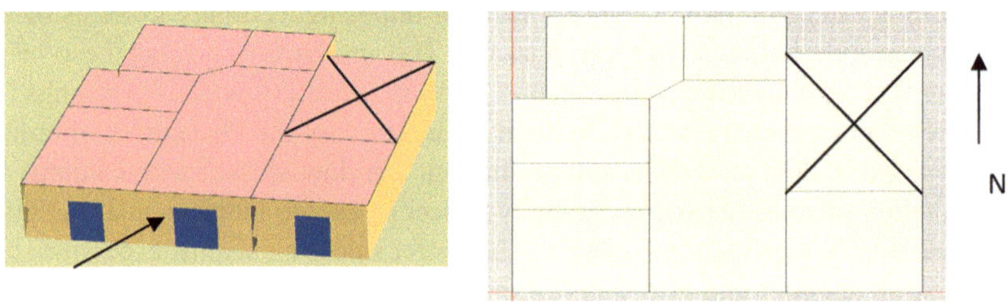

Fig. 56.1 Sketch diagram of a single flat model with total area 168 m2 and height 3.5 m

Table 56.1 *U* factor of different building materials used for constructing the flat

Construction	Description	*U* factor ($W/m^2\,°C$)
Roof	Adiabatic roof	0.114
Ceiling	Gypsum board ceiling, radiant barrier	4.229
Floor	Adiabatic floor	0.183
Internal floor	Adiabatic floor	0.183
Interior wall	partition	2.196

Table 56.2 Construction summary

Name	Area (m^2)	*U* factor ($W/m^2\,°C$)	Absorptance	Type	Category
Partition	162	2.19	0.3	Partitions	Light
Hollow clay brick wall	141	2.31	0.5	Walls	Light
Adiabatic roof	168	0.11	0.7	Roofs	Light
Adiabatic floor	168	0.18	0.0	Floors	All

Table 56.3 Illustration of the characteristic of different glazing types

	U factor ($W/m^2\,°C$)	Solar coefficient	Visible light transmission
Single clear	6.172	0.95	0.881
Opaque	0.02	0.01	0.1
Single blue 6 mm	6.172	0.71	0.57
Single gray	6.172	0.69	0.431
Single bronze	6.172	0.71	0.534
Double clear	2.742	0.81	0.781
Double clear argon	2.561	0.81	0.781
Double blue	2.742	0.57	0.505
Double blue argon	2.561	0.56	0.505
Double green	2.74	0.57	0.664
Double green Argon	2.561	0.57	0.644
Double gray	2.742	0.54	0.381
Double bronze	2.561	0.54	0.381
Double gray	2.742	0.57	0.473
Double bronze Argon	2.612	0.72	0.618
Internal shaded	6.172	0.95	0.881
External shaded	6.172	0.95	0.881
Gray tinted	2.351	0.2	0.13
Double gray tinted	2.742	0.54	0.381

The wall was constructed from hollow clay brick and has a *U* factor of 2.314 $W/m^2\,°C$. The first step in the investigation was to construct the flat using the Visual DOE and verify the base (actual or existing) case and its electrical consumption. The second step was to estimate the effect of using another glazing type instead of the base case or existing glazing on the cooling electrical power which was consumed in the ventilation and air-conditioning (HVAC) system and also the electrical power of the fan to achieve the thermal comfort for the occupants. Table 56.3 illustrates the characteristic of the different glazing types used in this investigation.

56.3 Results and Discussion

As shown in Table 56.4 for Aswan, the higher percentage value of saving electrical power consumption in cooling process was 0.55 when the clear glass was replaced by the opaque glass. The use of the gray-, blue-, and bronze-colored glass reduced

Table 56.4 Electrical usage summary for Aswan and Alexandria and their saving percentage of the cooling and the fan

Glass type	Aswan				Alexandria			
	Cooling	Saving	Fan	Saving	cooling	Saving	fan	Saving
Base case	3092	0	924	0	936	0	362	0
Opaque	1400	0.55	400	0.57	102	0.89	66	0.82
Single blue	2629	0.15	786	0.15	560	0.4	240	0.34
Single gray	2591	0.16	775	0.16	532	0.43	231	0.36
Single bronze	2632	0.15	787	0.15	562	0.4	241	0.33
Double clear	2931	0.05	856	0.07	893	0.045	345	0.047
Double clear + argon	2944	0.04	859	0.07	889	0.05	352	0.027
Double blue	2443	0.21	701	0.24	553	0.41	207	0.43
Double blue + argon	2445	0.21	700	0.24	563	0.4	209	0.42
Double green	2456	0.21	705	0.24	562	0.4	210	0.42
Double green + argon	2458	0.21	704	0.24	573	0.39	212	0.41
Double gray	2403	0.22	689	0.25	522	0.44	197	0.46
Double gray + argon	2404	0.22	688	0.26	531	0.43	199	0.45
Double bronze	2447	0.21	702	0.24	556	0.41	208	0.43
Double bronze argon	2748	0.11	798	0.14	814	0.13	293	0.19
Internal shading	2716	0.12	809	0.13	631	0.32	260	0.28
External shading	2209	0.28	665	0.28	311	0.67	143	0.61
Bronze tinted	2473	0.2	714	0.23	637	0.32	233	0.36
Green tinted	2215	0.28	630	0.32	515	0.45	185	0.49
Gray tinted	1719	0.44	483	0.48	152	0.84	81	0.77

the electrical power consumption for cooling process by 0.16, 0.15, and 0.15, respectively. This means that any of the above colored glass can be used instead of the clear-glassed type. The results show that the percentage of the electrical power saving for the double-glassed type in case of with or without argon was 0.21 except for the double bronze with argon which has 0.11. Only according to its installation cost that the owner or the occupant can make a decision for what type can be used for the higher value of saving electrical power. Finally, the results indicated that the external shading of the clear glass type can save 0.28 of the base case electrical power consumption for the cooling process which can be a more effective method to reduce the power consumption with low installation cost because it does not need to replace the clear glassed type. The last comparison was occurred between the colored tinted glassed (bronze, green, and gray) which have a percentage of 0.2, 0.28, and 0.44 of the electrical power consumption, respectively, in which means that the more suitable and effective to reduce the electrical power consumption in the cooling process is the tinted gray glass, but it still compromised between its cost and the power saving. From Table 56.4, it can be obtained that the same trend occurred for the reduction in the electrical power consumption for the fan process. For Alexandria, the same trend occurred for the selected colored glazed windows. The opaque glass has a higher saving capacity with a value of 0.89 of the base case electrical consumption for cooling process and has a value of 0.82 of the base case electrical power consumption for the fan. The gray tinted glass type has the second opportunity with a value of 0.84 and 0.77 for the cooling and the fan consumption, respectively. The last one is the external shading in which the value of the saving electrical power consumption for the cooling and the fan was 0.67 and 0.61, respectively. All the double-colored glass with or without argon have a value of 0.45 of the base case electrical power consumption for cooling and fan except the double bronze with argon which has a value of 0.13 for cooling and 0.19 for the fan.

56.4 Conclusion

There are many strategies for saving the electrical power consumption in the cooling process, the first one is using the external shading in which the installation cost was lower than the other types, and also this method does not need to replace the clear glass type. The second type of strategy is the use of the colored glass types instead of the clear glass. A decision will be taken according to the compromise between the saving percentage value and its cost. The recommendations for further work was to study the effect of using the shading with the colored glass types single or double with or without argon.

References

1. Erell E, Etzion Y, Carlstrom N, Sandberg M, Molina J, Maestre I, Maldonado E, Leal V, Gutschker O (2004) "SOLVENT": development of a reversible solar-screen glazing system. Energy Build 36:467–480
2. Lorenz W (1998) Design guidelines for a glazing with a seasonally dependent solar transmittance. Solar Energy 63(2):79–96
3. Chunying L, Chow T (2011) Water-filled double reflective window and it's year-round performance. Procedia Environ Sci 11:1039–1047
4. Wong H, Liping W, Chandra AN, Pandey AR, Xiaolin W (2005) Effects of double glazed facade on energy consumption, thermal comfort and condensation for a typical office building in Singapore. Energy Build 37:563–572
5. Carlos EO, Capeluto IG (2008) Strategic decision-making for intelligent buildings: comparative impact of passive design strategies and active features in a hot climate. Build Environ 43:1829–1839
6. Chow T, Li CY, Lin Z (2011) The function of solar absorbity windows as water-heating device. Build Environ 46:459–467

Chapter 57
Energy Savings and Environmental Benefits from Solar Window Film for Buildings in Kurdistan of Iraq

Kamil M. Yousif

Abstract The advances in thin film coatings for window glass products provide a means of substantially reducing heat gain without proportionally reducing daylight transmittance.

Households in Kurdistan of Iraq pay rather high electricity bills in summer due to the excessive use of fans, evaporative coolers and air-conditioners, since the solar radiation intensity in Kurdistan of Iraq (e.g. Duhok city) at midnoon in summer season exceeds 1000 W/m^2, and the air temperature goes as high as 45 °C. this leads to, the rise in the use of air-conditioning in residential buildings with the associated serious impact on the environment. This could lead to high levels of carbon dioxide (CO_2) and other greenhouse gas (GHG) emissions. As environmental concern grows, a more-efficient energy conversion and utilization technologies become cost-effective. Saving money on energy bills is attractive to businesses, industries and individuals alike. Regulations and standards worldwide tend to embody detailed energy analysis of buildings.

In this chapter, two wooden cabinets were built as a building simulation. Measurements were made for two wooden cabinets, one with solar control film coating on the window glass and the other without. The results of an experimental investigation aimed at assessing the performance of thin film windows in the wooden cabinets are presented. The work is performed under real weather conditions. The data are integrated with spectrophotometric measurements.

In order to decrease the solar heat gain, window films may be a good choice against the solar heat gain caused by the windows in the building. The findings showed that the solar film coating could cut down energy expenditures for air-conditioned buildings. Concerning the environmental benefits, the reduction of CO_2 emission by reducing the electrical energy consumption using window film was calculated. Briefly, in this study, it was found that, the window film works smarter and greener to save energy costs.

K. M. Yousif (✉)
Dept of Environmental Sciences, Zakho University, Zakho, Iraq
e-mail: kamy22002@yahoo.co.uk

A. Sayigh (ed.), *Renewable Energy in the Service of Mankind Vol I*,
DOI 10.1007/978-3-319-17777-9_57

Keywords Solar heat gain · Reduction of carbon dioxide (CO_2) · Transmittance for solar control film · Solar control window films

57.1 Theoretical Background

Energy security and climate change are significant issues globally. As we move into the future, we will need to adapt our energy consumption to the availability of renewable sources. A critical part of this adaptation will be reducing the energy consumption by buildings.

If we are to mitigate climate change, CO_2 emission, and secure our future energy resources with minimum economic impacts, we must change the way in which we design, procure and operate buildings. Science and engineering help us to design and construct high-performance buildings which are comfortable and functional, yet use natural resources efficiently and minimize environmental impacts of their construction and operation.

Heat gain is the rate at which energy is transferred to or generated within a space. The origin of solar heat gain (SHG) is the direct and diffused radiation coming from the sun and the sky (or reflected from the ground and other surfaces). Some radiation is directly transmitted through the glazing to the building interior, and some may be absorbed in the glazing and indirectly admitted to the inside. Some radiation absorbed by the frame will also contribute to the overall window SHG factor [1]. Other thermal (nonsolar) heat transfer effects are included in the overall heat transfer coefficient—U value of the window. Heat gains occur in many forms, for example, solar radiation passing through glazing and other openings, heat conduction with convection and radiation from the inner surfaces into the space. Other forms of heat gain are given by Kalogirou [2]. SHG through fenestration is considered as the largest provider to building envelope cooling load and the most important parameter for the overall thermal transfer value determinations. The wall and roof constructions affect the thermal performance of buildings in which high overall heat transfer coefficient (U value) yields to more heat flow by conduction inside the building through walls and roof. Thick construction or insulation (lower U value) reduces the heat gain which in turn reduces the indoor air temperature inside the buildings.

It would be advantageous to reduce the energy consumption of buildings to a more environmentally friendly level and to a more energy cost-efficient level by using solar control window films (SCWF). The optical and thermal properties of the SCWF were investigated by many researchers [3–6].

Yin et al. [7] developed a procedure to estimate the energy performance of SCWF as an energy-saving retrofit to existing buildings in Shanghai, China. An equation was used by Alnaser [8] to model the SHG in a car after exposure to the sun for time interval *(t)*. That equation was modified and an approximated equation was used by Yousif et al. [9] in order to fit the case of SHG for a room in a building. Yousif [10] discussed the advantages of using SCWF to reduce the energy consumption of buildings in Iraq.

For this study, two identical wooden cabinets—A (without the use of SCWF; WOSCWF) and B (windows with the use of SCWF; WSCWF) were built as a building simulation.

When the SCWF is applied to the windows, it will act to reduce the solar radiation that will enter the wooden cabinet *(room) through* the window. This will cause the SHG of the wooden cabinet *(room) or* building (SHG_1) to be reduced to some lower value (SHG_2), and the difference in SHG, ΔSHG, will be given by:

$$\Delta SHG = SHG_1 - SHG_2 \tag{57.1}$$

Then the difference in SHG, ΔSHG, in the wooden cabinet A (WOSCWF) and wooden cabinet B (WSCWF) can be estimated as:

$$\Delta SHG = (\tau_b - \tau_a) I A_g + \rho_{air} V_i C_{air} / t\{(\Delta T_b - \Delta T_f)\} + \varepsilon \sigma A_i \left(T_{fb}^{\,4} - T_{ff}^{\,4}\right), \tag{57.2}$$

where τ_b is transmissivity of the window glass for cabinet A. τ_a is the transmissivity of the window glass for cabinet B. I is the solar radiation (W/m^2), A_g is the area of the glass windows (m^2), ρ_{air} is the air density (kg/m^3) inside a wooden cabinet, V_i is the interior volume of the cabinet (m^3), ΔT is the temperature difference between the outside and inside air temperature of the cabinet, C_{air} is the air-specific heat capacity (J/kg K), ε is the emissivity, σ ($=5.6697 \times 10^{-8}$ W/m^2K^4) is the Stefan–Boltzmann constant, A_i is the cabinet interior area. The temperature difference ΔT_b will be given by:

$$\Delta T_b = T_{fb} - T_{ib}, \tag{57.3a}$$

where T_{fb} and T_{ib} are the final temperature and the initial temperature in the cabinet A, respectively.

The temperature ΔT_f will be given by:

$$\Delta T_f = T_{ff} - T_{if}, \tag{57.3b}$$

where T_{ff} and T_{if} are the final temperature and the initial temperature in the cabinet B, respectively.

57.1.1 Experimentation

Measurements were made for the two identical wooden cabinets A and B, and they were built and used for the study as a building simulation. Each of the internal dimensions, 120 cm height, 70 cm length, and 40 cm width, are shown in Fig. 57.1, with a window-to-wall ratio (WWR) of around 0.178.

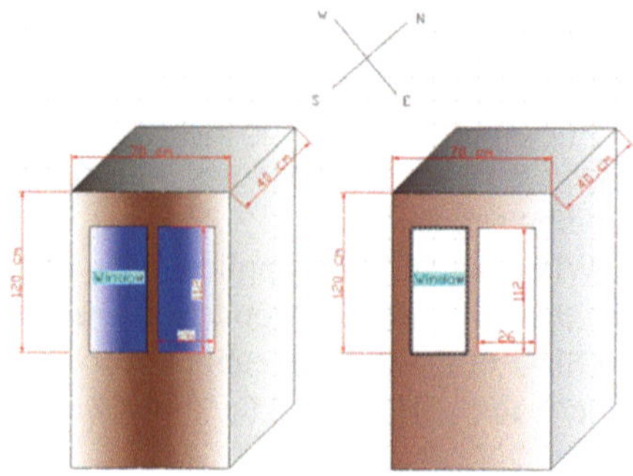

Fig. 57.1 The test wooden cabinets

The wooden cabinet B is with solar control film coating on the window glass and the other cabinet without, that is cabinet A. Both cabinets consist of two sections, upper and lower. The sections were separated from each other by a wooden shelf. There are two glass windows through each of the front and rear faces of the upper part. Each glass window is 26 cm wide, 112 cm high and 0.3 cm thick. The front side of the cabinet is directed to the south. The solar radiation intensity and other weather elements were measured inside and outside the cabinet for 10 h from 8 a.m. to 6 p.m. local time. The solar radiation intensity inside both wooden cabinets was measured during all the 10 h outside and inside both cabinets using solar radiation intensity meter. The temperature and relative humidity were also measured for the same cases and period using digital thermometer and simple hygrometer. The measurements of the wind speed were made by means of an anemometer. Readings of the solar radiation intensity, temperature, the wind speed and humidity were recorded every half an hour. Because of the semi-darkness inside the cabinet after applying the window film, a small aperture was made in the cabinet to make the task of taking the readings of the measurement devices easy. Field measurement was conducted between June and early September 2013 in Duhok city, Iraq. The measurements of the absorbance and transmittance of the window glass as well as the SCWF in the regions of UV, visible (VIS), and near infrared (NIR) were made by means of a (Perkin Elmer, Lambda 25) spectrophotometer.

57.2 Results and Discussion

Average indoor hourly temperatures for both cabinets A and B, together with average outdoor hourly temperatures, confirm that in summer, temperatures reach values in excess of comfort levels. The maximum temperature is 46 °C (for outside the cabinet) observed at 2 p.m. While the highest temperature is 68.5 °C inside cabinet A (without solar window films) at 2.30 p.m. With solar window films, significant reductions in the temperatures are observed (maximum temperature = 57 °C). Concerning the indoor and outdoor relative humidity, it was found that there was no

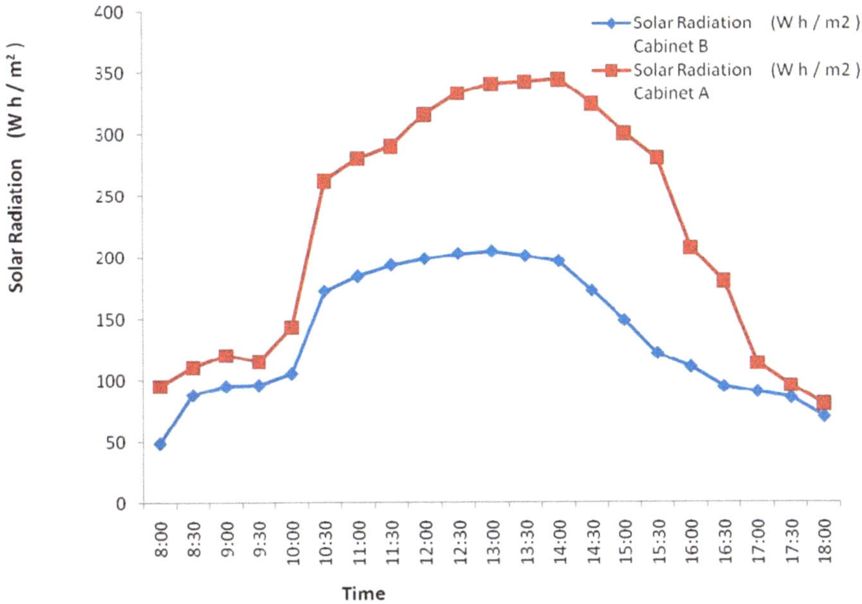

Fig. 57.2 Average daily interior solar heat profiles for the two cabinets A and B

noticeable change in the humidity both indoor and outdoor or with and without the use of SCWF. Wind speed was in the range 0.5–3.2 m/s.

The transmittance for solar control film in the UV–Vis–NIR region indicates that there is no light transmitted in the UV region. The transmittance for both glass and solar control film together is less than 8 % at wavelength of 400–650 nm. Then it increases exponentially to about 58 % at a wavelength of 750. It was about 57 % (wavelength of between 800 and 1000 nm). The transmittance for glass is 0.87 %.

To study the average performance of the film coating, mean daily solar heat profile based on the 20 sets of daily data was determined and is plotted in Fig. 57.2. Significant solar heat reductions are observed between 11 a.m. and 3 p.m. indicating that there was a clear sky and no cloud. It can be seen that the measured solar heat data for the two cabinets have very similar trends and peak at the same time (i.e. at 2 p.m.). In general, the features are quite similar for both curves in the morning and the evening, the solar radiation data are mainly diffused of low values, and the solar heat transmittance becomes smaller. The solar heat reduction ranges from 3 to 15 % at those periods. While increase in heat during noon to more than 50 % are due to direct component of solar radiation of high values, the solar heat transmittance of the solar control film depends strongly on the amount of solar heat received. The areas under the two curves in Fig. 57.2 indicate the average daily amount of solar heat received by the two cabinets, and the difference is the solar heat rejected by the film coating.

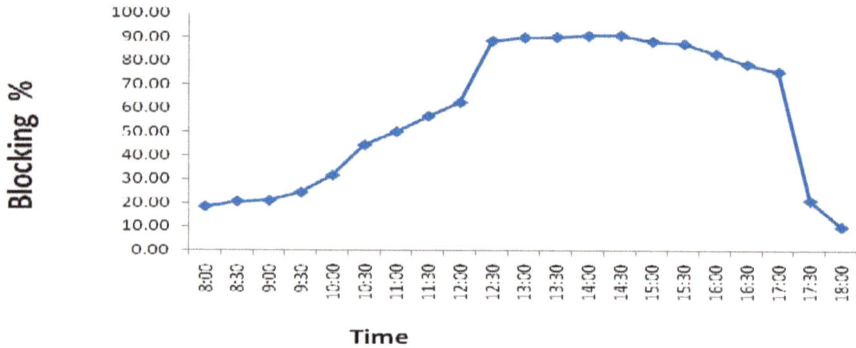

Fig. 57.3 The blocking of the solar radiation for solar control film

It was found that the measured solar heat varies from one day to another, but similar patterns can be found for the two curves indicating that the same amount of solar radiation falls on the facades of the two cabinets A and B.

Some other measurements show low solar heat, indicating a cloudy sky, which is dominated by diffuse solar component. It means that the solar heat reduction ability of the film coating is directly proportional to the amount of solar radiation received and depends on the types of solar radiation. Since the two cabinets A and B face the south, total solar radiation (pure direct and diffuse component) of solar heat would be received in the noon. Therefore, the performance of the film coating in a sunny sky at noon would be much better than those recorded in other few measurement times of cloudy sky.

Blocking of the solar radiation (or solar heat reduction percentages) for SCWF is shown in Fig. 57.3. This graph highlights the performance of the SCWF. The results show that the thin film is blocking between 10 and 90 % of the solar radiation, depending on the time of day, and on weather conditions, for example clouds, haze, etc.

When the time of day is between 12:30 and 4 p.m., a thin film is blocking about 90 % of the solar radiation. Large reduction in solar radiation usually represents a sunny sky with a large portion of direct component.

57.2.1 Calculations of Solar (Heat) Gain

Equation (57.2) was used to calculate the reduction in the SHG in the wooden cabinet under test. Following are the calculations of reduction in SHG.

The difference between the transmissivity $\Delta\tau$ of the window glazing of the test cabinet under test before and after applying the film is given by:

$$\Delta\tau_b - \Delta\tau_a = 0.87 - 0.57 = 0.30$$

Therefore, the reduction in the amount of SHG due to $\Delta\tau$ of the window glazing caused by window film application is give by the amount of $\Delta\tau_b - \Delta\tau_a$.I. A_g is the first term of the Eq. (57.2). The average value of the solar radiation (W/m^2) Ī is 900 W/m^2 and the total area of the glazing is 1.16 m^2.

Thus, the value of the first term of Eq. (57.2) is $0.30 \times 900 \times 1.16 = 313.2$ W.

The value of the second term of Eq. (57.2) is 0.0579 W. The value of the third term of Eq. (57.2) is 138.25 W. Total difference in solar (heat) gain will be:

$$\Delta P_{gain} = 313.2 + 0.0579 + 138.25 = 451.51 \text{ W}$$

The ratio of the glazing area to the total area $Ag/A_i = 0.178$, and this led the first term of Eq. (57.2) to be dominant. The second term is almost negligible.

Accordingly, in a large-scale practical case, if a window film with similar properties is applied to the windows of classrooms of a three-storey (floors) school along the south perimeter zone at Duhok (with dimensions: width = 5 m, height = 3 m, length = 7 m) containing two windows in each classroom, of total net glazing area of 8 m^2, (e.g. Safen Secondary School), and by considering that the temperature difference ΔT_b and ΔT_f of Eqs. (57.3a) and (57.3b) have the same values for both cases (i.e. the test cabinet and the case of classrooms), the results will be as follows:

The value of the first term of Eq. (57.2) is $(\Delta\tau_b - \Delta\tau_a)$.I. $A_g = 0.30 \times 900 \times 8 = 2160$ W

The value of the second term of Eq. (57.2) is 18.1 W. And the value of the third term of Eq. (57.2) is 6134.84. The ratio of the glazing area to the total area A/A_i is 0.16. This makes the third term of Eq. 57.2 to be dominant.

Total difference in solar (heat) gain (in case of classrooms) will be:

$$\Delta P_{gain} = 2160 + 18.1 + 6134.84 = 8321.94 \text{ W}$$

This causes the reduction in SHG (in case of classrooms) to be almost $\Delta P_{gain} = 8312.94$ W.

For the classrooms of a three-storey school along the south, there are five classrooms on each floor; this causes the reduction in SHG to be almost $\Delta P_{gain} = 8312.94 \times 5 \times 3 = 124694.1$ W.

57.2.2 Calculations of the Reduction in CO_2 Emissions

The total energy consumption of heating, ventilation and air-conditioning (HVAC) systems will be reduced by the total difference in solar (heat) gain.

In case of the test cabinet, ($\Delta P_{gain} = 451.51$ W), which saves about 1648.01 kWh/year.

The estimated reduction in CO_2 is 1.17 t (for the case study—the test cabinets). This was calculated on a factor of 0.71 kg of CO_2 for every kWh. (See Ref. 11).

Likewise, in the school's case:

Total difference in solar (heat) gain ($\Delta P_{gain} = 124694.1$ W) which saves about 455133.465 kWh/year. The estimated reduction in CO_2 emissions is about 323.14476 t/year. And for 100 school (or similar class rooms or a laboratory in the university) in Duhok, the estimated reduction in CO_2 emissions will be about 32314.476 t/year.

57.3 Conclusions

In this research, two wooden cabinets were built as a building simulation in Dohuk, Kurdistan of Iraq. Two wooden cabinets were measured, one with solar control film coating on the window glass and the other without. This chapter presents an investigation of the performance of a solar control film based on a clear window glass in a cabinet. The study is generalized in terms of a large-scale practical case, if a window film of similar properties is applied to the windows of a school classroom or similar classrooms or the laboratory in the university at Duhok.

Using a SCWF for a building reduces the fuel consumption in the building and subsequently the CO_2 and other greenhouse gas (GHG) emissions. Further, the advantages in using such films are reducing the glare, heat and UV. It was found that SCWF can block a significant amount of the sun's heat. This could lead to reducing the load on the air conditioner for the buildings.

Calculations of reduction in SHG for a small-scale test cell (wooden cabinet), and for 15 classrooms in a school in Duhok, indicated that the total difference in solar (heat) gain is about 451.51 and 124694.1 W, respectively.

Calculations of the estimated reduction in CO_2 emissions/year due to the savings in electrical energy by window films was about 1.17 t in case of a small wooden cabinet, and in case of 100 schools (of 15 classrooms or similar classrooms or laboratory in university) in Duhok is about 32314.476 t/year.

It was observed and confirmed that solar control thin film's blocking depend on the time of day, and on weather conditions, for example, haze, clouds, etc.

References

1. Cengel Yunus A (2007) Heating and mass transfer, Chapter 12. Mac-GraW Hill, Pennsylvania State University
2. Kalogirou SA (2009) Solar space heating and cooling solar energy engineering. Academic Press, Boston, pp 315–389
3. Nostell P (2000) Preparation and optical characterisation of antireflection coatings and reflector materials for solar energy systems. Dissertation for the Degree Doctor of Philosophy, Acta Universitatis Upsaliensis, Uppsala, Sweden
4. Rubin MD (1982) Solar optical properties of windows. Energy Res 6:123–133
5. Danny HW Li, Joseph C Lam, Chris CS Lau, Huan TW (2004) Lighting and energy performance of solar film coating in air-conditioned cellular offices. Renew Energy 29:921–937

6. Roos A, Polato P, van Nijnatten PA, Hutchins MG, Olive F, Anderson C (2000) Angular dependent optical properties of low-e and solar control windows—simulation versus measurements. Solar Energy 69(Suppl. 6):5–26
7. Yina R, Xub P, Shenb P (2012) Case study: energy savings from solar window film in two commercial buildings in Shanghai. Energy Build 45:132–140
8. Alnaser WE (2007) The solar gain in automobiles in Bahrain and its negative impacts. JAAUBAS 4(Suppl.):317–324
9. Yousif Kamil M, Jundi Kh Yousif, Manaf A Mahammed (2012) Potential of using window films to support protection of the environment in Kurdistan of Iraq. Proceedings of World Renewable Energy Forum 2012 (WREF2012), Denver, Colorado, USA
10. Yousif Kamil M (2012) Control of solar heat gain to reduce the energy consumption of buildings in Iraq. Proceedings of World Renewable Energy Forum 2012 (WREF2012), Denver, Colorado, USA
11. Aboulnaga Mohsen M, Al-Ali Najeeb Mohammed (2008) Low carbon and sustainable buildings in Dubai to Combat global warming, counterbalance climate change, and for a better future. In: Sayigh A (ed) World renewable energy congress (WREC-X). Scottish Enterprise Energy Team, Scotland, pp 566–575

Chapter 58
Research on Daylighting Introduction of Commercial Buildings in Different Climate Zone of China

Hong Jin and Xin-xin Li

Abstract This chapter chose Beijing, Harbin, Kunming, Shanghai, and Shenzhen, the key cities of different climate zones in China and studied the lighting energy-saving effect of daylight in large commercial buildings. Based on interior Pedestrian Street that is a typical composition pattern, this study uses the DAYSIM analysis to affect the trends of different daylighting system on energy saving of artificial lighting in different opening ratio, different orientation and different climate zone.

Keywords Commercial buildings · Daylight · Energy consumption of lighting · Climate zone

58.1 Introduction

With the enhancement of environmental awareness, building energy consumption and comfort is receiving increasing attention. Many countries made standard and illegal to save energy and improve the quality of environment. Nowadays, architecture design is not only about function, shape, and aesthetics alone, but also refers to how to create a better interior environment and decrease the energy consumption of lighting, air-conditioning systems, ventilation, and hot water by using natural resources in terms of site planning, volumes, windows design, shading pattern, and building materials [1].

According to statistics, the number of power consumption of large public buildings in China account for 22 % of the total energy use. The change conveys the intended meaning commercial buildings is 1.5–2 times in China than in Japan per square meter, 10–20 times more than residence, 2 times more than hotels and offices in China. A survey on energy consumption of 38 commercial buildings in Beijing, Shanghai, Guangzhou, Chongqing, Hangzhou, and Nanjing was carried out. The results shows that the average power consumption is 244 kWh/m^2

X.-x. Li (✉) · H. Jin
School of Architecture, Harbin Institute of Technology, 150001 Harbin, China
e-mail: 417361741@qq.com

A. Sayigh (ed.), *Renewable Energy in the Service of Mankind Vol I*,
DOI 10.1007/978-3-319-17777-9_58

Table 58.1 The typical cities and the analysis mode

	City	Climate zone	Sky conditions (lux)
1	Harbin	Severe cold zone	7186
2	Beijing	Cold regions	7800
3	Shanghai	Hot-summer and cold-winter zone	8659
4	Kunming	Mild zone	9165
5	Shenzhen	Hot-summer and warm-winter zone	9343

including 30 ~ 50 % air-conditioning energy consumption and 30 ~ 50 % lighting energy consumption. According to literature survey for large commercial buildings in Beijing, energy consumption of air-conditioning, lighting, and elevators is 50, 40 and 10 %, respectively [2]. Lighting accounted for 30 ~ 50 % of the total power consumption in Chongqing shopping malls [3] and 39 % in American shopping malls [4]. This ratio is 55 % in Holland [5] and 24 % in Brazil, particularly in non-air-conditioned shopping mall, which is as high as 70 % [6]. In building energy consumption, lighting holds a large proportion, so it have a vast of potential energy.

The use of daylighting requires special attention in Green Lighting Engineering. Not only does Daylighting put users in touch with nature, but it is also an important means of sustainable building from the angles of energy and health. However, at present quantitative studies on energy consumption in commercial building are infrequent. According to reasonable daylighting way which is fixed by the quality and site of function, this chapter analyzes the effect for daylighting openings on the indoor light environment in different climate zones using DAYSIM software. Daylighting is combined with artificial lighting as a premise, and then the effect for daylighting on lighting energy consumption is analyzed.

58.2 Research Objects

This chapter selects five typical cities in different climate zones of China to carry out the research, respectively are Harbin, Beijing, Shanghai, Kunming, and Shenzhen, and the sky conditions of each zone is shown in Table 58.1. The analysis mode is the commercial buildings with interior pedestrian street, as shown in Fig. 58.1, which has three stories, each story about 5875 m^2 and 117.5 m long, 50 m wide and 5.4 m high.

According to the function and space design, using the characteristics of natural daylight and the complexity of daylight opening open, this chapter sets six cases based on the mode of analysis, the daylight opening position and form as shown in Table 58.2 and the perspective of each case as shown in Fig. 58.2.

The different orientation of daylight opening will have great influence on the building's heat gain from solar radiation and heat dissipation, and thus affect the buildings energy consumption, the statistic of opening ratio for each case as shown in Table 58.3.

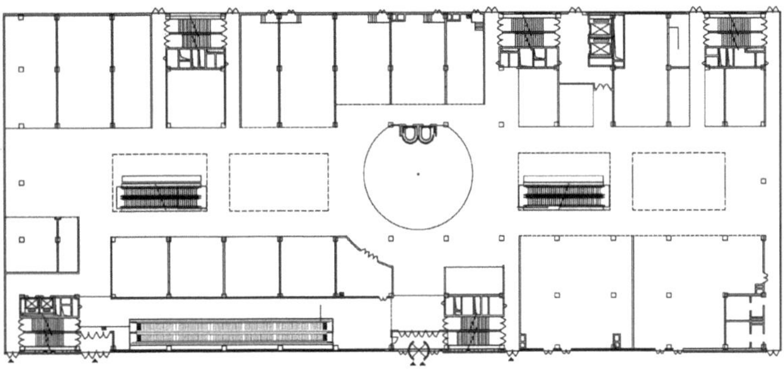

Fig. 58.1 The floor plan

Table 58.2 The setting of daylight opening

Daylight position	Entrance	Stores of first floor	Other stories	Auxiliary space	Atrium	Traffic space of top floor
Daylight form	Glass door	French windows	High side window	Side window	Skylight	Skylight
case 1	✓					
case 2	✓			✓		
case 3	✓	✓		✓		
case 4	✓	✓		✓	✓	
case 5	✓	✓		✓	✓	✓
case 6	✓	✓	✓	✓	✓	✓

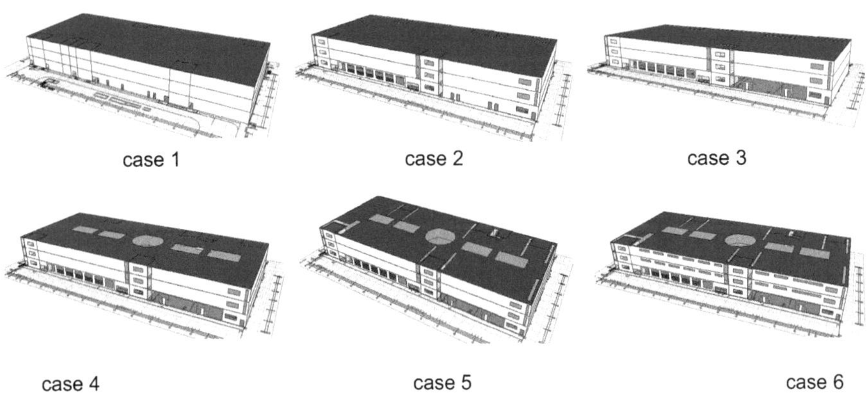

Fig. 58.2 The perspective of each case

In order to enhance the heat preservation and insulation ability, the doubleglazed lowE timberframe for which the heat transfer coefficient is 2.26 W/m^2K, solar heat gain coefficient is 0.75, and visible light transmission is 0.639 was selected as the type of window in the simulation.

Table 58.3 The statistic of opening ratio

	Opening ratio (%)			
	South	North	Roof	Total
Case 1	2.87	2.41	0.00	0.89
Case 2	16.59	8.29	0.00	4.19
Case 3	22.65	20.12	0.00	7.20
Case 4	22.65	20.12	12.01	13.44
Case 5	22.65	20.12	16.66	15.86
Case 6	28.10	24.85	16.66	17.58

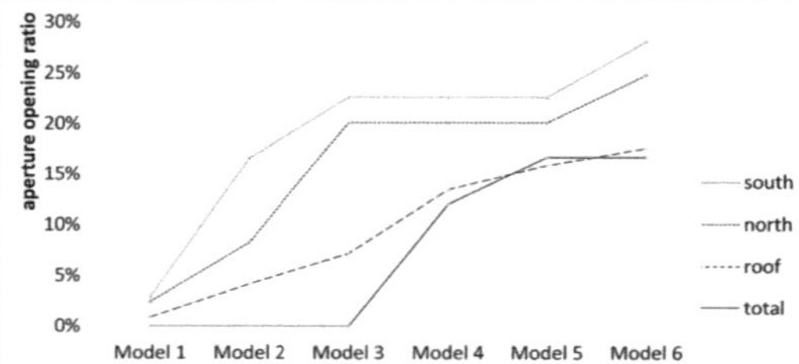

58.3 Research Methods

58.3.1 Simulation Software

The daylight factor only aims at the whole cloudy day model, so it is not appropriate for a comprehensive evaluation of building daylight performance. The daylight autonomy (DA) and maximum daylight autonomy (DAmax) are the dynamic evaluation index. Though setting the minimum illumination, it can calculate the percentage of time that meets the lighting requirements in the yearly work time [7]. In order to study the influence of daylight opening setting on lighting electricity consumption, it applies the Ecotect software to set the relative parameters and calculation grid, and then import the mode information into the DAYSIM software to calculate the all year illuminance value of each grid point between 9:00 and 21:00 every day. The value will be automatically converted to the DA.

58.3.2 Relative Indexes

Daylight autonomy (DA): The daylighting quantity of some point in a building is defined as the whole year working time the point only relies on natural light can meet the percentage of time that minimum design illumination requirement.

Continuous daylight autonomy (DAcon): It is a new concept put forwarded by Roger, that when the daylighting illuminance values are less than the specified minimum design illuminance values, it adopts the weighing coefficient method to inspect the dissatisfied degree.

Maximum daylight autonomy (DAmax): It is similar to DA, but the specified value of maximum daylighting quantity is 10 times the minimum design illuminance values, that is the upper threshold value.

Annum light exposure: The acceptable amount of visible light for the whole year at some point indoor [8].

58.3.3 Parameters Setting

According to the field survey for the indoor physical environment of large commercial buildings, when the illuminance value of a public area reaches 350 lx the subjective evaluation is moderate, so the fiducial illuminance value was set as 350 lx. It selects business hours that are from 9:00 to 21:00 every day as the simulation time, takes 5 min as the calculation step, and hourly calculates the DA of each grid point in the model.

The lighting power density (LPD) is the needed light power per square meter to meet the illuminance value, which is generally considered in the light design and calculated by the light power divided by the area of the space. According to the

Chinese national standard, the lighting power density of stores should be between 9 and 16 W/m^2. In this study, set the lighting power density as 12 and the area of the zone as 15.

58.4 Simulation Results and Analysis

58.4.1 Lighting Energy Consumption

Energy consumption of primary lighting is 2018.1 kWh per year by simulation calculation.

Table 58.4 shows the results of models that is DA, DAcon, DAmax, Annum light exposure, and lighting energy consumption in different climate zone. With the increase of daylighting opening ratio, lighting energy consumption is gradually decreasing, the most obvious in the Shenzhen area.

58.4.2 Analysis of Influence on Energy Consumption

Figure 58.3 shows the trends of lighting consumption with the opening ratio. It can be seen that the effective energy saved in Shenzhen is best, Beijing is second, Harbin and Shanghai is third, the worst is Kunming. Furthermore, using regression analysis, the relationship between opening ratio and lighting energy consumption was found as shown in Figs. 58.4–58.8, the results help to direct the design.

This chapter compared the results of opening growth rate of different orientations as shown Table 58.5 with lighting energy consumption as shown in Fig. 58.3. It can be seen from the slope of energy consumption that a normal size window located on the roof could drive down lighting energy consumption, second is at the south of the building, the worst is at the north of the building. Strip windows in all directions have little impact on lighting energy saving.

58.5 Conclusions

Based on the functional space, the use of daylighting, the direction of opening, and climate zone of commercial building, this chapter studied the assessments of lighting energy saving through quantitative analysis by DAYSIM and ECOTECT. As a top energy user, this is a time of GREEN which requires architects to deal with more complex problems. Rapid, human and material saving, high reliability method and tools should be adopted to assess the project in order to make a responsible decision to owner, environment, laws and standards before the building is completed.

Table 58.4 Annual daylighting and lighting situations

Place	Sky conditions (lux)	Number of model	DA (%)	DAcon (%)	DAmax (%)	Annum light exposure (Luxh)	Lighting energy consumption (kWh)
Beijing	7800	Case 1	2.26	5.61	0.18	446705890.17	1972.57
		Case 2	18.13	31.89	2.25	3669611115.33	1652.28
		Case 3	27.24	38.91	3.45	5370784007.17	1468.31
		Case 4	67.96	72.55	22.13	23314206326.83	646.61
		Case 5	68.76	72.96	26.04	27271883174.67	630.51
		Case 6	70.49	73.88	29.32	30265131981.00	595.49
Harbin	7186	Case 1	1.92	4.88	0.17	372883336.00	1979.28
		Case 2	16.24	28.87	1.99	3201549359.17	1690.40
		Case 3	24.21	35.27	2.96	4647501615.17	1529.57
		Case 4	62.35	67.02	18.50	19541751997.83	759.80
		Case 5	63.19	67.44	21.93	22863721477.33	742.82
		Case 6	64.96	68.30	24.83	25409896439.67	707.19
Kunming	9165	Case 1	1.93	4.83	0.15	301764074.50	1979.18
		Case 2	14.45	26.33	1.54	2413483782.50	1726.53
		Case 3	22.95	32.61	2.79	3727851936.67	1555.03
		Case 4	59.30	63.35	20.97	19282195903.83	821.39
		Case 5	60.03	63.74	24.28	22510031605.83	806.69
		Case 6	61.34	64.39	26.54	24794960058.17	780.11
Shanghai	8659	Case 1	1.77	4.66	0.13	331808668.33	1982.31
		Case 2	14.43	26.58	1.44	2636565101.33	1726.88
		Case 3	22.76	33.54	2.44	3977561176.17	1558.74
		Case 4	62.43	67.82	20.35	20348277316.50	758.19
		Case 5	63.40	68.39	23.79	23961607842.67	738.68
		Case 6	64.88	69.23	25.94	26115248233.67	708.77

Table 58.4 (continued)

Place	Sky conditions (lux)	Number of model	DA (%)	DAcon (%)	DAmax (%)	Annum light exposure (Luxh)	Lighting energy consumption (kWh)
Shenzhen	9343	Case 1	2.45	6.19	0.19	392621045.83	1968.57
		Case 2	18.40	33.31	1.89	3054783040.17	1646.81
		Case 3	29.93	41.60	3.54	4740340266.67	1414.14
		Case 4	75.74	78.17	28.43	25783715352.50	489.59
		Case 5	76.18	78.36	32.98	29806819441.00	480.77
		Case 6	77.31	78.81	36.06	32966516304.17	457.97

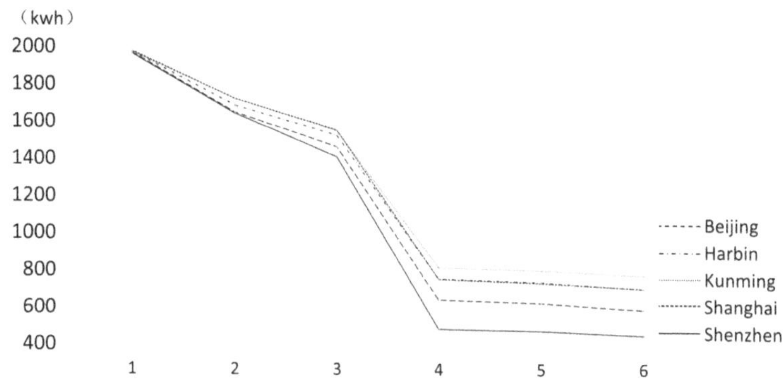

Fig. 58.3 Lighting energy consumption changes in different areas

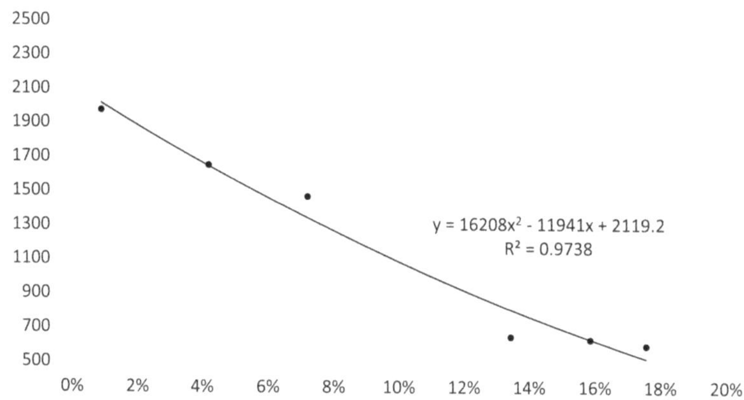

Fig. 58.4 Changes of lighting energy consumption with the opening rate (Beijing)

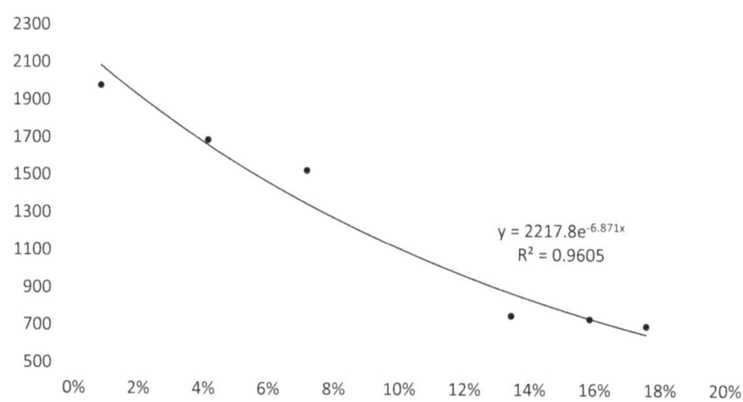

Fig. 58.5 Changes of lighting energy consumption with the opening rate (Harbin)

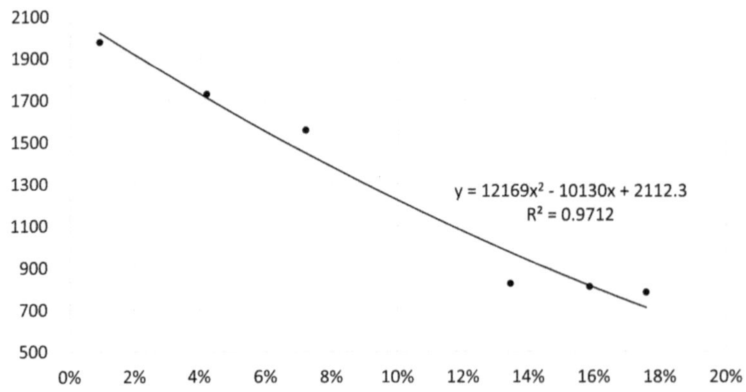

Fig. 58.6 Changes of lighting energy consumption with the opening rate (Kunming)

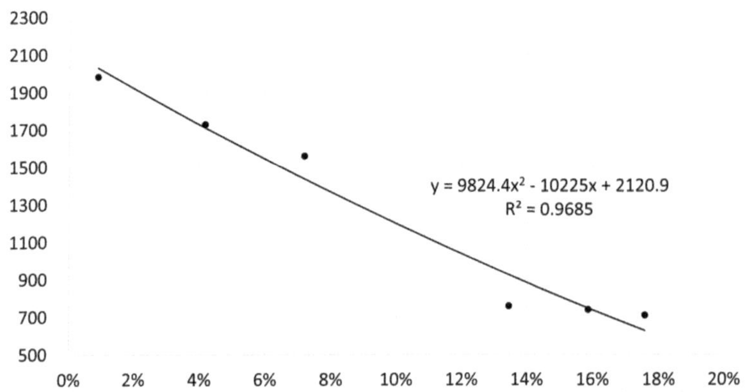

Fig. 58.7 Changes of lighting energy consumption with the opening rate (Shanghai)

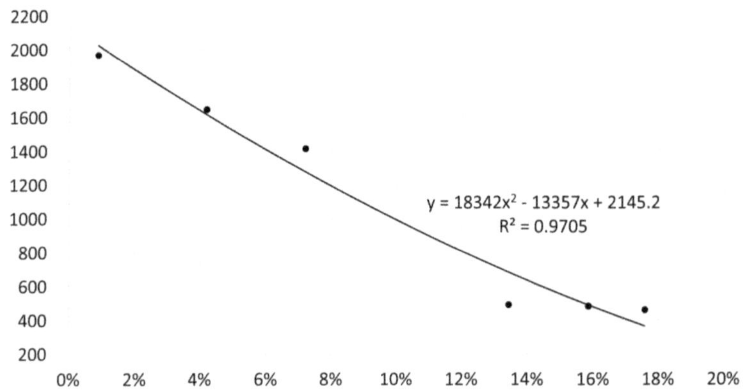

Fig. 58.8 Changes of lighting energy consumption with the opening rate (Shenzhen)

Table 58.5 The growth rate of daylighting opening

	Case 1 (%)	Case 2 (%)	Case 3 (%)	Case 4 (%)	Case 5 (%)	Case 6 (%)
South	2.87%	13.72%	6.06%	0.00%	0.00%	5.45%
North	2.41%	5.87%	11.83%	0.00%	0.00%	4.74%
Roof	0.00%	0.00%	0.00%	12.01%	4.65%	0.00%

References

1. Wangmengxin (2012) The application of Eco—architecture—aided design software
2. Xuezhifeng J (2004) Situation of energy using and analysis on energy-saving potential for large commercial buildings in Beijing. HVAC 34(9):8–10
3. Wangyouwei, Wangqingqin C (2012) Green building report
4. EIA (1941) Energy end-use intensities in commercial buildings. Energy Information Administration, US Department of Energy, Washigton
5. Sliepenbeek W, Van Broekhoven L (1995) In: Evaluation of stimev, the al-l Dutch utility-sponsered lighting rebate programs. Right Light Three, Third European Conference on Energy-Efficient Lighting. Proceedings. Vol. I. Presented papers. England
6. PROCEL (1993) Handbook of energy savings in public and commercial buildings
7. Yunpeng (2010) Lighting environment simulation in the building. China Architecture & Building Press, Beijing
8. Konglingyu (2012) Study of daylighting metrics based on luminance. Tianjin University, Tianjin

Chapter 59
Reducing Canadian Greenhouse Energy Costs Using Highly Insulating Glazing

William David Lubitz

Abstract This study explores the potential to significantly reduce heating costs in northern climates by adding additional layers and types of glazing to simulate greenhouse-like structures with highly insulating roofs that still allow appreciable light transmission. Representative typical meteorological year (TMY) data are used with a one-dimensional multilayer heat transfer model to explore the potential of insulated glazing panels for increasing solar gain. The model was validated against interior temperature and exterior weather data collected at a small unheated greenhouse in winter conditions. It was found that it is possible to reduce winter heating energy costs by using thicker-than-normal insulated glazing, although in practice, energy savings will be offset by added costs of the insulated glazing and additional measures would be needed to minimize summer heat gain. However, the potential for reducing winter heating energy costs suggests potential for highly insulating, semitransparent cladding.

Keywords Passive solar · Greenhouse · Model · Transparent insulation · Low-energy architecture

59.1 Introduction

Energy costs, primarily for heating and ventilation, are the second highest operating costs for commercial vegetable greenhouses in Ontario, Canada (after labor). Large commercial greenhouses can cover many hectares, essentially acting as large-scale, ventilated, plant-containing passive solar collectors. Most greenhouses are clad in only one or two layers of glass or plastic glazing to minimize structural cost and ensure high solar radiation transmissivity to the plants inside [1]. However, this light cladding also results in low thermal performance during cool periods.

Transparent insulation materials have been explored by many researchers for several decades [6]. Conventional glass or plastic glazing can be supplemented with

W. D. Lubitz (✉)
School of Engineering, University of Guelph, Guelph, ON, Canada
e-mail: wlubitz@uoguelph.ca

A. Sayigh (ed.), *Renewable Energy in the Service of Mankind Vol I*,
DOI 10.1007/978-3-319-17777-9_59

a three-dimensional structure consisting of many narrow channels that will suppress internal air movement and therefore reduce heat transfer [2]. If the structure is made from transparent materials, typically plastic films, it is possible to produce an insulating structure that also transmits light. It has been shown repeatedly that wall or roof panels can be constructed in several ways to combine some degree of light transmission with very good insulating properties. Wong et al. [10] note that insulating qualities comparable to conventional opaque materials can be achieved with greater than 50 % transmissivity. Despite promising results for a range of concepts, including transparent honeycombs, aerogel, multiple planar layers of material, and the related concept of vacuum-insulated panels, there are few cases of successful commercialization of these materials. Wong et al. [10] suggest one reason is that there is little cost information available on transparent insulating materials, making financial projections difficult.

Significant heat transfer occurs through transparent insulating materials by both conduction and radiation modes. While conduction heat transfer can be experimentally measured or simulated using conventional approaches, it is difficult to accurately characterize radiative heat transfer through, and within, a honeycomb system. Analytical models have been developed by a number of researchers to predict the thermal performance of specific configurations of transparent honeycomb materials [4, 11]; however, there is no known widely applicable thermal model available to engineers or architects considering incorporating transparent insulating materials into building designs.

Hollands et al. [3] measured heat transfer across panels consisting of an expanded plastic 40 mm thick honeycomb with 9.5 mm cells and a 9 mm air gap between plates simulating glass. The panels were inclined at 45 ° to the horizontal. Heat transfer coefficients were in the order of 2 W/m^2 K, with lower values associated with lower temperatures and lower temperature difference. Transmittance of the honeycomb was only 0.98 for zero incidence angle, reducing to 0.90 at a 45° incidence angle. Youngberg [11] experimentally investigated the conduction and radiation heat transfer characteristics of several honeycomb material samples of different overall and wall thicknesses and found similar results. Adding honeycomb adjacent to conventional polyethylene greenhouse glazing was found to reduce heat transfer by 18–26 %, with the degree depending on the thickness and configuration of the honeycomb.

Zhang et al. [12] experimentally measured heat transfer coefficients across several glazings in operating greenhouses during both summer and winter conditions and found heat transfer coefficients between 2.9 W/m^2 K (for the best two-layer polyethylene glazing with an antifog treatment) and 3.4 W/m^2 K. It was also observed that thermal screens reduced night heat losses by 23–24 %.

These values suggest that thin honeycombs could be utilized to reduce heat loss in greenhouses with minimal loss of transmittance. It may also be possible to use thicker transparent insulating panels to further reduce heat loss, at the cost of reducing transmittance of incoming solar radiation. This study used available heat transfer information on transparent insulation materials with a simplified greenhouse thermal energy model to investigate the impact that changing types and levels of transparent insulated glazing might have on a passively solar-heated interior space. The model is first validated against measurements taken in an existing greenhouse, then used with typical meteorological year (TMY) data to investigate greenhouse performance over a longer period.

59.2 Case Study

A measurement campaign was conducted at the greenhouse at the Guelph Centre for Urban Organic Farming (GCUOF), in the University of Guelph campus (43.53858 °N,−80.22297 °E). The greenhouse is 14.6 m long, 9.1 m wide, and 5.5 m tall, with polyethylene glazing and polycarbonate-glazed end walls containing sliding doors (Fig. 59.1). The floor is earth, with the surface a combination of light wood chips, plant covers, and well-worked soil. The greenhouse is exposed to winter solar radiation, but is sheltered from prevailing winds by rows of pine trees to the northwest and west.

One minute interval data were collected between the evening of March 4, 2014 and the afternoon of March 10, 2014. Two Pace Scientific XR5-SE-M dataloggers with synchronized clocks were used to record data. Outside temperature, total horizontal solar irradiance, and wind speed and direction (at 2 m above ground) were measured at a point 7 m southeast of the greenhouse. Temperatures were simultaneously measured at a central point inside the greenhouse at 3.34, 2.40, and 0.20 m above ground, at 0.05 m below ground, and 0.05 m above ground within a bed of dispersed kale plants covered by a semipermeable row cover fabric. Ventilated radiation shields were used for air temperatures. The outside ground was covered by snow approximately 0.3 m deep. No supplemental heating or lighting was applied and the greenhouse vents remained closed throughout the data collection period.

Figure 59.2 shows the measured data from 0:00 March 5, 2014 through 12:00 March 10, 2014. There was no measureable precipitation during this period, and frequent checks found that radiation sensors were free of contamination. Measured wind speeds were calm to low (Fig. 59.2), averaging 0.46 m/s. Interior air tempera-

Fig. 59.1 Guelph Centre for Urban Organic Farming (GCUOF) greenhouse. March 5, 2014

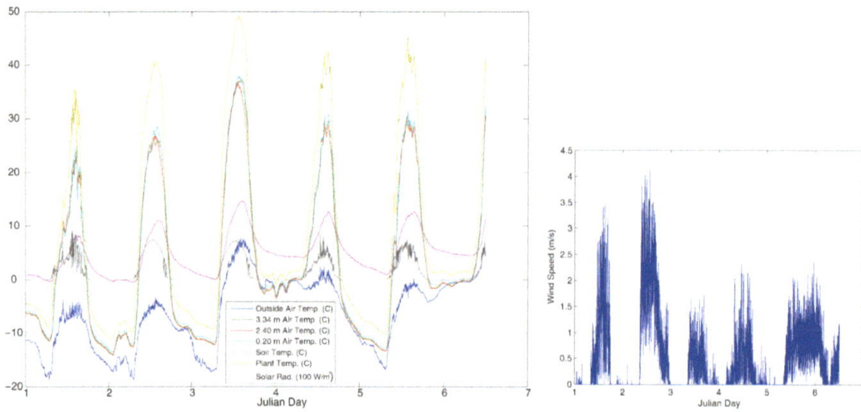

Fig. 59.2 Measured Guelph Centre for Urban Organic Farming (GCUOF) greenhouse data

tures at all three levels were similar throughout, suggesting there was little stratification within the greenhouse.

The greenhouse exhibited very little thermal inertia, suggesting minimal heat storage was occurring in the ground, likely due to a range of low-density surfaces such as plant covers, wood chips, and loose, dry, well-worked earth. Interior temperatures in the order of 30 °C above outside air temperature occurred during midday periods, while at night, interior air temperatures were generally only a few degrees above outside temperature. The data also suggest air leakage has a significant impact on interior temperature, even with all vents closed. For example, before midnight on the evening of March 6 (Fig. 59.2, at Julian Day = 3.0), interior air temperature is very similar to outside air temperature, while after midnight, a greater temperature difference occurred once winds became calm, although a change in sky condition may also have occurred at this time and contributed to reduced heat loss from the greenhouse.

59.3 Model

A one-dimensional, three-layered (glazing, air space, and ground layer) lumped capacitance model was utilized, with heat transfer between the layers by convection and radiation. Polyethylene glazing was assumed and a thin layer of soil was used to represent the ground layer. Neither plants nor moisture was included in the model. Convection heat transfer relationships were utilized from Roy et al. [7], while material properties are predominantly from Zhang et al. [13] and Incropera and Dewitt [5]. The model was forward differenced in time, with a short 6 s time step utilized to minimize drift due to nonlinearities.

Two variations of the model were used: Model A included a glazing layer with both convection and radiation processes on both the inside and outside surfaces.

Table 59.1 Heat transfer rate equations

Equation	Description	Model
$I_1 = G\ a_1$	Solar irradiance absorbed by glazing	A
$I_3 = G\ \tau_1\ (1-\alpha_3)$	Solar irradiance absorbed by ground	A, B
$I_{3ref} = G\ \tau_1\ \alpha_3\ a_1$	Ground-reflected solar rad. absorbed by glazing	A
$Q_{h1} = (7.2 + 3.84\ V)(T_1 - T_a)$	Convection, glazing to outside air	A
$Q_{h2} = 8(\|T_2 - T_1\|)^{0.33}(T_2 - T_1)$	Convection, inside air to glazing	A
$Q_{h3} = 3.4(\|T_3 - T_2\|)^{0.33}(T_3 - T_2)$	Convection, ground to inside air	A, B
$Q_g = U(T_2 - T_a)$	Convection across and conduction through glazing, inside air to outside	B
$Q_{r1} = \sigma(e_1(T_1)^4 - e_{sky}(T_{rs})^4)$	Infrared (IR) radiation, glazing to sky	A
$Q_{r3} = \sigma((1-\tau_{1IR})(e_3(T_3)^4 - e_1(T_1)^4)$	IR radiation, ground to glazing	A
$Q_{rg} = \sigma(1-\tau_{1IR})(e_3(T_3)^4 - e_{sky}(T_{rs})^4)$	IR radiation, ground to sky through glazing	B

The glazing layer also received reflected visible light from the ground surface. This model required specifying optical and thermal properties of the glazing and the glazing layer temperature (T_1), which was solved for in the model. However, this model required too much information about the glazing layer to be useful for comparisons of possible glazing treatments which would be defined based on solar radiation transmissivity (τ_1), infrared (IR) radiation transmissivity (τ_{1IR}), and heat transfer coefficient (U). Therefore, Model B was formulated with the glazing layer specified using only these three variables. Radiation exchange occurs between the ground and sky through the glazing layer, but radiation exchange between the glazing layer and other elements is not modeled, since it is included in the overall heat transfer coefficient.

The sets of equations solved to determine rate of heat transfer between layers (and the sky) is given in Table 59.1 for both models. Absolute temperatures (in K) are T, heat transfer rate per area (W/m^2) is Q, and I is irradiance per area (W/m^2). Numeric subscripts denote layer number. T_a is outside air temperature and T_{rs} is radiant sky temperature (explained below). Wind speed in m/s is V and $\sigma = 5.67 \times 10^{-8}$ W/(m^2 K^4) is the Stephen–Boltzmann constant. Remaining variables are defined in Table 59.2. The change in temperature of each layer during the time step was then found by dividing the net heat transfer into a layer over the time step duration by the product of density, heat capacity, and layer thickness.

Significant night energy losses can occur from greenhouses due to radiation exchange between the greenhouse and the night sky. Particularly on clear nights, the radiant sky temperature (the effective temperature of the sky for radiative heat transfer purposes) is often lower than ambient air temperature, and under the right conditions, the resulting radiative heat transfer from the greenhouse to the sky can reduce the interior greenhouse temperature to 1–2° below the outside air temperature [1]. Wang and Boulard [9] report radiant sky temperature T_{rs} is

$$T_{rs} = ST_a + 0.0552(1 - S)T_a^{1.5} \tag{59.1}$$

Table 59.2 Simulation parameters

Parameter	Value	Variable	Parameter	Value	Variable
Sky			*Layer 2—air*		
Emissivity	0.74	e_{sky}	Depth	4.2 m	dx_2
Layer 1—glazing			Heat capacity	1007 J/(kg K)	c_2
Thickness	0.5 mm	dx_1	Density	1.1614 kg/m³	ρ_2
Heat capacity	2000 J/(kg K)	c_1	*Layer 3—ground*		
Density	950 kg/m³	ρ_1	Depth	1 mm	dx_3
Transmissivity, solar rad	0.6	τ_1	Heat capacity	1840 J/(kg K)	c_3
Absorptivity, solar rad	0.2	a_1	Density	2050 kg/m³	ρ_3
Reflectivity, solar rad	0.2	α_1	Reflectivity	0.2	α_3
Transmissivity, Infrared (IR)	0.45	τ_{1IR}	Emissivity	0.94	e_3
Emissivity, IR	0.40	e_{1IR}			
Heat transfer coefficient	3.4 W/(m² K)	U			

where S is cloud-covered fraction of the sky and T_a is outside air temperature.

Air leakage rates in closed greenhouses range from 0.5 to 5 air changes per hour [8]. In practice, some component of air exchange will be driven by ambient wind speeds. The GCUOF greenhouse poly covering was very tight, but cracks were observed around doors and vents. An air change correction was applied at each time step, in which a portion of interior air was replaced with air at outside air temperature. The rate of air changes (AC) per hour was approximated as 1 AC/h plus measured wind speed in m/s. (This relation has not been extensively tested.)

The models were implemented in MATLAB, and used to predict the GCUOF greenhouse conditions based on the measured solar irradiance *(G)* and outside air temperature *(T_a)*. Since sky condition was not measured, Fig. 59.3 shows the predicted interior air temperature as a function of time, assuming both $S=0$ and $S=1$ in Eq. 59.1. An overcast condition is approximated when $S=1$, which is equivalent to assuming that the radiant sky temperature is equal to the outside air temperature. The opposite extreme of $S=0$ assumes a clear sky condition, when the radiant sky cooling effect would be greatest. (During the measurement period, cloud conditions were variable, ranging from clear to overcast.) Figure 59.3 shows substantial differences in predicted interior temperature depending on the sky condition with both models, highlighting that radiant sky cooling effect should be included, if possible, even in simplified greenhouse simulations.

Model A reproduced interior temperature with reasonable accuracy due to explicit modeling of glazing layer conditions. The Model B simulations reproduced the observed interior air temperature less accurately (Fig. 59.3) and added a lag between interior and exterior air temperatures because high glazing temperature, and the resulting radiative and convective heat transfer, is not directly simulated. However, predicted daily temperature ranges remained similar in the $S=1$ implementation,

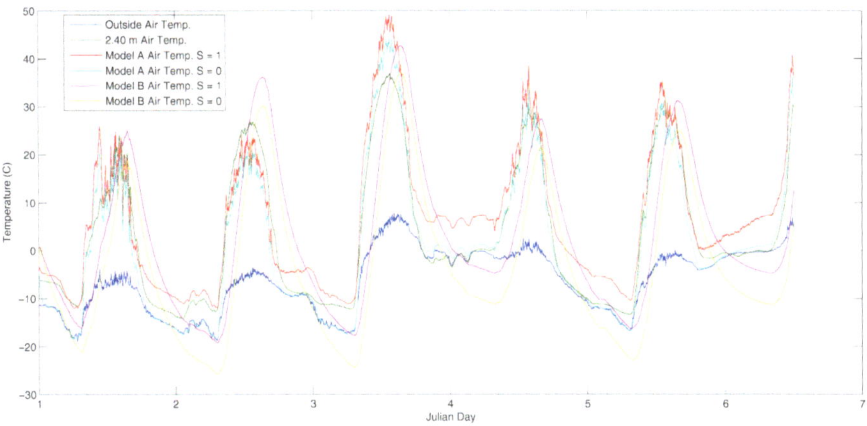

Fig. 59.3 Greenhouse interior air temperatures from Model A and Model B, compared to measured temperatures

and this formulation allows the glazing to be defined with the three variables of heat transfer coefficient, transmissivity to solar radiation, and transmissivity to IR radiation. Model B with $S=1$ will be used in the subsequent simulation exercise to explore the effects of different glazing treatments (since most results in the literature give heat transfer coefficients and transmissivities).

59.4 Simulations

Annual simulations of greenhouse interior temperatures were conducted using a modified version of Model B in which additional soil layers were included. This had the effect of incorporating thermal mass due to heating and cooling of the soil column, and hence, assumed good thermal coupling between the ground surface and the soil (which was not present in the GCUOF greenhouse). TMY data for Windsor, Ontario, Canada, from Environment Canada were utilized. Global horizontal irradiance (G) and air temperature (T_a) from this dataset were used as inputs to Model B. The model was configured to simulate 2 years of greenhouse operation (using the same TMY twice). Only the second year of the simulation was used for results. The first year of the simulation was effectively used only to provide appropriate initial conditions for the second year simulation.

A very simple control system was utilized. No heat input from lighting or occupants was considered. It was assumed that sufficient ventilation was always achievable to maintain interior air temperature at 3 °C above the outside air temperature or 28 °C, whichever was higher. Supplemental heat was added to interior air during cool periods to always maintain a minimum interior air temperature of 15 °C. Although, this greatly oversimplifies greenhouse ventilation and does not provide insight into energy costs of ventilation (which could be substantial with this

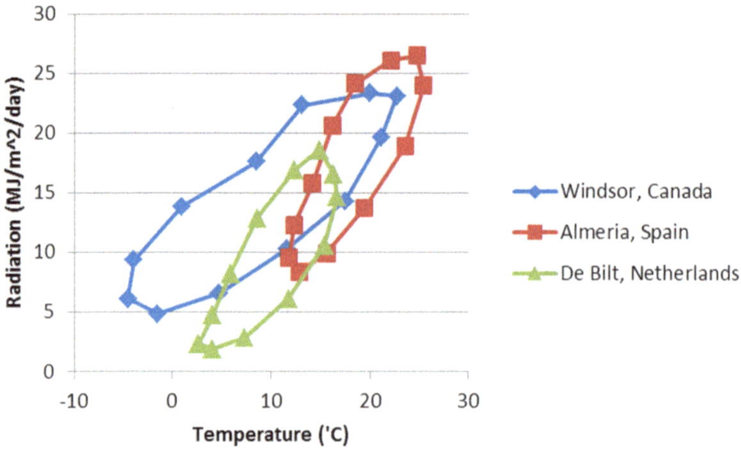

Fig. 59.4 Monthly average temperature and solar radiation for three locations

approach), it was felt this was a reasonable simplification since the goal of this work is the energy cost of maintaining reasonable winter temperatures.

Annual average outside air temperature in the Windsor TMY is 9.2 °C, with a maximum of 33.3 °C and a minimum of −20.6 °C. The monthly average temperature and solar irradiation (Fig. 59.4, with data for two European regions known for commercial greenhouse production from Castilla [1]), ranges widely during the year. In practice, both winter heating and summer cooling are significant issues.

An important question is whether there is significant value in using very thick transparent insulating panels. Compared to conventional greenhouse glazing, such material would have a small U value, but would also be expected to have reduced transmissivity due to the extra depth of glazing materials. These simulations were conducted for U values between 0.01 W/m^{2}K, which is the thermal conductivity of approximately 2 m of still air (considered as an ideal case for limiting conduction heat transfer) and 10 W/m^{2}K (which would correspond to a poor, conductive, single layer glazing).

Figure 59.5 shows the amount of heat required to be added to the greenhouse to maintain a minimum interior air temperature of 15 °C for a year. Results are shown for a range of U and τ, and with and without air leakage between the greenhouse and the outside (with the latter case being an ideal situation of a completely sealed building envelope). It was always assumed that τ_{IR} was 75 % of τ.

There are three main pathways for heat loss to the outside in a greenhouse: (1) conduction, (2) radiation, and (3) air exchange due to air leakage. Figure 59.5 shows that greatly reducing heat loss through a single pathway, such as significantly reducing U, will have diminishing returns because heat will continue to be lost through the other paths. The effect of reducing U (effectively assuming a more insulating glazing) has the greatest effect when U is large. For the conditions simulated here, there is little advantage to reducing U below approximately 0.5 W/m^2 K, even if it could be done without reducing τ. Interestingly, this is true whether air leakage

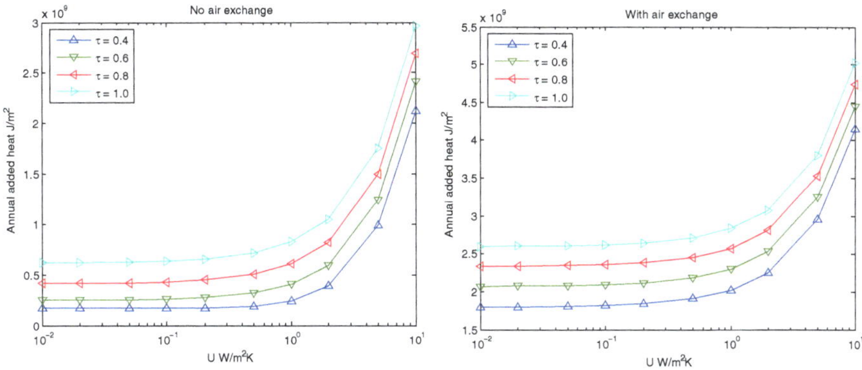

Fig. 59.5 Annual additional heat required to maintain interior air temperature of at least 15 °C

is present or not. Figure 59.5 also shows that reducing transmissivity results in reduced heating needs, since in this climate there is little solar gain during the coldest periods, and much of the radiative heat transfer is in the form of interior heat radiating to the outside. Hollands et al. [3] and Youngberg [11] found heat transfer coefficients in the order of 2 W/m^2K for centimeter thick transparent insulating panels. Figure 59.5 suggests that additional improvement may be possible through careful design of a thicker transparent insulating panel to achieve a heat transfer coefficient closer to 0.5 W/m^2 K, particularly, if this could be done while minimizing material and manufacturing costs and designing a configuration that would still allow significant light transmission.

It is important to note that this was only a simple, location-specific simulation exercise, for the purpose of trying to determine if there is practical value in reducing both heat transfer and light admittance through the use of very thick transparent insulating panels. These simulations have neither included occupants or plants within the greenhouse, nor attempted to optimize for the needs of plants. Light levels within a low transmissivity greenhouse would be inadequate to support plant growth in winter and would likely reduce plant growth during other seasons. This would be less important for human-occupied passively solar heated structures than it is for greenhouses.

59.5 Conclusions

This study has found or confirmed several important observations related to passive solar heating of greenhouses and other buildings. Heat transfer between glazing, outside air, and interior air should be explicitly modeled, if possible. Glazing temperature and thermal properties can significantly impact predicted interior air temperatures. The energy performance of a passive solar structure depends on optimizing conduction and radiation heat transfer through an envelope, while minimizing air leakage and also optimizing thermal mass within the structure. Simulations with a simplified model suggest there is a potential for designing thicker-than-normal transparent insulating panels in order to reduce heat loss, although, there will be a practical limit to the amount heat loss can be reduced through this process, and the benefits of a panel with greater insulating properties will only be achievable in a building carefully designed to minimize air leakage and radiation heat losses as well.

Acknowledgements The enthusiastic support of Martin Rhonda and Martha Gay Scroggins, of GCUOF, is gratefully acknowledged. Equipment used in this study was funded by the Canada Foundation for Innovation.

References

1. Castilla N (2013). Greenhouse technology and management. 2nd edn. CABI International
2. Charters WWS, Peterson LF (1972) Free convection suppression using honeycomb cellular materials. Sol Energy 13(4):353–361
3. Hollands KGT, Iynkaran K, Ford C, Platzer WJ (1992) Manufacture, solar transmission, and heat transfer characteristics of large-celled honeycomb transparent insulation. Sol Energy 49(5):381–385
4. Hum JEY, Hollands KGT, Wright JL (2004) Analytical model for the thermal conductance of double-compound honeycomb transparent insulation, with validation. Sol Energy 76(1):85–91
5. Incropera FP, DeWitt DP (1996). Introduction to heat transfer. 3rd edn. Wiley, New York
6. Kaushika ND, Sumathy K (2003) Solar transparent insulation materials: a review. Renew Sustain Energy Rev 7(4):317–351
7. Roy JC, Boulard T, Kittas C, Wang S (2002) PA—precision agriculture: convective and ventilation transfers in greenhouses, part 1: the greenhouse considered as a perfectly stirred tank. Biosyst Eng 83(1):1–20
8. USDA (2014). Virtual grower. Version 3 manual. U. S. Depart. Agriculture, Agricultural Research Service. http://www.ars.usda.gov/services/software/download.htm?softwareid=309. Accessed 3 July 2015
9. Wang S, Boulard T (2000) Predicting the microclimate in a naturally ventilated plastic house in a Mediterranean climate. J Agric Eng Res 75(1):27–38
10. Wong IL, Eames PC, Perera RS (2007) A review of transparent insulation systems and the evaluation of payback period for building applications. Sol Energy 81(9):1058–1071
11. Youngberg JE (2000). Solar transmittance and heat transfer through polypropylene honeycombs in a greenhouse. Master's thesis. University of Waterloo, Waterloo, Canada
12. Zhang Y, Gauthier L, De Halleux D, Dansereau B, Gosselin A (1996) Effect of covering materials on energy consumption and greenhouse microclimate. Agric For Meteorology 82(1):227–244
13. Zhang Y, Mahrer Y, Margolin M (1997) Predicting the microclimate inside a greenhouse: an application of a one-dimensional numerical model in an unheated greenhouse. Agric For Meteorology 86(3):291–297

Chapter 60
Self-Sufficient Prefabricated Modular Housing: Passive Systems Integrated

Alberto García Marín, Jorge Barrios Corpa, Javier Terrados Cepeda, Juan de la Casa Higueras and Jorge Aguilera Tejero

Abstract In recent years, the construction of efficient and sustainable housing has experienced a significant development, and it is thought that in the near future these criteria will be required in most new constructions. In this context, the architectural project includes the bioclimatic strategies to promote the energy saving and comfort. Spatial organization, zoning, and variability of form factor will be considered. By means of balancing plant-volume model, it can be adapted to a particular climate and remake the popular architectural tradition with nowadays technology (e.g., the Andalusian patio). The materials used should have fast regeneration, low environmental impact, and high durability. Ceramic or wood promote passive conditioning. Besides, they can be combined with new components such as phase change material that increases the thermal inertia. Its use to current constructive systems can be adapted in a flexible way to different places or uses. This research is based on the housing prototype "Patio 2.12" presented to the international competition Solar Decathlon Europe 2012, awarded with the second prize in the general ranking, the first prize in energy efficiency and energy balance, and the second prize in sustainability and innovation. Different architectural models are experienced that integrate the use of solar photovoltaics (PV) and "passive systems": The technological patio like a "buffer space" is an intermediate damper space of outdoor climate conditions that allows the control of solar incidence and the generation of drafts. It is constructed in two layers: the first one is glazed and also mobile, in order to open it or close it, and the second one consists of adjustable blades that provide shade. In winter, it causes a greenhouse effect, and during summer, it favors ventilation. Passive cooling by evapotranspiration or "jug effect" can cause the envelope to become an "active material"; in this case, the ceramic skin is breathable and incorporates an irrigation system that humidifies the ceramic pieces. When the water gets evaporated, the system absorbs the energy, cooling the cavity wall; the cool air is then recirculated into rooms through automated gates. Passive ventilation by solar chimney works by heating a volcanic rock on the roof, which generates a convection current that ventilates the module. It is not only to acclimatize the inside of the room but also to regulate the air quality. This chapter presents a study of an "open industrialized system"

A. G. Marín (✉) · J. B. Corpa · J. T. Cepeda · J. de la C. Higueras · J. A. Tejero
School of Architecture, University of Malaga, Plaza de El Ejido 29071, Málaga, Spain
e-mail: albertogm@uma.es

A. Sayigh (ed.), *Renewable Energy in the Service of Mankind Vol I*,
DOI 10.1007/978-3-319-17777-9_60

that comes from a spatial and constructive prefabricated modular system, based on compatible and interchangeable elements. The production in factory and the dry construction system used are advisable from a sustainable point of view because the quantity of waste generated during the manufacturing, assembly, and disassembly processes is minimized. Due to the utilization of supports over the ground, it is a construction without trace on landscape. Reuse of the living module responds to the economic sustainability claimed by nowadays society. The innovations presented can be applied in social housing, substandard housing, rehabilitation, eco-touristic housing, and the exportation of manufactured housing emergency.

Keywords Prefabrication · Passive systems · Self-sufficient

60.1 Patio 2.12

60.1.1 The House

House Patio 2.12 is a proposal for a self-sufficient prefabricated modular house using renewable energies. It is a project that develops a type of construction without a landscape footprint; it combines low energy consumption on its fabrication with the cultural identity where it is inserted. The main features are to set a *flexible housing* type and its *technological patio* that plays multiple functions in the house. The house is generated by the addition of living modules to a variable space—the patio, allowing a new type of house prefabrication. The *living modules* are fixed prefabricated spaces whose dimensions allow to be transported by road, incorporating all the housing systems and facilities. They are based on the medium-scale prefabrication concept. The *patio* is an intermediate space, covered by a pergola, which is used for living, thermal regulation, and energy generation. It is a *technological patio* that uses the new technologies and is based on the traditional knowledge about passive cooling and spatial comfort of the Mediterranean house.

This research is based on the *housing prototype "Patio 2.12"* presented to the international competition *Solar Decathlon Europe 2012,* awarded with the *second prize* in the general ranking, the first prize in energy efficiency and energy balance, and the second prize in sustainability and innovation (Figs. 60.1, 60.2 and 60.3).

The prototype is based on three main components which add up to create a complete house. These components are totally premanufactured in a workshop and assembled at the user's will, thus allowing for a higher user's personification, normally missing in premanufactured housing. The three components are: skin, closets, and isles:

- The *skin* is completely prefabricated in a workshop. It includes all the materials needed to create the comfort necessary in a house. It is made up of the exterior and interior envelopes and the bearing structure and is prepared to have any type of system installed in it.

Fig. 60.1 Prototype "Patio 2.12"

Fig. 60.2 Constructive and spatial layer

- The *closets* are also premanufactured and are thought to contain the necessary elements for any system to work properly. The lighting points, sockets, air conditioning grilles, etc., can be found inside of them.
- The *isles* help qualify the space by giving it a use (kitchen and bathroom). These are prefabricated "furniture" containing everything needed to qualify a space. These concepts, closets, and isles can also be used in any other place, for example, building's refurbishment.

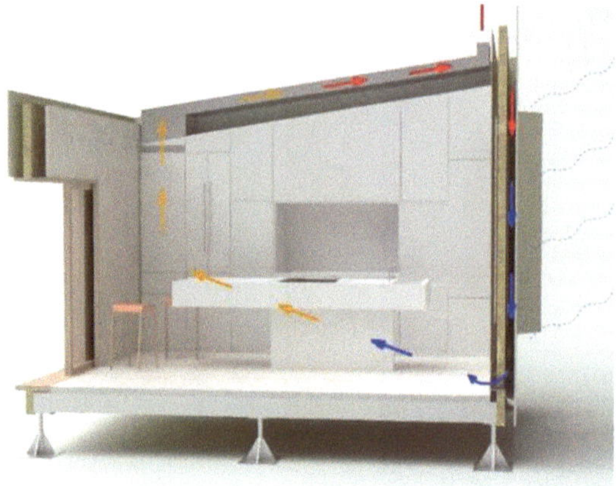

Fig. 60.3 Passive cooling and ventilation. Kitchen module

60.1.2 Industrialization: Medium-Scale Prefabrication (Complete Modules)

Medium-scale prefabrication is the basis for this new model, which allows the living modules to be made completely in a factory, to be taken later on to its final destination by road transportation.

Traditional construction materials are technologically reinterpreted and applied to the prototype, for example, ceramic and timber. They are suitable from an environmental perspective because they improve the passive performance of the spaces and need little energy consumption for the construction.

The constructive system of the envelope consists of a *balloon frame* timber structure, with ceramic claddings to the outside and timber sandwich panels filled with insulation inside. Each module is autonomous from the structural perspective and connected to the ground through a reduced number of supports.

The *assembly process* is very simple: First, the living modules are set in place. The foundation is minimal, thanks to the reduced weight of the modules; the plots are based on few small cubes of concrete (Figs. 60.4, 60.5 and 60.6).

Second, the patio floor is independent from the living modules and is connected to the ground through a support system similar to the module's one. Therefore, the house floor is elevated, ventilated, and insulated. The pergola's structure of the patio rests on the living modules. Last, all the systems are connected using the systems ring found underneath the patio's air chamber.

Once the prototype has been assembled, it will be completed with the implementation of exterior ramps, carpets, and water repositories.

60.1.3 Sustainability, Ecology, and Energy Efficiency

The prototype achieves the best architectural *integration* of *photovoltaic* (PV) solar system through the house shape. PV panels are integrated in the modules (Figs. 60.7

Fig. 60.4 Constructive system, balloom frame

Fig. 60.5 Road transport, habitable modules

and 60.8), placed on the roof to create a ventilated air gap on little supports. The living module's roofs have the appropiate inclination to get the higher efficiency. Therefore, the PV system has a double function: roof cover and electricity generation.

The innovative systems of this prototype decrease the need of air-conditioning during a great part of the year. This is possible through the use of different *bioclimatic* strategies: passive and semi-passive systems.

60.1.3.1 Passive and Semi-Passive Conditioning Strategies

1. **The technological patio as a "buffer space"**

An intermediate space damper of outdoor climate conditions allows control of solar incidence and generating drafts. It is constructed in two layers, a glazed and mobile,

Fig. 60.6 Minimal foundation, plots

Fig. 60.7 Photovoltaic solar system

that can be opened or closed and the other layer consisting of adjustable blades that provide shade (Figs. 60.9, 60.10 and 60.11). In winter, it causes a greenhouse effect, and in summer, it favors a cross ventilation. The patio is a thermal regulator, with adjustable systems that allow a different behavior in summer or winter, during day or night.

- *Winter* During winter days, the patio turns into a greenhouse (closing the glazing and opening the shading arbor). The glazing captures the sunlight, and the heated air is directed into the rooms. By night, both the patio and the rooms remain closed in order to decrease the energy loss through the envelope and to use the patio as a thermal transition between the interior and exterior.

Fig. 60.8 Photovoltaic panels, integration

Fig. 60.9 Patio, buffer space

- *Summer* During summertime, the glazing remains open, favoring the air circulation inside the house. The movable arbor is oriented to shade the patio.
- *Form factor* The patio allows varying the house form factor to adapt to any climate and different hours of each season (Fig. 60.11).

2. **The envelope**

- *High insulation capacity* The envelope has been designed with a very low transmittance. The prototype industrialization and its exhaustive construction control have allowed a precise assembly between its different components in order to assure minimal infiltrations towards the interior, so that the energetic waste of air-conditioning is minimized.

Fig. 60.10 LED lighting technology

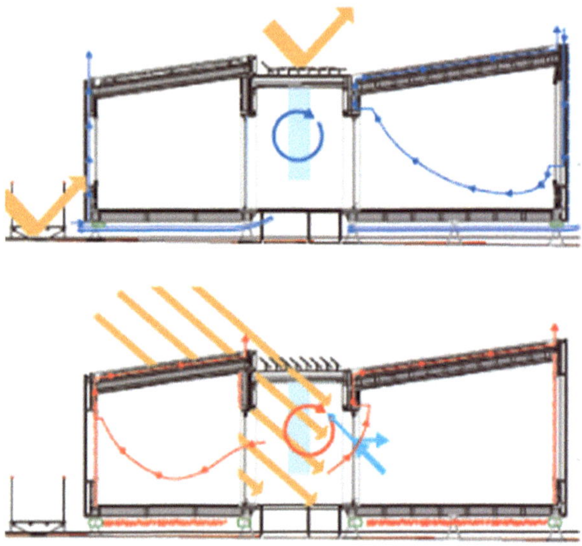

Fig. 60.11 Patio, thermal regulator

- *Ventilated exterior walls and reflective insulation* The use of ventilated ceramic curtain walls and a sheet of reflective insulation allows to considerably decrease the temperature at the exterior side of the envelope during summertime (Fig. 60.13).
- *Passive cooling* Envelope evapotranspiration and "Botijo" effect (Fig. 60.14)—the prototype ceramic finish includes a system of capillary irrigation that allows a natural cooling of the envelope, thanks to a process of evaporation of the water inside the ceramic material. This way the air gap in the facades can be cooled almost 10 °C below the exterior temperature. This cool air is channeled to the interior space through motorized locate over the roof in every living module.
- *Passive ventilation by solar chimney* It works by heating a volcanic rock on deck, which generates a convection current that ventilates the module. It is not

Fig. 60.12 Ceramic ventilated envelope

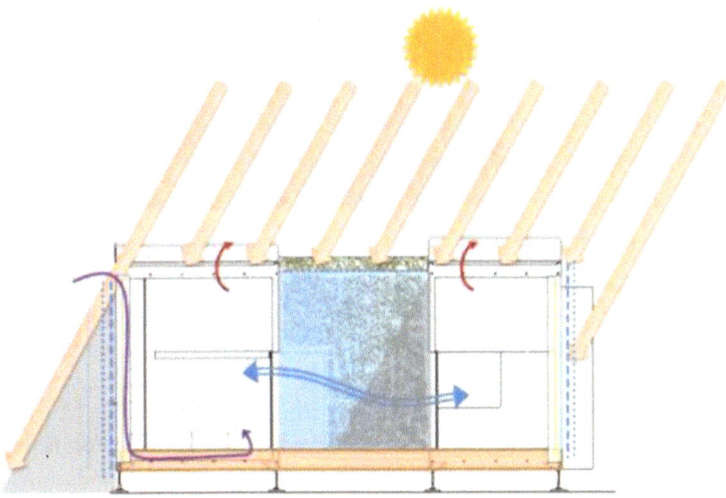

Fig. 60.13 Interior comfort

only to acclimatize the inside of the room but also to regulate the air quality (Fig. 60.12).

3. **Zoning**

A clear zoning reduces the full conditioned spaces while the patio can provide another usable living space with a minor energy demand. It means energy saving in the house operation.

Fig. 60.14 "Botijo" effect

60.1.3.2 Natural Lighting Optimization, Materials, and Maintenance

Windows size and location are designed to take the maximum advantage of natural lighting and to control the solar thermal gains. All windows include a system of thermal bridge interruption and low-emissivity triple glass panes (8 + 12 + 4 + 12 + 8), so that a transmittance of 0.7 W/m^2k is achieved. The solar radiation is controlled by the automation "artificial vine" at the patio.

While designing the Patio 2.12 prototype, we have tried to use a higher number of reusable, renewable, recycled materials, with the lowest incorporated energy.

The house is designed in such a way that almost no maintenance is required, except for mechanical systems. PV solar panels have a life cycle up to 35 years.

60.1.3.3 Sustainability in the Industrialization Proposal

The prototype develops an innovate concept in prefabrication, based in habitable modules, completely factory-made, and with a suitable size for road transportation. Every module is big enough to hold not only a room but also a group of compatible uses, like bedroom-bath, kitchen-dining, or living room-studio. This allows composing different kinds of programs, starting from a livable unit for one person in a single module.

In this field, sustainability is understood as the capability of the product to adapt to different situations with minor changes and efforts. This concept implies a double virtue: It is easier to market it, and its reutilization is more feasible when a change is needed.

There are five lines of adaptation that can mix and give an ample range of possibilities:

- *Adaptation to different budgets* Each of the main elements that compose the house have different versions, from a basic one to an expensive one, depending on the number of "extras."
- *Adaptation to different programs* Users can choose the number and arrangement of the required modules, according to their needs. The patio then works as an "elastic" in-between element that links and connects the chosen layout of modules.
- *Adaptation to different building typologies* The prefab modules could gather in different ways for multiple possibilities. They can be also used for the recycling obsolete structures by inserting them into the available skeleton construction.
- *Adaptation to different locations* The prototype energy self-sufficiency allows using in special situations such as temporary shelters, humanitarian catastrophes, or places with no urban services such as farming areas or mountain spots.
- *Adaptation to the other spaces* The way the isles and storage walls are designed in the prototype, for instance, for refurbishing obsolete residences or hotel facilities.

60.1.4 Habitability and Comfort

The integration of industrialization in the process of architectural design should not prevent the creation of pleasant and comfortable spaces, where there is a thermal, acoustic, and light balance. The walls are energy accumulators; the transmission of heat through the wall in variable conditions is affected by the thermal inertia, which is a direct function of their weight. The thermal and acoustic comfort is a subjective and personal sensation that makes us feel good in the environment where we are—make the difference between a house and a home (Figs. 60.15 and 60.16).

60.2 Industrialized Prototype: Highly Efficient Tourist Housing

As another alternative of Patio 2.12, we propose a prototype of temporary housing for tourism. It consists of two bedroom modules with bathroom and a living room/dining room module with kitchen. The three prefabricated modules (Figs. 60.17 and

Fig. 60.15 Natural lighting

Fig. 60.16 Prefabricated isles

60.18) are related through the patio [1], that is, a versatile space that contains the access. The housing modules are fixed spaces, with a dimension of 3.0 × 6.5 m, which allows the use of road transportation. These bring and incorporate all necessary equipment and facilities. They are places where compatible uses are developed: relax–eat, sleep–grooming.

This application is actually a set of elements of the original project—this is possible because it is an open and flexible industrial system, which can be extended or reduced by adding or removing modules and patio pieces. In this particular case, one of the four prefabricated modules has been deleted, and the patio has been reduced. It remains the same system but adapted to a different use.

The PV system (Figs. 60.19 and 60.20) generates electricity from solar energy, and it can incorporate a backup system for emergencies. The system will be located in a place where there is no access to conventional power grid. This is the concept of "energetic self-sufficiency." It can be connected to the general grid, with other

Fig. 60.17 Tourist housing

Fig. 60.18 Flexible industrial system

houses to form a micro-generation grid or incorporate a generator. The Island System is the fundamental component and an innovation in renewable energy used at homes. It feeds a complete equipment: high-efficiency appliances, conditioning system, etc.

The prototypes can be recycled or reused once their housing mission has finished. The possible adaptation to other uses responds to economic sustainability [2] demanded by today's society.

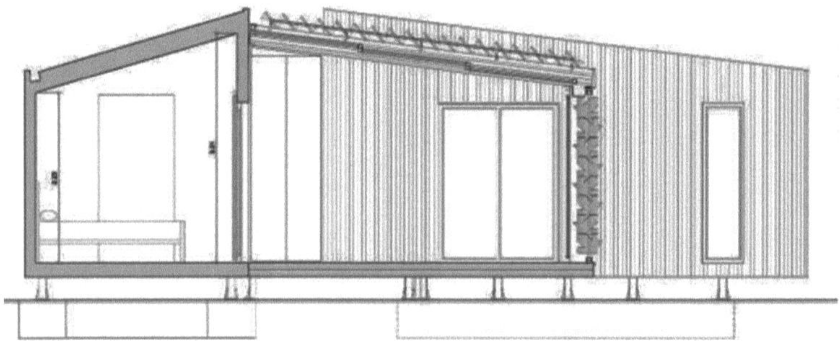

Fig. 60.19 Industrialized prototype

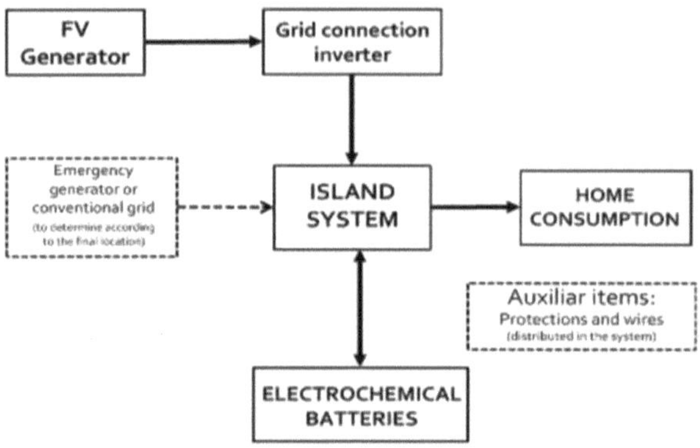

Fig. 60.20 Photovoltaic system

60.3 Industrialized Prototype: Emergency Housing

This research deals with various systems of emergency housing, self-sufficient housing industrialized with a very low cost, including renewable energies. Space requirements are set at a minimum housing for four people: living area with kitchen, two bedrooms, and a bathroom. The prototype can optionally incorporate a front porch (Fig. 60.21).

We start with a basic premise that we call "open industrialized building" [3] based on compatible, interchangeable, and renewable resources. For items that are interchangeable, we must take into account the factor of compatibility of the pieces. The kit house [4] concept refers to the sale of packages of precut pieces that allow user input, possibly by self-build parts with dry mounting following a "manual"—structural kit, surround kit + woodworking kit, facilities kit + PV kit.

The main advantage is that this house is designed to be sent anywhere in the world, exporting the packaged components contemplated by road in trucks or by sea in containers.

The prototype consists of a single compact volume with an optimal ratio between the volume of the interior space and the surface contact with the outside. Thus, the

Fig. 60.21 Constructive process

Fig. 60.22 Emergency housing

Fig. 60.23 Photovoltaic kit

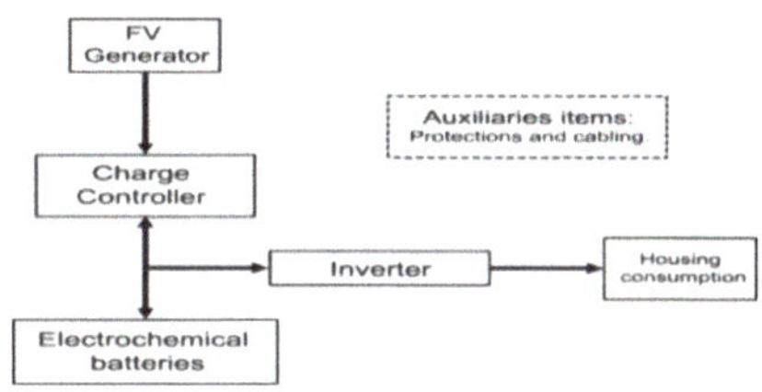

maximum performance is achieved at the lowest cost. The structure (Fig. 60.21) is formed by standard linear elements prefabricated in a workshop—a metallic system, reminiscent of the balloon frame.

By the "PV kit" (Figs. 60.22 and 60.23), we obtain a self-sufficient house, which generates its own electricity through a solar panel, an inverter, and a battery. The installation includes elements for storing and supplying electrical energy when required. It will feed point lighting (light-emitting diode (LED) technology) and small appliances.

The prototype is composed of sustainable materials (steel structure and wood surround oriented strand boards (OSB), sandwich panels, plates or ceramic, Onducober, etc.), considering the energy balance and the life cycle. Sustainability is extended to the use of materials of rapid regeneration, low environmental impact, and high durability (for possible reuse) along with the implementation of recycling solutions.

60.4 Conclusion

The construction sector, due to its size and branches, is the slowest to incorporate innovations offered by contemporary industry. Industrialized architecture linked to residential use, which has the largest impact on the construction sector by volume, is still anchored in tradition with great difficulties in applying existing systems and components on the market.

A restructuring of the construction industry, where the architectural design allows creating more effective linkages between technologies production and assembly, is required. This requires developing a methodology of teamwork that provides the basis for greater coordination between architecture and industry and all those involved in the construction process agents.

Acknowlegments The authors would like to thank the FEDER of the European Union for the financial support via the project "Sistemas Industrializados de Vivienda Eficiente de bajo coste con Energías Renovables (Proyecto SIVER)" of the "Programa operativo FEDER de Andalucía 2007–2013." We also thank all Agency of Public Works of Andalusia Regional Government staff and researchers for their dedication and professionalism.

References

1. Blaser W (1997) Patios. 5000 años de evolución desde la antigüedad hasta nuestros días. Gustavo Gili, Barcelona
2. Ruiz Larrea C, Prieto E, Gómez A (2008) Arquitectura, Industria y Sostenibilidad. Informes de la Construcción 60(512):35–45
3. Águila A (2006) La industrialización de la edificación de viviendas. Mairea Libros, Madrid
4. Cook P (1971) Arquitectura: planeamiento y acción. Ediciones Nueva Visión, Madrid

Chapter 61
The Road to Integrated Design Process of Net-Zero Energy Solar House

Mona Azarbayjani, Ben Futrell and Valentina Cecchi

Abstract *An integrated design process* encourages holistic collaboration of the design team to optimize solutions in an iterative development of a building design. This chapter discusses the design process of a net-zero energy solar-powered house developed for the 2013 solar decathlon competition to promote high-performance design while using traditional passive strategies. The University of North Carolina Charlotte (UNC Charlotte) entry tries to effectively integrate building systems' components to minimize energy consumption while optimizing comfort. To that end, each component delivers multiple functions.

Being part of a large and varied team, seeing a project from the preliminary design phase to construction and commissioning, the students were provided a true multidisciplinary hands-on opportunity. The opportunity proved to strengthen their technical skills, acquired in the regular curriculum, via integration of theoretical knowledge and practical experience. Moreover, organized in a multidisciplinary format, students were then able to share their strengths across disciplines and contribute to a synthesis of process and product. The greatest learning experience for the students occurred in the *integrated design process*—across engineering disciplines, architecture, and business—the student team members also learned how to raise funds, procure materials and construction equipment, and how to interact with one another. *This chapter reports a way of effectively designing and iteratively modeling process to inform the design decisions.*

Keywords Energy codes · Ventilation code · GPRS · Green building code

M. Azarbayjani (✉) · B. Futrell · V. Cecchi
Integrated Design Labs, Energy Performance Laboratory, School of Architecture, College of Arts + Architecture, University of North Carolina at Charlotte, 9201 University City Blvd, Charlotte, NC 28223, USA
e-mail: mazarbay@uncc.edu

A. Sayigh (ed.), *Renewable Energy in the Service of Mankind Vol I*,
DOI 10.1007/978-3-319-17777-9_61

61.1 Introduction

UrbanEden house was one of the 19 houses built through an international competition sponsored by US Department of Energy (DOE)-Solar Decathlon. The challenge was to design and build an affordable, innovative, and fully functional energy-efficient home that addressed each team's local geographic setting. The home was then to be deconstructed, transported, and reconstructed at the Marine Corp Air Station El Toro, a closed airstrip that is now part of Orange County Great Park in Irvine, CA.

The 19 participant houses were evaluated during the competition week. There were 10 contests. Each contest was worth a maximum of 100 points, for a competition total of 1000 points. These contests included juried contests (engineering, architecture, market appeal, communications, and affordability), and measured contests (comfort zone, hot water, appliances, home entertainment, and energy balance).

In this chapter, the road map to this experience from its initial concept to the real construction is presented. This study illustrates that an integrated engineering-assisted design is feasible and helpful at the very beginning stage of a project. To develop an economical viable solar house, concerns about affordability and aesthetic become relevant. No longer can expensive technologies that have limited relevance to the budget-driven residential construction industry in the USA be the main driver to reach performance goals. Therefore, our strategy was to achieve attractive, low-cost high-energy balance performance (Fig. 61.1).

The name of the project was derived by combining its two key environmental influences: "Urban," representing Charlotte's urban, densely populated geographical makeup and "Eden," aptly symbolizing a quiet and lush garden. "The whole concept was to build a house that brings the garden into the city." Students and faculty, from different colleges took the original design submission and began shaping concept into reality. This mix of academic backgrounds provided a wide range of ideas and collaboration. It would not have been feasible without a cross-disciplinary approach and that will lead into best results.

Fig. 61.1 The solar village on the great park at Irvine, California. (Credit: solardecathlon.org)

Our team was composed of experienced faculty, in addition to enthusiastic faculty and students. This project brought together a diverse team of faculty, scientific staff, and students to both lead and organize the project. Although support is being drawn from many colleges within the University, much of the project direction stems from the Colleges of Arts and Architecture (architecture, graphic design, and communication), Engineering (electrical engineering, computer science, building technology, project management, and civil and mechanical engineering), and Business (business administration, finance, and marketing).

Our first task was to publicize the Solar Decathlon project to the student body through a lecture series and invite students from architecture, engineering, and business. Approximately 100 students from architecture, engineering, and business have been regularly involved in research, planning, and strategy meetings. Fifty of these students were in attendance at a single design charrette in the spring. About 20 undergraduate and graduate architecture students took the results of this collective work and generated the conceptual design materials represented in the proposal. The next task was to meet with the administration, including the deans and the department chairs to solicit their support and initiate the advisory board. Student and faculty representatives from these varying backgrounds worked together to provide a well-rounded perspective of the project goals, so that integrated design, energy, affordability, and marketing goals and considerations were met.

61.2 The Design Process

The students involved with the Solar Decathlon project drew all of their strengths and skills to troubleshoot, communicate, dream, and build an original solar-powered house. The competition challenged them to think about energy in new ways and to consider the impacts it has on their everyday lives. With the expertise of our team and the intelligence and will of our students, this project helped advance the research and development of energy efficiency and energy production technologies. The students fully understood, from the outset, that the house must be powered entirely by the sun (Fig. 61.2).

Fig. 61.2 Design charrette

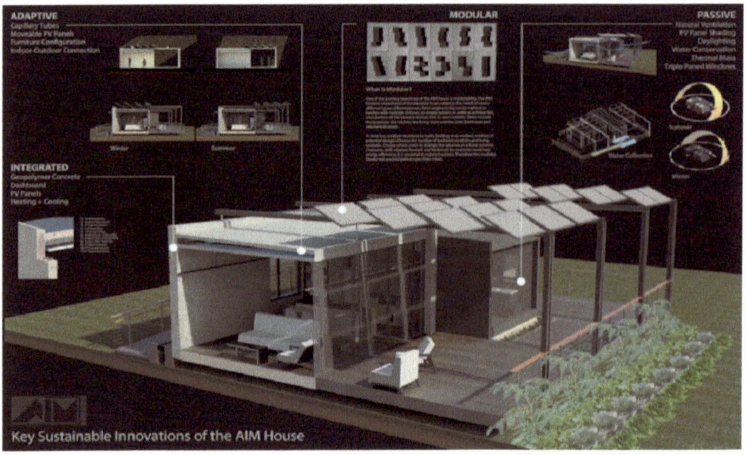

Fig. 61.3 Key sustainable features

61.2.1 Design Objectives

The overall objective of the competition was to design and build a maximum 1000 SF solar powered house. The design and construction of a net-zero energy solar house incorporated both passive and active strategies. The UrbanEden house focused on incorporating innovative passive strategies and developing new material, integrated with heating and cooling systems of the house, allows to provide a better comfort for the inhabitants while reducing the negative impact on the environment (Fig. 61.3).

61.2.2 Design Team and Design Process

Several design charretes were used during the beginning of design process to generate multiple concept and ideas. Several students from different levels and disciplines came together voluntarily to initiate the design ideas. First charrette comprised 50 architecture students. Each team created multiple preliminary designs which were then reviewed by the entire group. Through this process, the team collaboratively selected a single-floor plan design.

61.2.3 Selection of the Final Design

After the charrette series and preliminary design stage, the students finalized the house's concept in summer 2012. The integrated studio in the summer comprised 20 architecture students and about 8 engineering students who worked together

Fig.61.4 Selected design, *Left*: plan drawing, *Right*: rendered model

Fig. 61.5 Selected design concepts

to finalize the concept of the house, basic structure, and the materiality under the supervision of the faculty mentors and advisors (Figs. 61.4, 61.5).

61.2.4 Net-zero Energy Performance Parameter of the Design

A thorough energy analysis was executed to understand the energy consumption implications of each aspect of the design. To achieve net-zero energy performance, the design of the house itself was first analyzed. Before any energy modeling or climate analysis was done, an overall design theme was created based on one simple idea: the house is continuously transferring energy to and from the environment and it is our job to precisely regulate this energy exchange to naturally maintain comfort. This fundamental idea was behind all of the team's passive strategies and techniques and led to the highly integrated and efficient design of the house. The team's approach started from basic climactic principles: the main source of energy on planet earth is the sun, and in the northern hemisphere during the winter, most of its radiation is received on the southern facade of buildings. In Charlotte, in the winter, south-facing windows utilize this heat. In the summer, this solar gain is not desirable and can be easily avoided by overhangs on the roof, which are long enough to block the summer sun. The strategic placement of windows also allows the house to be autonomously daylit during the daytime.

The traditional passive solar strategy of thermal mass storage was also an integral element of our strategy to achieve net-zero energy. Since a massive object can absorb large amounts of heat without rising much in temperature, it is able to build up heat energy and still maintain its temperature gradient with the outside environment. It effectively stores energy in the form of heat and will release it once the ambient temperature drops. This simple fact is exploited in the UrbanEden house by using thick concrete slabs for the walls. The concrete used in the UrbanEden house, is custom-made, high-density fly ash based concrete. High-density material, such as concrete, absorbs heat when the temperature around it is higher and gives off heat when the surrounding temperature is lower. Our geopolymer walls store and release solar heat in the winter and pull heat out of the air in the summer, reducing temperature fluctuation in the house.

An innovative passive/active hybrid cooling system features roof mounted heat exchangers connected to small plastic tubes embedded in the concrete walls. On a hot day, flowing cool water removes the heat absorbed by the walls and releases it into the evening sky. This process is through integration of capillary tubes that are1/8th of an inch in diameter. These tubes are places close together and imbedded 3inches deep inside of the walls.

The house was primarily designed for Charlotte's climate. Manual J load calculations and BEOpt energy simulations were used to inform design decisions. The house energy and load modeling analysis was done independently of the design of the capillary tubes and roof heat exchangers system. Our approach was to optimize the design of the house and its heating, ventilation, and air conditioning (HVAC) systems, per se, and treat the capillary tube/roof heat exchanger system as a "bonus" to the energy performance of the house. Any cooling load met by the capillary tube/roof heat exchanger system would remove this load from the mini-split heat pump. Assumptions and calculation results are displayed in Table 61.1.

To meet the code the UrbanEden house requires about 23 CFM of ventilation. Supplying this ventilation and recovering some of the lost energy is energy recovery ventilation (ERV) which will run at 50 CFM. The required rate of ventilation is doubled because the return duct is located in the bathroom supplying its 50 CFM

Table 61.1 Significant parameters of Building Energy Optimization (BEOpt) energy model

Energy model parameter	Setting
Wall insulation	R-30
Floor insulation	R-30
Roof insulation	R-55
Continuous ventilation (based on ASHRAE 62.2)	25 CFM
Water heater	Heat pump water heater
Window shading summer	90%
Window shading winter	5%
Infiltration (based on Passivhaus standards)	0.03 ACH
Window U-value	0.2
Window SHGC	0.5

ASHRAE American Society of heating, Refrigeration and Air Conditioning, *U-value* opposite of R value, rate of heat transfer, *SHGC* Solar Heat Gain Coefficient, *ACH* Air Change per Hour, *R-30* thermal resistance, *CFM* cubic feet per minute exhaust requirement. This heat loss of this ventilation is calculated to equal 1.7 kBtu/hr. The UrbanEden house is designed to achieve an average annual infiltration rate of 0.03 air changes per hour. In that case the effect on the load is negligible and is calculated to be 115 Btu/hr. Since this parameter is not certain, it is modeled at a couple different values.

After simple calculations, it is easily seen that the windows are the critical factor in the design. They take up an area of about 400 sq ft which is a large fraction of the total wall area. The windows have an R-value that is about 10 times less than the walls. 75 % of the windows are on the southern wall where most of the desirable solar radiation is directed. This design works well for passively meeting winter heating loads but posed a problem of increasing the summer cooling loads. The solution to this problem was found in two places, the window construction and, more importantly, the window shading. Shading provides an easy way to block undesired solar gain and can cut the cooling load in half. The total peak heat loss through all of the windows is 1.468 kBtu/hr. This is even less than the load contributed by the insulation and ventilation showing that with high-quality windows the loads can be kept down while still retrieving plenty of sunlight.

BEOpt used to model the house, contains optimization tools that allows one to optimize one or several design factors, such as window area or wall insulation, for annual energy performance. Table 61.1 shows the significant input parameters of the UrbanEden house BEOpt model. Figure 61.6 compare the performance of the UrbanEden house against a similar "typical," or "code," house.

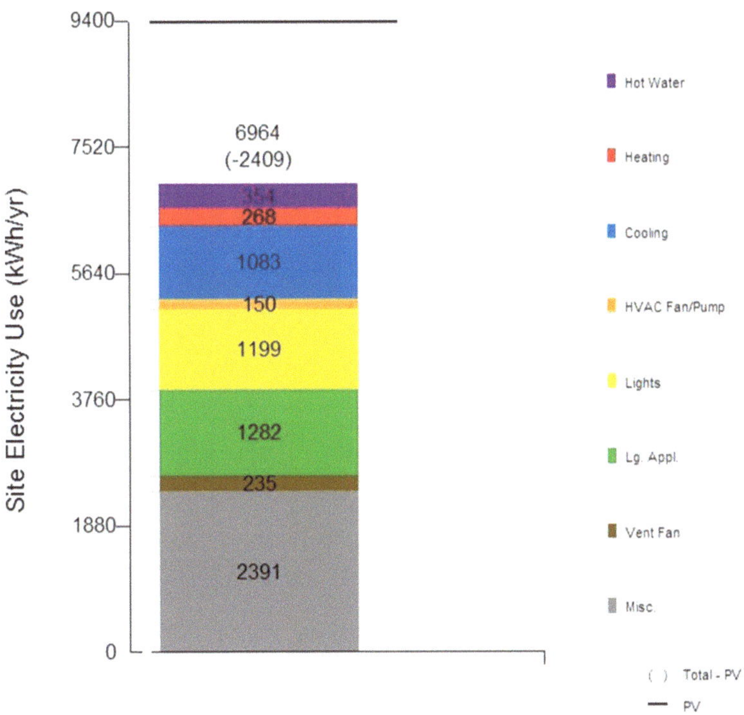

Fig. 61.6 Annual end energy consumption by end use

61.3 Strategy for Project Execution

The studio in the fall semester 2012 then worked in different work groups broken up by area of focus: interior design, exterior (envelope), structure and material development, estimating the cost, and exterior landscape design. The preliminary design was then refined, reviewed, and construction drawings were created. The groups created material lists, estimates, and construction schedules. This construction packet was submitted to the DOE for review and comment through the process. At the end, the students had created around 100 pages of construction documents, a BIM model, and 600 pages project manual.

Once the project and all the construction documents were approved by DOE, and the permits were gathered for state construction, the students started to construct the house on a site located on campus near engineering departments. The students themselves built much of the house except the concrete walls and floors, which were poured in a local plant, in addition the steel manufacturer built and erected the steel structure. Students worked with industry professionals to build and commission the home in preparation for the competition in Irvine California. While the student construction team were working on the home, other team members were working on finalizing the website, brochures, talking points and tour presentations, dinner party menus and logistical plans (Fig. 61.7).

Once the home was completed on campus, the teams hosted multiple open houses for the public before packing the home up for transportation. The house was taken apart and shipped to the competition site in California. Students along with faculty mentor, construction and project managers rebuilt the home in 7 days. They, then, competed in five juried and five measured contests while hosting open houses and dinner parties over the course of 10 days. Then, the home was disassembled in 4 days and shipped back to North Carolina.

Fig. 61.7 Construction on campus

Fig. 61.8 UrbanEden house completed on the site. (Photos:Solardecathlonorg)

61.3.1 Construction and Transportation

The core crew under the supervision of construction and project managers planned, scheduled and controlled the procurement, material deliveries, and the construction.

After over a year of countless planning meetings, design work refinements and securing donor commitments, UrbanEden was ready for construction. Four months later, UrbanEden had become the long-awaited, tangible reality. Days later, the steel beams, concrete foundation, solar panels, decking—every piece of UrbanEden—was painstakingly disassembled by the team. Using two large forklifts, everything was slowly loaded onto five flatbed trailers and two box trucks. As with any large-scale logistics effort, there were challenges (Fig. 61.8).

61.4 Roadmap for the Future

The ultimate goal of the UrbanEden was the establishment of a living lab, solar energy research center. The house will reside permanently on UNC Charlotte's campus, near the EPIC building. Several interdisciplinary research projects and classwork will emerge in this "living" laboratory. Many plans to work with a team "to collaborate on measurements, verification and monitoring of the house when it's reassembled." Graduate students pursue their thesis, studying the UrbanEden's integrated systems.

61.5 Outcomes of Solar Decathlon

Stretching every last watt of electricity that is generated by solar–electric panels or saved by passive heating and cooling features, the students treated energy as a precious commodity; they strived for innovation through the use of high-tech materials and ingenious applications of design elements. During the process, the students learned how to collect supplies, talk to subcontractors, raise funds, meet budgets, communicate across disciplines, and build their solar house. This problem-based and active learning model reinforced and advanced the kinds of nontraditional education known to deepen learning and build professional development skills.

With integrated business, construction management, and communication expertise, we committed to strong internal and external communication efforts. We anticipated a highly student-driven problem-based learning process that would be mission driven and life-changing for student participants as they begin their careers. Through the design and creation of our decathlon entry, our focus was on both student and community education. Our House is a bridge into the future for students. It demonstrated the inspiration and guidance the University can bring to the community and the contributions we can make to advancing sustainable technologies and practices for businesses and municipal agencies in our region.

Upon completion, thousands of visitors toured the competing houses for 8 days. UrbanEden team members proudly introduced their entry, demonstrating all of its sustainable features from a revolutionary geo-polymer concrete to flexible rooftop solar panels. Its lush, 7 ft tall "L" shaped vertical garden was a particular crowd favorite.

61.6 Conclusion

This chapter tried to present the story of the journey of Solar Decathlon project. To accomplish the project, students not only represented their individual discipline but also they mostly integrated in different areas to accomplish the goal of the project. Our 2013 Solar Decathlon effort was very successful and enjoyed tremendous campus support. Just making it to the competition site in California and competing was a huge accomplishment. Finishing third place in engineering and winning the People's Choice Award, showed that people liked the house and felt it was comfortable and livable.

With integrated business, construction management, and communication expertise, we committed to strong internal and external communication efforts, diligent scheduling and project management, skilled budgeting and fund-raising efforts, and creative technical expertise that had engineers and architects working in tandem to attend to every detail of design. It was a highly student-driven problem-based learning process that was mission driven and life-changing for student participants as they begin their careers. The faculty team and advisors committed and successfully and strategically integrated the project into the curricula and they guided and supported our student team throughout the process. Through the design and creation of our decathlon entry, our focus was on both student and community education. Our House is a bridge into the future for students. It demonstrated the inspiration and guidance the University can bring to the community and the contributions we can make to advancing sustainable technologies and practices for businesses and municipal agencies in our region.

Acknowledgments The authors would like to thank all the advisors, mentors, faculty, and the students who helped design, build, and transport the "UrbanEden House." We would like to express our gratitude to the University of North Carolina at Charlotte for giving us the financial support.

Further Readings

1. Azarbayjani M, Cecchi BT et al (2014) Lessons learned from an interdisciplinary collabortion in design and construction of a net-zero energy solar house. Proceedings of the ASME 2014 8th International Conference on Energy Sustainability, 29-July 2, 2014, Boston, Massachusetts
2. Cecchi V, Ebong A, Azarbayjani M et al (2013) Collaborative research and education in the design and building of a net-zero energy solar powered house—testimony of a solar decathlon entry. American Society of Engineering Education: ASEE, The 2014 International Forum June 14, 2014 Indiana Convention Center, Indianapolis, IN
3. European Solar Decathlon—expanding the global reach of zero energy homes through collegiate competitions (2009) 34th IEEE Photovoltaic Specialists Conference (PVSC), 7–12 June 2009, Philadelphia, PA
4. Holt E, Loss B, Shaurette M (2012) Students involvement in the solar decathlon competition: giving context to the classroom experience 48th ASC Annual International Conference Proceedings 2013 solar decathlon. http://www.solardecathlon.org. Accessed 13 July 2015
5. Warner C, Farrar-Nagy S, Wassmer M, Stafford B, King R, Vega Sanchez S, Rodriguez Ubias E, Cronemberger J, Serra Maria-Tome J (2010) "The 2009 Department of Energy Solar Decathlon and the 2010 European Solar Decathlon—expanding the global reach of zero energy homes through collegiate competitions", *Conference Record of the 34th IEEE Photovoltaic Specialists Conference*, 2009.

Chapter 62
Describing Native Architectural Features of Kandovan, a Sustainable Village with Rock Architecture

Navid Nahi and Maryam Singery

Abstract The purpose of the present chapter is to investigate the unique features of Kandovan as well as its rural elements playing a role in the traditional texture of the village as sustainable elements. The village is located 62 km from Tabriz, Iran, in a region with moderate mountain climate. The village of Kandovan is a specimen of the most primitive type of human life and its coexistence with nature. The village texture consists of a series of pointed cones with manmade and natural cavities emerging on the surface. These cones have sporadically extended over the steep mountain skirts toward the peak of the mountain in a condensed and scattered manner. The building of the houses in Kandovan are of the rock-made architecture and date back to the seventh century (in the Hijri solar calendar) and in cases to even as early as pre-Islamic eras.

The rocky architecture of Kandovan represents man's struggle to harness and tame the nature to satisfy his needs. The structure and texture of the cone-shaped habitats of the village, made of INEMBERITE and ALTHER, formed with the passing of time by the combination of the volcanic materials from the Sahand with the mud resulting from flood-causing rainfalls. The residential space in the Kandovan village is carved off the stone mass and rock Karans (rock-shells) create a firm crust around the space like a body, protecting the residents from natural phenomena such as rain, snow, wind, cold, and heat. In addition, the shape of the cones and their positioning has a substantial effect on increasing a sense of comfort and convenience for the residents through regulating environmental conditions like heat, cold, humidity, and light. The village Kandovan is listed as one of the three rocky villages of the world, the other two being the Kappadokia in Turky and Dakota in the USA. Kandovan is different from other rocky villages, in that life is still present and the residents of the village still continue to live their day-to-day life with the natural features of the village without making fundamental changes in the structure of their habitats.

N. Nahi (✉)
Department of Architecture, East Azarbaijan Science and Research Branch,
Islamic Azad University, Tabriz, Iran
e-mail: nahi.navid@yahoo.com

M. Singery
Department of Architecture, Tabriz Branch,
Islamic Azad University, Tabriz, Iran
e-mail: singeri@iaut.ac.ir

A. Sayigh (ed.), *Renewable Energy in the Service of Mankind Vol I*,
DOI 10.1007/978-3-319-17777-9_62

The methodology of the present chapter is based on library research method and field observations, and describes and analyzes the structural and architectural elements existing as sustainable elements in the texture of the village. Notable among the elements are the integrated and interconnected rural texture; interior and exterior design of the houses; access means; apparatus; ventilation; cooling and heating system; materials, types, and typology of the Karans; and the rocky architecture. Kandovan, with its high potential in aptly using sustainable architecture may be fit as a model for various sustainable structures around the world.

Keywords Village · Kandovan · Sustainable architecture · Rocky architecture

62.1 Introduction

The historical village of Kandovan is a tourist attraction and is located on the grass-covered skirts of the Sahand mountains, on the extension of Vidahr (Oskoochay) valley stretched toward Sahand heights, on the right of Mount Sahand and in terms of national divisioning is a district of Oskoo parish. This unique historical village with its uniquely natural architecture (rocky architecture) and the presence of light mineral water that has curing properties is renowned worldwide ([1], p. 353).

"What immediately occurs to the mind of any beholder, is the long-sustainability of the houses, and the agedness, and the intact traditional of the living from the far earlier ancestors of the now-living who carved out the heart of the stone making the rocks their roof" ([1], p. 275). This village is a live example of a type of combination and mixing of novel forms and proportions to which man has adapted his life in the best possible way. In the environment, corporal factors of human habitat is selected out of purely natural factors and a wondrous combination is made between the residential body and human behavior with the two, in fact, have influenced each other.

The present study attempts to explore the sustainability elements of Kandovan through a structural and climatic of investigation of this historical village. In the research, first a study of the structural patterns (Karans) and identification of their types are presented along with the natural characteristics of the village (Karan) and then the chapter sets out to identify the representation of the sustainable elements in this historical compound by analyzing climatic and physical issues of the village texture presented in the form of tables.

62.1.1 *Interior Elements of the Karans and the Proportions Between Spaces*

Architectural elements inside the Karans are formed with two types of use namely residential and public. In the former case, the interior space is divided in proportion to the particular needs and functions. Abbas Rafi'ie, a Kandivani landlord, describes the use of the interior "the main space of the Karan is used as both a living room and a bedroom and guest room at the same time. Other interior spaces include Yuk Yeri (closet), Astana (entrance), Matbakh (kitchen) and Sandoogh Khana or Gazana (storage)."

62.1.1.1 Terminology used in Relation to Karan Spaces

Yuk Yeri: a relatively large and flat space in the main chamber wall used to keep blankets, mattresses, rugs, and clothes.

Astana: the entrance of a Karan used in large Karans for dishwashing, laundry, and even bathing of the residents.

Matbakh or kitchen: It is also outside the main chamber and in the adjacent room.

SAndooghkhana or Gazana (fuel and log storage): Small storehouse inside a Karan that is used to keep the weekly supplies of groceries such as wheat, barley, legumes, and flour.

Odunlookh: It is used to store cattle manure and wood logs next to some Karans.

62.2 Etymology of Kandovan Naming

As regards the naming of Kandovan there are several accounts some of which lack scientific and historical ground, and are mostly personal opinions. However, some of them sound acceptable which are mentioned here: Kandovan is made of two parts "Kand" meaning village in Turkish and "van" a suffix making a place name. Some people believe that is derived from the Kandeno Kandegan or honey bee house which has to do with the mild climate of the region and growing honey bees, a common practice from the early times. Or it might be that, hollow trees are similar to hive-like (Kandies) piles of legumes and are probably derived from this word. According to another theory, Kandovan is the same as Kandivan or Kandevan, a three-morpheme compound made up of Kandi + Ev + An which mean "self," "house," and location noun-,aking suffix, respectively. Thus, the words mean "a house of one's own" ([1], p. 353).

62.2.1 Typology of Karans

Karans can be classified into two categories in terms of shape and the number of floors:

A) Shape: in Kandovan, tens of various types of Karan are seen which include multiplex, duplex (paired), conic, and spindle-shaped. These Karans and cones are made of Inemberite and lahar with a height reaching to over 60 m.

Fig. 62.1 A two-storied Karan. (Photo:Nahi)

Fig. 62.2 Three-storied Karan. (Photo:Nahi)

B) Number of floors: the number of floors and the form of the interior spaces of the Karans depends on the volume and shape. Karans are generally two-storied but in case three/four-storied Karans are also found. The connection between top-floor Karans is made possible through marvelous stairs that are integral parts of the Karan body (Shaterian 1387, p. 382) and there is no access to top floors from inside. The only internal connection between the floors is seen in the late Mo'tamid's Karan where the interior staircase connects the ground-floor Astana to the basement (storehouse) ([2], p. 4) (Figs. 62.1, 62.2).

The dimensions of the Karans are relatively small due to safety and static considerations and in cases where there has been a need for larger spaces, such as in case of mosques and stall-feeds, thick pillars are carved out of rock in the center of the Karans to serve in the stability of the ceilings ([3], p. 26).

62.2.2 Climatic Features of Kandovan Region

1. Altitude from the sea level: The most important contributor to climatic variety in the region can be said to be its altitude and in this regard the presence of high heights, their geographical orientation, and altitude difference from the sea level are more influential in determining climatic conditions than other factors. The Oskoo region, due to its considerable altitude from the sea level (over 1500 m) and the impact of the high-pressure center, has a moderate climate in the summer and cold in winter. Winters in the Sahand mountains are very cold and summers are moderate and as we move westward the coldness of the air decreases.
2. Latitude: Oskoo is located on the 38 northern latitude, in the north of Iran. This causes a difference of about 13° in the angle of sunlight when compared to the angle in the south of Iran. Apart from other factors, the direction of the general angle of sunlight leads to a decline in the energy reception in the region ([1], p. 77).

62.2.3 Climatic Analysis of the Site of the Historical Village of Kandovan in Oskoo Region

A. The Form and Positioning of Buildings to Increase the Amount of Sunlight Used The diagram of sun trajectory shows that the only face of the houses that is desirably exposed to sun is the southern face. In summers, sunlight in western and eastern directions causes the houses to heat up. The sun trajectory in the winter is much shorter and lower than the trajectory in the summer Fig. 62.3.

B. The Shape and Positioning of the Buildings to Decrease Wind Turbulence in Winters In general, as seen in Fig. 5 (wind blow diagram), the dominant winds in the region blow in southwestern direction in fall and spring and in western and northwestern directions in winters and summers. The compression of the form in the plan and cross section, is the first rule in decreasing the area of the surface in contact to wind flow ([4], pp. 1210–121).
In mountains-skirt regions, avoiding winter cold and summer heat, that almost occur during one third of the year, is very important. Thus, heat exchange through building walls should be minimized whereas the sunlight use in cold times and cool air flow in hot times should be maximized (Tahbaz, Mansoore p. 173) (Table 62.1, Table 62.2, Table 62.3).

Given the types of Karans under study in the above tables, the plan example 3 possesses higher functionality due to its wide area when compared to other types

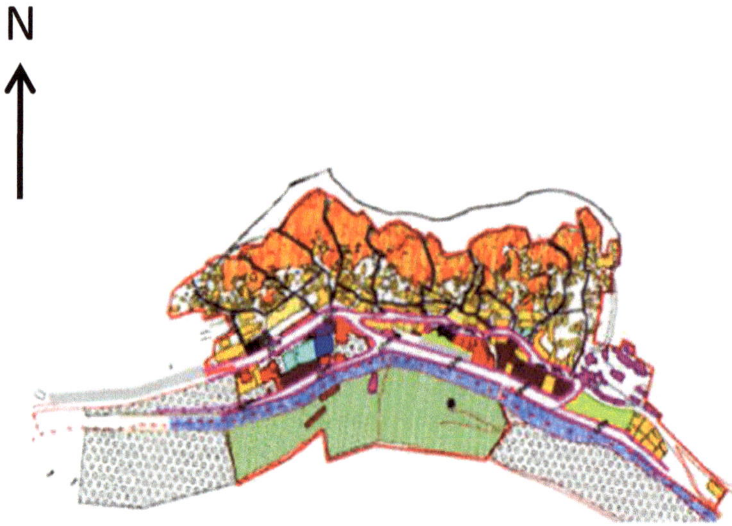

Fig. 62.3 The shape and position of the building to increase the amount of sunlight used in the winter (from [4])

Table 62.1 Typology of Karans in terms of plan and function. (Source: nahi)

Plan example 1	Function	Spatial structure		Plan form	Floor number
	First floor, stall-feed	Stall-feed		Tetragonal, nearly square	Two-storied
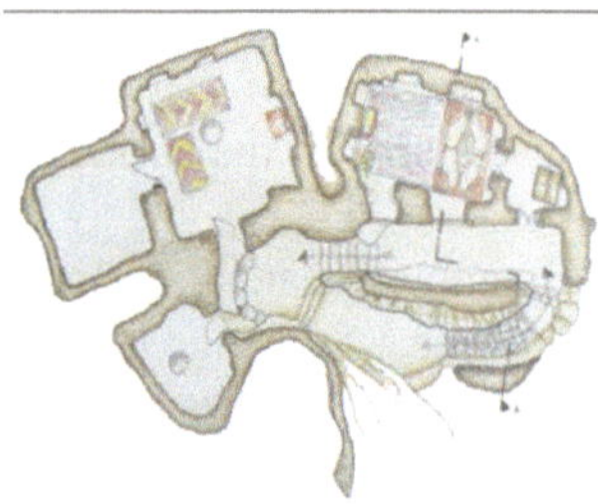	Second floor, residential	◘	Bedroom	Tetragonal, nearly square	
		○	Lobby and entrance	Combination of rectangle and ellipse forms	
		△	Living room	Tetragonal and nearly rectangular	

Table 62.2 Typology of Karans in terms of plan and function. (Source: nahi)

Plan example 2	Function	Spatial structure		Plan form	Floor number
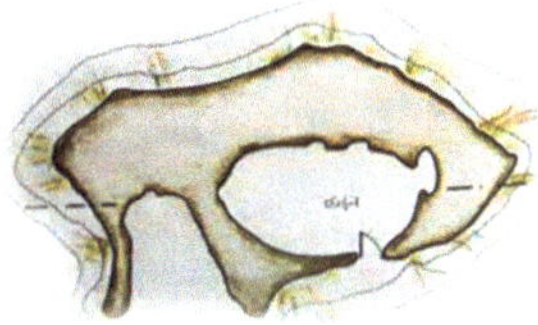	First floor, stall-feed	Stall-feed		Ellipse form	One-storied
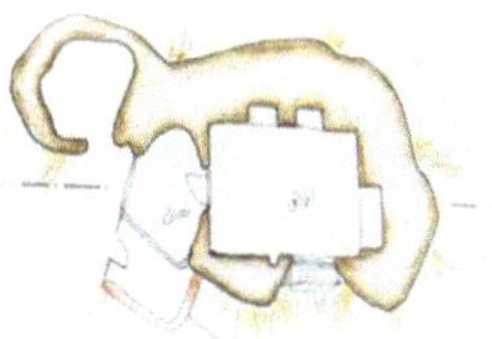	Residential	◘	Bedroom	Tetragonal, nearly rectangle	
		○	Lobby and entrance	Combination of rectangle and ellipse forms	
		△	Living room	Tetragonal and nearly rectangular	
		◇	Store	Combination of uneven eclipse and square	
			Kitchen	Tetrago-nal, nearly rectangular	

and in addition, the wide area makes additional natural ventilation possible. Also, the partitioning of spaces and fields are made in a desirable manner. The connective cycle is in a regular and hierarchical manner and the user will not have to have access from some exterior spaces.

Due to the specific conic form, most Karans have several floors which are not interconnected from inside. The idea behind the emergence of this type of architecture is that at times carving the rock, given its particular material, is more reasonable and wiser than building walls, roofs, etc. Meanwhile, the thick bodies of rock provide good protection against cold and the steep slope greatly removes water from rain and snow. In terms of lighting and ventilation, tall pointed Karans suffer more due to their lower cross section (Zargar 1383, p. 238).

The typological study of rock houses in Kandovan suggests that in Karans are generally two-storied and in some cases three- or even four-storied which are built based on the necessary process of following nature and form. The access to top floors of the Karans is provided externally through elegant staircases made of the Karan body. The ground floor of most houses is used as stall-feed and the second, third, and fourth floors are used as residential part of the house. In some cases the fourth floor is also used as a storage room.

Table 62.4 represents a randomized form of the typical Karans in Kandovan village that is evaluated (in terms of usage area and facing ecological limitations) in a comparative study above (Table 62.1). As mentioned before, the number of floors and the interior space of Karans depend on the volume and shape. Karans are

Table 62.3 Typology of Karans in terms of plan and function. (Source: nahi)

Plan example 2	Function	Spatial structure		Plan form	Floor number
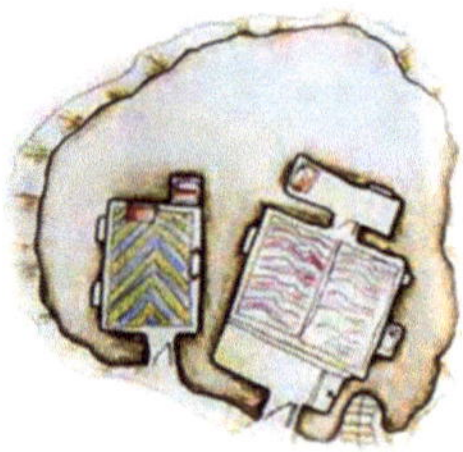	First floor, stall-feed		Stall-feed	Main space: tetragonal, nearly rectangular	Two-storied
	Residen-tial	◘	Bedroom	Main space: Tetragonal, nearly rectangle	
		○	Lobby and entrance	Formal combina-tion: Tetragonal, nearly rectangle	
		△	Living room	Tetragonal and nearly rectangular	
		◇	Store	Combination of uneven eclipse and square	
			Bathroom	Tetragonal, nearly rectangular	
			Kitchen	Tetragonal, nearly rectangular	
			Stairway	Unidirectional narrow	
			Balcony	Tetragonal, nearly square	

generally two-storied. In the conducted investigation, Karan form 3 is proposed as the ideal form because of the increase in used area and partitioning spaces as well as better resistance against ecological limitations.

As the table above (Table 62.5) indicates, stable residence of human in this historical village is the sign of a type of ecological balance between human activities and his surrounding environment. Residential complexes located in mountainous regions built on steep grounds have condensed textures to minimize thermal exchange with open space. Houses are generally built with their backside to the mountain and facing the sun, and on the warm slope in order to make optimal use of sunlight heat ([5], p. 128).

Because of the high thickness of Karans, they make good resources to store, transmit, and delay heat, a feature that plays a great role in heat and cold control. The walls of the Karans (Toofs) are hard and serve as insulators; thus, cool in the summer and warm in the winter.

Thermal capacity acts as a source of heat/cold saver, regulator, and controller of heat/cold fluctuations and in the summer, the heat saved during the day can be released and exhausted at night as ventilation.

Table 62.4 Climatic and area usage investigation of the Karan forms. (Source: nahi)

		Advantages	Disadvantages
Form 1—single Karan	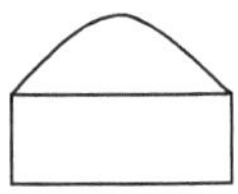	Karan form with high arc radius increases efficiency in a floor	Mixing and overlapping of rooms (private and service)—problems with heating the house due to low height—slow flow of rain/snow waters due to the dome-like structure—creating a light opening in lower floors is difficult due to the thickness of Karans
Form 2—conic Karans		Karan forms with small arc radius and large height provides usage of floors and partitioning of fields—facilitation of watershed removal—contribution to heating system of the residential area often placed on the first floor of each Karan	Decrease in the useful area in each floor in terms of usage and service—more interruption in the view
Form 3—duplex (doubled) Karan	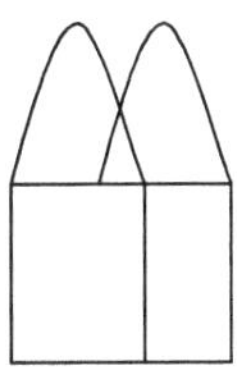	Double Karan form with high arc radius provides usage of the floors and partitioning of the spaces leading to increased used area in each floor—contribution to residential space heating system placed on the first floor—increase in view and possibility of view usage such as balcony	Slow transfer of rain and snow fall compared to form 2

In conclusion, it is noteworthy that the use of the early sustainable systems without adapting them to today's needs and conditions is not desirable and sensible. Using sustainable systems of the past lead to savings on seasonal and nonrenewable energy use.

62.3 Conclusion

- Thermal flow between the interior and exterior of the Karans is not much perceptible due to the thickness of the walls and as a result, the interior space remains cool in the summer and warm in the winter. Because of the peculiar positioning of the Karans beside each other, the surfaces of the cones break the wind and the placement of doors and windows inside the wall bulk prevents wind, rain, and snow from penetrating inside. Given the hard and polished durable conic rocks,

Table 62.5 Climatic and physical analysis of Karans. (Source: nahi)

Analysis of Karans				
Ventilation	Doors and windows	Material	Mostly wooden	
		Position	Apparatus on top floors depending on increased diameter in the ground floor—deep ledged windows due to: 1-prevention of rain, snow and storm penetration 2–better suction and regulation of winds	

Table 62.5 (continued)

Analysis of Karans				
		Shape and dimensions	Small shape and dimensions	
		Light exposure extent	In principle, wider and less lighted lower floors are used as stall-feed and top floors are used as residential section and the only apparatus and balcony is found in the southern face of the Karan	

Table 62.5 (continued)

Analysis of Karans		
Heating and cooling systems	The presence of thick walls with high thermal capacity to store heat for suitable heating and cooling—the existence of Matbakh (kitchen) in the main space to help heating—housing cattle in the lowest floor to make use of floor heating	
Materials	Larha and Imbrite formed through mixing with the volcanic materials from the Sahand	
Shape and structure	Conic (spindle-like)	

there is no need for ceiling covers and the rocks protect the Karan from any kind of climatic phenomena despite the harsh climatic conditions, particularly in the winter.

- Alleyways are narrow in these areas and are formed after ground typography. Since in these regions, the majority of the buildings are backside to the mountain making it impossible to use lateral ventilation, unidirectional ventilation is used instead. Rocky complexes located on the slope surface of the mountain have close condensed textures to minimize thermal exchange with open space. Houses are generally built backside to the mountain and facing the sun on the sun-warmed slope of the mountain to make optimal use of sun heat.
- The largest areas of windows are placed on the southern faces of the Karans to make maximal use of natural light and outside heat during the day, whereas there are no openings on the northern face.
- One of the reasons why the lower floors are not residential is that light openings do not exist on these floors, thus making lower floors deprived of sufficient light and in turn these floors are damp and humid. In addition, algal growth can be seen in such spaces.
- The residents of Kandovan region, having recognized the comfortable region, have established their homes on the first and second floors and have allocated the ground floor to stall-feed.

References

1. Moghimi-e Oskooyi H (1385) Oskoo: From the Urumyieh Lake Beach to Sahand Climax, 1st ed. pp. 101–106. 361–375, Oskoo Municipality, Tabriz [Moghimi-e Oskooyi H (1385) Oskoo az Sahel Dayacheye Orumyie ta Gholleye Sahand. Chape Aval, pp. 101–106. 361–375, Shahrdarie Oskoo, Tabriz]
2. Rezvani A (1378) Kandovan: The representation of man's adaptation to nature. Geogr Educ Dev J [Rezvani A (1378) Kandovan (Namayeshe Entebaghe Ensan ba Tabi'at). Nashreye Roshde Amoozeshe Joghrafia]
3. Biri-e Oskooyi B (1380) M.A. Thesis (*The Kandovan Tourist Complex*), Architecture and Urban Construction School, Shahid Beheshti University, 1380, Thesis No. 5664 [Biri-e Oskooyi B (1380) M.A. Thesis (*Majmoo'eye Tooristye Kandovan*), Daneshkade Me'mari va Shahrsazi, Daneshgahe Shahid Beheshti, No. 5664]
4. Watson D (1384) *Climatic Design: Theoretical and Practical Principles of Energy Use in Buildings*, Vahid Ghobadian & Mohammad Feiz Mahdavi Translators, Tehran: Tehran University, Pulications and Press Institute, 7th. Ed. [Watson D (1384) Tarahyie Eghlimi: Osoole Nazari va Ejrayie Karborde Energy dar Sakhteman. Tarjomeye Dr. Vahid Ghobadian—Mohammad Feiz Mahdavi—Tehran: Daneshgah Tehran, Moassesye Entesharat va Chap, Chape Haftom]
5. Tahbaz M (1392) *Climatic Knowledge, Design, and Architecture*, Shahid Beheshti University, Shahid Beheshti University Press Center. 1st. ed. [Tahbaz M (1392) *Daneshe Eghlimi, Tarahi va Me'mari*, Daneshgahe Shahid Beheshti, Markaze Chap va Entesharat, Chape Avval]
6. Aftahi S, Shervanitabar E, Mohammadreza F (1372) Village Project. Architecture and Urban Construction School, Shahid Beheshti University [Aftahi S, Shervanitabar E, Mohammadreza F (1372) Projhey-e Roosta, Daneshkade Me'mari va Shahrsazi, Daneshgahe Shahid Beheshti]

7. Haji Ebrahinm Zargar A (1383) An Introduction to Understanding Iranian Rural Architecture, Tehran, Shahid Beheshti University, Shahid Beheshthi University Press Center, 1st. ed. [Haji Ebrahinm Zargar A (1383) Daramadi bar Shenakhte Me'marie Roostayie Iran, Tehran, Daneshgahe Shahid Beheshti, Markaze Chap va Entesharat, Chape Aval]
8. Kasmayi M (1382) Climate and architecture, 1st ed. Baztab Publications, Tehran [Kasmayi M (1382) Eghlim va Me'mari. Entesharate Baztab, Chape Avval, Tehran]
9. Mojtahedzadeh G (1353) Kandovan. Art and Architecture Magazine, vol. 37–38, Eighth year, pp. 154–149 [Mojtahedzadeh G (1353) Kandovan. Majaleye Honar va Me'mari, jelde 37-38, Sale Hashtom, pp. 149–154, 1353]
10. Sayigh A (2014) Sustainability, Energy and architectecture Case Studies in Realizing Green Buildings. Academic Press is an imprint of Elsevier The Boulevard, Langford Lane, Kidlington, Oxford OX5 1 GB, UK 225 Wyman Street, Waltham, MA 02451, USA First edition 2014.
11. Azari R, Singery M (2013) Sustainable building and relationship with humans and nature. In: Sayigh A (ed) Sustainability, energy and architecture: case studies in realizing green buildings, Chap. 14. Elsevier Science Ltd., pp. 387–399
12. Singery M (2006) Practical principle build permanent building in Tabriz. World Renewable Energy Congress IX, Florence

Chapter 63
Investigating the Effect of Climatic Factors on the Spatial Structure of Old Texture of Yazd City: A Specimen of a Sustainable Urban Texture

Navid Nahi and Maryam Singery

Abstract Sixty-one percent of Iran's area in located in arid and super-arid climate. Possessing 2.1 % of lands, 4.2 % desert phenomena without vegetation, and 8.3 % desert regions of the world, Iran is one the hottest and driest countries. Most of the regions in Iran are faced with severe limitations in vegetation and water resources. Nevertheless, such areas have housed several residences, since a long time ago, where man has dwelled because of necessity. The adverse climate of these regions has always caused numerous difficulties for the living of the residents and the residents have had to come up with solutions to resist and have ultimately managed to reduce the adverse irritating climatic aspects, providing the conditions for comfort and peace. The major difficulties that has evoked the residents of hot and dry regions to seek solutions include intense summer heat, severe winter cold, burning sunlight, high day and low night time temperatures in the environment, sandstorm blows full of dust, water shortage, vegetation scarcity, air dryness, and earthquake threat.

Among the habitats that have managed to keep their residents safe against the harsh climatic conditions and overcome the above mentioned difficulties, is the city of Yazd. Yazd is ranked the first as an adobe city and the second, next to Venice, Italy, as a historical city in the world. Being located in the heart of Iranian deserts with a long history, Yazd has won the title "Desert Capital." The ancient texture of the city, an exemplar of a live desert city, may provide a model for sustainable urban designs, as the old texture of the city is still in optimal use after several years and at times, it has even demonstrated a better performance over the modern texture of the city which enjoys the technology and materials of the day. It has also been optimally responding to the needs of the residents. The integrated architecture of Yazd has taken advantage of all the available potentials in every single structural element

N. Nahi (✉)
Department of Architecture, East Azarbaijan Science and Research Branch, Islamic Azad University, Tabriz, Iran
e-mail: nahi.navid@yahoo.com

M. Singery
Department of Architecture, Tabriz Branch, Islamic Azad University, Tabriz, Iran
e-mail: singeri@iaut.ac.ir

A. Sayigh (ed.), *Renewable Energy in the Service of Mankind Vol I*,
DOI 10.1007/978-3-319-17777-9_63

in the texture city and buildings including the materials, forms, space planning, and has provided environmental comfort.

The purpose of the present research is to investigate the effect of climatic factors on the spatial structure of the ancient texture of Yazd. The study is carried out in a descriptive-analytic design and has described and analyzed the urban elements that serve as sustainable elements in the traditional texture of the city through library (review) and field observations. Namely among the elements are the integrated and connected urban texture, independent urban neighborhoods, optimal orientation toward sun, conducting and controlling winds to provide thermal comfort, semiopen squares, narrow passages with tall walls, typology and form of the roof covers, Sabats, courtyards, the application of native materials, and additional elements that play a significant role in stabilizing the city structure.

Keywords Climatic factors · City spatial structure · Traditional texture of Yazd city · Sustainable city

63.1 Introduction

Providing thermal comfort in human living environments is one the major goals of climate-compatible architecture. Throughout the history, man has always attempted to adapt his shelter for living with the surrounding environment to create a suitable condition for his living and, indeed, geographical and climatic conditions are also directly involved in the formation of this living space.

As Iran possesses four different climates, namely, hot and humid, hot and dry, moderate and humid, and cold, various climatically adapted architectures (particularly in native house designs) have emerged. In such human-made spaces native materials have been used, which present the least environmentally adverse effects and decrease energy consumption, leading to the sustainability of eco-environment and decreased durability of the buildings. Therefore, in the present discussion, it has been attempted to investigate Iranian native architecture in Yazd region, in terms of its environmental sustainability and energy consumption savings and to briefly evaluate the downsides of modern architecture in each climate in terms of the use of materials incompatible to regional conditions, inappropriate building designs, lack of savings on energy consumption, and environmental destruction.

Given the geographical situation of Yazd, a hot and arid climate rules the region. Little falls of rain and snow coupled with extreme evaporation, very small relative humidity along with extreme heat, and sharp fluctuations in temperature go hand in hand to produce a condition in which the region experiences very hot summers and very cold winters. Under the harsh conditions of Iranian central deserts, providing for the comfort of the residents depends highly upon the skills to create buildings and spaces in conformity with the ecological environment and climatic conditions. Climatic comfort in the region has been made possible in three ways: wind and shadow use, water and water use management, and sunlight effect reduction [2].

Probing the research problem is focused on a population in Yazd region and to delimit the research for better and more accurate investigation, the patterns of shadowing in the context of natural texture is addressed through which sustainable climatic elements in the region could be identified and studied.

In addition to describing the sustainable elements in the traditional architecture of Yazd region, the present study also investigates the factors influencing shadowing with a view of the phenomenal appearance of valuable and sustainable representations that predominate architecture in Yazd region.

63.2 Orientation

Orientation (?) is the basic principle taken into account in the construction of cities in hot and arid climate. The orientation of the texture and passages, given the need to avoid the hot afternoon sunlight in summers as well as hot and stormy north–southerly winds, represents a small deviation. Thus, the orientation of the pieces and buildings are along the northeast–southwest direction. This direction is almost along the Ghibla direction and normally, the halls of the houses are located with their back to the Ghibla. The way the buildings are situated with this orientation enables maximal use of mild winds and sunlight protection.

Since the environmental temperature is lower during the hours before noon than that during afternoon hours, and as a result, there is more winter need for sun heat during morning hours, a rotation to the east (25 °) can provide a better use of the sun heat. Therefore, in these buildings, the setting of the rooms around the house yard is in a way such that summer rooms face opposite to and away from the summertime afternoon sunlight. In this case, maximum shadow is enjoyed during the feverous hours of the summer. Winter rooms are sunny and oriented in the reverse direction to the summer rooms [4].

The most appropriate direction to establish the main front of the building is the one that receives minimum heat in hot periods and maximum heat in cold periods, while remaining protected against adverse winds of cold seasons and enjoying suitable winds in hot seasons.

63.3 Condensed Texture

The formation of condensed texture to prevent extreme heat and cold of the summer and winter, respectively, and from adverse winds to penetrate the residential texture, is the second principle in the sustainability of central Iran cities. Condensation of urban texture has formed to reduce the area of the surfaces exposed to sun and in particular, generate shadow.

Condensation results in the city complex to appear as a condensed mass and minimum exposed and sunward area to be foreseen, this in turn, has resulted in each house to be able to conserve necessary amounts of energy for a long time inside. In such a condensed texture, passages and yards are placed in the heart of the texture as tracks and holes, respectively. In this condensed texture, thanks to the reduction in surfaces in contact with open air, the penetration of extreme summer heat and winter cold is considerably lowered. The texture is also protected against the adverse whipping winds of desert [9].

63.4 Organic and Roofed Passages

Because of the existence of the tall low-width walls of the Sabat, shadow stations 3 have formed in the alleyways enabling the pedestrian to remain in the shadow while passing from time to time. Shadow-sun temperature difference, the result of the alternative shadows by the Sabats, causes a cool wind to move from the shadow to sunny areas all the time. Shadow stations enable the pedestrian to reach the next station before the saturation of thermal budget 4 and loss of tolerance to take sunlight. Upon the arrival in the next shadow station the pedestrian exhausts the body heat excess, enabled to pass through the sunlight [9].

Non-straight, winding, and roofed (with Sabats) passages are formed to protect the pedestrians. These organic passages, on one hand, prevent annoying winds and on the other hand, provide maximum shadow due to their large depth [4] (Fig. 63.1).

A characteristic of passages in hot and dry climates is the proportion between the height and width of the passage such that the passage remains shadow-fallen during hot summer afternoons. The height is created through placing relatively high bumps on the rooftops of the adjacent plucks.

Fig. 63.1 Dummy caption

63.5 Shadow-Sun Area Control

A further issue in the use of sun heat, is taking into account the areas of sun and shadow when adjusting the distance between buildings as in open spaces, passages, streets and yards; this determines the way full–empty spaces are arranged. In this regard, preventing the formation of all-time-shadow during cold periods and reducing such areas is desirable. The distance between the buildings should be selected such that the shadow by the southern building does not interrupt the sunlight to the northern building. This sun shield may be controlled for distancing the buildings by drawing the shadow visor on the sun trajectory map [4].

63.6 Wind Channeling and Control to Provide Thermal Comfort

Like sun energy, wind energy plays a determining role in climate compatible design. Whenever natural wind is present around a building, the characteristics of the wind should be identified to make use of desirable winds in arranging for comfort inside the building. When natural winds are not present, local winds may be produced using pressure rules and the vacuum suction by the temperature differences on the surface and in the height to provide the desirable cool and ventilation. Natural ventilation is created based on three phenomena: wind rate, wind direction, and pressure difference.

63.7 Wind-Induced Pressure and Vacuum Regions

When wind comes to a barrier along its path, parallel layers of air diverge their main path. Air particles exert pressure on the surface against the wind before they diverge. The pressure generates a force along the direction in which air layers move, called positive pressure. Behind the barrier, a vacuum region is thus formed, called negative pressure or suction. Air, then, is sucked from positive-pressure regions into negative-pressure regions. The region behind the barrier on the opposite side to the wind, has negative pressure and circulating flows and is called wind shadow or tail region.

63.8 Generating Local Winds

In cases when the use of local and regional natural wind is impossible, it is possible to generate local winds by causing temperature difference between horizontal and vertical surfaces of air to make it flow. Causing air flow along the horizontal surface can be achieved by sun-shadow temperature difference. This way, air flows from the shadow region—which has lower temperature—into the sunny region—which has higher temperature and lower pressure.

Generating local winds along the vertical direction is a method that operates by benefiting from the ascension of warm air and exhaustion through a wind sucker placed on the roof and the created vacuum generates air flow in any given area. Badkhor or Badgir (wind sucker) is a good method to improve the ventilation performance.

63.9 Using High-Thermal-Capacity Materials

In hot and arid climates the application of materials with high thermal capacities, which have high resistance against heat, has a considerable effect on the comfort of the residents. In this type of climate, the construction of houses with adobe and mud is a means of adaption to the harsh desert environment. Clay and its derivatives, in addition to high thermal resistance, reflect large amounts of sun energy because of their bright color and accordingly, keep the building cool in the summer and warm in the winter [4].

The amount of energy consumed in a house has an effect on the usable life cycle of the house. Our demand for construction materials with high availability plays a significant role in our sustainability and ability to keep and use planet earth sound and fair. The demand for modern more efficient buildings is growing. The increase in the world population compared to minerals and metals, presents buildings as a major reductive factor with respect to these limited resources [1].

63.10 Introversion

The creation of central yards to provide cool desirable air for each residential unit is another principle in the construction of cities in hot and arid climates of Iran. The respiration of the condensed houses is made possible through these yards.

To be safe from the hot summer sun, in particular, in summer afternoons, the part of the house used during summer is built with its back to Ghible, that is to the south. This part is called Nasar in the local dialect and receives shadow and remains cool. The opposite side that faces Ghible receives winter sun—a desirable feature for this area of the house. Because of the narrowness and height of the room walls that surround it from four sides, is hollow and during hot hours of summer days fall in the shadow. On the other hand, at nights when it is cool, the cool air settles at the bottom and the yard remains cool until the sun come up in the sky [10] (Figs. 63.2 and 63.3).

Fig. 63.2 Yazd: Lariha House—the southern corner of the yard

Fig. 63.3 Yazd: Lariha House

63.11 The Combination of Dome Rooftop and Arc

Domed rooftops always stay exposed to wind blow as a result of their protrusion. This is effective in lowering the heat that the roof has received from the sunlight. Since the intensity of the sunlight is uneven over the roof top, three parts of dome or cylinder shaped roofs receive less heat as compared to the other part and this influences the reduction of the temperature under the dome, especially if the dome is equipped with a stem [10].

Having studied the determining factors in the sustainability of Yazd traditional texture and the solutions extracted thereupon, the following table is proposed as a deductive documentation based on what went before (Table 63.1):

63.12 Conclusion

The study of the city of Yazd in terms of body and spatial structure indicates that the city substantially conforms to new scientific findings and the characteristics of a sustainable city. In the present chapter, the importance of shadow generation in

Table 63.1 Adaptive comparison of sustainable city features with city construction principles used in the ancient textures of Yazd (by the author). (Source: nahi)

Organized urban principles of Yazd ancient texture	Sustainable features in the city texture architecture				
	Consideration of ecological factors (adjusting the penetration of outside critical weather conditions)	Optimized energy consumption and adjustment for nonrenewable energy resources	Preservation of renewable natural resources	Improving the sense of liveliness and collectivism in the life quality and society	Prevention of eco environmental pollutions
Orientation	❖	❖	❖	❖	❖
Condensed texture	❖	❖	❖		
Topped and organic passages	❖	❖			
High-thermal-capacity materials	❖	❖	❖		❖
Introversion	❖	❖	❖	❖	❖

Yazd region (with hot and arid climate) was presented as one of the most useful techniques to reduce temperature during hot periods, given the large number of sunny hours and intense sunlight.

The appropriate orientation of the buildings, in many cases, can bring about considerable savings without imposing excessive costs on the users or the need for the availability of advanced technologies. These techniques in the native architecture of Yazd region has been experienced and it is necessary to make practical use of them to achieve sustainable architecture.

Compression causes the city complex to appear as condensed texture with minimum area exposed to the sun and thus enables each house to preserve the necessary energy for a long time. Because of the decreased areas exposed to open air, the extreme penetration of severe heat and cold in the summer and winter, respectively, is reduced which is considered as one of the most important principles in the sustainability of the city.

The creation of organic, winding, and top covered passages to provide thermal comfort for humans, optimal use of energy resources, and further, the presence of Sabats in the passages leads to the gathering of the residents in neighborhoods and emphasizes the sense of cooperation, collectivism, and social stability.

References

1. Ali S (2014) Sustainability, energy and architecture case studies in realizing green buildings. Academic Press, Oxford
2. Dehghan Menshadi M (1385) Tose'ey-e Paydar dar Sayeh Roshan-hay-e Shahr. Entesharat-e Mafakher, Yazd. [Dehghan Menshadi, Mahdi. 1385. Sustainable Development in the Shades and Lights of the City, Mafakher Publications, Yazd.]

3. Kasmayi M (1382) Eghlim va Me'mari. Entesharat-e Baztab, Tehran (Chap-e Aval) [Kasmayi, Morteza. 1382. Climate and Architecture, Tehran: Baztab Publications, 1st. edn.]
4. Lashkari E, Khalaj M (1390) Osool-e Paydari-e Shahri dar Eghlim-e Garm va Khoshk-e Iran ba Ta'kid bar Shahrhay-e Kohan, Tehran: Ganj-e Honar, Chap-eAval. [Lashkari, Elham, M. Khalaj, 1390. Principles of Urban Sustainability in the Arid Climate of Iran: with an Emphasis on Ancient Cities, Tehran, Ganj-e Honar Publications, 1st. edn.]
5. McCarty B (1381) Neveshtey-e Mohandesin-e Moshaver, Badkhan (Molahezat-e Kalbadi-e Bad dar Sakhteman), Tarjome Mohammad Ahmadinezhad, 1381, Esfahan, Nashr-e Khak, Chap-e aval. [Battle Mac Carty Consulting Engineers Authors, Badkhan [Wind Chamber: Considerations of Wind in the building Body), Mohammad Ahmadinezhad Translator, 1381, Isfahan, Khak Publications, 1st. edn.]
6. Pirnia MK (1371) Ashnayi ba Me'mari-e Eslami-e Iran, Tadvin Gholam Hossein Me'marian, Entesharat-e Daneshgah-e Elm va San'at-e Iran, Tehran. [Pirnia, Mohammad Karim, Familiarization with Iran's Islamic Architecture, Edited by Ghholam hossein Me'marian, Science and Industry University Press, Iran, Tehran, 1371.]
7. Rahman A, Singery M (2013) Sustainable building and relationship with humans and nature. In: Ali S (ed) Sustainability, energy and architecture: case studies in realizing green buildings (Chap. 14, pp 387–399). Elsevier Science Ltd, Amsterdam
8. Singery M (2006) Practical principle build permanent building in tabriz. World Renewable Energy Congress IX, Florence
9. Tahbaz M (1392) Daneshe Eghlimi, Tarahi va Me'mari, Daneshgahe Shahid Beheshti, Markaze Chap va Entesharat, Chape Avval. [Tahbaz, Mansooreh. 1392. Climatic Knowledge, Design, and Architecture, Shahid Beheshti University, Shahid Beheshti University Press Center. 1st. edn.]
10. Tavassoli M (1381) Sakhte Shahr va Me'mari dar Eghlim-e Garm va Khoshk-e Iran. Payam, Tehran (Chape Aval) [Tavassoli, Mahmood. 1381. Urban Construction and Architecture in Arid Climate of Iran, Tehran: Payam. 1st. edn.]
11. Tavassoli M (1368) Tarahi-e Shahri dar Baft-e Ghadim-e Shahr-e Yazd. Markaz-e Motale'at va Tahghighat-e Shahrsazi va Me'mari-e Vezarat-e Maskan va Shahrsazi, Tehran (Chap-e Aval) [Tavassoli, Mahmood. 1368. Urban Design in the Old Texture of the City of Yazd, Tehran: Urban Construction and Architecture Research and Studies Center of the Ministry of Housing and Urban Construction, 1st. edn.]

Chapter 64
Enable Environmental Policies for Eco-Industrial Growth: A Voluntary Government Tool for Local Productive Areas in Tuscany (Italy)

Paola Gallo

Abstract The Regional Government of Tuscany in Italy, approved a regional regulation to establish a new voluntary approach at the regional level for the development of "industrial parks." For the first time in European environmental and industrial policies, the "industrial ecology" principles are applied with an approach that allows an area to achieve an official environmental certification. In the literature, we find mainly experiences that rely on command and control policies to implement industrial symbiosis concepts. As opposed to those practices, this chapter presents and analyzes a new method to stimulate the creation and dissemination of eco-industrial parks based on a voluntary approach.

The "ecologically equipped productive area" has been introduced in the Italian Legislative Order by the D.Lgs. n. 112/1998, Bassanini law, which expects that "Italian Regions discipline, with their own laws, industrial areas and ecologically equipped areas, provided with infrastructures and systems necessary to ensure the protection of the health, safety and environment".

Ecologically equipped productive areas have to be planned, realized, and managed on the basis of "ecoefficiency" criteria, in order to ensure an integrated system of management of environmental aspects, reduction and prevention of air, water, and soil pollution, the protection of the health and safety as well as a widespread environmental improvement of territory. The question is to organize the productive site so as to favor the individual settled firms on realizing their own environmental objectives, both economically and technically.

In the case under discussion, the certification scheme is fully voluntary and the certification process is coordinated by a Regional Authority; Regional law of Tuscany (L.R. Toscana n. 61 22/12/2003) establishes priority objective to develop a new concept of industrial areas, characterized by quality management systems and infrastructures for protection of health, safety, and environment, obtained by local authorities assessment methods that integrate economical, social, and environmental issues.

P. Gallo (✉)
Department of Architecture DIDA, University of Florence, Via S. Niccolò 93, 50125 Florence, Italy
e-mail: paola.gallo@unifi.it

A. Sayigh (ed.), *Renewable Energy in the Service of Mankind Vol I*,
DOI 10.1007/978-3-319-17777-9_64

This chapter describes the criteria and requirements that an area must meet in order to obtain the qualification, and the roles and responsibilities of all involved actors. The criteria, inspired by industrial symbiosis concepts, are related to planning, infrastructure, and management-related issues.

Keywords Environmental policy · Eco-industrial parks · Voluntary environmental certification

64.1 Introduction

An appropriate change in industrial policy is required in order to respect and protect the environment and the welfare and that still need to ensure economic growth. The European concept of new industrial competitiveness is expressed in the economic abilities on a sustainable basis to ensure its population living standards and high growth and high employment rates (Lisbon 2000).

The eco-efficient industrial area is made by technical and management requirements that aim to minimize and manage, in integrated way, ecological footprint in order to start a knowledge process about legislation, economic and social aspects, and technical and planning requirements, in order to identify a model of sustainable productive area compatible with the local industrial reality. It is believed that such eco-efficient industrial area initiatives could bring great environmental, economic, and social benefits as a contribution to ecologically sustainable industrial local development.

According to Italian law, in force since 1998, an industrial ecology seeks to find the appropriate balance between environmental, economic, and social needs of a system, so some regions have disciplined on that meaning the ecologically equipped industrial area (APEA—Aree Produttive Ecologicamente Attrezzate), it is an innovative productive area developed and managed as a real estate development enterprise and seeking high environmental, economic, and social benefits as well as business excellence.

Below, we describe the key component of APEA model in Tuscany Region, including information about planning, the main environmental challenges, creation of resources management plans, and provision of supporting policies. Moreover, the document provides details on, which industrial areas should be called APEA, with description of indicators and criteria to reach the Regional qualify of APEA.

64.2 APEA: Description and Features of New Model

APEA is an acronym that means ecologically equipped industrial areas, and it expresses public administrations' purpose to join environmental policy and economical aspects.

The APEA has been introduced in the Italian Legislative Order by D.Lgs. n. 112/1998, Bassanini law, which expects that Italian Regions discipline, with their own laws, industrial areas and ecologically equipped areas, provided with infrastructures and systems necessary to ensure the protection of the health, safety, and environment.

Industrial areas ecologically equipped have to be planned, realized, and managed on the basis of "eco-efficiency" criteria, in order to ensure an integrated system of environmental aspects management, reduction and prevention of air, water, and soil pollution, the protection of the health and safety as well as a widespread environmental improvement of territory. The question is to organize the productive site so as to favor the individual settled firms on realizing their own environmental objectives, both economically and technically.

The goal of an APEA is to improve the economic performance of the participating companies while minimizing their environmental impacts. Components of this approach include green design of area infrastructure and plants (new or retrofitted); cleaner production, pollution prevention; energy efficiency and inter-company partnering. An APEA also seeks benefits for neighboring communities to assure that the net impact of its development is positive.

Some regions together with planners and local communities have used the term APEA in a relatively loose fashion. To be a real industrial area ecologically equipped a development must be more than:

- A single by-product exchange or network of exchanges;
- A recycling business cluster;
- A collection of environmental technology companies;
- A collection of companies making "green" products;
- An industrial area designed around a single environmental theme (i.e., a solar energy driven area);
- An area with environmentally friendly infrastructure or construction.

Although many of these concepts may be included within an APEA, the vision for a fully developed of this model needs to be more comprehensive.

The main objectives of APEA model are as follows:

1. The planning of new industrial estates, technologically and environmentally equipped in order to represent a strategic asset for local development.
2. The transformation and conversion of existing industrial estates, through technological and management's actions. This will contribute toward ongoing business efforts to increase competitiveness in these areas.

Traditionally, industrial settlement developers consider environmental factors only as required by regulations and permitting requirements. Developers of an APEA have to make a deeper assessment of potential sites in locating the area. It is important to consider the characteristics of the local and regional ecosystem, the site's suitability for industrial development, and potential constraints on the pattern of development. This ecological evaluation complements the usual evaluation of transportation, infrastructure, zoning, and other human systems.

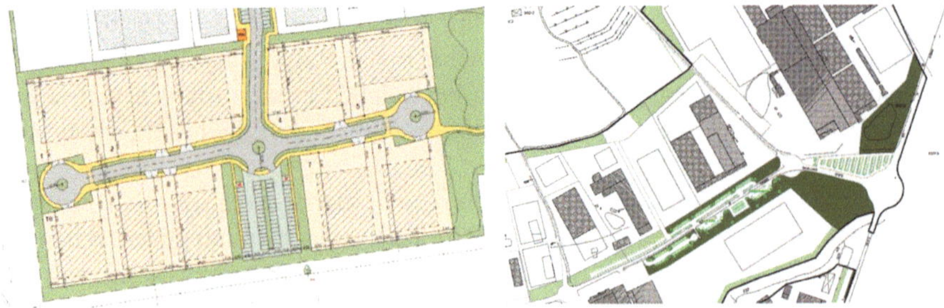

Fig. 64.1 Examples of projects of APEA

The industrial area ecologically equipped is characterized by common infrastructures and services, managed by a single entity that pursues environmental performances that positively influences final quality of total area. This new perspective, activated through cluster typical mechanism, allows combining a sustainable productive development with enterprise competitiveness improvement (Fig. 64.1).

In this way the APEA concept enable real estate developers, industrialists, policy makers, regulators, investors, and communities to collaborate in the vital search for sustainable development.

APEA is intended to reduce their environmental impact and their consumption of resources for industrial production to a minimum.

The ecological and environmental development of APEA areas are accompanied by growth in the competitiveness of the production system, offering the companies which locate here economies of scale, jointly-used infrastructure and services, shared environmental management, and a reduction in the costs of water and energy supplies.

The "APEA" will be entrusted to a managerial company responsible not only for planning the integrated services and improving production cycles, but also for the planning and development of avant-garde systems and infrastructure, for the right environmental management of the area, with the involvement of all the companies operating here to assist in the attaining of the objectives and, finally, a dialogue with local bodies and communities.

In fact APEA model permits environmental and economic advantages for enterprises, through adoption of common infrastructures (collective waste platform, energy production plant from renewable fonts, collective water treatment plant, collection of rain water, landscape mitigations) and management solutions (common emergency management, centralized management of green areas and common spaces, purchase groups for energy supply, waste recycle stock exchange, centralized logistics, environmental training area Environmental Management System). Advantages are as follows:

- Energy and water consumption reduction
- Waste treatment costs reduction
- Costs reduction (energy, water, matter)
- Maintenance costs reduction

- Administrative semplifications and incentives for enterprises
- Safety conditions improvement
- Imagine improvement

In addition, some common business services may be shared by firms in industrial areas ecologically equipped: these may include shared waste management, training, purchasing, emergency management teams, environmental information systems, and other support services. Such industrial cost sharing could help APEA members achieve greater economic efficiency through their collaboration.

Small and medium size firms often have a problem in gaining access to information, consultation, and know-how. This integrative approach can support such enterprises in overcoming these barriers and gaining access to investments they may require to improve performance.

64.3 An Innovative Model for the Planning of Industrial Areas in Tuscany

Tuscan Region intend the APEA as an innovative productive area developed and managed as a real estate development enterprise and seeking high environmental, economic, and social benefits as well as business excellence.

The Tuscan Region is an important stakeholder in this commitment and plays a key role in promoting APEA development in its territory, through more aspects like decision making, creating policies, issuing laws and regulations, organizing pilot activities, providing financial incentives, encouraging innovations in technology and systems, fostering new markets, and promoting both education and academic research partnership.

After national guideline, the regional law of Tuscany (since L.R. Toscana n. 61 22/12/2003) establishes as priority objective to develop a new concept of industrial areas, characterized by quality management systems and infrastructures for protection of health, safety, and environment, obtained by local authorities assessment methods that integrate economical, social, and environmental issues.

On the basis of these law information, a new Regulation (R.T. 2 dicembre 2009, n. 74) clarifies and updates the APEA concept as: "an industrial, craft and mixed use areas, included in multifunctional contests, equipped with pollution and emission control system; APEA are characterized by an integrated and unitary management of infrastructure, services to protect environment, security and health of operators and communities" (Art. 2).

In this way, the region promotes, through this new regulation, an innovative model in relation to planning, design and promotion of industrial areas equipped with innovative technological infrastructure, which conform to current and future expected environmental standards.

This legislative document, realized in scientific collaboration with the University of Florence, Architecture Technology Department (TAeD) and S.Anna Superior

School of Pisa (SSSUP), enhances the relationships between different actors—including municipalities, businesses, and the local community—and aims to optimize the sustainable use of resources in industrial areas.

The work, lasted 3 years, aims first of all to define the main features of APEA as follow:

- Sustainable urban planning and design of technological and mobility networks.
- Implementation of sinergies between enterprises, through a unit management of centralized technological systems, common spaces, and common services.
- Closed production cycle that aims at the reuse of waste streams and industrial symbiosis.
- Provision of barriers and other systems for the reduction of any kind of pollution.
- Use of renewable or low impact energy sources.
- Setting up of ecological platforms for waste collection, for water treatment, etc.

The Regulation makes difference between new industrial areas and restoration of existing ones, and gives to decision making bodies (Region, Provinces, and Municipalities) specific skills in APEA planning and management, including regional financings to promote APEA diffusion on the Tuscany territory.

The Regulation explores strategies and method through which managers of existing. Productive areas can gain the right to call their properties APEA. A complessive vision of area and a strategic planning process, drive site managers and their tenants to evaluate the benefits of participating in a regional APEA network and by product exchange as well as other means of improving their performance.

The team has elaborated guidelines of these processes and resources to support new industrial areas and existing ones, improving the environmental, social and economic performance of companies at each scale, through new services offered by APEA management, like:

- An integrated resource recovery system
- A system for encouraging and managing the exchange of by-product between companies
- Training and services in all aspects of eco-industrial development
- A single management/coordinating unit
- Public sector support in R&D, policy development, access to investment, and information management.

Moreover, the document also explains procedures for checks and acceptance and for performance assessment.

This Regulation, in detail, offers a rich menu of individual facilities, and shared support services, design options, including ideas for site and infrastructure design; moreover, also covers strategies for achieving environmental performance and management.

Several basic strategies are fundamental to developing an APEA, therefore, the Tuscany Regulament identifies the following as key strategies:

Management Entity APEA members will need a management system that maintains their cohesiveness without compromising their autonomy. Except where external regulation or property covenants are involved, the community will depend on voluntary participation in any common initiatives.

An effective management entity covers a role of primary importance in every productive industrial environment so, the Tuscan Regulation provides for the identification of a Legal Entity (named in the Regulament as Soggetto Gestore) with specific role and functions for the improvement of productive areas, intend as organisms that can grow with the needs and developments decided and implemented with the actors involved: users, operators of the settlement, local government, businesses, and service of the municipality.

Infrastructural Envelope The companies in APEA need a range of general services indirectly related to their production systems. These include governmental relations, dining facilities, purchasing of common supplies, information access, and many others. By acting in common to procure these services, they can reduce indirect operation costs (especially important for smaller companies). By coordinating satisfaction of these tenant needs, the APEA management company can increase its revenues. Sharing services will increase opportunities for communication among employees of different companies and build the community spirit of APEA.

64.4 The Qualification of APEA: The Criteria

A full evaluation framework for an ecologically equipped area combines economic, technical, social, and environmental objectives into a whole system. This means that APEA project can seek a design that optimizes objectives in these four domains as a whole, not separately. Clearly articulated objectives in each area, agreed by project stakeholders, will be essential. With this clarity, site managers and public administrations will be better able to determine the trade-offs among the objectives in all four domains, economic and environmental objectives, social and environmental, or any other pair of domains.

The Tuscany Regulation establishes requirements to qualify industrial area environmentally equipped and foresee a score system points in order to evaluate them: each criteria have a specific score to add in order to reach the APEA qualify. There are two kinds of requirements and they are as follows:

1. Minimum requirements: their satisfaction is necessary to obtain APEA status;
2. Flexible requirements: it is possible to choose requirements that are functional and compatible with the territory, to obtain threshold necessary to obtain APEA status.

Several basic strategies are fundamental to developing an APEA; individually, each adds value and together they form a whole greater than the sum of its parts so, the criteria of Tuscany Regulation to satisfy in order to reach APEA status are articulated in:

- Urban, about planning and design of industrial areas ecologically equipped
- Infrastructural, about innovative technologies and services
- Management, about organizational requirements.

In detail, urban and infrastructural criteria provides with technical requirements directed to diminish and to manage the pressures on environment in an integrated way, applied to buildings, industrial facilities and common areas, bought in, have the ambition and the aim of transforming the entire area in a body to serve its users. A body, which is in its integration, becomes a tool for its users by creating and providing content for environmental sustainability: living healthy, active safety, passive safety, comfort, but also socialization and connection services. Infrastructures (for sustainable mobility, energy saving and production, for water management, lighting, waste management, access control, the web server, WiFi access points, video surveillance, irrigation, etc.) will be centralized, and they are characterized by simplicity available to all actors involved. The APEA planning aims that buildings and infrastructure are designed optimizing the efficient use of resources and minimizing pollution generation. It is essential to minimize ecosystem impacts by careful site preparation and environmentally sensitive construction practices. The whole area will be designed to be durable, maintainable, and readily reconfigured to adapt to change. At the end of its life, materials and systems can be easily reused or recycled.

The realization of industrial areas ecologically equipped will be a tool for local governments and for the entire areas to support the economic and social development, which, since the implementation phases will generate jobs, and opportunities for the construction industry, and support socioeconomic area.

64.4.1 APEA Qualify Criteria

An APEA planning, calls for asking new questions within the context of traditional industrial development processes. Developing any industrial areas ecologically equipped requires several rounds of planning and design. The team should test project feasibility in greater detail with each stage. The project must satisfy financial, economic development, public planning/zoning, environmental, and technical criteria at each step. The APEA team will follow the traditional process, while considering new design options in each phase of project planning.

64.4.1.1 Urban Criteria

Criteria for the APEA localization, are finalized to warrant a complete efficiency of urban and environmental systems, and they must (Fig. 64.2):

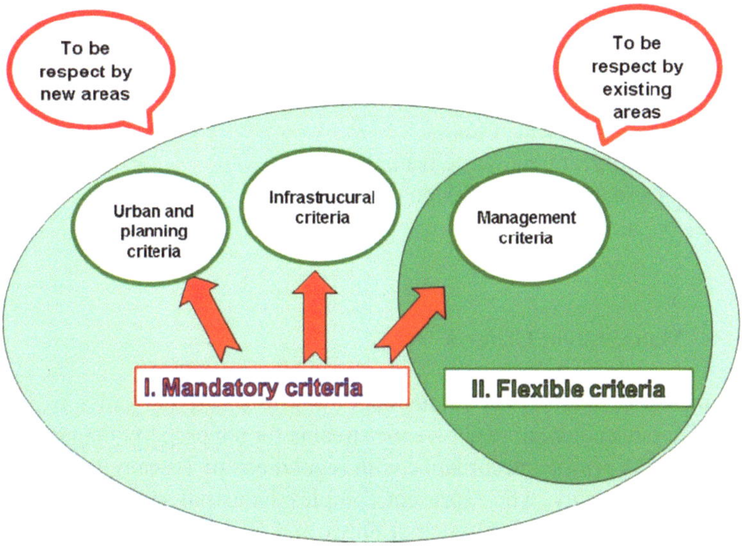

Fig. 64.2 Choice of criteria

- Privilege reuse and fulfillment of existing industrial areas and buildings;
- Evaluate insediative system efficiency and functionality, about standard infrastructures, social and economical local factors that can warrant actual and future stability of enterprises;
- Evaluate accessibility to major communication routes, fostering rail transport and transport intermodality;
- Realize urban area programs in spite of single isolated initiatives;
- Assess the presence of environmental, historical, and urban bond, monumental, archeological, geological constraints, protected areas like natural parks, etc.
- Evaluate environment quality and liveability and promote innovative instruments.

64.4.1.2 Infrastructural Criteria

APEA promotes environmentally improved performance, concerning industrial ecology and environmental sustainability.

In respect of energy and environmental aspects, APEA are:

- Equipped with infrastructures and networks coordinating system, to improve integrated prevention by air, water, and soil pollution, and suitable instruments to make a constant emission monitoring;
- Realized in geological and environmental safety;
- Provided with system to maximize energy efficiency (cogeneration systems, renewable energies use, heat waste recovery, etc.);

- Equipped by the presence of:

a. Environmental data station detection
b. Waste management systems
c. Water safe management systems
d. Collecting and treatment waste water system
e. Collecting and treatment emission system
f. Production and distribution energy system.

64.4.1.3 Management Criteria

The management of an APEA entails both traditional and innovative responsibilities and generates potential new revenue streams for the property manager.

It is an APEA priority, accordance with regulations of Tuscany Region, the creation of a single entity, who represent complex industrial settlement enterprises, able to act as interface with Municipal Entity and other public stakeholders, to arrange actuative planning program, to promote contracted formulas with Municipality, to manage action settlement through infrastructural and services creation, and to arrange for maintenance and management of common services and facilities. Management entity can be represented by Municipal Entity, Industrial Development Unions, mixed public-private societies, etc.

64.5 Conclusion

APEA model allows organizing productive settlements to facilitate single enterprises, economically and technically to reach their environmental goals, both prescriptive ones (emission control, energy safe, reduction of their own pollution, etc.) and voluntaries ones (Emas or other environmental management systems adherence).

APEA model is compatible with Tuscan Region productive and industrial structure, characterized by local systems highly specialized in production sector (paper mills, tanneries, steel, textile) and by small and medium enterprises presence. In fact the model:

- Facilitates SME to reach an improvement of their own environmental performance, through common infrastructural and services equipment, characterized by high quality, impossible to achieve and to manage individually
- Allows control and reduction of cumulative environmental impact, generated by all the enterprises of the settlement
- Facilitates and exempts enterprise by obtaining environmental permission when issuing and renewing
- Applies postulate of pollution prevention, precaution, and reduction

The work is the result of a synergy between University, enterprises, and public entity: this cooperation allows to demonstrate how scientific research contribution is necessary to definition of new models and mode of contemporary and future of lifestyle.

References

1. Allenby BR, Richards DJ (1994) The greening industrial ecosystem. National Academy Press, Washington
2. Eilering JAM, Vermeulen WJV (2004) Eco-industrial parks: toward industrial symbiosis and utility sharing in practice. Prog Ind Ecol 1(1/2/3)
3. European Council (2000) European Council Conclusions, Lisbon, 23–24 March
4. Gallo P, Carletti P (2000) Principles and guidelines for a sustainable environmental development. In: Steemers K, Yannas S (a cura di) (eds) Proceedings of PLEA 17th International conference on passive and low energy architecture London, 3–5 July. James & James. Science publishers) Ltd, London
5. Iraldo F (2002) Ambiente, impresa e distretti industriali. Franco Angeli, Milano
6. L'Architettura Naturale N 23 (2004) Produrre sostenibile. Edicom Edizioni, Monfalcone
7. L. R. Toscana n. 61 22/12/2003 - Norme in materia di autorizzazione integrata ambientale. Modifiche alla HYPERLINK "http://jtest.ittig.cnr.it/cocoon/regioneToscana/urnResolver.xql?css=&urn=urn:nir:regione.toscana:legge:1998-11-03;79" \o "URN = urn:nir:regione.toscana:legge:1998-11-03;79" legge regionale 3 novembre 1998 n. 79. Aree produttive ecologicamente attrezzate. Modifiche alla HYPERLINK "http://jtest.ittig.cnr.it/cocoon/regioneToscana/urnResolver.xql?css=&urn=urn:nir:regione.toscana:legge:1998-12-01;87" \o "URN = urn:nir:regione.toscana:legge:1998-12-01;87" legge regionale 1 dicembre 1998, n. 87

Chapter 65
Smart Envelope for Nearly Zero Energy Schools. The Case Study of Vallisneri Secondary School in Lucca

Rosa Romano

Abstract The research SE*NZES analyses the evolution of smart façade systems in the area of design and industrial production, in order to investigate the technological, functional and qualitative standards of dynamic façades and to evaluate the energy performance of the school building envelope as a dynamic system that interacts with the indoor and outdoor environment. The study focuses on dynamic envelopes for the refurbishment of the school buildings, analysing the evolution of façade systems in terms of building construction, energy and structural performance and durability.

The research project has been developed with the aim of spreading the research topics identified by Cordis and European Directives 2010/31 and 2013/27, developing a smart façade system, achieving the reduction of the energy requirement in the building sector by 2020 and realizing zero energy public buildings.

The refurbishment of the existing school building stock requires ground-breaking strategies in order to meet targets for improved indoor comfort in the classrooms and to reduce energy use and greenhouse gas emissions. One of the most important components to be addressed with an innovative approach is the building envelope, which has to develop into an active rather than a passive element, fulfilling more functions than just the separation of the outer space from the interior with insulation.

The concept of the adaptable envelope is twofold. On the one hand, it has been designed to accommodate further modifications (such as future renovation or technology upgrades). On the other hand, the envelope is able to adapt to a dynamic and intricate environment by measuring and processing multisource information (e.g. outdoor and indoor environment conditions, occupancy, behaviour of users and envelope performance) in order to respond to the building occupant's instructions and evolving environmental conditions at an appropriate time and to an appropriate extent.

The main topic of research has been to develop an innovative and smart façade system that will be possible to integrate in the retrofitting of existing school buildings. In this smart envelope, it will be possible to integrate

R. Romano (✉)
Department of Architecture DIDA, University of Florence, Florence, Italy
e-mail: rosa.romano@unifi.it

A. Sayigh (ed.), *Renewable Energy in the Service of Mankind Vol I*,
DOI 10.1007/978-3-319-17777-9_65

- Advanced materials or technologies for energy generation and storage
- Smart and performance insulation materials
- Reflecting materials and cool coatings
- Adaptable innovative precast solutions
- Materials with enhanced acoustic properties
- Materials with improved fire resistance
- Automated blinds or movable sun barriers with interrelated issues of summer overheating, air-tightness and natural light use
- Innovative sensing systems to control and optimize the real-time performance of the envelope

In this chapter, we introduce the characteristics of smart façade for school buildings, analysing the technological solutions and the energy performance to reduce energy needs of this type of building. We will be observing, in particular, the case study of the refurbishment of one school building: the "Vallisneri Secondary School in Lucca", where we have designed and built an innovative façade system. The project focused on the adoption of a precast façade in order to control solar radiation in summer and heat loss in winter, reduce the heat losses between the opaque envelopes and increase the natural ventilation inside the school building.

Keywords Smart façade · Sustainable school buildings · Energy saving · Renewable energy

65.1 Introduction

The UE school buildings stock, built up in 1960/1980, is actually one of the most energy-consuming building sectors, affected by decay. The continuous state of emergency in the Italian education process penalized the school building sector: buildings with construction faults (improperly realized) or affected by decay, are actually structures unable to satisfy spatial, lighting, structural and energy requirements. The Italian situation is far away from the North European trend, where school retrofitting programmes, like Innoschool [1] or BSF [2] aim towards constructive, technological, spatial and environmental improvement, acknowledging good influences on students' performance. Important activities to increase energy performance of the school buildings are carried out in Europe and in the USA. The UK and Sweden are the main European countries where an effective policy on school energy efficiency has been implemented, by Eco-Schools Network and a *Guide about Energy Consumption—Saving Energy in Schools*.

However, it is necessary in Europe and in Italy to invest more strongly in school buildings, by channeling efforts towards supporting the life quality of constructions for education, energy efficiency and public funding saving, placing the end users' requirements at the heart of the design and decision-making process.

These goals are in line with Horizon 2020 priorities, Society Challenges sector, related to:

- Secure source of energy
- Resources and raw materials saving
- Inclusive, innovative and safe societies

The assessment of the building component in a dynamic condition is promoted by the European Commission which during the past years invested in several projects to develop new building components using nanotechnologies, with the aim to optimize the insulating behaviour of the opaque and transparent building components, and in the meantime, to control the relative attenuation factor in order to reduce the energy consumption in the building. For example, the European funding call "Nanosciences, Nanotechnologies, Materials and New Production Technologies" under the FP7, encourages the development of new building technologies for energy efficient building components. In accordance with this strategy, the European Commission produced Directive 2010/31, providing guidelines to European countries to reduce energy use even more, considering, in the most appropriate way, the specific regional climatic conditions. This approach has moved the focus from the winter needs, usually investigated under static conditions, to the concept of nearly zero energy building that requires the integration of renewable energy in order to reduce energy consumption during the summer time also.

The environmental incidence of the research project SE*NZES is evident and strong because it focuses on the experimentation of energy-saving techniques, integration of innovative materials and renewable energies in the smart envelopes for reducing costs and consumptions in school buildings.

65.2 Objectives

The main goal of the SE*NZES research is to design and develop innovative smart façade systems for the energy and structural refurbishment of the school buildings built between the 1970s and the 1980s in southern Europe.

A real problem in the South European countries, at the moment, is to reduce energy loss for cooling, particularly in nonresidential buildings. This is possible by projecting envelopes that can change their configuration and energy performances during the year. At the moment, in Europe, it is necessary to define a new concept of nearly zero energy buildings, also for the Mediterranean area, and develop innovative envelope systems. However, the most important research on sustainable façade systems has been carried out by the research centre located in northeastern Europe (Germany, Austria, Finland, etc.), in order to study the problem related to thermal insulation in the winter months only. The façades, developed for this type of climate, are always built in glass or other light and transparent materials; so it is very important, as previously mentioned, to develop research in order to design and realize innovative façade systems for South Europe, studying the energy performance for the hot months and founding innovative solutions to improve the inertial value also.

At the same time, the EU energy policy encourages member states to start converting building stock into nearly zero energy buildings (nZEB), and public authorities to adopt exemplary actions nZEB achieved by combining high-energy efficiency and renewable energy sources. Low energy buildings are becoming the standard for new buildings in north-central European countries. However, there are several barriers that hinder school building renovation from achieving nZEB levels in Mediterranean climates, for instance

- Insufficient knowledge about existing technologies that can be used to renovate buildings into nZEB in Mediterranean climate;
- Inefficient tools for public sector to achieve nZEB renovation.

In addition, Mediterranean zones are characterized by high diversity climatic conditions (on the coast, with high population density, winters are mild and summers are relatively hot and humid, whereas inner areas are generally drier and less temperate), so they need a specific approach, different from that used for colder climates.

The research SE*NZES on dynamic façade systems for school buildings can give an input to achieve these EU goals, because the dynamic façades guarantee a variability of performance during the year and allow more layers in the one façade system in order to improve: thermal insulation and solar gains in winter; inertial value and shading in summer. The analysis of the state of art shows that the research on dynamic façade systems, moreover, is not very diffuse in Europe, where only a few examples of this type of façade were built.

In order to achieve these objectives, we have designed and realized the project of a new façade system of Vallisneri Secondary School in Lucca, as a first example of a type of innovative façade system that can be used in an energy retrofit of an existing school building located in the Mediterranean area (Fig. 65.1).

65.3 The Retrofitting of the High School Vallisneri Secondary School in Lucca Towards an nZEB

The retrofitting of Vallisneri Secondary School has been developed with the aim of integrating different needs of performance upgrading in existing buildings towards a sustainable construction, proposing a model of a new nZEB in Mediterranean areas, using appropriate policies tailored on local climate conditions. From 31st December 2018, Italy will have zero energy public buildings, but this definition is still too focused on continental climatic conditions and poorly adapted to Mediterranean areas.

The project of the retrofitting of the school building of Vallisneri has been a laboratory to promote technologies to improve energy performance of the building envelope and indoor comfort in the classrooms, integrated with the general

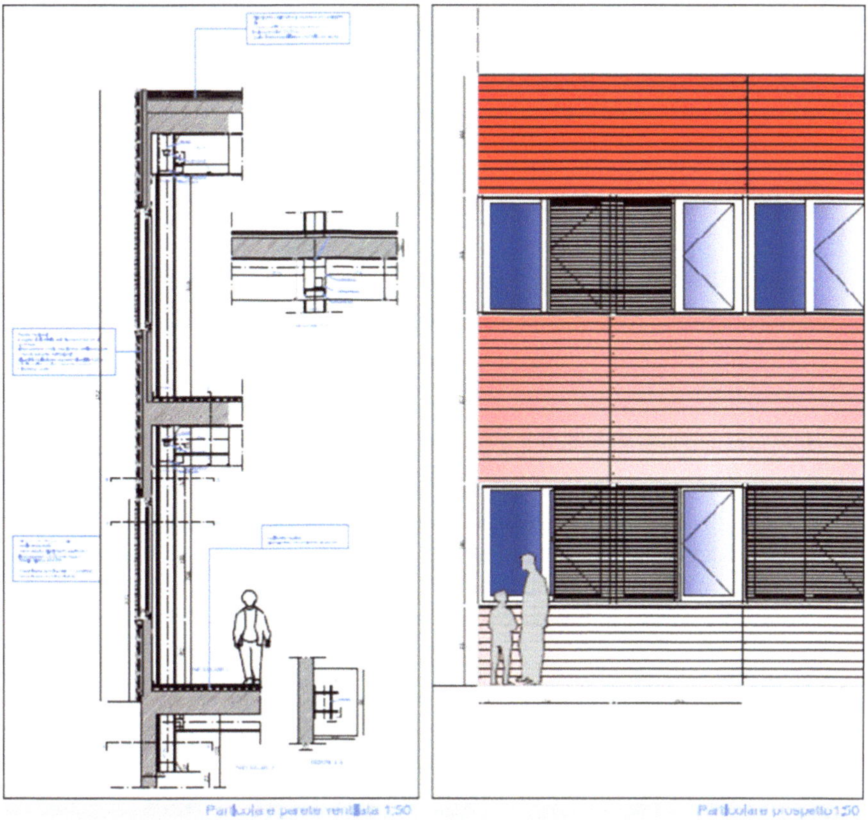

Fig. 65.1 The project of the new façade system for Vallisneri Secondary School

upgrading of the anti-seismic resistance of the building, according to the ultimate codes introduced in Italy.

The new envelope design with high-energy performance has reduced the energy consumption for heating and cooling, and the energy consumption for the air exchange system that is necessary to guarantee a high indoor air quality in the classrooms.

The envelope of the school buildings has been developed as a dynamic environmental filter, which can adjust the airflows, the solar radiation and the heat flows.

The skin of the building became a technological system that can produce thermal and electric energy that will be possible to use for the metabolism of the building.

The project of the retrofitting of Vallisneri Secondary School has been promoted and financed by a public administration (the Lucca province) and shows that it is possible to perform a total requalification action of an existing building at a lower cost rather than demolishing and rebuilding the same volume, upgrading at the same time, the structure and the energy performances.

Fig. 65.2 Vallisneri Secondary School before and after the energy retrofit

The Vallisneri school is divided into two different buildings built at two different times (Fig. 65.2):

- A building with structure in c.a. and envelope in brick, built in 1960; it has three floors and a plant in the form of an "H".
- A building with a steel structure and prefabricated panels, it was built in 1970 to increase the surface of the pre-existent school. It has two floors and a plant in the form of a "C".

The retrofit action has been carried out in the building built in 1970 which had many structural and energy problems, because the steel structure was inadequate to fulfil the criteria set by the Italian seismic legislation and the envelope (realized with a sandwich panel of 10.00 cm of thickness) did not guarantee adequate thermal insulation during all months of the year (U: 1.8 W/sqK).

In detail, the retrofit project was concerned with

a. The adaptation to the seismic legislation and standards, with:
   - Reinforcement and insertion of metal plates on some columns and beams
   - Insertion of wind bracing in the external surface of the existing floors that did not have the welded steel mesh into the c.a. slab
   - Introduction of new connecting beams into the internal surface of the floors
   - Reinforcement of rods and knots
   - Verifying the stability of the existing partitions in brick and effectiveness of the bond to the existing floors

b. The increasing of energy performances and indoor comfort, with:
   - Total replacement of the existing façade
   - Realization of a new heating system
   - Integration of a new lighting system
   - Maintenance of the existing photovoltaic (PV) system on the roof of the building

The project to improve the energy performances of the building and indoor comfort in the classrooms has been developed with the design of an innovative façade system that can also be used in the other retrofit projects of the school buildings.

65.4 The New Façade System

The old façade, which was realized with sandwich panels, had many problems with maintenance and thermal and acoustic insulation. The new façade system has been designed as a prefabricated module, so as to decrease the time and cost of realization and to guarantee good hygrothermal performance during the months of the year when the school is open (Fig. 65.3).

The new façade system has been designed as a modular façade with opaque and transparent partitions where it is possible to integrate

- Thermal insulation, to reduce thermal loss during the winter months but also in spring and autumn, when the temperature in Italy is very high;
- A shading device, to regulate the natural lighting into the classroom and to reduce glare and overheating phenomena;
- An air vent, to increase the air change and the natural ventilation inside the buildings.

The opaque and transparent panels were made in a factory in 3 weeks and were brought to the site where they were integrated into the structural grid (realized with steel pillars and beams with a thickness of 60.00 mm) in 2 weeks (Fig. 65.4).

The opaque panels are made up of various layers: aluminium sheet, air gap (8.00 cm), aluminium sheet, plasterboard (1.5 cm), two insulation panels of fibreglass (8.00 cm), two insulation panels of rock wool (8.00 cm), plasterboard (1.5 cm), and aluminium sheet. They have a thermal transmittance of 0.25 W/sqK and, thanks to the presence of an air gap and plasterboard, guarantee good performance in the intermediate seasons also.

Fig. 65.3 Phase of construction of the façade

Fig. 65.4 The classrooms before and after the energy retrofit

The transparent module is made with a frame with thermal break and double-glazing.

The frame is made of aluminium extrusion 6060, according to the UNI EN 12020, UNI EN 573 and UNI EN 515. The aluminium frame is insulated with the interposition between the bearing section (located inside the wall) and the mounting face of a lath with low thermal conductivity, according to the UNI 10680. The main characteristic of this system is the low value of thermal transmittance: 1.2 W/sqK. All windows have double typology of opening: swing door and turn and tilt, so as to facilitate cleaning and increase natural ventilation.

The aluminium shading device is located on the external surface of the transparent module; the shading device is rotated to reflect direct solar radiation and reduce the overheating and glare phenomena inside the classrooms.

Finally, the air vent is integrated into the frame of the transparent module. We have chosen a simple typology of air vent with manual opening.

65.5 Conclusion

The new system of façade was built and integrated into the structure of the existing school building during the winter of 2012. The possibility of realizing the technological system in the factory and, consequently, to reduce the time of construction has allowed to reduce the costs of retrofit and discomfort of the pupils.

After the replacement of the envelope, the image of the school building was completely changed and the energy consumption for heating and lighting decreased by about half.

The Vallisneri project shows that it is possible to do an effective retrofit action with the objective of really increasing the energy performance and indoor comfort of the existing school buildings. The possibility of developing a project of a precast façade module, in fact, allows to analyse during the design phase exactly what type of solution to employ and choose the ones that are the best to achieve the goal to realize an nZEB, according to the EU legislations.

In the coming months, it will be possible to analyse the real performance of the façade system and to test its capacity to decrease the energy consumption of the building. At the same time, we are working on the design of other envelope solutions for the school buildings as expected from the research program SE*NZES, to produce a catalogue of technological solutions that can be used in the retrofit action in the coming years.

References

1. The goal of InnoSchool was to develop the Future School Concept: a set of research-based good practices, processes, models and designs, and recommendations for their successful combinations in the Future School. InnoSchool consists of four co-projects: InnoArch—Places and Spaces for Learning; InnoEdu—Education with Innovation; InnoPlay—Innovative Playful Learning Environments; InnoServe—Service Innovations for the Future School. These co-projects conducted research and developmental work on the schools from their different scientific perspectives: architecture and urban planning; playful learning environments; education and the teaching–studying–learning process; and process and innovation management. The synergy between the four co-projects was achieved through theoretical dialogue that applied a joint Concept Design framework, and through collaborative empirical case studies on selected pilot schools. http://innoschool.tkk.fi/
2. Building Schools for the Future (BSF) is a program for the rebuilding or renovation of every secondary school in England over the next 10–15 years. It was launched by Tony Blair in February 2004, as the successor to the Private Finance Initiative (PFI). Under BSF, the government promised £ 45 bn to rebuild or refurbish every secondary school and half of all primaries in England by 2020

Chapter 66
A Comparison of Computational Simulation and Physical Measurement of Solar Radiation and Photovoltaic Outputs for Residential Dwellings

Stephen Pretlove and Patrick R. Osborne

Abstract By 2016, new residential buildings in the UK will have to be 'net zero carbon' to comply with proposed changes to Part L of the Government's Building Regulations. Approved document Part L of the Building Regulations requires energy use and generation, and the resulting carbon emissions, to be quantified using the Government's standard assessment procedure (SAP) model. To achieve a zero carbon dwelling, on-site renewable technologies must usually be incorporated into the design. Since the introduction of the feed-in tariff (FIT), in April 2010, photovoltaic (PV) systems have been seen as one of the most cost effective methods of achieving the higher levels of the code for sustainable homes (CfSH), the route to meet zero carbon emissions in domestic buildings. The quantification of energy generation for CfSH certification, comes directly from the SAP model, where the methodology used to justify the use of PV systems is crude in its prediction of shading and utilises simplified rules-of-thumb to predict how shading will affect energy generation. This chapter compares the prediction methods currently available to designers (SAP) against real data collected on live building projects in South West London. Included in this study is the physical measurement of solar radiation where PV panels will be installed at a later date, together with the measured outputs from two recent domestic PV installations that are benefiting from the FIT initiative. For both of these, in terms of solar radiation and electricity production, comparisons are made between actual measurements and predicted results. The results of this study show that the methodology provided in the SAP 2005 and 2009 models for determining the available energy at inverter output (kWh/year) for solar PV systems is crude and inaccurate, particularly in locations where there is significant shading from external obstructions, and particularly where an evaluation of the overshading is required. The SAP methodology for quantifying the shading coefficient is crude and there are little guideline provided. A novel technique for quantifying

S. Pretlove (✉)
School of Architecture & Landscape, Kingston University London,
Knights Park, Grange Road KT1 2QJ, UK
e-mail: s.pretlove@kingston.ac.uk

P. R. Osborne
Lee Evans Partnership, 175–185 Grays Inn Road, London WC1X 8EU, UK
e-mail: patrick.osborne@lee-evans.co.uk

A. Sayigh (ed.), *Renewable Energy in the Service of Mankind Vol I*,
DOI 10.1007/978-3-319-17777-9_66

overshading has been tested in this study and the results indicate that a more robust method is required. The methodology being proposed is in line with more comprehensive approaches that have been adopted by other organisations, in particular new guidelines and methodologies published and recommended by the microgeneration certification scheme (MCS). Proposals by the BRE for improvements to the SAP 2012 model, for the calculation of incident solar radiation, and the energy generated by a solar PV system, are generally positive but the determination of overshading in any particular location remains crude and difficult to quantify.

Keywords Solar radiation · Photovoltaic · Shading · Building regulations · SAP · Sustainability

66.1 Introduction

In 2007, the UK Government set, and has subsequently maintained, an ambitious target of ensuring all new homes will be net carbon neutral by 2016 [8]. The route to achieve this is through the CfSH [7], which categorises sustainability into nine broad areas; energy and carbon dioxide emissions; water; materials; surface water run-off; waste; pollution; health and wellbeing; management; and ecology. To achieve the highest level, and the status of 'zero carbon', the net carbon emissions must be zero. The definition of a zero carbon building has been disputed for some time. When the CfSH was first published in 2006, 'zero carbon' homes were defined as those where all CO_2 emissions were mitigated on site or by directly connected infrastructure. Subsequently, and after some debate, it became apparent that mitigating all CO_2 emissions on site would not be practically achievable for the vast majority of new dwellings. In 2008, the government redefined 'zero carbon' homes as those where a hierarchical approach should be adopted as follows [21]:

1. Ensure an energy efficient approach to building design
2. Reduced CO_2 emissions on-site via low and zero carbon technologies
3. Mitigate the remaining carbon emissions with a selection of 'allowable solutions'

In reality, the real energy consumption, and carbon emissions, in a dwelling are likely to be more than that predicted by the SAP model (BRE 2010), and alternative systems, such as the national home energy rating (NHER) system developed by national energy services [15], account for these unregulated emissions, and indicate that they can be a significant proportion of the total energy consumed, and carbon emissions produced, in a typical dwelling. In 2009, the housing minister announced that the 'carbon compliance level would be set at a 70 % reduction in regulated [that predicted by the SAP model] CO_2 emissions' [21].

The method used for calculating a building's CO_2 emissions is that described in the Government's SAP which utilises basic building geometry to calculate heat losses, energy consumption and carbon emissions. A building will always consume a certain amount of electricity, so to achieve a zero carbon rating, this must be offset; most commonly with on-site renewables.

Photovoltaic (PV) systems can currently benefit from a government funded FIT that pays the owner a tariff for every kWh that they generate, even if the owner then consumes that energy. Higher level tariffs are available if the owner sends excess energy back into the national grid. In effect, the owner gets free energy and they are paid for generating it. For this reason, PV systems have been recognised as a cost effective way of meeting the higher levels of the CfSH. Since the introduction of the FIT system in April 2010, and up until December 2012, over 340,000 domestic registered FIT PV systems (representing 99% of all FIT domestic technologies) have been installed, with a generating capacity of over 1100 MW and an average installation capacity of 3.2 kW [17].

There exists a wide range of tariffs available, currently and historically, based upon the technology adopted, the size of the installation, and more recently, on whether installed on an existing occupied building, a new building or as a stand-alone system not attached to a building. For a PV system installed on a new building, and less than 4.0 kW, the original generation tariff in April 2010 was 36.1 p for each kWh generated and this figure has progressively been reduced to the current tariff (from April 2013) of 14.9 p for each kWh generated. Additional export tariffs are available where the energy generated is sent back to the National Grid [18].

Just prior to this study being carried out, and towards the end of 2011, the UK Government attempted to reduce the tariff rates for solar PV significantly, with limited public consultation, but this was challenged and tested in the courts and was deemed to be unlawful to implement at that time. This delayed the onset of what was to become a considerable reduction in tariff which led to a significant increase in the number of domestic installations, caused by a general rush to install systems at the higher tariff rate. Subsequently, and inevitably, in March 2012 the government reduced the tariffs by approximately 50% which had an impact on the take up of this technology in the domestic market.

This study evaluates the solar PV prediction methods currently available to designers (SAP) against real data collected on live building projects in South West London. It includes the physical measurement of solar radiation where PV panels will be installed at a later date, together with the measured outputs from two recent real domestic PV installations that are benefiting from the FIT initiative. Comparisons have been made between actual measurements and predicted results and the main focus of the study is the quantification of shading. A number of previous studies have examined the impact of shading on the performance of PV installations but the vast majority of these studies have examined the impact of shading on the performance of the PV modules themselves, and the resulting reduction in electrical output, rather than an assessment of the general impact of shading on incident solar radiation and the electrical output of individual domestic system installations [1, 20]. Work by Murphy et al. [14] evaluated the simulation of PV installations using the SAP 2005 model and concluded that whilst there is less scope for errors when compared to more detailed models such as PVSYST, specific PV system data cannot be modelled, nor can site location be specified which leads to the SAP model overpredicting PV output for all UK locations North of Sheffield, and underpredicting PV output for locations further South. It has been estimated that there could be a potential output difference of up to 35%.

66.2 Methodology

This study was designed to test the prediction of solar radiation and PV electrical outputs against real measured data. There were two key areas of investigation, the first being the physical measurement of solar radiation on a site where an energy efficient dwelling is being built which has significant shading from trees and surrounding buildings and has been designed with a large PV array installed horizontally on the roof. The second investigation examines the effectiveness of two completed PV installations and compares a year's worth of collected electrical energy data against the SAP predictions that were used originally to justify their installation. These two systems were registered for the Government's FIT initiative and so were installed by a MCS accredited installer. The justification and eligibility of the systems under this initiative were determined using the SAP calculation methodology, which is a basic requirement under this scheme.

The measured solar radiation and electrical outputs in both investigations took place during the same one year period between September 2011 and October 2012. Both of these have also been compared to the predictions of solar radiation and electrical output that have been generated using the government's standard protocol for determining solar radiation, using the current SAP model (2009).

66.3 Prediction of Solar Irradiation Using the SAP Model

The SAP is the government's approved method for calculating a dwelling's energy consumption and carbon emissions. Both the earlier (2005) and current (2009) versions of SAP are relevant in this study because there exists a significant number of dwellings that were granted planning approval when the 2005 model was current in addition to those that have subsequently been approved in the 2009 version of the model. Both versions of SAP use a simple method of calculation to predict the energy output from PV installations although there are variations between the two versions. The available energy at inverter output (kWh/year) is determined using the following equation:

$$0.8 \times \mathrm{kWp} \times S \times Z_{pv},$$

where the kWp is the installed peak power for the PV modules, S is the annual solar radiation (kWh/m^2) (depending on orientation and tilt), and Z_{PV} is the over-shading factor provided in Appendix H of the SAP documentation [3, 4, 6].

Table 66.1 shows the annual solar radiation data (S) that is used for these variables and indicates those values used in the SAP 2009 model together with those values (*in brackets*) that are used in the earlier SAP 2005 model. It is interesting to note that these solar radiation figures have increased by approximately 3 % from the 2005 version of the SAP model and the 2009 version.

In SAP 2005, the annual solar radiation figure from Table 66.1 is used in the calculation of annual energy output for the PV installation. In the current SAP 2009

Table 66.1 Annual solar radiation, (S) kWh/m² [3, 4]

Tilt of collector	Orientation of collector				
	South	SE/SW	E/W	NE/NW	North
Horizontal	961 *(933)*				
30°	1073 *(1042)*	1027 *(997)*	913 *(886)*	785 *(762)*	730 *(709)*
45°	1054 *(1023)*	997 *(968)*	854 *(829)*	686 *(666)*	640 *(621)*
60°	989 *(960)*	927 *(900)*	776 *(753)*	597 *(580)*	500 *(485)*
Vertical	746 *(724)*	705 *(684)*	582 *(565)*	440 *(427)*	371 *(360)*

Table 66.2 Ratio of monthly solar radiation to annual average solar radiation [4]

Tilt	Jan	Feb	Mar	Apr	May	Jun	Jul	Aug	Sep	Oct	Nov	Dec	Sum
Horiz	0.24	0.50	0.86	1.37	1.74	1.84	1.78	1.50	1.06	0.63	0.31	0.19	*12*
30°	0.35	0.63	0.92	1.30	1.58	1.68	1.62	1.39	1.08	0.74	0.43	0.29	*12*
45°	0.39	0.69	0.95	1.27	1.52	1.61	1.55	1.34	1.08	0.79	0.48	0.33	*12*
60°	0.44	0.74	0.97	1.24	1.45	1.54	1.48	1.30	1.09	0.84	0.53	0.37	*12*
Vert	0.58	0.92	1.05	1.15	1.25	1.33	1.28	1.15	1.10	0.99	0.69	0.50	*12*

Table 66.3 Overshading factor, (Zpv) [3, 4]

Overshading	% of sky blocked by obstacles	Overshading factor
Heavy	>80	0.5
Significant	>60–80	0.65
Modest	20–60	0.8
None or very little	<20	1.0

Note Overshading must be assessed separately for solar panels, taking account of the tilt of the collector. Usually there is less overshading of a solar collector compared to overshading of windows for solar gain.

model, the most significant development is the determination of monthly available energy output which is calculated using the data in Table 66.2. This shows the ratio of monthly solar radiation to annual averages for different collector tilts from horizontal through to vertical that are used in this calculation. In addition, the later SAP 2009 model allows more than one type of PV installation to be accounted for.

Table 66.3 shows the overshading factors applied to the calculation and is the same in the SAP 2005 model and the SAP 2009 model.

Estimating and selecting an appropriate level of shading is clearly critical and has a significant impact on the resulting predictions of annual overshading. Despite numerous sources quoting this information, including the most recent SAP 2012 guidance documentation [6], there is very limited advice on how to select an appropriate figure for the overshading factor. Quantifying the percentage of sky blocked by obstacles can only be estimated in most cases, where some shading exists. This is fundamentally important in this calculation since the appropriate selection of an overshading factor can change the predicted solar radiation by a factor of up to 2. It is also worth pointing out that the results from the SAP calculations are used as the basis for determining credits that are available from the current CfSH [7], and for the calculation methodology adopted by the government for FIT payments.

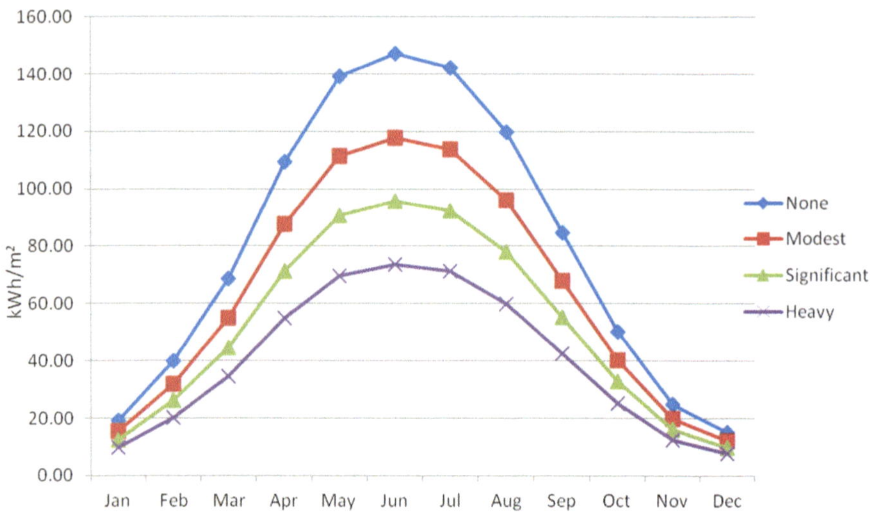

Fig. 66.1 Monthly solar radiation (kWh/m^2) predicted by SAP 2009 for a horizontal solar collector under the four standard categories of shading

Trees are often ignored in line with the guidance suggested as their shape is difficult to predict [10]. However, installing PV panels in heavily wooded areas can reduce direct solar gains, and therefore energy production, dramatically.

This study has evaluated two alternative methods for more accurately predicting the overshading factor; (a) taking panoramic images from the location and superimposing them onto a sunpath diagram for the relevant latitude, and (b) using draft MCS guidelines [12] for quantifying overshading.

Figure 66.1 shows the predictions for a typical horizontal solar collector based on an annual solar radiation figure of 961 kWh/m^2, and for each of the shading options provided in the SAP 2009 model.

66.4 Measurement of Solar Radiation and PV Output

This section describes the two case study investigations: (a) case study 1, where solar radiation was measured at three locations on a site where a significant solar PV installation is being planned, and (b) case study 2, where electrical output was measured at two new domestic solar PV installations benefiting from the government's FIT initiative.

66.4.1 Case Study 1: Measurement of Solar Radiation

This part of the study relates to the design and construction of a CfSH Level 5 domestic dwelling, in Teddington. The dwelling comprises 400 m^2 of floor area

Fig. 66.2 Architectural visualisation of case study 1

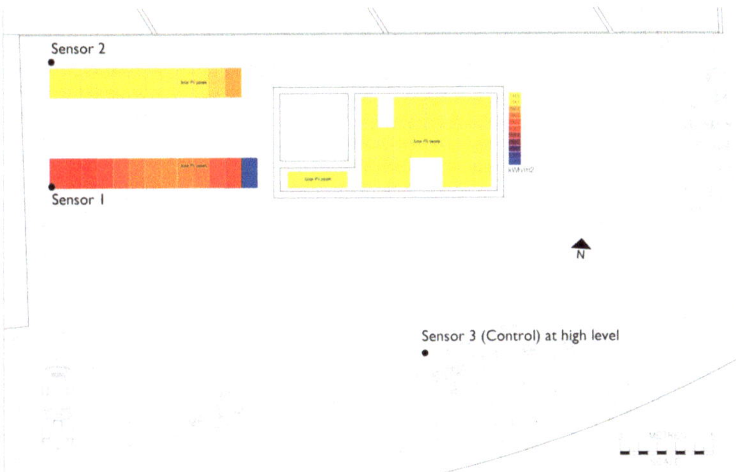

Fig. 66.3 predicted annual radiation (kWh/m^2) on the roof of Case Study 1

over three levels including a basement, ground floor and first, is highly insulated, airtight and energy efficient. In order to achieve CfSH Level 5 the dwelling design incorporates 70 m^2 of photovoltaic panels, installed horizontally on the roof of the building with an installed capacity of 12.0 kW$_p$.

At a very early stage in the design process an assessment was carried out, using Autodesk Ecotect [2], on the incidence of solar radiation on the roof area of the building in order to evaluate the effectiveness and location of the panels being proposed. Figure 66.3 shows the results of this assessment and indicates that there is some variability in the predicted solar radiation across the roof area.

The Ecotect predictions of the relative incidence of radiation on various parts of the roof were then used to determine the most appropriate locations to set up solar irradiance sensors and data logging equipment in order to measure solar radiation for a whole year. The monitoring equipment used for this study was provided by omni instruments [16] and included three SP-Lite pyranometers for measuring

Table 66.4 Average monthly solar radiation (kWh/m²) at each sensor location

	Jan	Feb	Mar	Apr	May	Jun	Jul	Aug	Sep	Oct	Nov	Dec	*Total*
S 1	9.2	28.4	37.5	50.0	66.4	61.9	70.2	61.0	42.0	26.4	14.2	10.5	*477.8*
S 2	10.3	30.2	48.4	63.9	97.1	82.1	93.3	78.9	53.5	30.5	16.2	12.3	*616.6*
S 3 (c)	12.9	30.5	58.1	89.1	126.3	106.5	124.3	102.6	71.0	42.6	20.1	14.8	*798.8*

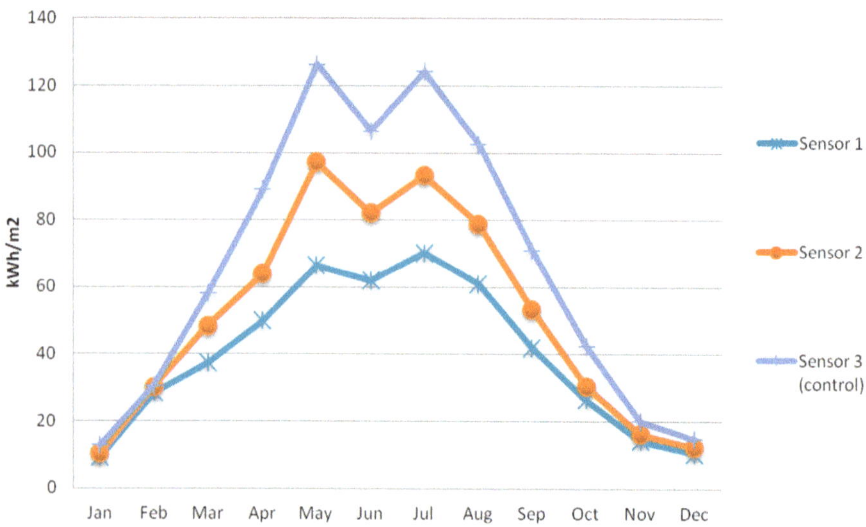

Fig. 66.4 Average monthly horizontal solar radiation (kWh/m²) at case study 1

solar radiation connected to an RME1 analogue Ethernet I/P module with internal memory and battery backup, with data retrieval via a secure website and with online graphing and data analysis.

As shown in Fig. 66.3, of the three sensors being used, Sensor 3 was positioned as high as possible in the centre of the site representing the control (most unshaded location) and the other two, Sensors 1 and 2, were positioned in two locations relevant to the position of the roof of the proposed building so that they were measuring incident solar radiation at appropriate positions and accounted for surrounding shading. At each of these three locations, the incident solar radiation was measured every 15 min and the data was collected for a whole year for the period September 2011 to August 2012.

Each sensor took a reading of the solar radiation being received (W/m²) every 15 min and this data was downloaded each month and modified to units of kWh/m².

Table 66.4 shows the average monthly solar radiation (kWh/m²) measured at each of the three sensor locations, September 2011 to August 2012. Figure 66.4 also shows this data graphically.

It is clear from the data collected that there was a significant reduction in the available solar radiation during the month of June 2012. Other reliable sources of solar radiation data, from Heathrow, UK, [11] during this period were evaluated and showed similar trends in this geographical location.

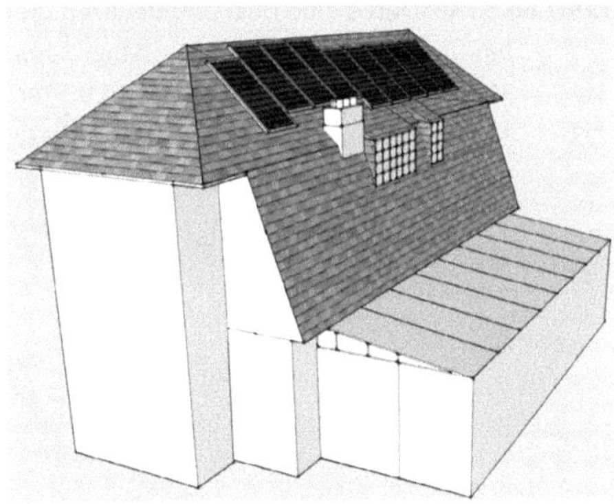

Fig. 66.5 Dwelling 1 representation of the PV installation

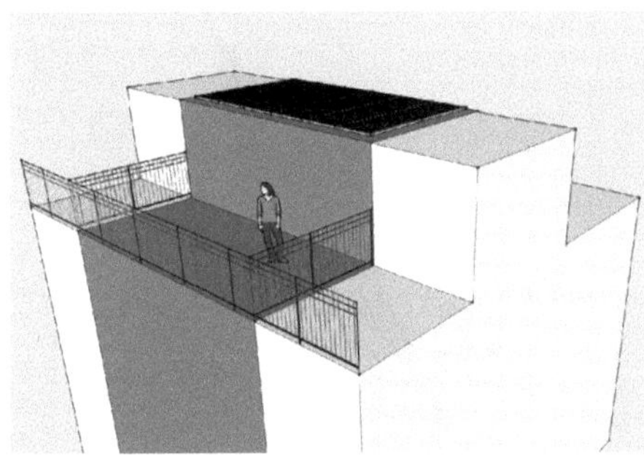

Fig. 66.6 Dwelling 2 representation of the PV installations

66.4.2 Case Study 2: Measurement of Electrical Output from Solar PV Installations

Two domestic solar PV systems, installed under the government's FIT initiative have been investigated in this study. They were installed by a recently accredited MCS installer who justified the installation in terms of installed load, costs and payback periods in line with the guidelines published by MCS at the time.

The first system, Dwelling 1, relates to the design, supply, installation and commissioning of a 2.25 kW_p PV system in Epsom, incorporating nine Sanyo HIT H series 250W PV panels with a Fronius IG20 2.0 kW inverter system.

The PV panels were installed at an inclination of 30° and were predominantly south facing in orientation. An on-site assessment of shading at this height and location led to the installation being classified as 'none or very little', using the methodology described in the SAP 2009 model documentation (Fig. 66.5).

The second PV system, Dwelling 2, was installed horizontally on the rooftop of a dwelling, comprising a 2.88 kW_p system, installed horizontally and incorporating twelve UpSolar 240 W poly panels with a Fronius IG30 2.6 kW inverter. On-site assessments of shading at the height of the installation led to a classification of 'none or very little' shading as described by the SAP 2009 model (Fig. 66.6).

During the period of monitoring the output electrical energy was measured for both of the solar PV installations and the results are shown in Table 66.5 and Fig. 66.7.

Table 66.5 Measured electrical outputs from the two PV installations

	Jan	Feb	Mar	Apr	May	Jun	Jul	Aug	Sep	Oct	Nov	Dec
D1	39.6	108.9	142.4	200.8	286.9	266.2	305.7	257.6	184.2	110.2	55.9	45.8
D2	32.0	88.1	164.8	232.4	332.0	332.7	382.0	321.9	170.7	102.1	51.8	37.1

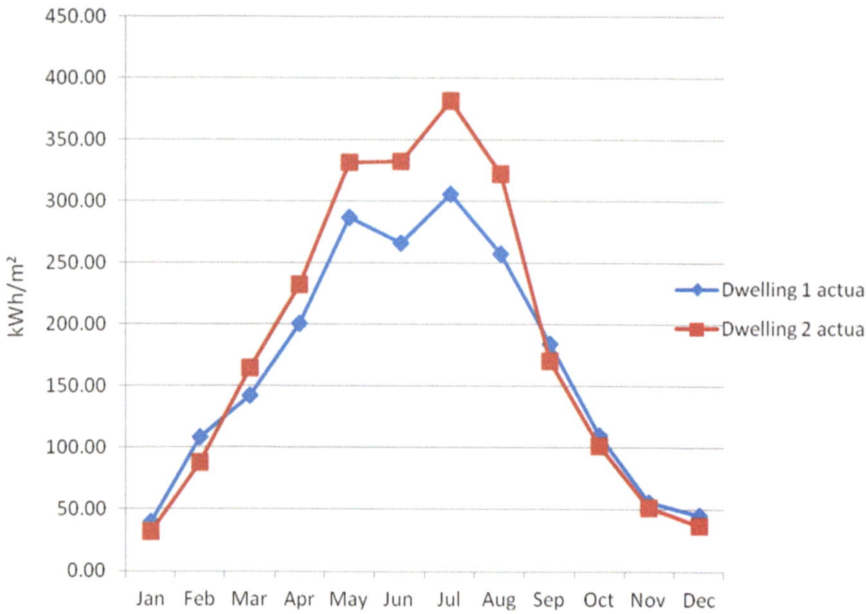

Fig. 66.7 Measured electrical outputs from the two PV installations

It is interesting to note that in this location there was a reduction in measured PV output during the month of June 2012, which coincides with the reduction in the measured solar radiation from case study 1 during the same period, and the Heathrow meteorological data.

66.5 Comparison of Measured and Predicted Results

This section compares the SAP model predictions of solar radiation with measurements for both the case study sites.

66.5.1 Case Study 1

In order to estimate the overshading factor (Zpv) for each of the three sensor locations for case study 1, the SAP methodology was used. Standard assessment procedure (SAP) provides this information (see Table 66.3), but, as previously discussed, there are very limited guidelines on how the overshading factors can be determined

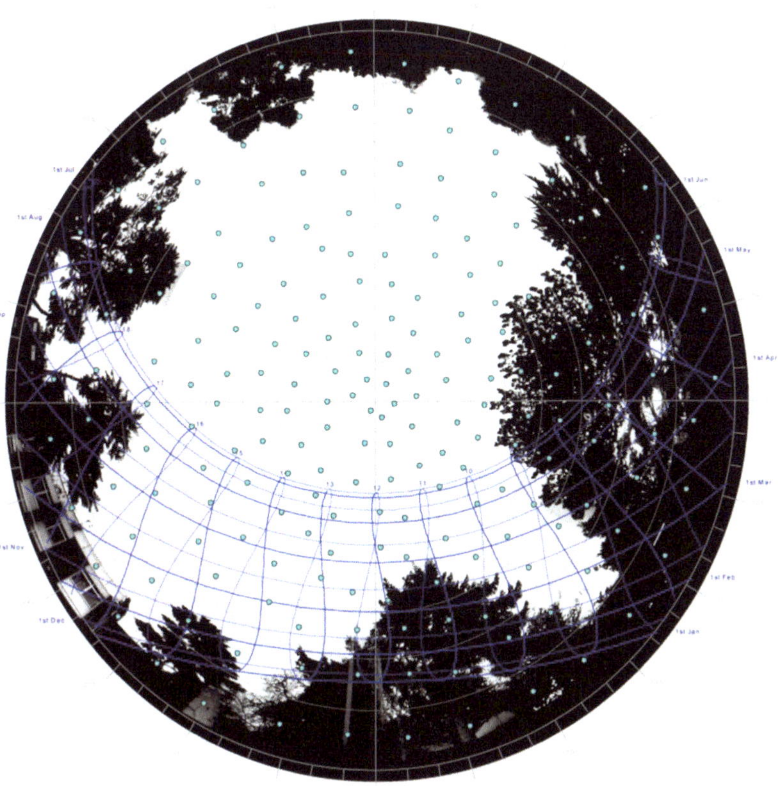

Fig. 66.8 Panoramic image of visible sky from sensor 3 (control)

because it is difficult to quantify the percentage of the sky which is blocked by obstacles, as required.

In this study, to determine overshading factors, a novel method was tested which superimposed a panoramic (fisheye) image of the sky into stereographic sunpath diagrams, for each of the three sensor locations, for the latitude of 51 °N generated using Autodesk Ecotect's solar tool [2]. This allowed the visible percentage of the sunpath to be determined for each of the sensors with a reasonable degree of accuracy. In each of the sensor locations, the vast majority of the shaded areas are represented by evergreen trees. Figure 66.8 shows the panoramic sky image for sensor 3 (control) superimposed onto the sunpath diagram. This was used to determine the percentage of the sunpath which was shaded, which in turn was used to select an appropriate overshading factor for the SAP calculation.

A similar image was generated for all three of the sensor locations on site, and Table 66.6, shows the results of the determination of appropriate overshading factors for each location.

It is important to differentiate between how much of the overall sky (direct and diffuse component) is shaded and how much of the sunpath area of the sky (direct component) is shaded. Guidelines published by the MCS [12] show clearly

Table 66.6 Overshading classifications for three sensor locations

Location	Sunpath shaded dots (out of maximum of 55)	Percentage of sunpath shaded (%)	SAP overshading classification, as shown in Table 66.3
Sensor 1	44	80	On the cusp of HEAVY and SIGNIFICANT shading (shading factor of 0.5 and 0.65)
Sensor 2	33	60	On the cusp of SIGNIFICANT and MOD-EST shading (shading factor of 0.65 and 0.8)
Sensor 3 (control)	18	33	MODEST (shading factor of 0.8)

that when assessing overshading factors only the sunpath area of the sky should be considered. It is also clear that the areas of the sunpath before 0800 and after 1400 should be ignored because the sun is insufficiently high in the sky to have an impact. Analysis of the panoramic sky images agree with this methodology. If the whole sky is used to quantify the percentage of the sky which is shaded, then the resulting overshading factors are underestimated resulting in predictions of solar incidence being overestimated by a significant degree.

The results in Table 66.6 also show that when quantified, the overshading factors for both sensors 1 and 2 sit right on the cusp of two different categories which leads to confusion about which category is most appropriate and could have a significant impact on the predictions of the available energy. This leads to the question of whether a more robust methodology for determining shading coefficients should be considered, particularly on sites where there is a lot of external shading from both surrounding buildings and evergreen trees.

Figure 66.9 shows the monitored and predicted results for case study 1. In comparing the measured and predicted data it is clear that in this study the measured conditions at sensors 1, 2 and 3 most appropriately fit overshading classifications of heavy, significant and modest respectively as defined by the SAP model. However, although the trends for the predictions are a reasonable fit, had the sunlight levels not dipped in June, the predictions of energy generation appear slightly low. Whilst, there is reasonable agreement between the measured and predicted overshading factors, it does not detract from the fact that for two of the sensor locations (1 and 2) the classification of shading sits on the cusp of two shading coefficients as defined by the SAP methodology, and in theory, either could have been adopted, resulting in potentially significant differences in predicted energy generation. Superimposing the panoramic image of the sky, onto the appropriate sunpath diagram results in an accurate quantification of the actual overshading coefficients (Zpv). This can then be applied to the fundamental equation for predicting solar irradiation in the SAP model.

66.5.2 Case Study 2

The assumptions that were made for the overshading of the two case study 2 dwellings were described in Sect. 4.2. In both cases the overshading on site was assessed as ‘none or very little’ and so the overshading factor in both cases was 1.0 (there

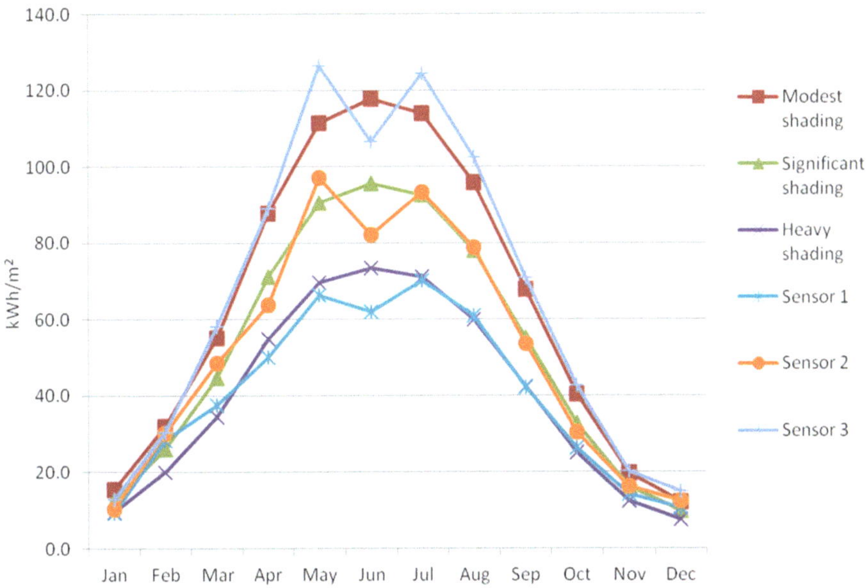

Fig. 66.9 Monitored and predicted results for case study 1

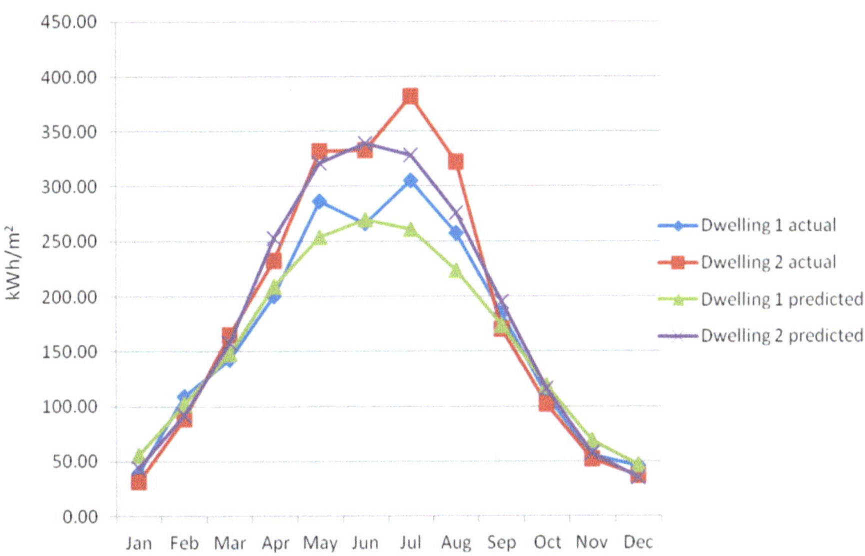

Fig. 66.10 Monitored and predicted data for the two PV installations

was no reduction in PV output predicted as a result). Figure 66.10 below shows the measured and predicted electrical output (kWh/m^2) for the two dwellings.

Whilst the overall trends for the predictions are a good fit, again, had the sunlight levels not dipped in June, the predictions of electrical generation appear low which

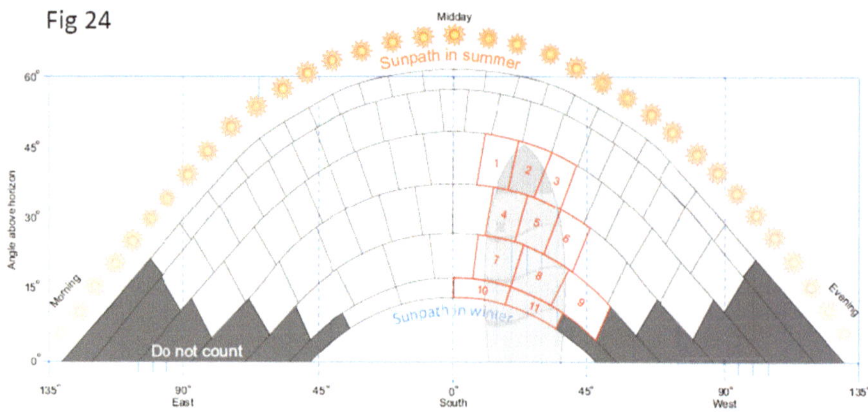

Fig. 66.11 MCS guidelines for determining shading factors [13]

is of particular concern since it was assumed that there was no overshading in the calculations.

66.6 MCS Guidelines for PV Installations

For some considerable time there have been concerns regarding the predictions of electrical outputs from domestic solar PV installations. In July 2011, Which? [19] investigated MCS accredited solar PV installers and found that 8 out of 12 companies underestimated the time it would take for PV systems to pay for themselves. They also found that the majority of companies failed to take into account solar shading where it was obvious and where it would make panel installations questionable. As a result, the MCS recently published comprehensive updated guidelines for registered installers [13]. These guidelines now include a revised methodology for quantifying PV system performance and a new system for estimating annual electricity generated which accounts for geographical location, PV array orientation and inclination to an accuracy of 5 and 1 ° respectively, and shading factors quantified using a sunpath diagram.

For shading factors, the MCS guidelines deal separately with situations where shading objects are up to 10 m away and where they are further than 10 m away from the array, although the basic methodology is similar. Figure 66.11 shows the sunpath diagram, published by the MCS, used for this purpose.

The method described in the MCS guidelines require you to (a) stand as near as possible to the proposed location of the array, (b) looking due south draw a line showing the uppermost edge of any objects visible on the horizon onto the diagram, (c) count the number of segments that have been touched or are within the line (in this case 11 segments), (d) multiply the total number of affected segments by 0.01, and (e) deduct this from 1 to arrive at the shading factor.

In this case:

$$1-(11\times 0.01)=0.89$$

The methodology adopted in these new guidelines is a significant improvement, both on what has been previously published by the MCS and also on the crude methodology currently used in the SAP calculation. It is also similar to the novel approach which we have adopted in this study, in the sense that it requires a more comprehensive visual assessment at the location of the solar PV installation and with specific reference to the relative position of the sun. However, there are a number of issues associated with the ease with which the MCS system can be applied. It is not clear how someone can position themselves (rather than a fisheye lens camera) easily at the specific location of a proposed PV array, particularly if the building has not yet been built. It is also not clear how the shading diagram is actually used in practice in terms of considering orientation and inclination of objects that could be some considerable distance away from the observer. It is worth also noting that the MCS guidelines make it very clear that this method is not intended to be as accurate as more sophisticated methods such as those used in software packages, but that it is estimated that this method will yield results within 10% of actual energy yield for most systems [13].

A number of professional and commercial organisations have responded to the need for more accurate methods of determining solar shading, and produce a variety of tools, both physical and digital, that enable the shading coefficients to be measured on site with a greater level of accuracy. Simple tools are available in the market that provides visual systems incorporating solar acetates similar to that shown in Fig. 66.11, and more complex systems, such as the solar pathfinder (www.solarpathfinder.com), are available which incorporates a transparent dome, positioned horizontally, which is then photographed from above, superimposing an image of the sky and shaded areas on a sunpath diagram which is positioned underneath. In addition, a number of more complex digital handheld tools are available, such as that developed by solar design (www.solardesign.co.uk), which incorporates a digital fisheye camera for quick and accurate assessments of shading in any location, similar to the novel technique used in this study, and described in Sect. 66.5.1.

66.7 Conclusions and Discussion

This study set out to investigate, and compare, various systems for predicting and measuring solar radiation and solar PV outputs for residential dwellings in the UK. As the UK building regulations move steadily towards the requirement for zero carbon homes, and since the introduction of the government's FIT initiative, solar PV installations have been seen as one of the main methods for achieving this. In 2012, the UK Government stated that 4 million homes in the UK will be powered by the sun within 8 years [9]. At the moment, approximately 10% of this figure has been achieved, and with FIT rates progressively and regularly reducing since their introduction in 2010, this is looking like a target that is going to be difficult to meet.

The study evaluated the solar PV prediction methodology available to designers at that time; the government's SAP (SAP 2009, 2011a), and has compared those predictions against real data collected on live projects in the South West London region of the UK. This has included the physical measurement of solar radiation on a site where a future solar PV installation is being planned and also the monitoring of the electrical output of two domestic solar PV installations compared to the predictions which justified and quantified their likely performance. Key to this study is the evaluation of shading from surrounding obstructions, including buildings and trees, and the comparison between assumptions that are made in predicting this shading and the reality of shading on site.

A considerable amount of monitoring took place during this project, including the measurement of solar radiation at a specific site and the measurement of electrical energy generation on another over one full year. Both of these datasets were then compared with the predictions of the standard models used to quantify solar radiation and electrical energy output. Most predictive models use standardised (theoretical and empirical) input data that has been developed over many years and it is possible that the measured conditions for this particular period is different to that incorporated into the predictive models. The study has identified some anomalies in the measured data but these have been compared to other weather station data collected by the Met Office locally and there is good agreement.

The SAP methodology and calculation for quantifying the available energy at inverter output (kWh/year) is relatively crude on a number of levels, although it is acknowledged that any increase in complexity of this methodology would risk it being used less by the designers who will make most use of it. The calculation assumes, in all cases, that there is a loss of 20 % and it is not clear on what basis this figure is used, particularly with the advances in efficiency of solar PV electrical generation over the last few years. Predictions of the annual solar radiation (S) were also crude and increased by approximately 3 % between the publication of the SAP 2005 and 2009 models. A technical paper published by the BRE in 2011 [5] set out proposed changes to the calculation of solar radiation (S) and in the SAP 2012 model [6] the methodology has been significantly improved, and adopted, so that the calculation is capable of estimating the solar radiation for 'the applicable climate and orientation and tilt of the PV'.

The determination of the overshading coefficient (Zpv) has the most significant impact on predicting available energy at inverter output, and quantifying this for any solar PV installation has been the main focus of this study. This is acknowledged by Murphy et al. [14] who state 'further work is needed to assess the impact of shading. It is known that shading can have a devastating effect on PV performance, and it is unclear how the basic categories in SAP can address this'.

The first case study evaluated the available solar radiation on a site where a future solar PV installation was proposed. Solar radiation was monitored over a full year and this was compared to the predictions of the SAP model. Comparisons of the measured and predicted results show good agreement, but only where the appropriate SAP overshading classifications are assumed. For this case study, and for each of the three locations monitored, selecting an appropriate overshading classification from SAP was not straightforward and there was little guidance provided. As a result a novel technique was adopted, which more accurately quantified the percentage of the sunpath that was blocked by obstacles. This novel technique required superimposing a fisheye image of the sky from each location onto a sunpath diagram and quantifying the proportion of this that was shaded. Comparisons between this novel method and the SAP method showed that in two out of three cases the overshading was on the cusp of two SAP classifications which suggests there is significant scope for error and inaccuracy in the model predictions. Recently published guidance from the MCS includes a new methodology for quantifying overshading for solar PV installations and this development supports the findings of this study. It also implies that the SAP model is not currently sufficiently detailed enough to be used to support the installation of solar PV, particularly where there is significant, and quantifiable, external shading.

The second case study compared measured electrical output to SAP predicted outputs for two domestic solar PV installations. In both cases the installations were located in positions where the overshading was classified, correctly, as 'none or very little' and, therefore, the overshading issue was essentially removed from the analysis. In both dwellings the measured and predicted outputs show reasonable agreement, although the results suggest that even with 'none or very little' shading the predictions of electrical generation were slightly low compared to the measured outputs, particularly during the summer. It is not possible to say for sure whether this was caused by the SAP assumptions for PV system efficiency or for actual solar radiation during this period.

The study has shown that the current methodology used by the SAP models to determine electrical output of solar PV systems is, in significant areas, crude, that various input parameters are open to interpretation and there is potential for errors. The BRE have made some significant improvements in new and current version of the SAP 2012 model and these will account for regional variations in solar irradiance and improved calculations for PV outputs for any given orientation and tilt. However, no changes have been put in place for the determination of the overshading coefficient (Zpv), and the methodology for this remains as before, which is a lost opportunity.

Acknowledgements This work was supported by the UK Government's Technology Strategy Board (TSB), under Knowledge Transfer Partnership (KTP) number 8351.

References

1. Alonso-Garcia MC, Ruiz JM, Herrmann W (2006) Computer simulation of shading effects in photovoltaic arrays. Renew Energy 31:1986–1993
2. Autodesk (2010) Ecotect analysis 2011. www.autodesk.com. Accessed 10 April 2014
3. Building Research Establishment (BRE) (2009) The government's standard assessment procedure for energy rating of dwellings. 2005 edition, Revision 3, Published on behalf of DECC, Crown Copyright

4. Building Research Establishment (BRE) (2011a) The government's standard assessment procedure for energy rating of dwellings. 2009 edition, Published on behalf of DECC, Crown Copyright
5. Building Research Establishment (BRE) (2011b) Changes to the calculation of solar radiation in SAP 2012, report reference STP11/SR01, Henderson, J., 24/11/11, Crown Copyright
6. Building Research Establishment (BRE) (2014) The government's standard assessment procedure for energy rating of dwellings, 2012 edition, Published on behalf of DECC, Crown Copyright
7. Department for Communities and Local Government (DCLG) (2010) Code for sustainable homes technical guide, November 2010, ISBN 978 1 85946 331 4
8. Department for Communities and Local Government (DCLG) (2013) Next steps to zero carbon homes—allowable solutions, Consultation document, August 2013, ISBN 978 1 4098 3978 1
9. Harvey F (2012, February 9) Greg Barker: 4M homes will be solar-powered by 2020, The guardian. www.theguardian.com. Accessed 10 April 2014
10. Littlefair PJ (2009) Site layout planning for daylight and sunlight: a guide to good practice, IHS BRE Press, Report BR209, ISBN 978-1-86081-041-1
11. Meteorological Office (2014) Station data, Heathrow, 1948–2013 (online) www.metoffice.gov.uk/climate/uk/stationdata/heathrowdata.txt. Accessed 10 April 1214
12. MCS (Microgeneration Certification Scheme) (2011) Guide to the installation of PV systems, 3rd Edition, Draft for comment, (online) www.microgenerationcertification.org/. Accessed 23 Oct 2013
13. MCS (Microgeneration Certification Scheme) (2012) Guide to the installation of photovoltaic systems, ISBN 978-0-9574827-0-8
14. Murphy GB, Kummert M, Anderson B, Counsell J (2009, July) A comparison of the UK standard assessment procedure (SAP) and detailed simulation of building-integrated renewable energy systems. Paper presented at Building Simulation 2009, the eleventh International IBPSA conference, Glasgow, Scotland, July 2009
15. National Energy Services (NES) (2013) NHER plan assessor, software, NES Ltd, Davy Avenue, Milton Keynes, MK5 8NA
16. Omni Instruments (2011) Omni instruments Ltd (online) www.omniinstruments.co.uk. Accessed 22 July 2011
17. Ownergy plc (2013a) Feed-in tariffs (online) www.fitariffs.co.uk/statistics/. Accessed 23 Oct 2013
18. Ownergy plc (2013b) Table C: historical feed-in tariff starting levels (online) www.fitariffs.co.uk/library/regulation/TariffTableC.pdf. Accessed 23 Oct 2013
19. Which? (12 February 2013) Solar panel savings now more accurate, (online) http://www.which.co.uk/news/2013/02/solar-panel-savings-now-more-accurate-310285/. Accessed 10 April 2014
20. Woyte A, Nijs J, Belmans R (2003) Partial shadowing of photovoltaic arrays with different system configurations: literature review and field test results. Solar Energy 74:217–233
21. Zero Carbon Hub (2009) Defining a fabric energy efficiency standard for zero carbon homes, Task Group Recommendations, November 2009

Chapter 67
Courtyards: Optimum Use as Means of Providing Daylight into Adjacent Zones

Maitha M. Bin Dalmouk and Khaled A. Al-Sallal

Abstract The aim of the study is to investigate daylighting performance in spaces adjacent to courtyards in the climatic conditions of the UAE (latitude 25°N longitude 55°E). Two cases were simulated mainly to understand how the space generally performed in terms of lighting performance. The two cases comprised of the same universal adjacent space but in one case the courtyard size was doubled while maintaining the same height for both cases. The Desktop RADIANCE 2.0 program (DR) was used to evaluate the lighting performance of the adjacent space under clear sky conditions. Horizontal illuminance levels, diversity and uniformity of illuminance, and Spatial Daylight Autonomy (sDA) were the lighting parameters used. These parameters were compared to the standard values or recommended ranges specified in lighting codes and handbooks such as the Chartered Institution of Building Services Engineers (CIBSE) and the Illuminating Engineering Society of North America (IESNA). The study revealed that although both models provided sufficient amount of illuminance level into the space it failed in the other aspect of diversity and uniformity, some suggestions were highlighted to be tested at a later stage.

Keywords Daylighting · Courtyard · Adjacent space

67.1 Introduction

The main incentive on conducting this study is the revival of the courtyard design configuration in newly constructed structure in the region, through the implementations of Green Building Policies, as well as the newly introduced the Estidama rating system in Abu Dhabi Emirate and other green rating/codes in other emirates.

Courtyards in the region have been mainly used in traditional houses and some small-scale schools. The function of many of the traditional houses has been changed to museums or art galleries. There is not enough evidence of their performance with regards to environmental aspects especially Daylight. The majority

K. A. Al-Sallal (✉) · M. M. Bin Dalmouk
Department of Architectural Engineering, UAE University, Al-Ain, United Arab Emirates
e-mail: k.sallal@uaeu.ac.ae

A. Sayigh (ed.), *Renewable Energy in the Service of Mankind Vol I*,
DOI 10.1007/978-3-319-17777-9_67

of studies carried out on the functionality of courtyards studied mainly its thermal performance, energy performance and shading, and exposure within the courtyard primacies [1–3]. Studies on adequate daylight availability to adjacent spaces to a courtyard are not covered thoroughly.

The courtyard houses residents in warm climate utilized the courtyard to serve as cool air collector at night and a shade provider in the daytime [4]. As a central space in buildings, direct link to the outdoor environment and light-enhancing tool to the adjacent spaces, the courtyard properties affects the daylight entering to the adjacent spaces. Therefore, some questions arise:

- How do the different sizes of the courtyard affect the daylight performance of the spaces adjacent to the courtyard?
- What are the effects of geographic location or climate upon courtyard lighting performance?
- How does the opening to wall ratio of the wall connecting the courtyard to the adjacent space affects the daylight performance in the space?

Therefore, the analysis of this paper attempts to investigate the daylight behavior of the zones adjacent to the courtyard in buildings under the UAE climatic conditions with two different courtyard sizes, in a single story general use building. At the current stage we focused only on understanding the problem and its boundaries and specify the involved variables that are significant to the problem, by using simple cases to understand the problem first. We have also started to generate some possible solution. This was carried out through conducting a comparative analysis of the daylight performance in the zones adjacent to the courtyard using Desktop RADIANCE program (DR) [5]. The outcomes of the study will be used to further investigate the above raised questions.

67.2 Background

The use of artificial lighting is considered a key problem that can lead to excessive energy usage as it affects cooling and heating loads requirements of the buildings. As a result, many papers and research studies reveal substantial opportunities for improving energy efficiency in the building by incorporating courtyards into buildings as daylight-enhancing techniques to bring light into the interior and consequently minimize space conditioning and lighting loads [2, 6–8]. Many researchers have studied courtyard as an effective design strategy to lower the energy demands of buildings and encourage further awareness of energy-conscious design [1–3, 6]. Most of the studied researches tackled the importance of courtyard in admitting daylight into adjacent zones thus reducing energy consumption [6], but never highlighted the appropriate amount of daylight admitted and how the size of the courtyard may affect the stated issue.

A study conducted by Ahmed S. Muhaisen suggests general guidelines for efficient courtyard design in the four main climatic regions [3]. In his research he studied how different proportions of rectangular courtyard affects the shading and exposure conditions of the internal envelope of the four proportions in the defined regions, and concluded that the shading and exposure of the internal envelope highly depends on three elements the courtyards form proportions, locations, and available climatic conditions.

Another study on the thermal performance of the courtyard was done by Abdelsalam Aldawoud [1]. In this research the energy performance of a square courtyard was examined under the following conditions; conditions of different factors of climate, height, glazing type, and glazing percentage. Using DOE 2.1E as the energy simulation program and weather files of four cities that represents the main climatic regions of the world were used, what was noticed from the analysis of this research is that in general, the open courtyard building exhibits a better energy performance in hot-dry and hot-humid climates.

Effect of courtyard proportions on solar heat gain and energy requirement in Rome is a research done by both Ahmed S. Muhaisen and Mohamed B Gadi [6]. In this research different methods such as the use of light shades of colors for the external surfaces to reduce the solar radiation absorption in summer and the use of shading devices and improving the thermal properties of the external walls and roof as means of increasing the building utilization of solar heat gain. What was found in this study is that deeper courtyard forms were the most preferable for the purpose of reducing the cooling load during summer seasons and heating load during winter. In terms of shading, the self-shading of the courtyard building indicated that obtaining solar radiation in winter was more critical than avoiding it in summer.

The daylighting metrics calculated in this study were: daylight factor (DF) (%), illuminance level (Lux), light distribution and uniformity, and diversity of illuminance. DF is the ratio of the interior illuminance level compared to the exterior horizontal illuminance level. The DF recommended by the Chartered Institution of Building Services Engineers (CIBSE) [9] for general room lighting is 5 % when electric lighting is not normally to be used and 2 % when it is normally to be used. Illuminance level is the visible energy (light) incident on a surface, measured in Lux. Illuminating Engineering Society of North America (IESNA) has made recommendations for the illuminance level for various tasks and building types [10]. According to IESNA Lighting Handbook, illuminance levels within general use are categorized according to the level of task in the room, 100–1000 Lux was chosen for ambient light for the above case [10]. The light distribution and uniformity can be described as the relative illuminance at different points in space. Eye fatigue is caused if the light level differs too much in a space. For the uniformity of light distribution, CIBSE Code for Lighting was used to test the performance of both courtyards [9]. In the CIBSE Code for Lighting, the uniformity of illuminance is measured using the ratio of minimum illuminance to average illuminance, which should not be less than 0.7. For uniformity of illuminance higher value indicates better uniformity. Diversity of illuminance is another metric that was used, sometimes referred as illuminance gradient and it is the ratio of the maximum to minimum illuminance level of the area 0.5 m off the wall perimeter. This ratio should not exceed 5 to 1. The lower the value the better uniformity it yields.

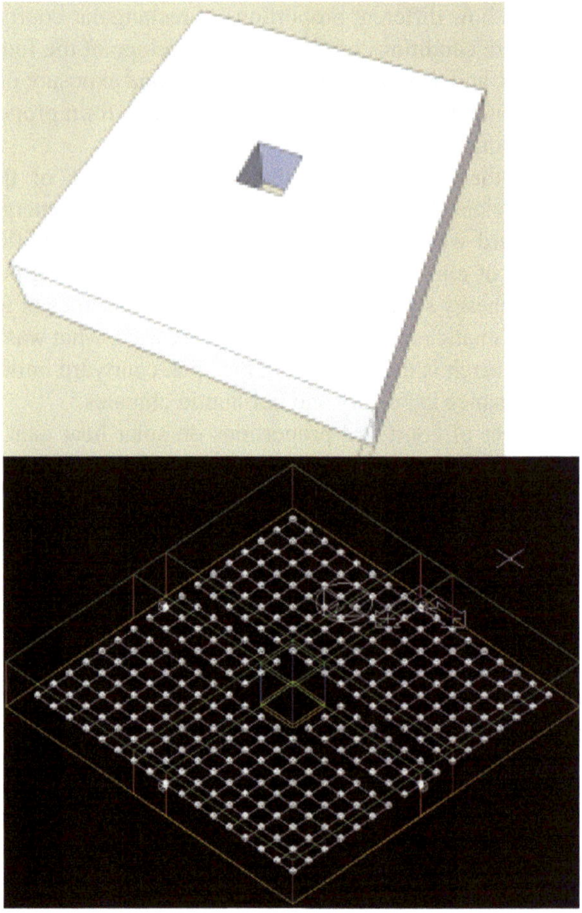

Fig. 67.1 Courtyard model A

67.3 Method

To investigate the daylight behavior in the zones adjacent to the courtyard the research was divided to stages, at this stage model A and model B (Figs. 67.1, 67.2) were defined as per Table 67.1 below. Daylight simulations and measurements using Desktop Radiance 2.0 Beta (DR) were conducted in the adjacent zone on summer solstice (June 21), equinoxes (March 21/September 21), and winter solstice (December 21) at 9:00 a.m., 12:00 p.m., and 3:00 p.m. The dates were chosen to see the variation of the lighting performance at extreme sub-angles. DR is used for detailed lighting calculations and visual quality analysis at the specified time of the year. The daylighting components that were tackled covered DF, illuminance level, light distribution and uniformity, diversity of illuminance, and finally general view of the space at the extreme times. The main function of the adjacent zones was considered as that to general function. The location tested was that of Dubai, UAE.

Fig. 67.2 Courtyard model B

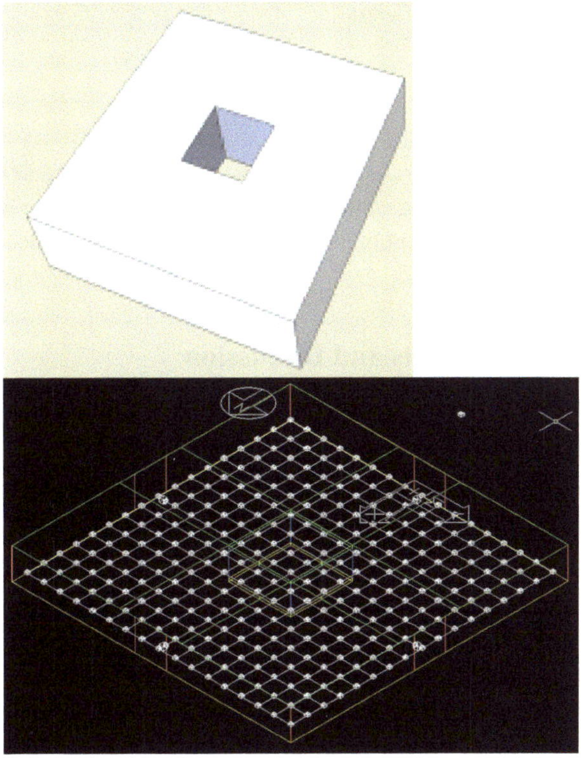

For the purpose of simulating the zones adjacent to the courtyard the surfaces reflectance values were defined based on the recommendations specified by IESNA Lighting Handbook and other suggested values [10–12], for ceiling the minimum reflectance value is 85 %, walls to have 50 % or greater, and floors of not less than 30 % (Table 67.2). The glazing was specified as per Table 67.3.

Table 67.1 Simulated models characteristics

	Model A			Model B		
	Length (m)	Width (m)	Height (m)	Length (m)	Width (m)	Height (m)
Courtyard	4	4	4	8	8	8
Courtyard Glazing	4	4	3.4	8	8	7.4
Space	36	36	4	36	36	4

Table 67.2 Reflectance values of materials used in DR

	Material name	Reflectance (%)
Floor	Gray 3k309 (IES091)	30
Walls	Beige 2k222 (IES091)	52
Ceiling	White	85.77

Table 67.3 Characteristics of the glazing (glass clear) used in DR

	Value
Transmittance (%)	84.10
Reflectance (%)	7.8
Thickness	12.50

Wide-ranging simulation cases were conducted to investigate in detail the luminous environment in the adjacent zones in both model A and model B.

67.4 Results and Discussion

Daylight Factor (DF) The results showed that the DF for both courtyard models A and B does not meet CIBSE recommended level of 5 % for daylight use only. Model B exhibited a DF of almost 4 % in all directions which is suitable for the 2 % DF recommended for the use of daylight with the aid of artificial lighting (Figs. 67.3, 67.4).

Daylight Trend An array of sensors were placed in each space to simulate the illuminance level at the specified times, the sensors were located within a 0.5 m off the wall parameters with a distance of 2 m between the sensors, a contour map was generated for each simulated time and then was compiled as a single contour map for the whole year at the extreme daylight time. The contour map of model A showed that the courtyard exhibits almost symmetrical illuminance map over the adjacent zone area (see Fig. 67.5), the maximum depth of recommended daylight covers most of the space, except for 2 m to the East, 1 m to both North and West, and 6 m

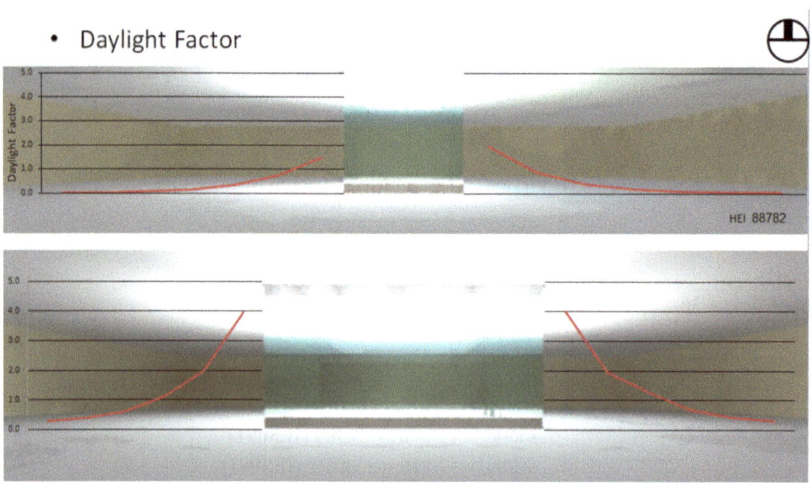

Fig. 67.3 Daylight factor (*DF*) in model A (*above*) and model B (*below*) on the East–West axis

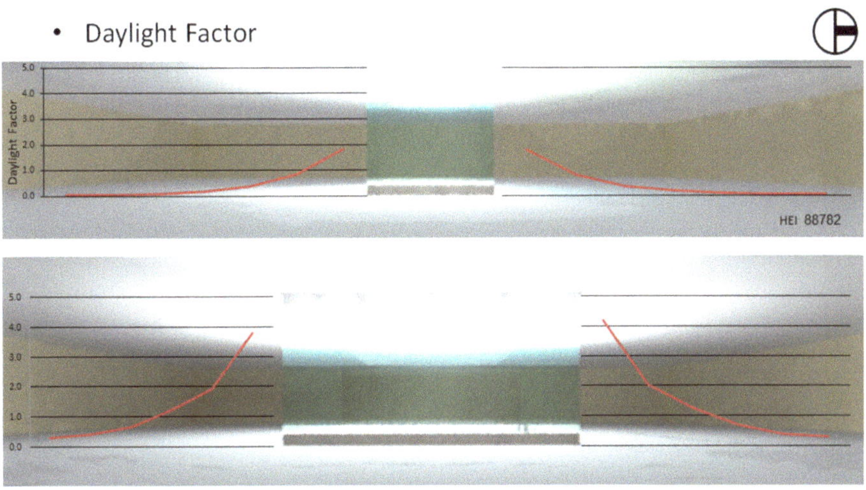

Fig. 67.4 Daylight factor (*DF*) in model A (*above*) and model B (*below*) on the North–South axis

to the South. 38 % is the area that receives adequate illuminance level; the rest is either over 1000 lx or under 100 lx. What can be visualized from the contour map of model A is that the area near the courtyard receives very high illuminance level, much higher than the recommended 1000 lx for general use. Model B also showed symmetry when it came to the illuminance map, but the (Fig. 67.6 map was wider spread over the space. Almost 50 % of the area received the recommended level of illuminance (100–1000 lx) and covers the parameter of the whole space except for small areas near the corners. What can also be visualized in model B is that the area near the courtyard receives very high illuminance levels. It spans almost 8 m from each side, this results that the area near the courtyard cannot be utilized for the defined function. In general light in both models penetrated well near the courtyard and in the room center. Model A provided limited light into the back spaces, model

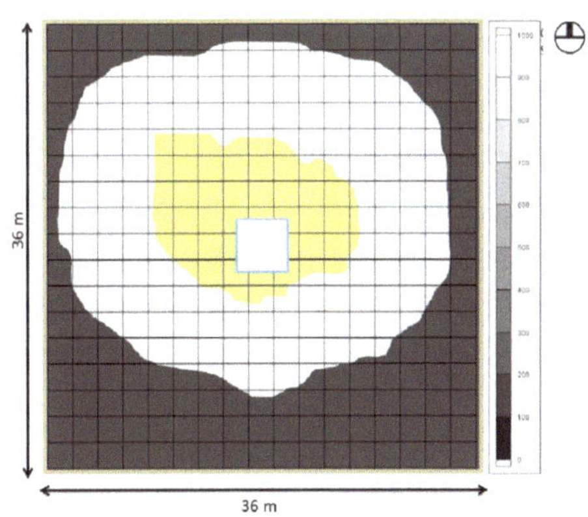

Fig. 67.5 Illuminance contour map of model A over the extreme daylight times

Table 67.4 Diversity[a] of illuminance for model A and model B respectively

		Courtyard model A	Courtyard model B
21-Mar	9:00 a.m.	7124	7124
	12:00 p.m.	336	336
	3:00 p.m.	8263	1439
21-Jun	9:00 a.m.	6982	1361
	12:00 p.m.	248	74
	3:00 p.m.	285	74
21-Sep	9:00 a.m.	8960	1438
	12:00 p.m.	344	109
	3:00 p.m.	7964	1728
21-Dec	9:00 a.m.	6162	1594
	12:00 p.m.	13548	1182
	3:00 p.m.	12349	1706

[a] Diversity should not exceed 5:1

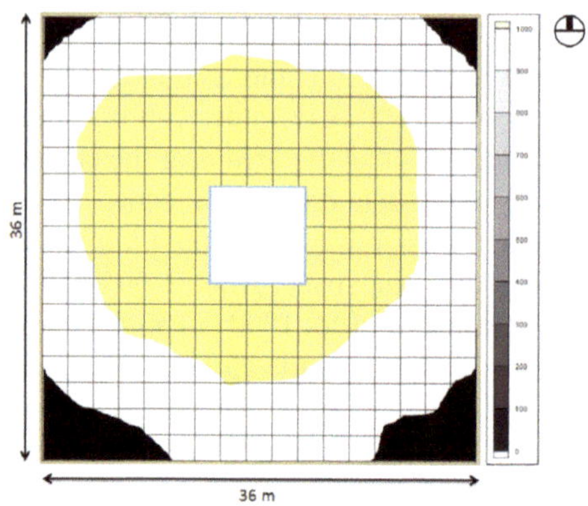

Fig. 67.6 Illuminance Contour Map of model **A** over the extreme daylight times

B although extended further, but was having a larger area with higher than the recommended illuminance levels.

Diversity of Illuminance Referring to Table 67.4 one can see that both models exceeded the recommended value assigned by CIBSE (should not exceed the ratio of 5:1) [9], except for December. Model B performed better than model A, but still higher than the recommended values.

Uniformity of Illuminance Both models A and B failed to avoid excessive variations of horizontal illuminance across the interior space of the adjacent zone. This excessive variation will lead to uneven distribution of light across the space and poor visual quality and visual discomfort and annoyance to the users, filtering might be a good solution to this problem (Table 67.5).

Table 67.5 Uniformity[a] of illuminance for model A and model B respectively

		Courtyard model A	Courtyard model B
21-Mar	9:00 a.m.	0.02	0.02
	12:00 p.m.	0.04	0.04
	3:00 p.m.	0.01	0.03
21-Jun	9:00 a.m.	0.01	0.03
	12:00 p.m.	0.04	0.08
	3:00 p.m.	0.04	0.08
21-Sep	9:00 a.m.	0.01	0.09
	12:00 p.m.	0.04	0.08
	3:00 p.m.	0.01	0.02
21-Dec	9:00 a.m.	0.01	0.01
	12:00 p.m.	0.01	0.04
	3:00 p.m.	0.01	0.02

[a] Uniformity should not fall beyond 0.7

Spatial Daylight Autonomy (sDA) It is defined as the percentage of space or building area that meets a given DA value for a set time period. For example, an sDA300/50 value indicates the percentage of the space that meets or exceeds 300 lx for 50 % of all the annual hours between 8 a.m. and 6 p.m. In the space illustrated above, sDA300/50 = 66 %.

67.5 Conclusion

Simulating the above models using DR helped to examine them better in terms of daylighting performance, and gave a direction on how to proceed further with this study. Generally, what was concluded is that both models although provided adequate amount of the recommended illuminance level into the space (38 and 45.6 %) it failed in the other aspect of diversity and uniformity. Some alternative solutions can be specified and could be tested at a later stage to verify if they solve the given problem, such solutions include; filtering using trees, vines, or some means of shading devices, both in the courtyard and inside the space. The use of light shelves might be another solution here. Changing the courtyard size and geometry, and reducing the glazing area in the courtyard could result in better results. Increasing the building height might give a more unified light level into the space. Such solutions will prevent the presence of concentrated light level at a specific area. Further investigations of the above suggestions are recommended to fine-tune the results and to achieve an optimum courtyard that will enable sufficient lighting into the space without having any problems.

References

1. Aldawoud A (2008) Thermal performance of courtyard buildings. Energy Build 40:906–910
2. Aldawoud A, Clark R (2008) Comparative analysis of energy performance between courtyard and atrium in buildings. Energy Build 40:209–214
3. Muhaisen A (2006) Shading simulation of the courtyard form in different climatic regions. Build Environ 4:1731–1741
4. Safarzadeh H, Bahadori M (2005) Passive cooling effects of courtyards. Build Environ 40: 89–104
5. Lawrence Berkeley National Laboratory Energy (2000) Technologies Division Building Technologies Department. Desktop RADIANCE 2.0 Beta. http://radsite.lbl.gov/deskrad/help.htm
6. Muhaisen A, Gadi M (2006) Effect of courtyard proportions on solar heat gain and energy requirement in the temperate climate of Rome. Build Environ 41:245–253
7. Al-Mumin A (2001) Suitability of sunken courtyards in the desert climate of Kuwait. Energy Build 33:103–111
8. Li D, Lam C, Wong S (2005) Daylighting and its effects on peak load determination. Energy 30:1817–1831
9. CIBSE (1994) CIBSE Code for interior lighting. Chartered Institution of Building Service Engineers, London
10. Rea M (Ed) (1999) Lighting handbook, 9th edn. Illuminating Engineering Society of North America, New York
11. Baker N, Fanchiotti A, Steemers K (1993) Daylighting in architecture a European reference book, 1st edn. James and James, London
12. Steefy G (2008) Architectural lighting design, 3rd edn. Wiley, Hoboken

Chapter 68
Earth Construction: The Mechanical Properties of *Adobe* with the Addition of *Laponite*

Francesca Scalisi and Cesare Sposito

Abstract The contribution describes testing of compression strength, flexural strength and abrasion resistance of *adobe* made up of soil, water and sand (AS), soil, water, sand and straw (ASP), soil, water, sand and laponite nanoparticles (ASN). Embodied energy in materials presents an increasingly high percentage of the energy spent in the whole life cycle of a building. The same applies for carbon dioxide (CO_2). Therefore, the development of new sustainable construction materials with lower embodied energy and lower CO_2 emissions is needed.

The use in construction of the brick made from soil, water and sand or straw, called *adobe*, boasts a millenary tradition and in recent years there has been renewed interest in a material readily available and ecofriendly. Earth is a building material that is able to act perfectly in balance with the environment: earth lends itself to achievements accessible to any manufacturing organization and is also a resource available in most geographical contexts. It allows one to manufacture products suited to pursue energy conservation and comfort in different climatic regions. The use of *adobe* presents: reduction of embodied energy and CO_2 at component level; improvement of insulation properties; reduction of the total costs compared to existing solutions.

Keywords Earth · Nanoparticles · Architecture · Low energy

Paragraphs with the initials F.S. by Francesca Scalisi; paragraphs with the initials C.S. by Cesare Sposito.

F. Scalisi (✉) · C. Sposito
Department of Architettura, University of Palermo, Palermo, Italy
e-mail: francesca.scalisi@unipa.it

C. Sposito
e-mail: cesare.sposito@unipa.it

A. Sayigh (ed.), *Renewable Energy in the Service of Mankind Vol I*,
DOI 10.1007/978-3-319-17777-9_68

68.1 Introduction

The construction sector represents one of the major consumers of energy, consequently emitting a great quantity of CO_2 into the atmosphere. This situation concerns both the consumption of energy required in the production phase of the materials to be utilized in construction and the consumption of energy in the actual deployment phase of the building. It is well-known that winter heating and summer cooling inside a building brings about an enormous consumption of energy.

In the field of construction the saving of energy needs to be shaped by the use of sustainable materials that will contribute to insulating the building adequately and reducing energy consumption.

The utilization of earth blocks in construction dates back over a thousand years and in the last few years there has been a renewed interest in this easily acquired and ecofriendly material.

Making earth blocks does not demand a great amount of energy. In fact, in contrast to bricks it does not require a baking stage, their being dried out simply in a natural way. It is precisely in the baking stage that ordinary bricks emit large quantities of CO_2; by building a house of 100 m^2 with earth blocks instead of bricks one avoids expelling 20 t of CO_2 into the atmosphere.

It is a safe and natural prime material available in great quantities in the natural world and is completely recyclable.

In the operational phase it enhances the comfort of the building and contributes to a saving of energy, since:

- It regulates the humidity because it can maintain a constant level of humidity in the atmosphere of around 50 %.
- It regulates temperature in the home; heat produced by man, from electrical appliances and lighting, on entering through windows is absorbed by the ground, which then restores it again when necessary (e.g. in the evening). This enhances home comfort and boosts energy saving considerably.
- It protects from high temperatures; thanks to its specific heat and substantial size, it prevents heat from entering. At night it cools, and in the morning, when temperatures rise, it once again absorbs a great quantity of heat, thus reducing the ambient temperature during the day.
- It protects from noise; it has excellent values as regards to acoustic cutbacks, since it is an "elastic" material and absorbs noise, by preventing it from passing through.
- It protects from electromagnetic pollution; 15 cm of earth cut out 99 % of electromagnetic waves, a value that is higher than that of all other building materials.
- It purifies the air; domestic odours are conveyed by water vapour. Earth, by absorbing water vapour, functions as a natural filter and purifies the air.
- It prevents condensation; precisely because it can absorb humidity, it prevents condensation from forming on walls (including interstitial ones).

Earth blocks may represent the material of the future because they can satisfy the increasingly important requisites of environmental sustainability, energy performance, positive energy balance, healthiness of abode, disposability. Therefore it becomes fundamental to improve the performance of this material, especially from the point of view of mechanical resistance, which is decidedly inferior to that of bricks. laponite, a synthetic layered silicate, at nanometric dimensions, represents a stabilizer compatible with earth, given the predominant presence of SiO_2, and can improve the mechanical resistance of earth blocks [1].

(F. S.)

68.2 Preparing Adobe

The sample realized is of the adobe type, an earth brick shaped by hand in a mould, without being compressed, and left to dry under natural conditions. The mixture used to produce traditional adobe is: soil and water with sand or straw as a stabilizer [2, 3]. The materials used to make the samples are: soil, sand, water, straw and laponite nanoparticles (Fig. 68.1). Soil was taken from the Roccasieli quarry, situated in the municipality of Motta S. Anastasia in Sicily. The sand utilized was lavic and of basaltic origin from Nicolosi in Sicily. The straw is from Nicolosi. The laponite RD is a synthetic layered silicate, supplied by Rockwood Additives Ltd. as a white powder and used without further purification. It is composed of rigid disk-shaped crystals with a well-defined thickness of 1 nm and a diameter of 25 nm.

In the first phase, the samples produced were of two types: the AS sample, made up of 31.54 % soil, 30 % sand and 38.46 % water; the ASP sample made up of 27.12 % soil, 30 % sand, 41.38 % water and 1.5 % straw (Fig. 68.2). The sizes of the samples were 40 × 40 × 160 mm, for the flexion and abrasion tests, and 50 × 50 × 50 mm for the compression test. The specimens were observed through the *scanning electron microscope* (*SEM*), which showed a graph of the elements and their distribution for each type of sample analyzed (Figs. 68.3, 68.4, 68.5).

(C. S.)

Fig. 68.1 The materials used to make the samples. From *left*: soil, sand and straw

Fig. 68.2 The samples: ASP (*left*) and AS (*right*)

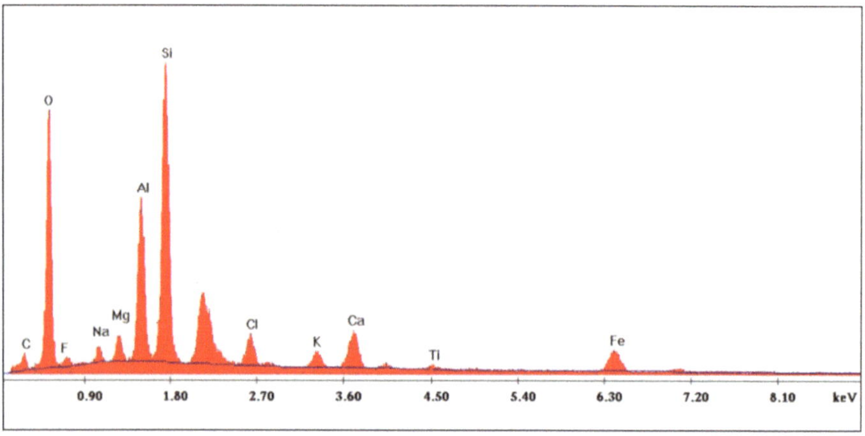

Fig. 68.3 SEM 100.000x chemical composition of AS sample

68.3 Testing of Compression Strength of the AS and ASP Samples

In the absence of regulations regarding earth bricks, for the administration of compression reference was made to the NORMA UNI EN 772–1:2011 the title being *Testing Methods For Masonry Elements—Part 1: Determining Compression Resistance.* Before carrying out the tests the samples were compacted to constant mass, at a temperature of 20±2 °C and relative humidity of 65±5 % [4–7]. The compression resistance demanded for an *adobe* is about 2 MPa. The samples made up of soil, water and sand (AS) showed an average compression resistance of 3.6 MPa, whilst those composed of soil, water, sand and straw (ASP) showed an average compression resistance of 2.8 MPa (Table 68.1; Fig. 68.6).

(F. S.)

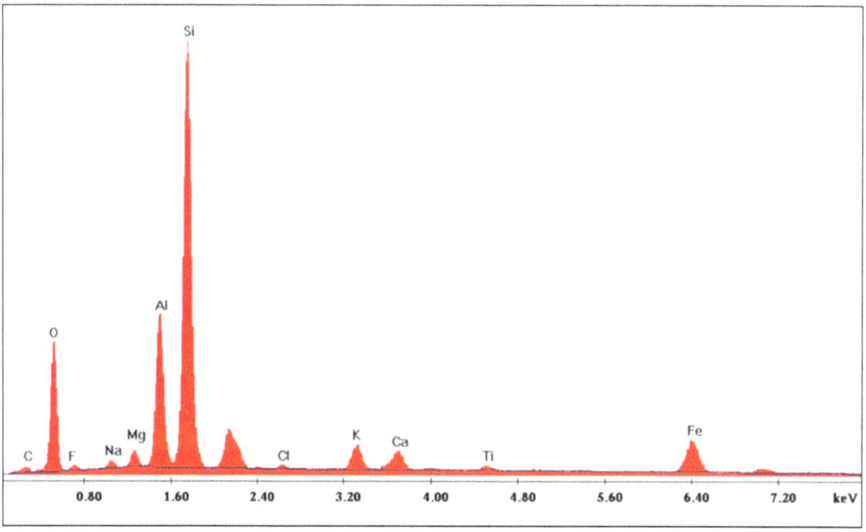

Fig. 68.4 SEM 100.000x chemical composition of ASP sample

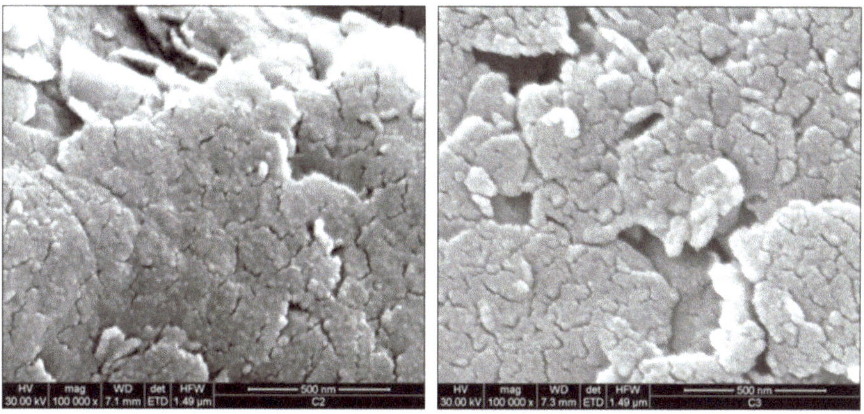

Fig. 68.5 SEM 100.000x of samples AS (*left*) and ASP (*right*)

Table 68.1 Test results of compression strength of the AS and ASP samples

Sample	Compression strength (MPa)	Sample	Compression strength (MPa)
AS/1	4.0	ASP/1	2.9
AS/2	3.0	ASP/2	2.7
AS/3	3.2	ASP/3	2.4
AS/4	3.4	ASP/4	2.7
AS/5	4.6	ASP/5	2.8
AS/6	3.4	ASP/6	3.0
Average compression strength 3.6 MPa		Average compression strength 2.8 MPa	

Fig. 68.6 The samples, AS (*left*) and ASP (*right*), to be tested for compression strength

68.4 Testing of Flexural Strength of the AS and ASP Samples

In the absence of regulations regarding earthen bricks, for the administration of the flexion tests, reference was made to the NORMA 12372:2007 with the title *Testing methods for natural stone—Determining flexion resistance under concentrated load.* Before carrying out the tests the samples were compacted to constant mass, at a temperature of 20 ± 2 °C and relative humidity of 65 ± 5 % [4–7]. The flexion resistance demanded for an *adobe* is about 0.4 MPa. The samples composed of soil, sand and water (AS) present an average flexion resistance of 1.7 MPa, whilst those composed of soil, sand, water and straw (ASP) present an average flexion resistance of 1.02 MPa (Table 68.2; Fig. 68.7).

(C. S.)

Table 68.2 Test results of flexural strength of the samples AS and ASP

Sample	Flexural strength (MPa)	Sample	Flexural strength (MPa)
AS/1	1.8	ASP/1	1.2
AS/2	1.8	ASP/2	0.9
AS/3	1.6	ASP/3	1.2
AS/4	1.6	ASP/4	0.9
AS/5	1.7	ASP/5	0.9
Average flexural strength 1.7 MPa		Average flexural strength 1.02 MPa	

Fig. 68.7 The ASP sample to be tested for flexural strength

68.5 Testing Abrasion Resistance of the AS and ASP Samples

The resistance-to-abrasion tests were carried out on three specimens for each typology, with the use of a metallic bristle brush. The evaluation of the test was carried out by quantifying the actual weight of the material removed, through the difference between the initial weight of each sample and the weight subsequent to the abrasive action. The procedure entailed the initial weighing of each sample using precision electronic scales and the subsequent rubbing of one side of the sample for one minute with a brush loaded with a weight of 3 kg; at the end of each test, the respective specimens were again weighed to determine the difference with the initial weight and, consequently, the amount of material removed. The AS samples registered a lower average weight for the material removed than that of the ASP, 2.96 g as against 3.93 g (Table 68.3).

(C. S.)

Table 68.3 Test results of abrasion resistance of the AS and ASP samples

Sample	Initial weight (g)	Final weight (g)	Amount of material removed (g)
AS/19	504.14	500.62	3.52
AS/26	502.00	498.68	3.32
AS/30	496.58	494.54	2.04
Average weight of material removed 2.96 g			
ASP/13	463.51	459.34	4.17
ASP/28	464.52	460.61	3.91
ASP/30	463.55	459.84	3.71
Average weight of material removed 3.93 g			

68.6 Testing of Compression Strength, Flexural Strength and Abrasion Resistance of ASN Samples

The tests carried out show how AS samples have greater resistance to compression, flexion and abrasion than the ASP samples. This is due to the presence of straw, which renders the mixture less compact. On the basis of these results, ASN samples were realized with the addition of a small amount of laponite. The ASN samples comprised 25 % sand, 38.46 % water, 31.54 % soil and 5 % laponite nanoparticles. The laponite nanoparticles and the ASN sample were observed through the SEM, which showed a graph of the elements and their distribution for each type of sample analyzed (Figs. 68.8, 68.9, 68.10). The ASN samples were subjected to compression, flexion and abrasion tests. The samples composed of soil, sand water and laponite nanoparticles (ASN) showed an average compression resistance of 4.7 MPa, an average flexion resistance of 2.42 MPa (Table 68.4) and an average weight for the material removed of 0.85 g (Table 68.5) (Fig. 68.11).

(F. S.)

68.7 Conclusions

The use of laponite nanoparticles in adobe bricks has brought an increase in compression, flexion and abrasion resistance when compared to traditional bricks. As regards resistance to compression, there was a 30 % increase over the AS samples,

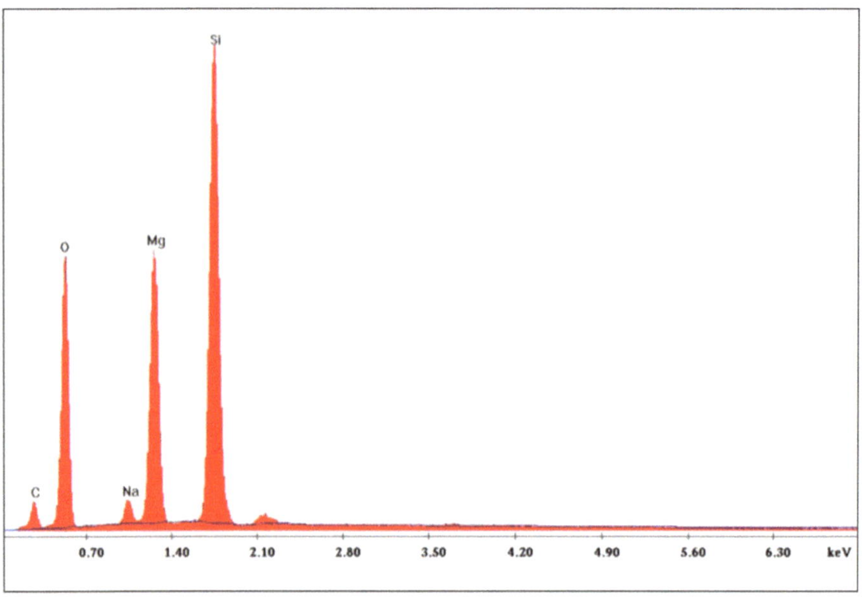

Fig. 68.8 SEM 100.000x chemical composition of laponite

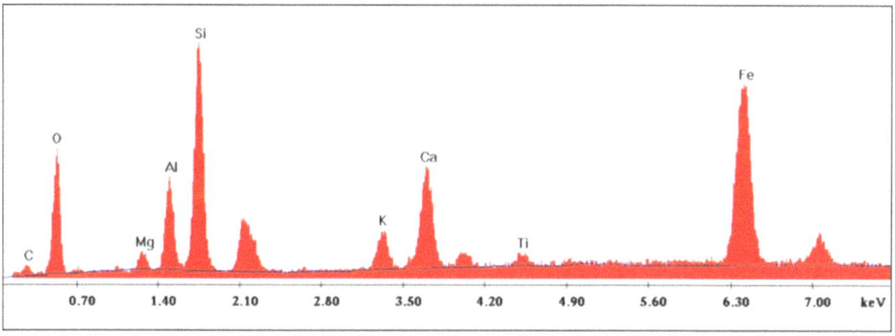

Fig. 68.9 SEM 100.000x chemical composition of ASN sample

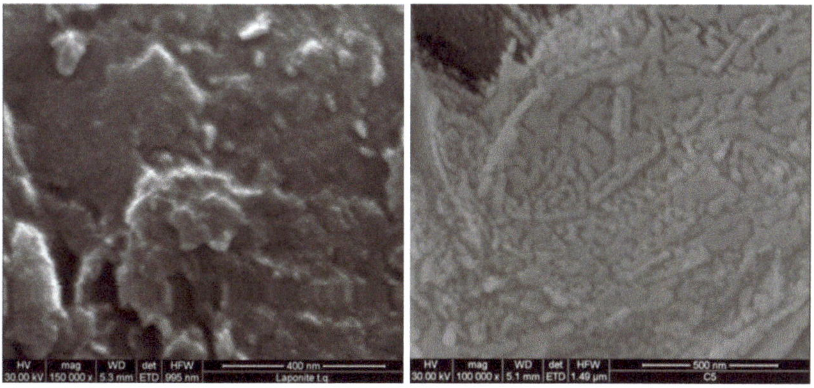

Fig. 68.10 SEM 150.000x of laponite (*left*) and SEM 100.000 of ASN sample (*right*)

Table 68.4 Test results of compression and flexural strength of the ASN samples

Sample	Compression strength (MPa)	Sample	Flexural strength (MPa)
ASN/1	4.1	ASN/1	1.9
ASN/2	5.0	ASN/2	2.2
ASN/3	4.4	ASN/3	2.3
ASN/4	4.7	ASN/4	2.4
ASN/5	5.3	ASN/5	2.8
ASN/6	4.7	Average flexural strength 2.42 MPa	
Average compression strength 4.7 MPa			

Table 68.5 Test results of abrasion resistance of the ASN samples

Sample	Initial weight (g)	Final weight (g)	Amount of material removed (g)
ASN/1	506.08	505.45	0.63
ASN/2	515.99	515.01	0.98
ASN/3	509.23	508.30	0.93
Average weight of material removed 0.85 g			

Fig. 68.11 The ASN sample to be tested for abrasion resistance

and 45 % over those made from soil, water, sand and straw (even though the latter also showed lower resistance than bricks made from earth, water and sand). As regards resistance to flexural strength, there was a 43 % increase over the AS samples and 60 % over those made from soil, water, sand and straw (even though the latter also showed lower resistance than bricks made from earth, water and sand). As for resistance to abrasion, the performance of the ASN samples was 3 times greater than the AS samples and 4 times greater than the ASP samples.

(F. S.)

References

1. Ashby MF, Ferreira PJ, Schodek DL (2009) Nanomaterials, nanotechnologies and design. Elsevier, Oxford
2. Fontaine L, Anger R (2009) Batir en terre. Du grain de sable à l'architecture. Belin, Paris
3. Rael R (2008) Earth architecture. Princeton Architectural Press, New York
4. Walker P (2002) The Australian earth building handbook, HB 195-2002. Standards Australia, Sydney
5. Morel JC, Pkla A, Walker P (2007) Compressive strength testing of compressed earth blocks. Constr Build Mater 21(2):303–309
6. Oliver M, Mesbah A, El Gharbi Z, Morel JC (1997) Mode opératoire pour la réalisation d'essais de résistance sur blocs de terre comprimée. Materials and Structures 30 (9): 515–517
7. Varum H, Costa A, Pereira H, Almeida J, Silveira D (2007) Mechanical characterization of adobe masonry walls. In: Proceedings of the 13th conference of sociedade portuguesa de materiais and IV international materials symposium—a materials science forum [CD-ROM], Porto, Portugal; April 1–4

Chapter 69
The Development of Renewable Energy Applications in Buildings in Greece During The Last Decade

Nikos Papamanolis

Abstract Greece is a country rich in renewable energy sources (RES). A consequence of this is that the country performs well in the field of renewable energy applications. The share of renewable energies in the national production of primary energy is over 20 %. The bulk of the energy generated by the country's RES is consumed in the building sector and covers mainly heating and hot water production needs. The forms of renewable energy that are found in applications in Greek buildings include solar, wind, geothermal and biomass energy. The quantity of energy that these energy sources represent increased by more than 70 % during the last decade for which statistics are available (2002–2012). This work presents and attempts to interpret the data for the development of renewable energy applications in buildings in Greece during this decade.

Keywords Buildings · Greece · Renewable energy · Renewable energy applications

69.1 Introduction

Greece, with an overall land area of approximately 132,000 km^2, is situated in the southeastern part of Europe. While four-fifths of the country consists of mountainous terrain, Greece also has numerous islands and a coastline of over 15,000 km in length. The country's population, according to the last census, stands at about 10.8 million, the bulk of which (i.e. about 61.2 %) lives in urban areas [1].

The climate in Greece is predominantly of the Mediterranean type, with mild and rainy winters, relatively warm and dry summers and sunshine almost all year round. The coldest months are January and February, with mean minimum temperatures ranging between 5 and 10 °C near the coasts and 0–5 °C in inland areas, and lower values in the northern part of the country. The warm season lasts from April until September. The warmest period is in July and August, with mean maximum

N. Papamanolis (✉)
School of Architecture, Technical University of Crete,
K4 Building, University Campus, 73100, Chania, Greece
e-mail: npapama@arch.tuc.gr

A. Sayigh (ed.), *Renewable Energy in the Service of Mankind Vol I*,
DOI 10.1007/978-3-319-17777-9_69

temperatures in the range of 29–35 °C [2]. The climatic conditions in Greece result in most of the buildings being heated for at least 4 months of the year and being cooled for another 4 months [3], which creates a series of challenges in their design and construction.

During the last few years there has been an increased interest in the use of renewable energy systems in buildings and, more generally, the technologies that exploit the country's climate conditions to improve the energy efficiency of buildings. This interest can be partly explained by the fact that since 2010 new laws have been implemented in response to Directive 2002/91/EC on the Energy Performance of Buildings [4]. In February 2013, this legislation was supplemented by the incorporation into national law of Directive 2010/31/EU, which revises and improves the previous directive on building energy performance [5].

The present study identifies and examines the development over the last decade of the renewable energy applications used in Greece, as well as the applications in the country's buildings that take advantage of the climatic conditions. It identifies and analyzes the factors that have shaped and influenced this development, with special reference to the impact they have on the behaviour patterns of these buildings. The emergence of factors that play a positive or negative role in the growth of RES applications in the country's buildings is expected to assist in the proper management of these applications.

69.2 Renewable Energy in Greece

The share of renewable energy sources (RES) in the national energy balance is approximately 21.7 %, at the level of the total domestic production of primary energy [6]. The primary production of renewable energies increased from 58,319 TJ in 2002 to 95,230 TJ in 2012 (i.e. an increase of approximately 63.3 %), where biomass accounted for the largest proportion (53.3 %), followed by hydropower (16.6 %), wind energy (14.6 %), solar energy (14.5 %) and geothermal energy (1.0 %) (Fig. 69.1) [7]. All of these renewable energy forms registered an increase over the decade 2002–2012. The greatest increase occurred in the case of geothermal energy production, with an increase of 1523 % (from 56 to 909 TJ), followed by an increase of 491 % in wind energy production (from 2344 to 13,860 TJ), an increase of 234 % in solar energy production (from 4140 to 13,818 TJ), an increase of 57 % in hydropower production (from 10,080 to 15,851 TJ) and, finally, an increase of 22 % in biomass production (from 41,699 to 50,792 TJ).

In 2012, the share of renewable energy in gross final energy consumption in Greece reached 15.1 %, surpassing for the first time the average for the EU-28 countries as a whole (14.1 %) [8]. This development represents an important step towards fulfilling the target for 2020, which has been set at 18 % for Greece, as measured against the indicative trajectory defined in Annex I Part B of Directive 2009/28/EC [9].

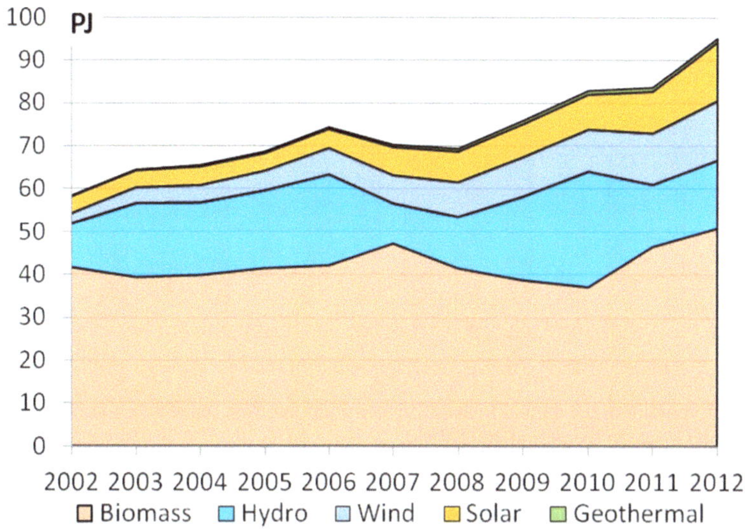

Fig. 69.1 Evolution of the annual primary production of renewable energy in Greece in the period 2002–2012

According to the latest Eurostat statistics, in the period 2004–2012 Greece recorded the fourth largest increase in gross final energy consumption from RES (from 7.2% in 2004 to 15.1% in 2012), behind Sweden (from 38.7 to 51%), Denmark (from 14.5 to 26%) and Austria (from 22.7 to 32.1%) [8].

69.3 Renewable Energy in the Building Sector in Greece

According to 2012 data, the total amount of hydropower, wind and photovoltaic (pv) energy generated in Greece was used for electricity production. The amount of finally consumed energy generated by other renewables and waste came to 64,435 TJ. Approximately 76.4% of this amount (i.e. 49,239 TJ) was directed to the building sector, with biomass representing the largest percentage of this total (83.1% or 40,945 TJ), followed by solar thermal energy (15.6% or 7678 TJ) and geothermal energy (1.3% or 616 TJ) [7]. Collectively, this energy consumption served mainly heating and hot water production needs and represented approximately 16.5% of total energy consumption in the Greek building sector, a percentage that was approximately 6.5% points greater than that of the EU-28 as a whole [6].

The bar chart in Fig. 69.2 shows the fluctuating shares of each renewable energy source in the final energy consumption of the Greek building sector during the decade 2002–2012. The chart includes the amounts of electrical energy generated by hydropower, solar pv and wind energy and consumed in the building sector, these amounts being calculated on the basis of the annual proportions of the total electri-

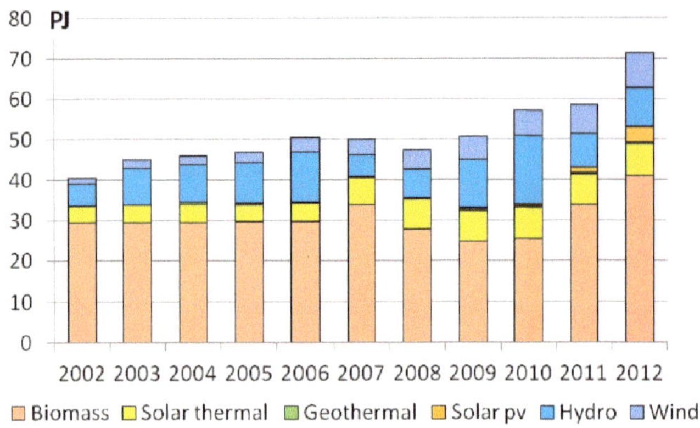

Fig. 69.2 Evolution of the annual final energy consumption from renewable sources in the building sector in Greece in the period 2002–2012

cal energy generated by the respective renewable sources and the share of the building sector in consumptions of electrical energy [6, 7, 10].

Of the six forms of RES that appear in this chart, three—biomass, hydropower and solar thermal—could be considered as traditional since they have contributed to the country's energy system for several decades (solar thermal is the most recent of these energy forms, having contributed to the country's energy system since the 1970s). The other three—i.e. geothermal, solar pv and wind energy—are the most recent energy forms of all, with wind energy being added to the system as recently as the late 1990s, while records for the other two have been kept only since 2004.

The total share of RES in the energy balance of buildings in Greece increased from 40,347 TJ in 2002 to 71,274 TJ in 2012 (i.e. an increase of 76.7%), while each individual form of RES also displayed an upward tendency over the same period. More specifically, of the traditional forms of RES, biomass increased from 29,447 TJ in 2002 to 40,945 TJ in 2012 (an increase of 39.0%), hydropower increased from 5494 TJ in 2002 to 9785 TJ in 2012 (an increase of 77.6%) and solar thermal energy increased from 4129 TJ in 2002 to 7678 TJ in 2012 (an increase of 86.0%). Biomass, mostly in the form of firewood and charcoal, clearly has the largest share in the consumption of RES in buildings in Greece. Hydropower comes second throughout the decade, except for 2007, when it was overtaken by solar thermal energy. Solar thermal energy had a constantly greater share than wind power, except in the final year of the decade (2012), when it was overtaken by the latter. The last two positions in the table are occupied by the most recent forms of RES, with solar pv registering a large increase in the last 2 years of the decade (2011 and 2012) and clearly overtaking geothermal energy.

The upward tendency in the use of RES in recent years has partly been helped by the new legislation governing the energy performance of buildings that has been in force since October 2010. This particular legislative framework replaced the earlier one of 1979 concerning the energy and environmental performance of buildings, which was primarily based on a study of thermal insulation and its compliance with relevant directives of a rather general nature [10]. Since the beginning of 2011, the new legislation has been combined with a package of actions like "Energy Efficiency at Home", "Building the Future" and others which, through a series of measures and incentives, have aimed to improve the energy performance of buildings in the country and to support the effort to achieve a 20% energy saving by 2020 [11, 12].

Under these arrangements, a strong endowment is being provided to apply renewable sources and exploit the climatic conditions in buildings. The contribution of these arrangements is both direct, through the specific requirements they introduce (e.g. the obligation that passive solar systems, as well as heating/cooling/electricity production systems utilising RES and combined heat and power (CHP), must be incorporated in the heating/cooling specification study submitted in the building licensing procedure), and indirect, through the increased standards for energy performance that they prescribe and the recommendations that are being provided for improving the energy performance of audited buildings [13].

69.4 The Main Characteristics of the Variability of RES Applications in Buildings in Greece

In the overall upward tendency of the shares of RES in the building sector, transitory fluctuations may be observed in the shares of some of these energy sources. These fluctuations are most obvious in the cases of biomass and hydropower. More specifically:

- The consumption of biomass during the first few years of the period under examination was more or less constant. Following a fluctuation in the middle of the period (an increase in 2007 followed by a decrease in the following two years), from 2010 onwards it registered a strong increase. The consumptions of biomass at the beginning of the decade represent the culmination of a downward trend that had been in evidence since the early 1990s, which was caused primarily by a change in the way buildings were heated in the countryside [14]. This trend has been reversed in recent years when the economic crisis that has affected the country, combined with increased prices and taxes imposed on alternative fuels, particularly oil and natural gas, has caused citizens to turn to firewood as a more economical solution [15]. At the same time, the production of firewood in Greece over the last decade has fallen markedly. From 1,093,000 m^3 in 2002 it dropped to 857,180 m^3 in 2011 [16]. This means that the increased demand that was recorded in the last few years of the decade was covered by imports. As a matter of fact, imports of firewood, mainly from neighbouring countries and particularly Bulgaria, displayed a strong increase over the decade (from 146,700 m^3 in 2002 to 389,200 m^3 in 2012) [16].
- Hydroelectric energy registered sharp decreases in the 2-year periods 2006–2007 and 2010–2011. Given that hydropower capacity remained relatively stable during the period under examination (from 2379 MW in 2002 to 2447 MW in 2012

[17]), these changes are a result of the low rainfall recorded in the country during the winters of the periods in question. In fact, as relevant press releases issued by the Public Power Corporation have revealed, the water stocks in the reservoirs of the hydroelectric power plants were down in the periods concerned [17, 18].

- Solar thermal energy displays a considerable increase in 2007, with periods of relatively stable values on either side of this year. In the last few years, the considerable drop in the construction of new buildings in Greece as a result of the economic crisis has had a negative impact on the solar thermal collector market. At the same time, however, there has been a swing in favour of the installation of new solar thermal collectors to replace electric and oil heating systems, as a more economical solution [17].
- In the case of the other forms of RES, the year-on-year increases are almost constant. If we exclude geothermal energy, which made slow progress, wind energy and, to an even greater extent, solar pv energy registered a clear upward trend. This trend can be accounted for by the growth of the respective technologies and their infiltration into the market as a result of the endowment they received from the state. It is characteristic that photovoltaics, in the context of the more general measures adopted to promote the use of RES, received strong support from the state [18]. Indeed, this fact led to the attainment of the relevant national targets set for 2020, 7 years earlier than projected [18].

69.5 Conclusions

During the period 2002–2012, RES applications in buildings in Greece registered a remarkable increase, both overall and in each individual category. Already the country's ranking among the EU countries in respect of the use of RES in the building sector is quite satisfactory, in keeping with its rich stock of RES.

In the last few years, the application of RES in Greek buildings has been significantly affected by two factors:

a. The incorporation into Greek national law of Directives 2002/91/EC and 2010/31/EU on the energy performance of buildings, combined with measures adopted by the Greek state to support and promote RES applications.
b. The economic crisis, which since about 2009 has curbed construction activity and also many other activities and initiatives that could have had an impact on the utilisation of RES for the improvement of the energy efficiency of buildings in Greece.

The effects of these two opposing factors appear for the moment to balance each other out. In any case, the desirable thing is that the positive progress of RES applications in Greek buildings should be maintained in the future and, if possible, strengthened.

References

1. Hellenic Statistical Authority (2012) Announcement of the results of the 2011 Population Census for the Resident Population, 28 Dec. http://www.statistics.gr/portal/page/portal/ESYE/BUCKET/General/A1602_SAM01_DT_DC_00_2011_02_F_EN.pdf
2. Hellenic National Meteorological Service (2013) The climate of Greece. http://www.hnms.gr/hnms/english/climatology/climatology_html?
3. Papakostas K, Kyriakis N (2005) Heating and cooling degree-hours for Athens and Thessaloniki, Greece. Renew Energy 30(12):1873–1880
4. Directive 2002/91/EC of the European Parliament and of the Council of 16 December 2002 on the energy performance of buildings. Official Journal L 001, 04/01/2003, 65–71. http://eur-lex.europa.eu/LexUriServ/LexUriServ.do?uri=CELEX:32002L0091:EN:HTML
5. Directive 2010/31/EU of the European Parliament and of the Council of 19 May 2010 on the energy performance of buildings. Official Journal L 153, 18/06/2010, 13–35. http://eur-lex.europa.eu/LexUriServ/LexUriServ.do?uri=OJ:L:2010:153:0013:01:EN:HTML
6. Eurostat, Supply, transformation, consumption—all products—annual data [nrg_100a] Last update: 17-02-2014, http://appsso.eurostat.ec.europa.eu/nui/show.do?dataset=nrg_100a&lang=en
7. Eurostat, Supply, transformation, consumption—renewable energies—annual data [nrg_107a] Last update: 11-02-2014, http://appsso.eurostat.ec.europa.eu/nui/show.do?dataset= nrg_107a&lang=en
8. Eurostat, Share of renewable energy in gross final energy consumption [t2020_31]. http://news.in.gr/science-technology/article/?aid=1231302035
9. Directive 2009/28/EC of the European Parliament and of the Council of 23 April 2009 on the promotion of the use of energy from renewable sources and amending and subsequently repealing Directives 2001/77/EC and 2003/30/EC. http://europa.eu/legislation_summaries/energy/renewable_energy/en0009_en.htm
10. Hellenic Ministry of Environment (1979) Planning and Public Works, Thermal Insulation Code for Buildings, Decree-Law 1/6/1979, The Hellenic Official Gazette, 362D
11. Hellenic Ministry of Environment (2011) Energy and climate change, Press release for the Program "Building the Future" (in Greek). http://www.ypeka.gr/Default.aspx?tabid=362&sni[524]=637&locale=en-US&language=el-GR
12. Hellenic Ministry of Environment, Energy and Climate Change, Energy Efficiency at Household Buildings, http://exoikonomisi.ypeka.gr/Default.aspx?tabid=629&locale=en-US&language=el-GR
13. Papamanolis N, Tsitoura M (2013) The use of renewable energies in the new legislative framework for decreasing energy consumption in buildings in Greece, In: Proceedings of the 4th International Conference on RES, Nicosia, Cyprus, June 2013, p 6
14. Hellenic Ministry of Environment (2012) Energy and climate change, activity report of forest services in Greece for 2010, p 66 (in Greek). http://www.ypeka.gr/LinkClick.aspx?fileticket=fjsA7Em8z0A%3D&tabid=588&language=el-GR
15. Zafeiriou E, Arabatzis G, Koutroumanidis T (2011) The fuelwood market in Greece: an empirical approach. Renew Sustain Energy Rev 15(6):3008–3018
16. Eurostat, Roundwood, fuelwood and other basic products [for_basic]. http://appsso.eurostat.ec.europa.eu/nui/show.do?dataset=for_basic&lang=en
17. European Solar Thermal Industry Federation (ESTIF) (2013) Solar thermal markets in Europe, Trends and Market Statistics 2012; p 20. http://www.estif.org/fileadmin/estif/content/market_data/downloads/Solar_Thermal_M%20arkets%202012.pdf
18. Ministry for the Environment (2010) Energy and Climate Change, National Renewable Energy Action Plan in the Scope of Directive 2009/28/EC, p 109, http://www.ypeka.gr/LinkClick.aspx?fileticket=CEYdUkQ719k%3d&tabid=37

Chapter 70
Preliminary Results Concerning the Thermal Comfort in a Romanian Passive House

Ruxandra Crutescu, Ioana Udrea, Ilinca Nastase, Cristiana Croitoru and Viorel Badescu

Abstract Energy saving is a recognized priority worldwide. The goal for the construction industry is to conceive and operate buildings with low energy consumption. However, sometimes energy saving can cause discomfort to the occupants. A field survey was done in a naturally ventilated passive office building during summer. Comfort parameters were measured and occupants were distributed comfort questionnaires consisting of questions related to their thermal sensation and preference.

This chapter compares the experimental results with the subjective responses. It analyzes the relationship between thermal preference and sensation of the occupants which depends on the floor where the occupants work. The survey results are compared with the comfort ASHRAE standard 55 and are in agreement with the standard.

Keywords Passive building · Thermal comfort · Adaptive · Low energy

---

I. Udrea (✉) · V. Badescu
Faculty of Mechanical Engineering and Mechatronics, Thermodynamics Department, Polytechnic University of Bucharest, Spl. Independentei 313, 060042, Bucharest, Romania
e-mail: udreaioana@yahoo.com

R. Crutescu
Faculty of Architecture, Spiru Haret University, Bucharest, Romania
e-mail: pasivhausoffice@gmail.com

I. Nastase · C. Croitoru
Building Services Department, Technical University of Civil Engineering in Bucharest, Avenue Pache Protopopescu, 020396, Bucharest, Romania
e-mail: ilinca.nastase@gmail.com

V. Badescu
e-mail: badescu@theta.termo.pub.ro

C. Croitoru
e-mail: cristianaverona@yahoo.com

A. Sayigh (ed.), *Renewable Energy in the Service of Mankind Vol I*,
DOI 10.1007/978-3-319-17777-9_70

70.1 Introduction

In environments where thermal loads are considered non-transient over short time intervals, the thermal comfort theory embraces two major approaches: the classic or Fanger's model and the adaptive approach. According to Fanger's model [11], human comfort depends on the quantitative, combined influence of six parameters (air temperature, mean radiant temperature, water vapour pressure, air velocity, closing level and metabolic rate). The adaptive model relates indoor design temperatures or acceptable temperature ranges to outdoor meteorological or climatological parameters (mean outdoor effective air temperature, mean outdoor air temperature) [8].

The adaptive comfort model is used for naturally ventilated buildings, taking into account the internal conditions; internal operative temperature is reported to monthly mean of outside air temperature. It is not necessary to fix the type of clothing to be worn, the humidity or air-speed limits when this model is used.

Current comfort standards such as ASHRAE standard 55 [1] and EN 15215 [6] propose designs for operative temperatures in offices, based on heat balance and the adaptive thermal comfort model. However, there is no assurance that the results of comfort studies performed in climate-controlled chambers, offices, or university classrooms reflect the thermal sensations and preferences of the people working in a passive office building like the one surveyed. Furthermore, the daily environment of buildings with special characteristics such as offices, hospitals, universities, and schools are different to an extent, which suggests that occupants probably adapt to different thermal conditions. While poor thermal comfort is just one of many indoor climate problems, it generates the greatest number of workers' complaints.

In the studies about hospital buildings [15, 16], [18], [7], it is demonstrated that the thermal preferences of medical supplies differ from patients' preferences and that actual comfort standards are not suitably adapted. In [19], the atypical thermal responses of school pupils are described. Other adaptive studies in universities include [5]. Dissatisfaction about thermal comfort of offices as regulated by current standards is demonstrated by Huizenga et al. (2006). The main conclusion in all these cases is that each office has specific traits and that standards must be changed in order to suit each.

The passive building is a special category, low energy consumption building that fulfils special requirements according to the Passivhaus Institute [17]. There are few studies in the literature about adaptive comfort in passive buildings.

The aim of this chapter is to analyze thermal comfort in a passive building through measurements, questionnaire surveys and the adaptive approach. The adaptive approach compares the survey results to predictions made by one of the current comfort standards, namely ASHRAE standard 55.

70.2 Method and Procedure

70.2.1 Questionnaire about Thermal Comfort

The field study presented here included questionnaire surveys and simultaneous measurements of the environmental variables in relation with thermal comfort. The questionnaires were distributed between 19 July and 19 August 2013 in various working places situated in the passive building Amvic.

In this survey, 142 questionnaires were filled out over 16 days. On average, the time taken to complete the questionnaire was about 2–3 min. The information requested by the questionnaire was: date and time of answering, the floor and office number, the working place (POST) and information about the person who had completed the questionnaire (name, age and sex). For the assessment of thermal sensation, the subjects chose an option on the ASHRAE 7-point rating scale; they also chose what thermal preference they had at the time of completion and had to answer questions about the suitability of the thermal environment and about local thermal discomfort.

The questionnaire included a checklist with clothing items for people to choose from. They had to specify the activity they had been doing in the 15 min before they filled the questionnaire.

70.2.2 Amvic Building Presentation

Amvic is a passive office building, located in Bragadiru (latitude 44.4°N), a small town 10 km south of Bucharest, Romania. It has a ground floor and four levels with offices. The building was inaugurated in February 2009 [13]

The main façade has a southeast orientation 292 (Figs. 70.1 and 70.2). A Neopor and Styropor factory is situated on the northwest side of the Amvic building. On the northeast side, there is a garage. The Amvic building system was used for the supporting structure. The foundations are continuous, made of reinforced concrete while the structural walls are made of reinforced concrete from which filling forms made of two panels of expanded polystyrene (EPS) connected by polypropylene webs have been made. Twenty cm of extruded polystyrene (XPS), which is put on the internal and external parts of foundations [9], has been used. External walls are provided with low-emissivity triple-pane glazing windows with reduced overall heat transfer coefficient U and high solar transmission factor.

The main functions of the Amvic office building are listed hereafter. On the ground floor, there is a wide open space where the sales department and the secretary's office are located. In a separated area, there is a service room. The first, second and third floor are wide open office areas. On the top floor, there are five apartments. Amvic PH is well documented [2–4] and has been monitored for a relatively long time (2009–2013).

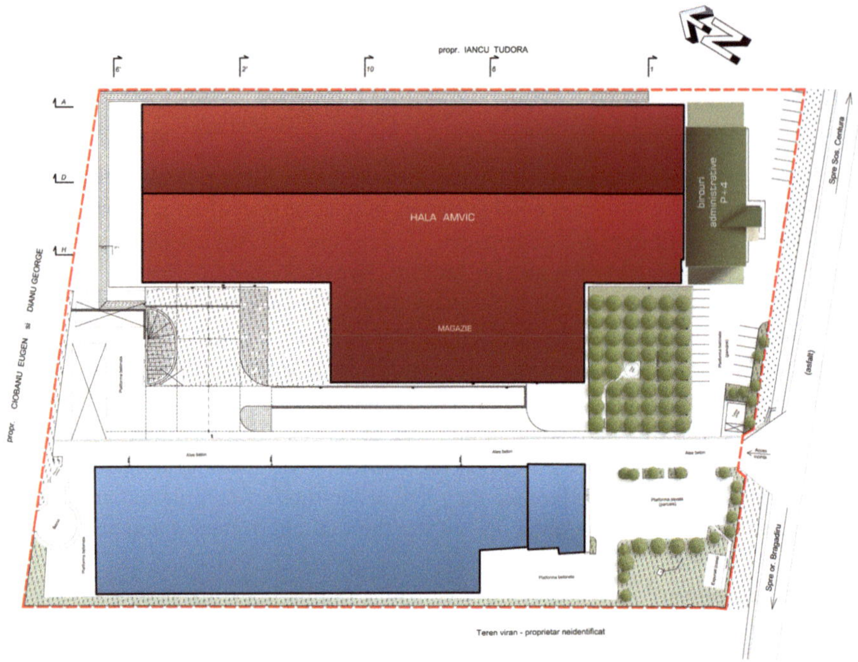

Fig. 70.1 Site plan of Amvic passive office building in Bragadiru, Ilfov, Romania

Fig. 70.2 Amvic office building. South façade

The internal heat sources and sinks in the Amvic building are grouped as follows. First, there is the heat released by residents and employees, electrical appliances, room lighting, solar energy storage, auxiliary power, cooking stoves and heating distribution pipes. Second, there are heat gains or losses due to domestic hot water pipes and fresh water pipes. Third, there is the heat gain or loss due to water condensation and evaporation. Details are given in [3, 4]. The heating and ventilation system implemented in the Amvic office building contains: (1) a registry-type ground heat exchanger; during winter, it warms up the air coming in from the outdoors, and during the summer cools it down; (2) a high-efficiency heat-recovery heat exchanger built into the ventilation system; during the cold season, the air going out warms

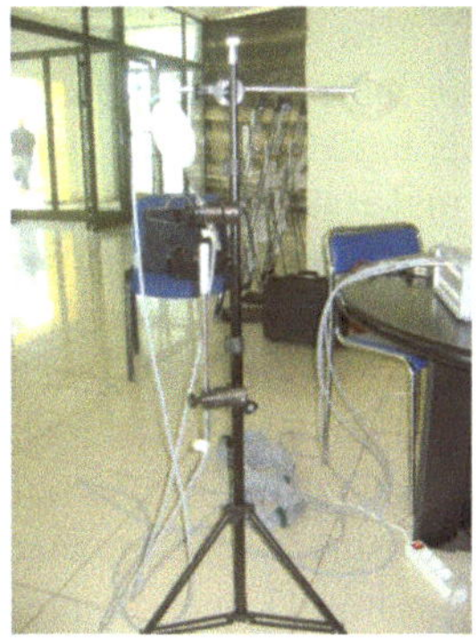

Fig. 70.3 The thermal measurement set-up on its stand inside the Amvic building

up the fresh air coming in; (3) two water-to-water heat pumps used to warm up the fresh air entering the building in the cold season, or during bad weather or nights, and to cool down the fresh air in summer. The cold reservoir for the heat pumps is the dip soil beneath the building.

Amvic is a passive building. This has been well demonstrated by the use of the Passive House Planning Package (PHPP) [12], a software designed for passive buildings. The performance of the Amvic building has been simulated in many locations all over the world [2, 20]

In July 2013, when the present thermal comfort survey was carried out, some interior changes were made. The arrangement of the ground floor was the same, an open space office, a toilette area and a technical room. On the first, second and third floors, the open space offices were divided into several separate rooms to allow people more privacy and concentration at work. The partition walls were made of tempered glass and double plated plasterboard. The partition near the staircase was made of double plated plasterboard with mineral wool insulation. The toilette area remained unchanged at all levels of the building. At that time, in July 2013, the third floor was under renovation. On the fourth floor, in the residential zone, there was a two-room apartment, four studios, a dining room and a small technical room, used for the solar collector storage tanks. In December 2012, the owner of the building changed, and consequently changes were made in the colours of the façades and the logos. The new owner is Arcon Amvic.

70.2.3 Measurement Surveys

The thermal comfort parameters were evaluated with the ComfortSense device, a system for measurement according to International Standard ISO 7730 (Fig. 70.3). The ComfortSense system consists of a main frame with build in A/D converter and USB 2.0 interface. A personal computer (PC) may be connected with this main frame and the measurement probes which are positioned on the system stand. The operator can use the PC to communicate with the whole set-up at the same time. The measurement is set up using a PC and application software.

The measurement was intended to determine the comfort level for each worker, or to be precise, for each workstation in the office building. The statistical analysis is based on a large number of independent measurements. One measurement was undertaken at each workstation given on a list.

The measurement procedure consisted of placing chairs taken from listed workstations near the measurement stand. The procedure took 5 min and the equipment was moved to the next workstation on the list. The measurements were undertaken in July, which was a characteristically warm month. The ComfortSense measurements at the Amvic building took place on July 9 and July 11, 2013. On July 9, measurements were made for the ground floor, first floor and a part of second floor. On July 11, measurements were made for the rest of the second floor and the fourth floor. Measurements could not be carried out on the third floor due to ongoing renovations.

The measurements of thermal comfort parameters were undertaken in a total of 56 posts. The estimated metabolic rate was 1.2 met and the clothing insulation was 0.38 clo. The time taken for each measurement was 3 min. The positioning heights of the probes on the ComfortSense measurement stand were: for the operative temperature probe—109 cm, for the humidity measuring probe—89 cm and for the draft probe—120 cm (with one exception, the reception area, where the occupants stood—180 cm). (See Fig. 70.3). The inclination angle for the operative temperature probe was 30° from vertical, when it simulated a sitting person, as in the case of the Amvic measurements, where the majority of occupants work at their desk in a sitting position. According to recommendations, and the observation of Fig. 70.1, the position of the humidity probe is vertical and the position of the draft probe is inclined to 30° from vertical.

The measured and processed parameters resulting from the measurement were operative temperature, relative humidity, temperature, velocity, turbulence intensity, draught rate, predicted mean vote, predicted percentage of dissatisfaction and mean radiant temperature. For each measurement, the system provides the mean parameter values and instantaneous distributions of value measured by the sensors of velocity and temperature.

At the time of the measurements, the ventilation was operating with fresh air provided by the ground heat exchanger, which was ejected through the ceiling diffusers. The ventilation system was functional on the first and second floors as well. On the fourth floor, the ventilation system functioned in the two-room apartment, but the ceiling diffusers were shut off in the other areas of the floor.

70.3 Results

Figure 70.4 shows the distribution of the thermal sensation vote (TSV) from the questionnaires in comparison with the predicted mean vote (PMV), obtained from calculation with measured data. We can observe an increase towards 'warm' from both TSV and PMV values from lower to upper levels. The difference between TSV

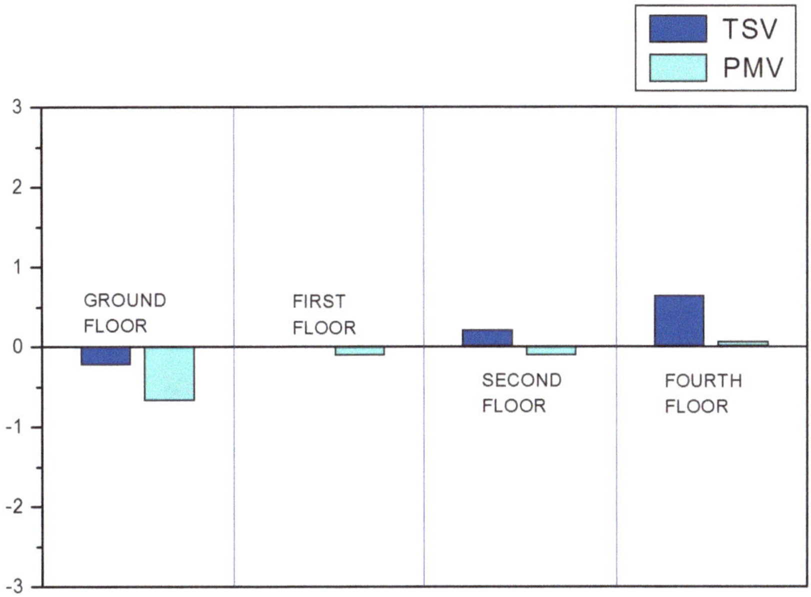

Fig. 70.4 Distribution of TSV comparatively with distribution of PMV on building levels for days with measured dates

and PMV is greater on the ground floor, TSV (−0.24) and PMV (−0.68) and on the fourth floor, TSV (0.64) and PMV (0.07). On the fourth floor, the adaptive comportment is not accomplished (this means that values are closer to comfort-zero value for TSV than PMV) because the majority of questionnaires was completed in the hottest room, which was the kitchen. Near it is a little room where the storage tanks for solar collectors are kept.

The comparative distribution of TSVs and thermal preference votes (TPVs) on building levels can be seen in Fig. 70.5. In this case too, the natural tendency to have a distribution of temperatures rising from low to high levels is demonstrated.

Figure 70.6 shows the distribution of TSVs in relation to thermal preference for the entire survey period. As can be seen, the majority of people voting for a specific thermal sensation preferred conditions which would give them a neutral temperature, as would be expected. However, it must be said that, for example, in the category of people who felt ‘a bit cold’, there were people who preferred temperatures ‘warmer’ and preferred ‘no change’, rather than only ‘a bit warmer’ as expected. The number of votes for ‘warmer’ (3) and ‘no change’ (5) were bigger, as compared to (12) for ‘a bit warmer’. This showed that the subjective behaviour of people does not correspond with expectations.

The occupants were asked whether they considered the thermal environment acceptable or not during the survey. The intention was to see the repartition of TSV and TPV for those who considered the thermal environment acceptable (see Fig. 70.6). We can see a concentration on ‘colder’ for TPV than the almost symmetrical distribution of TSV around ‘OK’.

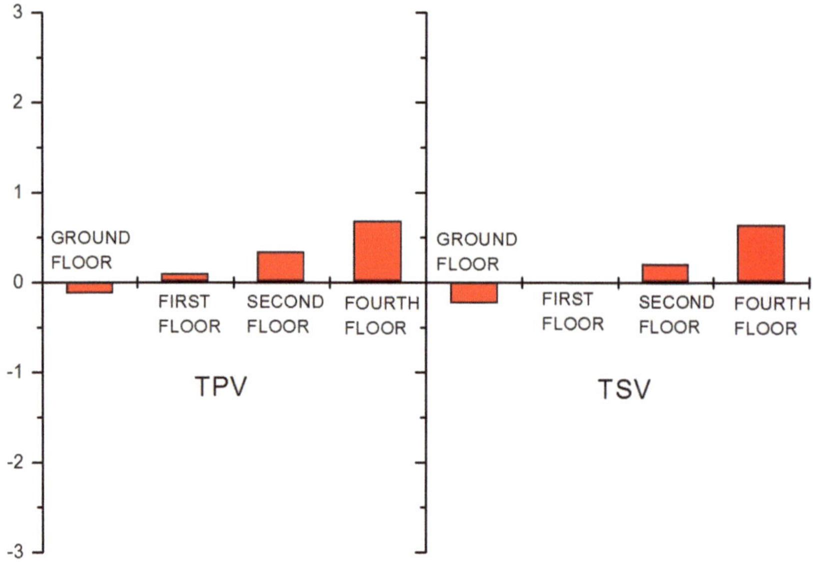

Fig. 70.5 Distribution of TSV and of TPV levels for the entire survey period

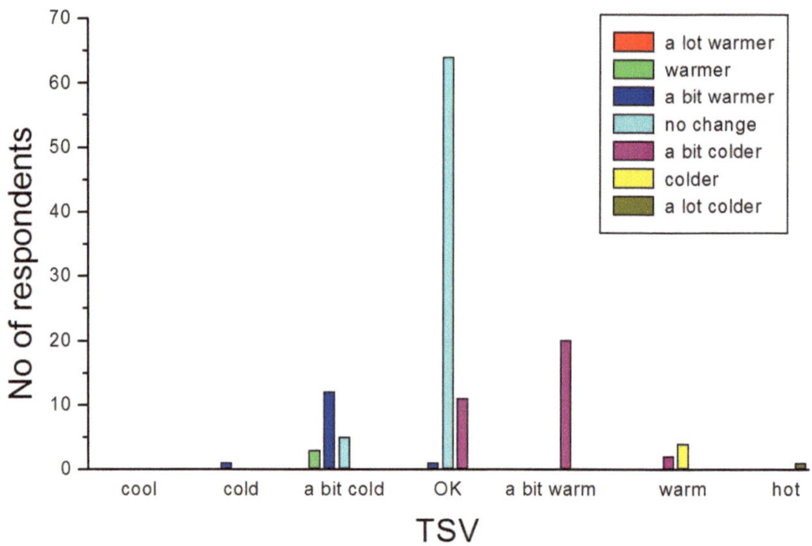

Fig. 70.6 TPV by TSV

Figure 70.8, and Fig. 70.7 show the repartition of TSVs and TPVs for those who did not feel any local thermal discomfort. The tendency is the same with a slight accentuation of warmth on TSVs.

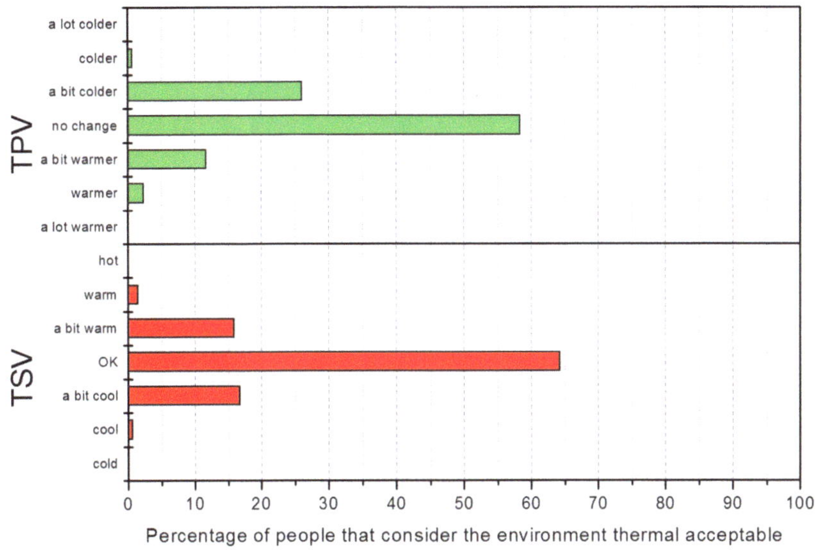

Fig. 70.7 Percentage of people considering the environment thermally acceptable as per TSV and TPV

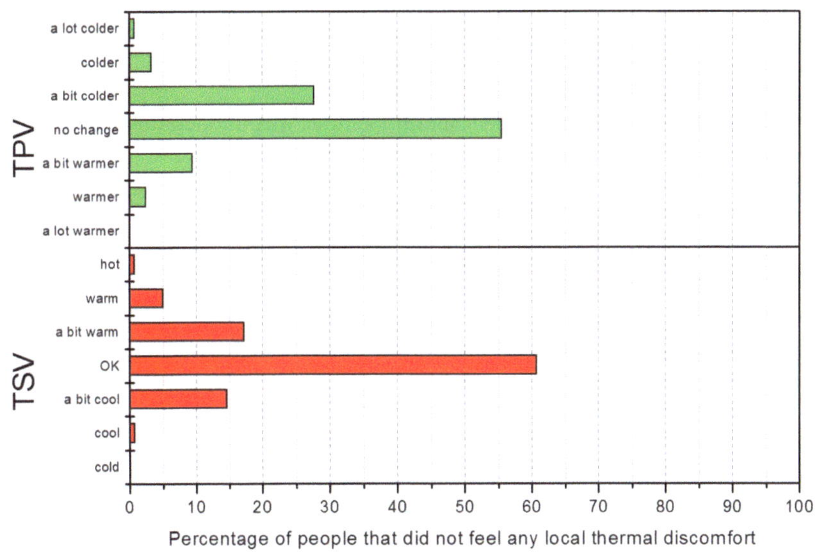

Fig. 70.8 Percentage of people feeling comfortable as per TSV and TPV

Figure 70.9 represented the average TSVs and TPVs for each day of the survey related to internal temperature. It is observed that the values didn't follow the linear distribution because of two points, corresponding to 1st and 6th August 2013. The explanation for this discrepancy was that the questionnaires in those days were

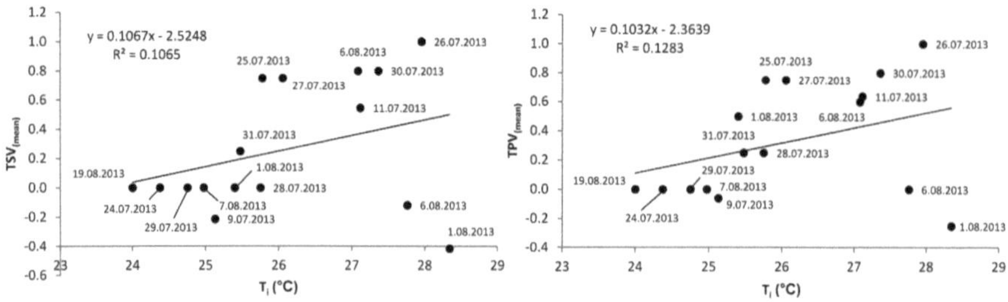

Fig. 70.9 Mean TSVs per survey (TSV(mean)) and mean TPVs per survey (TPV(mean)) against the interior temperature

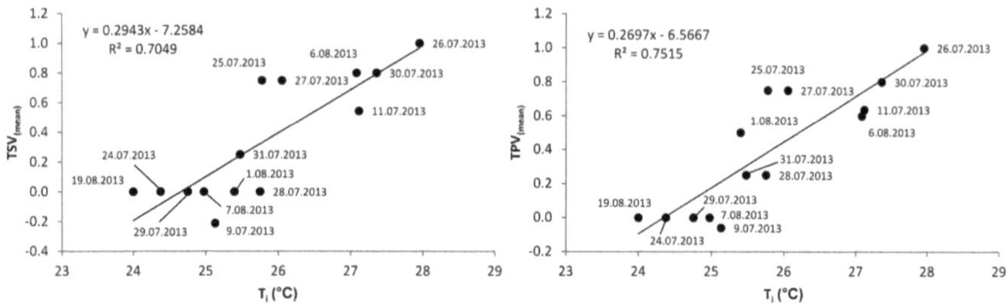

Fig. 70.10 Mean TSVs per survey (TSV(mean)) and mean TPVs per survey (TPV(mean)) against the interior temperature after excluding some dates

sometimes given when a cooling system tip splitter was in operation (It was a new endowment of the building that occurred after 11 July 2013).

After removing the points corresponding to 1st and 6th August 2013, the graph in Fig. 70.10 was obtained. The remaining points have a clear linear aspect.

For 2 days, 9th and 11th July 2013, for which we had measured values for various studied posts, an evaluation had been carried out according to ASHRAE standard 55 specifications.

Daily real-time weather data from EnergyPlus Energy Simulation Software, Real-Time Weather Data [10] was used. The meteo data file obtained for the requested period contained hourly values of atmospheric temperature. The daily average of all hourly values was provided.

As observed in Fig. 70.11, both 9th and 11th July 2013 belong to the gray band in the middle of the picture. This corresponds to 90 % acceptability, which me ans a high level of comfort, according to the ASHRAE standard 55. The point corresponding to 11 July 2013 is situated at a higher level of indoor temperature than the point corresponding to 9 July 2013. This is due to the fact that the measurements performed on 11 July 2013 were undertaken on the ‘hot’ levels (i.e. the second and fourth floors) while the measurements undertaken on 9 July 2013, were carried out on the ground floor, first floor and second floor.

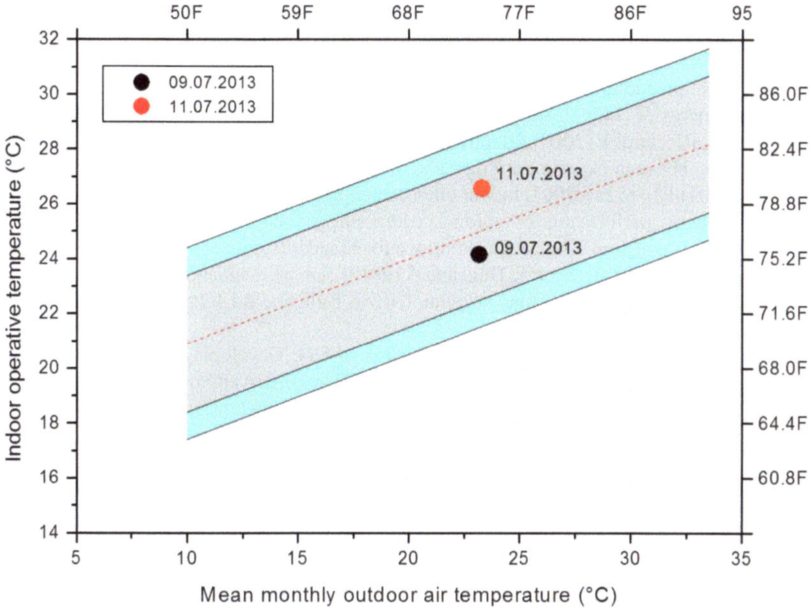

Fig. 70.11 Operative temperatures of the surveys, plotted on the ASHRAE standard 55 diagram for acceptable indoor temperatures

70.4 Conclusions

The distribution of TSVs comparatively with the distribution of PMVs on building levels for days with measured dates shows value for both TSV and PMV in interval (−1,+1). The same can be observed from the distribution of TSV and TPV levels for the entire survey period.

The ASHRAE standard 55 evaluates buildings in category I (90 % acceptability) and the results are in agreement with that standard.

One can conclude that passive buildings are buildings with a good comfort comportment. The measurements should be made in the winter too.

Acknowledgements Part of this work has been funded by the Sectoral Operational Programme Human Resources Development 2007-2013 of the Ministry of European Funds through the Financial Agreement POSDRU/159/1.5/S/138963

References

1. ASHRAE (2010), ANSI/ASHRAE standard 55-2010. In: Thermal Environmental Conditions for Human Occupancy. American Society of Heating, Refrigerating and Air-Conditioning Engineers, Inc., Atlanta, 2010 ASHRAE, ISSN 1041–2336
2. Badescu V, Rotar N (2012) Implementation of the German Passivhaus Concept in Southeast Europe. J Energy Eng 138(3):146–162. ISSN 0733–9402

3. Badescu V, Laaser N, Crutescu R (2010) Warm season cooling requirements for passive buildings in Southeastern Europe (Romania). Energy 35:3284–3300
4. Badescu V, Laaser N, Crutescu R, Crutescu M, Dobrovicescu A, Tsatsaronis G (2011) Modeling, validation and time-dependent simulation of the first large passive building in Romania. Renew Energy 36:142–157
5. Buratti A, Ricciardi P (2009) Adaptive analysis of thermal comfort in university classrooms: correlation between experimental data and mathematical models Build Environ 44:674–687
6. CEN (2007) EN 15251:2007. Indoor environmental input parameters for design and assessment of energy performance of buildings addressing indoor air quality, thermal environment, lighting and acoustics. European Committee for Standardization, Brussels, p 2007
7. Croitoru C, Vartires A, Bode F, Dogeanu A (2013) Survey evaluation of the indoor environment quality in a large Romanian hospital, INCAS Bulletin, Vol 4/2013, pp 45–52; December 2013
8. de Dear R, Brager G, Cooper D (1997) Ashrae rp-884; Developing and adaptive model of thermal comfort and preference. Technical report, The American Society of Heating. Refrigeration and Air-Conditioning Engineers, Inc., and Environmental Analytics, editor. Atlanta
9. Dimitriu S, Badescu V, Hera D, Crutescu M, Crutescu RG, Costache N, Ivan G, Drughean L, Ilie A, Iordache F, Iordache V, Marinescu M, Isvoranu D, Cazacu M, Budea S (2008) *Cladirea administrativa pasiva Amvic—o realizare pentru viitor*. Presented to the conference on energy efficiency and buildings, Asociatia Auditorilor Energetici pentru Cladiri, Bucuresti, 27–28 November 2008 (CD of the Proceedings, ISBN 978-973-85936–10, Editura ARTECNO Bucuresti)
10. Energy (2013) EnergyPlus energy simulation software, real-time weather data, http://apps1.eere.energy.gov/buildings/energyplus/weatherdata_download.cfm. Accesed 8 Dec 2013
11. Fanger PO (1970) Thermal comfort. Analysis and applications in environmental engineering. Published in the United States by McGraw-Hill Book Company 1972. Originally presented as the author's thesis, Danmarks Teknishe Højskole 1970
12. Feist W, Puger R, Kaufmann B, Schnieders J, Kah O (2007) Passive house planning package 2007, Specification for quality approved passive Hauses. Passivehouse Institute, Darmstadt
13. Hera D, Drughean L, Ilie A, Crutescu R (2008) *Climatizarea unei case pasive cu functiune mixta*, a XV-a Conferinta; Confort, Eficienta, Conservarea energiei si Protectia mediului, 26–27 noiembrie 2008
14. Huizenga C., Abbaszadeh S., Zagreus L., Arens E. *A., (2006), Air quality and thermal comfort in* office buildings: Results of a large indoor environmental quality survey, Indoor Environmental Quality (IEQ), Center for the Built Environment, Center for Environmental Design Research, UC Berkeley
15. Khodakarami J, Nasrollahi N (2012) Thermal comfort in hospitals—a literature review. Renew Sustain Energy Rev 16:4071–4077
16. Meldaho M, Beyer P, Hensen J, Siqueira L (2005) The thermal comfort, the indoor environment control, and the energy consumption in three types of operating rooms. In: Building simulation, 9th International IBPSA Conference
17. PHI (2014) Passivhaus Institut, The independent institute for outstanding energy efficiency in buildings. http://passiv.de/en/. Accessed March 2014
18. Skoog J, Fransson N, Jagemar L (2005) Thermal environment in Swedish hospitals summer and winter measurements. Energy Build 37:872–877
19. Teli D, Jentsch MF, James Patrick AB, Bahaj AbuBakr S (2012) Field study on thermal comfort in a UK primary school. Proceedings of 7th Windsor conference: The changing context of comfort in an unpredictable world Cumberland Lodge, Windsor, UK, 12–15 April 2012. London: Network for Comfort and Energy Use in Buildings. http://nceub.org.uk
20. Udrea I, Rotar N, Badescu V (2013) Implementation of the German Passivhaus concept in the Southern Hemisphere. Acta Technica Napocensis: Civil Eng Archit 56(4):15–21

Chapter 71
Towards a Comprehensive Approach to Sustainable Urban Planning: Integrated Estimation of Housing Electricity Consumption and Photovoltaic Generation Potential Using the web-based framework iGUESS®

Alessio Mastrucci, Christian Braun, Olivier Baume, Francesca Stazi and Ulrich Leopold

Abstract Cities play a key role in the mitigation of carbon emission, however there is a need for comprehensive tools able to support sustainable urban planning. The Smart City Energy platform powered by the web-based framework iGUESS® is a geospatial platform for decision support giving access to a series of calculation modules and providing visualization of energy related maps for cities.

This chapter presents an exemplary application of iGUESS ® for the combined assessment of housing electricity consumption and photovoltaic (PV) generation potential at the urban scale. The study provided relevant results for a test case study in Rotterdam city (Netherlands) for the integration of PV and assessment of the energy potential.

Keywords Web-based spatial decision support system · Housing stock · Photovoltaic · Urban scale

71.1 Introduction

In the run of carbon emission mitigation, cities play a major role as they are responsible for about 70 % of the overall primary energy consumption in Europe [5]. Local authorities can significantly contribute to tackle this issue by developing sustainable

A. Mastrucci (✉) · C. Braun · O. Baume · U. Leopold
Luxembourg Institute of Science and Technology (LIST) - Environmental Research and Innovation (ERIN) Department, 41 rue du Brill, L-4422 Belvaux, Luxembourg
e-mail: alessio.mastrucci@list.lu

A. Mastrucci · F. Stazi
Dipartimento di Ingegneria Civile, Edile e Architettura—Università Politecnica delle Marche, Via Brecce Bianche 12, 60131 Ancona, Italy

A. Sayigh (ed.), *Renewable Energy in the Service of Mankind Vol I*,
DOI 10.1007/978-3-319-17777-9_71

urban plans at the city scale. However, there is the need of more comprehensive geospatial data and tools for the development, implementation and monitoring of sustainable planning and policies.

The Smart City Energy platform powered by the web-based framework iGUESS®, developed as a part of the European MUSIC project[1], is a geospatial decision support platform based on an open source software [4]. iGUESS® provides access to the distributed data sources and modelling tools through a single web-based interface. Cities can use this system to create energy-related maps, to identify the opportunities for balancing energy sources and sinks and to assess the effects of energy savings and energy efficiency measures in the city.

This chapter presents an exemplary application of the Smart City Energy platform for the combined assessment of housing electricity consumption and photovoltaic (PV) generation potential for the case study of Rotterdam city (Netherlands). Building stock energy demand modelling (Keirstead et al, 2012; Swan et al, 2009) and integration of PV systems at the urban scale (Amado and Poggi, 2014; Jakubiec and Reinhart, 2013; Theodoridou et al, 2012) have been the focus of many recent studies [1, 6, 15] and is challenging, due to a series of aspects, such as the influence of occupants' behaviour, the matching of demand and production profiles and the lack of measured data for validation [13].

The approach introduced in this study integrates two distinct components: a statistical model to estimate the electricity demand of dwellings at the urban scale using measured electricity consumption and information about buildings and households; a model for the prediction of PV production based on a detailed solar irradiation calculation of roof surfaces at city scale using LiDAR data. The energy matching between demand and production on-site is further analyzed through hourly profiles to determine the energy fraction on-site covered by PV. Results are finally displayed on maps at the urban scale to provide the support for sustainable urban planning.

71.2 Materials and Methods

71.2.1 The iGUESS® framework

iGUESS® is a geospatial decision support system based on an open source software. The tool is organized into a series of components providing the main structure of the user interface [4]:

The *Data Manager* allows to connect and organize the remote spatial datasets, which represent the basic inputs and outputs of calculation modules.

The *Modules* constitute of a series of routines that calculate new spatial datasets from a set of inputs and are run in a dedicated server remotely invoked through

[1] The MUSIC Project. Mitigation of CO_2 emissions in Urban areas: Solutions for Innovative Cities. http://www.themusicproject.eu.

the OGC Web Processing Service (WPS). Results of calculation indicate the potential of a particular planning action (scenario), identifying the areas of the city where its employment is more convenient. At the current state, the models include the estimation at the urban scale of the following aspects: solar irradiation and PV potential; shallow geothermal potential; building stock energy consumption and energy savings potential; urban heat island; green roofs suitability; fuel poverty.

The *Scenarios Tool* allows the exploration of different urban development alternatives starting from a base scenario that characterizes current energy flows in the city. A set of indicators, such as CO2 emissions, can then be used to characterize the scenario outcome.

The *WebGIS* provides a mapping interface to visualize all the geospatial datasets distributed across different sources. Users can compare the outputs of alternate module configurations and can visualize the spatial interaction between alternative planning actions.

The architectural iGUESS® coordinates several distributed services, each dedicated to a specific function: (i) a database management system hosting the iGUESS® database where the resulting datasets and corresponding metadata to remote data services are stored; (ii) a web mapping server providing spatial data over the World Wide Web using the Open Geospatial Consortium (OGC) standards Web Map Service (WMS), Web Feature Service (WFS) and Web Coverage Service (WCS) (iii) a web server driving the user interface of iGUESS®; (iv) a web processing server in which modules are wrapped by services compliant to the Web Processing Service (WPS) standard, enabling to automatically publish outputs to the map server and register them in iGUESS®.

71.2.2 Statistical Model for Electricity Consumption

The model for the prediction of electricity consumption of dwellings at the urban scale is a statistical approach [9] based on Geographical Information Systems (GIS) and implemented as a module into iGUESS®. Measured electricity consumption values available at an aggregated scale (e.g. postcode areas) are disaggregated to single dwellings using multiple linear regression and depending on building and household features. The equation at the basis of the model is Eq. (71.1):

$$y = \beta_0 + x_{floor}\beta_{floor} + x_{occup}\beta_{occup} + \Sigma_i x_{type,i}\beta_{type,i} + \varepsilon \qquad (71.1)$$

where y represents measured yearly average electricity consumption per dwelling at postcode level, x_{floor} the average floor surface of dwellings, x_{occup} the average number of occupants, $x_{type,i}$ the percentage of dwellings per type in the postcode, β_0, β_{floor}, β_{occup}, $\beta_{type,i}$ the regression coefficients and εa random error term. The model is fitted using ordinary least squares technique. The prediction for the Rotterdam housing stock was made on a sample of postcode areas across the whole city providing a good accuracy [9].

Table 71.1 Input data for PV potential calculation in Rotterdam

PV module type	Overall system efficiency estimated	Economic payback assumed	PV production needed for covering costs
	(%)	*(years)*	*(KWh/m²a)*
Monocrystalline	13.6	10	126
Multicrystalline	11.9	10	99

71.2.3 Model for Solar Irradiation and PV Potential Prediction

To predict solar irradiation under real sky conditions, a model on the basis of GRASS GIS has been developed. GRASS GIS contains a model called r.sun, which is based on the work undertaken for development of European Solar Radiation Atlas [10–12].

By using a digital surface model (DSM) and several atmospheric parameters, such as cloud cover, atmospheric turbidity and albedo, the model calculates solar irradiation maps for a given time of the day. It also incorporates shadowing effects of the surrounding topography, for example, cast shadows by neighbouring buildings or even smaller roof structures such as dormers or chimneys. To further enhance the estimation of solar irradiation, the DSM has been used to characterize the rooftops and delineate suitable patches for installations of PV panel according to the roof profile. This means that a net roof surface was calculated which just represents unobstructed roof areas not being disturbed by dormers or chimneys. Furthermore, if a flat rooftop was found by the characterization model, it was assumed that a PV panel with optimal inclination and exposure could be placed. This would lead to a better estimation of the real amount of produced electricity.

The resulting PV potential was then calculated per identified roof patch. With given economic parameters, like envisioned panel life time, panel parameters, (efficiency, costs and maintenance rates) as well as payback tariffs a roof patch can be characterized for economic suitability and the electricity output can be estimated (Table 71.1).

71.2.4 Matching Electricity Demand and PV Production

The matching capability between on-site electricity production and building electricity demand was assessed using two indices according to a relevant literature [3]: the *on-site energy fraction* represents the proportion of the local load covered by the on-site energy production; the *on-site energy matching* is the proportion of on-site energy generation used by the building and system and not dumped or exported.

The output of calculation of electricity demand and PV potential in Smart City Energy platform are currently given at a yearly time resolution. However, a high

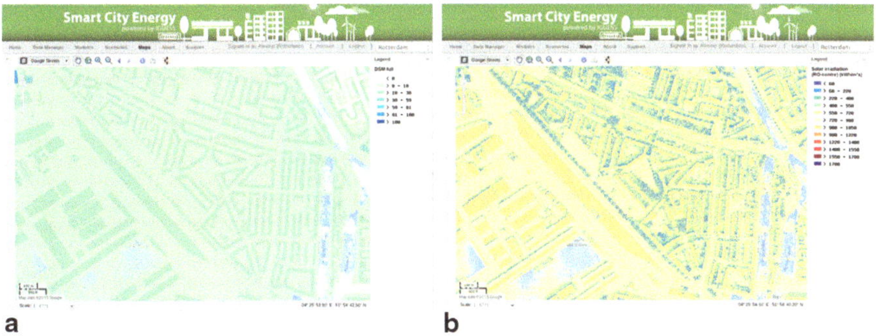

Fig. 71.1 Digital surface model (*DSM*) (**a**) and solar irradiation calculated (**b**) for Bospolder—Rotterdam visualized in the web-based Smart City Energy Platform powered by iGUESS®

time resolution has to be considered to correctly match the on-site electricity demand and PV production in order to avoid significant error.

At present, the measured electricity consumption data are available only at the yearly resolution for Rotterdam. European standard load profiles [8] were adapted to the Rotterdam electricity consumption to disaggregate yearly results at an hourly resolution at the urban scale.

The module for solar irradiation is run at high temporal resolution; however results are now available only as yearly values. An extension of the module is foreseen to provide hourly profiles and further perform the energy matching. In order to preliminarily generate results for a test neighbourhood in Rotterdam city, solar irradiation profiles at the hourly resolution were generated from the weather data [2] and then used to disaggregate yearly values.

71.2.5 Case Study

The district Bospolder situated in the North-West side of Rotterdam City was assumed as case study to test the methodology. The neighbourhood is mainly residential and it was largely built in the early 20th century and after the 1980s. Bospolder accounts for a total of 3120 dwellings mainly of type row house, maisonette, "portiek" flat and "galerij" flat.

The spatial database owned by the municipality of Rotterdam was the main source of information for buildings, including building footprints, building use, year of construction, housing type (detached, semidetached, row house, maisonette, flat "galerij", flat "portiek"), floor surface of dwellings, number of occupants and reference to the postcode area. A database of yearly electricity consumption records aggregated at the postcode area level has been provided by the local energy supplier and constitutes the input of the statistical model.

High resolution LiDAR data were used to generate a DSM for Rotterdam as an input to the solar irradiation model (Fig. 71.1). Additional weather data, such as

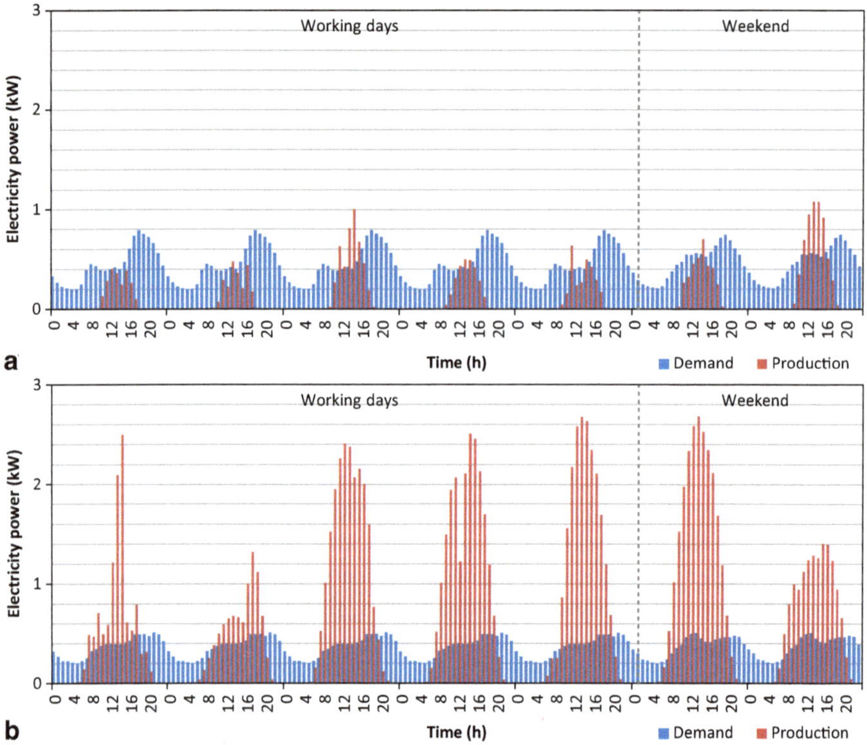

Fig. 71.2 Profiles of electricity demand and production from PV estimated for an exemplary building of Bospolder—Rotterdam for typical weeks in winter (**a**) and in summer (**b**)

cloud cover and ratio of diffuse to global irradiation were made available from a local weather station. Information about the average efficiency and investment cost of different types of PV panels were provided by local experts.

71.3 Results

71.3.1 Results at the Building Level

The profiles of electricity demand and potential PV production were generated for every residential building in the Bospolder area. As an example, Fig. 71.2 shows predicted weekly profiles in winter and summer (hourly time-step) for a two-storeys building with floor surface of 100 m^2. The roof area suitable for PV installation is 25 m^2 south exposed with slope 37° and the PV technology assumed is multicrystalline.

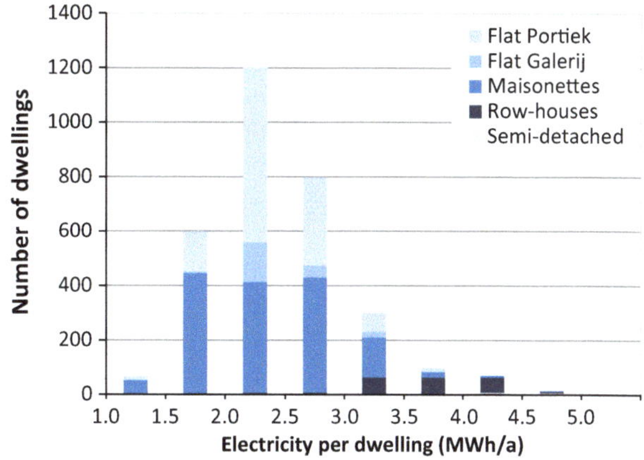

Fig. 71.3 Distribution of electricity consumption predicted for the housing stock of Bospolder—Rotterdam

The amount of electricity demand is higher in winter than in the other seasons, due especially to lower availability of daylight, and the peaks in the morning and in the evening are more accentuated. However, the amount of solar irradiation for this exposure is lower in winter and maximum for the months from April to August. As a consequence, in a typical week in winter most of the PV production is used to cover the local demand (82 %) but only a limited part of the demand matches with PV production (26 %). In summer, a significant part of the electricity demand is covered by the PV system (63 %); however, a large part of the production is fed into the grid (69 %). On a yearly base, the predicted electricity consumption for this building is 3.47 MWh and the PV potential 3.08 MWh. The part of the electricity produced matching the on-site demand is 1.32 MWh, corresponding to an on-site energy matching of 43 % and on-site energy fraction of 38 %.

71.3.2 *Results at the Neighbourhood Level*

The distribution of yearly electricity consumption of households predicted by the statistical model for Bospolder is showed in Fig. 71.3. The typical range of consumption for flats "portiek" and "galerij" is between 1.5–3.0 MWh/a, while for maisonettes the range is wider due to different floor surface of dwellings and number of occupants. Row houses and semidetached houses are more consuming, between 3.0–4.5 MWh/a. The average consumption in Bospolder is estimated as 2.49 MWh/a per dwelling and the total predicted electricity consumption of residential buildings is 7577 MWh/a.

The predicted electricity consumption values of dwellings were aggregated at the building level across the whole neighbourhood and matched with calculated PV potential given by modules of type monocrystalline and multicrystalline. Results for the whole Bospolder neighbourhood are reported in Table 71.2.

Table 71.2 Results of calculation for the whole Bospolder neighbourhood—Rotterdam: Roof area suitable for PV modules, PV production and PV coverage ratio on electricity consumption of residential buildings

PV module type	Suitable Roof area (m^2)	PV production total (MWh/a)	PV production matching on-site demand (MWh/a)	On-site energy matching (%)	On-site energy fraction (%)
Monocrystalline	26,495	3596	1472	41	19
Multicrystalline	44,534	5038	2133	42	28

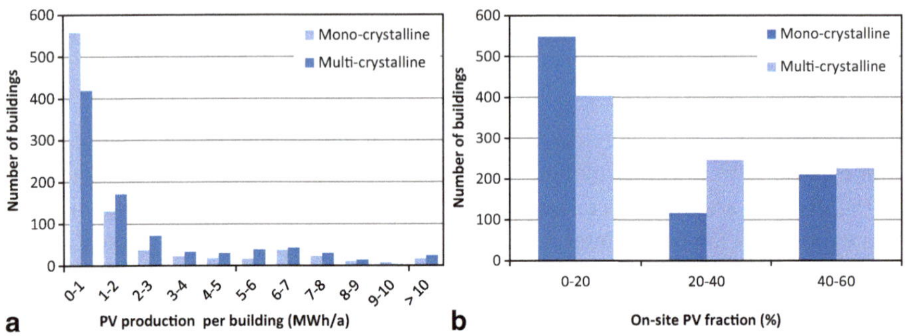

Fig. 71.4 Distribution of PV production matching on-site demand (**a**) and PV on-site energy fraction (**b**) predicted at the building level assuming different type of PV modules for the housing stock of Bospolder—Rotterdam

The rooftop area of residential buildings suitable for installation of PV modules ranges from 26,495 to 44,534 m^2 based on the PV module type assumed and on the economic feasibility of installation. The total PV potential is higher for monocrystalline modules (5038 MWh/a), despite their lower system efficiency, due to the larger suitable rooftop area. The part of PV production matching the energy demand of buildings is also higher for multicrystalline modules (2133 MWh/a) than for monocrystalline (1472 MWh/a), leading to an on-site energy fraction of respectively 28% and 19%. The on-site energy matching is similar for the two technologies (41–42%), meaning that the ratio between energy covering on-site demand and energy fed to the greed is comparable. Figure 71.4a shows the distribution of PV production matching on-site demand assuming different module types. The on-site PV fraction is between 0–20% for a significant part of the building stock and it is generally higher for multicrystalline modules (Fig. 71.4b).

Results were finally visualized as maps after having aggregated them on a regular spatial grid of *50 x 50 m* to prevent from the data protection issues. Figure 71.5 shows the spatial distribution of electricity consumption and PV potential by installation of multicrystalline panels across Bospolder.

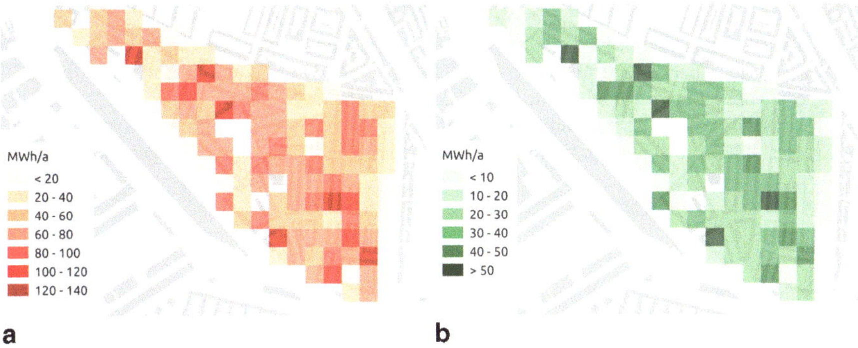

Fig. 71.5 Electricity consumption (**a**) and total PV production matching the on-site demand (**b**) of residential buildings in Bospolder—Rotterdam visualized on a grid 50 x 50 m

71.4 Conclusions

The study presented an exemplary application of the web-based Smart City Energy Platform powered by iGUESS®, to assess the electricity demand and PV potential at the urban scale.

Relevant results were provided for the test case neighbourhood of Bospolder in Rotterdam in order to promote the implementation of local renewable energy sources. The PV production potential in Bospolder is expected to cover 19–28 % of the electricity demand on-site. However, this represents 40 % of the total production, while the remaining part would be fed into the grid in absence of storage systems.

A series of limitations and potential improvements of the approach were identified. In particular, the lack of measured energy consumption data at high time resolution limits the development of demand profiles for the neighbourhood. Therefore, the influence of household type and occupants' behaviour should be further investigated. The selection of rooftops suitable for PV is currently made considering the largest surface available. However, the possible installation of other systems should be considered, for example, solar thermal modules or green roofs. Finally, uncertainty analysis and validation of the model should be carefully addressed.

The study showed the potential of embedding the models into iGUESS®, as it allows to remotely gather data from different sources, run calculations for electricity consumption and PV potential prediction and display results as maps using a unique web-based environment. A further development of the web-based platform will make it possible to combine results obtained in output from the two modules in an automated way and analyze the energy matching of electricity demand and PV production profiles. This approach can be further applied to other European cities provided the input data availability. The study contributes to move a step forward in the development of integrated geospatial modelling tools to support local authorities and stakeholders in sustainable urban planning at the city scale.

Acknowledgments This work has been funded by: the Public Research Centre "Henri Tudor", Luxembourg which was merged into the Luxembourg Institute of Science and Technology (LIST) on the 1st of January 2015; the EU INTERREG IVB NWE Project 165F MUSIC; Università Politecnica delle Marche, Italy by the Postdoc grant of Alessio Mastrucci.

The authors would like to thank the Rotterdam City Council for supporting this study.

Special thanks to Dipl.-Ing. Oliver O'Nagy for the valuable support and discussion for the PV analysis.

References

1. Amado M, Poggi F (2014) Solar urban planning: a parametric approach. Energy Procedia 48:1539–1548
2. ASHRAE—American Society of Heating, Refrigerating and Air-Conditioning Engineers (2001) IWEC- WMO#062400—Europe—original source data (Weather data for Amsterdam)
3. Cao S, Sirén K (2014) Impact of simulation time-resolution on the matching of PV production and household electric demand. Appl Energy 128:192–208
4. De Sousa L, Eykamp C, Leopold U, Baume O, Braun C (2012) iGUESS—a web based system integrating urban energy planning and assessment modelling for multi-scale spatial decision making. Proceedings International Environmental, Modelling and Software Society (iEMSs), Leipzig, Germany
5. EIFER European Institute for Energy Research Eifer 2. 0, 2012. http://www.eifer.uni-karlsruhe.de/IMG/pdf/120417_ eifer_broschuere_ansicht_web.pdf (access date: August 2014)
6. Jakubiec JA, Reinhart CF (2013) A method for predicting city-wide electricity gains from photovoltaic panels based on LiDAR and GIS data combined with hourly Daysim simulations. Sol Energy 93:127–143
7. Keirstead J, Jennings M, Sivakumar A (2012) A review of urban energy system models: approaches, challenges and opportunities. Renew Sustain Energy Rev 16(6):3847–3866
8. Knight I, Kreutzer N, Manning M, Swinton M, Ribberink H (2007) European and Canadian non-HVAC electric and DHW load profiles for use in simulating the performance of residential cogeneration systems. Technical report, Annex 42 of the international energy agency energy conservation in buildings and community systems programme. www.ecbcs.org
9. Mastrucci A, Baume O, Stazi F, Leopold U (2014) Estimating energy savings for the residential building stock of an entire city: a GIS-based statistical downscaling approach applied to Rotterdam. Energy Build 75:358–367
10. Page J, Albuisson M, Wald L (2001) The European solar radiation atlas: a valuable digital tool. Sol Energy 71:81–83
11. Rigollier Ch, Bauer O, Wald L (2000) On the clear sky model of the ESRA—European solar radiation atlas—with respect to the Heliosat method. Solar Energy 68:33–48
12. Scharmer K, Greif J (eds) (2000) The European solar radiation atlas, Vol 2 database and exploitation software. Les Presses de l' École des Mines, Paris
13. Strzalka A, Alam N, Duminil E, Coors V, Eicker U (2012) Large scale integration of photovoltaics in cities. Appl Energy 93:413–421
14. Swan L, Ugursal V (2009) Modeling of end-use energy consumption in the residential sector: a review of Modeling techniques. Renew Sustain Energy Rev 13:1819–1835
15. Theodoridou I, Karteris M, Mallinis G, Papadopoulos AM, Hegger M (2012) Assessment of retrofitting measures and solar systems' potential in urban areas using geographical information systems: application to a Mediterranean city. Renew Sustain Energy Rev 16:6239–6261

Chapter 72
Energy-Efficient Lighting by LED

Helmut F. O. Mueller and Francesco Sasso

Abstract Light emitting diodes (LED) is the light source of the future because of energy efficiency, performance, and life-time. Due to their small dimension and high flux many innovative solutions for non-glare and controlled light distribution are developed for various applications. This chapter deals with a joint R&D project (R&D project, funded 2012–14 by AiF: Entwicklung flacher, durchsichtiger und gleichzeitig selbst-leuchtender Elemente nach dem Prinzip der Lichtleitung mit gezielter Lichtauskopplung durch Mikrostrukturen. Research team: Green Building R&D, JO GmbH & CO. KG., RIF Institut für Forschung und Transfer, Temicon GmbH, TU Dortmund Faculty of Electro- and Information Technology) for edge lit and highly transparent light guiding plates with controlled single-sided light emission. The transparency allows for looking through the plate and is not affected by the specific micro-structure for light emission on one surface of the light guide.

The work plan, including optimization of microstructures by ray-tracing simulation, development of manufacturing technology for large scale production, and system design for different applications, is carried out by an integrated team from university and industry. The main objectives are economic manufacturing technologies, high quality and low energy lighting and architectural system integration.

Keywords LED · Light guide · Micro structure · System design · Applications

72.1 Controlled Light Emission of Transparent Panes by Micro Structures

The functional principle in Fig. 72.1a shows the LED edge lighting of a highly transparent, light guiding plate for example, made of PMMA (polymethylmethacrylate). The light emission by only one surface of the guide is generated by a specific micro

H. F. O. Mueller (✉) · F. Sasso
Green Building R&D, Duesseldorf, Germany
e-mail: helmut.mueller@greenbuilding-rd.com

F. Sasso
e-mail: francesco.sasso@greenbuilding-rd.com

A. Sayigh (ed.), *Renewable Energy in the Service of Mankind Vol I*,
DOI 10.1007/978-3-319-17777-9_72

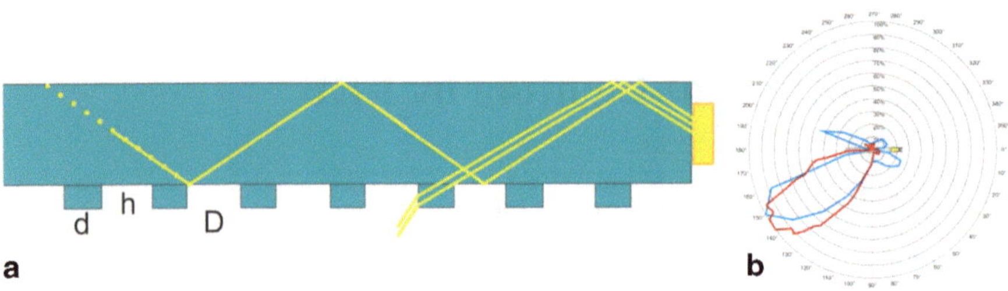

Fig. 72.1 **a** Principle of edge lighting a transparent plate by total internal reflection (*TIR*) and one-sided light emission by a plane cylindrical micro structure. Parameters: cylinder-diameter d, -height h, -axial distance D. **b** Measured (*blue*) und simulated (*red*) light distribution curves, cylindrical structure, diameter d 50 µm, distance D 150 µm. (A. Neyer, TU Dortmund [2])

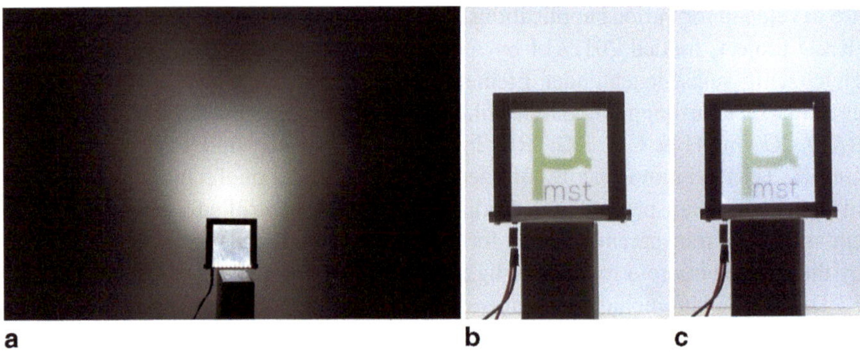

Fig. 72.2 **a** Sample of transparent pane, light emitted away from the eye to the background **b** Sample with background picture, LED switched out, **c** LED switched on. (Neyer, TU Dortmund [2])

structure. Form, geometry and position of the micro structure determine the angles of light distribution and whether the light is emitted by one or two surfaces. The task of this project is a one-sided emission in controlled angles, that is perpendicular or inclined, in order to achieve transparent building components with defined light distribution curves for various applications. Figure 72.1b shows exemplary simulation and measurement results of a specific cylindrical structure given in Fig. 72.1a. The coincidence of calculated and measured values is relatively high and demonstrates, that one-sided emission can be achieved to nearly 100 % [1].

The objective of high transparency is solved by flat micro structures with lateral dimensions near to the resolution limit of the human eye. Simulations showed, that optical structures of cylinders with flat top covers, diameter 50–100 µm and the ratio of pitch (D) to diameter (d) not smaller than 3, are well suited for the task of transparency and one-side light emission [1]. The transparent pane can be looked through from both sides, with one exception: the eye is exposed to light from the emitting surface and the luminance of this surface is higher than that of the background. Figure 72.2 shows the sample of a transparent pane, light emitted away from the eye to the background.

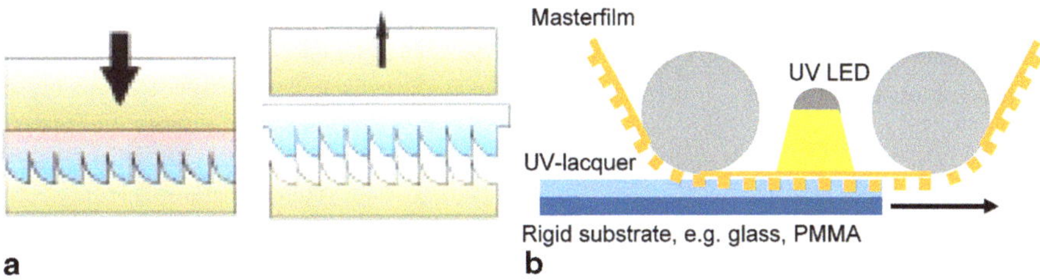

Fig. 72.3 Large-scale manufacturing of micro structures **a** hot embossing **b** UV embossing

72.2 Manufacturing

The manufacturing of light guiding plates (LGP) with optical micro structures for extracting the light is state of the art for TV and monitor backlight units (BLU) [3] or for edge-lit acrylic light-guide plates with interior micro-diffusers [4]. But these manufacturing techniques, for example by burning ceramic colour dots into glass [5], cannot be used for products with one-sided light extraction and distortion free view perpendicular to the plate.

First prototypes of 100 x 100 mm with 50 µm high cylindrical structures were produced by silicon casting (PDMS, polydimethylsiloxane), the master being manufactured by photolithographic technology [1]. For larger structures up to 500 µm forms were produced by cutting in brass.

The transfer to large-scale manufacturing can be realized by two different technologies, hot embossing and UV embossing (Fig. 72.3). Hot embossing uses flat form tools of metal, which are heated, to manufacture structures >500 µm in thermoplastic materials like PMMA. The pulsed plate to plate process is used for flat plates.

UV embossing uses cylindrical tools of metal or synthetic films with metal layers to form a thin lacquer film, which is dried by UV radiation. The substrate of the lacquer can be flat plates or films, thus allowing for a continuous plate to plate or roll to roll production process.

First experiences show, that cylindrical structures are difficult to manufacture by hot embossing in the necessary accuracy of shape and optical surfaces, even with cylinder shafts inclined up to 10°, while cones and pyramids are suited well for this embossing technique.

72.3 Architectural Integration

The thin and transparent light guiding plates are predestined for integration in building elements, like windows, skylights, transparent partitions and balustrades. Other applications are glass show-cases, shop windows or safety glazing for pieces of fine art. Additionally, they can be used as extremely thin and transparent luminaires.

Table 72.1 Areas of application of transparent LED lighting systems

Direction of light emission	Area of application	Reference	Exemplary sketch
Directed towards observer	Transparent component	Window partition balustrade roof, ceiling	
	Room lighting	Ceiling wall task lighting	
	External lighting	Streetlight facade lighting	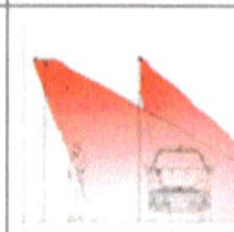
Directed away from observer	Object lighting	Poster, picture show case shop window reading lamp	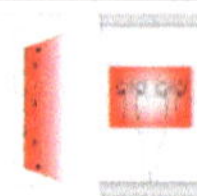
	Room lighting	Indirect lighting (ceiling, wall)	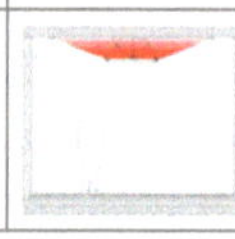

Table 72.1 gives a systematic survey of possible applications, tabulated according to direction of light emission and area of application. The overview demonstrates the wide functional and creative variety of the LED lighting system.

The integration of artificial lighting in daylight openings like windows or glazed areas of partitions allows for similar lighting conditions by day and night, thus increasing the variety of utilizing and furnishing rooms. Last not least, a combination of innovative sunlight redirecting systems [6] and LED lighting can be realized in clerestory windows and skylights in order to increase lighting- and cost efficiency.

72.3.1 Case Studies for Office Lighting

The lighting of typical office rooms with core and window zones was designed in order to compare the advanced LED lighting with standard solutions [7,8]. Figure 72.4 shows the location of transparent LED light guide panels in the upper areas of windows and partitions. The resulting illuminance in task level is shown by false colours in the plan.

For comparison a reference lighting system with fluorescent lamps in suspended luminaires was chosen. It generates a similar mean illuminance of 500 lx as the LED systems. Figure 72.5 shows the significant difference of brightness distribution on room surfaces (luminance) between the two lighting systems. The uniformity of luminance of the LED system is striking in comparison to the reference system with significantly higher contrasts. The transparent light guides, emitting the LED light in an angle of about 40 ° downwards, can be looked through.

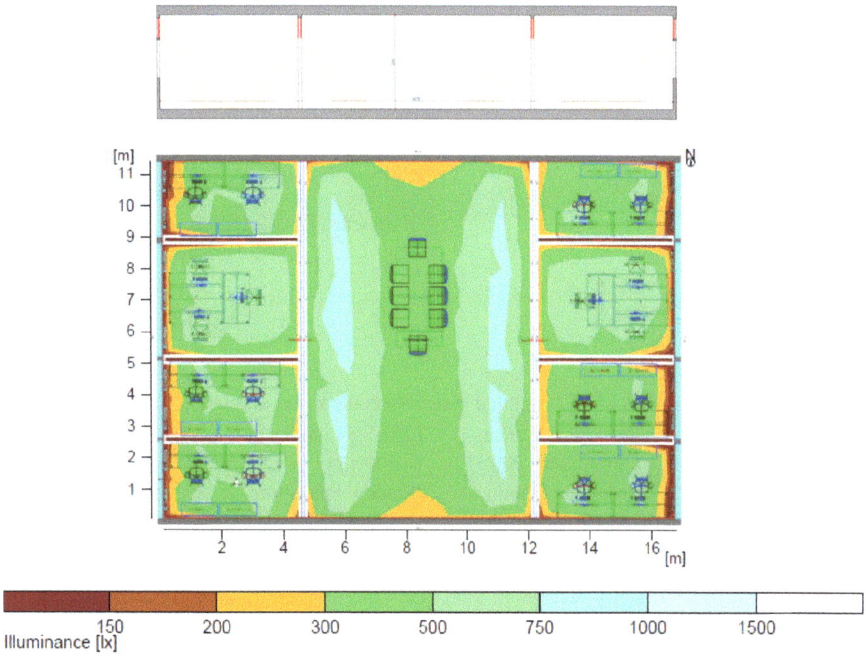

Fig. 72.4 Vertical section with location of transparent LED light guide systems in upper areas of windows and partitions and plan with illuminance in task level. Simulation by RELUX

Fig. 72.5 Comparison of luminance distribution on room surfaces for transparent LED light guides (*left*) and reference system with suspended luminaires and fluorescent lamps (*right*)

An additional case was developed with LED task lights complementing the general room illumination by LED light guides in clerestories (Fig. 72.6). The room has a basic illumination of 300 lx and the task area of 500 lx. This specific lighting system has ergonomic advantages, as direct or indirect glare is avoided, for example by light sources or contrasts on monitors or room surfaces. The highly transparent task luminaire is located directly above the monitor, distributing the light evenly on task area.

Fig. 72.6 Luminance distribution by a combined LED lighting system of task luminaires above monitors and integrated luminaires in upper areas of windows and partitions

Table 72.2 Characteristics of LED lighting system in comparison to reference system

	Reference system (fluorescent lamps)	LED light guides in clearstorys	LED light guides in clearstory + task-light
Description	Suspended lumin. Length 1500 mm	Window + partition L: 1100, H: 75 mm	2. + task lighting L: 500 mm, H: 150 mm
Luminous flux (per luminaire)	6600 Im	4000 Im	4000 Im (core area) 3000 Im (window area) 540 Im (desk area)
Wattage (per luminaire)	79 W	44 W	44 W (core area) 33 W (window area) 7 W (desk area)
Luminous efficacy (per luminaire)	75 Im/W	88 Im/W	88 Im/W (core area) 91 Im/W (window a.) 83 Im/W (desk area)
Maintenance factor (per luminaire)	56%	74%	74%
Mean illuminance	500 lx	478 lx	585 lx (desk area) 422 lx (periphery)
Specific wattage per room	15 W/m^2	12 W/m^2	10 W/m^2

All lighting simulations were carried out by RELUX. In contrast to standard luminaires there still is a demand of specification data for the advanced LED luminaires in the programme library. The most significant data of luminaires and room illumination are given in Table 72.2, comparing the LED lighting system with micro structures and the conventional reference system. Given approximately the same mean illuminance, the uniformity of luminance and the energy efficiency (specific wattage per room) are better for the LED light guide system than for the reference one with fluorescent lamps. The higher energy efficiency is mainly caused by the higher luminous efficacy of the LED and the luminaire (no losses by total internal reflection TIR in light guide), and the higher maintenance factor (long lifetime of LED, low degree of pollution due to enclosure of LED). Consequently, the operating costs for energy and maintenance are reduced.

72.4 Conclusions

The development of micro structured LED lighting systems yields manifold solutions of architectural integration and interior design. Transparent building components can additionally perform as controlled light emitters without glare, making conventional luminaires redundant. Hybrid systems for artificial- and day-lighting provide advantages of multifunctional configuration. The new product development has passed the phases of prototype manufacturing and testing successfully and principals of large-scale manufacturing are resolved. The main advantages of LED lighting by transparent light guides in comparison to luminaires with fluorescent lamps as well as OLEDS are obvious: economy as to investment and operation costs, long lifetime and high quality of room illumination.

References

1. Tengler; Carsten F, Jakubowski M, Neyer A (2013) High transparent light guiding plate for single-sided light emission, poster abstract (Micro and nano engineering) MNE 2013, London
2. Bericht des AiF-Forschungsvorhabens (2012–2014) Entwicklung flacher, durchsichtiger und gleichzeitig selbst-leuchtender Elemente nach dem Prinzip der Lichtleitung mit gezielter Lichtauskopplung durch Mikrostrukturen. RIF, TU Dortmund, 2012–2014
3. Feng D, Yan Y, Yang X, Feng GJD, Ying, Fan SJ (2005) Novel integrated light-guide plates for liquid crystal display backlight. J Opt, Opt. A: Pure Appl Opt 7:111–117
4. PLEXIGLAS® LED for edge lighting (2012). http://www.plexiglas-shop.com/DE/en/details.htm
5. Thomas Emde (2014), http://www.thomasemde.de
6. Klammt S, Müller HFO, Neyer A (2012) Mikrostrukturiertes Tageslichtlenksystem mit hoher Effizienz: Licht 2012, Berlin
7. Sasso F (2014) Design and energy efficiency of microstructured LED lighting systems. Master Thesis, A.B.I.T.A., Universität Florenz
8. Müller H, Sasso F (2014) LED-Beleuchtung mittels transparenter Bauteile. xia intelligente architektur, 87, Alexander Koch GmbH, Leinfelden-Echterdingen, 05–2014

Part VI
Wind Energy

Chapter 73
Power System Performance of Offshore Wind in the UK in 2030

P. Higgins and A. M. Foley

Abstract At least 34% of the UK's power must come from renewable energy sources to meet planned European Union targets in 2030. Wind power will provide the majority of this renewable electricity with an estimated 36 GW offshore and 21 GW onshore. The success of the Crown Estate's leasing rounds 1 and 2 in offshore wind has meant the UK is now one of the world leaders in offshore wind power development. Leasing round 3 will see offshore wind in the UK surpass 36 GW of installed capacity. This is a significant increase from the current installed offshore wind capacity of 3.6 GW. This research investigates the power system performance of offshore wind power in the UK in 2030.

Keywords Dispatch · Unit commitment · Offshore wind power · Wind curtailment · Wind constrained

73.1 Introduction

The United Kingdom (UK) has identified wind power as the main source of renewable energy to achieve the European Union's (EU) 2020 renewable energy targets. In 2012, 11.3% of electricity generated in the UK was produced from renewable generators [1] and is expected to increase to over 50% in 2030 [2]. The UK government has acknowledged that offshore wind is the key renewable energy source and has put plans in place to develop over 48 GW of offshore wind via the Crown Estates leasing round 3 [3]. Considering the recent government announcement to increase offshore wind power subsidies from £ 135 to £ 140/MWh until 2019 and the Crown Estate's leasing round 3, offshore wind power is likely to become a significant source of electricity beyond 2020 [4].

P. Higgins (✉) · A. M. Foley
School of Mechanical and Aerospace Engineering, Queen's University Belfast, Belfast BT9 5AH, UK
e-mail: phiggins14@qub.ac.uk

A. M. Foley
e-mail: a.foley@qub.ac.uk

A. Sayigh (ed.), *Renewable Energy in the Service of Mankind Vol I*,
DOI 10.1007/978-3-319-17777-9_73

National Grid [2] showed that 57 GW of wind generation is required by 2030 and in the 'Future Scenarios' study 36 GW comes from offshore wind. This is a sizeable increase from the 3.6 GW currently in operation. A tenfold increase in offshore wind power will have a significant impact on the operation and management of the electricity market in the UK. The variability of wind power and the response of conventional generation at short notice are crucial to system balancing [5]. Understanding the interaction of both will identify the high levels of renewable generation that the grid can incorporate in the future. Swinand et al. [6] showed that wind generation in the British Electricity Trading and Transmission Arrangement (BETTA) has substantial impacts on balancing costs. The flexibility of power systems with large wind energy penetrations has been shown to be beneficial to reacting to wind curtailment [7]. The economic benefit from wind energy is reliant on reducing wind curtailment and wind constraints. In 2012, the amount of wind curtailed in the UK was 206 GWh which had an estimated value of £ 7 million [8].

Edmund et al. [9] investigated the optimum amount of wind power the British electricity system can accommodate and the impact additional storage and interconnector capacity has on the wind. This study concluded that increased interconnection reduces the amount of unused wind and decreases carbon dioxide (CO_2) production. De Decker et al. [10] also demonstrated that interconnection to other EU countries is crucial to the deployment of offshore wind power. Denny et al. [11] examined the single electricity market (SEM) of the Republic of Ireland and Northern Ireland interconnected to a simplified BETTA test system with large quantities of wind power. The work showed in the event of a transmission constraint between Scotland and England, and then the SEM would act as a transmission corridor. The proposed development of large amounts of renewable energy, predominately onshore and offshore wind, in Scotland has the potential to create further grid constraints in the already congested interconnectors between Scotland and England. The lack of interconnection between the BETTA and Europe was one of the main limitations of the analysis. The BETTA is heavily interconnected with Europe and therefore an accurate representation of the BETTA requires EU interconnection. Deane et al. [12] developed a model of North West Europe and identified congestion on the interconnector between the BETTA and France as a problem. Curtis et al. [13] analysed the impact different electricity policies in the UK and Ireland could have on the operation of the SEM and BETTA. A higher price of carbon in the BETTA will result in carbon leakage from the SEM into the BETTA. Curtis et al. [13] described 'carbon leakage' as 'a gross outflow usually arising when competitive advantage is lost compared with foreign counties and production occurs in the foreign countries satisfying domestic demand'.

The aim of this study is to investigate the performance of the UK power system in 2030 with a large penetration of offshore wind. The impact that the location offshore wind can have on the power system is also analysed, as well as the difference in carbon prices between the EU and the BETTA. This investigation is performed by studying the impact of large offshore wind power penetrations in an unconstrained unit commitment (UUC) test system of the BETTA. The chapter is divided into four sections, including the introduction. Section 73.2 discusses the methodology and test system used in this research. Section 73.3 presents the results and analysis. Section 73.4 provides a brief discussion and conclusion.

73.2 Methodology and Test System

For the purpose of this research energy exemplar's PLEXOS for power systems version 6.208R03 was used [14]. The optimiser used was the Xpress solver [15]. The parameters for the Xpress solver were set at a relative gap of 0.1 % with mixed integer linear programming. The aim of the test system is to determine a dispatch schedule at the lowest cost conditional to a number of constraints. The test system minimises the objective function shown in Eq. (73.1) [16].

$$\min \sum_{t \in T} c_{jt}.U_{jt} + n_{jt}.V_{jt} + .m_{jt}.P_{jt} + vl.use_t + vl.usr_t \tag{73.1}$$

where t indexes time periods in chronological order $t = 1, \ldots, T$, c_{jt} is the start cost of unit j in period t, U_{jt} is a binary quantity representing if unit j has started the period before t, n_{jt} is the no load cost of unit j, $V_j t$ is a binary quantity representing the generating status of unit j, m_{jt} is the production cost of unit j, P_{jt} is the power output of unit j, vl is the penalty for loss load, use_t is the unserved energy of unit j and usr_t is the reserve energy not met by unit j. The objective function is subjected to constraints such as maximum generation, minimum stable generation, minimum up and down times and ramp rates.

In the test system generating/transmission parameters such as maximum and minimum ramp up/down time, minimum stable level, start-up cost, forced outage rate, wind input data, operational and maintenance costs and fuel costs are inputted for the power system to minimise the objective function. The test system for this research is optimised for the year 2030 at hour intervals. The test system is an UUC model of the BETTA as a wholesale electricity market. The UUC test system will have a perfect forecast with a 24 h look ahead. This means that the system operator will know, 24 h in advance of the trading period, the amount of wind available and predicted demand. The wholesale electricity market will be a mandatory pool with generators bidding in their short run marginal cost[1] (SRMC). The generators are organised by their SRMC and the cheapest are scheduled for dispatching. The system marginal price[2] (SMP) is determined by the scheduled generator with the most expensive SRMC. Operating reserves and network security constraints are not included in this UUC test system. Further work will develop a security constraint unit commitment model that will include reserves and network constraints.

The 2030 BETTA test system has interconnector capacity to simplified electricity markets of France, Belgium, the Netherlands and the SEM, as shown in Fig. 73.1.

[1] SRMC is the variable cost associated with providing one more MW of generation.

[2] SMP is the price at which 1 MWh of electricity is sold in any given trading period [17].

Fig. 73.1 Test system

Table 73.1 Interconnector capacity

Interconnector	Maximum capacity (MW)
East–West (SEM—BETTA)	1500
Moyle (SEM—BETTA)	400
FR–BETTA (France—BETTA)	2000
BEL–BETTA (Belgium—BETTA)	1000
NL–BETTA (Netherlands—BETTA)	1000

SEM single electricity market, *BETTA* British Electricity Trading and Transmission Arrangement

The French market is modelled as gas and nuclear generators. The Belgium and Netherlands markets are modelled with gas and coal generators. The SEM is modelled as a coal, gas, offshore wind and onshore wind generators.

The maximum capacity and ramp rates for each interconnector with the BETTA are shown in Table 73.1 [18–20]. In 2014, the capacity on the Scotland England interconnector (Scot–Eng line) was 3100 MW [21].

Table 73.2 Installed capacity per fuel type

Generator	Installed capacity (MW)
Gas (CCGT)	39,347
Gas (OCGT)	529
CHP	2141
Coal	6015
Oil	445
Nuclear	13,072
Hydro	1122
Pumped storage	2744
Offshore wind	39,128
Onshore wind	12,784
Renewables	8328
Biomass	5434
Total	*131,089*

CCGT combined cycle gas turbine, *OCGT* open cycle gas turbine

The thermal generation units for the 2030 test system are based on data from a number of sources [18, 22, 23]. The total generation capacity per fuel type is shown in Table 73.2. The plant efficiencies, ramp rates, heat rates, minimum on and off times, and minimum stable levels were included [24, 25]. Onshore and offshore wind was separated in zones as defined by national grid's 7 year statement study zones [22]. The capacity factor[3] for each wind zone was determined from ref [26]. The national grid wind generation time series for 2011 was scaled accordingly to achieve the 2030 targets [22], as seen in Table 73.2. This time series data was used as BETTA onshore wind. The capacity factor of each zone is different therefore the wind data files were adjusted accordingly using the same procedures as applied in [27]. Foley et al. [28] showed that the BETTA onshore wind data time lagged the SEM onshore wind data by 2 h. The SEM offshore wind data proceeds SEM onshore by 1 h. Offshore wind in the BETTA is time adjusted by an hour depending on whether the zone is to the east or west of Great Britain. Any zone with offshore wind to the east time lags BETTA onshore wind by an hour and any zone to the west proceeds BETTA onshore wind by an hour.

Datasets for 2030 forecasted fuel and carbon prices were obtained from [29]. Table 73.3 illustrates the fuel costs applied to the test system. The high fuel costs in the SEM are due to fuel transportation costs.

This research applied three different scenarios that might impact offshore wind generation. The first scenario examines the line capacity between Scotland and England and the location of the offshore wind capacity in the BETTA. The current line capacity between Scotland and England is 3100 MW. Scenario one increases the line capacity from 3400 MW to 6400 MW and examines the impact the line capacity has on the SMP and dispatch down of wind power in the BETTA. The dispatch down of wind power is unused wind power due to curtailment and

[3] Capacity factor is the ratio of a power plant's generation to available capacity.

Table 73.3 Fuel costs

Fuel type	Price (£/GJ)
BETTA and EU coal	3.62
BETTA and EU gas	9.23
BETTA and EU nuclear	0.5
BETTA oil	18.55
BETTA biogas	3.7
BETTA biomass	4.72
SEM coal	3.98
SEM gas	9.38

BETTA British Electricity Trading and Transmission Arrangement, *SEM* single electricity market, *EU* European Union

Table 73.4 Offshore wind capacity scenarios

Scotland (GW)	England (GW)	BETTA (GW)
4	35	39
6.5	32.5	39
9	30	39
11.5	27.5	39
14	25	39

constraint effects. Wind that is 'curtailed' refers to unused wind generation due to system services such system stability requirements, operating reserve requirements, a system nonsynchronous penetration limit and periods of very high wind generation but low demand levels. Wind that is 'constrained' is due to transmission network limitations such as overloading of transmission lines. The location of the offshore wind capacity is examined in correlation to the line capacity. The installed capacities of offshore wind in Scotland and England are varied while the BETTA offshore wind capacity remains constant at 39 GW, as shown in Table 73.4. The first scenario will show how much wind is curtailed or constrained due to the line capacity as well as the most beneficial location for deploying offshore wind capacity.

In 2030, the BETTA could have a significant amount of interconnection to the EU market and the policy selections the UK government takes on carbon pricing could impact on the operation of the BETTA. The second scenario investigates the impact the EU price of carbon has on offshore wind generation, SMP and thermal generation in the BETTA market. The BETTA carbon price will remain constant at £ 90/tCO_2 and two different EU carbon prices will be analysed £ 90/tCO_2 and 70/tCO_2.

The third scenario analyses the impact the location, an extra 5 GW of offshore wind, would have on the BETTA. An extra 5 GW of offshore wind capacity will be added to Scotland resulting in a 14 GW in Scotland and 30 GW in England. The results from this setup will be compared to a test system with 9 GW of offshore wind capacity in Scotland and 35 GW in England. The impact nuclear generation

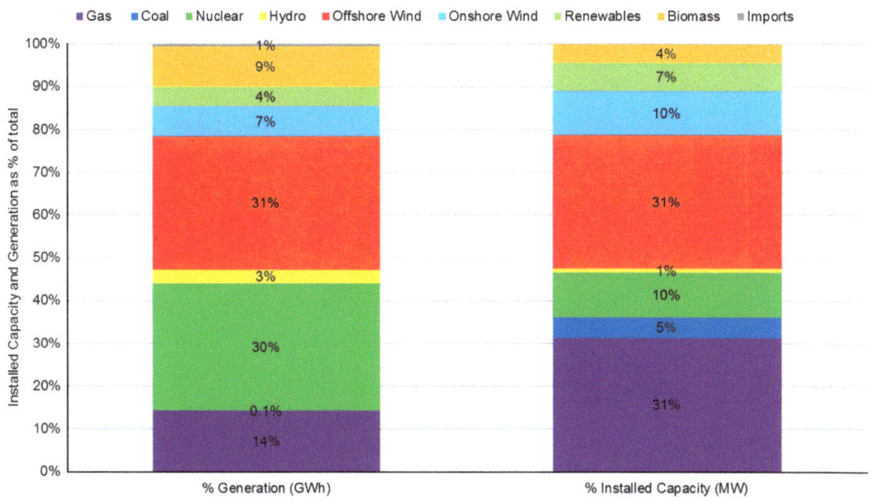

Fig. 73.2 Generation and installed capacity

has on the generation portfolio is also examined. The BETTA with 5 GW of nuclear generation capacity rather than 14 GW is investigated.

73.3 Preliminary Results and Analysis

Figure 73.2 illustrates the percentage of installed capacity and generation of each fuel type. Offshore wind and gas power plants provide the majority of the installed generation capacity with 30 % each with nuclear and onshore wind providing 10 % each. As expected the nuclear generating units have an average capacity factor of 91 %. The onshore and offshore wind should have average capacity factors of 27.2 and 30.9 %, however due to the dispatch down of wind power the average onshore and offshore wind capacity factors are 19 and 27 % respectively.

Gas can be broken into base load, mid-merit and peaking plants to determine the capacity factors. For this study the base load, mid-merit and peaking plants are determined by the number of hours operating, that is base load is greater than 6221 h/year, mid-merit operate from 911 to 6220 h/year and peaking plants operate less than 910 h a year. The average capacity factor of the mid-merit gas units in 2030 is 26 %. This is considerably smaller than the 60 % capacity factor of the mid-merit gas units of 2014 [30]. However, this is to be expected due to the large amount of wind generation in the BETTA, in 2030. The standard deviation of the generation for each fuel type was calculated and offshore wind and gas generation have the highest standard deviation of all the generating units. From these standard deviation results it is possible to determine that the gas units are required to accommodate the large variations in offshore wind generation because of their ramping capabilities.

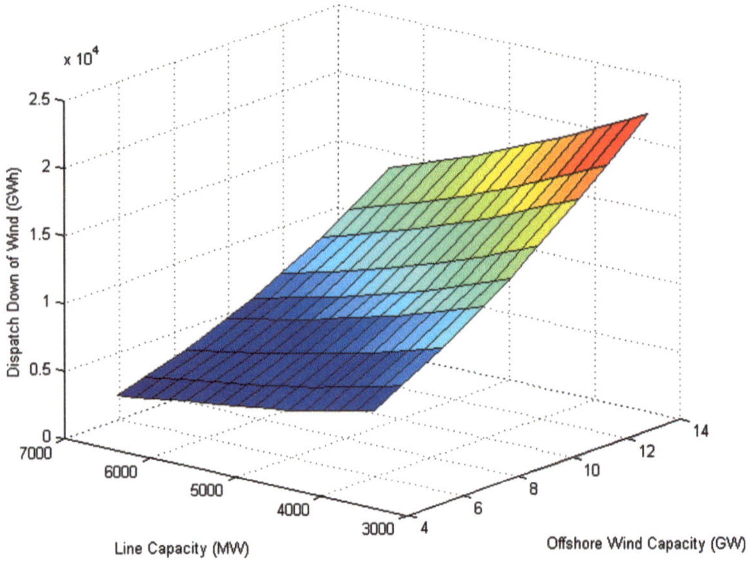

Fig. 73.3 Impact of the Scot–Eng line capacity and offshore wind capacity in Scotland on the dispatch down of wind power

73.3.1 Locational Effects of Offshore Wind Capacity

The deployment of wind generation on the BETTA is reliant on the ability of the power system to transport the electricity to areas of demand. In Scotland, the deployment of offshore wind generation is restricted by the demand in Scotland and the ability to export the electricity to the SEM (Moyle interconnector) and to the rest of the BETTA (Scot–Eng line).

As expected an increase in the Scot–Eng line capacity results in a reduction of dispatched downwind for all the different offshore wind location scenarios. An increase in the line capacity from 3400 to 6400 MW results in a reduction of wind dispatched down of 43 and 37 % for the 4 GW and 14 GW scenarios respectively. Figure 73.3 illustrates the impact the max capacity of the transmission line and the installed offshore wind capacity in Scotland can have on the dispatch down of wind power in the BETTA. The scenario with a line capacity of 3400 MW and an installed offshore wind capacity of 4 GW has 6423 GWh of dispatched down wind. This is a significant increase from 2012 levels of 206 GWh. In 2012, the percentage of wind dispatched down to the total amount of wind generated was 1 % whereas in the UUC 2030 test system it has risen to 4.95 %. This figure would be expected to increase if the reserves and network constraints were included in the test system. A high percentage of dispatched downwind represents lost benefits from wind energy. The breakdown of the 6423 GWh of dispatched down wind is 2565 GWh (40 %) from wind constrained and 3857 GWh (60 %) from wind curtailed.

Increasing the offshore wind capacity in Scotland to 14 GW and keeping the line capacity at 3400 MW results in dispatch down of wind power increasing to

Fig. 73.4 Dispatch down of wind power in Scotland

21,969 GWh with 18,052 GWh (82 %) from wind constraint and 3916 GWh (18 %) from wind curtailment. The increase in offshore wind in Scotland results in increased wind constrained. The wind that is being constrained is due to the line capacity of the Scot–Eng line. For approximately 48 % (4193 h) of the year the flow in the line is at its max capacity of 3400 MW. An increase in the Scot–Eng line to 6400 MW capacity compared to 3400 MW reduces the dispatch down of wind power from 15 to 9 % with the majority coming from constrained wind, 18,052 to 9103 GWh. The flow in the line reaches max capacity for only 28 % (2433 h) of the year when the line capacity is increased to 6400 MW. The Scot–Eng line capacity has the greatest impact on wind constrained.

The reason for the relatively little change in wind curtailment for increases in line capacity is due to the periods when curtailment occur. At certain time periods throughout the day, the line capacity is not reached but wind is dispatched down. At 05.00 on the 08/08/2030, the total available Scottish onshore and offshore wind generation and load demand are 9314 and 2245 MW respectively. The onshore and offshore wind is dispatched down by 5371 to 3943 MW. The flow on the Scot–Eng line is 2126 MW even though the available capacity is 6400 MW, as shown in Fig. 73.4. At this period the SEM has a low load demand which is being met by wind generation in the SEM, imports from Scotland through the Moyle interconnector and 509 MW through the East West interconnector. There is available capacity

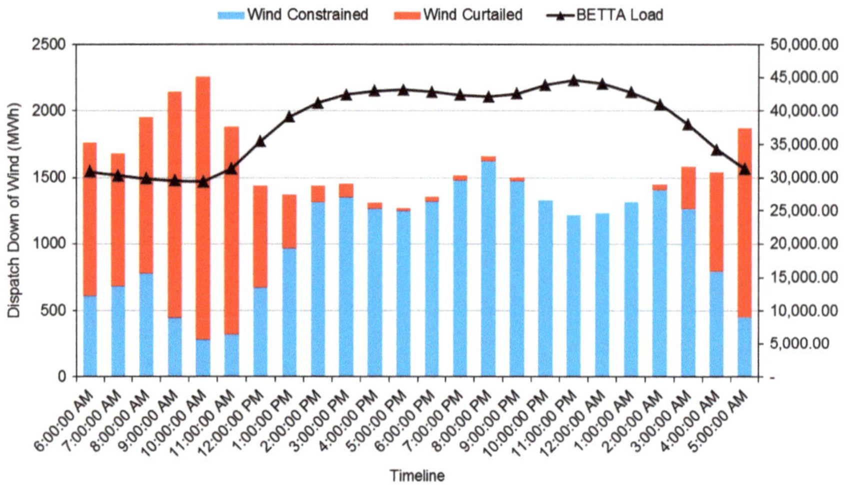

Fig. 73.5 2030 daily average dispatch down of wind power

on the East West interconnector; however the SEM load demand is not high enough to require more. The French nuclear power plants are operating at their minimum stable level which is generating approximately 69 MW higher than the French load. In addition, the price of electricity is cheaper in France than the BETTA and hence the small import flow of 69 MW from France to the BETTA. Both interconnectors to Belgium and the Netherlands are operating at full capacity. In England, the load demand is being met by onshore wind, offshore wind and nuclear generation. As a result of the low load demand in the BETTA and SEM the offshore wind generation in Scotland has to be curtailed.

Figure 73.5 illustrates the profile and breakdown of the daily average dispatch down of wind power for the scenario with 14 GW of offshore wind in Scotland and a Scot–Eng line capacity of 6400 MW. As expected, the dispatch down is lowest during periods of high demand (12.00 p.m. to 12.00 a.m.) and at its highest during low demand periods (12.00 a.m. to 12.00 p.m.). Wind constraint is the result of the majority of dispatch down during the high demand periods whereas wind curtailment is the largest component of the dispatch down during the low demand periods.

The impact from Scot–Eng line capacity and offshore wind capacity in Scotland on the SMP can be seen in Fig. 73.6. An increase of offshore wind capacity from 4 to 14 GW results in an average SMP increase of 6.1 % compared to an average decrease of 2.9 % due to a change of Scot–Eng line capacity from 3400 to 6400 MW. As discussed in the methodology section, the increase in offshore wind capacity in Scotland does not alter the total offshore wind capacity in the BETTA. The different offshore wind scenarios are the positioning of offshore wind capacity in Scotland and the rest of the BETTA. The dispatch down of wind power increases as more offshore wind is located in Scotland, which means less wind generation being sold onto the wholesale pool market. As a result, more imports and fewer exports accompanied with nuclear generation are required to meet the change in offshore

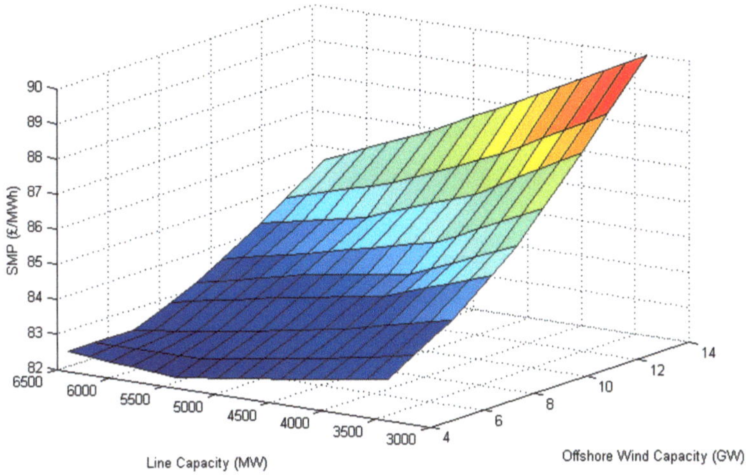

Fig. 73.6 Impact of Scot–Eng line capacity and offshore wind capacity in Scotland on the SMP

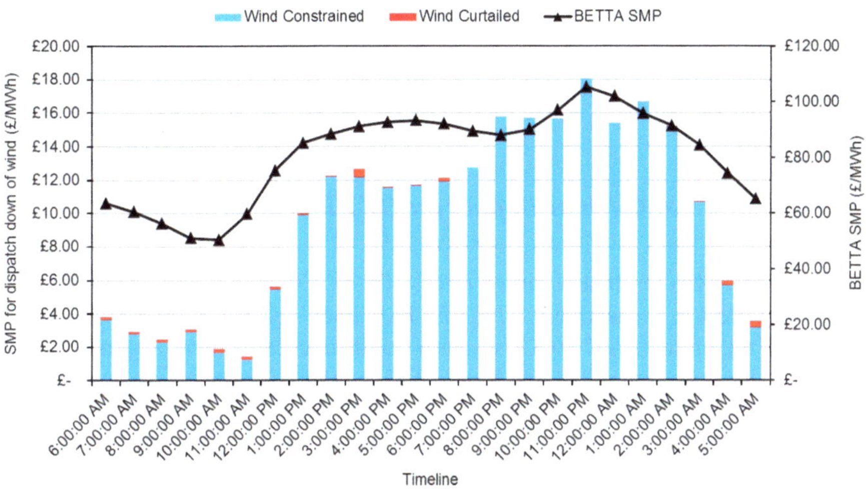

Fig. 73.7 2030 daily average SMP for periods of dispatched down wind power

wind generation. It is these changes that result in the SMP increasing as shown in Fig. 73.6.

The daily average SMP for the periods when wind is being dispatched down is shown in Fig. 73.7. The SMP is at its lowest during a.m. periods and at its highest during the p.m. periods as expected. Figure 73.7 shows on average at 11.00 a.m. the SMP is £ 60/MWh, however for days when dispatch down of wind power occurs at 11.00 a.m. the average SMP is below £ 2/MWh. It is illustrated that the SMP is negligible for periods when wind curtailment occurs. Whenever wind constrained occurs the SMP is significantly larger. Figure 73.5 highlights the periods from 12.00 a.m. to 12.00 p.m. as the highest for the dispatched down of wind. Since the

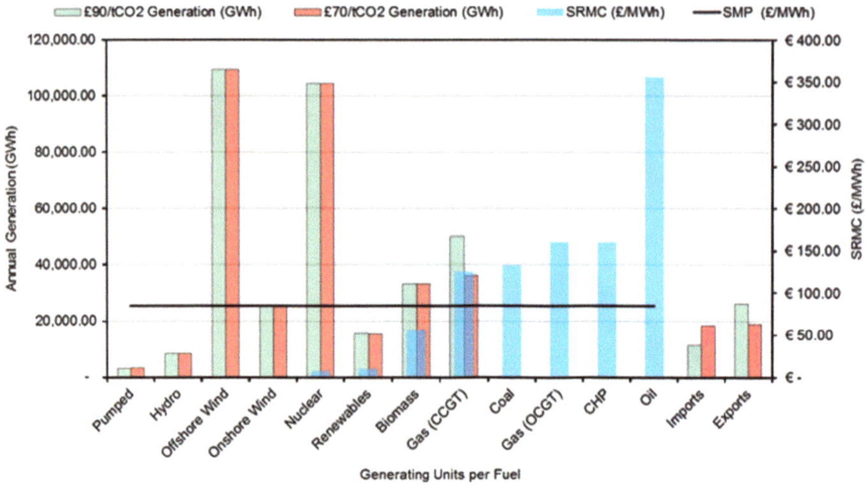

Fig. 73.8 EU price of carbon impact on BETTA generation

wind curtailment occurs at periods where the SMP is nearly zero the majority of lost savings to the BETTA wholesale market from cheaper wind electricity are due to periods where wind is constrained.

73.3.2 Price of Carbon Impact on BETTA Generation

The BETTA SMPs for the EU carbon prices of £ 90/tCO_2 and £ 70/tCO_2 are £ 84.49 and £82.45/MWh, respectively. The BETTA carbon price is £ 90/tCO_2 for both scenarios, therefore the £ 70/tCO_2 scenario results in cheaper thermal generation in the SEM, France, Netherland and Belgium. The main difference to the generation portfolio, in terms of GWh, when the EU carbon price decreases is generation from the gas CCGT units, as shown in Fig. 73.8. The decrease in gas CCGT generation of 13,774 GWh is offset by exports reducing by 7270 GWh and imports increasing by 6893 GWh.

As a result of the BETTA being modelled as a wholesale pool market and where the generators are scheduled on their SRMC, the cheaper renewables, nuclear and biomass units are scheduled ahead of the gas generating units. In 2012, approximately 30 % [1] of the electricity was generated from gas generators, however in 2030 it could fall to as low as 14 %, as shown in Fig. 73.2. The difference in carbon pricing policy was shown to have a negative impact on gas unit's generation. This decrease in generation results in a significant decrease in pool revenue[4] for the gas generators. The pool revenue for the gas generators when the EU carbon price is the same as the carbon price in the BETTA is £ 5.4 billion. It reduces by 27 % to £ 3.9 billion when the EU carbon price is £ 70/tCO_2. Gas generators are crucial to

[4] Pool revenue is the accumulation of SMP by generation for each generator for every period of the year.

Table 73.5 Location impact of an extra 5 GW of offshore wind

	Base (39 GW)	5 GW extra in Scotland	5 GW extra in England
Onshore and offshore wind generation (GWh)	134,336	139,757	146,232
Dispatch down of wind power (GWh)	11,082	19,604	12,267
Emission production (Tonne)	17,918,604	16,830,518	16,888,883
SMP (£/MWh)	£ 84.49	£ 83.33	£ 80.77

SMP system marginal price

supply the necessary reserves and system stability requirements. Therefore such financial impacts from different carbon pricing could prove difficult to attractive investment into gas generators in the BETTA.

73.3.3 Potential Changes to Generation Portfolio

The benefit of an extra 5 GW of offshore wind on top of the 39 GW depends on the location of the deployment, either in Scotland or England. Table 73.5 highlights the impact the deployment location of 5 GW of offshore wind can have on the onshore and offshore wind generation, dispatch down of wind, emission production and SMP in the BETTA. An extra 5 GW of offshore wind in Scotland will increase the total wind generation and dispatch down of wind power in the BETTA by 5421 and 8522 GWh respectively. Whereas, 5 GW of offshore wind in England increases the total wind generation and dispatch down of wind power by 11,896 and 1185 GWh respectively. The transmission line between Scotland and England (4400 MW) causes the dispatch down of wind power to increase as the line capacity is not large enough to accommodate the extra offshore wind generation. The majority of extra offshore wind generation in Scotland results in being dispatched down rather than sold to the market. For both scenarios, the SMP drops as more wind energy on the system replaces more expensive gas and nuclear generation. The deployment of extra offshore wind in England rather than Scotland results in greater reductions in SMP and emissions production.

The installed capacity of nuclear generation has a significant impact on the generation portfolio. The cheap price of electricity from nuclear power plants drives down the wholesale price of electricity and generation from other thermal generating units such as gas closed-cycle gas turbine(CCGTs) and coal. A reduction in nuclear generation capacity from 14 to 5 GW results in mid-merit gas generators increasing their capacity factors from 26 to 41 %. With increased capacity factors the gas generators also experience a 28 % rise in the number of start-ups throughout the year. The increased number of gas generation start-ups has resulted in the uplift component of the SMP increases from £ 1.79 to £ 4.49/MWh. The increase in uplift accompanied with an increase in shadow prices due to more expensive gas generation replacing cheaper nuclear generation results in the yearly average

SMP increasing from £ 84.49 to £ 104.15/MWh. The main benefit to decreasing the nuclear generation is the reduction in dispatch down of wind from 7.62 to 6.94 % of wind generated. The reduction of dispatch down of wind is critical to ensuring that the maximum benefit from onshore and offshore wind generation is achieved. However, this has to be achieved without a significant cost to the consumer.

73.4 Discussion and Conclusion

This chapter presented the preliminary findings of a study on the impact of a large penetration of offshore wind power on the UK power system in 2030. The initial finding indicates that a BETTA test system with 4400 MW of interconnector capacity can accommodate 52 GW of onshore and offshore wind capacity producing 145 TWh of electricity with only 6.35 % of dispatched down wind. An important factor in reducing the dispatch down of wind power is the line capacity of the interconnection between Scotland and England. The different location scenarios for offshore wind in the BETTA have an average reduction of 41.5 % of dispatch down of wind power due to an increase in line capacity from 3400 to 6400 MW. The location of offshore wind in the BETTA is critical to ensuring minimum dispatch down of wind power, SMP and emission production. From a customer's perspective the deployment of an extra 5 GW of offshore wind in England is more beneficial than in Scotland. However, from the developers' point of view, 5 GW of offshore wind in Scotland provides a greater return on investment [31].

The highest peaks of dispatch down of wind power occur during the night with periods of load demand (12.00 a.m.–12.00 p.m.). The majority of the dispatch down during the night period is wind curtailment. Since the test system does not include system and network services the wind curtailment is due to low load demand and excess wind generation. A smoother load curve with fewer reductions during the night could reduce the amount of wind curtailed and reduce emissions production. The greatest financial benefit to reducing the dispatch down of wind is in reducing wind constrained not wind curtailed. The results show that wind curtailment occurs at periods where the SMP is small whereas wind constrained occurs at periods with higher SMPs. It is for this reason that an increase in the Scot–Eng line capacity results in a 3 % decrease in the SMP.

Gas generation from CCGTs and open cycle gas turbines (OCGTs) have become the primary generators for balancing the significant installed capacity (30 %) of offshore wind in the BETTA. As a result their capacity factors have been dramatically reduced from 2014 levels. This significant reduction in capacity factor for the gas units could create financial difficulty for the owners. The proposed introduction of a 'target capacity measure' that will provide payments for generators that ensure the security of supply through reserves will be crucial for the profitability of gas units in 2030 [32].

The EU price of carbon has the greatest impact on gas units in the BETTA. An EU carbon price change from £ 90/tCO_2 to £ 70/tCO_2 results in a decrease in gas

CCGT generation of 13,774 GWh. This change in gas generation reduces the pool revenue gas generators receive by 27 %, down from £ 5.4 to £ 3.9 billion. Therefore, the impact future carbon pricing can have on the gas generators profitability and generation is significant. Considering the long project life of gas power plants the future carbon pricing policies need to be determined as soon as possible as future carbon pricing policies could impact on the financial return of gas generators in the BETTA in 2030.

Further research involves developing a reserve constrained unit commitment (RCUC) test system of the BETTA in 2030 and combining it with a RCUC of the SEM. A more accurate analysis of offshore wind on a combined BETTA and SEM market in 2030 will be performed.

Acknowledgments The authors acknowledge the support of Energy Exemplar through the academic license for PLEXOS, the Engineering and Physical Sciences Research Council for funding support and Dr. Paul Deane from University College Cork for some initial BETTA generator data.

References

1. Department of Energy and Climate Change DUKES (2013) Chap. 5: electricity. https://www.gov.uk/government/publications/electricity-chapter-5-digest-of-united-kingdom-energy-statistics-dukes. Accessed 6 Jan 2014
2. National Grid (2013) UK future energy scenarios. http://www2.nationalgrid.com/UK/Industry-information/Future-of-Energy/Future-Energy-Scenarios/. Accessed 03/02, 2014.
3. Crown Estate (2014) Offshore wind energy. http://www.thecrownestate.co.uk/energy-and-infrastructure/offshore-wind-energy/. Accessed 03 2014
4. Higgins P, Foley A (2014) The evolution of offshore wind power in the United Kingdom. Renew Sustain Energy Rev 37:599–612
5. Kubik ML, Coker PJ, Barlow JF, Hunt C (2012) A study into the accuracy of using meteorological wind data to estimate turbine generation output. Renew Energy 51:153–158
6. Swinand GP, Godel M (2012) Estimating the impact of wind generation on balancing costs in the GB electricity markets. In: Proceedings of the 9th international conference on the European energy market (EEM), IEEE, 10 May 2012
7. Finn P, Fitzpatrick C, Connolly D, Leahy M, Relihan L (2011) Facilitation of renewable electricity using price based appliance control in Ireland's electricity market. Energy 36:2952–2960
8. Wind Power Monthly (2013) UK wind curtailmetns fall dramatically. http://www.windpowermonthly.com/article/1185979/uk-wind-curtailments-fall-dramatically. Accessed 6 Dec 2013
9. Edmunds RK, Cockerill T, Foxon TJ, Ingham DB, Pourkashanian M (2014) Technical benefits of energy storage and electricity interconnections in future British power systems. Energy 2014:1–11
10. De Decker J, Kreutzkamp P, Woyte A, Dierckxsens C (2012) The impact of large-scale offshore electricity transmission: the European project offshoreGrid. In: Proceedings of the 9th international conference on the European energy market (EEM), IEEE, 10 May 2012
11. Denny E, Touhy A, Meibom P, Keane A, Flynn D, Mullane A et al (2010) The impact of increased interconnection on electricity systems with large penetrations of wind generation: a case study of Ireland and Great Britain. Energy Policy 38:6946–6954

12. Deane JP, Driscoll A, O'Gallachoir BP (2015) Quantifying the impacts of national renewable electricity ambitions using a North-West European Electricity Market Model. Renewable Sustainable Energy Reviews; renew. n. a. 80:604–609
13. Curtis J, Cosmo V Di, Deane P (2013) Climate policy, interconnection and carbon leakage: the effect of unilateral UK policy on electricity and GHG emissions in Ireland. Economic and Social Research Institute p 458. Available at: http://energyexemplar.com/wp-content/uploads/publications/Climate%20policy,%20interconnection%20and%20carbon%20leakage.pdf. Accessed 1 Oct 2014
14. Energy Exemplar (2013) http://energyexemplar.com/. Accessed 1 Jan 2013
15. FICO Xpress Optimiser (2013) http://www.fico.com/en/Products/DMTools/xpress-overview/Pages/Xpress-Optimizer.aspx. Accessed 02 March 2014
16. Deane P (2012) Improved modelling of pumped hydro energy storage. Department of Civil & Environmental Engineering and Environmental Research Institute, University College Cork
17. Single Electricity Market Operator (2012) http://www.sem-o.com/Pages/default.aspx. Accessed 01 Feb 2014
18. Redpoint Energy Ltd (2010) P229 cost benefit analysis: additional scenarios (1)
19. Ofgem (2013) Electricity capacity assessment report 2013 27/06;105/13
20. ENTSOE (2010) Evaluation of the ramping restriction in the energy market. Accessed 24 Nov 2010
21. National Grid (2012) Balancing Services Incentive Scheme (BSIS) 2011–2013 Methodology amendments.
22. National Grid (2013) Electricity ten year statement. http://www2.nationalgrid.com/UK/Industry-information/Future-of-Energy/Electricity-ten-year-statement/Current-statement/. Accessed 12 Jan 2013
23. Poyry Energy Consulting (2010) Options for low-carbon power sector flexibility to 2050. (v2_0)
24. DG Comp (2007) EU database product specification—Appendix I. (1)
25. National Grid (2011) Addendum to the Electricity SO Incentives Initial Proposals for 1st April 2011—modelling transmission constraints. 2010;BSIS 2011/13(1)
26. Department of Energy and Climate Change (2013) Renewable sources of energy: Chap. 6, digest of United Kingdom energy statistics (DUKES). 11/28, 2013; https://www.gov.uk/government/publications/renewable-sources-of-energy-chapter-6-digest-of-united-kingdom-energy-statistics-dukes. Accessed 1 Oct 2014
27. McGarrigle E, Deane P, Leahy P (2013) How much wind energy will be curtailed on the 2020 Irish power system? Renew Energy 55:544–553
28. Foley A, Leahy P, McKeogh E (2009) Wind energy integration and the Wales-Ireland interconnector. In: Proceedings of the IEEE PES/IAS conference on sustainable alternative energy; Valencia, Spain, pp 1–6
29. Department of Energy and Climate Change (2013) Updated energy and emissions projections 2013. 2013; 13D/231
30. Ofgem (2009) Project discovery energy market scenarios 09/10;122/09
31. Newbery D (2011) Reforming competitive electricity markets to meet environmental targets. Cambridge (1126)
32. Department of Energy and Climate Change (2012) Electricity market reform: Policy overview. https://www gov uk/government/uploads/system/uploads/attachment_data/file/65634/7090-electricity-market-reform-policy-overview-pdf. Accessed 10 March 2013

Chapter 74
Testing Operation and Control Functions of Wind Power Plant Control System by Hardware in-the-loop Simulation

Jong Yul Kim, Gyeong Hun Kim, Jin-Hong Jeon, Seul Ki Kim and Eung Sang Kim

Abstract The grid system operators in many countries require a grid code for a wind power generation system (WPGS) or a wind power plant (WPP). For satisfying the grid code, a wind power plant control system (WPPCS) is needed because it requires the possibility for controlling active and reactive power in point of common coupling (PCC). The WPPCS should have the ability to send out set points to all WPGSs, and each WPGS must be able to ensure set points from the WPPCS. This chapter presents a performance analysis method without a real WPP to evaluate a WPPCS using hardware-in-the-loop simulation (HILS) using a real-time digital simulator (RTDS). A HILS method for a WPPCS using a RTDS is presented, and performance of the WPPCS is analyzed. A 50-MW WPP which includes 5 MW WPGS is modeled and analyzed in RSCAD/RTDS. A prototype WPPCS is developed to control and operate a 50-MW class WPP, which is connected to the RTDS through the Distributed Network Protocol (DNP)3.0. The HILS results show the proposed method can be effectively utilized to validate control functions of WPPCSs.

Keywords Wind power plant control system(WPPCS) · Grid code · Hardware-in-the-loop simulation (HILS) · Real-time digital simulator(RTDS)

J. Y. Kim (✉) · G. H. Kim · J.-H. Jeon · S. K. Kim · E. S. Kim
Smart Distribution Research Center, Korea Electrotechnology Research Institute, Changwon, Republic of Korea
e-mail: jykim@keri.re.kr

G. H. Kim
e-mail: kgh1001@keri.re.kr

J. H. Jeon
e-mail: jhjeon@keri.re.kr

S. K. Kim
e-mail: blksheep@keri.re.kr

E. S. Kim
e-mail: eskim@keri.re.kr

A. Sayigh (ed.), *Renewable Energy in the Service of Mankind Vol I*,
DOI 10.1007/978-3-319-17777-9_74

74.1 Introduction

As fossil fuel energy has been exhausted and the global warming issue goes critical, renewable energy attracts more attention. Among renewable energy technologies, wind energy has been significantly increasing all over the world [1]. The increase in wind energy can give various benefits such as carbon dioxide (CO_2) reduction and peak demand cutting. However, high penetration of unregulated wind power plants (WPPs) has an adverse effect on the power system stability [2].

Hence, the power system operators in many countries such as Denmark, Germany, Spain, and so on require a grid code for a wind power generation system (WPGS) or WPP [3, 4]. According to general grid codes, the capability for controlling active power and reactive power in point of common coupling (PCC) should be needed, which is implemented by a wind power plant control system (WPPCS) [5]. For applying the WPPCS in a real WPP, the WPPCS has to be evaluated and validated before going to field. However, it is very difficult to test the WPPCS because of the security problem of the power system. The hardware-in-the-loop simulation (HILS) is a technique used to develop and test complex, real-time embedded systems. The HILS technique has merits of reduction of development cost and build-up time [6]. In particular, the real-time digital simulator (RTDS) is a special purpose computer designed to study electromagnetic transient phenomena in real-time. Electric power utilities have used the RTDS simulator for testing of controllers for high-voltage direct current (HVDC) systems, protective relays, and microgrid management systems [7, 8]. This chapter proposes an HILS system using a RTDS as a new method to test power control functionalities of the WPPCS. A 50-MW WPP, which is composed of ten 5-MW WPGSs, is modeled and simulated in an RSCAD/RTDS. A prototype WPPCS is developed to control and operate a 50-MW class WPP, which is connected to the RTDS through the Distributed Network Protocol (DNP) 3.0. The HILS results show the proposed method can be effectively utilized to validate control functions of the WPPCS.

74.2 Prototype WPPCS

74.2.1 Grid Code Requirements

Increase of wind energy in a power system affects the power balancing between generated and demanded power caused by unpredictable wind conditions. Unpredictable wind conditions can cause a significant variation in the system frequency that may affect the grid system stability [2]. Hence, the grid system operators in many countries require a grid code for a WPGS or a WPP. According to the grid codes, the technical requirements are defined for the connection and operation of WPPs connected to the transmission system [3, 4]. These requirements cover:

- Active power and frequency control
- Reactive power and voltage control
- Fault ride through (FRT) requirements
- Frequency and voltage operation range

The active power and frequency control requirements define the regulation of WPP's active power output and the frequency response to control their active power outputs with respect to frequency deviations. Regarding active power, various power regulating schemes such as power cap limitation, set point control, and power reserve are required. In addition, reactive power control capabilities are also required by the grid code. It is performed either by setting a reactive power value or power factor value. Further, reactive power control should be extended for controlling the voltage at the PCC or at the distant node.

74.2.2 *WPP Control Scheme*

A cooperative control scheme is used to satisfy the requirements of the grid code, which are active power and frequency, reactive power, and voltage [5]. The control scheme is composed of three steps, as shown in Fig. 74.1. At the first step, the active and reactive power references of a WPP ($P_{\text{wf\_ref}}$ and $Q_{\text{wf\_ref}}$) are produced in the reference generation module, then the active power reference of the WPP is limited by the ramp rate limit, and the power controllers (proportional integral, PI) compare the measured power in the PCC and the reference value ($P_{\text{wf\_ref}}$ and $Q_{\text{wf\_ref}}$) to obtain the error. This error generates the total required power ($P_{\text{PI\_out}}$ and $Q_{\text{PI\_out}}$). And the final active power reference of each WPGS is calculated by dispatch control and weight value (PF^{i}_{wt}) in (74.1).

$$P^{i}_{\text{wt\_ref}} = PF^{i}_{\text{wt}} \times P_{\text{PI\_out}},\; PF^{i}_{\text{wt}} = \frac{P^{i}_{\text{wt\_available}}}{P_{\text{wpp\_available}}} \tag{74.1}$$

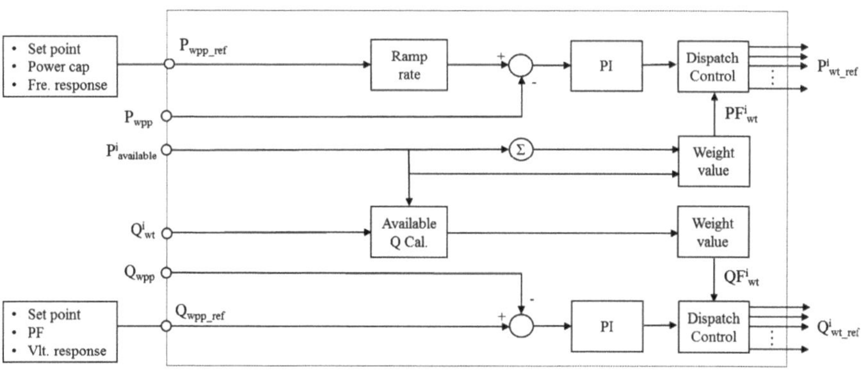

Fig. 74.1 Control blocks of a WPPCS. *PI* proportional integral

The reactive power reference of each WPGS is also calculated by dispatch control and weight value (QF^i_{wt}) in (74.2).

$$Q^i_{wt\_ref} = QF^i_{wt} \times Q_{PI\_out},\ QF^i_{wt} = \frac{Q^i_{wt\_available}}{Q_{wpp\_available}} \tag{74.2}$$

The available reactive power of the wind farm depends on the active power because of the current capacity of the back-to-back converter. The available reactive power of the wind farm is the sum of each available reactive power of the WPGSs, as follows in (74.3).

$$Q_{wpp\_available} = \sum Q^i_{wt\_av},\ Q^i_{wt\_available} = \sqrt{(S^i_{wt\_rated})^2 - (P^i_{wt\_ref})^2} \tag{74.3}$$

74.2.3 Development of Prototype WPPCS

The prototype WPPCS has been developed as shown in Fig. 74.2. The hardware platform of the WPPCS is based on six PCs, which are mounted in two racks. The operation software for the WPPCS is composed of a supervisory control and data acquisition database (SCADA/DB) module, human machine interface (HMI) module, and a control algorithm module. In the SCADA/DB module, DNP 3.0 is applied to communicate with each WPGS in every second. The HMI module gives

Fig. 74.2 Feature of a prototype WPPCS

the operating status of the WPP and selects WPPCS control parameters. The control algorithms to comply with the grid code are implemented by C++ language.

74.3 HILS System for Testing WPPCS

The configuration of HILS setup for testing a WPPCS is shown in Fig. 74.3. The WPPCS can connect to the real-time simulation model of the 50-MW WPP through the DNP 3.0 interface of the RTDS. The modeled WPP is simulated with a 50-μs time step in real time. The operation data such as the active (P_{wpp}) and reactive powers (Q_{wpp}) in the PCC, active and reactive powers of each WPGS (P^{i}_{wt} and Q^{i}_{wt}), values of pitch angles, and rotation speed are sent to the WPPCS. The WPPCS calculates each WPGS's active and reactive power commands ($P^{i}_{wt\_ref}$ and $Q^{i}_{wt\_ref}$) for controlling the total active and reactive power of WPP.

74.3.1 Configuration of WPP

The configuration of a WPP is depicted in Fig. 74.4, which is modeled in an RSCAD/RTDS. A total rated power of a WPP is 50 MW, and it has ten WPGSs, a substation, and distribution lines. Each 5 MW of variable speed WPGS consists of a permanent magnet synchronous generator (PMSG) and a back-to-back voltage source converter. The WPGS is connected to medium voltage (MV) at 33 kV. The MV is boosted up to 154 kV by using two 60-MW step-up transformers. The substation is connected to the PCC through a 20-km-long cable. Table 74.1 shows parameters of the 50-MW-modeled WPP.

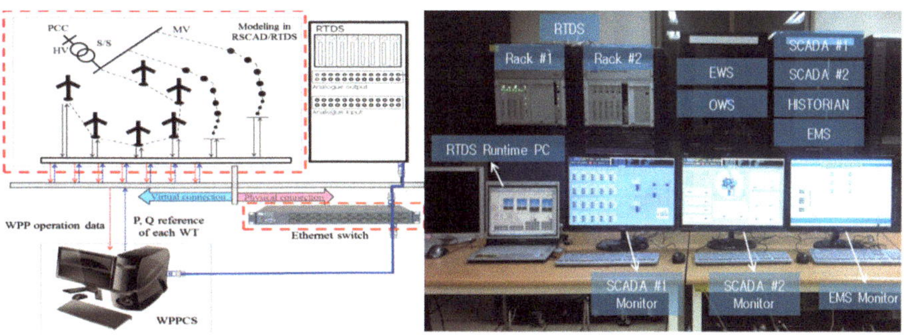

Fig. 74.3 Feature of an HILS. *PCC* point of common coupling, *HV* high voltage, *MV* medium voltage, *WPP* wind power plant, *WPPCS* wind power plant control system, *RTDS* real-time digital simulator, *SCADA* supervisory control and data acquisition

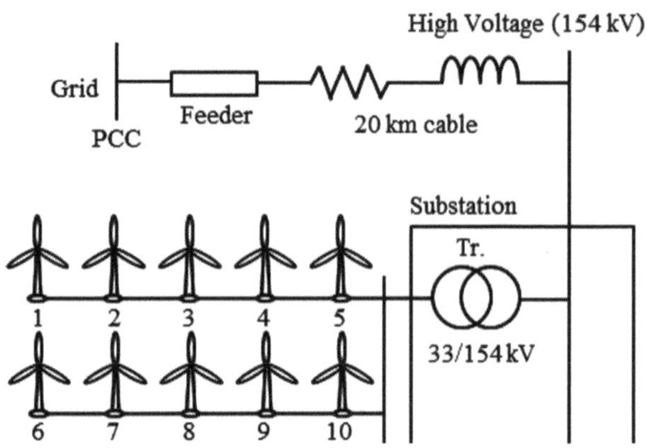

Fig. 74.4 Configuration of a WPP.*PCC* point of common coupling

74.3.2 Simplified Wind Turbine Model

In the case of the wind turbine model, the mechanical torque (T_m) of the turbine captured from the wind power can be calculated by:

$$T_m = \frac{1}{2}\rho\pi R^3 v^2 C_p(\lambda,\beta)/\omega, \tag{74.4}$$

where ρ is the air density (kg/m³), v is the wind velocity (m/s), R is the radius of the blade (m), ω is the angular velocity (rad/s), and C_p is the power coefficient [9].

The power coefficient is a function of tip speed ratio *(λ)* and pitch angle *(β)* of the blade (degree) as follows (74.5):

$$C_p(\lambda,\beta) = c_1\left(\frac{c_2}{\lambda_i} - c_3\beta - c_4\right)e^{\frac{-c_5}{\lambda_i}} + c_6\lambda, \tag{74.5}$$

Table 74.1 Parameters of 50-MW-modeled WPP

Type	Parameters	Values
Wind power plant	Number of wind turbines	10
	Distributed voltage	33 kV
	Transmission voltage	154 kV
Substation	Number of transformers	2
	Rated power	120 MVA
	Leakage inductance of transformers	0.05
	Voltage	33/154 kV
Feeder	Cable length	20 km
	Resistance	0.19 Ω/km
	Reactance	0.41 Ω/km

Table 74.2 Parameters of the 5-MW wind turbine model

Parameters	Values
Rated wind velocity	11.5 m/s
Radius of the blade length	60 m
Rotation speed	15 rpm
Maximum power coefficient	0.48
Optimal tip speed ratio	8.1
Air density	1.225 kg/m^3
Inertia of the rotor	117,000,000 kg m

$$\lambda_i = \frac{1}{\dfrac{1}{\lambda - 0.08\beta} - \dfrac{0.035}{\beta^3 + 1}}, \tag{74.6}$$

where c_1 is 0.5176, c_2 is 116, c_3 is 0.4, c_4 is 5, c_5 is 21, and c_6 is 0.0068 [10]. Parameters of the 5-MW wind turbine model are given in Table 74.2. The maximum power coefficient ($C_{\mathrm{p\_max}}$) is 0.48 when tip speed ratio is 8.1 (λ_{opt}). The controller of maximum power point tracking (MPPT) calculates the power reference (P_{mppt}), which is given by (74.7).

As in (74.7), the rotation speed of the wind turbine is used to determine the power reference of the MPPT. In the case of overrated power, the power reference is limited to 5 MW.

$$P_{\mathrm{mppt}} = \frac{1}{2}\frac{\rho\pi R^5 C_{\mathrm{p\_max}}}{\lambda_{\mathrm{opt}}^3}\omega^3 \tag{74.7}$$

The WPGS consists of the PMSG, whose stator winding is directly connected to a bidirectional frequency converter made up of two back-to-back insulated gate bipolar transistor (IGBT) bridge. However, there is a constraint of operation time and processor card [11] of the RTDS for implementing the real-time simulation. For this reason, a simplified model of the WPGS [12] is considered, as shown in Fig. 74.5, which includes a three-phase current source and wind turbine model. The three-phase current source of the modeled 5-MW WPGS is controlled by a power controller, as represented in Fig. 74.5. References of a DQ-axis current component frame are used for control of the active and reactive power, whose reference is received from the WPPCS. As shown in Fig. 74.5, the reference power of the MPPT control is limited by active power commands ($P_{\mathrm{ref\_wt}}$ and $P_{\mathrm{min\_wt}}$) from the WFC. The reactive power of the WPGS is controlled by reactive power commands ($Q_{\mathrm{ref\_wt}}$). If the active power command from the WPPCS is smaller than the reference power of the MPPT control, the output power of the wind turbine is limited. It causes an increase in the rotation speed of the wind turbine because of the difference in torque between the electrical output power and the mechanical power. In that case, the rotation speed of the wind turbine is regulated by the pitch angle controller.

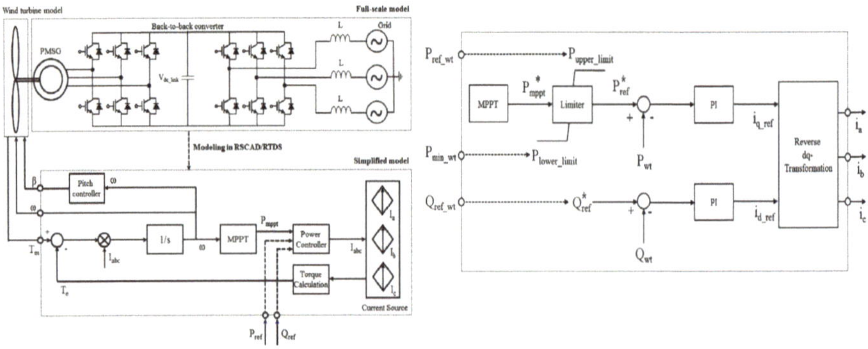

Fig. 74.5 Simplified 5-MW wind power generation system model in RSCAD/real-time digital simulator (RTDS). *PMSG* permanent magnet synchronous generator, *MPPT* maximum power point tracking, *PI* proportional integral

74.4 Test Results

Figure 74.6 shows the power output of a WPP under MPPT mode and set point control mode. In this case, all WPGSs operate in MPPT mode between 0 and 100 s. After 100 s, the operation mode of the WPP is changed to set point control mode with various power references, which are supposed to be required by the grid operator. As shown in Fig. 74.6, the active power of the WPP is controlled by its reference. Between 400 and 500 s, the active power output of the WPP cannot follow its reference 30 MW because the available power of the WPP is less than 30 MW.

Figure 74.7 represents the power output of the WPP in power reserve and frequency response mode. In the normal state, the power output of the WPP is reduced

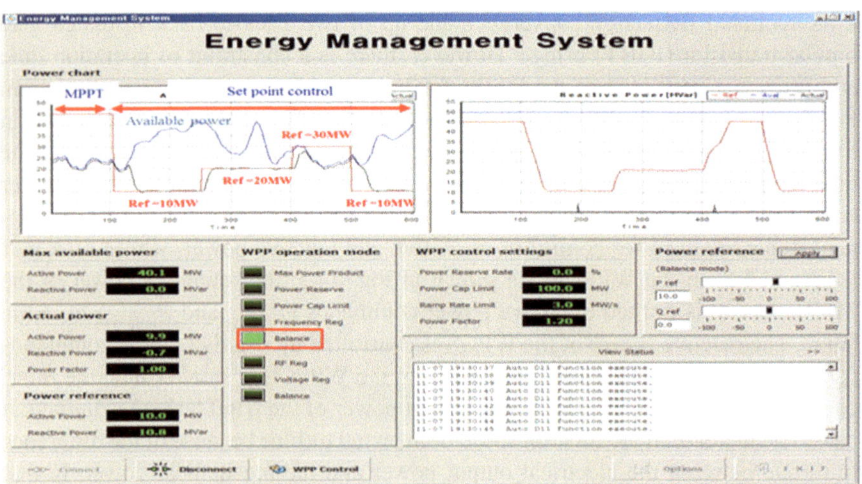

Fig. 74.6 Active power output of a wind power plant (WPP) under maximum power point tracking (MPPT) and set point control modes

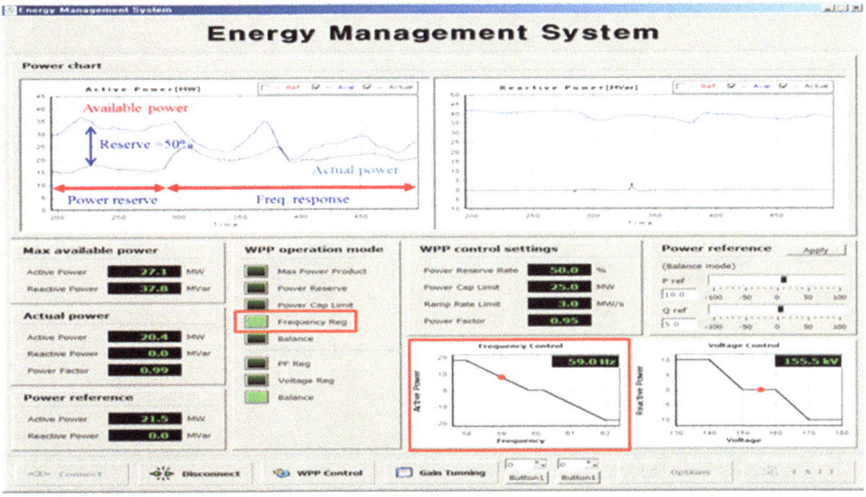

Fig. 74.7 Active power output of a wind power plant (WPP) under power reserve and frequency response modes

by 50 % of the available power for frequency response. During the first 80 s, the frequency is maintained at 60 Hz, and the power output of the WPP keeps 50 % of the reserve rate. At 280 s, the frequency is suddenly dropped to 59 Hz, and then the operation mode of the WPP is moved to the frequency response mode.

In this mode, the power output of the WPP is increased to regulate the frequency according to predefined frequency droop characteristic. In Fig. 74.8, the reactive power set point control and the voltage response modes are tested. In the first period, the operation mode of the WPP is the reactive power set point control mode. During this period, the reactive power output of the WPP is controlled by its reference, which is assigned from the system operator.

At 630 s, the voltage of the PCC is decreased to 145.6 kV, and then 4MVar of the reactive power is injected to the grid for regulating the voltage of the PCC according to predefined droop characteristic.

74.5 Conclusion

The HILS system using RTDS is proposed for testing operation and control functions of the WPPCS. The proposed HILS system emulates 50 MW of the WPP which consists of ten 5 MW of WPGSs and connects to the developed prototype WPPCS through the DNP 3.0. The prototype WPPCS for a 50-MW WPP has been developed with the active and reactive power control capabilities to comply with the grid code. The test results show that the various regulating functions of the WPPCS can be evaluated by the HILS environment.

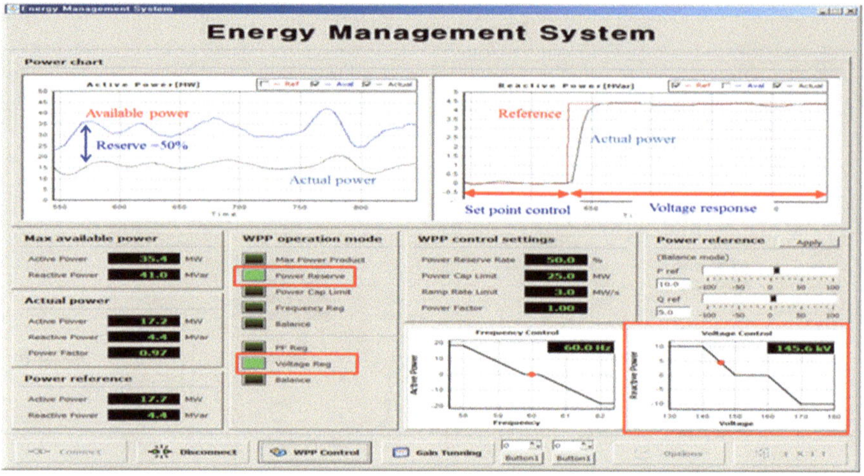

Fig. 74.8 Reactive power output of a wind power plant (WPP) under set point control and voltage response modes

References

1. Jeong H-G, Lee J-H, Lee K-B (2013) A 2nd order harmonic compensation method for wind power system using a PR controller. J Electr Eng Technol 8(3):507–515
2. Lee S-E, Won D-J, Chung II-Y (2012) Operation scheme for a wind farm to mitigate output power variation. J Electr Eng Technol 7(6):869–875
3. VDN (Aug. 2007) Transmission code 2007, Network and system rules of the german transmission
4. Energinet (Dec. 2004) Grid connection of wind turbines to networks with voltages above 100 kV, Regulation TF 3.2.5. Denmark
5. Altin M, Teodorescu R, Bak-Jensen B, Rodiriguez P, Iov F, Kjær PC (2010) Wind Power Plant Control—An Overview. In: Proceedings on the 9th International Workshop on Large-Scale Integration of Wind Power into Power Systems, GmbH
6. Lu B, Wu X, Figueroa H, Monti A (Apr 2007) A low cost real time hardware in the loop testing approach of power electronics controls. IEEE Trans Ind Electron 54(2):919–931
7. Bi T, Zhang Y, Xiao X, Forsyth P, Wierckx R (Dec. 2007) Large scale power system simulation and PMU testing using a real time digital simulator. IPEC 2007, Singapore
8. Jeon JH, Kim JY, Kim HM, Kim SK, Cho CH, Kim JM, Ahn JB, Nam KY (Dec. 2010) Development of hardware in the loop simulation system for testing operation and control functions of microgrid. IEEE Trans Power Electron 25(12):2919–2929
9. Heier S (1998)Grid integration of wind energy conversion systems. Wiley, pp 34–36
10. Rosadi M, Muyeen SM, Takahashi R, Tamura J (2011) Transient stability enhancement of variable speed permanent magnet wind generator using adaptive PI-Fuzzy controller. In publication of Powertech Conference, Trondheim, pp 1–6, Jun. 2011
11. Forsyth P, Maguire T, Kuffel R (2004) Real time digital simulation for control and protection system testing. In: Proc. 35th Annual IEEE Power Electron. Spec. Conference, pp 329–335, Jun. 2004
12. Kim G-H, Bae H-H, Kim S-Y (2010) ChulSang Hwang, Hyo-Guen Lee, Namwon Kim, Hyo-Rong Seo, Minwon Park, In-Keun Yu. Application of a Battery Energy Storage System for Power Quality Improvement of Jeju Power System. In publication of ICEMS 2010 Conference, Incheon, Korea, Oct. 2010

Chapter 75
Blade Element Momentum Theory and CFD Modeling as a Tool for Optimizing Wind Turbine Blade Design

K. Dogan and G. Martinopoulos

Abstract The present chapter focuses on the comparison of the capabilities of Blade Element Momentum Theory (BEMT) in relation to computational fluid dynamics (CFD) modeling as a tool for the design and performance optimization of a horizontal axial wind turbine (HAWT).

A generated blade is examined in different scales by BEMT with the use of the QBlade software. The same wind turbine blade design was then incorporated in a detailed 3D CFD model (in ANSYS CFX). Computations were performed and the results were compared to the ones produced with the BEMT method.

For the CFD modeling, a National Advisory Committee for Aeronautics (NACA) profile was initially validated through a two-dimensional analysis and flow field investigation regarding lift and drag coefficients for a variety of angles of attack (AoA). For computation reasons, a rotating domain was applied. The domain is discretized into 4,320,733 elements, most of which are tetrahedral, creating 781,582 nodes. An extra inflation layer is used on the turbine boundary and mesh density is higher in that vicinity.

After the validation of the two-dimensional analysis, the wind turbine blade design was incorporated in a detailed 3D CFD model and computations were performed and compared to the ones of the BEMT method. Detailed transition formulas were applied and compared and a mesh independent solution was achieved. Furthermore, pressure and velocity distribution on the blade are analyzed and Cp graphs were produced.

The shear stress transport (SST) turbulence model capability to simulate the flow around an airfoil in the pro stall region was verified, while, the angle of attack at which stall begins could also be predicted using CFD modeling.

The study revealed the superior performance and advantages of CFD modeling in relation to BEMT since CFD can take into account the 3D effects of actual flow around a turbine blade which cannot be obtained by BEMT methodology.

Keywords Wind energy · CFD modelling · BEM theory

G. Martinopoulos (✉) · K. Dogan
School of Science and Technology, International Hellenic University, 14th km Thessaloniki—Moudania, EL 57001 Thessaloniki, Greece
e-mail: g.martinopoulos@ihu.edu.gr

A. Sayigh (ed.), *Renewable Energy in the Service of Mankind Vol I*,
DOI 10.1007/978-3-319-17777-9_75

75.1 Introduction

Among renewable energy sources (RES), wind power is the leading form with a worldwide installed capacity of 318 GW that produced more than 662 TWh of electricity in 2013 [1]. In an attempt to further increase the utilization of RES, there is an ongoing effort in optimizing wind turbines. Towards this direction, the investigation of wind turbine aerodynamics is of significant importance.

Such an effort is presented in this chapter where the aerodynamic performance of a wind turbine is first investigated with various computational fluid dynamics (CFD) models that are available in the ANSYS software and then a comparison is made with the open source wind turbine calculation software QBlade, which is based on Blade Element Momentum Theory (BEMT) [2, 3]. The selection of both approaches (i.e., BEMT and CFD) was based on the different advantages and disadvantages that each approach presents.

BEMT analysis can provide a fast approach for the basic analysis and design of a wind turbine blade, whereas CFD can provide a 3D detailed view of the flow field development during the operation of a wind turbine, capturing secondary flow phenomena which cannot be resolved during BEMT analysis and are necessary for optimizing the wind turbine blade design, although computations are more time-consuming. For these reasons, a noticeable number of works is presented in international literature using BEMT method [4–6] or CFD computations [7–9] for the analysis of wind turbines' performance.

In the present chapter, initially a wind turbine is simulated in order to investigate which of the k-ε, k-ω, and SST turbulence models provide the best results, how much the result is dependent on the selected model and the model is validated against experimental data. After the optimum computational model is found, a comparison between the results from the BEMT together with CFD computations in ANSYS CFX software [10] was used to examine the aerodynamic performance of a wind turbine for various angles of attack (AoA). As a beginning, a blade length of 1 m is created which is then scaled up to 2, 4, 8, 16, 32, and 64 m. At this point, it must be mentionedregarding the BEMT calculations that the effect of the airfoil design is taken into account by the calculation of the lift and drag coefficients which is performed by the XFOIL design analysis code [11, 12] which is integrated in the open source software QBlade, developed by Department of Experimental Fluid Mechanics from the Berlin Technical University.

75.2 Comparison of Turbulence Models

There are a variety of turbulence models and depending on the application the selected model makes a difference. The selection of the correct turbulence model requires expertise and there are a number of studies in the literature that have investigated this issue, with the majority of them reaching the conclusion that the SST model provides the best results [13, 14] as it combines both near wall and far away conditions.

Table 75.1 Comparison of turbulence models

AoA (deg)	0°	5°	10°	15°
k-Epsilon model(L/D)	No-convergence	38.5	51	46
k-Omega model(L/D)	0	34	46	No-convergence
SST model(L/D)	0	45	45	29

In the current chapter, a wind turbine is simulated in order to investigate which of the k-ε, k-ω, and SST turbulence models provides the best results. A National Advisory Committee for Aeronautics 0012 (NACA 0012) profile with a chord length of 1 m is used, while the chosen wind speed is 7.25 m/s, and the lift/drag values for different AoA are given in Table 75.1.

The simulations were conducted in CFX ANSYS and for the two-dimensional analysis, a computational domain containing only a 2D cross section of the airfoil was created. The boundary conditions which were applied are shown in Fig. 75.1. At the inlet of the computational domain, the velocity components together with turbulence intensity value were provided. At the outlet of the computational domain, ambient static pressure was imposed. The bottom and top limits of the computational domain were given a translational periodic connection so as to use the same computational grid and setup for varying angles of inlet velocity without having to recreate the computational grid.

From the results presented in Table 75.1 as well as from Fig. 75.2, it is evident that the SST model provides the best overall results, whereas the turbulence model used has been found to have a profound impact, especially as the angle of attach increases.

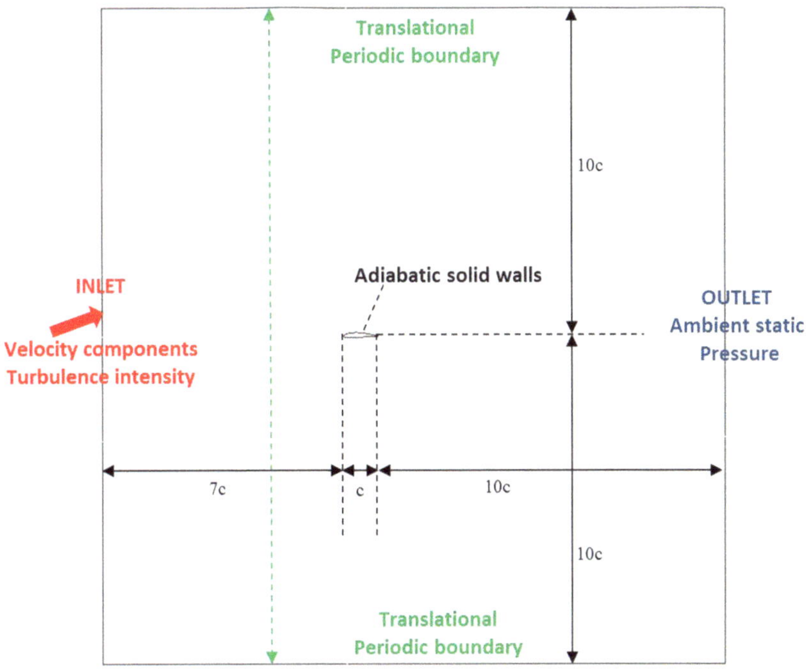

Fig. 75.1 Two-dimensonal computational fluid dynamics (2D CFD) model boundary conditions

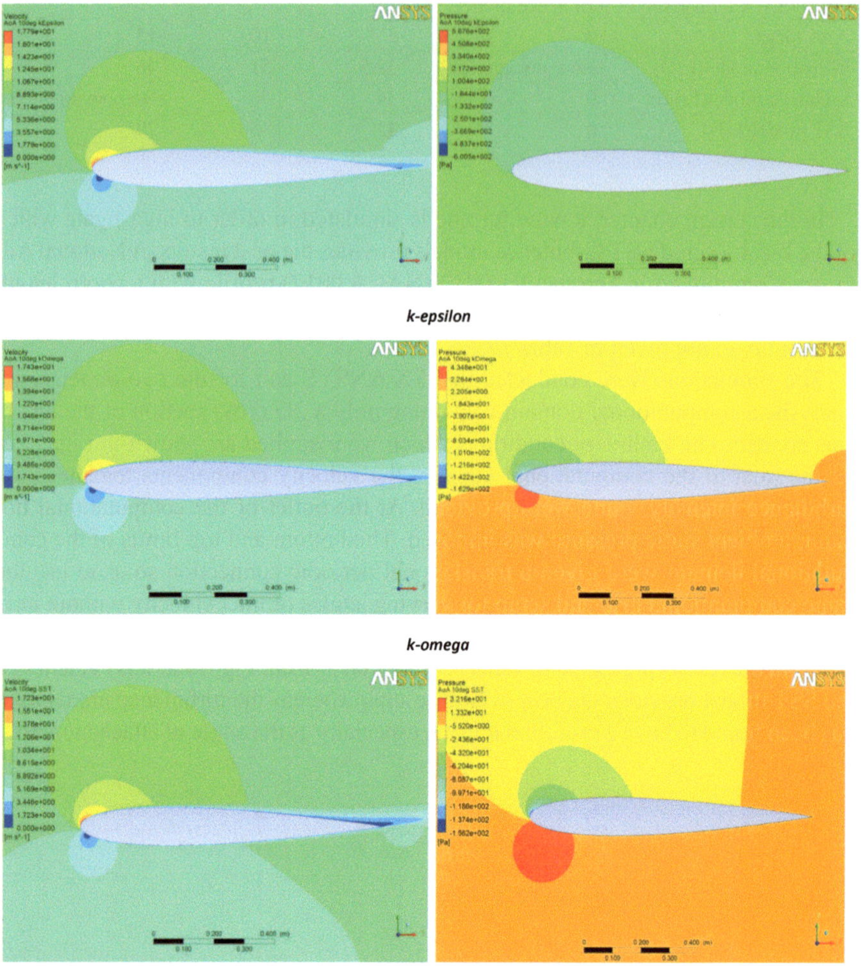

Fig. 75.2 Velocity *(left)* and pressure *(right)* contours for angle of attack (AoA) 10° for the three models investigated

Furthermore, for some cases all models are not able to provide a result, whereas in other cases models may need a significant number of iterations in order to converge, thus increasing central processing unit (CPU) overhead and time considerably.

75.3 Three-Dimensional Computations

At the next step, a 3D design of the wind turbine blade was performed using QBlade based on the airfoil polar values obtained through XFOIL calculations. The program Qblade is a free license wind turbine analysis software which makes use of

Table 75.2 Power coefficient for various rotor radiuses

Rotor radius (m)	Peak power (W)	Torque (Nm)	Rotational speed (1/min)	Power coefficient
1.00	394	5.5	680	0.566
2.01	1654	50	313	0.574
4.10	6857	418	157	0.570
8.06	26,209	3100	81	0.573
15.92	104,580	26,100	38	0.590
31.84	438,243	209,250	20	0.590
63.68	1,766,700	1,687,100	10	0.590

the BEMT. In the beginning, a blade length of 1 m is created which is then scaled up to 2, 4, 8, 16, 32, and 64 m. The main aim is to find out the power output of the design for a wind speed of 7.25 m/s which corresponds to the majority of Turkey's overall economic viable wind power capacity (almost 30 out of a total of 50 GW) [13]. The Qblade design optimization tool is available to create and optimize the blade geometry. The results are presented in Table 75.2.

First, a polar of the blade profile is created by using XFOIL which is incorporated in Qblade. The polar angle range is between −5 and +20°. Then, the polar is extrapolated to a full range of −180 to +180°. The blade consists of sections which are defined by their profile, position, chord, and twist. Its length is selected as 2 m and hub radius 0.05 m and through the Qblade, the geometry is optimized for the selected wind speed of 7.25 m/s. After creating the geometry in Qblade, it was imported to ANSYS CFX. The blade is connected to the hub by cylindrical profiles as pictured in Fig. 75.3.

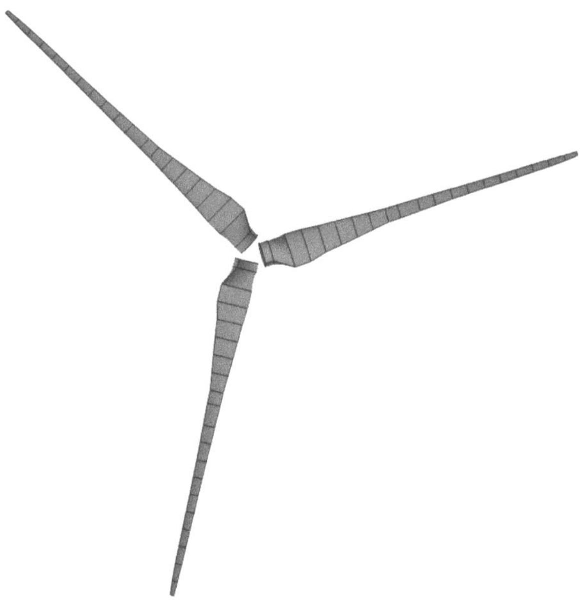

Fig. 75.3 A three-dimensional (3D) view of the rotor

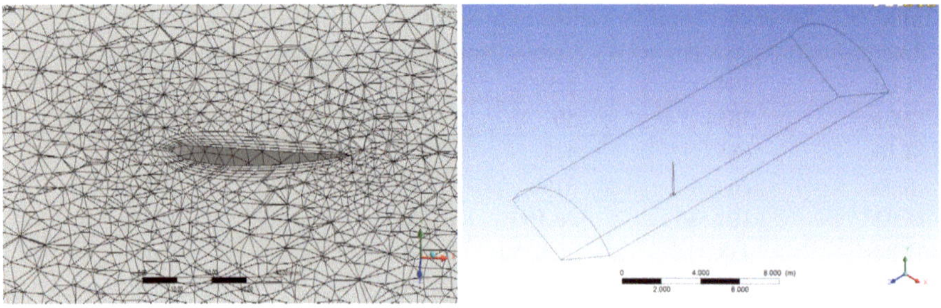

Fig. 75.4 Grid used for the three-dimensional (3D) simulations *(left)* and turbine simulation domain *(right)*

The model is created as a periodic one to save central processing unit (CPU) time. As the turbine under analysis is a three-blade one, a 120° part is simulated. The distance from the turbine to the inlet is 8 m, while it is 12 m to the outlet. The simulation is carried on in a rotating frame. The domain is discretized into 4,320,733 elements, most of which are tetrahedrals creating 781,582 nodes. An extra inflation layer is put on turbine boundary and mesh density is higher in the vicinity of it (Fig. 75.4).

The 3D CFD computations were performed only with the SST turbulence model as this was the model showing the best performance in the 2D computations and time economy. Special care was provided so that the outside boundaries of the computational domain were placed significantly far from the wind turbine surface.

At the inlet of the computational domain, the velocity components were provided (at the stationary frame of motion), together with turbulence intensity. The wind velocity was selected as 7.25 m/sec. At the outlet of the computational domain, ambient static pressure was imposed. The side boundaries of the computational domain were given a rotational periodic connection of 120° in order to incorporate the effect of the presence of the three wind turbine blades. At the wind turbine surface, a non-slip wall boundary condition was applied since the computations were performed as steady in a rotating frame of motion, in relation to which the wind turbine surface is considered stationary (Fig. 75.5).

From the CFD calculations, the maximum efficiency of the wind turbine is 42.2 %, including the hub area and 52.1 %, excluding the hub area. From the comparison with the BEMT it is evident that the BEMT overestimates peak power for the given wind speed, something that is in line with the literature [14–16]. The overestimation can be attributed to the overestimation of the lift coefficient calculated through XFOIL, the values of which are used by QBlade in BEMT method.

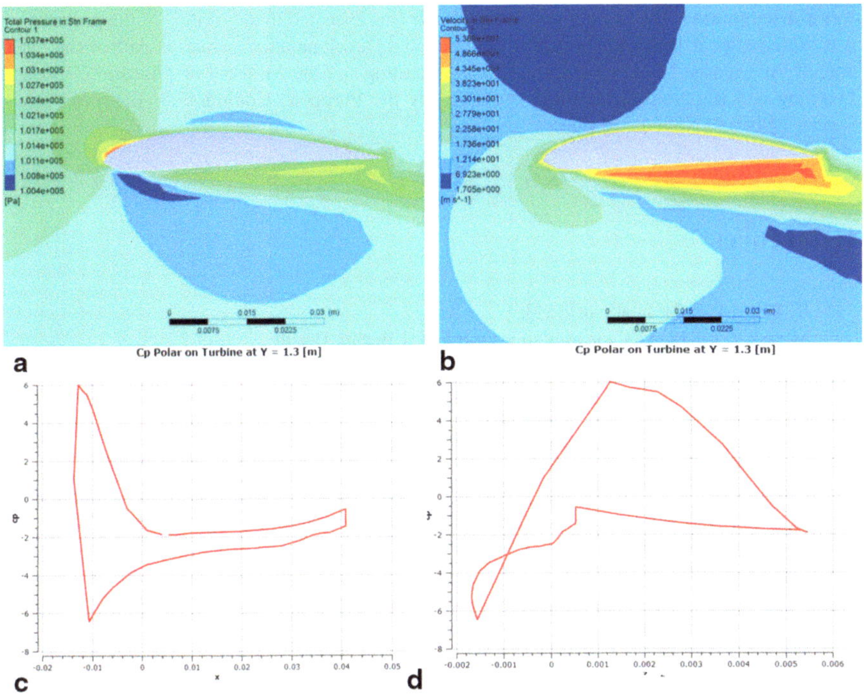

Fig. 75.5 Static pressure contour plots and velocity vectors for a 1.3 m cross section distance from the center of rotation and Cp vs X and Z axis respectively

75.4 Conclusions

In the context of the present work, the flow field and the performance of a wind turbine blade were investigated with the contribution of CFD computations and BEMT method calculations.

The most significant conclusions are that CFD computations can accurately compute the flow around an airfoil to be used for their optimization. There is also a significant correlation between the CFD model used and the results, with the best results coming from the use of the SST turbulence model.

Furthermore, CFD can provide a detailed inside view in the flow field which can be significant for optimization purposes, especially for 3D computations.

The comparison between XFOIL values and CFD results for the lift and drag coefficients reveals the weakness of XFOIL to perform satisfactorily for conditions where viscous effects are particularly dominant, i.e., close to stall conditions. In general, the results of XFOIL tend to overestimate the lift coefficient values.

The study verifies that the 3D effects of actual flow around a turbine blade must be taken into account and BEMT can only be a first step in blade design process.

Acknowledgments The present scientific chapter was executed in the context of the project entitled "International Hellenic University (Operation—Development)," which is part of the Operational Programme "Education and Lifelong Learning" of the Ministry of Education, Lifelong Learning and Religious Affairs and is funded by the European Commission (European Social Fund—ESF) and National resources.

References

1. IEA (2014) Solar heat worldwide
2. Marten D (2012) QBlade guidelines v 0.5. TU Berlin, Berlin
3. Marten D, Wendler J, Pechlivanoglou G, Nayeri CN, Paschereit CO (2013) QBlade: An open source tool for design and simulation of horizontal and vertical axis wind turbines. Int J Emerg Technol Adv Eng 3:264–269
4. Menter FR (1994) Two-equation eddy-viscosity turbulence models for engineering applications. AIAA-J 32(8):1598–1605
5. Launder BE, Reece GJ, Rodi W (1975) Progress in the developments of a Reynolds-stress turbulence closure. J. Fluid Mechanics 68:537–566
6. Tenguria N, Mittal ND, Ahmed S (2010) Investigation of blade performance of horizontal axis wind turbine based on blade element momentum theory (BEMT). Int J Eng 12:25–35
7. Rajakumar S, Ravindran D (2012) Iterative approach for optimizing coefficient of power, coefficient of lift and drag of wind turbine. Renew Energy 38:83–93
8. Yao J, Yuan W, Wang J, Xie J, Zhou H, Peng M, Sun Y (2012) Numerical simulation of aerodynamic performance for two dimensional wind turbine airfoils. Procedia Eng 31:80–86
9. Singh K, Ahmed MR, Zullah MA, Lee Y-H (2012) Design of low Reynolds number airfoil for small horizontal axis wind turbines. Renew Energy 42:66–76
10. Thumthae C, Chitsomboon T (2009) Optimal angle of attack for untwisted blade wind turbine. Renew Energy 34:1279–1284
11. ANSYS CFX Documentation Ansys Inc (2010)
12. Ostowari C, Naik D (1984) Post stall studies of untwisted varying aspect ratio blades with an NACA 4415 airfoil section-part I. Wind Eng 8:186–194
13. Turkish Wind Energy Association (2012) Wind business in Turkey
14. Ladson C, Aeronautics USN, Scientific, SA, Division TI (1988) Effects of independent variation of Mach and Reynolds numbers on the low-speed aerodynamic characteristics of the NACA 0012 airfoil section, NASA technical memorandum, National Aeronautics and Space Administration, Scientific and Technical Information Division
15. Deperrois A (2009) XFLR5—Analysis of foils and wings operating at low reynolds numbers
16. Drela M, Youngren H (2001) XFOIL 6.94 user guide. MIT Aero & Astro New York

Chapter 76
Integration of Wind Energy in Power System—Modelling of a Market Oriented Energy Concept

Yassin Bouyraaman, Jörg Bendfeld, Philipp Breymann and Stefan Krauter

Abstract Regenerative energy sources play a leading role in future energy generation. In this chapter, some commercialisation scenarios concerning offshore wind farms are discussed. The focus is set on energy-exchange trading combined with offering control reserve (secondary control reserve (SCR) or minute reserve (MR)) or offering only control reserve without participating in energy exchange trading. In other words, some alternative ways of operating a wind farm compared to Renewable Energy Act (EEG) commercialisation (feed-in tariff) is presented. First, a brief introduction is given, then the so-called power-to-gas storage system (P2G) is introduced, later, technical as well as economical results are presented and finally, a brief look into the future is made.

76.1 Introduction

Now, some technical and economical aspects should be addressed to justify the way of direct marketing of wind energy. First, we focus on the economical perspective, later, a technical point of view is discussed in more detail. As mentioned before, operating a wind farm on the basis of Renewable Energy Act (EEG) is the simplest and comfortable way, but as the EEG was introduced for subsidising renewable energy concepts, it is not surprising that EEG is running out sooner or later. Currently the EEG initial tariff for offshore wind farms is set to 15 cent/kWh, the base tariff is set to 3.5 cent/kWh. According to the most recent status, EEG compensation (initial tariff and base tariff) will be decreased by 7 % from 2018 on [1]. As a result of political discussions in Germany, this process could be accelerated, so from an economical point of view it can be finally said that alternative ways of commercialisation must be found.

In addition, there are, of course, technical aspects which have to be kept in mind. Due to an increasing portion of regenerative energy production, conventional power plants are squeezed out of the market. As a consequence, the overall energy pro-

Y. Bouyraaman (✉) · J. Bendfeld · P. Breymann · S. Krauter
Department of Sustainable Energy Concepts, University of Paderborn, Paderborn, Germany
e-mail: yassin.bouyraaman@uni-paderborn.de

A. Sayigh (ed.), *Renewable Energy in the Service of Mankind Vol I*,
DOI 10.1007/978-3-319-17777-9_76

duction is no longer consistent with the existing demand, thus the need for control reserve will be increasing. The amount of control reserve must be applied by regenerative energy sources due to a decreasing portion of conventional power plants. Obviously, storage systems are required for providing energy in the right quantities at all times independent of weather conditions. Concluding, there are economical as well as technical reasons for a direct marketing strategy and, furthermore, it is obvious that storage capacities are required because in contrast to EEG compensation, delivery commitments are entered.

76.2 System Concept

The main problem regarding the wind energy direct marketing for the wind farm operators or rather the balancing group managers (BGM) is the uncertainty of wind forecasts. These usually come from institutions or Transmission system operators (TSO) and are normally published at 6 p.m. for the complete following day and exists in 15 min time intervals [2]. This means that the forecasts are nearly 2 days old at the end of the day. With an almost 100 % accurate forecast, 100 % of energy supply of wind farm would be economically ideal, however, such forecasts do not reflect the reality. By direct marketing of wind energy in the power system, the TSO tries to correct the forecast error using control power; in this case, the wind farm operators or BGMs have to pay the used control power (correctional control reserve). If the operator obtains the EEG tariff, the TSO is obliged to pay by himself, because he must perform the balancing group to which the wind farm is attributed.

As an approach to solve the problem of forecast uncertainties, a combination of wind farm with storage system based on methane should be considered. This storage system consists of a methanation unit, a methane gas tank, and gas power plant. The basic idea of this combination is:

- *At forecast uncertainties that lead to a surplus of energy, this will be converted to methane and stored in the gas tank.*
- *At forecast uncertainties that cause a deficit of energy, the stored methane can be converted back to electrical energy to balance this deficit.*

The combination of the storage systems is shown in Fig. 76.1. To guarantee a stable mode of operation of such virtual power plant, a possible use of the methanation plant and the gas power plant is important at every time, which should not be restricted by the fill level limits of the methane tank.

As a result, this storage should never be filled completely, otherwise, the further excess of power cannot be achieved by the storage unit. Analog to this, the methane storage unit should never be completely empty, alternatively, the deficit of power cannot be balanced by the gas power plant.

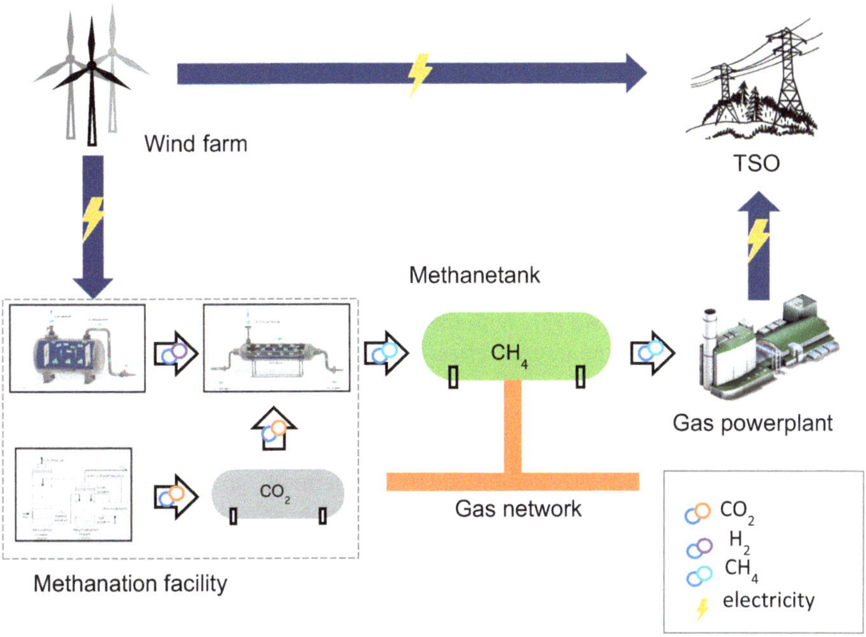

Fig. 76.1 Concept of virtual power station with connection to the gas network

76.3 Power-to-Gas Storage Technology (P2G)

As mentioned in the previous section, storage technologies are needed. Now the so-called P2G storage technology will be introduced. To do so, a brief technical description will be given and, furthermore, we discuss why this kind of energy storage might be an appropriate choice. As the name suggests, power is transformed into gas, more precisely into methane, also referred to as renewable gas or EE gas and it is deemed to be CO_2-neutral. The other way round, gas can not only be transformed into electricity but also be used as fuel for cars or as heat generator.

The transformation to synthetical natural gas is done in two steps. First, hydrogen is generated by electrolysis and second, methanation follows. The electrolysis works as follows:

$$2H_2O(l) \rightarrow 2H_2(g) + O_2(g) \quad \Delta H^{\circ} = 571,8\frac{kj}{mol}(25^{\circ}C)$$

(Endotherm reaction at 25 °C and atmospherical pressure [3])

The generated hydrogen can be used to produce methane (CH_4). Therefore, carbon dioxide CO_2 is needed. This extreme exotherm reaction, also known as sabatier reaction, looks as follows (25 °C) [4]:

$$4H_2(g) + CO_2(g) \rightarrow CH_4(g) + 2H_2O \quad \Delta H^{\circ} = 164{,}9\frac{kj}{mol}(25^{\circ}C)$$

The methanation based on the sabatier reaction is still under development. There are some research systems and the methanation achieves a system effectiveness of approximately 80 % at temperatures of about 250–500 °C [5]. Nowadays the largest P2G facilities reach all in all effectiveness of about 40 % and power outputs of 25 kW [6].

In the near future, facilities with of multiple power and system effectiveness of 60 % and more are planned [7]. With the further use of wasted heat and the usage of other CO_2 sources, the system effectiveness of a methanation-facility can be improved.

The main advantages of such a P2G storage system are the existing natural gas grid in Germany and also large storage capacities. Thus, the storage system and the power source can be spatially separated [8]. Furthermore, gas storages in the form of gas holders can be build at all locations, in addition, gas storages with extremely large capacities up to several billions of cubic metre can be realised in salt or granite formations.

Last, is the question of where to get the required CO_2gas from. There are several possibilities for the acquisition of CO_2 [9], for example, biogenies CO_2 generated with biomass fermentation, CO_2 extraction of the atmosphere, CO_2 as industry waste product, CO_2 as waste product of energy carrier combustion or CO_2 from recycling processes of power plants (Fig. 76.2).

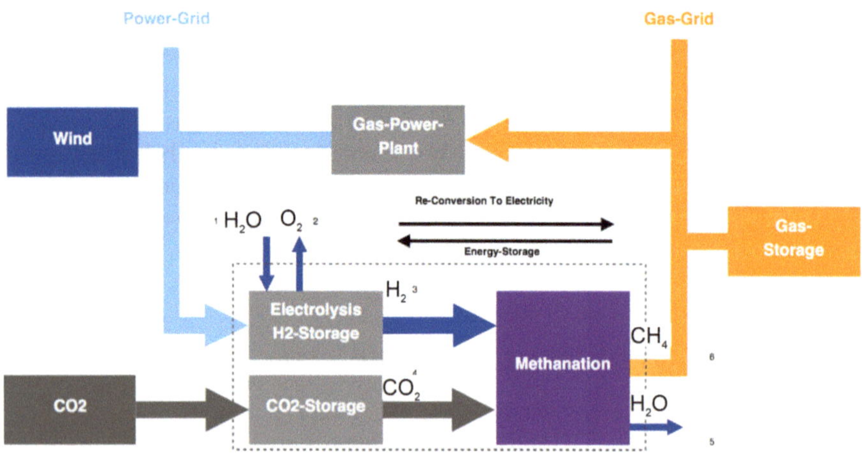

Fig. 76.2 Schematic of power-to-gas storage technology

76.4 Simulation Results

Now the technical results are presented. This section is separated into two subsections, the first one deals with direct marketing scenarios which are based on European Energy Exchange (EEX) trading combined with offering control reserve. Here 20 % of the wind farm performance are offered as maximum positive as well as negative control reserve.

For this chapter, an offshore wind farm with a maximum power output of 400 MW was evaluated. The gas storage capacity is set to 577.5 MWh, the input performance (performance of methanation facility) comes to 203 MW and the output performance (performance of gas turbine) amounts to 128 MW. All results are simulated with Matlab® based on original data of the forecasted and real power of the considered wind park. The prices for EEX trading or for control reserve are taken from 2013.

76.4.1 EEX Trading Combined with Secondary Control Reserve (SCR)

The following figure illustrates a 72 h time period. In case of this first simulation, this graphic is described in a detailed manner, for all other simulations, only the main differences will be pointed out. As can be seen in this figure, there is a stairs line which indicates the gas storage's filling level. The purple line indicates the power produced actually by the wind farm and the cyan line indicates the intraday schedule for EEX trading. The dark blue bar represents the called positive SCR, the normal blue bar stands for the called negative SCR and finally, the light blue bar indicates gas purchases which is due to the fact that the gas storage is completely filled and thus the energy overrun cannot be stored anymore. As just described, there are energy overruns as well as deficits. A deficit occurs if the EEX schedule plus all positive control reserve calls (SCR and MR) are greater than the really produced energy. On the other hand, an overrun occurs in case of a very low EEX schedule or when calling negative control reserve or even both at the same time. This (deficit/overrun) is indicated by the orange line in the figure. The gas storage system should handle both situations but obviously deficits or overruns might be too large (Fig. 76.3).

In case of an unmanageable deficit, further correctional control reserve have to be bought which is marked by the red bars. As mentioned before, overruns which cannot be stored due to a completely filled gas storage are purchased in the form of natural gas (light blue bar).

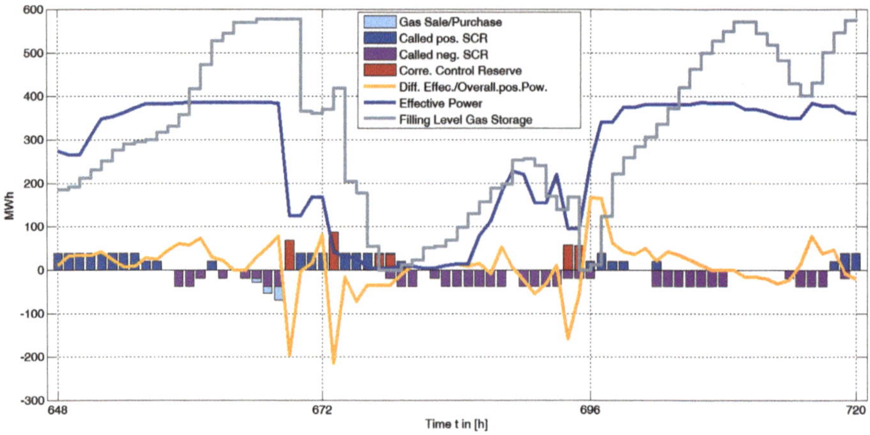

Fig. 76.3 Offering secondary control reserve (*SCR*) combined with European Energy Exchange (*EEX*) trading. Most times the wind farm and storage system can handle EEX schedule as well as control reserve calls, if not, further correctional control reserve is needed *(red bars)*

76.4.2 EEX Trading Combined with Minute Reserve (MR)

Now, EEX trading combined with MR is presented in more detail. The next figure gives an overview of the same 72-h time period again. Now we focus on the main differences compared to the previous scenario. As can be easily seen in the figure, there are not as much consecutive calls of MR as there were in case of SCR. The figure illustrates only a small time period but if one takes all 3.5 months into account, there are notable more negative than positive calls which makes offering MR more suitable for a wind farm because the gas storage is not as much stressed as it would be in case of SCR. Furthermore, the negative calls even support the gas storage filling level.

Besides, this fact there is another very important aspect which makes MR the more appropriate choice. In case of SCR, an offer has to be made for a time period of 1 week in advance, whereas in case of MR it has to be made only 1 day ahead. Thus the wind farm operator can adapt his offer with regard to weather conditions (purple line).

Till now two ways of direct marketing were discussed. Both of them are based on EEX trading and in addition, control reserve offers are made. Last, it should be discussed if only offering control reserve without taking part in EEX trading is a suitable way of direct marketing (Fig. 76.4).

76.4.3 Offering Control Reserve (SCR/MR without EEX trading)

When offering only control reserve there are some aspects which have to be considered. First, due to the fact that there are no EEX trading or even other delivery

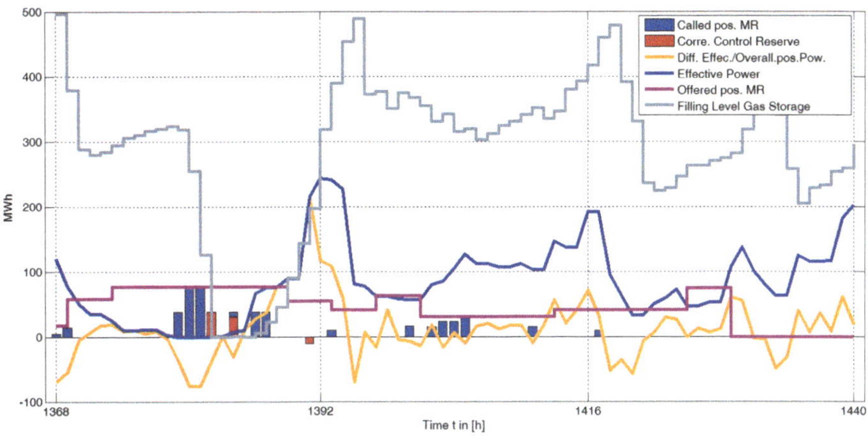

Fig. 76.4 Offering minute reserve (*MR*) combined with EEX trading. There are not as much consecutive MR calls as compared to secondary control reserve (*SCR*), thus the storage system is less stressed. Nevertheless, there are some correctional control reserve calls again

commitments, the amounts of offered control reserve can or must be increased to achieve appropriate results. Second, as mentioned before, in case of SCR, offers have to be made 1 week in advance which can lead to problems in case of long windless periods. As a consequence, only MR makes sense in this scenario. Third, as can be seen in both graphics above, most of the times there are no control reserve calls, thus there is no use for the produced wind energy. As a result, the wind farm performance has to be reduced, otherwise the energy overrun cannot be handled in case of an overstressed methanation facility.

Thus, the storage system is overstressed and tremendous amounts of correctional control reserve are required (red bars). In addition, gas sales have increased (light blue bars) and the wind farm performance is adapted to the demand which can be easily seen at hours 930–960.

Figures 76.5 and 76.6 illustrate the described behaviour. Concluding, it can be said that offering only control reserve currently does not make sense at all compared to EEX-based scenarios. Due to the required reduction of the wind farm performance the whole potential of the wind farm cannot be exhausted. Furthermore, great amounts of correctional control reserve may be caused when promised amounts of control reserve cannot be delivered. Nevertheless, the economical perspective of all introduced scenarios depends on several parameters like EEX prices, gas prices or the development of control reserve. Depending on the development of all these parameters, offering control reserve without EEX trading might be an appropriate choice in the future.

Before giving the economical overview, first some data concerning EEG commercialisation (feed-in tariff) are given for comparison purposes. In case of EEG commercialisation and at a price of 35 €/MWh this results in an overall turnover of 71 million €/year (Fig. 76.7).

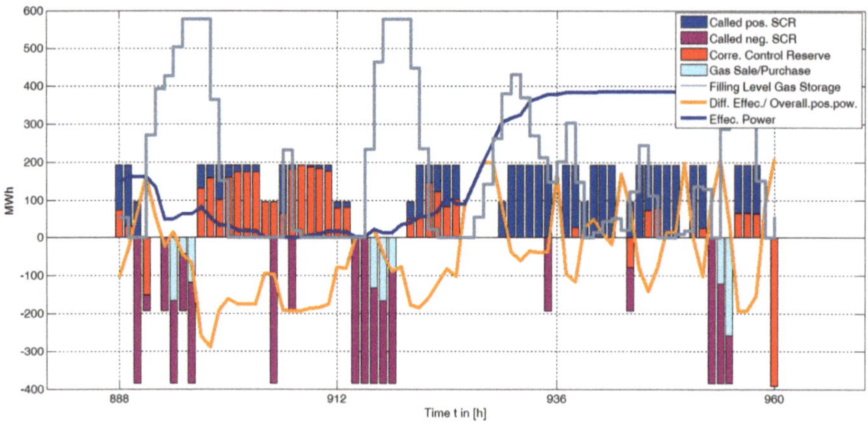

Fig. 76.5 Offering secondary control reserve *(SCR)*. There are lots of consecutive SCR calls even in windless periods *(blue line)*

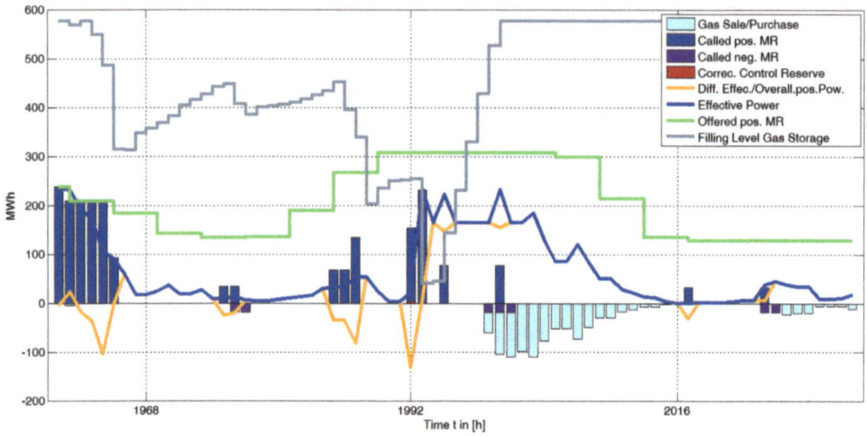

Fig. 76.6 Offering minute reserve (*MR*). There are not as much consecutive MR calls compared to secondary control reserve (*SCR*), thus the storage system is less stressed. Furthermore, the offer of minute reserve adapts itself *(purple line)* with regard to weather conditions (*blue line* shows the windfarm power output) and gas storage filling level *(grey stairs line)*

Fig. 76.7 Overview of the three ways of commercialisation. (Renewable Energy Act (EEG), EEX-based marketing combines with secondary control reserve (*SCR*) or minute reserve (*MR*) and non-EEX-based marketing—only offering SCR or MR)

In case of EEX trading combined with minute reserve the best results in an economical as well as technical manner are achieved. The overall turnover increases up to more than 100 million € (+40 %) and comparatively small amounts of correctional control reserve are caused. In other words, the wind farm and storage system operates in an appropriate way, neither great underperformance nor much overstressing.

EEX trading combined with SCR also achieves great increase in overall turnover (+35 %), but due the fact that no adaptions with regards to weather conditions can be made, the amount of called correctional control reserve is about five times greater as compared to EEX and MR.

As mentioned before, only offering control reserve (MR) without participating in EEX-trading does not make sense under the given circumstances but in case of decreasing EEX-prices and increasing amounts of control reserve and/or increasing prices of control reserve, this assessment may change in the future.

76.5 Conclusion

Concluding, it can be said that sooner or later there is no getting away from finding alternative marketing strategies for regenerative energy power plants. Direct marketing means that commitments towards energy exchanges or transmission system operators are entered, as a results highly capable storage technologies are required. The P2G storage technology currently is under development and there is still a lot of research to do, but it holds great potential especially with respect to the tremendous capacities which will be required in the future.

An offshore wind farm combined with a high performance storage system is capable to participate in energy exchange trading as well as control reserve market which opens up new economical perspectives. In case of offering control reserve, minute reserve is a more suitable choice as compared to SCR because offers have to be made only one day in advance. Nevertheless, at the end of the day, the development of energy exchange prices, gas prices and the progress in the control reserve market will determine the most appropriate way of direct marketing.

On the other hand, great investments have to be made and thus, financial risks also have to be taken. Nowadays, EEG marketing is still the most simple and low-risk method of operating a regenerative energy power plant, but as mentioned at the beginning of this chapter, EEG compensation will decrease annually by 7 %. The lower the EEG compensation, the higher is the incentive for regenerative power plant operators to strike new paths with respect to marketing strategies.

References

1. Act of renewable energy (EEG-Gesetz)
2. Website (2013) „Tennet". http://www.tennettso.de, path: "Tranzparenz">"Veröffentlichungen" > "Netzkennzahlen">"Tatsächliche und prognostizierte Windenergieeinspeisung"

3. Gerthsen T (2006) Chemie für den Maschinenbau 1. Universität Karlsruhe, Karlsruhe. ISBN: 978-3-86644-079-1
4. Halmann MM (1993) Chemical fixation of carbon dioxide–methodes for recycling CO_2 into useful products. CRC Press, Rehovot. ISBN:987-0-84934-428-2
5. Trost T, Jentsch M, Holzhammer U, Horn S (2012) Die Biogasanla gen als zukünftige CO_2-Produzenten für die Herstellung von erneuerbarem Methan. gwf Das Gas-und Wasserfach, Gas–Erdgas 153(3):172–179 (ISSN: 0016-4909)
6. Zentrum für Sonnenenergie- und Wasserstoff-Forschung Baden-Württemberg (ZWS) http://www.zsw-bw.de/uploads/media/pi-2012-ZSWIWESSolarFuel-Einweihung250kW-Anlage_01.pdf
7. EtoGas (2014) http://www.solar-fuel.net/;http://www.etogas.com. Accessed 17 March 2014
8. Bouyraaman Y, Bendfeld J (2013) optimization of offshore wind farm feed-in fluctuations via power to gas storage sytsem , Europe's premier wind energy event—EWEA 2013 annual event, Vienna, Austria, 4–7 Feb 2013
9. Sterner M, Jentsch M, Holzhammer U (2011) Energiewirt-schaftliche und ökologische Bewertung eines Windgas-Angebotes. Fraunhofer Institut für Windenergie und Energiesystemtechnik (IWES), Kassel

Chapter 77
Design and Experimental Validation of Thick Airfoils for Large Wind Turbines

Iva Hrgovan, Wen Zhong Shen, Wei Jun Zhu, Jesper Madsen and Rolf Hansen

Abstract In this chapter, two new airfoils with thickness to chord ratios of 30 and 36 % are presented, which were designed with an objective of good aerodynamic and structural features. Airfoil design is based on a direct method using shape perturbation function. The optimization algorithm is coupled with the viscous/inviscid flow solver XFOIL.

Wind tunnel tests for both airfoils were performed in the LM Wind Power low speed wind tunnel at Reynolds numbers of 1.5–6 million and various surface conditions. The results from the experiment confirm that the optimization algorithm is suited for airfoil design and the designed airfoils have top performance.

Keywords Wind turbine · Airfoil · Blade surface · Roughness · Wind tunnel

77.1 Introduction

The growing demand for power requires larger wind turbines. The latest installed turbine prototype has an 80 m long blade that weighs around 35 t [1]. These sizes present a challenge for the structural integrity of the blade. To deal with the increased weight, designers are using thick airfoils on larger parts of the blade, such as the inboard and the middle sections.

There are several airfoil families specially designed for modern wind turbines [2, 3]. However, thick airfoils from these series are designed to be used only at the inboard section of the blade where the structural performance is more important than the aerodynamic performance. Since the aerodynamic performance of thick airfoils is becoming important, the airfoils need to be improved to provide more power and to satisfy growing structural demands. Moreover, turbines operate in

I. Hrgovan (✉) · W. Z. Shen · W. J. Zhu
Department of Wind Energy, Technical University of Denmark, Kongens Lyngby, Denmark
e-mail: ivah@dtu.dk

J. Madsen · R. Hansen
LM Wind Power, Kolding, Denmark

A. Sayigh (ed.), *Renewable Energy in the Service of Mankind Vol I*,
DOI 10.1007/978-3-319-17777-9_77

environments filled with particles that either damage the blade surface or accumulate on the surface and change the flow conditions. This degradation causes a significant drop in annual energy production [4]. Therefore, airfoils need to be optimized to be less sensitive to roughness.

Using a direct method based on a shape perturbation function [5], we designed two new airfoils with thickness to chord ratios of 30 and 36%. The objective was to create thick airfoils with good aerodynamic features and low roughness sensitivity. The optimization algorithm was coupled with a viscous/inviscid flow solver XFOIL.

A wind tunnel experiment was performed in the LM Wind Power low speed wind tunnel. The results from the experiment for DTU-230 and DTU-236 are discussed in the chapter.

77.2 Airfoil Design

Inverse design methods have been used in airfoil design since 1980 [6]. They are used when a desired airfoil surface pressure distribution at a given operating condition is prescribed, which is then translated into the corresponding geometrical shape. Since the surface flow is taken only at a single design point, this method has limitations regarding multiple design points.

Alternatively, direct design methods are more flexible and allow multidisciplinary optimization with multiple constraints. A numerical optimization algorithm can be coupled with a general flow solver where the response parameters can be directly used as design objectives. In this study, a commercial gradient-based constrained optimization algorithm [7] was coupled with the XFOIL code [8].

77.2.1 Design Method

The new airfoil shapes were created by modifying an existing airfoil using a shape perturbation function. This method is based on adding smooth perturbations to an initial airfoil. The smooth perturbations Δy are a linear combination of base functions P_k as

$$\Delta y(i) = \sum_{k=1}^{N} \delta_k P_k(i)$$

Base functions can be chosen to be any set of smooth functions [9]. This study is based on the work of [5]. The airfoil is split to an upper and a lower side with leading edge and trailing edge points fixed at $x=0$ and $x=1$, and the corresponding perturbation functions can be written for the upper surface as

$$\Delta y_u(i) = \sum_{k=1}^{N} \delta_{ku} P_{ku}(i)$$

and for the lower surface as

$$\Delta y_l(i) = \sum_{k=1}^{N} \delta_{kl} P_{kl}(i)$$

The shape functions for the upper and lower surface along the x-coordinate are

$$P_{ku}(i) = \sin^{\xi}(\pi x_u(k,i)^{g(k)})$$

and

$$P_{kl}(i) = \sin^{\eta}\left(\pi x_l(k,i)^{g(k)}\right)$$

The subscripts u and l symbolize the upper and lower surface of the airfoil, k is the index of the shape modes and i is the index of x and y coordinates. Variable g corresponds to the location of maximum of the base function. This variable is defined by the designer and gives him more influence on the decision-making process.

77.2.2 *Optimization*

The optimization objective was to maximize lift coefficient and lift-to-drag ratio. The objective function was expressed as

$$obj = \min\left(\frac{1}{C_p}\right), C_p = f(C_L, \frac{L}{D})$$

The coefficients C_L and L/D were weighted for clean and rough conditions with more emphasis on the rough case.

The objective function was identified as a smooth non-linear constrained function so "fmincon" from the MATLAB optimization toolbox was chosen as the appropriate solver. The solver algorithm was set to "SQP".

As previously mentioned, the optimization algorithm was coupled with a flow solver XFOIL [8]. The Reynolds number was set to 3×10^6 and the accounted angles of attack varied between 0° and 10°. This set of angles of attack was chosen to keep the lift coefficient in the linear region, which made the calculations more reliable. Free transition was modelled e^N with the envelope method with $N=9$ as the default value. Transition was forced by fixing the upper and the lower transition points at 5%, and 10% chords from the leading edge.

In this study, the initial airfoil for the design method was DTU-LN221. This is an airfoil previously developed at DTU and was chosen as a good starting point.

The DTU-LN221 has a favourable after stall behaviour, and assures that the created airfoil will belong to the new airfoil family.

Response parameters from XFOIL were used either directly as design objectives (lift coefficient and lift/drag ratio) or they were bounded in a constraint function (boundary layer thickness or lift coefficient degradation in the forced case). Other constraints imposed came from structural requirements such as position of maximum thickness, skewness of the airfoil, trailing edge thickness, etc.

The weighting coefficients or amplitudes δ_{ku} and δ_{kl} from the shape functions were the design variables, and with the two additional power factors of ξ,η the total number of design points was $2N+2$[5].

The objective of this study was to create high performing, low noise wind turbine airfoils that can accommodate a more demanding internal structure. Structural integrity was indirectly expressed through a concept of effective thickness to chord ratio. This is defined by the value of moment of inertia of a spar cap on the actual blade compared to the one which is uniform over the same width at the maximum relative thickness [10].

77.2.3 Optimization Results

The resulting DTU-230 and DTU-236 airfoils are shown in Fig. 77.1 together with three other airfoils in the same new family of airfoils DTU-LN2xx. It can be noted that the new thick airfoils have a very good shape compatibility with the middle thickness airfoils.

In order to see the performance of the DTU-230 airfoil, comparisons with two commonly used wind turbine airfoils—NACA 63-430 and FFA-W3-301 are carried out. Figure 77.2 shows the lift coefficients and lift-to-drag ratio of the three airfoils versus angle of attack for both clean and rough case. The coefficients have been computed with XFOIL for an angle of attack range of 0°–20° with a step 0.5°, $Re=1.47\times10^6$, $N=9$. Rough surface cases were simulated with a fixed transition of

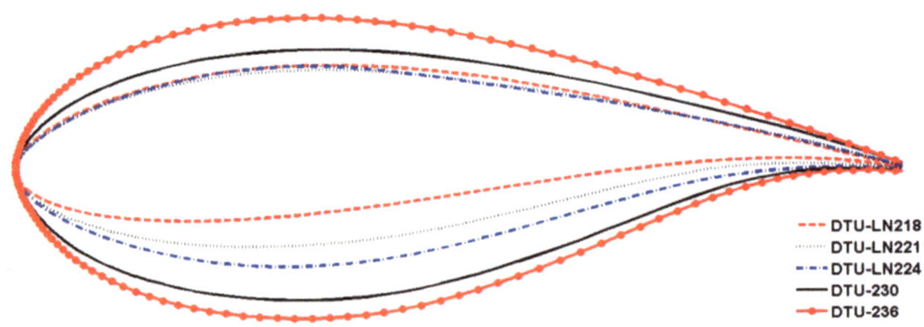

Fig. 77.1 New airfoil family DTU-LN2xx

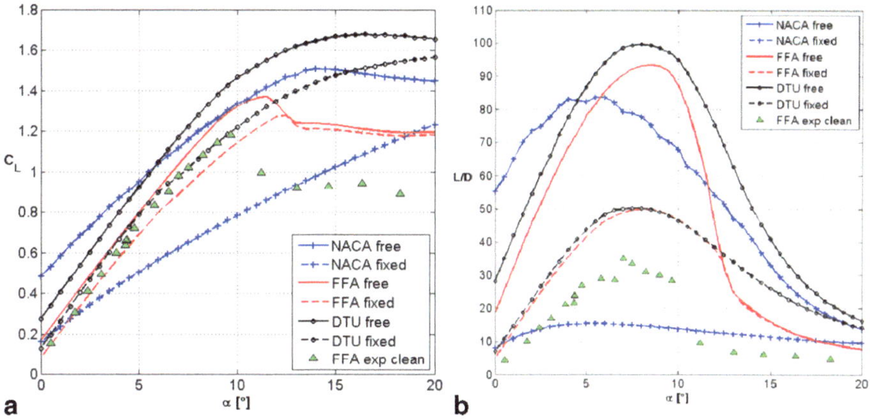

Fig. 77.2 Comparison of performance of airfoils NACA 63-430, FFA-W3-301 and DTU-230. *Left*: lift coefficients versus angle of attack; *Right*: lift to drag ratio versus angle of attack. The coefficients were computed with XFOIL for angle of attack range 0–20° with a step 0.5°, $Re = 1.47 \times 10^6$ with free and fixed transition (*5% suction, 10% pressure side*). The experimental data are for airfoil FFA-W3-301, clean surface, $Re = 1.47 \times 10^6$

5% chord on the suction side and 10% on the pressure side. Results are compared with experimental data for FFA-W3-301 measured in clean case at $Re = 1.47 \times 10^6$[3].

As can be noted, DTU-230 has higher maximum lift coefficient in both rough and clean cases. The difference between the maximum lift coefficients in the rough and the clean case, that is, roughness sensitivity is also quite good for DTU-230. The least roughness sensitive airfoil is FFA-W3-301; however, this merit was paid by a rather low lift. Comparison of lift-to-drag ratio over a set of angles of attack demonstrates that DTU-230 is the highest performing airfoil (Fig. 77.2, right).

77.3 The Experiment

77.3.1 The Wind Tunnel

The wind tunnel at LM Wind Power is a closed circuit, variable fan speed tunnel, with a cooling system to ensure constant flow temperature. The flow quality is very high with turbulence intensity levels around 0.1%. The tunnel has a contraction of 10–1 and is equipped with specially designed corner vanes, a honeycomb structure and three fine mesh screens. The width of the test section is 1.35 m and the height is 2.7 m. The model chord length of 0.9 m at the maximum wind speed of 105 m/s gives a chord Reynolds number of 6×10^6 and a Mach number of $M = 0.3$.

77.3.2 The Measurement Campaigns

The DTU-230 and DTU-236 airfoils were tested at several Reynolds numbers and in different surface configurations with different devices (such as vortex generators and Gurney flaps). The tested Reynolds numbers were $Re = 1.5, 3, 4, 5$ and 6×10^6. The airfoil performance in rough conditions was investigated using a standard zig-zag tape and bump tape. Data presented in this chapter are the measurements with zigzag tape of 0.4 mm thickness placed at 5 % chord on the suction side and 10 % chord on the pressure side.

77.4 Results and Discussions

77.4.1 DTU-230

Figure 77.3 shows a comparison of expected lift coefficients calculated with XFOIL and measured ones. The most noticeable result is that XFOIL over predicts lift and does not capture stall effects. The XFOIL was developed primarily for thin airfoils so this behaviour was expected. In the linear region, the difference is not that big but as soon as stall approaches, the difference between the calculated and measured data is greater. Therefore, the decision to limit the angle of attack to 10° in the optimization proves to be valid. The same effect can be observed when the zigzag tape is applied to the test model surface; only in this case, the difference is more pronounced already at lower angles of attack. The airfoil with "tripped" boundary

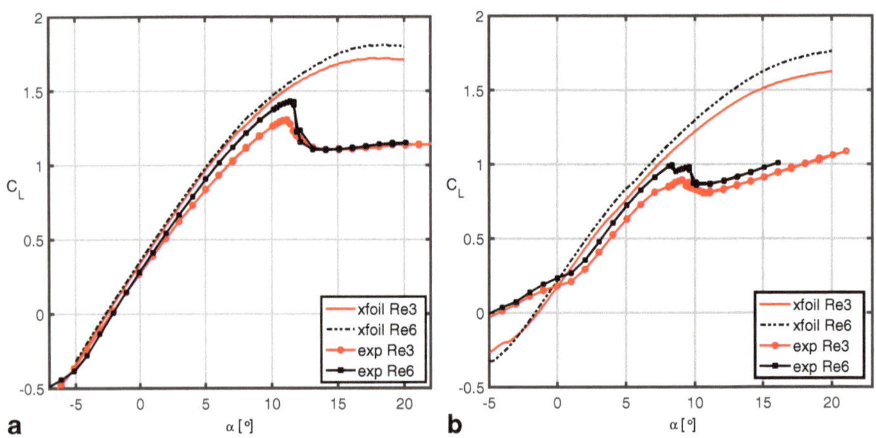

Fig. 77.3 DTU-230–Comparison of lift coefficients versus angles of attack computed with XFOIL and experimental data from LM wind tunnel at Reynolds numbers = 3 and 6 million. *Left*: free transition in XFOIL and clean surface of the test model; *Right*: forced transition in XFOIL and applied zigzag tape at 5 % chord on the suction side and 10 % chord on the pressure side of the test model

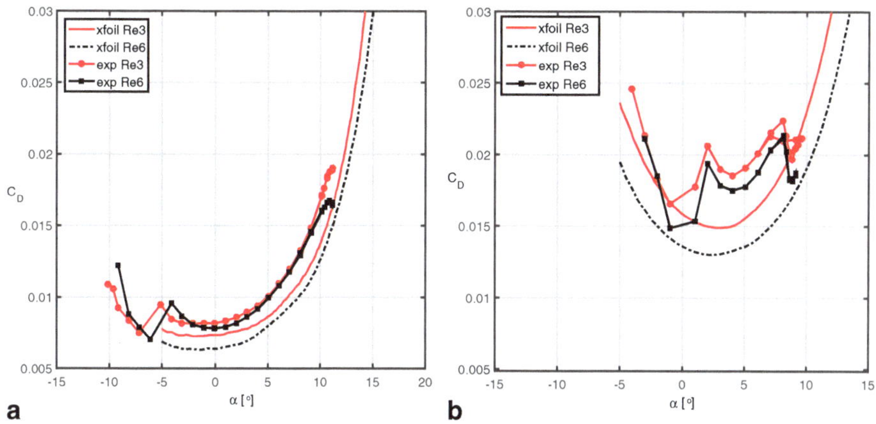

Fig. 77.4 DTU-230–Comparison of drag coefficients versus angles of attack computed with XFOIL and experimental data from LM Wind tunnel at Reynolds numbers = 3 and 6 million. *Left*: free transition in XFOIL and clean surface of the test model; *Right*: forced transition in XFOIL and applied zigzag tape at 5 % chord on the suction side and 10 % chord on the pressure side of the test model

layer stalls at around 8° so the range of angles of attack in optimization could have been even more conservative. The power coefficient in the objective function was a blended function of both the clean and the rough cases. Since the ranges of attack should be different for those cases, the power coefficient should have been also adjusted to that.

The obvious change of zero lift angles between "clean" and "tripped" airfoils was expected and is similar to that in cases with other thick airfoils [3].

When looking into drag coefficients (Fig. 77.4) the opposite can be observed: XFOIL is under predicting. Although drag coefficients were obtained from pressure taps on the airfoils, load cells and a wake rake, only the data from the wake rake are shown here. To protect the wake rake in stall conditions, it was removed from the wake just before the stall point.

The effect of increasing Reynolds number can be seen in both lift and drag coefficients—the lift increases and the drag decreases.

The overall properties of the DTU-230 are quite good. The aerodynamic characteristics are satisfactory and its geometrical simplicity and compatibility with the whole DTU-2xx family is an advantage for manufacturability and reliability. This ensures that the performance of the blade built with this airfoil will match the designed performance.

77.4.2 DTU-236

The DTU-236 airfoil was quite challenging to measure due to the large relative thickness and a high aspect ratio of the test model (aspect ratio = 1.5). To alleviate the problem of returning flow stall cells forming on the side edges of the airfoil, the

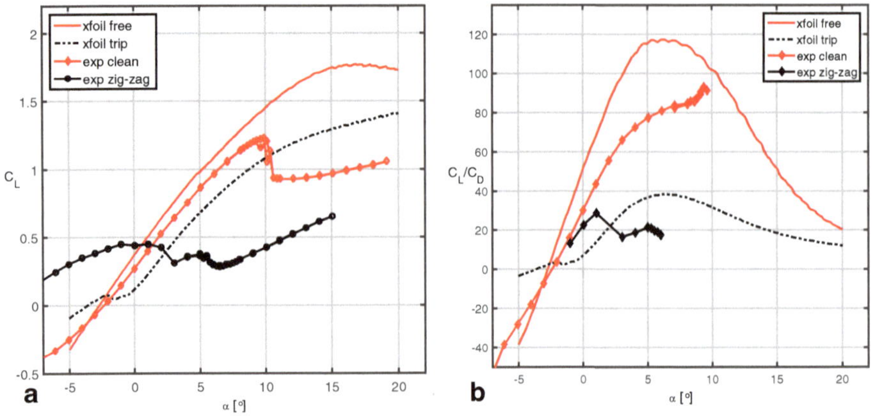

Fig. 77.5 DTU-236–Comparison of lift coefficients and lift-to-drag ratio versus angles of attack computed with XFOIL and experimental data from LM Wind tunnel at Reynolds number = 6 million; test model surface condition—clean and with applied zigzag tape at 5 % chord on the suction side and 10 % chord on the pressure side, corresponding free and forced (*trip*) boundary layer transitions in XFOIL. Only drag coefficients measured with a wake rake are included in the lift-to-drag ratio plot

wall boundary layers were energized using a blowing system built in the sidewalls of the test section. The results shown in Fig. 77.5 are only for the highest measured Reynolds number (6 million). As was the case with the DTU-230, the angle of attack was limited to 10° in the optimization. Given that the airfoil is stalled around 9° in the clean case, this limit should have been set to a lower angle. Only wake rake drag measurements are included in the lift-to-drag ratio plots. Since the airfoil stalled at around 9° and 6° in clean and tripped surface conditions, the data range is rather small.

Figure 77.5 also shows an even greater discrepancy in calculated and measured lift coefficients than with the DTU-230. This clearly proves XFOIL was not designed to handle such thick airfoils and should be used with precaution. Results for "tripped" calculations and data obtained while a zigzag tape was applied to the model surface indicate neither of the results should be trusted: XFOIL because it was not made for such cases and the experimental data should be verified in a tunnel with less blockage or with a smaller model. One possibility for this particular wind tunnel would be to use a model with a shorter chord line, for example choosing a length of 0.6 m. A smaller absolute thickness and larger aspect ratio would reduce the tunnel blockage but the maximum achievable Reynolds number in that case would be around 4 million. That is a compromise worth investigating.

Numerical flow solvers should receive more attention since neither the existing panel codes nor the computational fluid dynamics (CFD) codes have the ability to accurately capture stall and post stall behaviour of very thick airfoils compared to wind tunnel tests, even though the CFD performs a bit better as is shown in [11].

Given the high relative thickness and a small camber, the DTU-236 airfoil performs quite well. Rotational effects were not included in this analysis, but are assumed to lessen the roughness sensitivity [2].

77.5 Conclusion

The introduced design method is an effective tool to develop thick high performing wind turbine airfoils. The shape perturbation function was integrated in a numerical constrained optimization algorithm coupled with XFOIL. The results from the wind tunnel confirmed that using XFOIL in the linear region for DTU-230 was reasonable. However, for the very thick airfoils, using a CFD may have been a better option and should be considered in the future even though it is much more time-consuming.

Since the optimization process is rather fast, the design could be further improved with constraints that are more elaborate without significantly influencing the optimization performance.

References

1. Vestas Wind Systems A/S (2014) World's most powerful wind turbine now operational. http://www.vestas.com/en/media/news#!140128_nr_vws. Accessed 31 July 2014.
2. Timmer W, van Rooij R (2003) Summary of the delft university wind turbine dedicated airfoils. J Solar Energy Engg 125(4):488–496
3. Bertagnolio F, Sorensen N, Niels N, Johansen N, Fuglsang P (2001) Wind turbine airfoil catalogue. Tech.rep., Risø-R-1280(EN).
4. Sareen A, Chinmay SA, Selig MS (2013) Effects of leading edge erosion on wind turbine blade performance. Wind Energy. doi:10.1002/we.1649
5. Zhu WJ, Shen WZ (2013) Integrated airfoil and blade design method for large wind turbines, Proceeedings of the 2013 International Conference on Aerodynamics of Offshore Energy Systems and Wakes, Lyngby, Denmark, pp. 318–326
6. Eppler R, Somers DM (1980) A computer program for the design and analysis of low speed airfoils, NASA TM.80210.NASA Langley Research Center, Hampton
7. MATLAB (2012) version 8.00.783 (R2012b). The Math-Works Inc., Natick
8. Drela M (1989) Xfoil: an analysis and design system for low reynolds number airfoils. In: Mueller T (ed) Low reynolds number aerodynamics, vol 54 of lecture notes in engineering. Springer Berlin, Heidelberg, pp. 1–12.
9. Hager J, Eyi S, Lee K (1994) Two-point transonic airfoil design using optimization for improved off-design performance. J Aircraft 31(5):1143–1147
10. Jamieson P (2011) Innovation in wind turbine design, 1st edn. Wiley, West Sussex
11. Zahle F, Bak C, Sørensen N et al (2014) Design of the LRP airfoil series using 2D CFD. J Phys: Conference Series 524

Chapter 78
A Novel Topology for Enhancing the Low-Voltage Ride-Through Capability for Grid Connected Wind Turbine Generators

R. A. Ibrahim, M. S. Hamad, Y. G. Dessouky and B. W. Williams

Abstract Energy shortage and environmental pollution have led to the increasing demand of using renewable sources for electricity production. Currently, power generation from wind energy systems (WES) is of global significance and will continue to grow during the coming years leading to concerns about power system stability where wind farms replace conventional generating technologies that use fossil fuels as the primary energy source. One of these concerns is low-voltage ride-through (LVRT). In this chapter, a novel topology based on the use of magnetic amplifier for enhancing the low voltage ride through capability for grid connected permanent magnet synchronous generators (PMSG).

Keywords Component · Permanent magnet synchronous generator · Low-voltage ride-through · Wind energy grid connection · Magnetic amplifier · Renewable energy

List of Symbols

C_p	Performance coefficient of wind turbine
f	Supply frequency (Hz)
i_b	Boost converter inductor current (A)
i_{dg}, i_{qg}	*dq* PMSG stator current components (A)
i_a, i_b, i_c	Three phase grid currents (A)
i_d, i_q	*dq* axis components of grid currents (A)
i_d^*, i_q^*	*dq* reference grid current components (A)
i_s	PMSG supply current (A)
i_c	Magnetic amplifier control winding current (A)
I_{inv}	Inverter IGBT current (A)

Y. Dessouky (✉) · R. A. Ibrahim · M. S. Hamad
Electrical and Control Engineering Department, Arab Academy for Science and Technology & Maritime Transport, Alexandria, Egypt
e-mail: ygdessouky@yahoo.com

B. Williams
Electronics and Electrical Engineering Department, Strathclyde University, Glasgow, Scotland, UK

A. Sayigh (ed.), *Renewable Energy in the Service of Mankind Vol I*,
DOI 10.1007/978-3-319-17777-9_78

I_{rec}	Rectifier diode current (A)
L_f	Booster inductance (H)
L_d, L_q	PMSG *dq* axis inductances (H)
P_w	Power absorbed by windmill
P_{Grid}, Q_{Grid}	Grid active & reactive power (Watt, VAR)
P_{Gen}, Q_{Gen}	PMSG active & reactive power (Watt, VAR)
R_a	Stator armature resistance (Ω)
R	Windmill blade radius (m)
T_e	Electromagnetic torque (Nm)
V	Wind velocity (m/sec)
V_{inv}	Inverter IGBT voltage (V)
v_b	Input capacitor voltage of boost converter (V)
v_{dc}	DClink capacitor voltage (V)
V_{rec}	Rectifier diodes reverse voltage (V)
v_a, v_b, v_c	Three phase grid voltages (A)
β	Blade pitch angle (BPA)
δ	Load angle, between *q*-axis and phase "a" voltage, (rad)
φ_f	Permanent magnet flux linkage (Wb)
φ_d, φ_q	*dq* axis flux-linkages (Wb)
θ	Phase angle of grid voltage (degree)
λ	Tip-speed ratio (TSR)
ρ	Air density
ω	Electrical rotor angular speed (rad/s)

78.1 Introduction

The increasing consumption of the traditional fossil energy raises problems such as global warming, pollution, and energy security. Renewable energy sources are the most promising alternative energy in future. With the increasing penetration of wind energy and the improvement of its technology make it important to carefully reconsider the operating characteristics of wind farms, especially during grid fault conditions. Previously, disconnection of distributed generators (DGs) if one of the parameters (e.g., voltage, frequency, etc.) was beyond or below the permissible band was not a problem as the installed DG capacity was low. However, with installed megawatt (MW) of wind power, disconnection of the DG is no longer valid. Disconnection of many wind turbines (WT) during fault and evoked loss of a large share of generation can be dangerous for the stability of the power system operation. Thus, new connection requirements for DG units have been introduced by many system operators [1]. For these reasons, WT are expected to behave like conventional synchronous generators during voltage dips, remaining connected and supplying reactive power during and after the voltage dip. As an example, the low-voltage ride-through (LVRT) requirement for the UK electrical grid is shown in

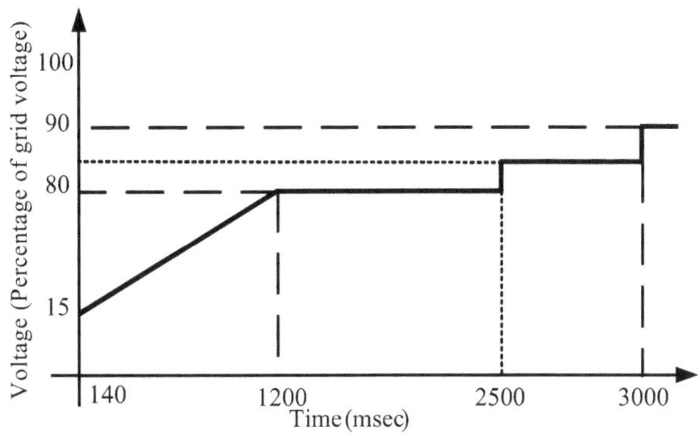

Fig. 78.1 British grid code for fault ride through

Fig. 78.1 [2], where the wind turbine has to remain connected to the network for grid faults that last up to 140 ms and upon the restoration of voltage to 90 % of nominal.

During normal operation of WT and with the absence of grid voltage dips, the wind energy conversion system (WECS) extracts maximum power from the wind regardless of wind speed. This function of the turbine controller is known as maximum power point tracking (MPPT) [3–5]. The maximum extracted power passes through many stages of power electronic conversions till finally being exported to the grid. When a fully rated converter WT is connected to an alternating current (AC) network and a network fault occurs, the WT direct current (DC) link voltages will rise rapidly as the WT grid side converters are prevented from transmitting all the maximum active power coming from the generators. Therefore, to maintain the WT's DC link voltages below their upper limit, the excess power has to be dissipated or the generator power has to be reduced.

Methods for LVRT capability enhancement and overcoming the rise in DC link voltage vary according to the turbine variables and utility grid variables. To enhance the LVRT capability for WT, blade pitch angle control can be used as in [6] through shifting the WT blades from an angle where it extracts maximum power to another angle of attack, thus reducing the amount of power being extracted from the wind and compensating the rise in the DC link voltage. Another suggested solution is to oversize the DC link capacitor to compensate for the DC link voltage rise during grid dips as in [7]. Active crowbar technique is considered one of the mostly used solutions for LVRT. By the insertion of a resistor in the DC link controlled by a power electronic switch, excess energy is dissipated and a balanced power flow is restored [8]. Other methods for LVRT improvement can be through de-loading of the wind turbine as demonstrated in [9] by the proper control of generator side converter, grid side converter control. The former controls the generator electromagnetic torque by controlling the generator armature current while the latter maintains the DC link voltage constant by controlling the grid side converter [10]. Reactive power and voltage compensation (shunt/series, shunt and series) has been one of the effective techniques in enhancing the LVRT capability on a larger scale such as wind farm level at the point of common coupling. Reactive power compensators employ the use of static Reactive power compensators (SVC), static synchronous compensator (STATCOM), magnetic energy recovery switch (MERS), dynamic voltage restorer (DVR), and thyristor controlled series capacitor (TCSC) [11–14].

This chapter proposes a novel topology that can be used to enhance the LVRT capability for permanent magnet synchronous generators (PMSG) grid connected WT. The topology is based on series voltage compensation using magnetic amplifiers. Magnetic amplifier has been known for years and is used in varaiety of applications such as position control. It is a device which acts as a variable inductance if inserted between a source of e.m.f and a load. Through proper control of its control winding current, magnetic amplifier will eventually act as a current limiter and a series voltage compensator at the event of transient voltage dip to enhance LVRT capability.

The basic principles of magnetic amplifiers design considerations are demonstarted to show the effectiveness of magnetic amplifier in reducing the DC link voltage rise as well as limiting the stresses on the power electronic converters in case of grid voltage dips.

78.2 System Configuration

78.2.1 System Modeling

The WECS employed for grid connection of PMSG wind turbine is found in Fig. 78.2 which consists of the drive train, PMSG, diode bridge rectifier, boost converter circuit and a voltage source inverter (VSI). As mentioned earlier, MPPT is one of the essential features of grid connected WT. The techniques used for PMSG

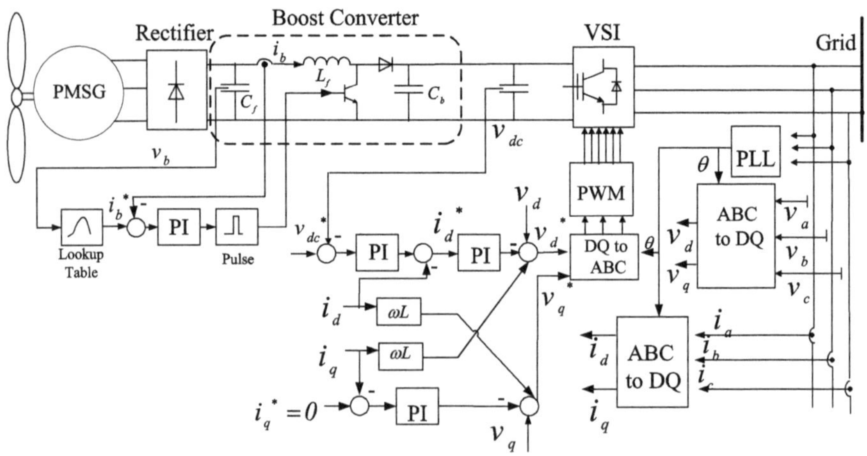

Fig. 78.2 Block diagram for system under investigation

Table 78.1 PMSG data [21]

Parameter	Value
Base power (MVA)	1.5
Base voltage (V)	$\frac{690}{\sqrt{3}}$
Base frequency (Hz)	11.5
Pole pairs of PMSG	40
PMSG inertia constant (pu)	0.5
Shaft stiffness (pu)	2
Rated generator torque (pu)	1
Rated generator line voltage (pu)	1
Generator inductance in the *d* frame (pu)	0.7
Generator inductance in the *q* frame (pu)	0.7
Generator stator resistance (pu)	0.01
Flux of the permanent magnets (pu)	0.9

WT are vast and diverse. One of the oldest approaches is the tip speed ratio (TSR) technique , where the turbine pitch angle is regulated according to the measured wind speed. TSR's main disadvantage lies in the cost of the anemometers used as well as its low accuracy in the presence of wind gusts [15]. Hill climbing search technique (HCST) is another method for MPPT, where the PMSG shaft speed is perturbed in small steps while observing the wind turbine mechanical power [16, 17]. The disadvantage of HCST lies in the slow responses to wind changes as well as oscillation around the optimum point making the HCST suitable for small-scale WT. Finally, optimum relationship based (ORB) MPPT, is the most popular technique used with large scale WT. Lookup tables are recorded with stored values of optimal generator speed and the corresponding maximum power (or maximum torque) at various wind velocities. ORB requires system pre-knowledge and many field tests to obtain accurate optimal relation between system parameters [18–20].

In the absence of LVRT enhancement methodology, grid voltage dips can cause a serious rise in the DC link voltage which might lead to the destruction of the power converters. To demonstrate this effect, Fig. 78.2 shows the block diagram being used in the simulation analysis of a 1.5 MW grid connected PMSG WT whose parameters can be found in Table 78.1. System description can be found in the following sections:

78.2.1.1 Wind Turbine Model

The power absorbed by windmill from air, P_w, is given by (1)

$$P_w = \frac{1}{2}\rho\pi R^2 V^3 C_p \tag{78.1}$$

When wind velocity is invariable, the power P_w is directly proportional to *Cp*.

$$\lambda = \frac{\omega R}{V} \tag{78.2}$$

$$C_p = 0.73\left(\frac{151}{\lambda_1} - 0.58\beta - 0.2\beta^{2.14} - 13.2\right)e^{\frac{-18.4}{\lambda_1}} \tag{78.3}$$

$$\lambda_1 = \frac{1}{\left(\frac{1}{\lambda - 0.02\beta} - \frac{0.08}{\beta^3 + 1}\right)} \tag{78.4}$$

As it can be noticed from (78.1)–(78.4), *Cp* varies as a function of both the TSR and the Blade pitch angle (BPA). For every fixed Blade pitch angle (BPA) there is a corresponding *Cp*–λ curve each has a peak *Cp* value corresponding to an optimum TSR. Maximum *Cp* is achieved at optimized rotational speed ω when β is equal to zero [10].

78.2.1.2 Permanent Magnet Synchronous Generator

Variable speed direct driven PMSG WT have gained increasing interest over the externally excited synchronous machine due to a significant reduction in magnet prices and magnetic material characteristics improvement. Variable-speed operation of a Wind turbine energy systems (WTES) has many advantages such as rapid response to wind speed and torque variation, causing less wear and tear on the tower and other components in the drive train. Variable-speed systems can also increase the production of the energy and reduce the fluctuation of the power injected into the grid [22]. A *dq* model of PMSG has been developed using Eqs (78.5)–(78.7) found in [23].

$$v_d = R_a i_d + L_d \frac{\partial i_d}{dt} - \omega L_q i_q \tag{78.5}$$

$$v_q = R_a i_q + L_q \frac{\partial i_q}{dt} + \omega \varphi_f + \omega L_d i_d \tag{78.6}$$

$$T_e = \frac{3}{2} \rho i_q ((L_q - L_d) i_d + \varphi_f) \tag{78.7}$$

78.2.1.3 Three Phase Diode Rectifier

For grid connected PMSG WT, generator's output voltage and frequency must be maintained within certain values to ensure proper grid interconnection. Since the PMSG output power is of fluctuating nature due to wind speed variability, power electronic converters are utilized to overcome such problems. Three phase diode bride rectifier has been used for AC–DC conversion of the generator output power.

78.2.1.4 Boost Converter

A boost converter has been added in the DC link to extract maximum power from the WT at any given wind speed as well as expand its scope of low speed operation. Using the ORB technique for MPPT, an optimal relation between booster converter inductor current i_b versus boost converter capacitor DC voltage v_b can be found [18]. This is true as v_b is proportional to the PMSG electromotive force, and the latter is proportional to the rotational speed. By sensing the DC voltage v_b and using the optimal relation between v_b and i_b, the chopper circuit extracts maximum power from the PMSG at any wind speed. Although this method requires many offline tests, it accounts for turbine effeciency and takes into consideration the total power converters losses (diode and chopper).

78.2.1.5 Three Phase Pulse Width Modulation Voltage Source Inverter (PWM VSI)

A current regulated PWM VSI maintains the output voltage of boost converter (DC link voltage) at a fixed value by balancing the input and output power of the DC link. The PWM VSI employs two control loops, the inner loop is for the grid current control while the outer loop is for the DC link voltage control. Decoupling the grid currents in the *dq* axis frames provides means for controlling the active grid current at unity power factor by setting the *q*-axis current to zero.

78.2.2 System Behavior for Voltage Dips and Low Voltage Ride—Through Capability

The system in Fig. 78.2 has been studied for several grid voltage dips of 30, 60, and 90 % at 0.4, 0.8, and 1.2 s to demonstrate the impact on the system elements and the simulation results are shown in Figs. 78.3, 78.4, 78.5, 78.6, 78.7, 78.8, 78.9. All sim-

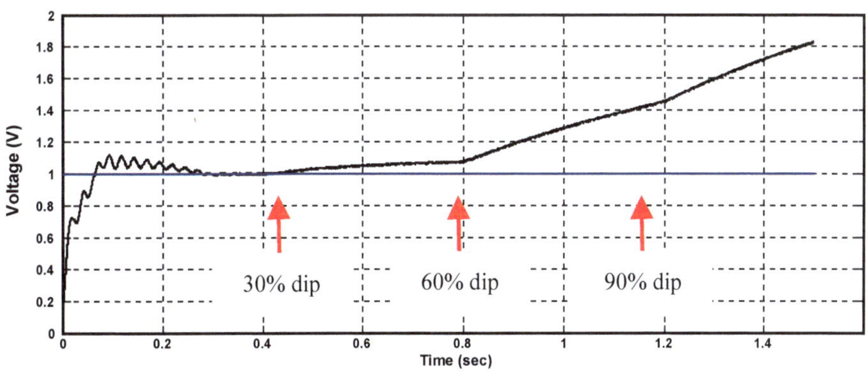

Fig. 78.3 DC link voltage for 30, 60, and 90 % voltage dip

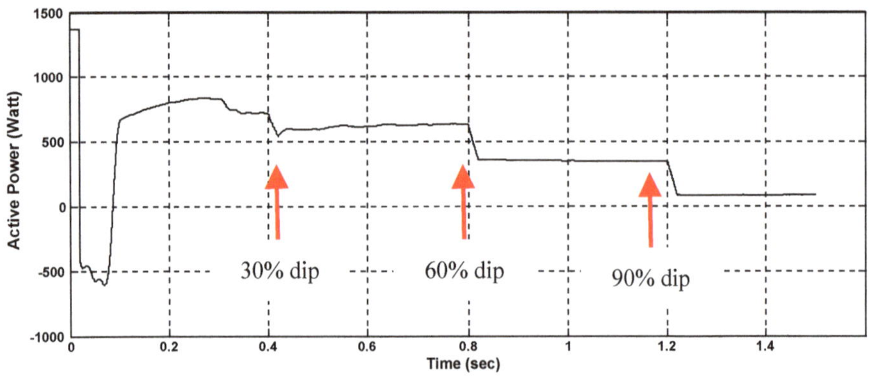

Fig. 78.4 PMSG active power for 30, 60, and 90 % voltage dip

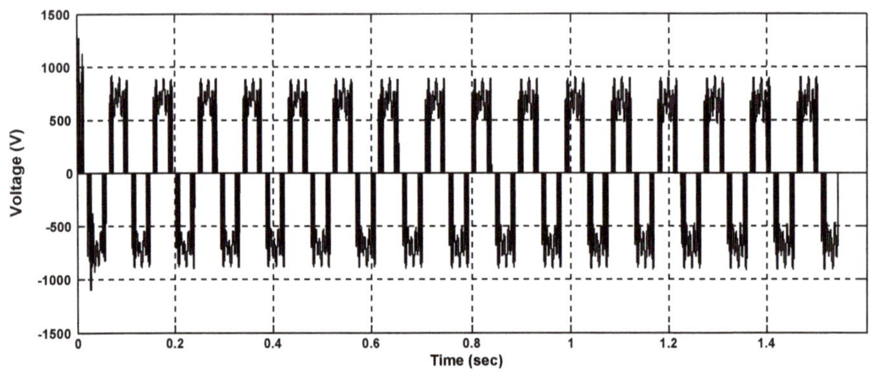

Fig. 78.5 PMSG terminal voltage for 30, 60, and 90 % voltage dip

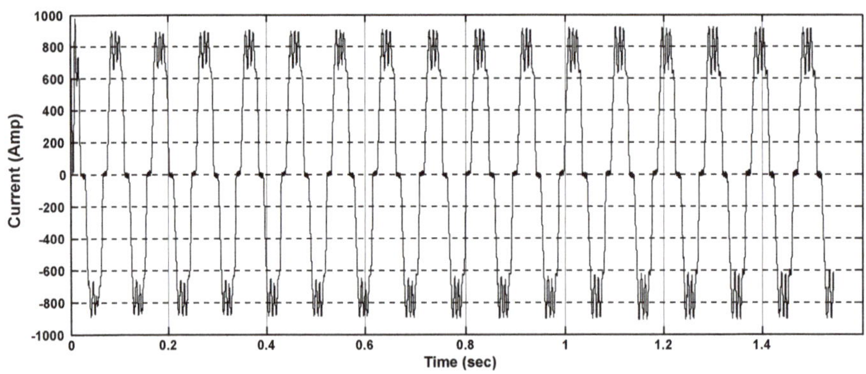

Fig. 78.6 PMSG supply current for 30, 60, and 90 % voltage dip

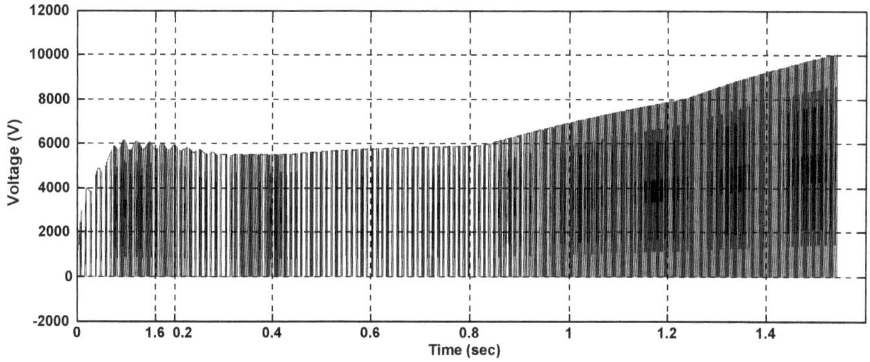

Fig. 78.7 IGBT switch voltage for 30, 60, and 90 % voltage dip

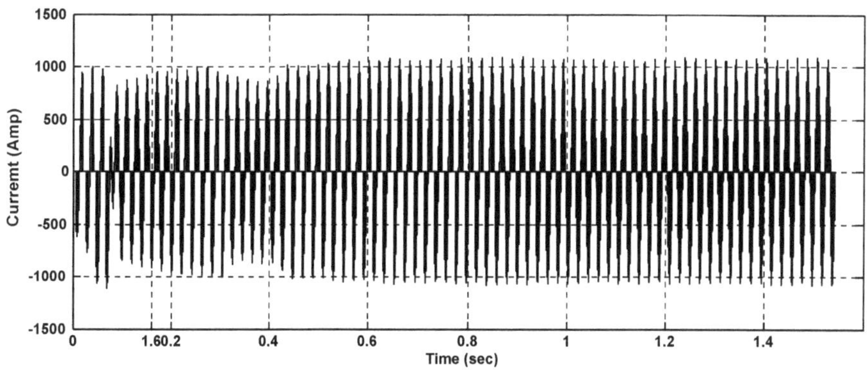

Fig. 78.8 IGBT switch current for 30, 60, and 90 % voltage dip

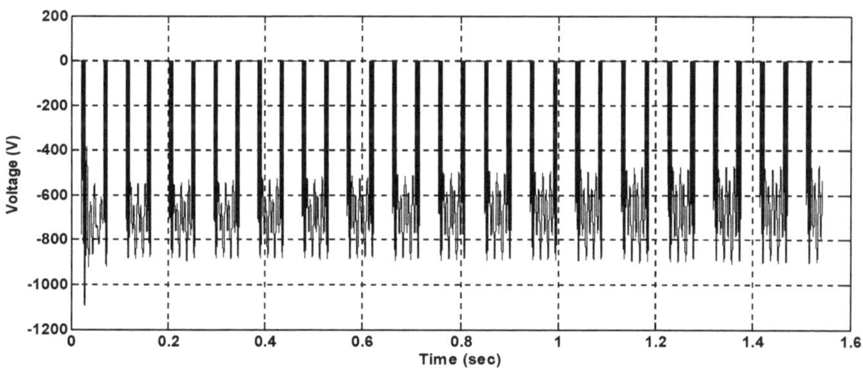

Fig. 78.9 Diode reverse voltage for 30, 60, and 90 % voltage dip

ulation results used in the analysis are carried out using Simulink–Matlab software. During the voltage dip incidence, the grid power will fall in response to the voltage dip while the PMSG will continue to extract maximum power from the wind turbine through the boost converter as seen from Figs. 78.5 and Fig. 78.6. Since the grid power has fallen to very low value, the power extracted from the wind will be transformed into stored energy in the DClink capacitor. As for Fig. 78.3 and Fig. 78.7, grid voltage dips cause a rapid rise in the DC link voltage, reaching almost to double the normal voltage which the capacitor can tolerate when a 90 % dip occurs. The rise level depends on the capacitor size, dip voltage duration, and network effective impedance, which can be more than just double the normal capacitor operating voltage. The DClink voltage rise accompanies more voltage stresses on the PWM VSI as seen in Fig. 78.7 while the diode bridge rectifier suffers from neither voltage nor current stress as found in Fig. 78.8 and Fig. 78.9.

78.3 Novel Topology for Low-Voltage Ride-Through Enhancement

As previously demonstrated, grid voltage dips can cause severe damages to the power converters due to the energy unbalance between the PMSG output power and grid electrical power which must be dissipated in the DClink capacitor accompanied by a rapid rise in the capacitor's voltage. A new topology is suggested in this chapter to enhance the LVRT capability. The topology is based on using magnetic amplifier as series voltage compensators to compensate the DClink voltage rise during voltage dips.

78.3.1 Magnetic Amplifiers Basic Principles

Magnetic amplifiers, also known as transductors, have been used in applications such as instrumentation, relays, position servo mechanism, and automatic battery chargers since 1916. It had established itself as a powerful tool as it is a rugged and trouble-free device. Magnetic amplifiers are known for their high efficiency (up to 90 %), reliability (long life, freedom from maintenance and reduction of spare parts inventory), ruggedness (shock and vibration resistance, high overload capability and freedom from effects of moisture), and needs no warm-up time. They have been used in vast applications, both in low and high power, one of which is voltage and frequency control. [24–27]. For years, the magnetic amplifiers were the topic of research and development such as in understanding its non-linear characteristics [25], steady state and transient behavior [26], possible circuit configurations for single and three-phase operation [27, 28] and its design for vast applications [29].

Fundamentally, the magnetic amplifier is a device whose input (control) signal is a voltage applied to a coil linking an iron cored system as shown in Fig. 78.10.

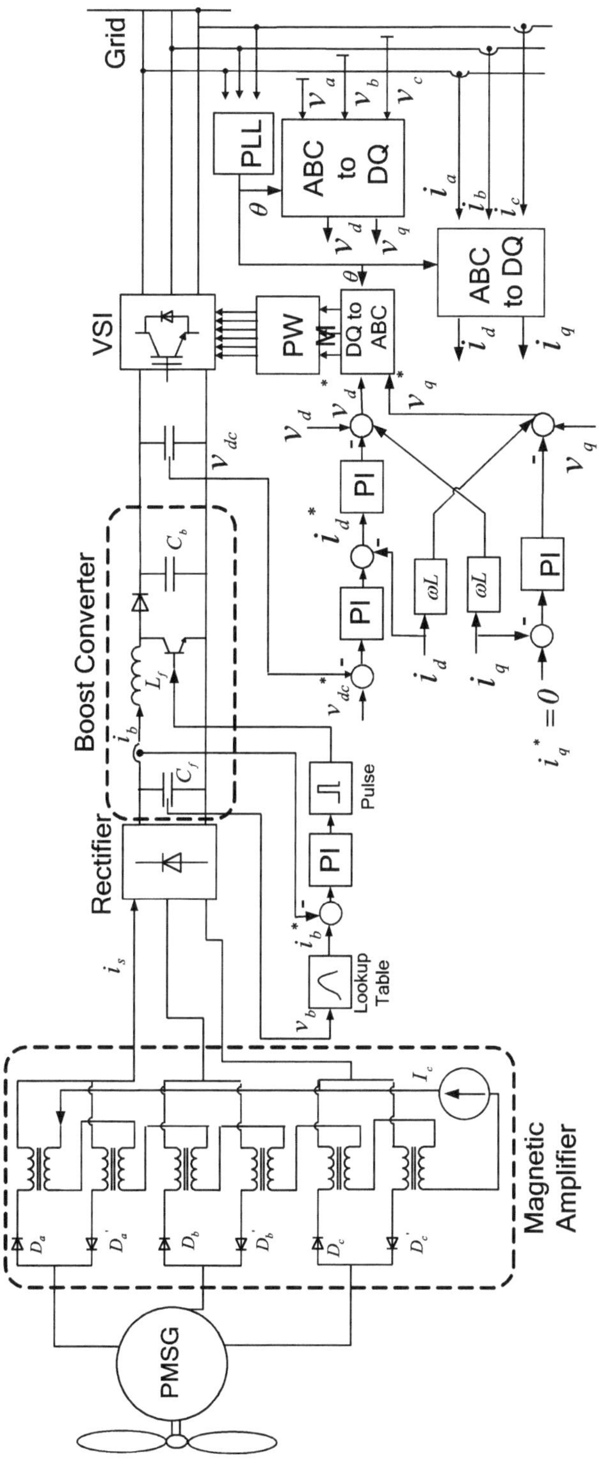

Fig. 78.10 PMSG WECS with magnetic amplifier

Other coil linking the system (main coil) carry the load current which is supplied from a low impedance source of alternating e.m.f. Variation of the control current signal changes the flux level in the core and modifies the operating conditions in the load circuit so that the load current is modulated according to the variations of the control signal. Changing the control current I_c demagnetizes the core and changes the effective impedance seen by the main coil. This is done by forcing the operating point away on the *B–H* curve in Fig. 78.10 from the saturation region at point A (effective impedance seen by the main coil is low) to the linear region at point B (effective impedance seen by the main coil is high). The basic magnetic amplifier consists of two cores, one being connected in opposite dot notation to the other. The reason for that is if one core is used only, the main and control coils are mutually coupled and an alternating voltage of frequency f will be induced in the control circuit. Thus an interaction will occur between the two circuits, so disrupting the amplification action.

On the other hand, if the coefficient of coupling is small, then a large signal current would be required to polarize the core and the power level at the control winding would be reasonably high [30]. Splitting the load winding into two coils connected in series opposition neutralizes the transformer action. Another modification can be made to the circuit is to include a diode in series with each of the two main windings. In this case, full utilization of both sides of the AC would give higher efficiency and good waveform as well as a considerable reduction in time delay that can reach up to half cycle of the supply frequency.

78.3.2 Magnetic Amplifiers and Enhancement of Low-Voltage Ride-Through Capability

After explaining the basic principle of operation of the magnetic amplifier, it is possible to use it as one method of series compensation. Changing the control winding current of the magnetic amplifier changes the main winding effective impedance, thus the magnetic amplifier acts as a variable inductance if inserted between a source of e.m.f and a load. To validate this idea, the magnetic amplifier have been used and tested for LVRT enhancement as seen in Fig. 78.11. In this configuration, each of the three phases of the PMSG is split into two windings as to avoid demagnetization of the magnetic amplifier in the negative half cycle of the supply voltage. Each winding is connected in series with a rectifying diode which helps the magnetic amplifier cores to reach saturation faster [30]. Six magnetic amplifiers have been used in the following analysis, each rating 500 VA, 100:45 turn and connected as seen in Fig. 78.11.

During normal operation with the absence of any grid faults, the magnetic amplifier is driven into saturation by the control winding current (controlled current source in Fig. 78.11) with minimum voltage drop exerted on its main winding terminal. The only source of voltage drop will be due to the diode forward voltage drop as well as the impedance of the main winding. To generate the control winding

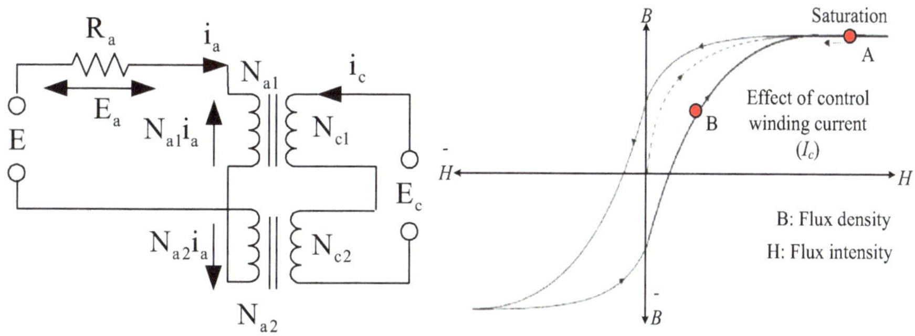

Fig. 78.11 Connection and theory of magnetic amplifier

current I_c, current controlled class "A" chopper, shown in Fig. 78.12, is used which activates as soon as the DC link voltage hits a predetermined threshold.

A *PI*-controller is used to control I_c at a given reference value shifting the operation from point "A" to point "B." At point "B," the reflected impedance on the main winding will be high causing high voltage drop across the magnetic amplifier main winding. Simulation results can be found in Figs. 78.15, 78.16, 78.17, 78.18 for a 20 kW, 24 poles PMSG grid connected system which validates the idea of using the magnetic amplifier in the enhancement of the LVRT capability for PMSG WT. When 90 % voltage dip occurs at 0.5 s, the control winding current controller senses the DC link voltage rise and activates the control current action.

As seen in Fig. 78.13, the DC link voltage is prevented from rising but with a system overshoot of 0.06 %. Overshoot problem can be avoided if magnetic amplifier is optimally designed. It is shown that the voltage stresses on the PWM VSI are

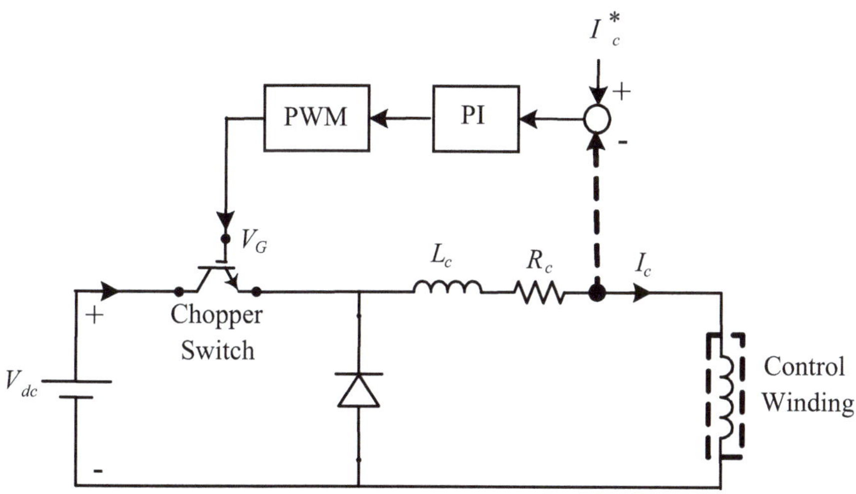

Fig. 78.12 Control winding current controller

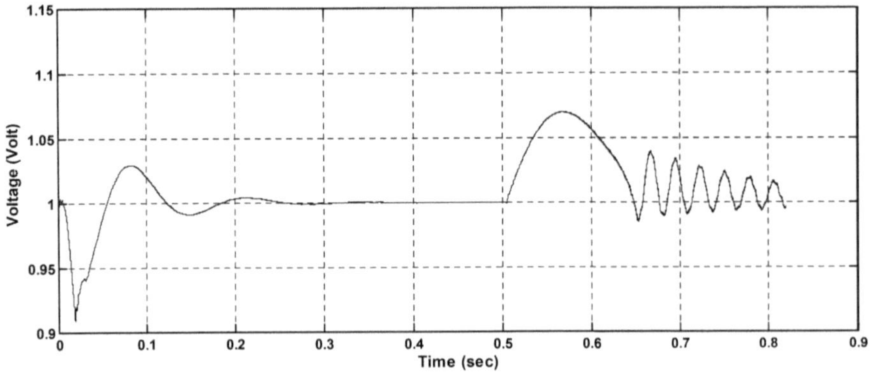

Fig. 78.13 DC link voltage at 90 % voltage dip with magnetic amplifier

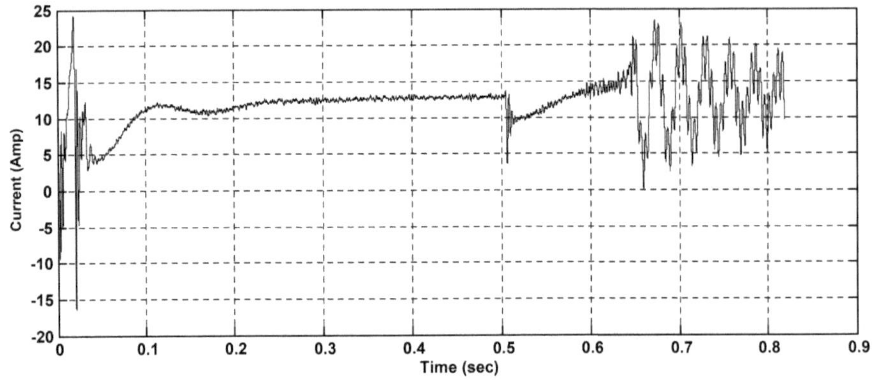

Fig. 78.14 DC link current at 90 % voltage dip with magnetic amplifier

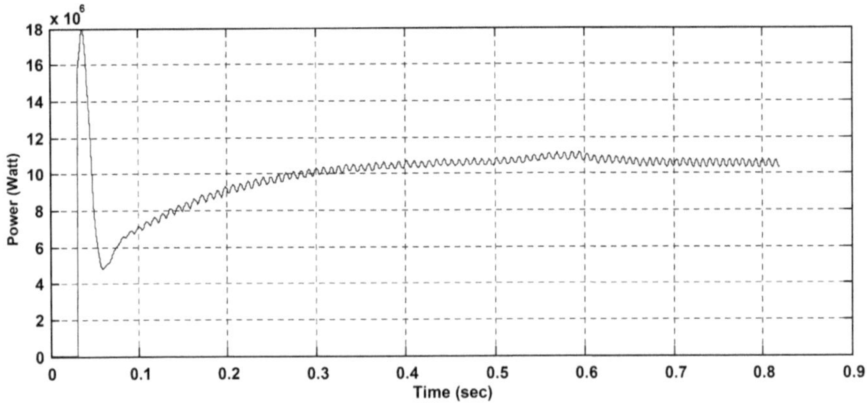

Fig. 78.15 PMSG active power at 90 % voltage dip with magnetic amplifier

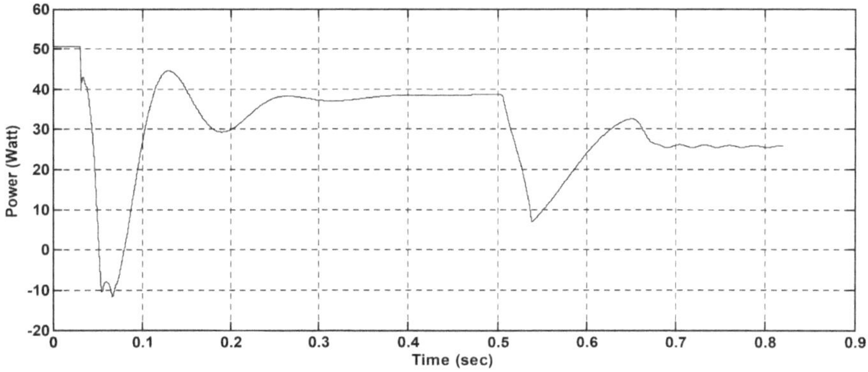

Fig. 78.16 Grid power at 90 % voltage dip with magnetic amplifier added

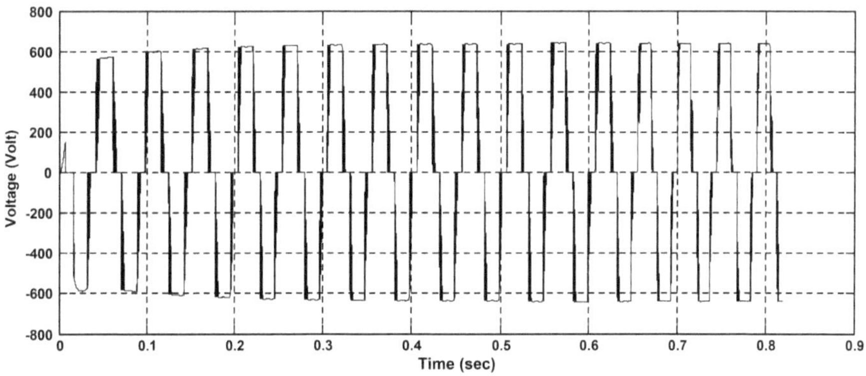

Fig. 78.17 PMSG terminal voltage at 90 % voltage dip with magnetic amplifier

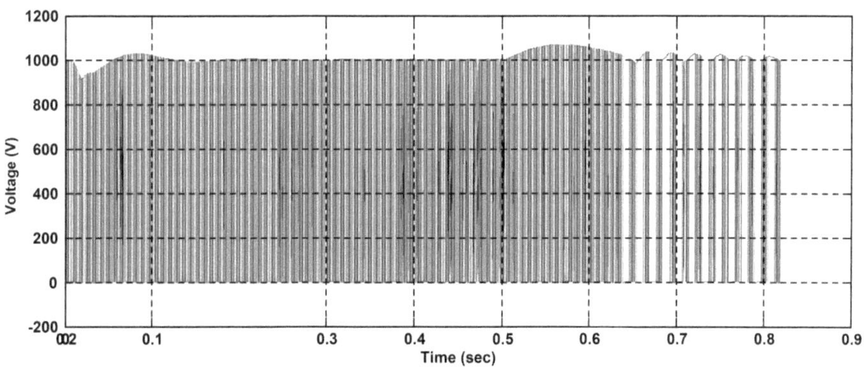

Fig. 78.18 IGBT switch voltage at 90 % voltage dip with magnetic amplifier

minimum as in Fig. 78.18 and that the magnetic amplifier was able to compensate for the power unbalance during the voltage dip. In the event of grid voltage restoration, magnetic amplifier control winding action is reversed to restore the system to its original state just as before the voltage dip (Fig. 78.14).

78.4 Conclusion

Enhancement of low voltage ride through capability for WT is becoming one of the most important issues related to the utilization and connection of WT to the electrical grid. Several techniques have been adopted in literatures that ensure that grid codes are being complied in the event of grid voltage dips which, not only ensure the stability of the electrical grid post the voltage dip, but also ensure the safety of wind farm during the dip event. A novel topology based on using magnetic amplifiers in the event of grid voltage dips is demonstrated. Simulation results show that magnetic amplifier can be used to enhance the LVRT capability for direct driven PMSG WES. The application extends for the use with DFIG wind energy systems as well.

References

1. B. M. Buchholz et al (2006) Dynamic simulation of renewable energy sources and requirements on fault ride through behavior. In: Power engineering society general meeting, 2006. IEEE, p 7
2. Tsili M, Papathanassiou S (2009) A review of grid code technical requirements for wind farms. IET Renew Power Gener 3:308
3. Pan CT, Juan YL (2010) A novel sensorless MPPT controller for a high-efficiency microscale wind power generation system. IEEE Trans Energy Convers 25:207–216
4. Morimoto S et al (2005) Sensorless output maximization control for variable-speed wind generation system using IPMSG. IEEE Trans Ind Appl 41:60–67
5. Cecconi V et al (2008) Active power maximizing for wind electrical energy generating systems moved by a modular multiple blade fixed pitch wind turbine, presented at the SPEED-AM 2008 international symposium on power electronics, electrical drives, automation and motion, 2008
6. Strachan NPW, Jovcic D (2007) Dynamic modeling, simulation and analysis of an offshore variable-speed directly-driven permanent-magnet wind energy conversion and storage system (WECSS), presented at the OCEANS 2007–Europe 2007
7. Conroy JF, Watson R (2007) Low-voltage ride-through of a full converter wind turbine with permanent magnet generator. IET Renew Power Gener 1:182
8. Zhang W et al (2008) Analysis of the by-pass resistance of an active crowbar for doubly-fed induction generator based wind turbines under grid faults, presented at the international conference on electrical machines and systems, 2008. ICEMS 2008
9. Ramtharan G et al (2009) Fault ride through of fully rated converter wind turbines with AC and DC transmission systems. IET Renew Power Gener 3:426
10. Xiao-ping Y et al (2009) Low voltage ride-through of directly driven wind turbine with permanent magnet synchronous generator, presented at the power and energy engineering conference, 2009. APPEEC 2009. Asia-Pacific 2009

11. Ali MH, Wu B (2010) Comparison of stabilization methods for fixed-speed wind generator systems. IEEE Trans Power Deliv 25:323–331
12. G. Wenming et al (2011) A survey on recent low voltage ride-through solutions of large scale wind farm, presented at the power and energy engineering conference (APPEEC), 2011 Asia-Pacific 2011
13. Abdel-Baqi O, Nasiri A (2011) Series voltage compensation for DFIG wind turbine. IEEE Trans Energy Convers 26:272–280
14. Raphael S, Massoud A (2011) Unified power flow controller for low voltage ride through capability of wind-based renewable energy grid-connected systems, presented at the 8th international multi-conference on systems, signals & devices 2011
15. Koutroulis E, Kalaitzakis K (2006) Design of a maximum power tracking system for wind-energy-conversion applications. IEEE Trans Ind Electron 53:486–494
16. Raza K et al (2008) A novel algorithm for fast and efficient maximum power point tracking of wind energy conversion systems. In: 18th international conference on electrical machines, 2008. ICEM 2008, pp 1–6
17. Buehring IK, Freris LL (1981) Control policies for wind-energy conversion systems. IEE Proc Gener Transm Distrib 128:253–261
18. Adam M et al (2007) Architecture complexity and energy efficiency of small wind turbines. IEEE Trans Ind Electron 54:660–670
19. Tan K, Islam S (2004) Optimum control strategies in energy conversion of PMSG wind turbine system without mechanical sensors. Energy Convers IEEE Trans 19:392–399
20. Ching-Tsai P, Yu-Ling J (2010) A novel sensorless MPPT controller for a high-efficiency microscale wind power generation system. IEEE Trans Energy Convers 25:207–216
21. Geng H et al (2011) Unified power control for PMSG-based WECS operating under different grid conditions. IEEE Trans Energy Convers 26:822–830
22. Chen Z et al (2009) A review of the state of the art of power electronics for wind turbines. IEEE Trans Power Electron 24:1859–1875
23. Krause PC et al (2002) Analysis of electric machinery and drive systems. Wiely, New York City
24. Lufcy CW (1955) A survey of magnetic amplifiers. Proceedings of the IRE 404–413
25. Feinberg R (1950) A review of transductor principles and applications. Power Eng Proc IEE Part II 97:628–644
26. Hudson CS (1950) A theory of the series transductor. Power Eng Proc IEE Part II 97:751–755
27. Parton KC et al (1981) Superconducting power-system transductor. IEE Proc Gener Transm Distrib 128:235–242
28. Lamm AU (1947) Some fundamentals of a theory of the transductor or magnetic amplifier. ATEE Trans 66:1078–1085
29. Shindo Y et al (2011) A magnetic amplifier using nanocrystalline soft magnetic material, presented at the 8th international conference on power electronics—ECCE Asia, May 30–June 3, 2011
30. Smith EHF (1966) The theory and design of magnetic amplifiers. Chapman and Hall, London

Chapter 79
Conceptual Design of Airborne Wind Turbines

Hossein Mirzaii and Liam Griggs

Abstract Airborne wind turbines (AWT) are a novel conceptual approach of enhancing the existing capabilities of capturing wind energy. AWT generate power from winds at altitudes of up to 1500 m, using a tethered structure with on board turbines, through a range of lift production methods. The chosen path for conceptual design was with a lighter than air (LTA) structure combined with a current small-capacity wind turbine. The chosen design approach was to use airship design principles, once the materials, gas, tether and turbine had either been designed or selected. Development of the design was carried out using a computational fluid dynamics (CFD) package, ANSYS CFX, to analyse the aerodynamic characteristics of the structure's shape. The design of AWT is still in the infancy phase and has numerous associated issues, with Civil Aviation Authority (CAA) regulations regarding airspace and tethering of balloons, and environmental conditions and impact. Through this research, the intention is to present further AWT designs, which utilise readily available small-capacity wind turbines, paving the way for additional research to be conducted.

Keywords Airborne wind turbine · High altitude wind energy · Computational fluid dynamics (CFD)

79.1 Introduction

The consequences of nonrenewable resources, such as coal, oil and gas, are increasing the need for viable renewable energy production. The limited availability and adverse environmental impacts associated with fossil fuels have been a driving force for the growth in renewable energy sector, primarily wind power. Conventional horizontal axis wind turbines (HAWT) are limited in their design, due to the tower being required to maintain the turbines at their operating height facing to the wind. The U.S. Department for Energy (DOE) published a report in which

H. Mirzaii (✉) · L. Griggs
Faculty of Science, Engineering and Computing, Kingston University, Roehampton Vale, Friars Avenue, SW15 3DW London, UK
e-mail: h.mirzaii@kingston.ac.uk

A. Sayigh (ed.), *Renewable Energy in the Service of Mankind Vol I*,
DOI 10.1007/978-3-319-17777-9_79

the weight percentage breakdown of a 1.5 MW HAWT was presented, with the tower accounting for 66.7 % of the total wind turbine weight and 67 % for a 4 MW HAWT [1].

High-altitude wind energy (HAWE) is the harnessing of wind energy at altitudes out of reach for conventional wind turbines. The use of airborne wind turbines (AWT) eliminates the height restriction imposed by the structural requirements of the tower, and has been conceptually explored through tethered kites, aircraft style structures and aerostatic airships. The most accomplished company to date are Altaeros Energies, who are the first to introduce a commercial application of an AWT in Alaska, with an lighter than air (LTA) design. Operating at 300 m the 30 kW AWT will produce enough power for 12 homes at half the cost of grid power; the design has been safety tested in winds up to 70 kmph and can be deployed from a shipping container in under 24 h [2]. Google X have also taken an interest in the AWT market, purchasing Makani Power in 2013. Makani Power's concept is an aircraft style structure with numerous turbines located across the span, and a capacity of 30 kW, although a 600 kW AWT is currently being developed [3]. The chosen design approach was a LTA structure due to the reduced limitations regarding the landing of the AWT in periods of low wind; additionally, LTA approaches have provided the most success to date.

The wind potentially available to these concepts at altitudes upwards of 1 km (outside of the earth's boundary layer) is more consistent and powerful. Due to the increased altitude, the available power increases by a factor of 8 when the wind speed doubles. Harnessing this wind will increase efficiency and reduce cost compared with conventional wind turbines. This in turn will reduce the dependency upon fossil fuels and meet energy targets set by governments and the European Commission.

The infrastructure and design of such concepts is not simple; however, the technology required for successfully achieving the goal already exists. The availability of technology has been proven by numerous conceptual designs, some of which have had scaled prototypes successfully tested at lower altitudes. Nevertheless, the research and development of AWT is still in its infancy and requires additional key issues to be addressed, including but not limited to, infrastructure, location, control and stability, material selection and cost reduction.

79.2 Design

The design discussed here was selected from a series of conceptual designs based on favourable potential properties of the design and the uniqueness in design compared with existing concepts. With the design similar to an airship shape the design principles for an airship were adopted, with the steps and equations required for the process taken from Khoury's book "Airship Technology" [4] and Konstantinov's journal publication "The Basics of Gas and Heat Airship Theory" [5].

79.2.1 Design Elements

The sizing of the concept required the weights of the structures elements to be determined. For this process the tether, envelope materials, turbine and gas were selected following a series of trade-off studies.

The tether materials required a low density with high-strength and high-weather resistance. The tether design consisted of three layers. The inner most layer of aluminium was selected over alternatives including copper and zinc, due to the density of aluminium being a third of the density of copper at 2.7 g/cm^3 and with comparable electrical resistivity [6]. The load bearing mid-layer of the tether was required to have a high specific strength, with a composite in a stranded rope configuration where strands are layered helically around an elastic and flexible core. The material selection determined Dyneema as the ideal choice with a strength 15 times that of steel and a hydrophobic property ideal for extreme environmental conditions [7]. The outer sheath material selected for flexibility and resilience with years of use in industry was polyurethane. Hwang et al. [8] designed a tether of a similar composition with a weight of 0.2 kg/m.

Maekawa et al. [9] conducted a series of tests using laminate compositions for stratospheric conditions, comparing the tensile strength and density of compositions before and after exposure to ultraviolet light. The optimum laminate composition used Vectran, a yarn spun from liquid crystal polymer [10], as the load-bearing element, Eval as a gas-retention film and polyurethane for the external environmental and weathering protection layer. Polyurethane was also selected as the material for the internal ballonet system, required for the pitch, altitude and pressure control of the structure. The Vectran laminate had a density of 0.192 g/m^3 with a thickness of 0.233 mm [9].

The lifting gas trade-off outcome was helium as the ideal choice, helium is the second lightest element after hydrogen which also ranked highly, however, due to the potential dangers associated with hydrogen this option was ruled out. The density of helium for the sizing was taken as 0.178 g/cm^3 [5]. The wind turbine trade-off was carried out for a selection of 10 kW wind turbines, with size, weight, maximum power output and wind speed as the criteria. The chosen wind turbine was from Ampair with a combined nacelle and blades weight of 480 kg and a high maximum operating wind speed [11].

79.2.2 Design

The initial stage of the design required the calculation of the weights required to be lifted by the design, with the turbine weight and tether being the initial known weights and additional weights provided by Khoury [4] including, pressure control pumps, maintenance access hatches and a load transfer bracket for the tether and structure contact point. The combined weight of the tether (assuming a maximum operating altitude of 1500 m), turbine blades and nacelle, and the additional features was 870 kg. With the initial weight value the volume of helium required to lift the weight could be calculated using Konstantinov's [5] equation, Eq. (79.1), with an assumed operational ceiling height (H) of 1500 m and where $\Delta\rho_0$ is the change in helium density with altitude.

$$H = 44300\left(1 - \frac{mg}{V\Delta\rho_0}\right)^{0.235} \tag{79.1}$$

With the initial volume of helium required known, further structural features stated in Khoury's [4] work could be calculated, the catenary, suspension system and nose reinforcement framework, which are calculated as a weight percentage dependent upon the volume of the envelope. The catenary and suspension system allow for the weight of the gondola, typically located below an airship, to be evenly distributed along the airships structure, this would be adapted to support the weight of the turbine nacelle internally. According to Konstaninov [5] the shape of an airship is initially based on a revolved ellipse, hence using Eqs. (79.2) and (79.3) the major (a) and minor (b) axes of an ellipse could be calculated. The fineness ratio (λ), a ratio of the major and minor axes, was taken as 3.75, based on historical data for nonrigid airships provided by Konstantinov [5].

$$a = \sqrt[3]{\frac{3V}{4\pi\lambda}} \tag{79.2}$$

$$V = 4\pi a^2 b \tag{79.3}$$

The surface area of the design was calculated using Eq. (79.4), with the surface area known the weight of the structures envelope could be calculated. Additionally, the weight of the ballonets required for the pitch and altitude control of the structure were also calculated. These weight were then included in the initial calculations and the volume of helium required recalculated, this iterative process was repeated until the variation between weights was negligible.

$$S = 9.719\pi\left(\frac{V}{10.917}\right)^{\frac{2}{3}} \tag{79.4}$$

The total weight of the structure was calculated as 1120 kg, with the final values for the major and minor axes, fineness ratio, volume and surface area shown in Table 79.1.

Table 79.1 Final overall dimensions of AWT concept

Overall dimensions				
a (m)	b (m)	λ	Volume (m^3)	Surface area(m^2)
4.6	17.33	3.77	1535.97	809.4

79.3 Experimental

79.3.1 Turbulence Model

The computational fluid dynamic (CFD) analysis of the designs was conducted in order to optimise the shape based on the potential flow conditions at the operational altitude. Voloshin et al. [12] carried out a comparison of turbulence models for airship steady-state CFD simulations. The comparison focused on the accuracy and resource consumption of four different turbulence models, with verification against wind tunnel data. The optimal turbulence model from the comparison, and the model chosen for this study, was Spallart Allmaras, a model developed specifically for the modelling of external flows within the aerospace industry [13].

79.3.2 Mesh Parameters

The domain for the structure was based on Hills [14] C-mesh, with a semi-circular inlet region and a rectangular section attached. The C-mesh was sized based off the length of the model being simulated, and revolved 360° around the body. The domain was tested with an unstructured coarse mesh to ensure the flow around the model was fully captured, the domain size satisfied this requirement and the use of the domain size was continued. The structure was meshed using a hybrid meshing technique, with prism layers around the body to create an inflation layer, calculated to ensure the boundary layer was fully captured. The remainder of the C-type mesh was unstructured, with tetrahedral elements inflating in size towards the far-wall regions of the domain.

79.3.3 Experimental Setup

The simulations were conducted at the maximum operating altitude (1500 m), with a velocity of 15 m/s, shown in Fig. 79.4a. The density, temperature and viscosity were also calculated for the operating altitude based on International Standard Atmosphere (ISA) data, these values were input into ANSYS CFX to replicate the actual environmental conditions at the operating altitude.

79.4 Results and Discussion

The development of designs required numerous iterations in order to establish the ideal change in shape of the structure, with variations in the shape of the nose and aft regions considered. The alteration of the structures shape in the development

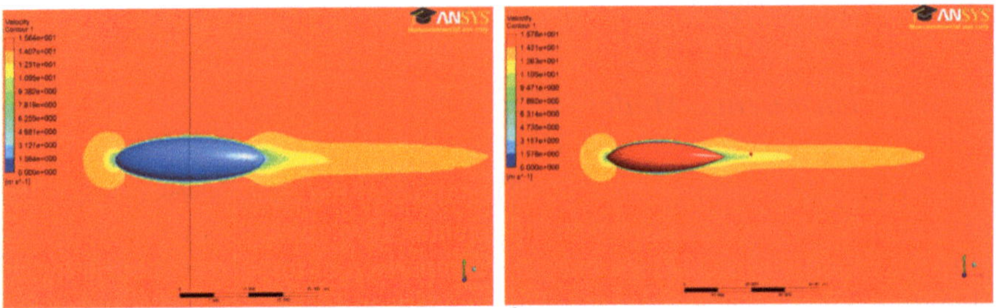

a b

Fig. 79.1 Velocity contour plots at 15 m/s. **a** Initial conceptual design ellipse shape. **b** Altered aft region of conceptual design

process would alter the sizing and weights previously calculated, however Khoury stated “it is almost impossible to calculate a detailed weight for the ‘final’ structure until it has been fully detailed (by which time it has probably been manufactured and weighed).” [4].

Figure 79.1a shows the initial elliptical profile structure, with two clear regions of disturbed flow, which would have significant drawbacks on the stability of the structure, due to the induced drag. The nose region of the structure is one that would be hard to alter to reduce the effect of the flow. The nose region is also a region of high pressure, shown in Fig. 79.2, hence the nose region is reinforced with nose reinforcement cap previously mentioned in the Sect. 79.2.2. The nose reinforcement prevents the structure from the possibility of deformation caused by the wind. The aim of altering the aft section was to move the separation that was occurring further along the structure and reduce the decrease in velocity. Figure 79.1b shows the chosen design, with an altered aft region, although the separation is occurring about the same point along the structure the near stagnation point that was occurring behind the structure has been eliminated. Figure 79.2 also shows how the pressure around the body has varied with the change in the aft section shape, the region of negative pressure covers a smaller region of the body. Additionally the increased pressure region behind the structure has been reduced.

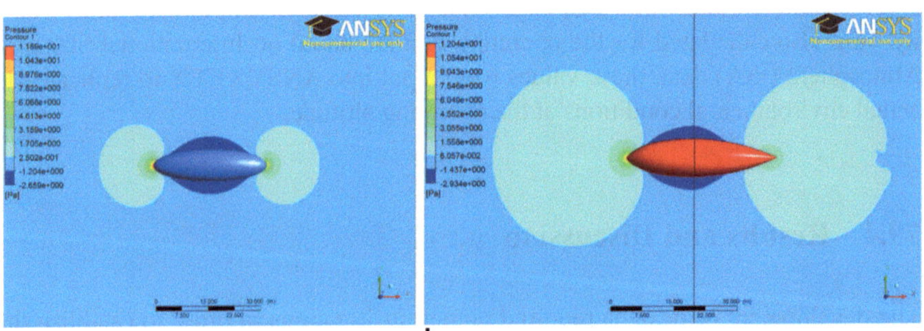

a b

Fig. 79.2 Pressure contour plots at 15 m/s. **a** Initial conceptual design ellipse shape. **b** Altered aft region of conceptual design

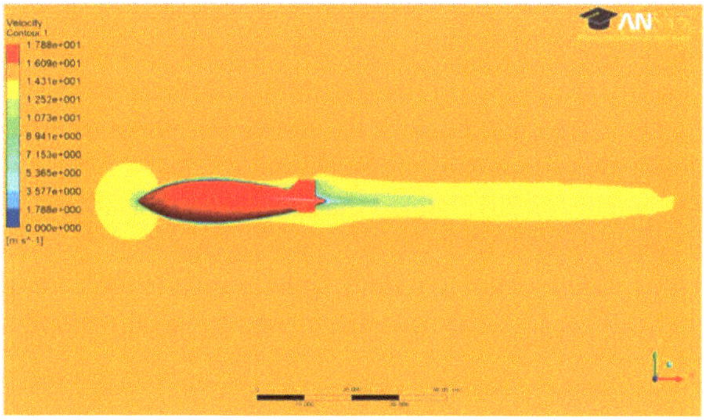

Fig. 79.3 Velocity contour plot at 15 m/s of finalised conceptual design shape with fins

The final conceptual design is shown in Fig. 79.3, with the addition of fins for stability and lateral control of the structure. The tail fins were sized from historical data in Konstantinov's work [5], where it was found the typical length of the tail was 10 % of the overall structure and a tip height equal to the radius of the structure. The fins were created in a Y formation, with a straight back and a front angle of 45° and a thin symmetrical aerofoil profile, for the simulations a NACA 0008 was chosen. The near stagnation region that was reduced in Fig. 79.1 has been reintroduced with the addition of fins, despite the drag resulting from the flow disruption the fins are an essential requirement for the stability of the AWT.

The operating conditions and wind energy available to AWT was also considered to explore their potential and feasibility for wind energy applications. The graphs in Fig. 79.4 shows a comparison between two sites in Scotland, the offshore site being a site highlighted for future development and ranked as the 27th highest average wind speed at 11.63 m/s, at a hub height of 100 m, in the world [15]. The onshore location considered was Whitelee wind farm, the largest onshore wind

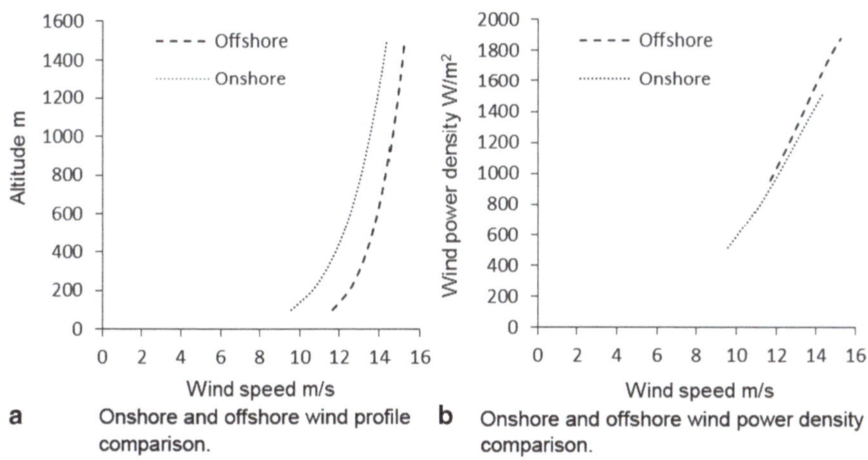

Fig. 79.4 Comparison of onshore and offshore wind turbine locations. **a** Comparison of onshore and offshore wind profiles. **b** Comparison of onshore and offshore wind power density

farm in the UK, with an average wind speed of 9.58 m/s at a hub height of 100 m [15]. Figure 79.4a shows the wind profiles for both onshore and offshore following similar trends, with little difference in wind speed at 1500 m. The power density for the two sites, Fig. 79.4b, shows the offshore location has a favourable power density profile, hence the selection of the site for future development. The Autumn Statement for 2013 stated that as of 2015 onshore subsidies would be cut, with an increase in offshore subsidies [16]. Assuming the same would apply to AWT the favourable location would definitely be offshore. An offshore location could hypothetically reduce the associated risks from failure and airspace restrictions.

79.5 Conclusion

In conclusion, the research and review has shown that the required technology for the advancement of AWT is readily available. The review of onshore and offshore locations has shown that operation at both sites could provide wind power density almost double that accessible to conventional HAWT. The estimated cost of the concept presented is primarily based on the lifting gas, materials and turbine required. The initial overall cost would be significantly higher due to the machinery and manufacturing supplies, however the increased wind power available and capacity factor of the system would allow for a shorter payback period. The main obstacle for the implementation of designs would come from the CAA, however, the CAA states that airspace can be zoned as a no fly zone if the reason benefits "society as whole" [17]. Additional issues include the stability and control of designs, with existing designs requiring complex control systems. Future work will aim to address these issues, with passive control measures being introduced to reduce the control system requirements. In addition, transient CFD simulations will be conducted, allowing for the effects of the rotating turbine to be modelled. The transient study will provide data on the stability effects to the system. Finite element analysis of the design in operating conditions and cost reduction methods through the safe use of hydrogen will also be investigated.

References

1. US Department of Energy (2008) 20% Wind energy by 2030. p 62
2. Tweed K (2014) World's highest wind turbine will hover abover Alaska. http://spectrum.ieee.org/energywise/energy/renewables/first-commercial-floating-wind-turbine-hovers-above-alaska. Accessed 1 April 2014
3. Makani (2013) Makani power Inc. http://www.google.com/makani/. Accessed 14 Dec 2013
4. Khoury G (2012) Airship technology, 2nd edn. Cambridge University Press, Cambridge
5. Konstantinov L (2003) The basics of gas and heat airship theory. AEROPLAST Inc, Kyiv, pp 1–21
6. Ashby M (2005) Materials selection in mechanical design, 3rd edn. Butterworth-Heinemann, Oxford

7. Ropes M (2014) Dyneema. http://www.marlowropes.com/dyneema.html. Accessed 28 Feb 2014
8. Hwang I, Kang W, Kim S (2011) An airborne cloidal wind turbine mounted using a tethered balloon. IJASS 12(4):P354–P359
9. Maekawa S, Nakadate M, Takegaki A (2007) Structures of the low-altitude stationary flight test vehicle. J Aircr 44(2):P662–P666
10. Fibre V (2014) Vectran. http://www.vectranfiber.com/. Accessed 28 Feb 2014
11. Ampair (2014). 10 kW westwind. Ampair. http://www.westwindturbines.co.uk/pages/index.asp?title=Overview_10kw&catID=199&subcatID=. Accessed 17 Jan 2014
12. Voloshin V, Chen Y, Calay R (2012) A comparison of turbulence models in airship steady-state CFD simulations. ArXiv e-prints, Cornell University. p 1–14
13. Aupoix B, Spalart P (2003) Extensions of the spalart-allmaras turbulence model to account for wall roughness. Int J Heat Fluid Flow 24(4):P454–P462
14. Hills J (2005) Numerical modelling of turbulent flow past an airfoil. Czech Technical University in Prague, Prague
15. 4COffshore (2014) 10 year global wind speed rankings. http://www.4coffshore.com/windfarms/windspeeds.aspx. Accessed 13 Feb 2014
16. Treasury HM (2013) Autumn statement 2013. Crown, London
17. CAA (2014) Air navigation: the order and the regulations CAP 393. Civil Aviation Authority

Chapter 80
Wind Shear Assessment Using Wind LiDAR Profiler and Sonic 3D Anemometer for Wind Energy Applications—Preliminary Results

Yoshiaki Sakagami, Pedro A. A. Santos, Reinaldo Haas, Júlio C. Passos and Frederico F. Taves

Abstract This chapter presents an assessment of wind shear up to 400 m at a coastal location. The vertical structure, temporal variations and influence of atmospheric stability were investigated. The results show four important findings using light detection and ranging (LiDAR) and dimensionless shear analysis. First, the top of surface layer can be considered around 200 m. Second, the wind speed decreases with height above 240 m from 11:00 to 16:00 local time. Third, the atmosphere can be considered to be stable from 10:00 to 14:00 and unstable from 19:00 to 01:00 according to the Monin–Obukhov similarity theory (MOST) theory. Fourth, the wind speed remains almost constant, with a height of up to 300 m in the first part of the night when the dimensionless shear is low. In addition, measurements taken with a sonic 3D anemometer confirmed the excellent correlation between the turbulent fluxes and the dimensionless shear. For strong winds, when the friction velocity is high, dimensionless shear varies almost linearly and at low friction velocity, it varies exponentially.

Keywords Wind speed profile · LiDAR · Wind shear · Turbulent fluxes · Atmospheric stability

---

Y. Sakagami (✉)
Department of Health and Service, Federal Institute of Santa Catarina, Florianópolis, Brazil
e-mail: yoshi@ifsc.edu.br

Y. Sakagami · P. A. A. Santos · J. C. Passos
Department of Mechanical Engineering, Federal University of Santa Catarina, Florianópolis, Brazil

R. Haas
Department of Physics, Federal University of Santa Catarina, Florianópolis, Brazil

F. F. Taves
Tractebel Energia S.A. (GDF Suez), Florianópolis, Brazil

A. Sayigh (ed.), *Renewable Energy in the Service of Mankind Vol I*,
DOI 10.1007/978-3-319-17777-9_80

80.1 Introduction

Recent developments in wind energy technology, with tall wind turbines reaching a height of up to 200 m, have increased the wind power capacity [1]. At this height, conventional means of measurement, such as a cup anemometer mounted on a tower, are no longer cost efficient and the wind flow can differ significantly from that of the theoretical logarithmic wind profile. As a consequence, financial risks increase because the performance and structural loads of the wind turbines may differ from the loads expected [1]. The characterization of wind flow in the atmosphere boundary layer is a complex task because it is dependent on numerous factors, including the topography of the terrain, surface roughness and atmospheric stability [2,3]. Two models are commonly used in wind energy applications to quantify vertical variations in the wind speed: the empirical power law and the semi-empirical logarithmic law profile [4]. However, these models are only valid under neutral conditions, which rarely occur in the atmosphere. Moreover, the parameterization of these models is site specific and limited to ideal conditions, such as flat and homogeneous surfaces [5]. Thus, a better understanding of wind profiles and other factors, such as turbulence and stability, will be of increasing importance to the wind energy industry over the next few years [6].

The aim of this study was to assess the wind shear up to 400 m using the dimensionless shear parameter based on the Monin–Obukhov similarity theory (MOST) [7]. The vertical structure, temporal variation and influence of friction velocity and sensible heat flux at 21 heights were analysed. This study uses data obtained from the Pedra do Sal experiment from September to November 2013. The site is located on the northeast coast of Brazil where meteorological conditions are strongly influenced by trade winds and sea breeze. The experiment is instrumented with a light detection and ranging (LiDAR) system with a range of up to 500 m height, a sonic 3D anemometer and a met-mast (100 m high) with five instrumented levels. All instruments are installed at the Pedra do Sal wind farm where there are 20 wind turbines of 55 m hub height, totaling 18 MW of installed power. The first section of this chapter presents a review of the theory of dimensionless shear and turbulent fluxes. In the methodology section, the experiment and instrument specifications are described and the assessment of wind shear in Pedra do Sal over three consecutive months is reported in the results section. Finally, important findings are discussed and conclusions are presented in the last section.

80.2 Theory Review

80.2.1 Wind Shear

Wind shear is simply defined as the local variation in the wind vector [8]. It is always present at the atmospheric boundary layer (ABL) because of flow interaction with the surface roughness. In the surface layer, the lowest 10 % of the ABL, vertical variations in the wind speed can be modelled using a semi-empirical logarithmic profile [8].

$$\frac{\partial \bar{U}(z)}{\partial z} = \frac{u_*}{\kappa z} \tag{80.1}$$

where, u_* is the friction velocity, κ is the von Kármán constant with a value of 0.4 and $\bar{U}$(z) is horizontal mean flow at height z. The logarithmic profile can then be rewritten, and the dimensionless shear (ϕ_m) is defined according to Monin and Obukhov [7].

$$\phi_m = \frac{\kappa z}{u_*} \frac{\partial \bar{U}(z)}{\partial z} \tag{80.2}$$

The derivative of the wind profile can be estimated by measuring the wind speed at two heights and using a logarithmic approximation [9].

$$\frac{\partial \bar{U}(z_m)}{\partial z} = \frac{(\bar{U}_2 - \bar{U}_1)}{z_m \ln(z_2 / z_1)} \tag{80.3}$$

where $z_m = \sqrt{z_2 z_1}$ is the geometric mean of the heights z_2 and z_1. Thus, by combining Eqs. 80.2 and 80.3, the dimensionless shear (Φ_m) can be expressed by,

$$\phi_m = \frac{\kappa}{u_*} \frac{(\bar{U}_2 - \bar{U}_1)}{\ln(z_2 / z_1)} \tag{80.4}$$

Φ_m is a parameter used to correct the wind profile according to the atmospheric stability. The atmosphere is considered stable when the stability parameter, known as the Obukhov length (L), is positive, which implies that the dimensionless shear increases with height [5]. On the other hand, the atmosphere is unstable when L is negative and Φ_m decreases with height. Many studies have been conducted in the past 50 years and some expressions have been suggested based on experimental data [5]. The formulations are always based on MOST, which considers turbulent fluxes to be constant with height within the surface layer [7].

80.2.2 Turbulent Fluxes

Wind shear is strongly affected by atmospheric stability, which is related to turbulent fluxes of momentum and heat. The eddy correlation method is used to calculate these fluxes, this being the most reliable and direct way to determine the turbulent fluxes in the atmosphere [7]. Turbulent fluxes are measured with a 3D sonic anemometer and the data are processed using Eddypro software [10], which is used to estimate the friction velocity (u_*) and heat fluxes (H_s).

The friction velocity is directly correlated to the intensity of the wind speed (mechanical stability). A strong wind has a large friction velocity, and consequently, strong wind shear. Using covariance of the standard deviation of the horizontal and vertical wind data, the friction velocity (u_*) can be calculated through the following equation:

$$u_* = \sqrt{\overline{u'w'}} \tag{80.5}$$

where, $\overline{u'w'}$ is the kinematic momentum with the x-axis oriented along the mean wind direction, and u and w are the horizontal and vertical wind speed components, respectively.

In addition, the sonic temperature is approximated to the potential virtual temperature (θ_v), and the sensible heat flux can then be calculated by covariance of the standard deviation of the vertical wind and potential virtual temperature:

$$H_s = \rho c_P \overline{u'\theta_v'} \tag{80.6}$$

where, c_p is the specific heat of dry air at constant pressure (1004 J kg^{-1} K^{-1}) and ρ is the air density.

Both fluxes simultaneously affect the wind shear and therefore, the method of bins is used to separate the fluxes at different intensities. In this procedure, the average and standard deviation of the dimensionless shear are calculated in bins of friction velocity and heat flux, which are centred on multiples of 0.05 m s^{-1} and 2 W m^{-2}, respectively. The standard uncertainty (Type A) is estimated as the standard deviation divided by the square root of the number of events of each bin. The expanded uncertainty is based on the standard uncertainty and coverage factor (k=2) for a confidence interval of 95.45 % [11]. Bins for which the number of occurrence is < 5 are not considered.

80.3 Methodology

80.3.1 Site Description

The experiment was started in June 2013 and preliminary data were collected from September to November 2013, which is considered the windy season in this region [12]. The Pedra do Sal experiment is located on a flat coastal area in the northeast of Brazil where 20 wind turbines of 900 kW are aligned with the local shoreline (Fig. 80.1a). The turbines are pitch controlled and spaced at intervals of approximately 170 m apart. Each turbine has 55 m of hub height and a rotor diameter of 42 m.

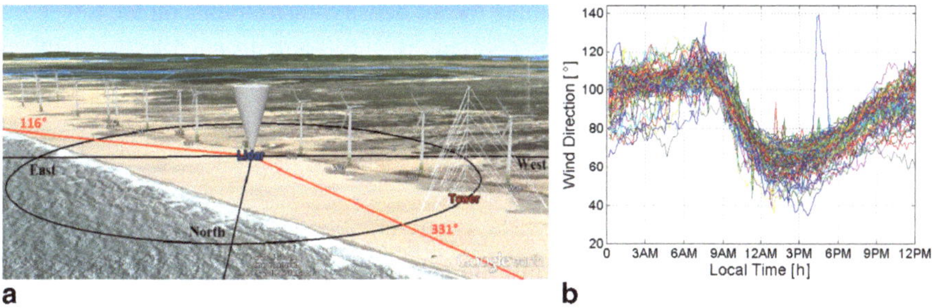

Fig. 80.1 **a** Picture showing LiDAR, tower and wind turbines (not scaled) aligned with shoreline at Pedra do Sal Wind Farm. *Black circle* indicates the distance of 565 m from LiDAR to tower, and *red lines* represent the sector free of wake from 331 to 116°. **b** Diurnal wind direction at Pedra do Sal from September to November 2013 (not filtered)

Local winds are influenced by the sea breeze and trade winds with a very persistent wind direction [13]. Figure 80.1b shows a small variation in the wind direction from 40° to 120° during three consecutive months. There was only one day on which the wind direction changed and this was because of a local storm. This high degree of steadiness is related to the permanence of an anticyclone in the South Atlantic Ocean, which tends to make the trade winds stronger in winter and weaker in summer [13]. Also, the air temperature, humidity and pressure show small diurnal variations. During this period, the average temperature registered was 25.5 °C, with a maximum of 27.6 °C and minimum of 22.4 °C. The relative humidity of the air ranged from 69 to 94 % and the air pressure from 1003.3 hPa to 1012.4 hPa. The average air density was estimated at 1.163 kg m^{-3} with a standard deviation of 0.003 kg m^{-3}.

80.3.2 Measurements

The LiDAR wind profiler, model Windcube8, with a range of up to 500 m and a meteorological tower (100 m high) with five instrumented levels is used in the experiment. The LiDAR system is installed around 300 m from the shoreline, 100 m in front of the turbines (upwind) and 565 m away from the tower. The first levels of the wind profile system are set at 40, 45, 50, 60 m and then every 20 m from 80 to 500 m. Horizontal and vertical components of the wind speed at each height are measured based on the velocity azimuth display (VAD) technique. Each measurement takes approximately 6 s and averaged data are stored at intervals of 10 min. A sonic 3D anemometer, model 81000, mounted on the tower at a height of 100 m, measures the components of the wind speed (u, v, w) and the sonic temperature at 20 Hz. In addition, a thermo-hygrometer at a height of 98 m and a barometer at a height of 13 m are mounted on the tower to measure the air temperature, relative humidity and air pressure, and estimate the air density. Instruments on the tower are operated according to the recommendations of the International Electrotechnical Commission (IEC) [4].

80.3.3 Dataset

The data collected are organized into three datasets, that is, those obtained from the LiDAR, tower and sonic 3D anemometer. The LiDAR and tower data represent the average values for 10 min periods while the turbulent fluxes are processed in 60 min intervals. A linear interpolation is applied to the flux data in order to format their time series interval of to 10 min. An algorithm was previously developed to ensure the dataset consistency, and two filters are used to find suspect data. The first filter excludes the wind sector which is affected by the turbine wake, as suggested by the IEC [14]. All data for a wind direction greater than 116° and lower than 331° are discarded. The second filter checks the LiDAR data available from 40 m to 400 m. If the data obtained at one level is $<100\,\%$, the profile within this interval is discarded. Wind data above 400 m is not considered because of the low amount of data available.

80.4 Results

The data collected during the Pedra do Sal experiment over a 3-month period were reduced to 1183 h of measurements after filtering, which represent 54.2 % of the total period. The dataset filtered by considering the wake sector represents only 1.3 % of the data filtered out. The poor aerosol content in the atmosphere and dust on the LiDAR lens affected the performance of the LiDAR. Thus, only 44.8 % of the LiDAR data is considered good, which significantly reduced the experimental database. The results are divided into two parts. The first part presents the vertical structure and diurnal variation in the wind shear from heights of 40–400 m, and the second part shows the influence of the friction velocity and sensible heat flux on the wind shear.

80.4.1 Vertical Structure and Diurnal Cycle

The wind speed at Pedra do Sal is strongly influenced by the trade winds and sea breeze, and thus the diurnal cycle is very regular in terms of the wind intensity (Fig. 80.2a) and direction (Fig. 80.1b) during this period. The trade winds always blow from the southeast in the Southern Hemisphere and the sea breeze turns the winds to the northeast, while the land breeze turns it to the southeast. At 14:00 (local

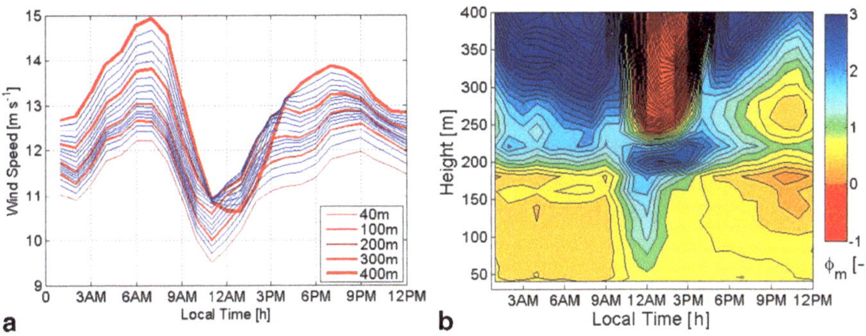

Fig. 80.2 Diurnal cycle of **a** wind profile and **b** dimensionless shear at Pedra do Sal from heights of 40 m to 400 m

time), the maximum temperature gradient of the land–sea surface and the maximum diurnal wind speed are expected to be reached [15]. However, the first maximum wind speed of the day is observed only at around 19:00. This occurs because the roughness of the ocean surface is low and the wind flow continues to accelerate until the land–sea temperature difference becomes negative, which happens at around 19:00 [15]. The reason for the second maximum wind speed at 06:00 is not clear. Preview analysis suggests the influence of a mountain range located in the southeast around 100 km away which diverts the trade winds and delays the decrease in the wind speed caused by the land breeze. The effect of this mountainous region can be observed in the data for the wind direction, which remains almost constant at around 100° during the night and starts turning to the northeast only when the sun rises and the diurnal cycle starts over again.

Figure 80.2b shows the diurnal dimensionless shear (Φ_m) at Pedra do Sal up to a height of 400 m, and thus describes the vertical structure of the wind flow. It can be observed that up to 200 m, the dimensionless shear varies from 0 to 2 in accordance with MOST theory. When Φ_m increases with height, the atmosphere is considered stable and when Φ_m decreases with height it is unstable [7]. Thus, in Fig. 80.2b it can be observed that the atmosphere is stable from 10:00 to 14:00 and unstable from 19:00to 01:00. Neutral conditions are reached at around 16:00 when Φ_m remains constant with height, which represents the transition from stable to unstable conditions. The boundary layer regime is different because stable conditions occur in the middle of the day and unstable conditions prevail at night. This shows that the site is very specific, which can be attributed to the dynamic regimes in coastal areas, where boundary layer structures are often found to differ significantly [16].

The stable conditions in the afternoon may be related to a warm advection current (land–sea) carrying warm air over a cooler sea surface when the wind flow crosses the shoreline [17]. The unstable conditions during the first part of the night are not yet well understood for this site. Nevertheless, it can be observed that Φ_m is very low up to 300 m at 23:00, and thus the wind speed is almost constant with height up to this level. Above 240 m, the wind flow cannot be described by MOST theory. Negative values for Φ_m are observed from 11:00 to 16:00, which means that the wind speed decreases with the height. This may be associated with a return flow above sea breeze, as explained by the Bjerknes theorem for sea breeze circulation [18]. Finally, there are no significant changes in the wind profile at the lowest levels measured by LiDAR. The possible influence of the internal boundary layer (IBL) is not observed, because LiDAR is located at about 300 m from the shoreline and the depth of the IBL is estimated to be around 30 m [19].

80.4.2 Stability Effect

In this section, the influence of atmospheric stability on the dimensionless shear is analysed considering the wind speed at heights of 80 and 40 m; thus, the average geometric height is considered to be 56.6 m. Figure 80.3a shows the variation in Φ_m with the friction velocity under neutral conditions where the heat flux is considered within an interval of ± 1 W m^{-2}.

It can be observed that a correction is necessary because the preview analysis shows an average dimensionless shear of 0.54, which is assumed to be constant and equal to one [7]. The measurements taken with the 3D sonic anemometer are usually the main cause of this difference [20], and a correction of −35 % is applied to the friction velocity, in agreement with $\Phi_m = 1$ under neutral conditions. After the friction velocity correction, Φ_m is compared with the heat flux at different friction velocities. Figure 80.3b shows that Φ_m varies almost linearly with heat flux when u_* is 0.4 m.s^{-1}. In this case, the influence of the heat flux on Φ_m is low because u_* is associated with strong winds and mechanical wind shear prevails. On the other hand, Φ_m varies exponentially with the heat flux when u_* is 0.2 m.s^{-1}. These results suggest that under calm wind conditions, Φ_m is strongly influenced by the heat flux mainly under stable conditions when the heat flux is negative. The marine atmospheric conditions are responsible for the narrow range of heat fluxes from −10 to 40 W·m^{-2}. However, strong stratification was observed with a Φ_m value of 1.8 under stable conditions.

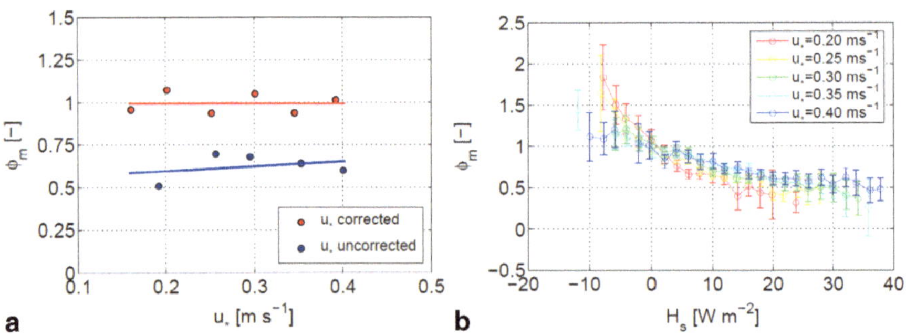

Fig. 80.3 **a** Variation in Φ_m with friction velocity under neutral conditions when the heat flux is zero. **b** Variation in Φ_m with heat flux at different friction velocities

80.5 Discussion and Conclusions

The results obtained from measurements taken in the Pedra do Sal experiment verify the excellent performance of the LiDAR wind profile and sonic 3D anemometer, with over 50 % of the data being available over three consecutive months. However, the turbulent flux measured by sonic anemometer presents a difference of −35 % in the u_* value, which is high for flux measurements, errors of ±20 % generally being considered acceptable so further analysis are still necessary [5]. It should also be noted that a correction factor related to the water vapour in the atmosphere was not applied to the sonic anemometer data and thus the atmosphere is considered to be dry in this study.

The method using dimensionless shear to assess the wind shear has the advantage that the results are in agreement with MOST theory, and it allows the atmospheric stability to be evaluated. However, it has the disadvantage that Φ_m is not actually the wind shear observed. The ratio (κ/u_*) in Eq. 80.2 can be viewed as a slope correction due to mechanical stability. Thus, Φ_m can be considered a coefficient of the wind shear normalized by u_*. Comparison of Φ_m with the friction velocity confirms that Φ_m does not change with u_* under neutral conditions (Fig. 80.3a). However, Φ_m is affected by u_* under non-neutral conditions. Figure 80.3b shows that Φ_m is correlated with the heat flux, but at high u_* the Φ_m value varies almost linearly, while at low u_* it varies exponentially. When the heat flux is negative, Φ_m varies substantially at different friction velocities and consequently, the uncertainty is higher under stable conditions.

The assessment of the vertical structure of Φ_m shows four important results. First, there is a substantial change in Φ_m at a height of around 200 m, which is probably the top of the surface layer where MOST theory is still valid. Second, above 240 m, negative values of Φ_m were observed, which is probably associated with the return flow above the sea breeze. Third, the variation in Φ_m with height within the surface layer allows the atmospheric stability to be evaluated. It is stable from 10:00 to 14:00 and unstable at night from 19:00 to 01:00 according to MOST theory. Fourth, Φ_m is low up to 300 m under unstable conditions during the night, when the wind speed remains almost constant with the height up to this level.

The experimental installations have the advantage of being located on flat terrain, which improves the analysis of the diurnal and vertical structure of the wind flow within the surface layer and the influence of turbulent fluxes. However, the site surface is not homogeneous because it is close to the shore. The diurnal variations in the wind verify the strong influence of the coastal conditions and the surrounding mountainous region. Therefore, this assessment is valid and specific to this site.

For wind energy applications, it is clear that atmospheric stability cannot be neglected at this site. The use of the logarithmic profile or power law without stability correction can lead to overestimation of the wind speed under unstable conditions, when Φ_m decreases with height, and underestimation under stable conditions, when Φ_m increases with height. This has a direct impact on the power generation estimates. In conclusion, the LiDAR profiler and the sonic 3D anemometer were found to be powerful instruments for assessing the vertical structure of the wind flow and the effects of atmospheric stability. Thus, they can collaborate to improve the understanding of the environmental conditions to which tall wind turbines may be exposed.

Acknowledgements This study was supported by the Brazilian Electricity Regulatory Agency (ANEEL) under R&D project 0403 − 0020/2011, funded by Tractebel Energia S.A. (GDF Suez). The authors also acknowledge support from the National Council for Scientific and Technological Development of Brazil (CNPq).

References

1. Platform EWET (2014) Strategic research agenda and market deployment strategy, Tech. rep., European Wind Energy Association
2. Wharton S, Lundquist JK (2012) Assessing atmospheric stability and its impacts on rotor-disk wind characteristics at an onshore wind farm. Wind Energy 15(4):525–546
3. Bunse U, Mellinghoff H, (2008) Influences of vertical wind profiles on power performance measurements, In: Proc. Dewi, DEWI, German Wind Energy Conference DEWEK, p 5
4. IEC (2005) Wind turbines—Part 1: design requirements 61400–1, 3rd edn. International Electrotechnical Commission, Geneva
5. Foken T (2006) 50 years of the Monin Obukhov similarity theory. Bound Layer Meteorol 119(3):431–447
6. Banta RM, Pichugina YL, Kelley ND, Hardesty RM, Brewer WA (2013) Wind energy meteorology: Insight into wind properties in the turbine-rotor layer of the atmosphere from high-resolution doppler LiDAR. Bull Am Meteorol Soc 94(6):883–902
7. Monin AS, Obukhov AM (1954) Basic laws of turbulent mixing in the surface layer of the atmosphere. Tr Akad Nauk SSSR Geophiz Inst 24:163–187
8. Stull RB (1988) An introduction to boundary layer meteorology. Kluwer, The Netherlands
9. Arya S (2001) Introduction to micrometeorology, international geophysics series. Academic Press, London
10. LI-COR (2012) EddyPro 4.0 & 4.1 Eddy Covariance Software, User's Guide & Reference, 4th edn. LI-COR Company, Lincoln
11. JCGM (2008) Evaluation of measurement data—guide to the expression of uncertainty in measurement, Tech. rep. Joint Committee for Guides in Metrology, Paris
12. Damas LB (2013) Análise de geração e aplicação do método mcp em dois parques eólicos em operação no Brasil Florianópolis 2013. Master's thesis, UFSC
13. Oliver JE (2005) Encyclopedia of world climatology, encyclopedia of earth sciences series. Springer, Netherlands
14. IEC (2005) Wind turbines Part 12–1: power performance measurements of electricity producing wind turbines technical report N ◦. IEC 61400-12-1, 1st Edition. International Electrotechnical Commission, Geneva
15. Haurwitz B (1947) Comments on the sea-breeze circulation. J Atmos Sci 4(1):1–8
16. N. R. Council (1992) Coastal meteorology: a review of the state of the science. The National Academies Press, Washington
17. Lange B, Larsen S, Hjstrup J, Barthelmie R (2004) The influence of thermal effects on the wind speed profile of the coastal marine boundary layer, Bound Layer Meteorol 112 (3)
18. Miller STK, Keim BD, Talbot RW, Mao H (2003) Sea breeze: structure, forecasting, and impacts, Rev Geophys 41 (3)
19. Garratt J (1990) The internal boundary layer a review. Bound Layer Meteorol 50(1–4): 171–203
20. Hogström U (1988) Non-dimensional wind and temperature profiles in the atmospheric surface layer: a re-evaluation. Bound Layer Meteorol 42(1–2):55–78

Chapter 81
Atmospheric Stability Effects on Small Wind Turbine Power Collection in a Complex Terrain

Pedro A. A. Santos, Yoshiaki Sakagami, Reinaldo Haas, Júlio C. Passos and Frederico F. Taves

Abstract The power performance assessment of a small wind turbine (SWT) based on atmospheric stability is reported herein. An experimental setup was used to study a 2.1 kW SWT in a suburban environment, where 1 Hz power collection data and 20 Hz turbulent flux measurements were obtained. The dataset consists of 4287 h of raw data covering a 6-month period. The measured International Electrotechnical Commission-based (IEC) power curve shows an average performance 30% above the manufacturer's curve, but with decreasing power output close to the rated wind speed. Values relating power collection increase at low wind speeds with high turbulence levels concur with previous studies. The Obukhov length was used as a stability parameter, and stability-dependent power curves were compared with the measured average. Above 8 m/s, unstable conditions were predominant and evidenced the decreasing power tendency, where turbulence intensity (TI) was unable to give consistent results. The results reported in this chapter validate an approach suitable for SWT assessment using a physical parameter as a classification criterion, which better explains the power collection behaviour close to rated conditions.

Keywords Small wind turbine · Turbulence · Atmospheric stability · Obukhov length

P. A. A. Santos (✉) · J. C. Passos
Mechanical Engineering Department, Federal University of Santa Catarina, Florianópolis, Brazil
e-mail: pedroasantos@lepten.ufsc.br; pedroazevedosantos@gmail.com

Y. Sakagami
Health and Science Department, Federal Institution of Santa Catarina, Florianópolis, Brazil

R. Haas
Department of Physics, Federal University of Santa Catarina, Florianópolis, Brazil

F. F. Taves
Tractebel Energia S.A., GDF Suez, Florianópolis, Brazil

A. Sayigh (ed.), *Renewable Energy in the Service of Mankind Vol I*,
DOI 10.1007/978-3-319-17777-9_81

81.1 Introduction

The small wind turbine (SWT) market is currently increasing with more reliable and cheaper technologies, reaching $ 3000/kW in the USA in 2011 [1]. Notably, most applications are still concentrated on stand-alone systems. However, the use of small wind grid-tied systems is increasing in countries with government incentives such as compensation support schemes. In 2012, the National Electrical Energy Agency (ANEEL) of Brazil released the regulatory framework for renewable energy distributed generation systems with up to 1 MW of installed capacity [2]. Net metering was chosen to be the first Brazilian compensation scheme. With this scenario, the appearance of small and medium wind turbines in Brazil will probably increase in the next few years.

Close to the final consumer, small wind grid-tied systems face major challenges related to the urban environment, characterized by a complex terrain, relatively low wind regimes and high levels of ambient turbulence [3, 4]. Field monitoring studies [5] show a mismatch between the manufacturer's power curve and data measured on site. Upstream turbulence present over a complex terrain is an important factor in relation to power curve data scattering and inaccuracies in energy prediction for small wind power applications. Numerical simulations [6] have been applied in order to evaluate the potential of built-mounted wind turbines. Such studies have provided interesting insights regarding the effects of turbine positioning, as well as predictions for near-obstacle wind profiles.

Experimental studies have also been conducted, with field measurements being taken to quantify ambient turbulence and evaluate the relative impact on SWT performance. Lubitz [7] reported experimental results for the variation in the power output based on turbulence intensity (TI), measuring a 2 % decrease in the power output for low TI (<0.14) between 4 and 7 m/s. Results are less consistent for high TI (>0.18) values. Sunderland et al. [8] performed power curve simulations for different TI levels, considering a neutrally stratified atmosphere. The authors also discuss the limits for TI as a parameter for analyzing urban environment applications, given the potential for unrealistic values during common episodes of gusts and low wind speeds. All of the above studies concur with the observation that, for lower wind speeds, high turbulence levels can improve the SWT power collection in urban environments. In contrast, turbulence tends to decrease the performance at rated wind speed.

The aim of this research was to advance beyond studies based on TI, introducing a physical parameter to evaluate the influence of turbulence on SWTs. Based on previous results obtained in studies on large wind turbines [9], the Obukhov length is used to assess the impact of atmospheric stability on SWT power collection.

81.2 Methodology

81.2.1 Test Site

The experimental setup is characterized as a suburban area on Santa Catarina Island, southern Brazil. Wind and power measurements were performed next to a two-storey building on a university campus, surrounded by vegetation of relatively homogeneous height at an elevation close to sea level and not far from the city airport. The selected site (Fig. 81.1) is a nonurbanized area but with a high associated roughness length around the measuring equipment, therefore considered to lie in complex terrain.

A Skystream 3.7 SWT equipped with a power transducer was used in this study. The certified rated power is 2.1 kW at 11 m/s, as established by the Small Wind Certification Council (SWCC) and thus, this 3-blade, downwind, grid-tied, horizontal axis wind turbine was installed at $19m_{agl}$ (above ground level) next to a two-storey building (27°41′1.40″S, 48°32′43.08″W). Operating since January 2013, all produced energy is supplied to the university laboratories. A met mast was placed on the rooftop of the same building, 22 m away from the SWT (27°41′1.00″S, 48°32′43.73″W), equipped with a First Class cup anemometer and a 3D ultrasonic anemometer. Appendix (A) details the equipment installed.

Considering specific conditions of the urban wind profile, the ultrasonic anemometer was placed twice as high as the mean building height in order to consider the inertial sub-layer properties of the conventional atmospheric surface sub-layer, where turbulent fluxes are considered to remain constant with height [10]. Measurements have been taken since November 2012 and 13,010 h of valid data are available (98.6% recovery rate). The met mast provided a representative wind

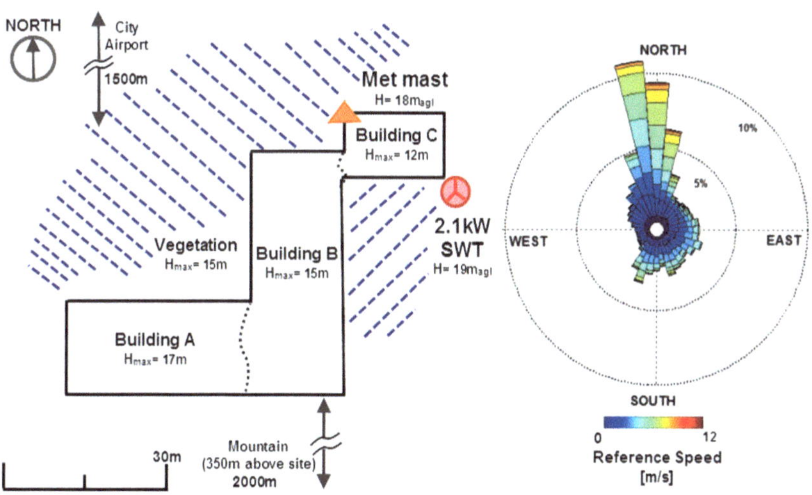

Fig. 81.1 Layout of experimental setup and wind rose at 17 m_{agl}

regime, with an average TI[1] of 0.32. The shape and scale Weibull parameters are 1.70 and 3.23 m/s at 17 m_{agl}, respectively. The prevailing wind directions are north and north-northwest (Fig. 81.1), with an average TI of 0.28. Contrarily, south winds (with direction between 120° and 240°) have a higher average TI of 0.36, characterizing the occurrence of gusty winds at the site.

81.2.2 Stability Parameter

With the use of a 3D ultrasonic anemometer, latitudinal, longitudinal and vertical wind speeds, as well as high frequency fluctuations from the mean (u', v' and w'), can provide detailed information on the eddy structures. Concerning the measurement of turbulent fluxes, the broadly known eddy covariance technique was applied [11]. The wind profile is affected by thermally produced/consumed turbulence as well as mechanically produced/dissipated turbulence. Practical stability classification schemes which rely on the Obukhov length (Eq. 81.1) consider a scale parameter of atmospheric mixing conditions within the surface sub-layer given by,

$$L = -\frac{u_*^3}{\left[\kappa\left(\frac{g}{\theta_v}\right)\left(\frac{H}{\rho c_p}\right)\right]} = -\frac{(\overline{u'w'}^2 + \overline{v'w'}^2)^{\frac{3}{4}} T_a \left(P_0/P\right)^{0.286}}{\kappa g \overline{w'\theta_v'}} [m]. \quad (81.1)$$

Where u_* is the friction velocity (m s^{-1}) derived from kinematic vertical momentum fluxes ($\overline{u'w'}$ and $\overline{v'w'}$, m^{2}s^{-2}), k is the von Karman constant (0.41) and g is acceleration due to gravity (9.81 m s^{-2}). Vertical sensible heat flux (H, W m^{-2}) is calculated from kinematic vertical sensible heat flux ($\overline{w'\theta_v'}$, m K s^{-1}), air density (ρ, kg m^{-3}) and specific heat at constant pressure (c_p, J kg^{-1} K^{-1}). The virtual potential temperature (θ_v, K) is evaluated based on the ambient temperature (T_a, K) with an atmospheric pressure (p, Pa) reference correction ($p_0 = 10^5\ Pa$). In contrast to TI, the Obukhov length is considered a physical stability parameter, for the sake of turbulent eddy assessment.

With a net upward heat flux ($\overline{w'\theta'} > 0$), unstable conditions, with low wind shear, are observed. Alternatively, a stable environment is characterized by high wind shear and a negative sensible heat flux ($\overline{w'\theta'} < 0$). Extreme values for L tend to indicate a neutrally stratified surface layer, where neither thermal nor mechanical turbulent components are dominant [12]. Wharton and Lundquist's [9] wind farm study reported strong seasonal and diurnal variations in the wind profile, which increases power production during stable conditions and reduces the output for unstable profiles. In this study, a stability classification similar to that of the aforementioned study was used.

[1] The statistical parameter given by the ratio of horizontal wind standard deviation and mean value $\left(\sigma_u/\overline{U}\right)$ [16].

81.2.3 Dataset

A 6-month period between December 2013 and May 2014 was selected for analysis. Ambient data (wind velocity and direction, atmospheric pressure and temperature) as well as power output were acquired at 1 Hz with average intervals of 1 min and 10 mins. Turbulent flux measurements were post-processed using EddyPro® software [13]. Rainy days were identified based on airport data and were not analyzed in order to avoid dropout events due to ultrasonic sensor obstruction. The 20 Hz data were averaged at 60 min intervals using linear detrending as well as double rotation to account for tendencies and tilt corrections. With a high associated roughness length, the vertical velocity fluctuation (w') tends to be overestimated due to the influence of the building, resulting in incoming winds with a high angle of attack [10]. Therefore, statistical filtering was applied with spike and angle of attack flags (Eddy Pro Quality Control, QC). As described by Vickers and Mahrt [14], nonstationary conditions can be identified in alongwind and crosswind components in order to filter conditions that would violate the quasi-stationary flow hypothesis of the Monin-Obukhov similarity theory.

Two distinct datasets, according to the filtering levels, are shown in Table 81.1 with respective sizes and recovery rates. For the stability-dependent dataset, all filters were used to favour a consistent stability parameter evaluation. The International Electrotechnical Commission-based (IEC) dataset was used for overall power curve calculation and thus filters related to the flux measurements were not applied.

81.3 Results and Discussion

81.3.1 IEC-Based Power Curve

Figure 81.2 shows the measured power curve for Skystream 3.7 based on IEC 61400-12-1, plotted from 1 min binned data with air density normalization, height adjustment (power law) and standard uncertainty error bars [15]. The SWT showed a 95.1 % operation time fraction over 4074 h, above the minimum IEC 61400-2 requirement of 90 % over 2500 h [16]. Cut-in wind speed was 3 m/s and valid filled bins reached up to 11 m/s.

Table 81.1 Filtered datasets collected within the analysis period (6 months)

Dataset	Filter procedure	Hours of data	Recovery rate (%)
IEC-based	Raw data	4287	98.2
	Power output	4074	93.3
Stability-dependent	Rain threshold	4061	93.0
	EddyPro QC	2570	81.7
	Steady flow	1503	29.2

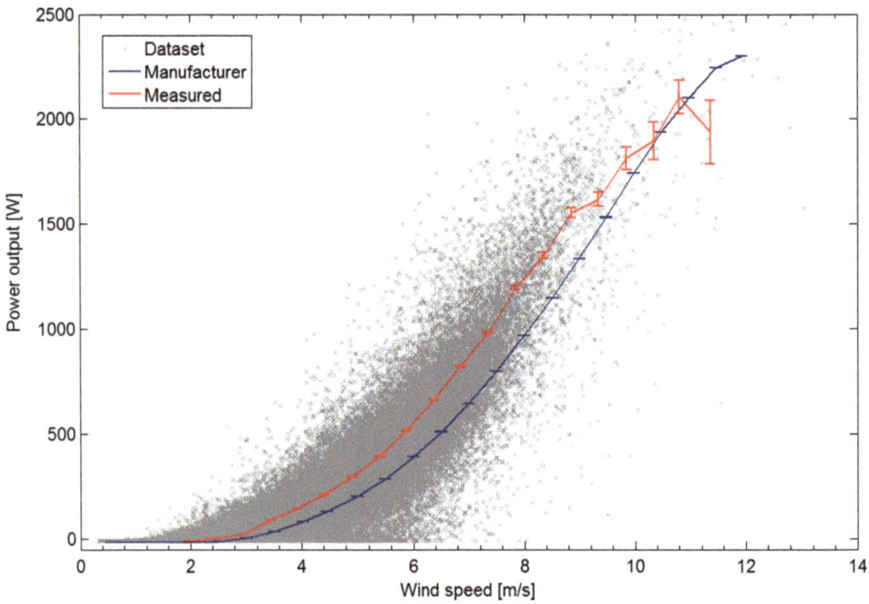

Fig. 82.2 Measured IEC-based power curve, with 1 min dataset and manufacturer agreement

The measured power curve shows an average performance 30 % above that expected according to the manufacturer's curve, although with power output values below the manufacturer's data for wind speeds close to the rated value. This result is in agreement with those obtained in previous studies [7, 8] and can be explained by higher average TI (0.32) at the site when compared with the manufacturer's test site (average TI of 0.09) [17].

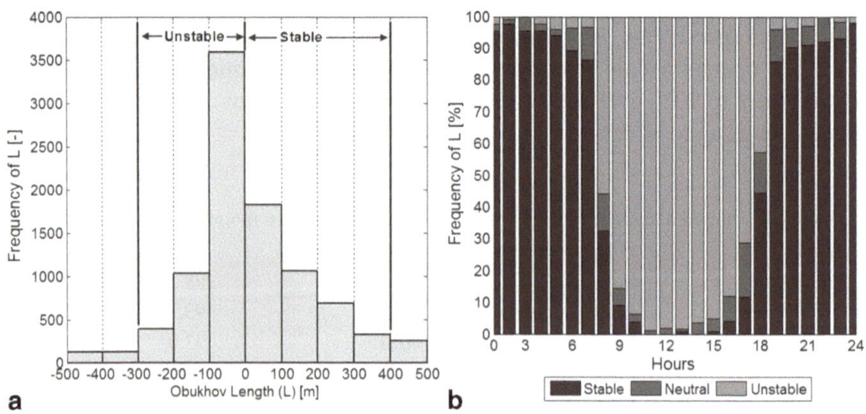

Fig. 81.3 Histogram of *L* with stability classification (**a**), stability diurnal regime (**b**)

81.3.2 Stability-Dependent Performance

Three stability classes were defined according to the Obukhov length: stable ($0 \leq L < 400$), near-neutral ($L \geq 400$) or ($L \leq -300$) an unstable ($-300 \leq L < 0$). Figure 81.3a shows the histogram for the stability-dependent dataset with stable, near-neutral and convective stability conditions in a 42:5:53 ratio. When the stability diurnal regime is analyzed (Fig. 81.3b), as expected, daytime has a predominantly unstable condition due to surface heating, in contrast to a stable surface layer at night.

Power output 1 min data were segregated into the above stability classes and respective power curves were constructed (Fig. 81.4a), with the IEC-based curve taken as the overall power curve. The dataset was also divided based on TI values as classification criteria, considering 17 % as the threshold between high and low turbulence. Figure 81.4b shows the power output deviation from overall values for each stability and TI class.

As in the work of Lubitz [7], no major power output deviation from the average was observed up to 8 m/s. An average increase of 4.9 % in power production, compared with the overall curve was observed under stable conditions (Fig. 81.4a) while unstable conditions led to a production 4.6 % below the average. Deviations up to 51 % were observed between stable and convective conditions. Close to the rated wind speed, the atmospheric stability parameter was able to provide evidence towards the observed tendency of a decrease in power collection (Fig. 81.4b). Unstable conditions were predominant at higher wind speeds and, as discussed in Sect. 2.2, they tend to transport vertical momentum more effectively over the

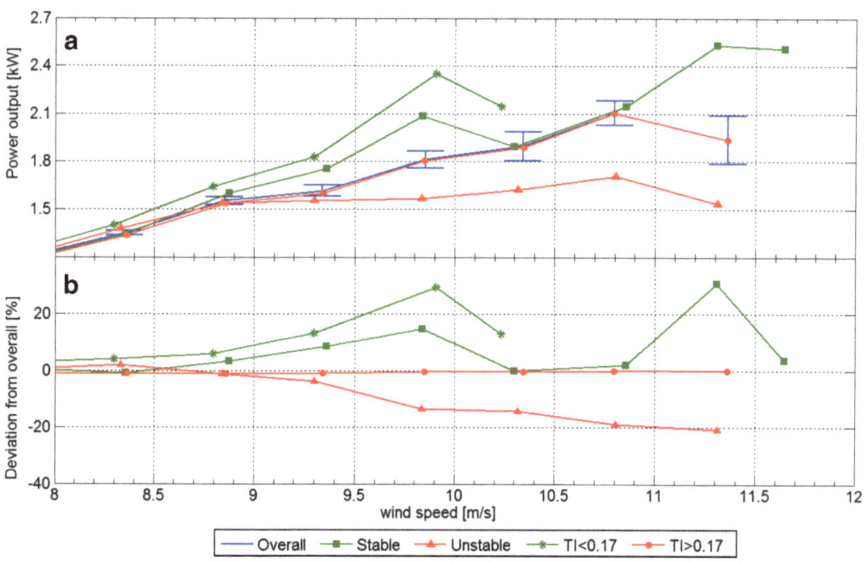

Fig. 81.4 Stability-dependent and TI-dependent power curves versus overall IEC-based (**a**), deviations from overall power curve for each stability and TI class (**b**)

surface layer, thus reducing wind shear. In contrast, the TI parameter could not be used to determine the overall pattern, whereas high wind speed bins tend towards lower standard deviations.

The aim of the data filtering procedure was to minimize distortions and unrealistic values, caused by flow interaction with high roughness obstacles. The experimental setup focused on capturing the conditions of the incoming near-building flow, but was not necessarily designed for optimal flux measurement sensor location. It should be noted that acquiring unbiased high resolution wind and power output data from sensors and wind turbines in urban environments is an arduous task. The instrumentation described is not commonly used for wind energy applications and thus, further studies can be conducted along with measurements for pollutant dispersion, urban thermal comfort and other usages that require fast-response sensors in the urban environment. Another interesting strategy used by Carpman [18] focused on other benchmark sites with a similar turbulence structure observed in urban environments, such as tall forest canopies.

The IEC standard [16] still lacks a procedure for SWT power performance assessment under high TI conditions. Power curve turbulence corrections have been presented [8, 19], but normalizing power output with atmospheric stability can be introduced to account for the observed deviations. Further research, already being carried out on large wind turbines, aims to establish a correlation between the results reported herein and parameters which can be calculated with instruments available in the wind energy sector, such as LiDAR wind profilers [20].

81.4 Conclusions

A stability-dependent SWT performance assessment is presented, using the Obukhov length as classification criteria. Based on data obtained from a well-instrumented experimental setup, site-specific power curves were constructed and the atmospheric stability was classified, shedding light on the relation between SWT power collection and turbulence.

Results are in agreement with those obtained in previous studies on TI, though clarifying the decreasing power collection phenomena close to rated wind speeds. The Obukhov length was found to be an appropriate stability parameter. The results indicate a consistent power collection dependence on stability conditions close to rated wind speeds. With a statistical nature, the TI parameter presents limitations in relation to explaining the measured pattern. Therefore, this study contributes as a complementary analysis, with TI as the representative parameter for lower wind speeds (<8 m/s in this study) and atmospheric stability accounting for conditions close to rated wind speed.

Further studies are required in order to obtain safe and viable small wind applications in complex terrain. The results of this study indicate that successful SWT grid-tied systems rely on more elaborated turbine tests and a better description of atmospheric conditions at the site, and for this the use of the procedure described herein is recommended.

Acknowledgments This study was supported by the National Electrical Energy Agency (ANEEL) under R&D project 0403–0020/2011, funded by Tractebel Energia S.A. (GDF Suez). Authors also acknowledge support from the Brazilian Research Council CNPq.

Appendix A. Experimental equipment

The main technical specifications of the experimental setup are:

Instrument	Height (m_{agl})	Scan (s)	Average (min)
Cup anemometer Thies P6101H Class S 0.5	15.6	1	1 and 10
Wind Vane Thies First Class P6200H	14.0	1	1 and 10
Ultrassonic anemometer RM Young 81000	17.2	0.05	60
Power transducer Ohio Semitronics PC5	18.9	1	1 and 10
Hygrothermal sensor Thies P6312P	14.3	1	10
Barometer P6330.2	9.0	1	10

A Campbell CR1000 data logger was used along with a NL115 module for the raw data storage and transmission. EddyPro® open source software was used for post-processing the 20 Hz data. The data collected were sufficient to meet IEC requirements [15, 16] and were arranged in a database which is available on request from the authors.

References

1. WWEA (2012) Small wind world report. World Wind Energy Association, Bonn
2. ANEEL (2012) Resoluçao Normativa 482 de 17/04/2012, Brasilia. Agencia Nacional de Energia Eletrica, Brazil
3. Smith J, Forsyth T, Sinclair K, Oteri F (2012) Built-environment wind turbine roadmap. National Renewable Energy Laboratory (NREL), Golden
4. Wood D (2011) Small wind turbines: analysis, design and application. Springer, London
5. Encraft (2008) Warwick wind trials: FInal Report, Encraft
6. Ledo L, Kosasih PB, Cooper P (2011) Roof mounting site analysis for micro-wind turbines. Renew Energy 36:1379–1391
7. Lubitz WD (2014) Impact of ambient turbulence on performance of a small wind turbine. Renew Energy 61:69–73
8. Sunderland K, Woolmington T, Blackledge J, Conlon M (2013) Small wind turbines in turbulent (urban) environments: a consideration of normal and Weibull distributions for power prediction. J Wind Eng Ind Aerodyn 121:70–81
9. Wharton S, Lundquist JK (2012) Assessing atmospheric stability and its impacts on rotor-disk wind characteristics at an onshore wind farm. Wind Energy 15:525–546
10. Lee X, Massman WJ, Law B (2005) Handbook of micrometeorology: a guide for surface flux measurement and analysis. Springer, Netherlands
11. Foken T, Nappo CJ (2008) Micrometeorology. Springer, Berlin
12. Stull RB (1988) An introduction to boundary layer meteorology. Springer, Netherlands
13. LI-COR (2013) EddyPro® 4.2 Help and user's guide. LI-COR, Inc, Lincoln

14. Vickers D, Mahrt L (1997) Quality control and flux sampling problems for tower and aircraft data. J Atmos Ocean Technol 14:512–526
15. IEC (2005) Wind turbines, part 12–1: power performance measurements of electricity producing wind turbines. International Eletrotechnical Commission, Geneva
16. IEC (2013) Wind turbines, part 2: design requirements for small wind turbines. International Eletrotechnical Commission, Geneva
17. WINDTEST (2009) Power performance measurement of the Skystream 3.7 at Kaiser-Wilhelm-Koog, Germany according to IEC 61400-12-1 and BWEA, WINDTEST Kaiser-Wilhelm-Koog GmbH
18. Carpman N (2011) Turbulence intensity in complex environments and its influence on small wind turbines. Dept. of Earth Sciences, Uppsala University, Uppsala
19. Trivellato F, Battisti L, Miori G (2012) The ideal power curve of small wind turbines from field data. J Wind Eng Ind Aerodyn 107–108:263–273
20. Wagner R, Courtney M, Gotschall J (2011) Accounting for the speed shear in wind turbine power performance measurement. J Wind Eng 14:993–1004

Chapter 82
Robust Design of Savonius Wind Turbine

Vishaal Dhamotharan, Ranjana Meena, Piyush Jadhav, Palaniappan Ramu and K. Arul Prakash

Abstract The Savonius wind turbine, a class of vertical axis wind turbine (VAWT), is simple and provides a better cost-benefit ratio. It works on the principle of differential drag and is effective in rooftop and ground mounting. Despite the advantages of Savonius wind rotors, they are not preferred due to their low aerodynamic performance levels. In order to address this, several experimental and numerical studies have been carried out in recent years. The primary aim of this work is to provide a simple methodology for the robust optimal design of the Savonius wind turbine. In the parameter design stage, the performance of the turbine is maximized using the traditional Taguchi method. An L_{27} orthogonal array is used considering five factors of three levels each, which affect C_p. Wind speed is considered to be the noise factor. Signal-noise ratio (SNR) metric is used to find the optimal settings for robust design. The aerodynamic performance of the turbine is investigated through dynamic computational fluid dynamics (CFD) models of the design sets. The numerical models used for the simulations are also discussed.

Keywords Small hydro-turbine · Contra-rotating rotors · Performance · Internal flow

---

V. Dhamotharan (✉) · R. Meena · P. Jadhav · P. Ramu
Department of Engineering Design, IIT Madras, Chennai, Tamilnadu, India
e-mail: visiitm2011@gmail.com

R. Meena
e-mail: ranjanameenaiitm@gmail.com

P. Jadhav
e-mail: piyushjadhav13@gmail.com

P. Ramu
e-mail: palramu@iitm.ac.in

K. A. Prakash
Department of Applied Mechanics, IIT Madras, Chennai, Tamilnadu, India
e-mail: arulk@iitm.ac.in

A. Sayigh (ed.), *Renewable Energy in the Service of Mankind Vol I*,
DOI 10.1007/978-3-319-17777-9_82

82.1 Introduction

Currently, a majority of energy requirements are fulfilled by conventional energy sources. As a consequence, resource depletion and environmental degradation occur at a rapid rate. To overcome the potential threat of running out of energy sources against the backdrop of increasing energy needs, efforts are made towards harnessing renewable energy sources. Wind is considered as one of the most important renewable sources of energy. Researchers have widely worked on developing effective power generation systems from wind energy. One of the integral parts of such a system is the wind turbine. The Savonius wind turbine is one of the simplest and cheapest wind rotors which was developed by Finnish engineer Sigurd Savonius in 1925 [1]. It features advantages like low starting torque, relatively low operating speed and ability to capture wind in all directions etc.

Researchers have investigated the effect of various parameters on the efficiency of the Savonius rotor. The parameters include overlap ratio, aspect ratio, end plates, curtains, separation gap between rotors, presence or absence of support shaft, tip speed ratio (λ), Reynolds number (*Re*) and other geometrical parameters [2–4]. The effects of the number of rotor blades, twisted blades and helical blades were also investigated [5].The performance is expressed in terms of static torque (T_s), torque (T), power (P), coefficient of torque (C_t) and coefficient of power (C_p). The literature consists of both numerical and experimental studies of the Savonius rotor. Qualitative analysis were carried out through pressure distribution plots and flow visualizations [6–7]. Recently, Kamoji et al. [8] investigated the effect of geometric parameters, aspect ratio and Reynolds number on the performance and reported a correlation between C_p, λ *and Re*.

Savonius rotors convert wind energy into mechanical energy, which can further be preserved in some form or used to perform work. Generators are employed to use the stored energy and are rated per their capacity. In the context of the Savonius turbine, the output is the rotations per minute (rpm) and the generator capacity is chosen according to the expected rpm. However, the rpm is dependent on wind speeds which vary and the generator can work effectively only for a band of rpm. Therefore, it is desirable to have a robust design of the turbine. That is, the rpm does not vary much in spite of variation in wind speed. The idea is to minimize the effect of the cause of variation without eliminating the cause itself. The cause of variation is the wind speed, and the design variables include geometric parameters of the turbine. The literature does not report much work in the robust design of the Savonius turbine except for [9] that studies horizontal axis wind turbines. This work focuses on an orthogonal array-based characterization of rpm. Signal-noise ratio (SNR) is used to find the optimal combination of design variables that will minimize the variation in the rpm subject to input variation in the wind speeds.

82.2 Model Details

82.2.1 CAD Model

A mono-stepped Bach-type [10] (also referred as modified design in literature) Savonius wind rotor is used for this study. The parametric details of the geometry used are presented in this section. By modifying certain parameters, the interest is to minimize the variation in performance subject to randomness in the wind speeds. Hence, these parameters need to be identified first. This is also referred to as parametric design. Coefficient of power, C_p, is a metric for the performance of the rotor. The C_p is influenced by the shape of the rotor as well as the Reynolds number. Five geometrical parameters are identified that completely define the cross-section of the rotor. These chosen parameters are the design variables for the robust design problem and are shown in Fig. 82.1, where d is the perpendicular distance between the blade extensions, r is the blade arc radius, θ is blade rotation angle depicting the angle by which the blades are rotated with respect to the arc centre and not the blade centre, φ is blade arc angle and D is the diameter of the blade. The diameter of the turbine is D_t which is not a design parameter and is required for the explanation of geometry. The design variable space is explored extensively, and combinations of the design variables are worked out such that only positive overlap ratios were possible.

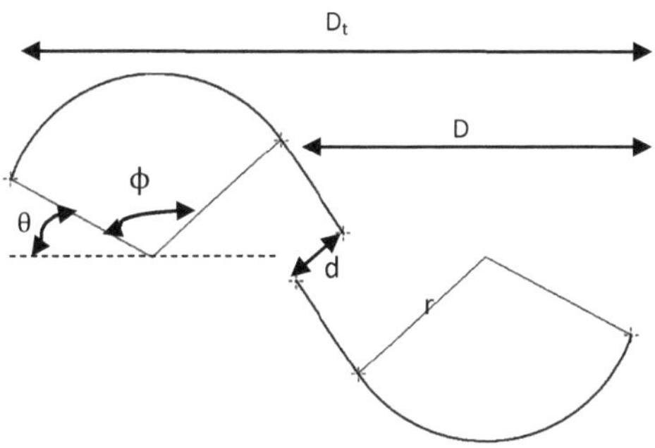

Blade arc radius	r
Blade arc angle	Φ
Blade rotation angle	θ
Diameter of the blade	D
Perpendicular distance	d
Turbine diameter	D_t

Fig. 82.1 Geometry of blade profile

82.2.2 Numerical Model

The computational fluid dynamics (CFD) model is developed based on the geometry described above. The outer domain of the CFD model is taken to be $5D_t$ in width and the inner domain is taken to be $2D_t$ diametrically as in Fig. 82.2. These

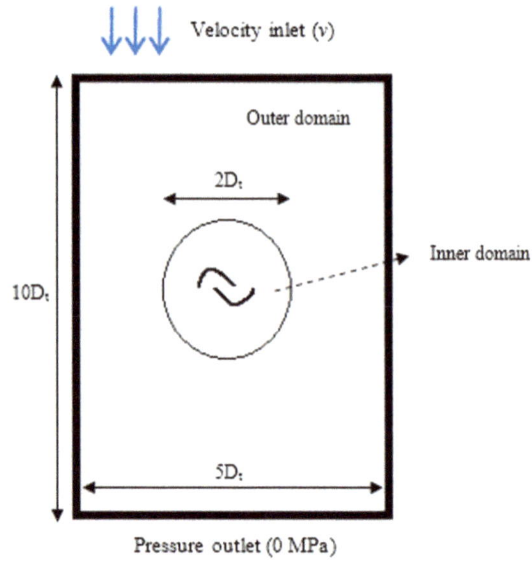

Fig. 82.2 Computational inner and outer domain of the model

Table 82.1 Mesh and solver details

Type of mesh element used	Quadrilateral elements
No of mesh elements	150,000
Meshing package	ANSYS ICEM
Solver model	K-ε
Algorithm	Coupled
Solver package	Fluent

are based on the upper limits of D_t. This size of the domain ensures that reflections of the fluid from domain walls do not affect the rotors. Care is taken to make the fidelity of the mesh high in the vicinity of rotor blades and coarse radially towards the outer domain. The exponential mesh law is followed with a ratio of 1:5 along the rotor blades to the domain boundaries. The mesh and solver model details are summarized in Table 82.1.

The mesh quality is well within the constrained limit required to qualify as a good mesh. A mesh convergence study was done by authors in [11], and the same mesh is used for all the simulations. The finite volume mesh as generated is shown in Fig. 82.3. In the robust design framework, new meshes were generated for each design and we have used K-ε. The advantage of K-ε is that it is proven to be stable and numerically robust. A Second-order interpolation method is used. Pressure and velocity distributions are calculated with a COUPLED analysis algorithm. The models were iterated till the values of the residuals converged to a value of 10^{-4}.

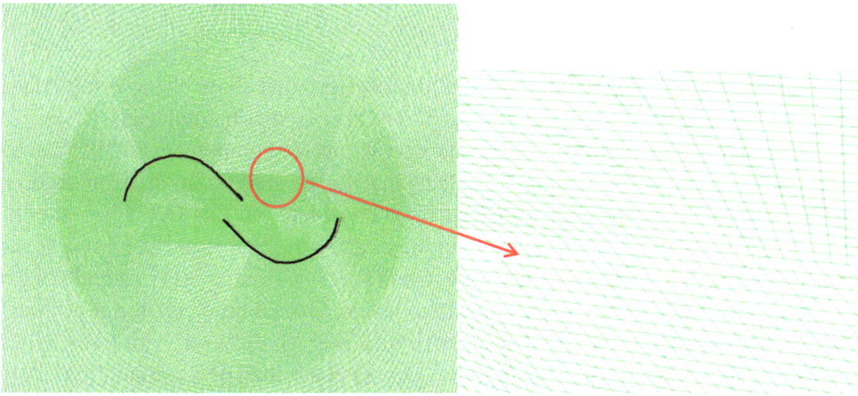

Fig. 82.3 Mesh model used for CFD study

82.3 Robust Design

The study uses the Taguchi method of orthogonal arrays to investigate the effect of different parameters on turbine performance. The objective is to maximize the C_p while minimizing its variations owing to the uncertainties in wind. This is done by identifying the optimal settings of control factors which would maximize C_p with minimum variation due to uncertainties in wind speed. The quality characteristic of interest here is larger the better (L-type), and hence the following SNR is used. Maximizing SNR is equivalent to minimizing the variance and improving the mean.

$$\frac{S}{N} = -10\log_{10}\left\{\frac{1}{n}\left[\sum_{i=1}^{n}\frac{1}{y_i^2}\right]\right\} \tag{82.1}$$

Where y_i is the response experiment, n is the number of runs. The numerical levels for the different design variables along with their SI units are specified in Table 82.2.

An important feature of the Savonius class turbine is that its performance is independent of the wind direction. Here, the only external factor which affects the performance of the turbine is wind velocity (v). In order to understand the sensitivity of the performance characteristic with respect to uncertainties in v, it has been considered as a noise factor in the study.

Table 82.2 Design parameters with their levels

	Design variable	Unit	Level 1	Level 2	Level 3
A	Perpendicular distance	mm	40*	50	60
B	Blade arc radius	mm	100	107.5	115*
C	Blade arc angle	deg	115	122.5*	130
D	Blade rotation angle	deg	7*	11	15
E	Blade diameter	mm	235	250	265*

* marks the levels which form the profile with maximum mean Cp among the 27 base profiles

82.3.1 Orthogonal Array

Orthogonal arrays (OA) are fractional factorial designs that allow one to compute the main and interaction effects via a minimum number of experimental trials. Choice of a suitable OA depends on the number of experiments (or computer runs) that can be conducted and the factor, interaction effects that can be studied. In order to account for the variation of wind, an external array is considered to this OA. Each experiment is realized at every occurrence from the outer array. Here, our outer array is the wind velocities and the three levels are 4, 6 and 8 m/s. The inner array is made of the design variables and each of them has three levels. From a time constraint perspective, we limit our study to 100 computer simulations. To this end, with three levels in the outer array, the best we can do is L_{27} OA. That is a total of 81 simulations. Any other choice of OA with three level factors would mean more simulations.

The parameter design for this study uses two matrices—the design parameter matrix and the noise matrix—as shown in Table 82.3. The fractional factorial design uses a standard L_{27} OA for the inner loop and a 3×1 array for each outer loop. A standard L_{27} OA for 5 factors and 3 levels contain 13 columns, allowing the investigation of five main and four interaction effects. For this study, only the interactions of perpendicular distance with blade arc angle, blade arc radius and blade rotation angle are considered. Table 82.3 shows both the inner array and the outer array. Each experiment in the inner array is run for the three elements in the outer array. Hence, 81 (27×3) runs are carried out for the experimental design.

82.4 Results and Discussions

As mentioned above, 81 numerical experiments are carried out in total with respect to the L_{27} OA. C_t values are extracted from CFD studies for each run and C_p for the different noise levels are evaluated using the Eqs. (82.2–82.6).

$$C_t = \frac{4T}{\left(\rho D^2 H v^2\right)} \tag{82.2}$$

$$\lambda = \frac{\omega r}{v} \tag{82.3}$$

$$C_p = \lambda x C_t = \frac{T\omega}{0.5\rho A v^2} \tag{82.4}$$

$$\frac{C_t}{Re^{0.3}} = -0.0107 \times \lambda + 0.0149 \tag{82.5}$$

$$Re = \frac{\rho \times D \times v}{\mu} \tag{82.6}$$

Where ω is angular velocity, A is cross-section area of the rotor, μ is fluid viscosity, H is the rotor height.

Table 82.3 L_{27} OA for inner loop and noise array for outer loop

Exp.			1		2		3	
v(m/s)			4		6		8	
	A(1)	B(2)	AXB(3)	C(5)	AXC(6)	D(8)	AXD(9)	E(11)
1	40	100	1	115	1	7	1	235
2	40	100	1	122.5	2	11	2	250
3	40	100	1	130	3	15	3	265
4	40	107.5	2	115	1	11	2	265
5	40	107.5	2	122.5	2	15	3	235
6	40	107.5	2	130	3	7	1	250
7	40	115	3	115	1	15	3	250
8	40	115	3	122.5	2	7	1	265
9	40	115	3	130	3	11	2	235
10	50	100	2	115	2	7	2	235
11	50	100	2	122.5	3	11	3	250
12	50	100	2	130	1	15	1	265
13	50	107.5	3	115	2	11	3	265
14	50	107.5	3	122.5	3	15	1	235
15	50	107.5	3	130	1	7	2	250
16	50	115	1	115	2	15	1	250
17	50	115	1	122.5	3	7	2	265
18	50	115	1	130	1	11	3	235
19	60	100	3	115	3	7	3	235
20	60	100	3	122.5	1	11	1	250
21	60	100	3	130	2	15	2	265
22	60	107.5	1	115	3	11	1	265
23	60	107.5	1	122.5	1	15	2	235
24	60	107.5	1	130	2	7	3	250
25	60	115	2	115	3	15	2	250
26	60	115	2	122.5	1	7	3	265
27	60	115	2	130	2	11	1	235

The results are then used to compute the SNR (η) values with respect to the L-type characteristic as defined in Eq. 82.1. The computed η values corresponding to each of the 27 combinations are indicated in Table 82.4. Table 82.5 provides the factor-wise average of SNR.

It is observed from Table 82.5 that the optimal combination of design settings A3, B2, C2, D1 and E2 offers a robust profile for the Savonius with the L-type quality characteristic. A corresponding CFD analysis was done on this profile and results are summarized in Table 82.6.

The optimum profile is observed to have a mean C_p of 0.287, same as the best geometry from L_{27}. It is also observed that there is a reduction in variance of C_p by 35 % in comparison to the best profile.

Table 82.4 Signal-to-noise ratios (SNR) for the 27 experiments

Expt.	η	Expt.	η	Expt.	η
1	− 11.952	10	− 11.604	19	− 11.29
2	− 11.465	11	− 11.565	20	− 11.637
3	− 12.278	12	− 12..385	21	− 12.547
4	− 11.189	13	− 11.093	22	− 11.178
5	− 11.53	14	− 11.424	23	− 11.515
6	− 11.309	15	− 11.385	24	− 11.466
7	− 11.48	16	− 11.795	25	− 11.36
8	− 10.915	17	− 11.185	26	− 11.076
9	− 11.627	18	− 11.381	27	− 11.407

Table 82.5 Factorwise SNR values

SNR	A	B	C	D	E
1	− 11.53	− 11.86	− 11.44	− 11.35	− 11.53
2	− 11.54	− 11.34	− 11.37	− 11.39	− 11.50
3	− 11.50	− 11.35	− 11.75	− 11.81	− 11.54

Table 82.6 Performance at the optimum design

Design variable	Perpendicular distance	Blade arc radius	Blade arc angle	Blade rotation angle	Blade diameter	Mean C_p	$\sigma(C_p)$	η
Optimum	60	107.5	122.5	7	250	0.287	0.014	− 10.88
Best of 27	40	115.0	122.5	7	265	0.286	0.022	− 10.91

82.5 SNR Analysis

One by-product of the robust design is to understand the effect of each factor on the objective. This can be observed by plotting the SNR for different factors as presented in Fig. 82.4.

From the SNR plots, it is observed that factor B has the largest effect on C_p among all the design parameters. Levels 2 and 3 have almost the same SNR values. But when the factor setting is changed to level 1, there is a large increase in the SNR value (of about -50 dB) which suggests that it can affect the performance of the turbine in a dramatic way. Factor D has the next largest effect on the performance of turbine. Similar to factor B, D has a difference of only about -4 dB between levels 1 and 2, and when the setting is changed to level 3, the SNR decreases by around 42 db. Factor C is the next most significant factor owing to the -31 dB or -38dB decrease in the S/N ratio value when the setting is changed from level 1 or 2 to level 3. Factors A and E are the least significant factors as they maintain roughly the same SNR values among all the three settings. Hence when the setting is changed from one level to another, the performance is not affected much.

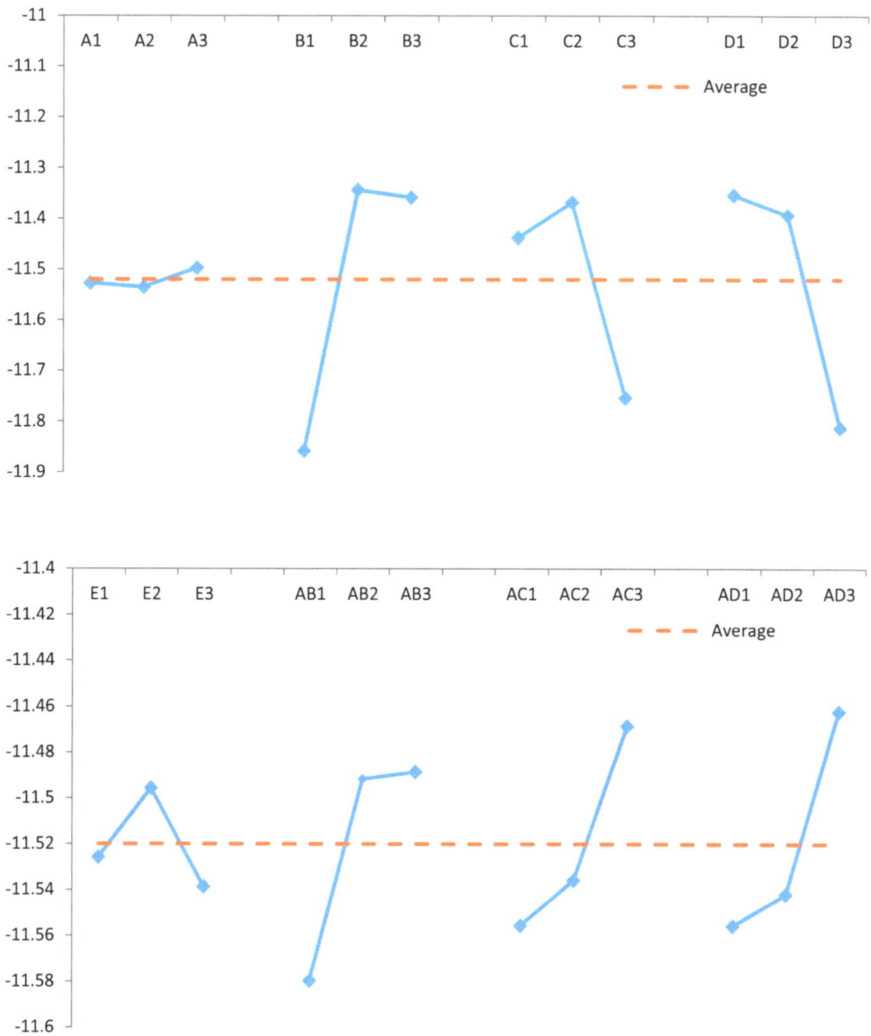

Fig. 82.4 Signal-to Noise Ratio response plots for the main effects and interactions

82.6 ANOVA

Though the behaviour of the model can be interpreted through SNR plots, a better understanding of the relative effects of different factors can be obtained through the ANOVA method.

Hence, the L27 OA design is subjected to an ANOVA analysis and the results are presented in Table 82.7. The F-statistic values or the P-values are used to identify the significance of the design variables. Lesser the P-value, the corresponding

Table 82.7 Results of ANOVA

	Parameter	Sum of squares (SS)	Mean of squares (MS)	Degrees of freedom (DoF)	F	P (Prob>F)	Rank
A	Perpendicular distance (d)	0.00714	0.00357	2	0.07	0.9358	8
B	Blade arc radius (r)	1.54516	0.77258	2	14.58	0.0145	1
C	Blade arc angle (φ)	0.76045	0.38022	2	7.18	0.0475	3
D	Blade rotation angle (θ)	1.16409	0.58204	2	10.99	0.0237	2
E	Blade diameter (D)	0.00872	0.00436	2	0.08	0.9225	7
AxB	d x r	0.05837	0.01459	4	0.28	0.8802	5
AxC	d x φ	0.14672	0.03668	4	0.69	0.6348	4
AxD	d x θ	0.04621	0.01155	4	0.22	0.9153	6
	Error	0.2119	0.05298	4			
	Total	3.94877		26			

parameter is more statistically important. The factors and the 3 two-factor interactions are then ranked according to the P-value starting from the smallest.

It is evident from Table 82.7, that blade arc radius with the least P value is more statistically important than its counterparts. The blade rotation angle stands at rank two with next lowest P-value, followed by blade arc angle which has a slightly higher P-value. These three factors having $P<0.05$ are considered the most significant parameters which should be taken care of while designing a blade profile. The other two factors, namely perpendicular distance and blade diameter have very high P-values and are the least significant. Among the interactions, the interaction of perpendicular distance with blade arc angle is observed to influence the performance more.

82.7 Summary

1. The present work provides an orthogonal array-based procedure for designing robust Savonius wind turbines under realistic conditions wherein wind velocity is considered an uncertainty.
2. The performance of the Savonius is maximized using the Taguchi method with 8 % increase in C_p and 38 % decrease in performance variation.
3. Blade arc radius is observed to have the most influence on the rotor performance, followed by blade rotation angle and blade arc angle. Hence, these parameters are required to be chosen wisely in the design stage.
4. Perpendicular distance and blade diameter are observed to be the least statistically significant parameters.

References

1. Savonius SJ (1931) The S-Rotor and its application. Mech Eng 53(5):333–338
2. Altan BD, Atılgan M (2008) An experimental and numerical study on the improvement of the performance of Savonius wind rotor. J Energy Convers Manage 49:3425–3432
3. Driss Z, Damak A, Karray S, Abid MS (2012) Experimental study of the internal overlap ratios effect on the performance of the Savonius wind rotor. J Eng Technol 1(1):15–21
4. Ushiyama I, Nagai H (1988) Optimum design configurations and performance of Savonius rotors. J Wind Eng Ind Aerodyn 12:59–75
5. Kamoji MA, Kedare SB, Prabhu SV (2009) Performance tests on helical Savonius rotors. Renew Energy 34:521–529
6. Fujisawa N (1992) On the torque mechanism of Savonius rotors. J Wind Eng Ind Aerodyn 40:277–292
7. Fujisawa N, Gotoh F (1992) Pressure measurements and flow visualization study of a Savonius rotor. J Wind Eng Ind Aerodyn 39:51–60
8. Kamoji MA, Kedare SB, Prabhu SV (2009) Experimental investigations on single stage modified Savonius rotor. J Appl Energy 86:1064–1073
9. Hu Y, Rao SS (2011) Robust design of horizontal axis wind turbines using Taguchi method. J Mech Design 133(11):111009 (15 pages)
10. Al-Bahadly I (2009) Building a wind turbine for rural home. Energy Sustain Dev 13(3): 159–165
11. Gole S, Meena R, Jadhav P, Vishaal D (2013) Numerical simulations towards optimal design of a Savonius wind turbine. International conference on Computer Aided Engineering (CAE-2013)

Chapter 83
Coupling Floating Wind Turbines with Large-Scale Air-Conditioning Systems Through Deep Sea Water Pumping: Case Studies of System Performance in European Deep Waters

Tonio Sant, Robert N. Farrugia and David Arroyo López-Carro

Abstract This chapter presents a novel concept for utilising offshore-floating wind turbines to concurrently exploit cold deep sea water (DSW) for large-scale cooling applications, as well as electricity production. This concept utilises wind turbine-driven pumps that extract cold DSW and pump it across a high-pressure pipeline to a land-based hydroelectric power station coupled to a centralised air-conditioning (AC) unit of a district cooling system. The wind-driven intermittent supply of cold water exiting the hydroelectric station is diverted to the condenser of the unit and mixed with sea surface water (SSW) to maintain a steady flow. The use of DSW lowers the condensation temperature of the refrigerant in the AC unit, resulting in an improved coefficient of performance for cooling conditions. This chapter investigates the potential application of the concept described above to four European deep offshore sites located in the Mediterranean Sea: the Greek Islands, Malta, the South of France and Spain. The numerical model used to compute the energy yield characteristics at these sites assumes a single offshore wind turbine system with modules for the wind turbine–pump assembly, the deep sea thermocline formations and the frictional/thermal losses across the pipeline and the hydroelectric turbine. Another module is dedicated to the thermodynamic refrigeration cycle for the AC unit. The study confirmed that although the losses incurred in transmitting energy from offshore wind turbines through the DSW pipeline are larger than those for conventional wind turbines relying on electrical cables, such losses may be partially or fully compensated for by the superior performance of the wind turbine-driven hydraulic pumps at high wind speeds and by the energy savings incurred in AC plants through DSW utilisation. The latter savings were found to be the highest for

T. Sant (✉) · R. N. Farrugia
Institute for Sustainable Energy and Department of Mechanical Engineering,
University of Malta, Msida, MSD 2080, Malta
e-mail: tonio.sant@um.edu.mt

R. N. Farrugia
e-mail: robert.n.farrugia@um.edu.mt

D. A. López-Carro
E.T.S. Ingenieros Industriales, Universidad de Valladolid, Paseo del Cauce, Valladolid, Spain
e-mail: darroyolc@gmail.com

A. Sayigh (ed.), *Renewable Energy in the Service of Mankind Vol I*,
DOI 10.1007/978-3-319-17777-9_83

the Greek Islands due to favourable wind conditions also prevailing during the hot summer months.

Keywords Wind turbine-driven pump · Hydraulic transmission · Deep sea water · Air-conditioning

83.1 Introduction

Floating wind turbine technology development has made significant progress in recent years [1]. However, this is still based on conventional approaches whereby the turbine only supplies electricity. Deep offshore sites having sea depths more than 150 m also offer an immense resource for renewable energy-based space cooling. The temperature of deep sea water (DSW) far below the surface deviates considerably from ambient temperatures observed in the onshore urban environment. New offshore floating technology capable of concurrently exploiting offshore wind resources in conjunction with thermal energy resources stored in deep seas may therefore offer the opportunity to reduce the costs of floating wind farms by maximising renewable energy exploitation in a deep sea context. This new concept, being referred to as *offshore wind and thermocline energy production* (*OWTEP*) concept, is currently being researched at the University of Malta. The concept is based on the application of hydraulic wind turbines consisting of wind turbine rotors directly coupled to positive-displacement pumps in lieu of electrical generators to supply pressurised sea water across a pipeline. The pipeline is linked to a land-based grid-connected hydroelectric power station consisting of a Pelton turbine connected to a generator enabling the conversion of fluid power into electricity. The DSW exiting the Pelton wheel is then diverted to a heat exchanger to exploit its thermal energy potential before flowing back to the open sea.

Although hydraulic wind turbines on a micro- or medium-scale have been used for many years primarily for water pumping applications in agricultural and flood control, the application of large-scale hydraulic turbines for offshore wind energy harvesting is still in its early stages of research and development. There has recently been renewed interest in such systems given the advantages these may potentially have over conventional systems [2–4]. The use of a hydraulic-based wind harvesting system, including hydraulic energy transmission, will not only simplify offshore wind technology and minimise the use of copper required in electrical systems but it may also facilitate the integration of offshore wind farms with other forms of renewables (such as wave and tidal energy) and hydro-storage facilities. Hydraulic-based systems have a higher power-to-weight ratio than electrical generator-based equivalents and were shown to present significant potential in reducing nacelle weights [2]. This advantage coupled with the capability of hydraulic machinery to operate at large torques offer possibilities for increasing the operating efficiencies of wind turbines at the higher wind speed classes.

Hydraulic wind turbine technologies will be readily applicable to the OWTEP concept introduced above. Computational modelling at the University of Malta of a single OWTEP turbine has shown how, through this concept, considerable amounts of renewable thermal energy can be supplied through DSW extraction, in addition to electricity generation [5, 6]. However, the higher energy transmission losses when compared to those encountered in electrical power transmission via cables presents a major drawback. Such losses result from fluid friction when pumping the pressurised DSW to the hydroelectric power station onshore. The resulting net electricity production for a conventional 126 m diameter, 5 MW wind turbine generator (WTG) was found to be considerably larger than for an OWTEP turbine with the same diameter and a hydraulic pump of equal power capacity. Such analysis was restricted to offshore sites in Maltese territorial waters in the Central Mediterranean Sea. Recent analysis has investigated the energy performance of a single OWTEP wind turbine coupled to a centralised air-conditioning (AC) unit of a district cooling system operating on a vapour compression cycle [7]. In this analysis involving an OWTEP-AC system, the DSW exiting from the hydroelectric power station of the OWTEP system was modelled to flow through the condenser of the AC unit. This was found to increase the coefficient of performance of the AC unit substantially, resulting in energy savings which partly offset the energy losses incurred in hydraulic energy transmission through the pipeline. The present chapter simulates the performance of single turbine OWTEP-AC systems operating in the following four Euro-Mediterranean offshore scenarios: the Greek Islands (Aegean Sea), Malta (Central Mediterranean), the South of France and Spain (West Mediterranean). The Mediterranean region has been specifically selected given its growing energy demand for cooling of buildings as a result of warm climatic conditions when compared to other European states. The analysis considered three different OWTEP-AC system configurations, each having the same wind turbine rotor diameter, but with a varying hydraulic pump capacity.

83.2 System Description

Figure 83.1 below illustrates the single wind turbine OWTEP system coupled to the centralised AC unit, making-up the baseline OWTEP-AC system that is being considered in this study. The offshore floating wind turbine-driven pump extracts DSW, supplying it at high pressure to a land-based hydroelectric power station consisting of a Pelton turbine coupled to a grid-connected electrical generator. The numerical model used in this study for the OWTEP system comprising the wind turbine-driven hydraulic pump, sea water pipeline and land-based hydroelectric power station is described in detail in [5]. Assuming steady-state performance at each timestamp, the model uses a time-series wind data set to generate a corresponding data set for the DSW flow rate and temperature (Q_{DSW} and T_{DSW}, see Fig. 83.1) and electrical power output. The model integrates sub-models for the wind turbine-pump characteristics, the pressure drops in the hydraulic power transmission to shore, the temperature rise

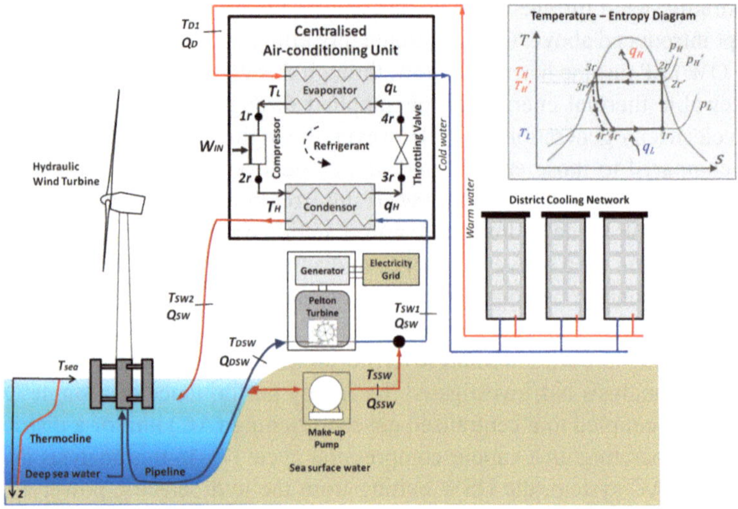

Fig. 83.1 The OWTEP wind turbine coupled to the large-scale AC unit for district cooling. The temperature–entropy diagram for the refrigeration cycle is included

in DSW as it flows through the pipeline and the energy performance of the Pelton wheel–electrical generator assembly. A simplified model for the thermocline formations is also included, taking into account the variation of sea surface water (SSW) temperature on a monthly basis while maintaining the deep sea temperature fixed, independent of season. The latter model is used by the heat transfer model to estimate the temperature of DSW at the pipeline exit (T_{DSW}). The temperature rise of the sea water across the Pelton turbine is ignored as this is found to be negligible [5].

As may be noted in Fig. 83.1, the DSW exiting the turbine flows through the condenser of the centralised AC unit that is being modelled to chill water circulating through a district cooling system. The AC unit operates on a vapour compression cycle with a refrigerant circulating internally to act as a cooling medium. The compressor is powered by the electricity grid. The thermodynamic cycle for the refrigeration process is shown in temperature–entropy (T–s) diagram in Fig. 83.1. In the present study it is assumed that the refrigerant exits the compressor as a saturated vapour, and hence, no superheating takes place. The coefficient of performance (COP_C) for the cycle is computed from:

$$COPc = \frac{q_L}{q_H - q_L} = \frac{q_L}{W_{IN}} = \frac{h_{1r} - h_{4r}}{h_{2r} - h_{1r}} \tag{83.1}$$

Where, q_L and q_H are the cooling and heating loads, while W_{IN} is the net input work from the compressor; h denotes the enthalpy of the refrigerant. The present study applies the same thermodynamic model for the performance of AC unit as that described in [7]. The heat extraction process by the sea water flowing through the condenser is modelled using the number of transfer units (NTU) method, with the sea water temperature rise ($T_{sw1} - T_{sw2}$) being fixed. Given that the flow of DSW is

Table 83.1 Offshore site coordinates and 5-year average wind speed and ambient temperature values

Site	Location	NCEP wind data coordinates latitude/longitude	Mean wind speed 10 m amsl (m/s)	Mean wind speed (extrapolated) 90 m amsl (m/s)	Mean ambient temperature (°C)
A	Greek Islands	37.14°N/26.25°E	6.73	8.56	18.25
B	Malta	35.0°N/15.0°E	6.35	7.32	18.97
C	South of France	42.86°N/5.63°E	7.60	9.68	14.79
D	Eastern Spain	39.05°N/0.0°E	5.97	7.61	16.48

intermittent as a result of the dependency on the wind driven hydraulic pump offshore, a make-up water pump is included in the OWTEP-AC system to supply SSW to compensate for any short fall in supply in DSW, thereby maintaining a steady flow of sea water through the condenser. The temperature of the SSW is unavoidably warmer than that pumped from offshore deep waters and the temperature T_{SW1} of the mixed water entering the condenser may be estimated from a simple energy conservation formula:

$$T_{SW1} = \frac{(T_{DSW})^* Q_{DSW} + (T_{SSW})^* Q_{SSW}}{Q_{DSW} + Q_{SSW}} \tag{83.2}$$

Where, T_{DSW} and Q_{DSW} are the temperature and flow rate of the DSW supplied by the offshore wind turbine to shore, while T_{SSW} and Q_{SSW} are the temperature and flow rate of the SSW. All temperatures are in Kelvin.

83.3 Description of Euro-Mediterranean Sites

Time series data for wind and ambient temperature covering the 5-year period 2008–2012 were used in conjunction with marine thermocline data and the OWTEP and AC unit numerical models outlined in Sect. 83.2, to estimate the performance of a single turbine OWTEP-AC system for the four sites in the Mediterranean. The wind turbines were assumed to be located at a distance of 10 km from the coast in an area having a sea depth of 200 m. Wind data consisted of six-hourly average estimates from the National Centers for Environmental Prediction (NCEP) Reanalysis II database [8]. The latter has a grid resolution of 5°, with the internodal distance being equal to 208 km. The coordinates of the grid nodes for which offshore wind data were extracted are given in Table 83.1. The offshore site locations are marked on the map shown in Fig. 83.2. Wind data was extrapolated from an elevation of 10 m above mean sea level (m.s.l) to an assumed rotor hub height of 90 m using a power lower, with the wind shear exponent equal to 0.11 for Site A (the Greek Islands), C (the South of France) and D (Eastern Spain). The resulting 5-year average wind speeds were found to be within the ranges for long-term wind speed predicted by the European Offshore Wind Atlas [9] for these sites. In the case of Site B (Malta), a wind shear exponent of 0.065 was applied given that the resulting 5-year average

Fig. 83.2 Location of the four Euro-Mediterranean sites investigated. (Map source: Google Earth 2013)

was within range of long-term average wind speeds derived from previous studies involving both the use of wind measurements and CFD [10] modelling.

Ambient temperature data for Sites A, C and D were also obtained from the NCEP Reanalysis II database, in this case for 2 m above m.s.l. It was assumed that such a temperature provides a realistic representation of the ambient temperatures experienced at coastal urban locations at these sites. However, in the case of Site B, the closest grid node located at 35.0°N/15.0°E is 96 km away from the nearest onshore location. Given the relatively small size of Malta, the ambient temperature at this node is unlikely to be representative of the true ambient temperature experienced in coastal urban areas. Therefore, METAR data integrated in the WindPro [11] software package was also used. Monthly-averaged temperature-depth (thermocline) profile data from the National Oceanography Data Centre (NODC) [12] was taken from the nearest data point on a 5° mesoscale modelling grid resolution. The monthly thermocline data obtained for one of the sites, Site D (Spain), is plotted in Fig. 83.3. While the temperature of the SSW changes considerably depending on season, the DSW temperature at depths below 160 m is stable. Similar trends were observed at the other three sites being investigated.

Figure 83.4 shows the monthly average values for the wind speeds derived from the NCEP Reanalysis II database for the 5-year period 2008–2012. Months 1 and 12 plotted on the x-axis denote January and December, respectively. The corresponding values for the ambient air, SSW and DSW temperature are also included. The monthly average sea surface temperatures and ambient air temperatures are relatively close during the summer months at all the four sites. On the other hand, the temperature differences between the SSW and DSW are considerable during the summer months, offering a potential renewable energy resource for cooling. Furthermore, it may be observed in Fig. 83.4 that, while the DSW temperatures are very close for all the four sites (approx 13–15 °C), considerable differences in the wind and ambient temperature exist. Both the South of France and the Greek Islands have a rich offshore wind resource; however, the summer ambient temperatures at the latter region are higher, hence demanding a larger AC load for cooling of buildings within urban areas. Malta and the Spanish site under consideration both have

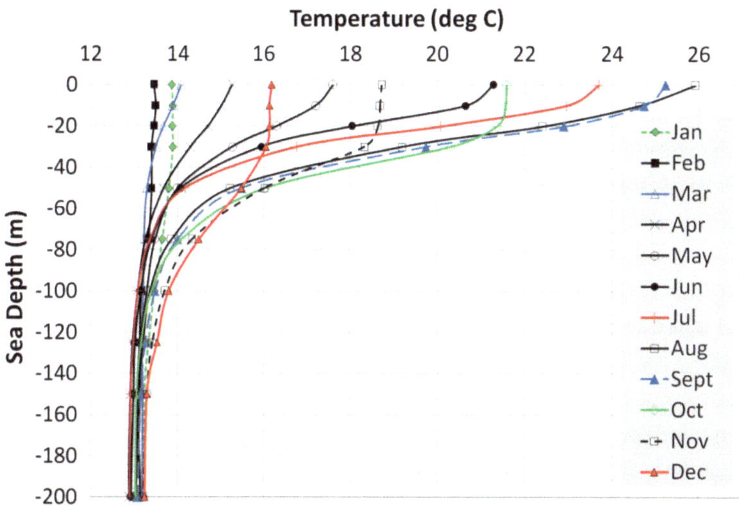

Fig. 83.3 Monthly-averaged thermoclines for site D (Eastern Spain)

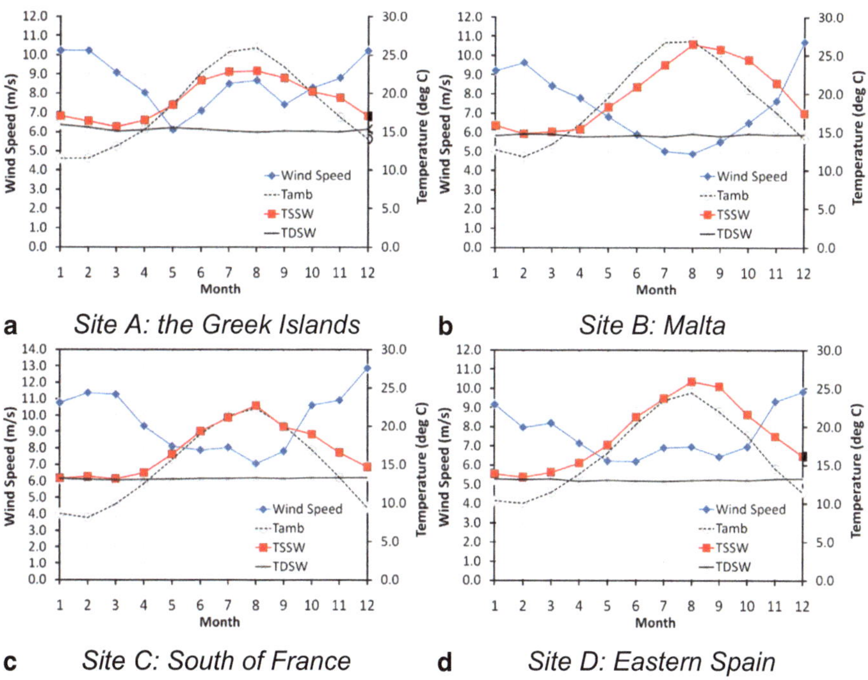

Fig. 83.4 Monthly average values for the wind speed at 90 m above m.s.l., ambient air temperature at 2 m above m.s.l., SSW temperature and DSW temperature at 200 m below sea level

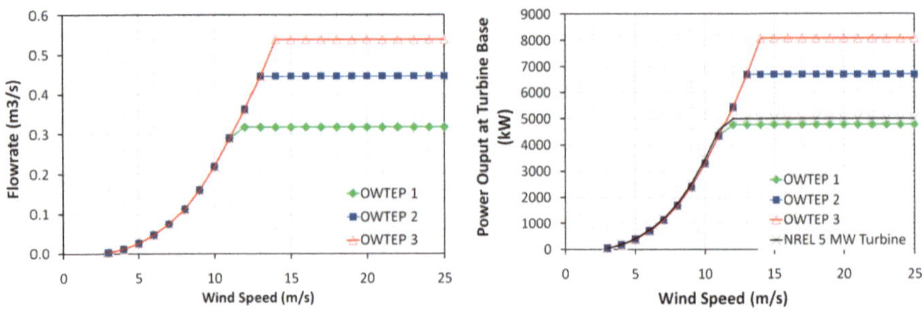

Fig. 83.5 Characteristics of the three modelled 126 m diameter OWTEP turbines. The pressure rise across the turbine pump is assumed constant with wind speed at 150 Bar

medium wind resources, but the ambient temperatures in Malta are higher given the more southerly position of the island in the Mediterranean.

83.4 Design Parameters for the Modelled OWTEP-AC Systems

Three different OWTEP hydraulic turbines were considered in the present investigation, each having a different capacity but with fixed rotor diameter and hub height of 126 and 90 m, respectively. The three OWTEP turbines, being referred to as OWTEP models 1, 2 and 3, were assumed to have the rotor design of the NREL 5 MW baseline conventional WTG [13], but individually coupled to a positive displacement pump of different rated volume flow rate. It was also assumed that the three hydraulic turbines operate at a constant efficiency (η_p) and pressure rise (Δp) with wind speed, equal to 90 % and 150 Bar, respectively. The water volume flow rate (Q) in m^3/s at each wind speed could then be derived for each wind speed from the following basic relationship:

$$Q = \frac{\eta_p \times P_{\text{shaft}}}{\Delta p} \tag{83.3}$$

where P_{shaft} is the rotor shaft power. The flow rate-wind speed and power-wind speed characteristics for the three OWTEP turbines considered in the study are presented in Fig. 83.5. The electrical power curve for the National Renewable Energy Laboratory (NREL) 5 MW baseline WTG is also included. One should note that the power shown in Fig. 83.5 for the three OWTEP models is hydraulic not electrical power. The increased rated volume flow rate capacity of the pumps results in a higher-rated wind speed, thus enabling the OWTEP hydraulic wind turbines to capture more energy at high wind speed conditions than a conventional WTG. The design parameter values common to the three OWTEP systems and the AC units to which they were coupled in the numerical analysis are presented in Table 83.2. The

Table 83.2 Fixed parameters for the OWTEP-AC unit systems

OWTEP system parameters		Parameters of district cooling AC unit	
Pipeline length	10 km	Refrigerant	R134a
Pipeline diameter	0.6 m	Evaporator temperature of refrigerant	6 °C
Pipeline thickness	5 mm	Condenser heat transfer coefficient	1000 W/m^2/K
Pipeline surface roughness	0.4 mm	Outlet temperature of district cooling Water ($T_{D,2}$)	12 °C
Pipeline thermal conductivity	0.2 W/mK	Temperature difference across condenser ($T_{SW2}-T_{SW1}$)	10 °C
Elevation of Pelton turbine above sea level	20 m		
Efficiency of hydroelectric power station	90 %		
Exit pressure of hydraulic turbine	10 Bar		
Wind turbine sea depth	200 m		
Thickness of upper layer	30 m		
Depth of thermocline (depth to stable temp region)	120 m		

Table 83.3 Condenser surface areas

OWTEP-AC units		Conventional WTG-AC units	
	Condenser surface area (m^2)		Condenser surface area (m^2)
OWTEP 1	1669	WTG 1	1669
OWTEP 2	2337	WTG 2	2337
OWTEP 3	3004	WTG 3	3004

temperature rise in the sea water flowing through the heat exchanger ($T_{SW2}-T_{SW1}$) (refer to Fig. 83.1) was maintained fixed. The sea water flow rate through the condenser was also kept fixed and set equal to the rated flow rate of the OWTEP turbine. Hence, the AC unit coupled to OWTEP models 1, 2 and 3 had a volume flow rate of sea water constant at 0.32, 0.45 and 0.54 m^3/s, that is, the OWTEP rated flow rates shown in Fig. 83.5. The surface area of the condenser was varied depending on the OWTEP turbine models, with the assumed values increasing with the turbine pump capacity. The values are given in Table 83.3. A steel pipeline was considered for each OWTEP system.

In order to be able to compare the performance characteristics of the OWTEP system to conventional offshore WTGs only producing electricity, the performance computations were repeated for the NREL 5 MW baseline WTG and AC units utilising solely SSW in the condenser. The electrical power curve in Fig. 83.5 was used to derive the electricity yield estimates for the 5 MW WTG located at the same offshore locations as the OWTEP turbines. It was assumed that the generated electricity is transmitted across an electrical cable to the onshore grid from which a fraction is consumed to operate the centralised AC unit compressor. The electrical energy losses associated with the cable transmission over the 10 km distance to shore were assumed to be equal to 6%. Three different conventional WTG-AC unit systems were modelled, WTG 1, WTG 2 and WTG 3, each having the same 5 MW WTG; however, supplying electricity to different AC units. The latter units were assumed to have the same design parameters, included the rated flow rate and condenser areas, of those used with OWTEP models 1, 2 and 3, respectively (see Table 83.3). The time series six-hourly data for wind and ambient temperature described in Sect. 83.2 were prescribed to the numerical models. Given that no thermal model for the district cooling system was considered in the present work, the temperature of the district system water at the evaporator intake of the AC units (T_{D1}, Fig. 83.1) was taken to be equal to the ambient temperature. Furthermore, each AC unit was only modelled to be operational when ambient temperatures exceeded 22 °C, the latter value being a realistic temperature to maintain thermal comfort inside habitable buildings. Time series of six-hourly values for the following variables were generated through the numerical simulations:

- Electricity generated and fed to the grid by the OWTEP and conventional WTG systems
- Electricity consumed by the compressor of the AC units
- AC unit COP_C
- Cooling load produced by the AC units

The above analysis was repeated for each of the three OWTEP-AC systems and the three WTG-AC systems, hence resulting in six different modelled systems. The annual/monthly average values for the above listed parameters over the 5-year period 2008–2012 were also determined. In the case of the COP_C computation, this was taken to be equal to the annual/monthly cooling load (q_L) produced divided by the annual/monthly electricity consumed by the compressor. A compressor efficiency equal to 80% was taken into account. The net (surplus) annual average electricity fed to the grid by the OWTEP and WTG system, but not consumed by the AC Unit, was also computed using the following equations:

$$E'_{\text{WTG}} = E_{\text{WTG}} - W_{\text{c,WTG}} \qquad E'_{\text{OWTEP}} = E_{\text{OWTEP}} - W_{\text{c,OWTEP}} \tag{83.4}$$

where E_{WTG} and E_{OWTEP} denote the net annual average electricity supplied to the grid by the OWTEP and WTG systems and $W_{\text{c,WTG}}$ and $W_{\text{c,OWTEP}}$ denote the gross annual average electricity consumed by the AC unit compressors, with the compressor energy losses being factored in. The overall performance of the three OWTEP-AC systems was compared quantitatively with the corresponding conventional WTG-AC systems by applying the following relations on the annual average values:

$$\Delta_1 = \frac{E_{\text{OWTEP}} - E_{\text{WTG}}}{E_{\text{WTG}}} \times 100 \qquad \Delta_2 = \frac{E'_{\text{OWTEP}} - E'_{\text{WTG}}}{E_{\text{WTG}}} \times 100 \qquad \Delta_3 = \frac{COP_{\text{c,OWTEP}} - COP_{\text{c,WTG}}}{COP_{\text{c,WTG}}} \times 100 \tag{83.5}$$

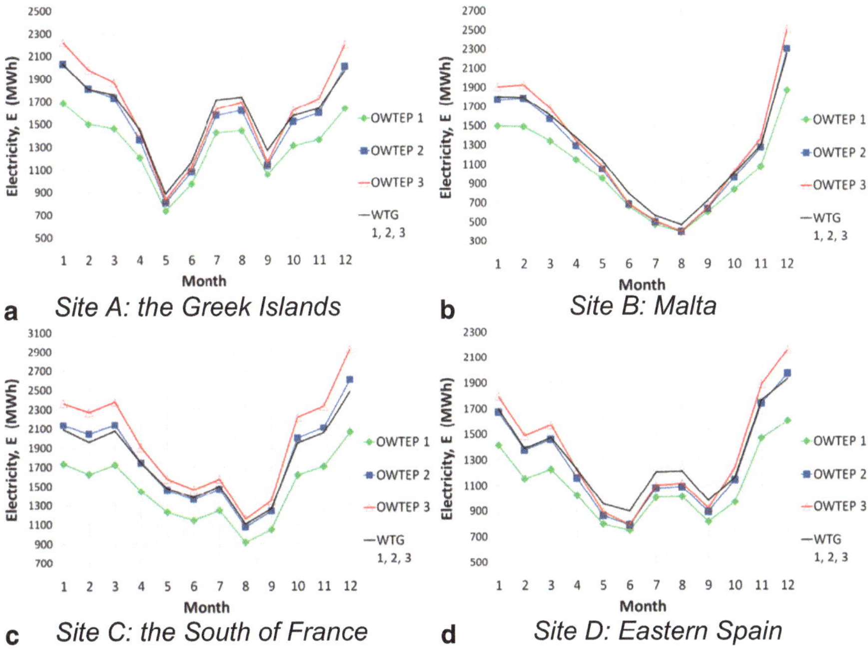

Fig. 83.6 Variation of the 5-year monthly average electricity production

83.5 Results and Discussion

Figure 83.6 plots the monthly average values for the net electricity produced by the three OWTEP models at the four offshore sites. The average values are based on the 5-year period 2008–2012. The net electricity fed to the grid by each OWTEP's hydroelectric station, *E*, is being plotted on the *y*-axis, while months 1 and 12 on the *x*-axis denote January and December, respectively. The corresponding values for the net electricity production for the conventional wind turbines (WTGs 1, 2, 3) are also included. The values for WTG 1, 2 and 3 are identical given that they are all based on the power curve of the NREL 5 MW turbine (see Fig. 83.1). It may be observed from the plots in Fig. 83.6 that the monthly electricity yield from OWTEP 1 is considerably lower than that for the conventional wind turbines at all four sites. This is primarily due to the higher energy losses incurred in transmitting fluid power hydraulically across the pipeline. Such losses were estimated to be in the range of 20–22 % of the gross hydraulic energy yield produced at the wind turbine base for the three OWTEP models. However, the increase in the rated wind speed and volume flow rate of the hydraulic turbines enables a high energy exploitation of wind energy at high wind conditions, thus offsetting the pipeline transmission losses. It was estimated that OWTEP 3 having a rated wind speed and volume flow rate of 14 and 0.54 m/s (see Fig. 83.5) has an energy yield that exceeds those from the 5 MW conventional WTG at all four sites.

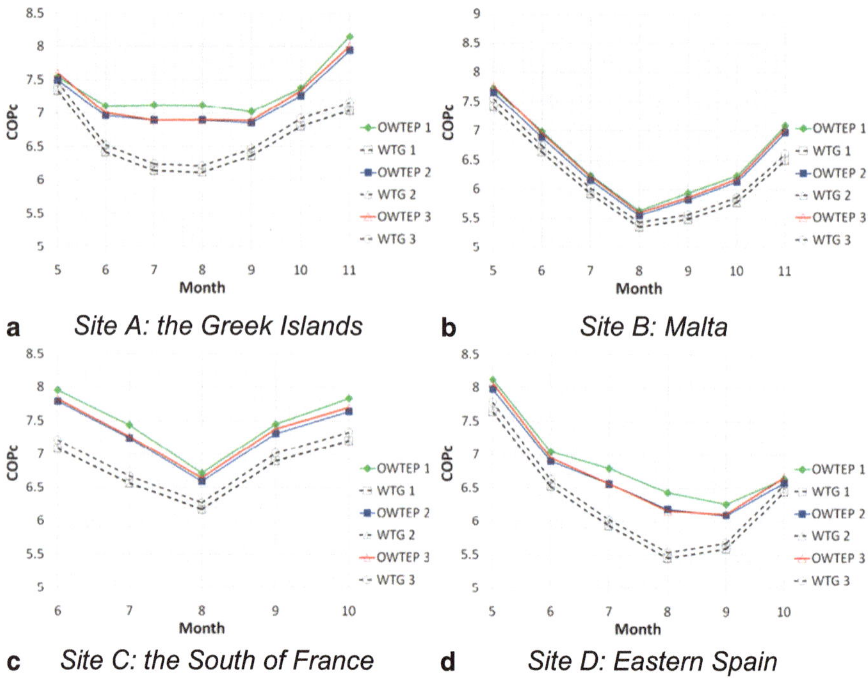

Fig. 83.7 Variation of the monthly average values of the *COPc* for the AC units coupled to the corresponding three OWTEP turbines and the three conventional WTG turbines

Figure 83.7 presents the monthly average predictions for the *COPc* of the AC units operated for the six different systems. The graphs in Fig. 83.7 clearly show how the *COPc* values for the AC units coupled to the OWTEP turbines are considerably higher than those utilising solely SSW with electricity consumed from the WTGs. The mixing of DSW from the OWTEP turbines with SSW in the condenser of the AC units lowers the condensation temperature of the refrigerant, improving the *COPc*. This implies savings in the electricity consumed by the AC systems to meet a specific cooling load. The monthly variation in the *COPc* is considerable at all sites, with the lowest values reached in the summer months. In the case of the conventional systems, this variation originates from the monthly variations in the SSW temperature. The lower SSW temperatures during the cooler periods (May, November), as may be seen in Fig. 83.4 have a beneficial influence on the *COPc*. In the case of the OWTEP-driven AC units, the *COPc* is also influenced by the wind speed that determines the DSW supply to the condenser. The lower wind speeds encountered during the summer period lower the flow rate of DSW from the OWTEP turbines.

OWTEP systems offer best potential at sites having good wind conditions coinciding with periods of the year when the ambient temperature is high. The Greek Islands experience favourable wind speeds during the summer months (see Fig. 83.4a), leaving a positive impact on the COP_C as noted in Fig. 83.7a. The same

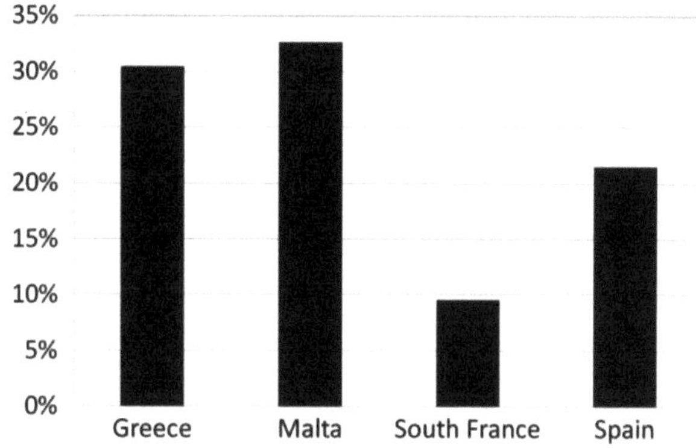

Fig. 83.8 Annual percentage operational hours of the AC unit at the four sites

phenomena are also predicted to occur to a lesser extent at Site C (Eastern Spain). No values for the COP_C could be plotted during some winter/autumn/spring months given that the ambient temperature is normally below the 22 °C limit. In such conditions the AC units would be inoperative and the OWTEP turbines would effectively only produce electricity. Figure 83.8 shows the percentage of operational hours for the AC units at the different sites. A correlation can be easily noted between the mean ambient temperature values for the four sites in Table 83.1 and the percentage operational hours in Fig. 83.8: Sites with high mean ambient temperature also have high operational demands from AC units.

The annual average values for the various performance parameters are tabulated in Table 83.4. These were used together with Eq. 83.5 to compute parameters Δ_1, Δ_2 and Δ_3 for the three OWTEP-AC units and to quantitatively assess the performance of such systems against conventional technology. The results for Δ_1, Δ_2 and Δ_3 are plotted in Fig. 83.9. Δ_1 represents the percentage deviation in the total net annual electricity produced by the OWTEP systems from those produced by the WTGs. Such a deviation is negative for OWTEP 1 system, at approximately −16% at all four sites, implying a lower efficiency than conventional WTGs of equal rotor diameter and power capacity when producing electricity. Such a deficiency is primarily due to the greater energy losses incurred in the hydraulic transmission pipeline, although the energy losses encountered in the hydraulic pump and Pelton turbine also have some contribution. This deviation is reduced by exploiting OWTEP's capability to supply cold DSW to improve the performance of the AC unit. This is quantified through parameter Δ_2 (Eq. 83.5). Figure 83.9 shows that in the case of OWTEP 1, $\Delta_2 > \Delta_1$. Differences between Δ_2 and Δ_1 are most noticeable in the Greek Islands as a result of the good wind conditions prevailing during the hot summer months. Differences are also noticeable for both Malta and Eastern Spain. On the other hand it can be noted that despite the fact that the South of France has very good wind conditions, the benefits of OWTEP with the assumed design parameters are predicted to be small for this region. This is primarily due to the lower ambient temperatures which result in reduced operational hours of $<10\%$ for the AC unit (see Fig 83.8).

Table 83.4 Estimated annual average values for various performance parameters based on the 5 year period 2008–2012

Greece						Malta					
	E	E′	W	QL	COPc		E	E′	W	QL	COPc
	GWh	*GWh*	*GWh*	*GWh*			*GWh*	*GWh*	*GWh*	*GWh*	
WTG 1	19.0	14.0	5.0	31.3	6.28	WTG 1	14.8	9.2	5.6	33.0	5.84
OWTEP 1	15.8	11.4	4.4	31.3	7.12	OWTEP 1	12.3	7.0	5.3	33.0	6.20
WTG 2	19.0	12.0	7.0	43.7	6.27	WTG 2	14.8	6.9	7.9	46.2	5.84
OWTEP 2	18.3	12.0	6.3	43.7	6.93	OWTEP 2	14.2	6.6	7.6	46.2	6.11
WTG 3	19.0	10.7	8.3	52.8	6.37	WTG 3	14.8	5.4	9.4	55.8	5.93
OWTEP 3	19.5	11.9	7.6	52.8	6.95	OWTEP 3	15.1	6.0	9.1	55.8	6.16
South of France						*Spain*					
	E	*E′*	*W*	*QL*	*COPc*		*E*	*E′*	*W*	*QL*	*COPc*
	GWh	*GWh*	*GWh*	*GWh*			*GWh*	*GWh*	*GWh*	*GWh*	
WTG 1	21.1	19.6	1.5	9.8	6.42	WTG 1	15.9	12.1	3.8	21.9	5.75
OWTEP 1	17.5	16.1	1.4	9.8	7.07	OWTEP 1	13.2	9.9	3.3	21.9	6.57
WTG 2	21.1	19.0	2.1	13.7	6.42	WTG 2	15.9	10.6	5.3	30.5	5.74
OWTEP 2	21.4	19.4	2.0	13.7	6.92	OWTEP 2	15.2	10.4	4.8	30.5	6.36
WTG 3	21.1	18.5	2.5	16.6	6.52	WTG 3	15.9	9.6	6.3	36.9	5.83
OWTEP 3	23.5	21.1	2.4	16.6	6.97	OWTEP 3	16.2	10.4	5.8	36.9	6.36

Figure 83.9 illustrates the positive impact resulting from the increase in the pump capacity of the 126 m rotor diameter hydraulic wind turbines on OWTEP system performance. Increasing the rated turbine pump capacity to 0.45 m^3/s, with a corresponding increased rated wind speed equal to 13 m/s (OWTEP 2, see Fig. 83.1), effectively reduces Δ_1 down to 4 % at all the four offshore sites. Further increases in pump capacity shifts Δ_1 to a positive value which, as can be seen in Fig. 83.9 (OWTEP 3), imply better performance than that of conventional turbines. The integration of the DSW in centralised AC systems with high capacity hydraulic turbine pumps contributes to further performance enhancement with the values of Δ_2 estimated to be around 2.5–4 % at sites A, B and D.

A major limitation of this study is the fact that wind and ambient temperatures are based on six-hourly values. Consequently, high temperature and wind speed occurrences spanning over a few hours but which result in better performance of the OWTEP system would not be captured in the six-hourly average values. This suggests that the values of Δ_1 and Δ_2 plotted in Fig. 83.9 are likely to be underestimated. Parameter Δ_3 quantifies the improvement in the annual average COP_C of the AC Unit as a result of mixing DSW from the OWTEP turbines with SSW in the condenser. It is noted in Fig. 83.9 that Δ_3 is significant, although it decreases for higher hydraulic turbine pumping capacities (OWTEP 2 and 3). The increased supplies of

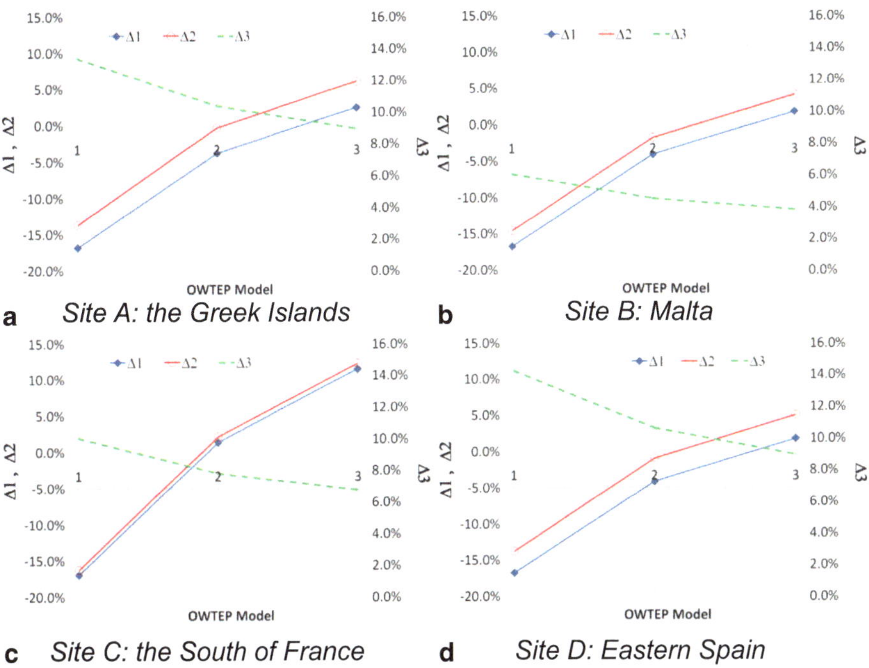

Fig. 83.9 Percentage variations in E, E′ and COP_C for the three OWTEP models

DSW resulting from the larger pump capacities of OWTEP 2 and 3 are only available at higher wind speeds that are less frequent. For the design parameters assumed in this study, the overall proportion of DSW flowing through the condenser of the corresponding AC units with respect to that of SSW was smaller for OWTEP 2 and OWTEP3. This results in slightly higher condenser temperatures, leaving a negative impact on the *COPc*.

83.6 Conclusion

This study presented a preliminary assessment of the potential of using offshore floating wind turbines in deep water to concurrently exploit cold DSW to improve the cooling performance of centralised AC systems while producing electricity. The proposed concept is most suited for countries having good wind conditions that coincide with periods of higher ambient temperature. Improved potential will also result at offshore sites having lower DSW temperatures.

It has been shown that although hydraulic energy transmission losses in the OWTEP system are considerable, such losses may be partially offset through the use of DSW in large-scale centralised AC units. Hydraulic wind turbines with larger pump capacities augment the energy yield at higher wind speeds, contributing to better performance of the OWTEP system. This would however demand for a more robust wind turbine design to sustain higher wind loads on the wind turbine structure. Future work will involve detailed design optimisation and cost analysis to examine whether the benefits of the proposed concept outweigh the added costs incurred for modifying wind turbine designs and for integrating OWTEP farms with district cooling systems.

References

1. European Wind Energy Association (EWEA) (2013) Deep water—The next step for offshore wind energy. July
2. Diepeveen N (2013) On the application of fluid power transmission in offshore wind turbines. Delft University of Technology, Delft, Ph. D. Thesis
3. Laguna AJ (2014) Fluid power network for centralised electricity generation in offshore wind farms. J Phys: Conf Ser—Sci Mak Torque Wind. 524:1–10. http://iopscience.iop.org/1742-6596/524/1/012075 (Copenhagen)
4. Jones JA, Bruce A, Lam AS (2013) Advanced performance hydraulic wind energy. IEEE Green Technologies Conference, Denver, April
5. Sant T, Farrugia RN (2013) Performance modelling of an offshore floating wind turbine-driven deep sea water extraction system for combined power and thermal energy production: a case study in a central Mediterranean context. In: Proceedings of the ASME 32nd international conference on ocean, offshore and Arctic engineering. Nantes, June. doi:10.1115/OMAE2013-10714
6. Buhagiar D, Sant T (2014) Steady-state analysis of a conceptual offshore wind turbine driven electricity and thermocline energy extraction plant. Renew Energy 68:853–867
7. Sant T, Buhagiar D, Farrugia RN (2014) Offshore floating wind turbine-driven deep sea water pumping for combined electrical power and district cooling. J Phys: Conf Ser—Sci Mak Torque Wind 524 (Copenhagen)
8. NCEP Reanalysis 2 data provided by the NOAA/OAR/ESRL PSD, Boulder, Colorado, USA. http://www.esrl.noaa.gov/psd/. Accessed May 2014
9. European wind resources map (2015). http://www.wasp.dk/Wind-Atlas/European-Wind-Atlas/European-Wind-Atlas_VI. Accessed June 2015
10. Farrugia RN, Sant T (2013) Mediterranean inshore wind resources: combining MCPs and CFD for marine resource quantification. Wind Eng 37(3):243–256
11. WindPRO Ver. 2.7 http://www.emd.dk/WindPRO/Frontpage, EMD International A/S Niels Jernes Vej 10, 9220 Aalborg Ø, Denmark
12. NODC (2012, August) NODC Live Access Server 7.3. http://data.nodc.noaa.gov/las/getUI.do. Accessed May 2014
13. Jonkman J, Butterfield S, Musial W, Scott G (2009) Definition of a 5-MW reference wind turbine for offshore system development. U.S. Department of Energy, National Renewable Energy Laboratory, Golden, CO, Technical Report

Chapter 84
Denmark Wind Energy Programme

Wen Zhong Shen

Abstract In this chapter, a summary of some ongoing wind energy projects in Denmark is given. The research topics comprise computational model development, wind turbine (WT) design, low-noise airfoil and blade design, control device development, wake modelling and wind farm layout optimization.

Keywords Wind turbine aerodynamics · Wind turbine aeroacoustics · Wind turbine design

84.1 Introduction

Wind energy is developing very fast in the world. To reduce CO_2 emission from fossil energy and alleviate global warming, governmental targets for developing wind energy are very ambitious in most countries. For example, in Denmark, electricity production is forecasted to reach 50% from wind energy in 2020 and 100% from renewable energy in 2050. This means there will be important installations of both onshore and offshore wind turbines (WTs) in the near future. The implementation of WTs is accompanied with challenges and research related to both onshore and offshore wind energy developments. For onshore wind energy development, the main challenge is the noise issue. For offshore wind energy, it is mostly to reduce the cost of energy as the cost of offshore wind energy is high as compared to onshore wind energy.

In Denmark, most of wind energy projects are funded by the following funding programmes: FP7 of the EU, the Energy Technology Development and Demonstration Programme (EUDP) of the Danish Energy Agency, the Danish High Technology Fund, and the Strategic Research programme (DSF) of the Danish Ministry of Research, Innovation and High Education. Each year, between 10 and 20 wind energy projects receive funding. The funded projects can be pure research, or pure product development and demonstration. They can also be both research and demonstration. The duration of the projects is between 2 and 6 years.

W. Z. Shen (✉)
Department of Wind Energy, Technical University of Denmark,
Nils Koppels Alle, Building 403, 2800 Lyngby, Denmark
e-mail: wzsh@dtu.dk

A. Sayigh (ed.), *Renewable Energy in the Service of Mankind Vol I*,
DOI 10.1007/978-3-319-17777-9_84

The ongoing wind energy projects in Denmark comprise the following topics: computational model development, offshore WT design, low-noise WT rotor design, control device development (such as trailing edge flaps, vortex generators, etc.), wake model development and wind farm layout optimization.

In this chapter, a summary of the above-mentioned ongoing research topics, especially in the projects coordinated by Technical University of Denmark (DTU) Wind Energy, are given. To illustrate, a number of selected results are presented.

84.2 Computational Models

In wind energy research, there are different types of computational methods that are currently being used for predicting the performance of WTs, from simple methods based on blade element momentum (BEM) theory [1] to fully 3D computational fluid dynamics (CFD) methods [2]. BEM is very simple and easy to run on a personal computer, but it has limitations and relies on the employed airfoil data which should be corrected for 3D rotational effects [3] and tip effects [4]. Fully 3D CFD is quite expensive and at the moment is often used for steady computations. Thus, intermediate methods are being developed. Among others, the Q^3UIC [5] and MIRAS [6] codes are the two new codes developed at DTU. The former is a quasi-3D steady/unsteady viscous–inviscid interaction code using strong coupling and is used to calculate the performance of airfoils. To illustrate its performance, computations are carried out for the NACA632xx airfoils at a Reynolds number of 3×10^6. From the comparisons presented in Fig. 84.1, it is seen that Q^3UIC predicts very well the performance for both lift and drag coefficients.

The latter method is based on a three-dimensional panel method using a surface distribution of quadrilateral singularities with a Neumann no-penetration condition. Viscous effects inside the boundary layer are taken into account through the coupling with the quasi-3D integral boundary layer solver Q^3UIC. A free-wake/mixed-wake model is employed to simulate the vorticity released by the blades in the wake. To illustrate its performance, computations are carried out for a 2.3 MW NM80 WT. In Fig. 84.2, the power and thrust coefficients are well predicted as compared to measurements.

84.3 Wind Turbine Design

The design of efficient and low-cost WTs is the focus of the current wind energy researches in many countries as offshore wind energy is still very expensive. One way to reduce the cost is to increase the size of WTs. Commercial single turbines

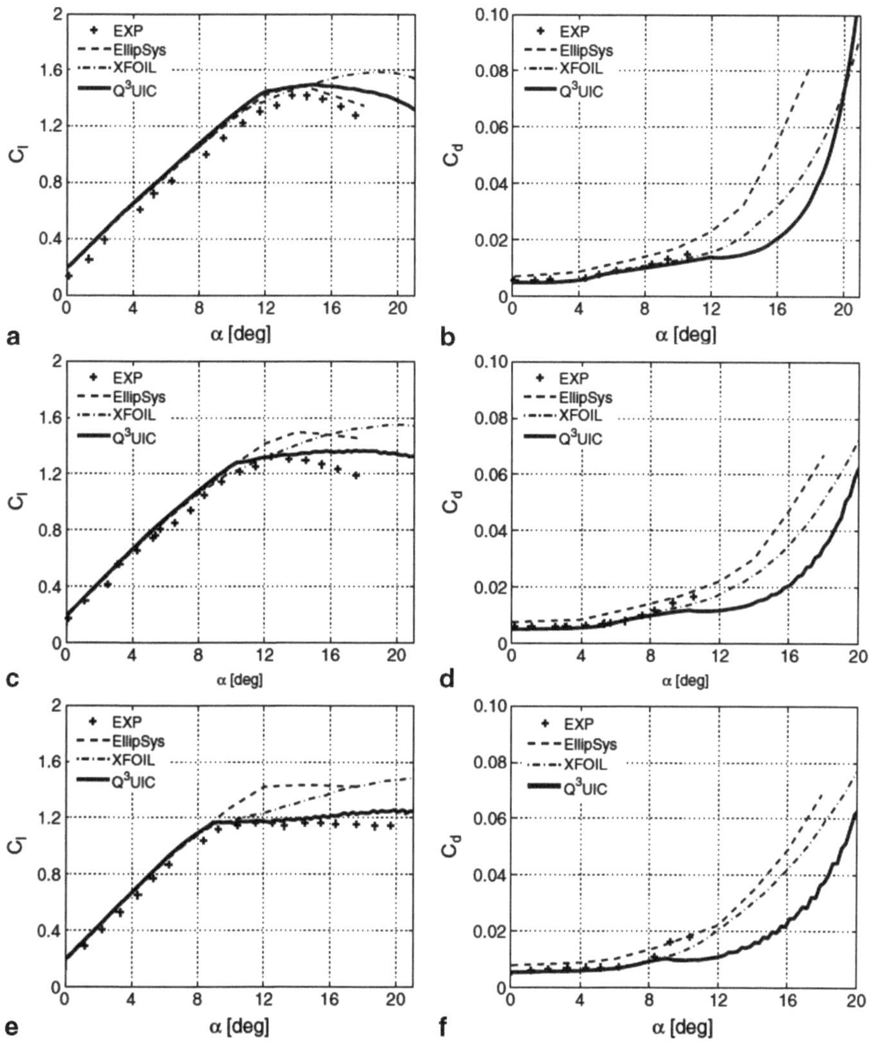

Fig. 84.1 Lift and drag coefficients of the NACA632xx airfoils at a Reynolds number of 3×10^6. (**a**, **b**) 15 %, (**c**, **d**) 18 %, and (**e**, **f**) 21 %

reach 8 MW [7] while the size of research turbines reaches 20 MW [8]. With this increase, many aerodynamic and structure problems may appear. To illustrate the activities, the surface pressure of a 20-MW turbine with a 130 m blade, which is designed using an integrated airfoil and blade design method developed in [8], is plotted in Fig. 84.3.

For onshore wind energy, most of wind energy developments are focusing on low-noise WT airfoil and blade design. In Fig. 84.4, the aerodynamic and

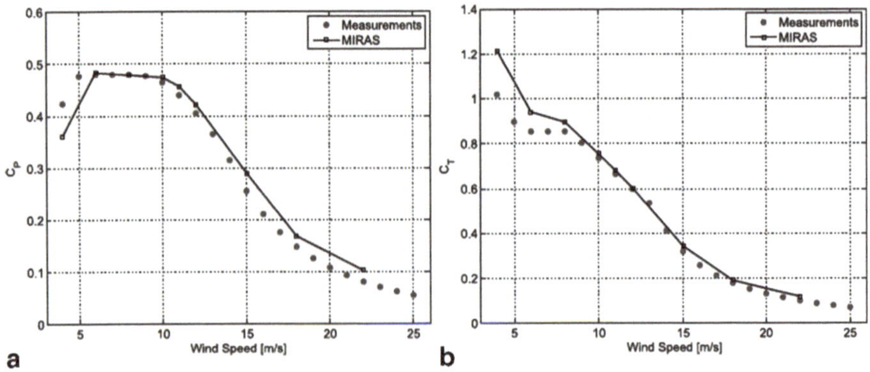

Fig. 84.2 Power and thrust coefficient of the 2.3 MW NM80 turbine

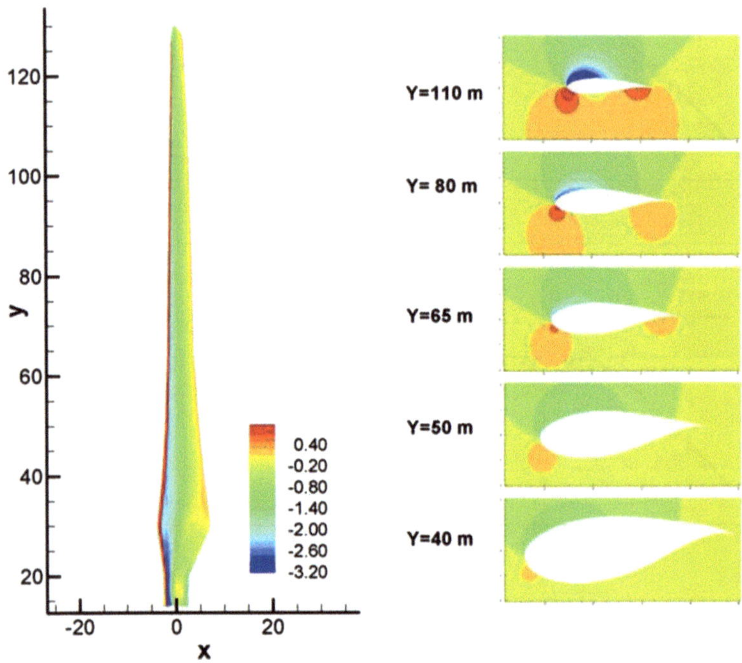

Fig. 84.3 Surface pressure and pressure contour at five different cross-sections

aeroacoustic performance of the low-noise CQU-DTU-LN118 airfoil [9] is shown. From the figure, it is seen that the CQU-DTU-LN118 airfoil has high aerodynamic performance and low noise emission than the NACA64618 airfoil which is used to construct modern WT blades.

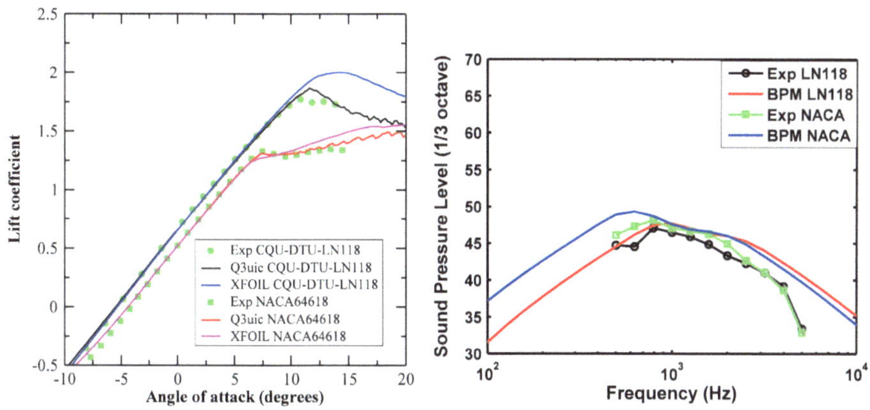

Fig. 84.4 Lift coefficient (*left*) and sound pressure level in 1/3 octave at a wind speed of 45 m/s and Cl=0.50 (*right*) for the CQU-DTU-LN118 and NACA64618 airfoil

84.4 Control Devices

Control devices are often used to optimize power output and alleviate loads on turbines. In recent researches, vortex generators [10] and trailing edge flaps [11–12] are the two main focuses. Due to the page limitation, it cannot be discussed further. For more details, the reader is referred to [10–12].

84.5 Wake Modelling

WTs are often clustered together into so-called wind farms. The prediction of performance of the WTs in wind farms is important for power predictions and/or wind farm control. The most used methods are the actuator disc and actuator line models implemented in a Navier–Stokes solver for large eddy simulation (LES). To illustrate, the actuator line model is applied to simulate the performance of the MEXICO rotor with a radius of 2.25 m [13]. In Fig. 84.5, the axial, radial and tangential velocity at a radial position of 1.848 m and 9 o'clock is plotted against particle image velocimetry (PIV) measurements [14]. Good agreement is seen. In Fig. 84.6, the flow past three 2 MW Tjæreborg turbines in tandem is simulated using the actuator line technique. From the figure, wake interactions are seen clearly but different when the separation distance between the three turbines is 4D and 6D (Fig. 84.7).

For wind farm predictions [15], the actuator disc model is applied to simulate the performance of the Lillegrund wind farm with 48 Siemens 2.3 MW turbines. Good agreement with the measurements is seen.

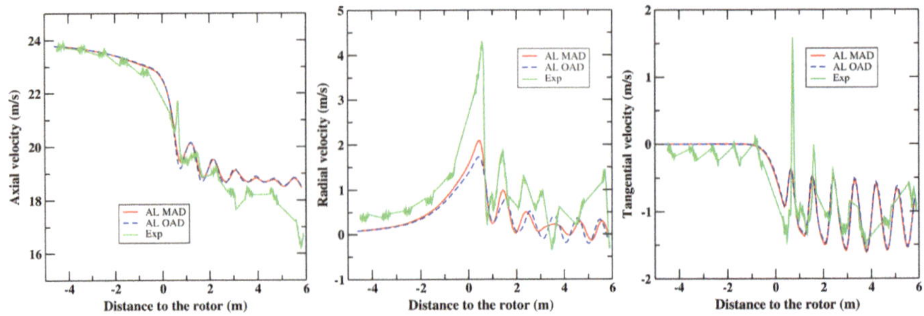

Fig. 84.5 Axial, radial and tangential velocity along the line in the axial direction at $r = 1.848$ m and 9 o'clock for the MEXICO rotor at a wind speed of 24 m/s

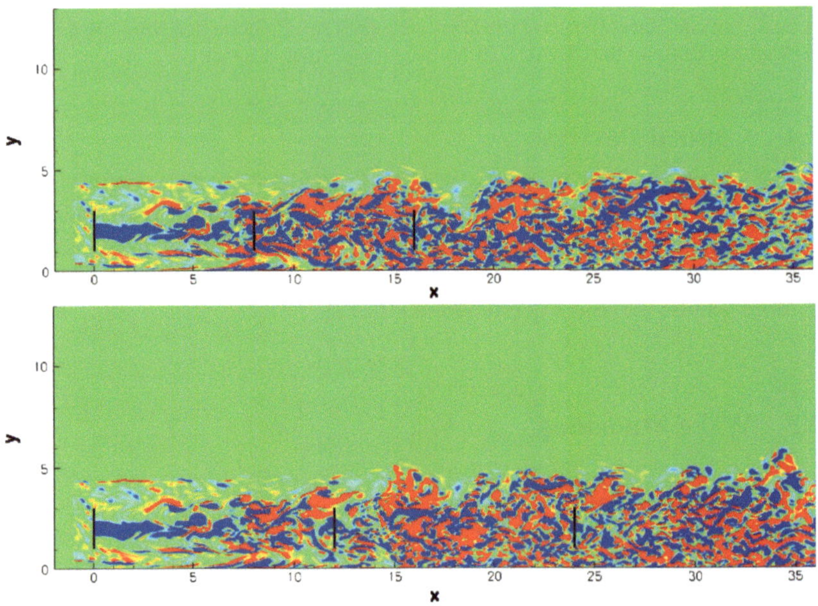

Fig. 84.6 Vorticity plot of three 2 MW Tjæreborg turbines in tandem at a hub wind speed of 10 m/s at separation distances of 4D (*up*) and 6D (*down*)

84.6 Wind Farm Layout Optimization

Wind farm layout optimization is an essential step for planning wind farms. The technical development has essentially two branches of optimization algorithm and wake model. At DTU, a new random search (RS) algorithm based on continuous formulation has been developed recently [16], which starts from an initial feasible layout and then improves the layout iteratively in the feasible solution space. It can serve both as a refinement tool to improve an initial design by expert guesses or other optimization methods, and as an optimization tool to find the optimal layout of wind farm with a given number of WTs. To illustrate the performance of the algorithm, optimization is

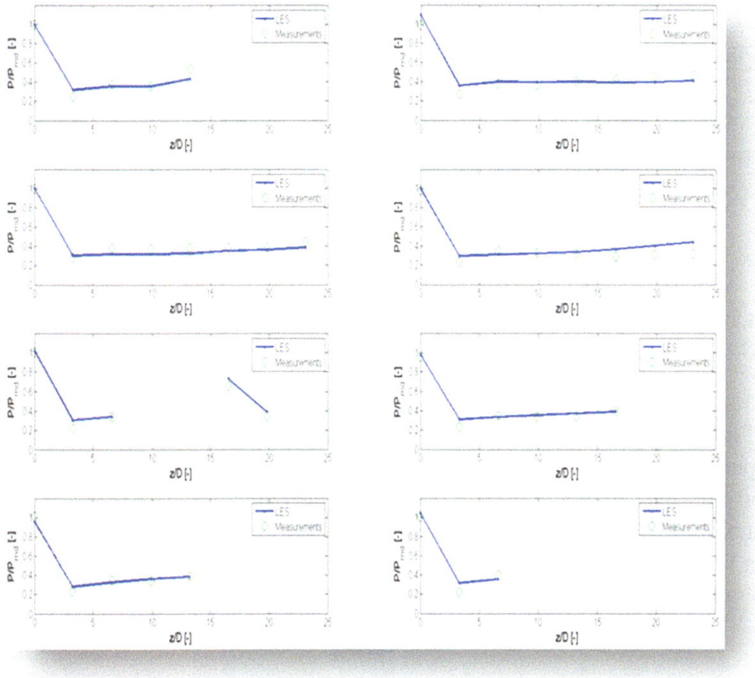

Fig. 84.7 Power performance of the 48 WTs in the lillegrund wind farm

made for the Danish Horns Rev 1 wind farm with 80 Vestas 2 MW turbines. From the figure, it is seen that slight improvements have been made (Fig. 84.8).

For installing WTs in complex terrain, CFD methods should be used because the wind resources in complex terrain are often difficult to determine. Different levels of complexity can be used. Steady Reynolds averaged Navier–Stokes (RANS)/

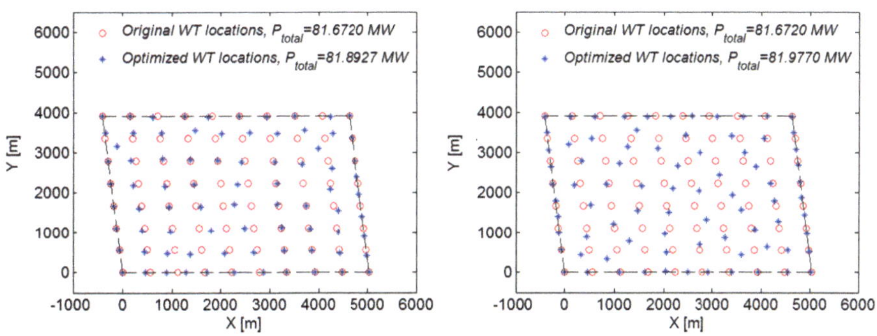

Fig. 84.8 Original and optimized layouts for the Horns Rev 1 wind farm with 80 Vestas 2 MW turbines started from the original layout (*left*) and from a random layout (*right*)

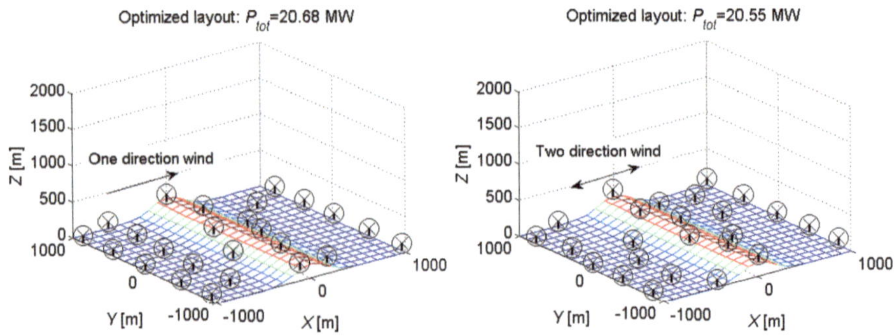

Fig. 84.9 Optimized layouts of the WF with 25 WTs with maximal 100,000 evaluations

time-averaged LES simulations for the complex terrain should be performed first. Based on the obtained results, layout optimization with improved wake models is then performed. To illustrate the development, wind farm layout optimization is carried out for a Gaussian hill of 200 m height and 200 m hill half-length at the mid-height of the hill. In Fig. 84.9, we see results of the optimization with 25 2 MW V80 Vestas WTs at a Weibull distribution with A = 10 and k = 2.3 (a mean wind speed of 8.9 m/s) for wind for one or two wind directions.

84.7 Concluding Remarks

This chapter gives an overview on some ongoing research activities in Denmark. In general, the research environment in wind energy and renewable energy in Denmark is very good. We hope such investment will continue in the way to increase the penetration of wind energy and renewable energy in the global energy consumptions.

Acknowledgements We gratefully acknowledge the support from the Danish Council for Strategic Research under the projects (09-067216, 10-094544 and 12-130590), the Energy Technology Development and Demonstration Program under projects (63011-0190, 64011-0094, 64012-0146 and 64013-0405), the Danish National Advanced Technology Foundation (028-2007-3) and EU-FP7 projects (EU-2012-308974).

References

1. Glauert H (1963) Airplane propellers. In: Durant WF (Ed) Aerodynamic theory. Dover, New York, 169–360
2. Sørensen NN, Michelsen JA, Schreck S (2002) Navier–Stokes predictions of the NREL phase VI rotor in the NASA Ames 80 ft x 120 ft wind tunnel. Wind Energy 5:151–169
3. Snel H, van Holten T (1995) Review of recent aerodynamic research on wind turbines with relevance to rotorcraft. AGARD Report CP-552, Chap. 7, pp 1–11
4. Shen WZ, Mikkelsen R, Sørensen JN, Bak Ch (2005) Tip loss corrections for wind turbine computations. Wind Energy 8:457–475

5. Garcia NR, Sørensen JN, Shen WZ (2013) A strong viscous–inviscid interaction model for rotating airfoils. Wind Energy, 28 pages, doi:10.1002/we.1677
6. Garcia NR, Sørensen JN, Shen WZ (2014) Validation of a three-dimensional viscous–inviscid interactive solver for wind turbine rotors. Renew Energy 70:78–92
7. http://www.sunwindenergy.com/news/dong-and-vestas-test-8-mw-turbine-together
8. Zhu WJ, Shen WZ, Sørensen JN (2014) Integrated airfoil and blade design method for large wind turbines. Renew Energy 70:172–183
9. Cheng JT, Zhu WJ, Fischer A, Garcia NR, Madsen J, Chen J, Shen WZ (2013) Design and validation of the high performance and low noise CQU-DTU-LN1 airfoils. Wind Energy. doi:10.1002/we.1668
10. Velte CM, Hansen MOL (2013) Investigation of flow behind vortex generators by stereo particle image velocimetry on a thick airfoil near stall. Wind Energy 16:775–785
11. Zhu WJ, Behrens T, Shen WZ, Sørensen JN (2013) Hybrid immersed boundary method for airfoils with a trailing-edge flap. AIAA J. 51:30–41
12. Fischer A, Madsen HA (2013) Load alleviation potential with trailing edge flaps for turbines in wake operation. Proceedings of the ICOWES conference, pp 102–112
13. Schepers J, Snel H (2007) Model experiments in controlled conditions. ECN-E-07-042, ECN
14. Shen WZ, Zhu WJ, Sørensen JN (2012) Actuator line/Navier-Stokes computations of flows past the MEXICO rotor: comparisons with detailed measurement. Wind Energy 15(5):811–825
15. Nilsson K, Ivanell S, Hansen KS, Mikkelsen R, Sørensen JN, Breton S, Henningson D (2014) Large-eddy simulation of the Lillegrund wind farm. Wind Energy. doi:10.1002/we.1707
16. Feng J, Shen WZ (2014) Solving the wind farm layout optimization problem using random search algorithm. Renewable Energy 78:182–192

Chapter 85
Development of Realistic Demand Side Management Strategies Using Artificial Neural Networks for the Production of Informative Wind Speed Prediction Signals

D. Zafirakis, K. Moustris, Ch. Maragkos, M. Stathopoulos and G. Tzanes

Abstract Urgency to achieve large-scale integration of the stochastic wind energy production calls among others for the employment of novel solutions including energy storage, upgrade of electricity grids, and application of effective demand side management. To this end, the problem of limited wind energy integration becomes even more severe in isolated island regions, owed to the weak character of local electricity grids. For this purpose, the current study emphasizes on the need for the production of adequately reliable forecasting wind speed signals that can, in turn, inform the development of appropriate energy, and especially demand side management strategies. In this context, we use artificial neural networks (ANN) and provide prediction of wind speed for three different island locations of the Aegean Sea, evaluating the wind speed signals for 1–10 h ahead. Our results demonstrate that wind speed predictions up to even 3 h ahead can sufficiently inform the development of appropriate energy management strategies, designating the potential of ANNs in the field of wind speed prediction.

Keywords Forecasting techniques · Wind speed · Artificial neural networks

85.1 Introduction

Achieving large-scale integration of the stochastic wind energy production [1] calls among others for the employment of novel solutions including energy storage [2], upgrade of electricity grids [3], and application of effective demand side management [4]. To this end, the problem of limited wind energy integration becomes even more severe in isolated island regions, owed to the weak character of local electricity grids [5]. Furthermore, due to the low levels of power generation diversity, such

D. Zafirakis (✉) · K. Moustris · C. Maragkos · M. Stathopoulos · G. Tzanes
Soft Energy Applications and Environmental Protection Laboratory,
Mechanical Engineering Department, Technological Educational Institute of Piraeus,
P.O. Box 41046, Athens 12201, Greece
e-mail: dzaf@teipir.gr

A. Sayigh (ed.), *Renewable Energy in the Service of Mankind Vol I*,
DOI 10.1007/978-3-319-17777-9_85

electricity grids have limited capacity in terms of energy management flexibility, while at the same time they rely heavily on the operation of diesel-based thermal power generation.

This results to increased electricity production costs and fast wear of the local thermal units, owed to the requirement for them to follow the intense variations of the local load demand. At the same time, increased wind energy contribution poses additional challenges to the operation of thermal units, introducing higher levels of uncertainty concerning the residual load demand to be covered. Considering this, optimum energy management is required in order to increase the local energy security and also allow for the improved operation of all participating power generators; but for this to be achieved, accurate prediction of power generation and electricity demand is also required in order to operate the system under optimal conditions. In this regard, although it is common for the load demand to follow a certain daily and often seasonal pattern, it is not valid for the stochastic nature of wind speed. In fact, the prediction of wind speed, even short-term, comprises a challenge for different forecasting methods [6, 7].

For this purpose, the current work emphasizes on the production of adequately reliable forecasting wind speed signals that can, in turn, inform the development of appropriate energy, and especially demand side management strategies. To this end, by using artificial neural networks (ANNs) [8–10] we develop a forecasting model for the short-term prediction of wind speed. The developed model is accordingly applied to an area of major interest, that is, the Aegean Sea, considering certain representative island regions of different quality wind potential. More precisely, we use a detailed dataset of time series for wind speed together with additional meteorological parameters for three different Aegean islands that can sufficiently support the development of an effective ANN forecasting model. The obtained application results demonstrate the potential of ANNs to provide accurate predictions up to 3 h ahead, independently of the wind regime each time examined.

85.2 Data and Methodology

ANNs are a branch of artificial intelligence developed in the 1950s aiming at imitating the biological brain architecture. They are a mathematical approach to the description of the human nervous system functions. Typical ANNs use very simple models of neurons. These ANN models retain only very rough characteristics of biological neurons of the human brain. ANNs are parallelly distributed systems made of many interconnected nonlinear processing elements. A renewal of scientific interest has grown exponentially since the last decade in the field of ANNs, mainly due to the availability of appropriate hardware that has made them convenient for fast data analysis and information processing. In fact, during the last two decades more and more scientists around the world have used ANNs modeling in many different scientific fields.

In this work, a special and well known kind of ANNs, the multilayer perceptron (MLP) is applied in order to predict—on an hourly basis—the wind speed of three different Aegean Sea islands. The MLP is the most commonly used type of ANNs.

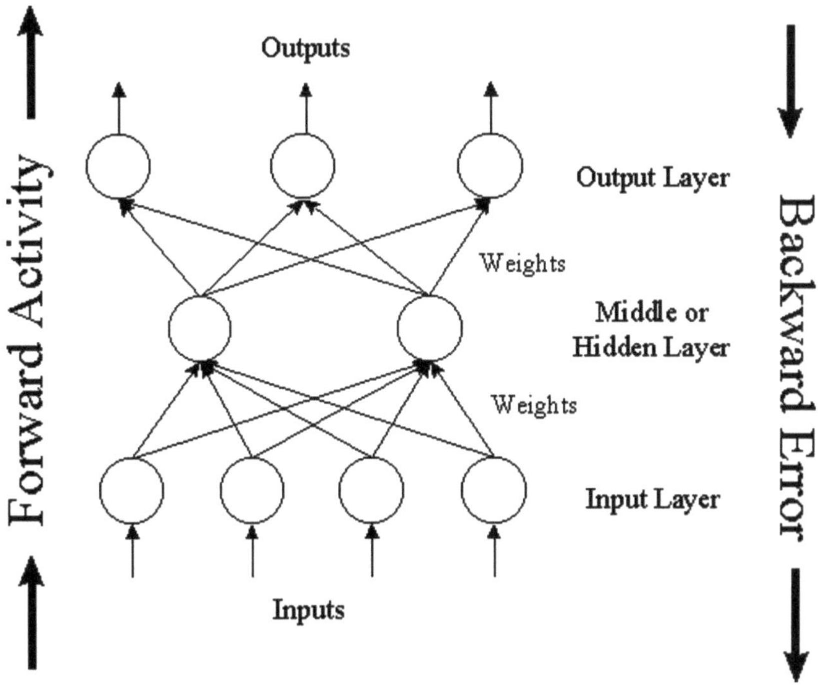

Fig. 85.1 Typical MLP feed-forward ANN structure

Its structure consists of processing elements and connections. Processing elements, which are called artificial neurons, are arranged in layers. The first layer is the input layer, one or more hidden layers follow, and the final layer is the output layer. An input layer serves as buffer that distributes input signals to the next layer, which is a hidden layer. Each neuron of the hidden layer communicates with all the neurons of the next hidden layer, if any, having in each connection a typical weight factor. Thus, each unit-artificial neuron in the hidden layer sums its input, processes it with a transfer function, and distributes the result to the output layer.

It is also possible that there are several hidden layers connected in the same fashion. The unit-artificial neurons in the output layer compute their output in a similar manner. Finally, the signal reaches the output layer, where the output value from the ANN is compared with the target value and an error is estimated. Thus, the values of weight factors are amended appropriately and the training cycle is repeated until the error is acceptable, depending on the application. As the data flow within the ANN from a layer to the next one without any return path, such kind of ANNs are defined as feed-forward ANNs. The structure of a feed-forward MLP ANN can be represented in Fig. 85.1.

Concerning the current work, a 3-year period (2011–2013) dataset of hourly wind speeds was used (Fig. 85.2) for three islands examined (i.e., the islands of Ikaria, Mykonos, and Naxos), corresponding to different quality wind potentials (Fig. 85.2). The first 2 years (2011–2012) were used for the training of ANNs that were accordingly asked to predict wind speed values of the last year (2013). Apart

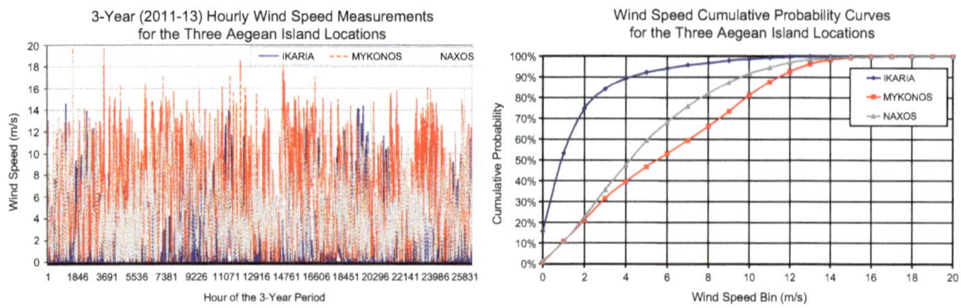

Fig. 85.2 Hourly wind speed measurements and wind speed cumulative probability for the three locations examined

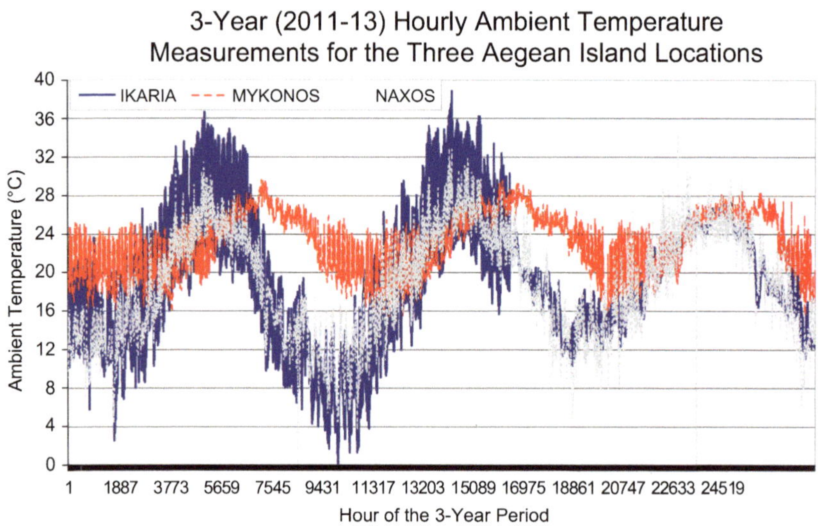

Fig. 85.3 Hourly measurements of ambient temperature for the areas of interest

from the hourly wind speed values, additional input data also included hourly information on ambient pressure, ambient temperature, and finally relative humidity of the local areas (Figs. 85.3 and 85.4). Furthermore, after a number of trials, input nodes (input layer) were educated on the basis of data corresponding to the 3-previous hours, this designated to be the optimum input signal for the prediction of hours-ahead wind speed values. The target value (output layer) was the wind speed value of the *t*-hour, extended from 1–10 h ahead.

85.3 Application Results

For the evaluation of the predictive ability of the developed ANN model, a number of statistical indices were calculated [9], specifically, the mean bias error (MBE), the coefficient of determination (R^2), and the index of agreement (IA). In this con-

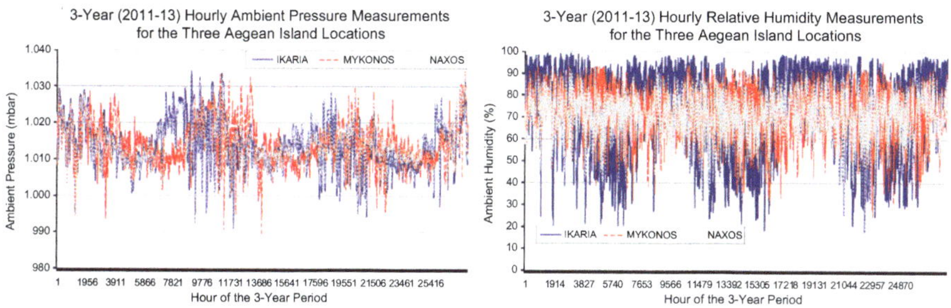

Fig. 85.4 Hourly measurements of ambient pressure and relative humidity for the areas of interest

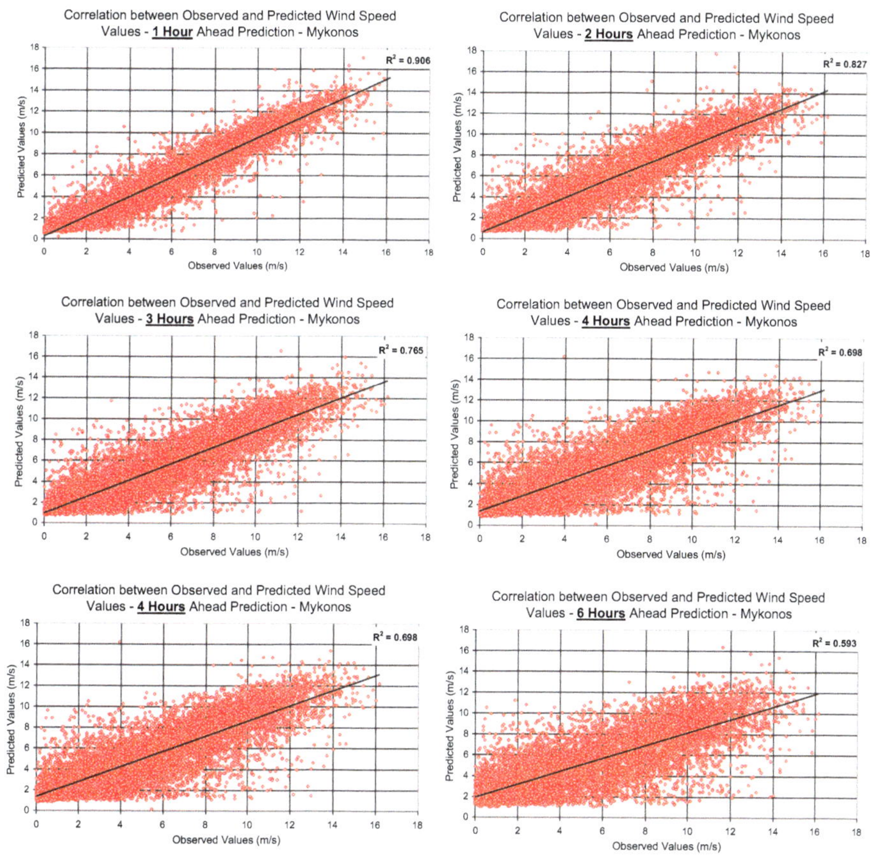

Fig. 85.5 Regression results for the area of Mykonos; prediction of 1–6 h ahead

text, in the following Figs. 85.5 and 85.6, the application results concerning the island of Mykonos are presented.

More precisely, correlation between observed and predicted wind speed values is presented for a prediction time horizon of 1–10 h ahead. According to the results obtained, the developed ANN model achieves to provide rather satisfactory predic-

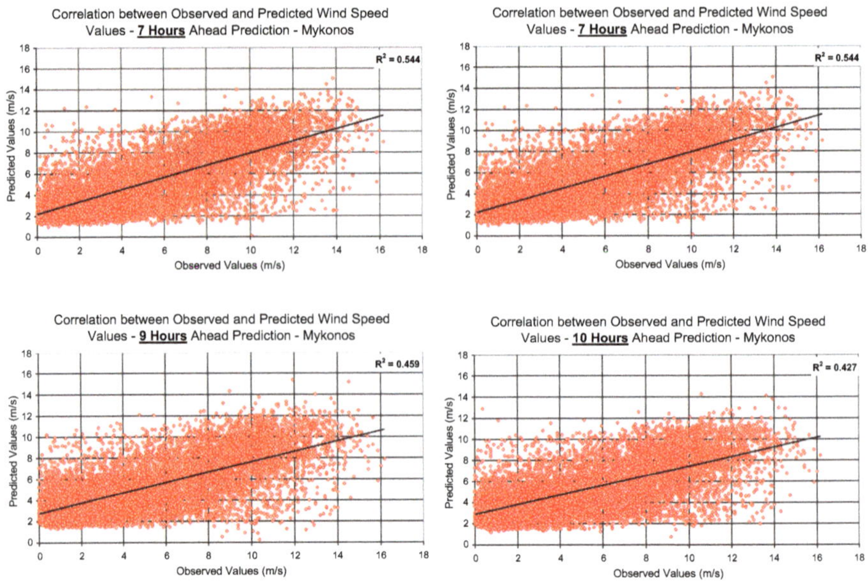

Fig. 85.6 Regression results for the area of Mykonos; prediction of 7–10 h ahead

tion up to 3 h ahead, with the respective 1 h ahead prediction being determined by $R^2 = 0.906$. On the other hand, when asking for a 10 h ahead prediction, the ANN model fails to provide satisfactory results, with the respective R^2 being equal to almost 0.427.

Overall results for all three locations are given in Table 85.1, where the values of MBE, IA, and R^2 are included in order to evaluate the predictive ability of the developed ANN model. The conclusion previously drawn concerning Mykonos is validated for the other two islands as well, reflecting the fact that the predicting ability of the developed ANN model is not affected by the wind regime examined. Furthermore, predicting ability of the developed ANN models is better illustrated in Fig. 85.7, where a typical weekly comparison is provided between observed and predicted wind speed values for the islands of Ikaria and Naxos, respectively.

Finally, by using a typical pitch-control wind power curve (cut-in wind speed of 4 m/s, cut-out wind speed of 25 m/s, and nominal power wind speed of 10 m/s), we estimate the respective annual capacity factor based on the observed and predicted wind speed values respectively. Results obtained for 1 h ahead prediction are given in Fig. 85.8, where as it may be seen, predicted-based values appear to be marginally lower than the corresponding observed-based ones. At the same time, difference between the observed and the predicted capacity factors seems to increase as the local wind potential quality improves with the greatest difference—in the order of 4 %—noted for the area of Mykonos.

Table 85.1 Values of statistical indices for the evaluation of ANN forecasting ability

	MBE (m/s)			IA			R^2		
	IKARIA	MYKONOS	NAXOS	IKARIA	MYKONOS	NAXOS	IKARIA	MYKONOS	NAXOS
10HA	0.027	−0.147	−0.004	0.718	0.703	0.679	0.447	0.427	0.397
9HA	0.001	−0.096	−0.007	0.759	0.741	0.719	0.489	0.459	0.435
8HA	−0.010	−0.176	−0.008	0.790	0.776	0.749	0.531	0.500	0.470
7HA	−0.017	−0.161	0.008	0.821	0.811	0.791	0.572	0.544	0.517
6HA	0.028	−0.181	0.006	0.852	0.842	0.824	0.608	0.593	0.563
5HA	0.019	−0.189	0.029	0.876	0.871	0.861	0.654	0.640	0.621
4HA	−0.034	−0.171	0.017	0.903	0.900	0.891	0.713	0.688	0.685
3HA	0.008	−0.213	0.013	0.928	0.927	0.920	0.773	0.765	0.751
2HA	−0.008	−0.223	0.015	0.952	0.949	0.947	0.832	0.827	0.822
1HA	−0.043	−0.141	0.015	0.976	0.975	0.976	0.912	0.906	0.910

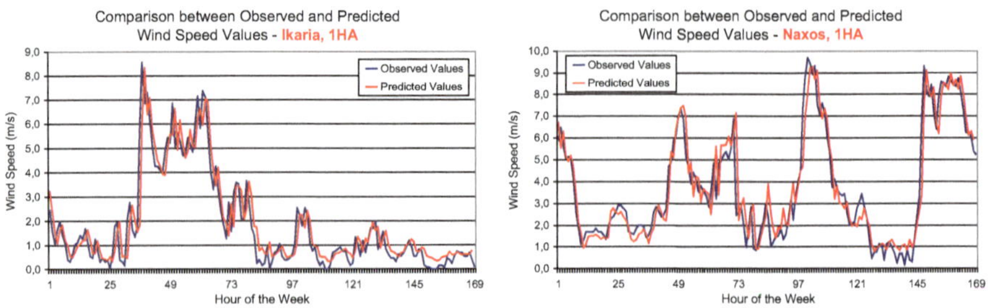

Fig. 85.7 Comparison between observed and predicted wind speed values for a typical week and the islands of Ikaria and Naxos (1 h ahead)

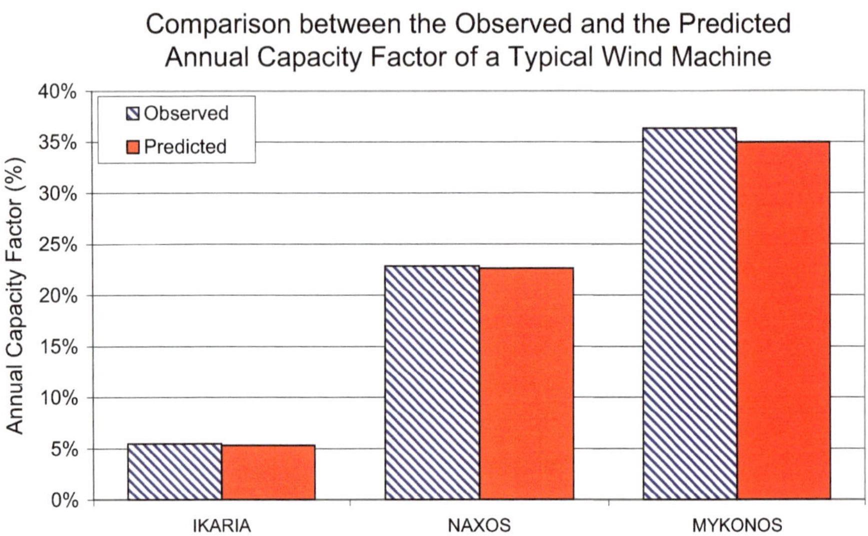

Fig. 85.8 Comparison between observed and predicted annual capacity factor for the operation of a typical wind machine in the three areas of investigation (1 h ahead)

85.4 Conclusions

Aiming to the production of reliable wind speed prediction signals, an ANN model was developed in the current study, able to provide satisfactory forecasting of wind speed up to 3 h ahead. To validate the developed model, three Aegean Sea areas of different quality wind potential were used as case studies, namely the island areas of Ikaria, Mykonos, and Naxos. According to our results, the ability of the developed ANN model to provide short-term wind speed prediction of sufficient accuracy was designated for all three areas examined, reflecting its applicability to other, different types of wind regimes as well.

References

1. Shafiullah GM, Amanullah MT, Shawkat ABM, Wolfs P (2013) Potential challenges of integrating large-scale wind energy into the power grid—a review. Renew Sustain Energy Rev 20:306–321
2. Zafirakis D, Chalvatzis KJ, Baiocchi G, Daskalakis G (2013) Modeling of financial incentives for investments in energy storage systems that promote the large-scale integration of wind energy. Appl Energy 105:138–154
3. Weigt H, Jeske T, Leuthold F, Hirschhausen C (2010) "Take the long way down": integration of large-scale North Sea wind using HVDC transmission. Energy Policy 38:3164–3173
4. Stathopoulos M, Zafirakis D, Kavadias K, Kaldellis JK (2014) The role of residential load-management in the support of RES—based power generation in remote electricity grids. Energy Procedia 46:281–286
5. Kaldellis JK, Zafirakis D (2007) Present situation and future prospects of electricity generation in Aegean Archipelago islands. Energy Policy 35:4623–4639
6. Petković D, Shamshirband S, Anuar NB, Saboohi H, Wahab AWA, Protić M, Zalnezhad E, Mirhashemi SMA (2014) An appraisal of wind speed distribution prediction by soft computing methodologies: a comparative study. Energy Convers Manage 84:133–139
7. Tascikaraoglu A, Uzunoglu M (2014) A review of combined approaches for prediction of short-term wind speed and power. Renew Sustain Energy Rev 34:243–254
8. Nastos PT, Paliatsos AG, Koukouletsos KV, Larissi IK, Moustris KP (2014) Artificial neural networks modeling for forecasting the maximum daily total precipitation at Athens, Greece. Atmos Res 144:141–150
9. Nastos PT, Moustris KP, Larissi IK, Paliatsos AG (2013) Rain intensity forecast using artificial neural networks in Athens, Greece. Atmos Res 119:153–160
10. Moustris K, Paliatsos AG, Bloutsos A, Nikolaidis K, Koronaki I, Kavadias K (2008) Use of neural networks for the creation of hourly global and diffuse solar irradiance data at representative locations in Greece. Renew Energy 33:928–932

Chapter 86
Comparative Study of Two Types of Wind Turbine Simulators for Wind Energy Conversion System

Gyeong-Hun Kim, Jin-Hong Jeon, Jong-Yul Kim, Jong-Bo Ahn, Chulsang Hwang and Eung-Sang Kim

Abstract Electricity generation using wind energy is becoming the fastest growing source of renewable energy. There are several types of wind turbine simulator (WTS) or emulator for evaluation of wind turbine generators, power electronic converters, and turbine control algorithms. These WTSs have each a different purpose, and have been used to serve the purpose.

It is difficult to construct a wind farm using a conventional WTS with motor–generator (MG) set because of its cost and volume. In this chapter, a WTS composed only of a power converter without an MG set is proposed, and the two types of WTSs for a wind energy conversion system are compared.

The conventional WTS with MG set consists of two 10 kW directly coupled permanent magnet synchronous generators, a motor driver, a power converter, and a microcontroller. The microcontroller calculates wind turbine torque or rotation speed for the motor driver. However, the proposed WTS with only a full power converter and without an MG set includes a rectifier which is connected to the grid and the microcontroller. The rectifier is used instead of the generator of the wind conversion system, and just supplies electric power to the grid-connected converter. The microcontroller for only the full power converter type wind simulator simulates status values of the wind turbine.

The experimental results show the output power and the rotating speed of two types of wind simulator under varied wind speed conditions. It is expected that the wind farm simulator can be designed using the proposed WTS. The results of the chapter can effectively be employed to design or construct wind farm simulators for testing wind farm controllers.

Keywords Wind power generation system · Wind turbine · Wind turbine emulator · Wind turbine simulator · Wind generator

---

G.-H. Kim (✉) · J.-H. Jeon · J.-Y. Kim · J.-B. Ahn · C. Hwang · E.-S. Kim
Smart Distribution Research Center, Korea Electrotechnology Research Institute, Changwon, Republic of Korea
e-mail: kgh1001@keri.re.kr; kgh1001gh@gmail.com

A. Sayigh (ed.), *Renewable Energy in the Service of Mankind Vol I*,
DOI 10.1007/978-3-319-17777-9_86

86.1 Introduction

Electricity generation using wind energy is becoming the fastest growing source of renewable energy. A total of 282 GW was installed in 2012, and there was an increase in installed cumulative capacity of wind power generation system (WPGS) of about 18.6 % compared to the previous year [1].

Increase of wind farm in power system affects the power balancing between generated and demanded power caused by unstable wind velocity [2]. Unstable wind velocity can cause significant variations in the system frequency that may affect grid system stability [3]. Hence, grid system operators in many countries such as Denmark, Germany, Spain, USA, China, and so on require a grid code for a WPGS or a wind farm [4, 5].

A wind farm controller (WFC) is needed to satisfy the grid code, because it requires the possibility for controlling active and reactive power in point of common coupling (PCC), continuous operation in a limited range, and rate limit operation. Before the WFC is applied to the wind farm, it is difficult to evaluate and validate the WFC because of security problems of the power system, cost, and development time. There are two methods which are used to test the WFC. One is hardware-in-the-loop simulation method. The other method uses a wind turbine simulator (WTS) to compose the wind farm.

The conventional WTS is shown in Fig. 86.1. It uses a motor (DC or AC motor), its driver, a generator, a coupler between the motor–generator (MG), and a microcontroller [6]. The WTS simulates the steady-state and dynamic behaviors of a modeled wind turbine in a controlled environment without reliance on natural wind resources and real wind turbines [7]. The WTS can be used for research applications to drive an electrical generator put a similar torque by a wind turbine for a given wind velocity [8].

However, it is difficult and expensive to construct a wind farm composed of this type of WTS, because it consists of an MG set. A WTS with only a power converter

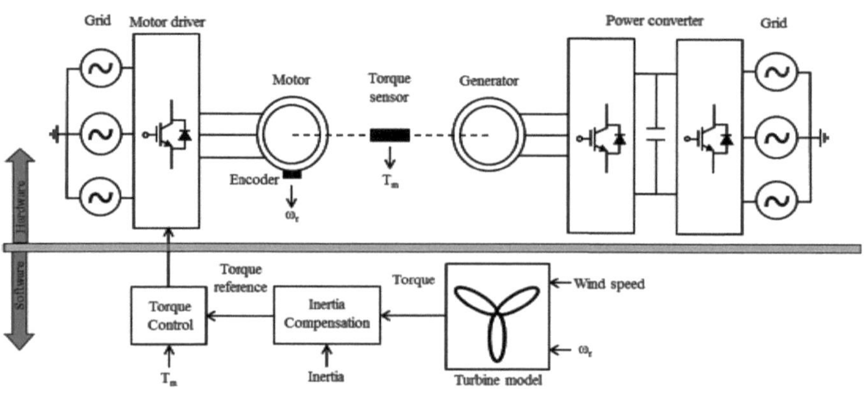

Fig. 86.1 Configuration of the conventional WTS

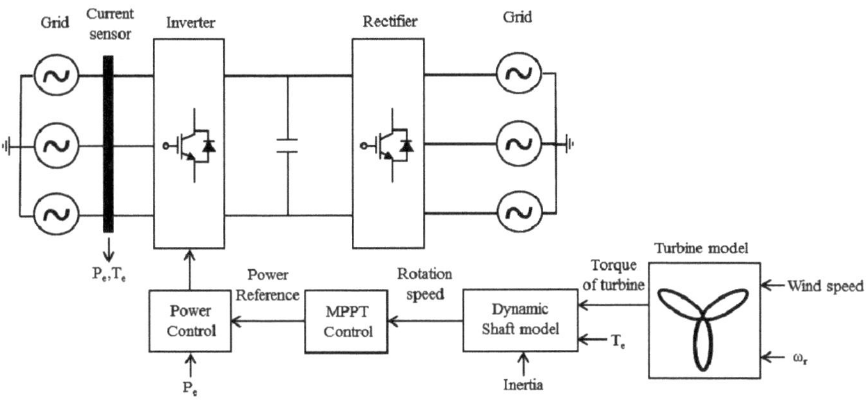

Fig. 86.2 Configuration of the WTS which has only a power converter.

is proposed in this chapter. As shown in Fig. 86.2, the proposed WTS consists of a grid connected inverter, a rectifier, and a wind turbine model.

The experiment results of the conventional WTS and the proposed WTS are compared. The output power and rotation speed of the two types of WTSs are similar as shown in Sect. 4. Therefore, it is expected that the wind farm simulator for testing and evaluating the WFC can be constructed using the proposed WTS.

86.2 Modeling of the Wind Turbine

Wind energy is the kinetic energy of the wind. The mechanical torque of the turbine captured from the wind energy can be calculated by

$$T_m = \frac{1}{2}\rho\pi R^3 v^2 C_p(\lambda,\beta)/\omega \quad (86.1)$$

where ρ is the air density (kg/m^3), v is wind velocity (m/s), R is the radius of blade (m), ω is angular velocity (rad/s), and C_p is power coefficient [9].

The power coefficient is a function of tip speed ratio (λ) as follows (2).

$$C_p(\lambda) = 0.3\ \sin\ (2.61 - 0.139\lambda) \quad (86.2)$$

The maximum power coefficient (C_{p_max}) is 0.3 when the tip speed ratio (λ_{opt}) is 7.5 [10]. A 10 kW wind turbine is modeled by (1), (2). Parameters of the modeled wind turbine are shown in Table 86.1 [11].

Table 86.1 Parameters of the 10 kW wind turbine

Parameters	Values
Rated wind velocity	11 m/s
Radius of the blade length	3.5 m
Rotation speed	200–400 rpm
Maximum power coefficient	0.3
Optimal tip speed ratio	7.5
Air density	1.197 kg/m^3
Inertia of the rotor	100 kgm^2

86.3 Power Converter Control

86.3.1 Power Converter Control for Wind Turbine Simulator with MG

A topology of the power converter for the WTS with the MG set is a back-to-back converter. It includes a grid-connected converter (GCC), DC-link capacitor, and generator-side converter (GSC). The control block diagram of the power converter is shown in Fig. 86.3. The GCC keeps the DC-link capacitor voltage to a set value. The reactive power of the grid is controlled to zero. The DC-link voltage is controlled by the q-axis current of grid, and the reactive power is controlled by the d-axis current of grid [12, 13]. The dq-axis currents of grid are implemented via the two proportional–integral (PI) controllers.

The three phase output of the permanent magnet synchronous generator (PMSG) is connected to the GSC. The active power reference of the GSC is calculated by a maximum power point tracking algorithm as follows (3).

$$P_{\max} = 0.5\frac{\rho\pi R^5 C_{P\_\max}}{\lambda_{opt}^3}\omega^3 \tag{86.3}$$

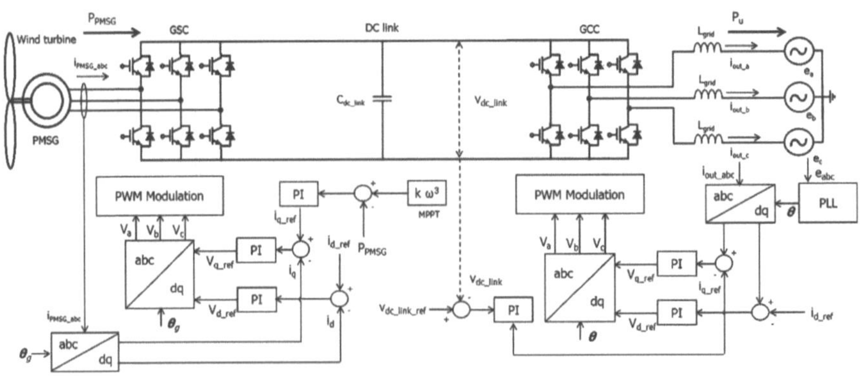

Fig. 86.3 Control block diagram of the power converter for WTS with MG

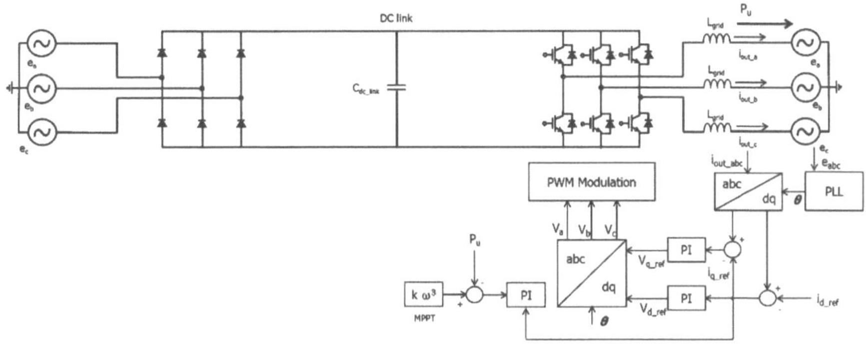

Fig. 86.4 Control block diagram of the power converter for WTS without MG

As a reference power as in (3) of the GSC is calculated from rotation speed *(ω)*, the GSC tracks the maximum power point according to the rotation speed variance [14]. The reactive power of the generator is controlled to zero for unity power factor operation.

The active power of the generator is controlled via the q-axis current of the generator output current, and the reactive power of the generator is controlled via the d-axis current. The dq-axis currents of the generator are also controlled by using a PI controller. The pulse width modulation (PWM) signal for the GSC is generated by results of inverse Park's transformation.

86.3.2 Power Converter Control for WTS without MG

A power converter for the WTS without MG includes the GCC, the DC-link capacitor, and a rectifier. The control block diagram of this type of power converter is shown in Fig. 86.4. In the case of the rectifier, there is no controller. It just supplies electric power to the GCC. The output power reference of GCC is calculated by (3). The reactive power of the grid is controlled to zero equally with the above. The dq-axis currents of the grid are also controlled by the two PI controllers. The output power is changed to electric torque, which is transmitted to the wind turbine model.

86.4 Experiment Results

The two types of WTS have been implemented for comparison as shown in Fig. 86.5. The output power and rotation speed are compared when the wind speed varies. Figure 86.6 depicts the reference wind speed applied to two types of simulators, which varies from 4 to 12 m/s. The output power of two types of WTS is compared in Fig. 86.7 under same wind speed. As shown in Fig. 86.7, the output power is approximately same.

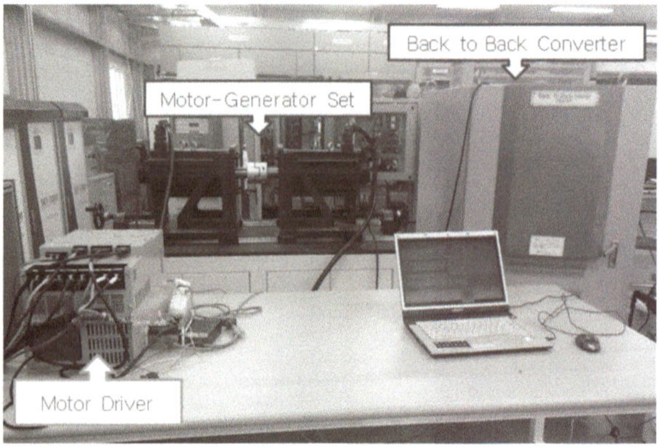

Fig. 86.5 Photograph of the experimental equipment

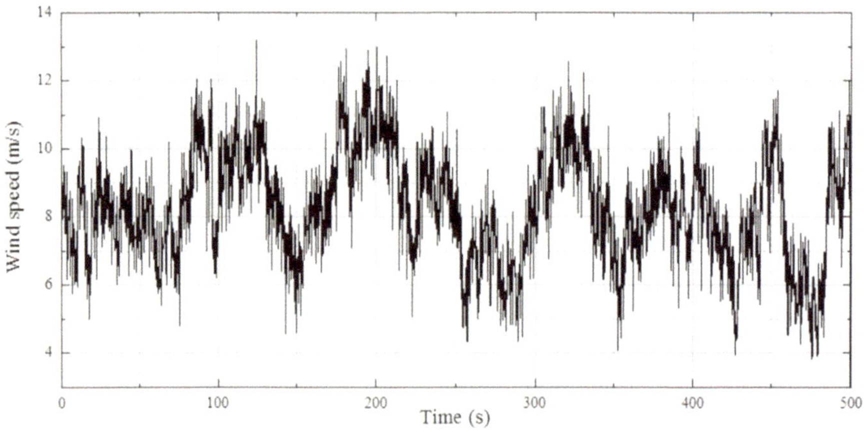

Fig. 86.6 Reference wind speed applied to two types of simulators

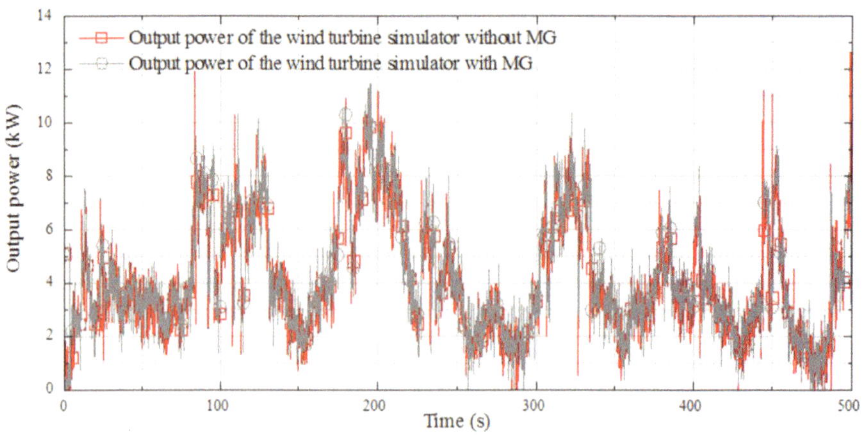

Fig. 86.7 Output power of the WTSs with and without MG

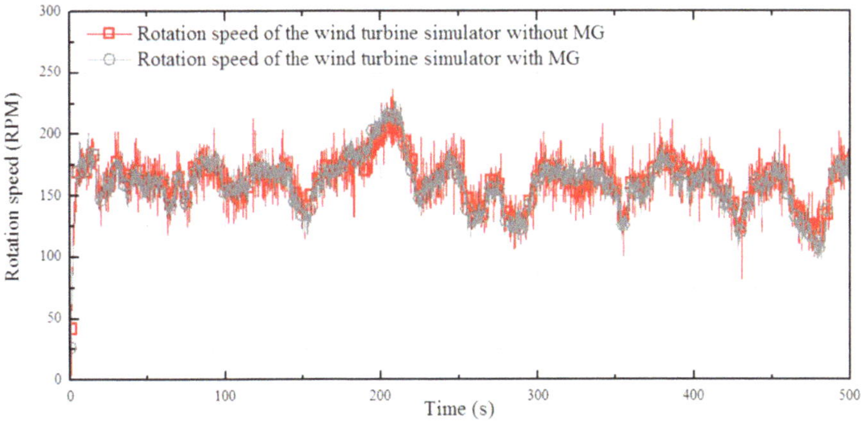

Fig. 86.8 Rotation speed of the WTSs with and without MG

The rotation speed of the wind turbine is shown in Fig. 86.8 at the same time. The rotation speeds of two types of WTSs change from 100 to 250 rpm, which are also same. Compared to the conventional WTS with MG, the proposed WTS without MG has same performance.

86.5 Conclusions

In this chapter, the two types of WTSs for the wind energy conversion system are studied. One of the WTSs uses an MG set. The other uses only a full power converter. Additionally, the study will investigate comparison with a real WPGS. The experimental results show the output power and the rotating speed of WTSs under varied wind speed conditions. Compared to the conventional WTS with MG, the proposed WTS without MG has same performance. If the dynamic performances of two types of WTSs are same, the wind farm simulator for testing and evaluating WFC can be constructed using the proposed WTS. The results of the chapter can effectively be employed to design wind farm simulators for WFC.

References

1. Fried L (2013) "Global wind statics 2012," *Global Wind Energy Council*
2. Sørensen P, Cutululis NA, Vigueras-Rodríguez A, Jensen LE, Hjerrild J, Donovan MH, Madsen H (2007) Power fluctuations from large wind farms. IEEE Trans Power Syst 22(3):958–965
3. Lee S-E, Won D-J, Il-Y C (2012) Operation scheme for a wind farm to mitigate output power variation. J Electr Eng Technol 7(6):869–875

4. Altin M, Ö G, Teodorescu R, Rodriguez P, Jensen B-B, Helle L (2010) Overview of recent grid codes for wind power integration, in publication of OPTIM 2010 Conference
5. Sim J-B, Kim K-C, Son R-W, Oh J-K (2012) Ride-through of PMSG wind power system under the distorted and unbalanced grid voltage dips. J Electr Eng Technol 7(6):898–904
6. Martinez F, Herrero LC, de Pablo S (2014) Open loop wind turbine emulator. Renew Energy 63:212–221
7. Kojabadi HM, Chang L (2004) Development of a novel wind turbine simulator for wind energy conversion systems using an inverter-controlled induction motor. IEEE Trans Energy Convers 19(3):547–552
8. Monfared M, Kojabadi HM, Rastegar H (2008) Static and dynamic wind turbine simulator using a converter controlled dc motor. Renew Energy 33:906–913
9. Heier S (1998) Grid integration of wind energy conversion systems. Wiley, Hoboken, pp 34–36
10. Arifujjaman M, Iqbal MT, Quaicoe JE, Khan MJ (2005) Modeling and control of a small wind turbine. Conf Electr Comput Eng 778–781
11. http://bergey.com/documents/2013/10/excel-10-spec-sheet_2013.pdf
12. Chinchillar M, Arnaltes S, Burgos JC (2006) Control of permanent magnet generators applied to variable-speed wind-energy systems connected to the grid. IEEE Trans Energy Convers 21(1):130–135
13. Kim GH, Kim KM, Sung HJ, Jung SJ, Kim S, Park M, Yu IK, Lee S (2012) Control scheme of a superconducting synchronous generator applied to a grid-connected wind power generation system. Conf Electr Comput Eng 1–4
14. Baroudi JA, Dinavahi V, Knight AM (2005) A review of power converter topologies for wind generators. In Proceedings of IEEE International Conference on Electrical Machines and Drives, San Antonio (TX), pp 458–465

Chapter 87
Wind Power Is the Last to Be Stored

Donald Swift-Hook

Abstract Pumped water storage had been widely used on power systems well before the advent of wind power. The economics rely on arbitrage between cheap night-time electricity and expensive daytime generation: a commercial store must buy electricity when it is cheap and sell when it is dear. As there is no solar power at night, this means solar power cannot be economically stored on a power system.

Pumped water storage systems can also help to cope with power system fluctuations but these control and balancing activities are quite distinct from energy storage. Even though the basic physical processes underlying any electrical engineering plant on a power system, such as capacitors and inductances, involve storage of energy, it would be wholly misleading to describe them as being for power system energy storage.

That would be like calling a school bus "a means of transporting fuel". A description that will be literally correct for a physicist may be totally misleading in a practical engineering context.

Storage cannot be economically installed in association with wind. Providing capacity for times of base load has little value because the maximum amount of spare plant is available then. Wind power is mainly for saving fuel, not providing capacity, and round-trip losses will only waste some of the fuel that has already been saved if such electricity is stored and regenerated. A store must buy at night and generate during the day to be economical, but wind power may need to generate at any time around the clock and sometimes all around the clock for many days on end. So wind power is contractually incompatible with storage.

Nevertheless, with storage already in place for other reasons, it might be thought that when a store is taking in power (typically during times of base load) and the wind happens to be blowing, wind is as likely as any other sort of power to be stored. Not so. The power being stored is that plant which would be switched off if storage ceased (e.g. because the store failed or filled). On a well-run power system which always uses the cheapest power generation available at any time, that will be the dearest power online. With no fuel cost, wind is always the cheapest online and so it will never be the one to be switched off. So wind is never stored unless the

D. Swift-Hook (✉)
Kingston University, Woking, Surrey GU21 4XX, UK
e-mail: donald@swift-hook.com

A. Sayigh (ed.), *Renewable Energy in the Service of Mankind Vol I*,
DOI 10.1007/978-3-319-17777-9_87

cheapest is also the dearest, and wind is the only plant still generating. Then wind is the last to be stored.

Keywords Save fuel · Wind and nuclear · Global warming · Store electricity · The energy losses in the storage

87.1 Introduction

Energy storage for electrical power systems has almost as long a history as power systems themselves [1]. The first public power supply was in Godalming, Surrey, England, 45 km from the county town of Kingston. It was a hydroelectric power system and was followed within a few months by steam plants in central London and New York and a rapid expansion—a positive explosion!—of both types of plant worldwide.

It was soon clear that a run-of-river hydropower system would frequently run out of water, and that to provide water storage by means of a dam was an essential, if costly, part of such a scheme.

Whatever size dam was built, it was relatively cheap to install extra water turbine generators to increase the generating capacity to use all the available water, so the system would again run out of water during times of peak demand.

Within a decade, pumped water systems were being installed to use cheap off-peak electricity to pump water backup above the damp to have it available to generate more valuable electricity at peak demand times.

87.2 Pumped Water Storage

Pumped water is the only system of bulk energy storage in practical use on power systems today. Figure 87.1 shows diagrammatically how such a system operates.

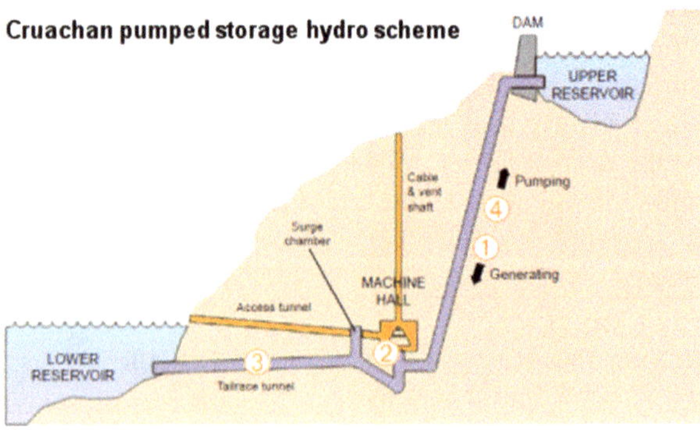

Fig. 87.1 Typical features of a modern pumped water storage facility

Fig. 87.2 The upper lake and dam at Cruachan, above Loch Awe in Scotland, a 400-MW reversible-pumped water storage scheme commissioned in 1965

Figure 87.2 shows the upper storage lake and dam of a 400 MW scheme at Cruachan that has been in operation in Scotland since 1965. This was followed in the UK by other schemes, most notably Dinorwig in Snowdonia, North Wales, which was commissioned in 1984 with a generating capacity of around 1800 MW (6×300 MW water turbo-generators).

87.3 Other Types of Storage

Two commercial scale demonstration projects using pumped air instead of pumped water were put into operation in the last century, a 290-MW scheme at Huntorf in Germany in 1978 and a smaller 110-MW plant 13 years later at McIntosh, Alabama. Neither of these proved economic, and they have not been replicated since then.

Studies continue (including recent experimental units at Gaines in Texas in 2012 generating 2 MW and at Seabrook, New Hampshire in 2013 generating 1.65 MW), but nothing on a commercial scale has materialised.

Many other methods of storage have been proposed for commercial use on power systems, most notably the fuel cell. This system was used in the first satellites and then on a much larger scale in the Apollo space missions (Fig. 87.3), but it has not proved commercially viable for Earth-bound applications.

Large banks of lead acid batteries have been used to power submarines silently for nearly 100 years. Today, huge efforts are going into the development of lithium ion batteries for electronic applications (because they have much higher energy density, much lower loss of charge and no memory effect) and they are making some inroads into vehicle applications. On power systems, batteries are used for some power system control and balancing purposes but they are far too expensive to consider for bulk energy storage.

In 1994, Ter-Gazarian [2] described many other methods of storage that have been proposed over the years based on many different physical principles, including various types of thermal energy storage, fly wheels, compressed air, various electrochemical systems, capacitor banks and superconducting coils.

Fig. 87.3 Alkaline fuel cells have been used by NASA in the Apollo missions *(shown here)* and in the space shuttles

Few methods have been added to this list since then, and, despite continuing research activities and development projects, no new practical systems have emerged for large-scale use. Pumped water is the only realistic method to be installed for electrical energy storage on power systems today.

87.4 Arbitrage

The economics of such a storage system rely on arbitrage.

Electricity is put into store (e.g. by pumping water) when it is cheap and regenerated when it is dear. If the extra value is sufficient to pay for the cost of the storage plant (pumps, etc.) as well as the round-trip loss of energy (in pumping and then regenerating), the system can be economic.

On most power systems, this arbitrage between cheap and expensive electricity is between night-time and daytime generation.

So, storing electricity is part of normal power system operation, and 5 % of all EU generation is associated with a pumped water storage system.

87.5 Solar Power

An immediate consequence is that solar power cannot be economically stored on a power system because there is no sunshine (by definition) at night and therefore no solar power, when power is cheap.

This conclusion could be invalidated if ever there was so much solar power generation installed that daytime generating costs fell below night-time generating costs, but such a scenario is very far off.

It is claimed that, if a consumer installs batteries (typically lead acid) alongside solar panels on the roof of their house, "demand and peak charges can be reduced if not eliminated, and independence from the utility, to some degree, can be achieved". This is wholly misleading.

Independence from a utility can be achieved readily enough: it is a simple matter of disconnection. Just switch off! But few people are prepared to accept the unavailability and loss of a continuous power supply which that involves. No individual can possibly provide a power system to match the redundancy and diversity which enable a large power system to provide high levels of reliability.

And independence is a bit like virginity. You cannot be *almost* disconnected; you either are or you are not, and there are fixed connection costs (and charges) to be met if you are not. It is not a matter of "to some degree": it is a definite and clear-cut on-off situation.

By the same token, demand and peak power costs are not reduced if they are covered privately but simply transferred, probably becoming part of the capital payment and installation cost. They are almost certainly increased pro rata in the process, because of the small scale involved.

If not—and this is perhaps most likely—they will just be dropped, if the peak power available is simply reduced and the consumer has to go without more often, as will then be necessary. We can expect a less reliable system to cost less.

87.6 Misleading Terminology

It is frequently asserted that the variability of wind makes power system energy storage essential, and often batteries and other systems, like capacitor banks, are proposed for this purpose. This is a misuse of terminology. What can be needed with high penetrations of wind power is not bulk energy storage but short-term balancing and control for some additional power system fluctuations, which only become significant for wind power capacity penetrations above 30 %.

There are several sources of confusion here. Firstly, these other types of plant (batteries etc.) can perform such control tasks without being able to cope with large-scale energy storage. Much shorter timescales are involved: just while the other plant is brought online or up to speed or equipment is reconnected. Such control functions are quite distinct from that of bulk energy storage. It is misleading to claim that units are for power system energy storage when they are only for power system control or balancing.

But secondly, pumped water systems that are installed for bulk energy storage purposes are also widely used to provide these power system control services. Their rapid response times are very suitable for providing spinning reserve and other essential power system control features. Dinorwig can go from full pumping (1.6 GW) to full generation (1.8 GW) in about 2 min, a swing of 3.4 GW which is well able to cope with the loss of two large nuclear plants on the UK power system almost simultaneously (as has happened).

Thirdly, like most other electrical engineering components and systems (such as capacitors, inductors and transmission lines), the physical processes underlying system balancing and control involve the physical process of energy storage. It is misleading, however, to call such control equipment "power system energy storage equipment" just as it would be misleading to describe a school bus as a "fuel transportation system". Fuel is certainly transported as an essential part of the operation of a school bus, but fuel transportation is not the aim or intention, and to describe it so would be wholly misleading in an engineering context.

Fourthly, batteries and other control devices can indeed be used for energy storage applications, but only on a much smaller scale, notably for transport and electronic applications. They are far too expensive to be used for power system energy storage.

So, widespread misuse of terminology is found in the fields of power system control and network balancing, confusingly and misleadingly claiming that storage devices are being developed and installed for such applications.

87.7 Storage vis-à-vis Wind Power

It has been pointed out that storage cannot be economically associated with solar power, but nor can it be with wind power. There are several obvious mistakes that are frequently made.

For instance, it is often suggested that storage would be valuable to transform intermittent wind into base-load generation. A little thought will show that it is during periods of base-load that most spare capacity is standing idle and is freely available for generation. The value of yet more capacity at such a time is clearly negligible, so the installation of storage or, indeed, any other type of power plant, cannot be justified economically then.

The times when capacity is needed are peak periods, but these are so brief that, again, storage cannot be justified then (unless it is competitive with cheap gas turbine peaking plant). Wind power itself cannot be justified on capacity grounds—it is to save fuel—and the same is true for storage.

There is an in-principle problem with storing electricity from a fuel-saver. Having saved a certain amount of fuel, to put the electricity into a store and then regenerate it involves losses, so less fuel will have been saved at the end of the storage process than without it and storage must always waste fuel. There is no variation in the cost of fuel between when the wind blows and does not blow to compensate for this loss and allow arbitrage. So storing wind energy must always expect to waste money as well as fuel.

Enthusiasts who claim that wind power needs storage to cope with its intermittency manage to ignore the fact that, with more than 300 GW of wind-power-generating capacity installed—and there is now more wind capacity operating than nuclear, both in the UK and the EU—no wind farm has ever had to install a storage plant alongside it.

If a store cannot cope with 10 days of wind it is not effectively tackling the wind storage problem. The classic work of van der Hoven (Fig. 87.4) showed three peaks in the wind frequency spectrum: one due to the local turbulence (1–2 min), a small diurnal (day to night) peak and a broad synoptic peak from 2 to 10 days when the

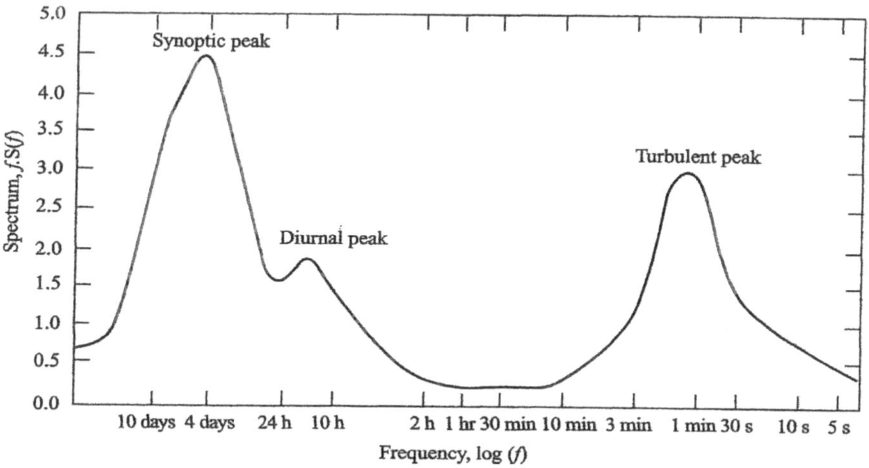

Fig. 87.4 The classical frequency spectrum measured by van der Hoven at Brookhaven National Labs on Long Island

wind blows for up to 10 days (as a weather pattern passes) and ceases for up to 10 days.

A store coping with 10 days of wind must be up to 20 times larger and more costly than a typical night-to-day store, making effective storage for wind power prohibitively expensive.

Swift-Hook [3] points out that storage and wind power are contractually incompatible. The wind only blows for a fraction of the time, typically around one third on average in the UK. Any associated storage would therefore find wind taking three times as long and therefore three times as expensive to store as nuclear or any other continuously available type of power would.

A store cannot afford to buy power around the clock but only when it is cheap (at night), and it can normally only take in power for a single night before it is full. Wind, on the other hand, needs to be able to sell power whenever the wind blows, which may be at any time around the clock and often all around the clock for many days on end. As already noted, a store with a much larger capacity (say 10 days) would be prohibitively (20 times more) expensive.

Contrary to frequent misunderstandings, the installation of storage cannot be justified on a power system to store wind energy.

87.8 Wind Is the Last to Be Stored

Although the installation of storage cannot be justified in conjunction with wind power, storage can have an important role to play on a power system, and it will often be in place regardless of any wind power that may also be in place.

It might then be thought that when electricity is being put into a store, typically during base-load periods, and the wind happens to be blowing at the same time, wind is as likely as any other source of power to be stored or perhaps even more likely. More careful consideration shows that this is not the case.

To discover which source of energy is being stored at any particular time when many different types of power plant are on line, we need to look at the state of the power system with and without storage. The difference between the two tells us which power is being stored.

The power plant, which would need to be switched off if storage ceased (because the store failed or became full) would be the power plant that was filling the store.

In a well-run power system which always uses the cheapest power available, the plant to be switched off would be the *dearest* generation on the system at the time. The "wind blows free" and has zero marginal cost, so wind will always be amongst the *cheapest* plants and never the dearest, with one or two provisos, see [3], unless the cheapest generation is also the dearest, that is to say wind is the only generation left on the system. In that case, it can be said that wind is the last to be stored.

87.9 Conclusions

Despite continuing research activities and other development projects, pumped water is the only system of bulk energy storage being installed on power systems today.

Energy storage is economically justified by arbitrage between cheap night-time and expensive daytime electricity.

Pumped water storage plants can also provide many control and balancing services such as spinning reserve, but these are quite distinct from storage. Battery and capacitor banks and other plants can also provide such control and balancing functions, but it is misleading and confusing to claim that they are for "power system energy storage".

With no solar power at night, storage of photovoltaic (PV) solar power is not economic on a power system.

Contrary to frequent misunderstandings, the installation of storage cannot be justified on a power system to store wind energy either. Wind power is to save fuel, and to store wind power would waste fuel on round-trip losses. Storage and wind power are contractually incompatible. Wind only blows for one third of the time, so it will take three times as long and cost three times as much to store wind compared with nuclear or any other continuously available type of power. The wind frequently blows for up to 10 days at a time, and a store of such capacity would be far too large and costly (20 times current costs) to be economic.

If storage is already installed on a well-run power system, wind is the last type of generation to be stored.

References

1. Swift-Hook DT (2010) Grid-connected intermittent renewables are the last to be stored. Renew Energy 35:1967
2. Ter-Gazarian A (1994) Energy storage for power systems. In: Swift-Hook DT (ed) IEE energy series, vol 6. Peter Peregrinus, London
3. Swift-Hook DT (2013) Wind energy really is the last to be stored and solar energy cannot be stored economically. Renew Energy 50:971–976

Index

A. Sayigh (ed.), *Renewable Energy in the Service of Mankind Vol I*,
DOI 10.1007/978-3-319-17777-9